Dyke Swarms: Keys for Geodynamic Interpretation

Rajesh K. Srivastava
Editor

Dyke Swarms: Keys for Geodynamic Interpretation

Foreword by Richard E. Ernst

Springer

Editor
Prof. Rajesh K. Srivastava
Banaras Hindu University
Department of Geology
Varanasi
India
rajeshgeolbhu@gmail.com

ISBN 978-3-642-12495-2 e-ISBN 978-3-642-12496-9
DOI 10.1007/978-3-642-12496-9
Springer Heidelberg Dordrecht London New York

Cover design: deblik, Berlin

Printed on acid-free paper

Springer is part of Springer Science+Business Media (www.springer.com)

Foreword

Only 25 years have passed since the initial conference that launched modern mafic dyke swarms studies. In 1985, Henry Halls gathered researchers from around the world to the Erindale Campus of the University of Toronto, Toronto, Canada for the first International Dyke Conference (IDC-1) that has led to subsequent gatherings in Australia (1990), Israel (1995), South Africa (2001), Finland (2005) and to the very successful 6th International dyke conference (IDC-6) recently held in Varanasi, India (February 4–7, 2010), organized by Rajesh K. Srivastava, and being celebrated in this volume. Book volumes from the prior dyke conferences were published in 1987 (Halls and Fahrig, Geol. Assoc. Canada Spec. Pap. 34), in 1990 (Parker et al., Balkema publisher), in 1995 (Baer and Heimann, Balkema publisher) and 2006 (Hanski et al., Taylor-Francis publisher).

Pioneers in Dyke Swarm Studies

In earlier days (prior to 1985) mafic dyke swarms were generally poorly understood and little appreciated by the broader geological community; they were typically conceived of as distractions to more important basement or supracrustal geology. However, some early interest in dykes came from paleomagnetism, and to some degree from geochronology and geophysical modeling (because of the simple tabular geometry of dykes). The work and inspiration of pioneering scientists opened new frontiers for dyke swarm studies. For instance, Walter Fahrig of the Geological Survey of Canada systematically (year-after-year starting in the early 1960s) distinguished the main swarms of the Canadian Shield, leading to the identification of the first radiating swarms. Another pioneer was Henry Halls (University of Toronto) who developed many key conceptual ideas that were summarized in his classic and influential paper, Halls (1982, Geoscience Canada). In 1985 Henry Halls also launched the series of International Dyke Conferences whose ongoing legacy we are celebrating in this volume. Elsewhere in the world, notable scientists were also making initial progress on characterizing the swarms of their regions (e.g., see regional papers in Halls and Fahrig, eds., 1987, Geol. Assoc. Canada Spec. Pap. 34).

For India, the history of dyke swarm research is reviewed in Srivastava et al. (in Srivastava et al., eds., 2008, Narosa publishers).

Through the vision of all these pioneering scientists, dyke swarms are now recognized as key indicators in geodynamic processes, the topic of the IDC-6 conference that has led to this volume.

Our early and continuing progress in dyke swarm studies has primarily come from the contributions of geochronology (especially the U-Pb method), paleomagnetism, and geochemistry which allow us to distinguish and characterize individual dykes, and group them into swarms, and from aeromagnetic maps which allow tracing of regional swarm distributions even under younger cover.

More recent progress derives from recognizing that regional-scale dyke swarms can be linked with other coeval intrusions, and volcanism, and related felsic magmatism, and collectively grouped into broader magmatic associations including Large Igneous Provinces (LIPs). This is leading to a prominent role for dyke swarms in efforts involving LIPs to reconstruct past supercontinents (e.g., www.supercontinent.org; Ernst and Bleeker, 2010, Can. J. Earth Sci., v. 47)

Dyke Swarm Myths

Recent studies have burst various long-standing myths about dyke swarms, and have resulted in important insights such as the following:

(1) *Trend matters*: Various trends of dykes in a single region (e.g. "massif") are sometimes assumed to all have the same age. However, from detailed studies, particularly in the Superior craton of the Canadian Precambrian Shield, it is now recognized that each distinct dyke swarm typically has a consistent linear trend or regional radiating pattern. Crosscutting regional dyke sets with different trends will almost certainly have different ages and belong to different swarms.

(2) *Dominant emplacement as Mode 1 cracks (i.e. normal to sigma 3)*: Dykes sets that have regional trends defining an acute angle have often been inferred to be coeval and to be emplaced along conjugate shear sets. However, in every proposed case, subsequent study has revealed that the two trends of dykes have different ages and so are unrelated. Thus we are confident that regional dyke swarms are typically emplaced along principal stress directions (i.e., parallel to the maximum compressive stress, sigma 1, and normal to the minimum compressive stress, sigma 3), and are not emplaced along shear stress directions, except perhaps locally.

(3) *Horizontal emplacement can be important*: It was automatically assumed that dykes were fed vertically from underlying source areas. However, starting in the 1980s, it became increasingly clear (based on magnetic fabric studies as well as modelling and other considerations) that dyke swarms can also be emplaced for long distances laterally in the crust into cratonic interiors (i.e. up to > 2,000 km).

A common style might be for vertical emplacement near the focus of a radiating swarm (above a mantle plume head) and lateral flow at all greater distances from the focus.

(4) *Each dyke is a unique event*: Each dyke in a swarm represents a distinct and independent emplacement event. This is evident from studies that have shown that pairs of long dykes (say 200 km long) that were only separated by a narrow gap (e.g., 10–20 km) could exhibit a consistent chemistry and paleomagnetic direction along the complete length of each dyke and yet be distinct in composition and paleomagnetic direction from that of the other dyke.

(5) *Radiating dyke swarms are the norm*: Work in the Superior craton of the Canadian Precambrian Shield has demonstrated that major Paleoproterozoic dyke swarms radiate into the interior of the Superior craton from discrete points along its margin. This observation is consistent with the identification of discrete mantle plume and associated breakup (or attempted breakup) events along the cratonic margin that collectively represent breakout of the Superior craton from a larger continent. As intraplate magmatic units from around the world continue to be precisely dated, such radiating dyke patterns are also becoming evident on other cratons (e.g. the Slave and Indian cratons). Similarly, the Phanerozoic record shows that LIPs are preferentially centred along the breakup margins within the Gondwana-Pangea supercontinent (in some cases the radiating swarms can be observed, but generally flood basalts obscure the plumbing system). It therefore seems likely that every major continental block through time will similarly host radiating dyke swarms (and their associated LIPs) which can be used to locate mantle plumes and help sort out the breakup history of Precambrian supercontinents.

LIPs and Their Dyke Swarms as a System

I would like to address a frontier issue here. As more and more dyke swarms around the world are precisely dated, it becomes increasingly clear that regional dyke swarms belong to LIP events. In their fullest context such LIP events consist of flood basalts, their plumbing systems of dykes, sills and layered intrusions, and also felsic magmatism produced from subcrustal magmatic underplating. LIPs are often associated with carbonatites and, in some cases, even with kimberlites. In addition, it is clear that LIPs are linked with both magmatic ore deposits (e.g. Ni-Cu-PGE, Cr, and Fe-Ti-V) and hydrothermal deposits resulting from the thermal pulse setting up fluid circulation patterns that redistribute and concentrate metals. In addition, LIPs can have a role in petroleum exploration: (1) oceanic LIPs (plateaus) can cause anoxia events resulting in significant production of black shales (oil source rocks) (e.g., Kerr et al. 1998, J. Geol. Soc., London, v. 155), (2) LIPs associated with breakup can cause a thermal pulse that affects the thermal maturation of oil, (3) sills can also provide structural traps for hydrocarbons. Similarly, dykes and sills can have an effect on ground water circulation and the geometry of aquifers.

For all these reasons, I expect many new insights to arise from looking at LIPs and their dyke swarms as integrated mass/thermal transfer systems that in a complicated way: (1) transport magma upward into the lithosphere via point sources such as plumes, and linear sources such as rifts and translithospheric faults, (2) redistribute the magma through the mantle lithosphere and crust and also to the surface, and (3) along the way the magma interacts with host rocks to potentially produce ores of various types, and affect hydrocarbon potential and regional aquifers.

Present IDC-6 Volume

The IDC-6 conference and the present volume are focused on "Dyke Swarms as Keys for Geodynamic Interpretation". This title accurately reflects the current 'state of the art', and represents a worthy continuation of the tradition of volumes produced after previous Dyke Swarm Conferences.

Given that IDC-6 was hosted in India then it is appropriate for a concentration of papers on Indian dyke swarms to appear in the current volume. Such an opportunity to shine the light on less well understood dykes of a particular region is the main reason that each IDC is hosted in a different country. Of the 30 papers in the current volume, 17 are from India of which 2 are on the classic 65 Ma Deccan flood basalt province and the remainder on older swarms. The other papers in the *Dykes of Gondwana* group address events in Antarctica, Australia, Brazil, Iran, Madagascar, Namibia, Oman, Pakistan, and South Africa. Most of the Gondwana papers are focussed on geochemistry and mostly on mafic units, but a few address felsic or intermediate dykes, and some address associated hydrothermal effects and one addresses U-Pb dating of kimberlites. The *Dykes of Laurasia* group contains five papers, two of which are on U-Pb geochronology in the UK and Greenland, one on melt emplacement in a deformational environment (Kazakhstan), and two of which assess patterns of crystallization in dykes (both from Russia).

We are in a golden era of mafic dyke swarm studies and need to continue to focus on understanding their regional distributions, and relationships to other coeval magmatism, and associated ore-deposits. Expanded U-Pb geochronology will be key to progress as well as asking questions about the context of dykes in the broader LIP systems.

Ottawa, Ontario, Canada Richard E. Ernst

Preface

Dyke Swarms: Keys for Geodynamic Interpretation

Although research on dyke rocks has been of interest to many geologists since eighteenth century, 25 years ago Henry Halls realized that systematic study of dykes, particularly mafic dyke swarms, may play an important role in understanding many geological problems. Dykes swarms are key elements to understanding geodynamic processes. Their study is particularly important for understanding Precambrian terrains. We know that Earth history is punctuated by events during which large volumes of mafic magmas were generated and emplaced by processes unrelated to "normal" sea-floor spreading and subduction. These are recognized as Large Igneous Provinces (e.g. Coffin and Eldholm, 2005; Bryan and Ernst, 2008). Study of dyke swarms helps to recognize Large Igneous Provinces (LIPs), particularly for the Precambrian period. Many LIPs can be linked to regional-scale uplift, continental rifting and breakup, and climatic shifts (e.g., Ernst et al., 2005 and references therein). In the Paleozoic and Proterozoic, LIPs are typically deeply eroded. They are represented by deep-level plumbing systems consisting of giant dyke swarms, sill provinces and layered intrusions. In the Archean the most promising LIP candidates are greenstone belts containing komatiites. Therefore, such studies are considered to be an important tool for paleocontinental reconstructions. LIPs (and their dyke swarms) may help to solve the reconstruction puzzle for pre-Pangean supercontinents back to 2600 Ma as discussed in recent publications (e.g. Bleeker and Ernst, 2006; Ernst and Srivastava, 2008; Ernst and Bleeker, 2010; Ernst et al., 2010) and showcased in a recent special issue of Precambrian Research on "Precambrian Large Igneous Provinces (LIPs) and their dyke swarms: new insights from high-precision geochronology integrated with paleomagnetism and geochemistry" (Srivastava et al., eds, 2010).

Recognizing the unrealized potential and importance of dyke swarm studies, Henry Halls started a series of International Dyke Conferences in 1985. The first conference was held at the University of Toronto, Mississauga Campus, Canada. It attracted over 120 scientists from 20 countries. A book "Mafic Dyke swarms" (Halls and Fahrig, 1987) was published based on the conference proceedings. Shortly after, an IGCP project, number 257, "Precambrian Dyke Swarms" was launched by Henry Halls, this kept the momentum going in terms of bringing dykes to the attention of the scientific community. Also important was special session organized by Henry Halls "Giant Radiating Dyke Swarms and Mantle Plumes" at the Geological Association of Canada annual meeting in 1991. During IGCP 257 the idea was borne and finally realized that an International Dyke Conference should be held every 5 years. An important feature of all IDC meetings is the associated field trips. The second IDC meeting (IDC-2) was held in Adelaide, Australia in September 1990 and was organized by Peter Rickwood, Dave Tucker and John Parker. It resulted in a proceedings volume (Parker et al., 1990). The third conference (IDC-3) was held in Jerusalem, Israel, in September 1995, where the principal organizer was Gideon Baer and Ariel Heimann and again an excellent proceedings volume was produced (Baer and Heimann, 1995). The fourth meeting (IDC-4) was held June 2001 in a game reserve in KwaZula, Natal, South Africa and organized by Mike Watkeys. IDC-5 was held in July–August 2005 in Rovaniemi, in the centre of Finland, and organized by Jouni Vuollo and Lauri Pesonen. A proccedings volume (Hanski et al., 2006) was subsequently published, including contributions from IDC-4. IDC-6 was held at Banaras Hindu University, Varanasi during 4th to 7th February 2010. Considering multifaceted importance of study of dykes, the Organizing Committee chose the overall theme "Dyke Swarms: Keys for Geodynamic Interpretation" with the following 12 sub-themes:

1. Regional maps of dyke swarms and related magmatic units
2. Emplacement mechanism of dykes
3. Petrology, geochemistry and petrogenesis of dykes
4. Geophysics of dykes with special reference to paleomagnetism, new aeromagnetic maps and remote sensing studies
5. Geochronology of dykes
6. Dykes as plumbing system for Large Igneous Provinces
7. Giant dyke swarm and Supercontinents
8. Alkaline dykes (including kimberlites, lamproites, lamprophyres and carbonatites)
9. Synplutonic mafic dykes
10. Dyke swarms and planetary bodies
11. Links to mineralization: e.g. U, PGE & Au, base metals, diamond, Ni-Cr-Co, laterite, etc.
12. Miscellaneous: any other research related to dykes

Out of 175 abstracts submitted for the conference, 168 were included in the Abstract Volume. 52 participants were from abroad, covering 15 countries, 86 participants were from different parts of India and 128 local participants attended the conference (see group photo). In addition to main conference, there were four field trips. These were

1. Mafic dyke swarms and synplutonic dykes emplaced within the Dharwar craton, Southern India (29th January – 4th February 2010)

2. Dyke Swarms of the Deccan Volcanic Province (29th January – 4th February 2010)
3. Wajrakarur Kimberlite Field, Eastern Dharwar Craton, Southern India (7th – 10th February 2010)
4. Mafic Intrusive Rocks from the Western Himalaya (7th – 12th February 2010)

Group photograph of IDC-6 participants

Continuing the tradition of a volume associated with the IDC, we herein offer the IDC6 proceedings volume "Dyke Swarms as Keys for Geodynamic Interpretation". A total 44 papers were submitted for the IDC-6 volume and after peer review 30 were accepted. All accepted manuscripts were submitted to Springer-Verlag at the end of August 2010. The manuscripts are divided into two parts i.e. "Dykes in Gondwana" and "Dykes in Laurasia" and are arranged in alphabetical order, country-wise.

Dykes in Gondwana

Twenty-five manuscripts fall in this group. **Sushchevskaya and Belyatsky** discuss geochemical and petrological characteristics of Mesozoic dykes from Schirmacher Oasis (East Antarctica) and suggest that plume-related magmatism within Antarctica shows the distribution of the Karoo-Maud plume, which is associated with splitting of the Gondwana continent and formation of the Indian Ocean. The time and spatial position of the dykes indicate eastward spreading of the plume material from Dronning Maud Land to the Schirmacher Oasis over at least 10 Ma. **Oliveira** studied the Late Archaean Uauá mafic dyke swarm of the São Francisco craton, Brazil. His studies are based on geological mapping and field relations which recognizes two dyke swarms in the Uauá block. Dyke branching suggests magma flow towards NE and this information is used to discuss potential centres of magma propagation in the ancient São Francisco-Congo craton.

There are five papers on the dykes of Dharwar craton, India. The Proterozoic mafic dyke swarms exposed in the Dharwar craton (Southern India) are of special interest because the craton has been a principal constituent of several ancient supercontinents (e.g. Rogers, 1996; Heaman, 2008; Srivastava et al., 2008). A number of reliable age determinations have been reported (e.g. Halls et al., 2007; French et al., 2008; French and Heaman, 2010). An integrated study (paleomagnetic, geochronological and geochemical) of Proterozoic dykes from Dharwar craton, Southern India by **Piispa et al.** supports the presence of at least two different E-W trending dyke swarms (~2370 and ~1890 Ma) in the Dharwar craton. These results are also consistent with the possibility that the Bastar and Dharwar cratons were amalgamated before ~2370 Ma. **Goutham et al.** provide integrated data on the Proterozoic mafic dykes located south of the southern margin of the Cuddapah Basin, India in two papers. In the first paper they provide the geochemistry of the mafic dykes and discuss their petrogenesis. They are interpreted as intraplate basalts with some crustal contamination or involvement of subduction modified lithospheric mantle (probably metasomatised). In the second paper they provide palaeomagnetism and Ar/Ar geochronology. The alternate interpretations are provided: Either the dykes record an Early Neoproterozoic continental rifting event or the magmatism and its magnetism are of Palaeo-Mesoproterozoic age and were overprinted during the Eastern Ghats (Grenville age) orogeny. **Mohanty** examined the palaeomagnetic database and Palaeoproterozoic evolution of the Indian and Western Australian cratons to determine the possible movement patterns for the assembly of Columbia (Nuna). He concluded that between 2500 to 1800 Ma the South Indian craton moved from high latitude to an equatorial position. Mafic dykes of ~2400 Ma age from the Western Australia and the South India provide evidence for the juxtaposition of the Yilgarn craton against the east coast of India and support the existence of a supercontinental block "SIWA" (an acronym for South India and Western Australia). Indian dykes of 2100–1800 Ma were possibly related to the breakup of SIWA at ~2000 Ma. **Sesha Sai** presented petrological and mineral chemical data on a picrite sill from the Peddakudala-Velpula area, in the southwestern part of the Proterozoic Cuddapah basin, Andhra Pradesh, India. This picrite body contains nearly 38% olivine (forestrite) by mode along with an enstatite-phlogophite-plagioclase-magnetite-chrome-spinel association, depleted HREE signature and the presence of mantle xenoliths of garnet-lherzolite composition.

Gold mineralization reported from the Wayanad, Southern Granulite Terrain, India occurs closely associated with probable mantle-derived quartz-carbonate dykes. **Pruseth** used field criteria, geochemistry of pyrite-hosted fluids and Rb-Sr geochronology of the quartz-carbonate dykes to constrain the source(s) for these auriferous veins.

The next 11 manuscripts cover different aspects of mafic dykes exposed in different cratons of Indian shield. **Lala et al.** presents rock and mineral geochemistry (including isotopes) and paleomagnetic data on the mafic dykes of the Rewa Basin, central India. It is suggested that these dykes are late stage intrusions of Deccan magma along the Narmada lineament and were possibly emplaced along intrabasinal faults within the Rewa and other Gondwana basins of eastern India. The tectonic significance and age of doleritic sill near Bandhalimal in the Singhora protobasin of Chhattisgarh Basin, Central India is the subject of **Sinha et al.'s** manuscript. Their observations confirm the existence of igneous/hydrothermal activity in the Chhattisgarh Basin which is consistent with the Rb/Sr age of sill. A concomitant thermal event in the basin at ~1100 Ma with related hydrothermal activity is thus proved by this communication. **Mishra et al.** discussed the genesis of the mafic-ultramafic intrusives and extrusives in relation to the basement gneiss of the Central Indian Tectonic Zone and suggested that the basement gneisses has been dissected by numerous mafic dykes, Padhar mafic–ultramafic complex and pillow lavas of Betul supra-crustal belt. They show depleted mantle model ages for the mafic volcanics of between 1951 and 2320 Ma and that for a pyroxenite from the PMUM of 2770 Ma. Based on the available age data they suggest that the basement gneiss and the PMUM rocks appear to have acted together as basement for the younger mafic magmatic dykes and flows. These basement rocks appear to have influenced the evolution of the younger volcanics and dykes through limited contamination.

Different generations of Precambrian mafic dyke swarms are well documented in different parts of the Archaean Bastar craton (Srivastava and Gautam, 2009). In the present communication **Gautam and Srivastava** present petrological and geochemical characteristics of the early Precambrian mafic dyke swarm from central part of Bastar craton. The authors conclude that there is a co-genetic relationship between these dyke samples and probably crystallized from a melt originated by ~25% melting of a depleted mantle source followed by 30–40% fractional crystallization. **Chakraborty et al.** studied mafic dykes that intrude the tremolite-zone siliceous dolomites of Palaeoproterozoic Mahakoshal Group of rocks of Central India which are exposed near Jabalpur, Madhya Pradesh, India. Fluid-rock interactions (in a partially open system) triggered substantial mass and fluid transfer across the contact between mafic dykes and the enclosing dolomitic marble. This mineralogical transformation of the mafic dykes changed the physical properties and allowed strain concentration preferentially along the dyke margins. The next paper by **Bose et al.** also deals with metamorphosed mafic dykes, in this case, exposed around Chilka Lake granulites, in the Eastern Ghats Belt, India. These mafic bodies were metamorphosed and deformed by tectonothermal events at ca. 800–500 Ma. **Ray et al.** have explored the petrological controls on rheological inversion of a suite of deformed mafic dykes from parts of the Chhotanagpur Granite Gneiss Complex of eastern India; they propose a viable mechanism for the development of fish-head boudins. Formation of the dyke swarms in this area is correlated with the ca.1.5 Ga extension of their gneissic basement. **Vallinayagam** studied acid dyke rocks from the Malani Igneous Suite (MIS), northern Peninsular India. Petrological and geochemical studies of MIS dyke rocks indicate that they contain rare metals, rare earths and radioactive mineral resources.

Hari and Swarnkar present the petrogenesis of basaltic and doleritic dykes from Kawant in the Chhotaudepur province of Deccan Traps. The results of mass balance calculations for these Deccan units suggest a generalized differentiation scheme from picrite to the most evolved rock that involved removal of olivine, pyroxene and Fe-Ti oxides in the proportion 44:50:6 with ~ 64% of the magma remaining. **Randive** studied xenoliths encountered in the lamprophyre and picrobasalt dykes of Bakhatgarh-Phulmal area, Madhya Pradesh, India. The petrographic and geochemical characteristics of these xenoliths indicate that these are crustal in origin were incorporated into the magma by disaggregation of the country rocks. These xenoliths represent the quartzo-feldspathic residues left in the host magma after to variable degrees of partial melting of basement rocks. **Ghose and Chatterjee** used textural evidence to demonstrate felsic melt emplacement at a microscopic scale in quenched host basalt. The inter-relationship between basalt, plagiogranite and late-granitoids as well as source of sulphide mineralsation were evaluated on the basis of petrography, field relationships, and limited geochemical data. This study indicates that the origin of the late-granitoid intrusives in basalt may be related to hydrothermal metamorphism and partial melting of basalt in the contact aureole of a magma chamber beneath an oceanic spreading center.

Torkian describe magma mingling processes in the Parishan pluton in Qorveh area, southeastern Sanandaj, Iran. He suggested that the mafic enclaves could have been formed by magma mingling process within the host granitoid magma. **Cucciniello et al.** use geochemical data to model the evolution of Cretaceous mafic and silicic dykes and spatially associated lavas in central-eastern coastal Madagascar. They suggest that magmatic evolution of the mafic dykes is dominated by fractional crystallization of plagioclase and clinopyroxene with minor olivine. The Vatomandry silicic rocks are the result of prolonged fractional crystallization from basalt parental magmas coupled with small amounts of crustal contamination. They also report a ^{40}Ar–^{39}Ar date for rhyolite from the Sakanila massif, southwest of Vatomandr to evaluate the age relationship between basalt and rhyolite in this silicic massif of the east coast.

Mafic dyke swarms are major components of the South Atlantic Large Igneous Province, which originated during the Cretaceous break up of western Gondwana. **Wiegand et al.** present magnetic fabric studies on these mafic dykes at a volcanic rifted margin in the Henties Bay – Outjo Dyke Swarm, NW Namibia. The dykes were emplaced in the NE-SW trending, Neoproterozoic Damara mobile belt. Their results point to magma ascent both far inland and close to the coast. Subhorizontal magma flow directions can be observed in each area. Since there is no lateral continuity in dyke trends because of segmentation, strictly horizontal transport over long distances is not possible. They presume a complex magma flow model, with multiple conduits of magma ascent for the Henties Bay-Outjo dyke swarm. **Python et al.** studied diopsidites and rodingites (which are two specific kinds of altered mafic dykes) that outcrop in the mantle section of the Oman ophiolite. Both result from alteration in the Ca-rich environment of the mantle or pre-existing gabbroic dykes. **Khan et al.** presented petrogenetic studies on mafic dykes of the Kohistan paleo-island arc-back-arc system, from the Himalayas of North Pakistan. Mafic dykes of the Jaglot Group show enriched MORB-type affinity whereas mafic dykes of the Chilas Complex and the tholeiitic dykes of the Kohistan Batholith give island arc type signatures. The authors suggest that all mafic dykes of the area are derived by the partial melting of depleted, heterogeneous mantle and enriched mantle sources during island arc, continental margin and back-arc tectonic settings. The last paper in this section is on the Monastery Kimberlite, South Africa. **Noyes et al.** investigated for the first time the feasibility of U-Pb ilmenite geochronology as a new tool for determining the age of kimberlite emplacement and formation of kimberlite indicator minerals. The U-Pb results for nine ilmenite fractions prepared from six megacrysts yield a date of 95.0±14.0 Ma, which is less precise but is in good agreement with the 90.1±0.7 Ma composite age determined by other radiometric techniques for the Monastery kimberlite. This study has also demonstrated that the large variations in ilmenite $^{238}U/^{204}Pb$ ratios are primary and indicates the feasibility of dating a single ilmenite megacryst. They also suggest that this could have importance for evaluating the provenance of ilmenite from indicator mineral suites during diamond exploration.

Dykes in Laurasia

There are five papers included in this section. **Hamilton and Pearson** provided a precise U-Pb age determination of 297.4±0.4 (21>(σ) Ma for baddeleyite for the Great Whin Dolerite Complex (GWDC), northeastern England. This new age places the most precise minimum constraint on the absolute age of the Permo-Carboniferous stratigraphic boundary in the UK and provides a maximum age of mineralization in the North Pennine orefield. The precise age determination of this large intrusive complex also permits improved understanding of the timing of structural evolution in the Northumberland basin. The reversed paleomagnetic polarity of the GWDC is in accord with intrusion during the Kiaman reversed superchron and provides firm geochronological support for establishment of this long-lived reversed interval before 297 Ma. **Halls et al.** presented data on the Melville Bugt dyke swarm of Greenland. They obtained paleomagnetic and geochemical data and U-Pb age for this NNW-trending dyke swarm. This study suggest that at 1.6 Ga Greenland lay in such a position to allow the Melville Bugt dyke swarm to trend towards the 1.5–1.6 Ga Fennoscandian rapakivi province, raising the conjecture that the dyke swarm was fed laterally from this magmatic centre. Satellite imagery from south-east Greenland shows several NNW-trending dykes that may represent a southerly continuation of the Melville Bugt swarm, necessary if the dykes are to have a source in Fennoscandia. **Dokukina and Konilov** discuss mafic melt emplacement during the shock deformation in the subvolcanic environment. They have taken Tastau volcanoplutonic ring complex, Eastern Kazakhstan as an example. On the basis of their study they propose the a synkinematic model of mafic melt fragmentation and mixing between magma and deforming rock at the hypabyssal levels in the crust (i.e. at about 3 km deep). The next two papers discuss chemical zonation in thin dykes. In the first paper **Chistyakova and Latypov** studied a 16 cm thick, fine-grained, dacitic dyke from Southern Urals that reveals a remarkable internal zonation with a systematic inward increase in normative An (100An/(An+Ab)), normative Opx, and whole-rock MgO and FeO, and a decrease in normative Pl content, whole-rock SiO_2, Na_2O, Ba and Sr. All these compositional trends indicate that the dyke becomes more primitive in composition inwards from its margin, the opposite to normal in situ fractionation. This suggests that, despite being fine-grained, the dacitic dyke should be interpreted as a cumulate that provides only indirect information on the parental magma composition. In the other paper they (**Latypov and Chistyakova**) present primary and secondary chemical zonation in the Vochelambina dolerite dyke, Kola Peninsula, Russia. This 140 cm thick dolerite dyke has an unexpected internal zonation that combines the features of both reverse and normal differentiation trends. These trends likely reflect increasing alteration caused by the concentration of post-magmatic fluids in the centre of the dyke (secondary zonation).

Future Perspectives

Richard Ernst in his Foreword rightly mention that we are in a golden era of mafic dyke swarm studies and that there is a need to focus on understanding their regional distributions, and relationships to other coeval magmatism, and associated ore-deposits. Most important part of future work would be robust U-Pb geochronology and paleomagnetic studies which certainly help to understand paleocontinental reconstructions, particularly during Precambrian.

During the concluding session of IDC-6 it was also recommended that, in the future, dykes should be studied in the context of global or regional geodynamics and they can be especially useful in deciphering paleo-plumes and related LIPs and as a tool for deciphering supercontinent reconstructions. M. Ramakrishnan emphasized the need to obtain expanded coverage of low altitude (<500 feet) aeromagnetic maps and the need for developing baddeleyite dating techniques through collaborative efforts. Henry Halls suggested, in the context of future Indian research on dykes, that a concerted program of study be carried out on the Godavari rift that separates the Dharwar and Bastar cratons. It was observed that this rift must be of major importance since it separates two Archean cratons with completely different ages and trends of dyke swarms.

The LIP record for a crustal block can be summarized as a barcode and the LIP barcodes of different blocks can be compared to determine which blocks have matching barcodes and were therefore probably nearest neighbors during the interval of matching (Bleeker and Ernst, 2006; Ernst et al., 2008). We need robust U-Pb geochronology of major dyke swarms to achieve this goal. In the context of Indian shield, very limited precise ages for mafic dykes are available (e.g. Halls et al.,

2007; French et al., 2008; French and Heaman, 2010). As high-precision U-Pb geochronology is increasingly applied to the multitude of poorly dated swarms of dolerite dykes and sills of India, it will be expected that barcode comparisons between the Indian Shield and other crustal blocks will become increasingly more robust for testing paleocontinental reconstructions.

References

Baer G, Heimann A (1995) Physics and Chemistry of Dykes. A.A. Balkema, Rotterdam

Bleeker W, Ernst RE (2006) Short-lived mantle generated magmatic events and their dyke swarms: The key unlocking earth's paleogeographic record back to 2.6 Ga. In: Hanski E, Mertanen S, Rämö T, Vuollo J (eds) Dyke Swarms -Time Markers of Crustal Evolution, Taylor & Francis Group, London: 3–26

Bryan SE, Ernst RE (2008) Revised definition of large igneous provinces (LIPs). Earth Sci Rev 86: 175–202

Coffin MF, Eldholm O (2005) Large igneous provinces. In: Selley RC, Cocks R, Plimer IR (eds) Encyclopedia of Geology, Elsevier, Oxford: 315–323

Ernst RE, Bleeker W (2010) Large igneous provinces (LIPs), giant dyke swarms, and mantle plumes: Significance for breakup events within Canada from 2.5 Ga to the present. Can J Earth Sci 47: 695–739

Ernst RE, Srivastava RK (2008) India's place in the proterozoic world: constraints from the large igneous provinces (LIP) record. In: Srivastava RK, Sivaji C, Chalapathi Rao NV (eds) Indian Dykes: Geochemistry, Geophysics, and Geochronology, Narosa Publ House Pvt. Ltd, New Delhi, India: 41–56

Ernst RE, Buchan KL, Campbell IH (2005) Frontiers in large igneous province research. Lithos, 79: 271–297

Ernst RE, Wingate MTD, Buchan KL, Li ZX (2008) Global record of 1600–700 Ma large igneous provinces (LIPs): Implications for the reconstruction of the proposed Nuna (Columbia) and Rodinia supercontinents Precamb Res 160: 159–178

Ernst RE, Srivastava RK, Bleeker W, Hamilton M (2010) Precambrian large igneous provinces (LIPs) and their dyke swarms: New insights from high-precision geochronology integrated with paleomagnetism and geochemistry. Precamb Res 183: vii–xi doi:10.1016/j.precamres.2010.09.001

French JE, Heaman LM (2010) Precise U–Pb dating of Paleoproterozoic mafic dyke swarms of the Dharwar craton, India: Implications for the existence of the Neoarchean supercraton Sclavia. Precamb Res 183: 416–441 doi:10.1016/j.precamres.201005.003

French JE, Heaman LM, Chacko T, Srivastava RK (2008) 1891-1883 Ma Southern Bastar-Cuddapah mafic igneous events, India: a newly recognised large igneous province. Precamb Res 160: 308–322

Halls HC, Fahrig WF (1987) Mafic dyke swarms. Geol Assoc Canada Spl Paper 34, 503p

Halls HC, Kumar A, Srinivasan R, Hamilton MA (2007) Paleomagnetism and U-Pb geochronology of eastern trending dykes in the Dharwar craton, India: Feldspar clouding, radiating dyke swarms and the position of India at 2.37 Ga. Precamb Res155: 47–68

Hanski E, Mertanen S, Ramo T, Vuollo J (2006) Dyke swarms: Time markers of crustal evolution. Taylor & Francis, London

Heaman LM (2008) Precambrian large igneous provinces: an overview of geochronology,origin and impact on earth evolution. J Geol Soc India 72: 15–34

Parker AJ, Rickwood PC, Tucker DH (1990) Mafic dykes and emplacement mechanism. A.A. Balkema, Rotterdam

Rogers JJW (1996) A history of continents in the past three billion years. J Geol 104: 91–107

Srivastava RK, Gautam GC (2009) Precambrian mafic magmatism in the Bastar craton, central India. J Geol Soc India 73: 52–72
Srivastava RK, Sivaji C, Chalapathi Rao NV (2008) Indian dykes: Geochemistry, geophysics and geochronology. Narosa Publ House Pvt. Ltd, New Delhi, India
Srivastava RK, Ernst RE, Bleeker W, Hamilton M (eds) (2010) Precambrian Large Igneous Provinces (LIPs) and their dyke swarms: new insights from high-precision geochronology integrated with paleomagnetism and geochemistry. Precamb Res Spl Issue 183/3, Elsevier, Amsterdam

Acknowledgements

This proceedings volume presents a collection of 30 papers (out of 44 submitted) which were presented during the 6th International Dyke Conference (IDC-6). At the foremost, I would like to thank all the participants of IDC-6 who made this event as a grand success. I am grateful to those who have taken interest in this proceedings volume and submitted their work for consideration. I wish to thank all the contributors to this volume for their full support and cooperation during its preparation by responding very promptly and efficiently. It is also a pleasure to acknowledge that many of the authors have chosen to submit contributions containing new data, including some significant findings, to this volume contrary to the solely review type papers that characterise many such endeavours.

Each paper was peer reviewed by one or two referees. I take this opportunity to profusely thank each of the reviewers for their ready response and for sparing their precious time in evaluating the manuscripts. Their list includes Talat Ahmad, B. V. Belyatsky, S. K. Bhowmik, S. K. Bhushan, M. E. Bickford, M. Bidyananda, N. V. Chalapathi Rao, N. Chatterjee, M. Deb, T. C. Devaraju, R. E. Ernst, N. C. Ghose, Dmitry Gladkochub, Saibal Gupta, Henry Halls, M. Jayananda, N. R. Karmalkar, Raymond Kent, Santosh Kumar, Andrew McCaig, Meenal Mishra, J. G. Meert, Peng Peng, P. C. Prabhakar, Marie Python, M. Ramakrishnan, H. M. Ramchandra, J. J. W. Rogers, S. J. Sangode, Rajesh Sharma, Hetu Sheth, S. F. Sethna, J. P. Shrivastava, D. C. Srivastava, D. V. Subba Rao, Jouni Vuollo, Miriam Wiegand and those who preferred to be anonymous.

I am grateful to Prof. D. P. Singh, Vice-Chancellor, Banaras Hindu University and Prof H. B. Srivastava, Head, Department of Geology and other University authorities for their unstinted support and constant encouragement.

All efforts were made to complete this volume within 6 months time from the last date of submission of the manuscripts (i.e. February 28, 2010). All accepted manuscripts were submitted to Springer-Verlag at the end of August 2010. I am grateful to Springer-Verlag, particularly Chris Bendall, Helen Rachner and Janet Sterritt for efficiently handling the final manuscripts and bringing out the volume in time.

My special thanks go to few people who helped and encouraged me a lot during organization of IDC-6, field-trips, and also towards the completion of this volume.

This includes Talat Ahmad, N. V. Chalapathi Rao, A. N. Dongre, R. A. Duraiswami, Richard Ernst, V. K. Gaur, G. C. Gautam, H. K. Gupta, K. R. Gupta, Henry Halls, M. Jayananda, N. R. Karmalkar, Alok Kumar, Santosh Kumar, Suresh Kumar, J. Mallikharjuna Rao, M. Mohanty, Shailesh Nayak, B. P. Radhakrishna, M. Ramakrishnan, D. K. Paul, R. H. Sawkar, Saurabh K. Singh, Anup K. Sinha, Ch. Sivaji, M. Srinivas, D. C. Srivastava, Hari B. Srivastava, V. C. Thakur and all the colleagues, research scholars and post-graduate students of the Geology Department, Banaras Hindu University.

All the events related to the IDC-6 were made possible by the financial support of several organizations including Department of Science & Technology, Ministry of Earth Sciences, Council of Scientific and Industrial Research, Oil & Natural Gas Corporation Ltd., Cameca & Gannon Dunkerley & Co. Ltd., Hutti Gold Mines Ltd., Banaras Hindu University, National Geophysical Research Institute, Board of Research in Nuclear Sciences, Buehler & Censico International Pvt. Ltd., Atomic Minerals Directorate for Exploration and Research, and Indian Association of Geochemists. The Geological Society of India is thanked for constant technical support since beginning.

Final thanks must go to my wife Ranjana and son Chinmaya for their enormous patience and constant support all these months.

Varanasi, India
September 27, 2010

Rajesh K. Srivastava

Contents

Contributors

T. Ahmad Department of Geology, Centre for Advance Studies, University of Delhi, Delhi 110007, India, tahmad001@yahoo.co.in

Shoji Arai Division of Earth and Environmental Sciences, Graduate School of Natural Science and Technology, Kanazawa University – Kakuma-machi, Kanazawa, 920-1192, Japan, ultrasa@kenroku.kanazawa-u.ac.jp

Boris Belyatsky VNIIOkeangeologiya Russia Research Institute of the Geology and Mineral Resources of the Ocean, St. Petersburg, 190121 Russia, bbelyatsky@mail.ru

Sankar Bose Department of Geology, Presidency College, Kolkata 700073, India, sankar.bose@gmail.com

Kasturi Chakraborty Geological Survey of India, Southern Region, Hyderabad, Andhra Pradesh, India, kasturi.c8@gmail.com

Supriya Chakraborty Department of Geology, Presidency College, Kolkata 700073, India, suppresi1985@gmail.com

Nilanjan Chatterjee Department of Earth, Atmospheric and Planetary Sciences, Massachusetts Institute of Technology, Cambridge, MA 02139, USA, nchat@mit.edu

S.Yu. Chistyakova Department of Geosciences, University of Oulu, FIN-90014 Oulu, Finland, sofya.chistyakova@oulu.fi

A.K. Choudhary Indian Institute of Technology, Roorkee 247667, Uttaranchal, India, akc.iitr@gmail.com

J. Conrad USGS, Menlo Park, CA, USA, jconrad@usgs.gov

R.A. Creaser Department of Earth and Atmospheric Sciences, University of Alberta, Edmonton, AB, Canada T6G 2E3, robert.creaser@ualberta.ca

C. Cucciniello Dipartimento di Scienze della Terra, Università di Napoli Federico II, Napoli, Italy, ciro.cucciniello@unina.it

Kaushik Das Department of Geology, Bengal Engineering and Science University, Howrah 711103, India, kaushik.met@gmail.com

K.C. Dass Department of Geology, Centre for Advance Studies, University of Delhi, Delhi 110007, India, kanhu65@gmail.com

S.W. Denyszyn Berkeley Geochronology Center, Berkeley, CA 94709, USA, sdenyszyn@bgc.org

T.C. Devaraju Department of Studies in Geology, Karnatak University, Dharwad 580001, Karnataka, India, tcdevaraju38@gmail.com

S.J. Devi Department of Geology, Centre for Advance Studies, University of Delhi, Delhi 110007, India, 1785jwellys@gmail.com

Ksenia Dokukina Geological Institute of Russian Academy Science, Moscow 119017, Russia, ksdokukina@gmail.com

Gulab C. Gautam Department of Geology, Centre of Advanced Study, Banaras Hindu University, Varanasi 221 005, India, gcgautam@gmail.com

Naresh C. Ghose G/608, Raheja Residency, Koramangala, Bangalore 560034, India, ghosenc2008@gmail.com

M.R. Goutham Dr. K.S. Krishnan Geomagnetic Research Laboratory, Indian Institute of Geomagnetism, Allahabad 221 505, India, gouthammr@rediffmail.com

Reinhard O. Greiling Institut für Angewandte Geowissenschaften, Karlsruhe Institute of Technology, 76187 Karlsruhe, FR Germany, er8@agk.uka.de

C. Grifa Dipartimento di Studi Geologici ed Ambientali, Università del Sannio, Benevento, Italy, celgrifa@unisannio.it

H.C. Halls Department of Geology, University of Toronto, Toronto, ON, Canada M5S 3B1; Department of Chemical and Physical Sciences, University of Toronto Mississauga, Mississauga, ON, Canada L5L 1C6, henry.halls@utoronto.ca

M.A. Hamilton Jack Satterly Geochronology Laboratory, Department of Geology, Earth Sciences Centre, Toronto, ON, Canada M5S 3B1, mahamilton@geology.utoronto.ca

K.R. Hari Government V.Y.T.P.G. Autonomous College, Durg, Chhattisgarh 491001, India, krharigeology@gmail.com

L.M. Heaman Department of Earth and Atmospheric Sciences, University of Alberta, Edmonton, AB, Canada T6G 2E3, larry.heaman@ualberta.ca

S.K. Jain Atomic Mineral Division, Department of Atomic Energy, Nagpur 440001, India, dr.guddu1962@rediffmail.com

T. Kaulina Geological Institute of the Kola Science Centre, Apatity, Russia, tatiana_kaulina@yahoo.com

Tahseenullah Khan Department of Earth and Environmental Sciences, Bahria University, Islamabad, Pakistan, bangash1444@hotmail.com

Alexander Konilov Institute of Experimental Mineralogy of the Russian Academy Science, Institutskaya ulitsa, Moscow 142432, Russia, konilov@iem.ac.ru

R. Krishnamurthi Department of Earth Sciences, Indian Institute of Technology, Roorkee 247667, India, krishnamurthi.iitr@gmail.com

S. Kumar Department of Geology, Centre for Advance Studies, University of Delhi, Delhi 110007, India, santoshgeology@gmail.com

Trisha Lala Department of Geology, Presidency College, Kolkata 700073, India, geoltrisha@gmail.com

R.M. Latypov Department of Geosciences, University of Oulu, FIN-90014 Oulu, Finland, rais.latypov@oulu.fi

M. Lingadevaru Department of Applied Geology, Kuvempu University, Shankaraghatta 577451, Karnataka, India, ml_devaru@yahoo.com

L. Melluso Dipartimento di Scienze della Terra, Università di Napoli Federico II, Napoli, Italy, melluso@unina.it

M. Mercurio Dipartimento di Studi Geologici ed Ambientali, Università del Sannio, Benevento, Italy, mamercur@unisannio.it

D.P. Miggins Denver Argon Geochronology Laboratory, USGS, Denver Federal Center, Denver, CO 80225, USA, dmiggins@usgs.gov

M.K. Mishra Department of Geology, Centre for Advance Studies, University of Delhi, Delhi 110007, India, geo_mukesh@yahoo.com

Hiroyuki Miura Department of Natural History Sciences, Hokkaido University, Sapporo 0600810, Japan, hiro@ep.sci.hokudai.ac.jp

S. Mohanty Department of Applied Geology, Indian School of Mines, Dhanbad 826004, India, mohantysp@yahoo.com

V. Morra Dipartimento di Scienze della Terra, Università di Napoli Federico II, Napoli, Italy, vimorra@unina.it

Mamuro Murata Department of Geosciences, Faculty of Science, Naruto University of Education, National University Corporation Naruto, Tokushima 772-8502, Japan, atarumm@naruto-u.ac.jp

K.S. Anantha Murthy Department of Applied Geology, Kuvempu University Shankaraghatta 577451, Karnataka, India, murthy_ask@yahoo.co.in

Kusum P. Naganath Atomic Mineral Division, Department of Atomic Energy, Banglore 560072, India, kusum_amd@rediffmail.com

A.K. Noyes Department of Earth and Atmospheric Sciences, University of Alberta, Edmonton, AB, Canada T6G 2E3, noyes_andreak@hotmail.com

Elson P. Oliveira Department of Geology, Institute of Geosciences, University of Campinas, Campinas, Brazil, elson@ige.unicamp.br

S.K. Patil Dr. K. S. Krishnan Geomagnetic Research Laboratory, Allahabad 221505, India, skpatil13@yahoo.com

D.K. Paul Department of Geology, Presidency College, Kolkata 700073, India, dalimpaul@yahoo.co.in

D.G. Pearson Department of Earth Sciences, Durham University, Durham DH1 3LE, UK, d.g.pearson@durham.ac.uk

L.J. Pesonen Division of Geophysics and Astronomy, University of Helsinki, FI-00014 Helsinki, Finland, lauri.pesonen@helsinki.fi

E.J. Piispa Department of Geological and Mining Engineering and Sciences, Michigan Technological University, Houghton, MI 49931, USA, ejpiispa@mtu.edu

J.D.A. Piper Geomagnetism Laboratory, Department of Earth and Ocean Sciences, University of Liverpool, Liverpool L69 7ZE, UK, sg04@liverpool.ac.uk

C.V.R.K. Prasad AMARARAMA, Vijayawada, India, prasadcvrk@yahoo.com

K.L. Pruseth Department of Geology & Geophysics, Indian Institute of Technology, Kharagpur-721302, India, klpruseth@gmail.com

Marie Python Division of Earth and Environmental Sciences, Graduate School of Natural Science and Technology, Kanazawa University – Kakuma-machi, Kanazawa, 920-1192, Japan; Institute for Geothermal Sciences, Graduate School of Science, Kyoto University, Beppu 874-0903, Japan, marie@bep.vgs.kyoto-u.ac.jp

K.R. Randive Department of Geology, Rashtrasant Tukadoji Maharaj Nagpur University, Nagpur, Maharashtra 440001, India, randive101@yahoo.co.in

V. Ravikant Indian Institute of Science Education and Research-Kolkata, Mohanpur 721452, India, ravikant.vadlamani@gmail.com

Sayan Ray Department of Geological Sciences, Jadavpur University, Kolkata 700032, India, ry.syn05@gmail.com

V. Damodara Reddy Department of Geology, Sri Venkateswara University, Tirupati, India

Hafiz Ur Rehman Department of Earth and Environmental Science, Faculty of Science, Kagoshima University, Kagoshima 890-0065, Japan, hzbangash@hotmail.com

Sanjoy Sanyal Department of Geological Sciences, Jadavpur University, Kolkata 700032, India, sanyal.sn@gmail.com

Pulak Sengupta Department of Geological Sciences, Jadavpur University, Kolkata 700032, India, pulaksg@gmail.com

V.V. Sesha Sai Petrology Division, Geological Survey of India, Shillong 793003, India, seshubb@yahoo.co.in

Tomoyuki Shibata Graduate School of Science, Institute for Geothermal Sciences, Kyoto University, Beppu 874-0903, Japan, tomo@bep.vgs.kyoto-u.ac.jp

D.K. Sinha Atomic Mineral Division, Department of Atomic Energy, Hyderabad 500016, India, dksjai@gmail.com

A.V. Smirnov Department of Geological and Mining Engineering and Sciences, Michigan Technological University, Houghton, MI 49931, USA, asmirnov@mtu.edu

Rajesh K. Srivastava Department of Geology, Centre of Advanced Study, Banaras Hindu University, Varanasi 221 005, India, rajeshgeolbhu@yahoo.com

K.V. Subbarao University Centre for Earth and Space Sciences, University of Hyderabad, Hyderabad 500 046, India, kvsubbarao_iitb@yahoo.co.in

Nadezhda Sushchevskaya Vernadsky Institute of Geochemistry and Analytical Chemistry, Russian Academy of Sciences, Moscow 119991, Russia, nadyas@geokhi.ru

Vikas Swarnkar Government V.Y.T.P.G. Autonomous College, Durg, Chhattisgarh 491001, India, vikash_geology@yahoo.co.in

Ashraf Torkian Department of Geology, Faculty of Sciences, Bu-Ali Sina University, Hamedan, Iran, a-torkian@basu.ac.ir

Robert B. Trumbull GFZ German Research Centre for Geosciences, 14473 Potsdam, FR Germany, bobby@gfz-potsdam.de

R.D. Tucker USGS, Reston, VA, USA, rtucker@usgs.gov

G. Vallinayagam Department of Geology, Kurukshetra University, Kurukshetra 136 119, India, gvallinayagam@rediffmail.com

S. Varghese Department of Earth Sciences, Indian Institute of Technology, Roorkee 247667, India, saajuvarghese@gmail.com

Tim Vietor NAGRA, CH-5430 Wettingen, Switzerland, tim_vietor@yahoo.com

M. Vincent CNRIT, Antananarivo, Madagascar, modvincent@yahoo.fr

J.N. Walsh Department of Geology, Royal Holloway, University of London, Egham, TW 20 OEX, UK, j.walsh@gl.rhul.ac.uk

Miriam Wiegand Institut für Angewandte Geowissenschaften, Karlsruhe Institute of Technology, Karlsruhe, Germany, miriam.wiegand@kit.edu

Masako Yoshikawa Graduate School of Science, Institute for Geothermal Sciences, Kyoto University, Beppu 874-0903, Japan, masako@bep.vgs.kyoto-u.ac.jp

Mohammad Zafar Department of Earth and Environmental Sciences, Bahria University, Islamabad, Pakistan, to_mzaf@yahoo.com

Part I
Dykes in Gondwana

Chapter 1
Geochemical and Petrological Characteristics of Mesozoic Dykes from Schirmacher Oasis (East Antarctica)

Nadezhda Sushchevskaya and Boris Belyatsky

Introduction

Continental flood basalts known from western Dronning Maud Land (DML) and Transantarctic Mountains are regarded as manifestations of the Karoo–Maud mantle superplume activities at about 180 Ma ago, which enabled subsequent Gondwanaland break-up (White and McKenzie, 1989). Unlike well-studied Karoo basalts in Africa, Antarctic large igneous province is hardly accessible and therefore much less studied. Superplume development in space and time, its geochemical character and nature, magma generation and crystallization conditions are most important attributes in understanding the superplume general evolution features. The rise of the spacious Karoo–Maud mantle plume intruded the lithosphere of central Gondwana (southeastern Africa – East Antarctica) and created a thermal anomaly (approximately 2,000 km across) in the upper mantle. The consequent dome-shaped down-warping and extension of the crust provoked intense volcanism on the continent and its margins and initiated riftogenesis in the Weddell Sea (Leitchenkov et al., 1996). The areas of the western Dronning Maud Land and its continental margin located in the central zone of the thermal anomaly (about 1,000 km across) appeared to be most strongly influenced by the plume. The Ferrar Province dolerites and basalts, which intruded the Paleozoic Beacon sedimentary formation along the Transantarctic Mountains, as well as sequences of Tasman Island and southeastern Australia, probably represent a specific manifestation of the Karoo–Maud plume as a consequence of melt "spill" from the heated mantle that formed a spacious dome-shaped structure in the Antarctic lithosphere (Leitchenkov and Masolov, 1997). The large mafic pluton extending along the coast of the Dronning Maud Land and marked by a high-amplitude magnetic anomaly (100 nT) can probably be considered

N. Sushchevskaya (✉)
Vernadsky Institute of Geochemistry and Analytical Chemistry, Russian Academy of Sciences, Moscow 119991, Russia
e-mail: nadyas@geokhi.ru

B. Belyatsky
VNIIOkeangeologiya Russia Research Institute of the Geology and Mineral Resources of the Ocean, St. Petersburg, 190121 Russia

R.K. Srivastava (ed.), *Dyke Swarms: Keys for Geodynamic Interpretation*,
DOI 10.1007/978-3-642-12496-9_1,

another result of this plume. In most cases, opening of a new ocean begins soon after termination of the main phase of plume-related magmatism (Hinz and Krause, 1982; White and McKenzie, 1989). If this is the case, such a process should have begun about 160–170 Ma ago (Middle Jurassic), according to age estimates of basalts and dolerite dykes in the Dronning Maud Land (Cox, 1992).

Petrological Features of Mafic Dykes

The basalts and dolerites outcropping on the area of 35 km^2 in the region of Schirmacher Oasis glacier which is situated close to the Princess Astrid coast in the central part of DML and extend in the E–W direction along the coastline within the territory covered with continental and shelf ice. The samples of these rocks were collected during Russian Antarctic Expeditions in 2002–2003 (Laiba et al., 2002). The Jurassic age of dyke bodies is determined by overlying sedimentary rocks. We studied the mineral phase compositions of dyke rocks from Schirmacher Oasis situated onshore of the central Dronning Maud Land. These olivine-bearing dolerites and gabbro-dolerites most likely represent Jurassic trap activities in Antarctica.

The mountainous Schirmacher Oasis is situated in the central part of DML and composes, together with the adjoining nunataks, the northernmost exposed segment of a large mountain chain extending through the whole region (Fig. 1.1). The oasis is a belt of hilly outcrops, which extend to 20 km in E–W direction at a maximum width of 4 km. The outcrops of the oasis are completely composed of Precambrian metamorphic rocks, which underwent at least two stages of metamorphic transformations, at about 1,000 and 500–450 Ma. The oldest rocks of Late Proterozoic age are alaskite metagranites, metadiorites, metadolerites, metamorphosed gabbronorites, and biotite granites. The Early Paleozoic epoch of granite formation produced three compositional series of pegmatites, diorite dykes, and pegmatoid granites. There is a separate group of Silurian dykes and small bodies of alkaline lamprophyres (Hoch and Tobschall, 1998; Hoch et al., 2001).

The complex of Mesozoic dolerites (basic rocks) occurs over the entire Schirmacher Oasis (Fig. 1.1). The dolerite dykes cut all of the known metamorphic sequences, metagranites and metabasic rocks, veins of pegmatite series, and alkaline lamprophyres occurring within the Oasis, which indicates that the dolerites are the youngest igneous rocks of the region. The dykes strike mainly NW–SE and NE–SW and dip 25–90°. The thickness of the dykes is 0.1–1.7 m, occasionally up to 8 m. Their lengths are up to 250–270 m and usually a few tens of meters. As for petrographic composition, the dykes are made up of olivine and olivine-free dolerites and gabbro-dolerites affected by secondary alteration of different range. The dolerites are porphyritic rocks, and the gabbro-dolerites are equigranular rocks with microdoleritic, microgabbroic, or vitrophyric groundmass textures. Olivine and clinopyroxene phenocrysts account for 10–25% of the rock by volume. Chill zones were observed in the thickest dykes. Concerning chemical composition, the majority of basalts and dolerites of the Schirmacher Oasis can be classified as weakly alkaline magnesian basalts with 0.6–1.6 wt% K_2O, 0.7–2.0 wt% TiO_2, and 10–17

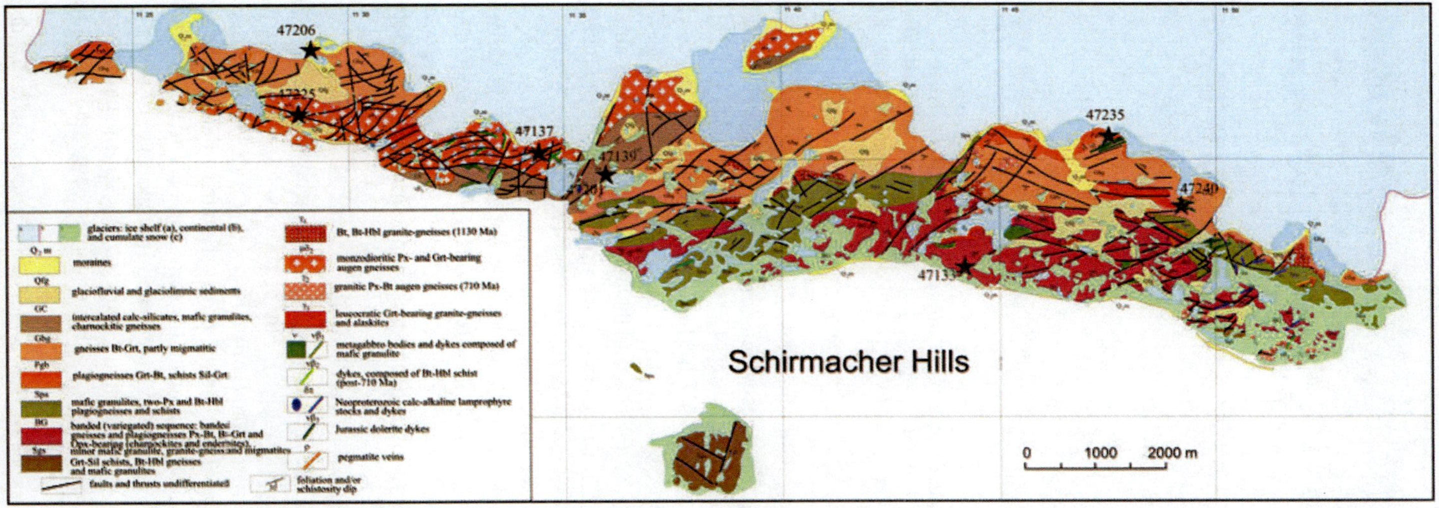

Fig. 1.1 Simplified geological map of the Schirmacher Oasis region (compiled by Laiba A.A. et al. (Sushchevskaya et al., 2009)) showing sampling sites (*asterisk*). The distribution of the ice cover is shown by light blue colors, and outcrops of old continental rocks (*granulites, gneisses, plagiogranites, schists*, etc.) are colored in accordance with insert legend. Tectonic dislocations, ancient dykes of lamprophyres, dolerites, and metagabbroids, as well as Jurassic olivine-bearing dolerites and pegmatite vein are shown as *line* of corresponded colors. Sample numbers are shown on the map

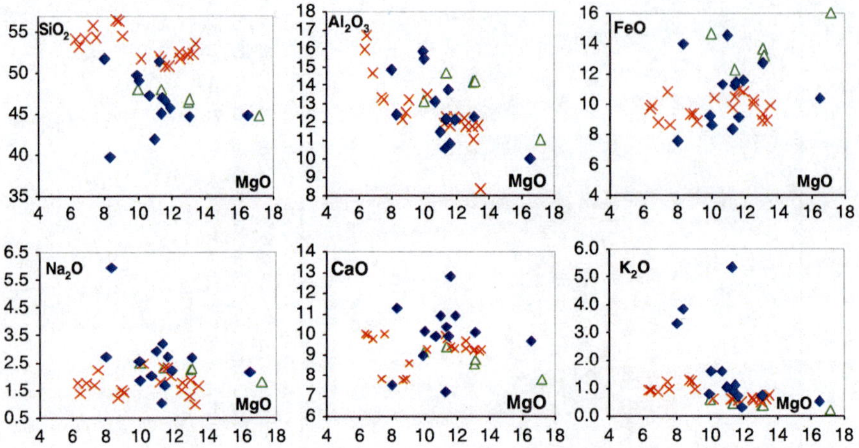

Fig. 1.2 Variations in major element contents for Schirmacher Oasis dolerites: rhombs – studied olivine-bearing dolerites (Table 1.1), triangles – dykes from the Muren massif, DML (Vuori and Luttinen, 2003), crosses – ancient anorthosite dykes from Schirmacher Oasis (Belyatsky et al., 2002)

wt% MgO. Figure 1.2 shows variations in TiO_2, K_2O, Na_2O, and SiO_2 as functions of MgO content for the dolerite samples which compositions are given in Table 1.1. Three samples from our collection (47240-2, 47235-19 and 47201-4) show elevated potassium contents, and sample 47240-2 is also depleted in silica and enriched in alkalis and titanium, which allows to classify it as an alkali basalt. Sample 47201-4 is a strongly altered amygdaloidal dolerite, and the presence of secondary amphibole and biotite in sample 47235-19 indicates that its composition cannot be regarded as primary. Two of the freshest olivine dolerite samples, 47225-7 and 47139-7, were selected for the investigation of major constituent minerals. The Mesozoic basites of the Schirmacher Oasis (Fig. 1.2) are identical in composition to the dolerites of the dyke complex of the Muren massif in western DML (Vuori and Luttinen, 2003) and slightly different from the Mesozoic basalts and dykes of ancient anorthosites occurring at the Schirmacher Oasis and often showing similar strikes (Belyatsky et al., 2002).

In general, the crystallization sequence of the basalts is magnesian olivine–plagioclase–clinopyroxene. This is supported by the analyses of minerals, in particular, more magnesian compositions of olivines compared to clinopyroxenes.

The olivine composition varies strongly from $Fo_{91.5}$ to Fo_{55} (~400 grains), which suggests the occurrence of crystal cumulation processes during melt differentiation. The most widespread olivine compositions fall within $Fo_{88–89}$ and are similar to the most magnesian olivines from the least differentiated Jurassic lavas of western DML (Sushchevskaya et al., 2009). A distinctive feature of the Schirmacher dolerites is the presence of Mg-rich olivines ($Fo_{90–91}$), which may indicate rapid magma ascent from the generation zones.

Variations in Mn and Ni as a function of forsterite content reflect a decrease in Ni and an increase in Mn contents during primary magma crystallization. From

Table 1.1 Major (vol%) element concentrations of basalts and dolerites from Schirmacher Oasis

component	47133-1	47137-20	47139-7	47201-4	47206-3	47225-6	47225-7	47235-19	47240-2	47232-3	47237-3	47252-2	47225-9
SiO_2	49.70	45.05	44.80	51.39	45.75	42.05	44.70	51.76	36.13	49.10	47.33	47.07	46.70
TiO_2	1.00	1.29	0.82	1.09	1.30	2.22	1.56	1.00	3.25	1.0	1.17	1.24	0.70
Al_2O_3	15.84	12.13	9.96	10.57	12.07	11.45	12.25	14.87	11.25	15.40	13.10	13.73	10.80
Fe_2O_3	10.18	12.51	11.56	9.31	12.87	16.13	14.15	8.39	14.09	9.57	12.52	12.74	10.10
MnO	0.16	0.19	0.36	0.15	0.19	0.21	0.24	0.13	0.23	0.15	0.17	0.18	0.17
MgO	9.91	11.34	16.49	11.25	11.89	10.95	13.05	8.01	7.56	10.00	10.65	11.45	11.60
CaO	8.95	10.35	9.68	7.21	10.88	10.89	10.06	7.50	10.21	10.10	9.88	9.87	12.80
Na_2O	2.56	3.18	2.14	1.04	2.21	2.94	2.68	2.72	5.39	1.84	2.01	1.67	2.73
K_2O	0.77	0.88	0.55	5.33	0.32	1.03	0.76	3.32	3.46	1.59	1.60	1.09	0.73
P_2O_5	0.18	0.30	0.20	0.91	0.18	0.31	0.25	0.51	0.64	0.34	0.28	0.17	0.28
Loss	0.88	2.62	2.62	1.16	2.07	1.57	0.28	1.30	7.35	0.60	0.80	0.20	3.13
Total	100.14	99.84	99.18	99.41	99.73	99.75	99.98	99.51	99.56	99.69	99.51	99.41	99.74

Note: Oxide concentrations determined by XRF (Vernadsky Institute, RAS), detection limits ranged from 0.001 to 0.02 wt%, RSD from 1.5 to 13%.

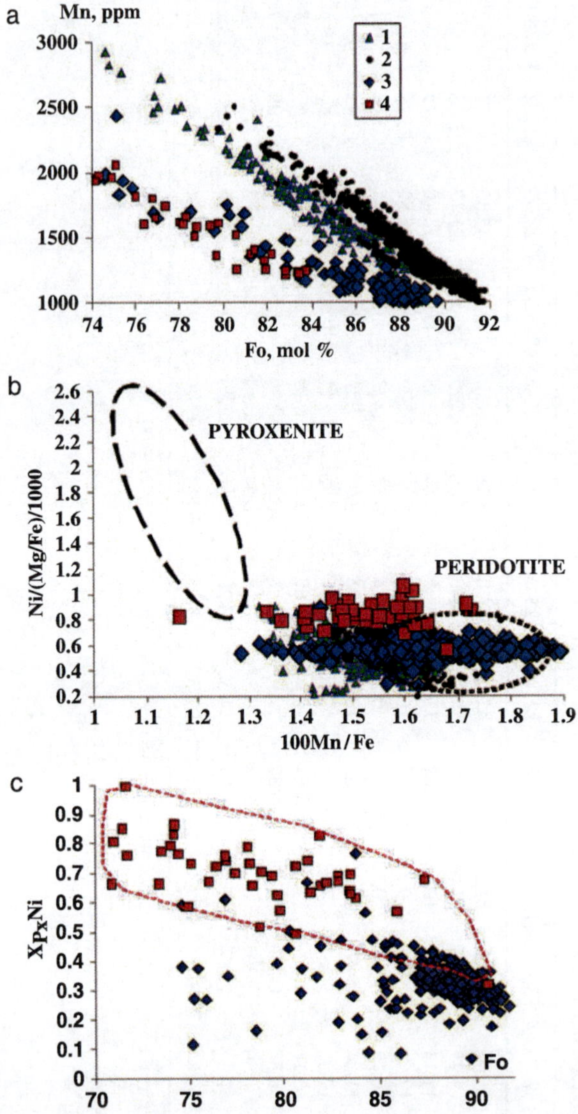

Fig. 1.3 Composition of olivines from Schirmacher Oasis dolerites: (**a**) – Mn vs Fo plot for olivine from Bouvet triple junction basalts (1), different MORB (2), sample 47139-7 Schirmacher Oasis (3) and sample 47225-7 Schirmacher Oasis (4); (**b**) – Ni/(Mg/Fe)/1,000 vs 100 Mn/Fe plot for initial composition of model mantle component sources (Sobolev et al., 2007); (**c**) – difference in olivines from samples 47339-7 and 47225-7 (denoted by stripped contour field) in $X_{Px}Ni$ index versus Fo plot

Fig. 1.3 can be inferred that the variability in Mn content is lower then for olivines from MORB and enriched Bouvet triple junction basalts (Migdisova et al., 2004). In addition, it can be clearly seen in the diagrams that the signatures of olivine compositions from the two studied samples form two independent trends of NiO and MnO variations with different slopes. The average NiO content in the most magnesian olivines is ~0.35 wt% ($N = 340$) for sample 47139-7, ~0.4 wt% ($N = 46$) for sample 47225-7, and ~0.5 wt% for olivines from the Vestfjella Mountains (Sushchevskaya et al., 2009). The differences in Ni content of the liquidus olivines are primarily related to different Ni contents in the initial melts. According to the model of Sobolev et al. (2007), an increase in NiO and decrease in MnO content in magnesian olivines indicate the presence of pyroxenite fragments in the peridotite source. Figure 1.3b shows differences in Ni and Mn concentration for equilibrium melts from pyroxenite and peridotite sources. The model was proposed for Hawaiian magmas and assumed that their plume source contained fragments of crustal eclogites, which melting began during the ascent of the heterogeneous mantle material at a depth of about 150 km (Sobolev et al., 2005). The produced melts reacted with the peridotite matrix that is resulted in the formation of pyroxenites with Ni-rich pyroxenes. Olivine is the main phenocryst in peridotite-derived melts while pyroxene is the main phenocryst in the pyroxenite-derived melts. These minerals have different predilections for rare elements accumulation. Relative contributions of peridotite and pyroxenite can be inferred from Ni and Mn concentrations in olivine phenocrysts crystallized from "hybrid" magmas. Melts with highest ratios Ni/(Mg/Fe)/1,000 and 100 Mn/Fe are more likely to be produced from the sources containing reaction pyroxenite (Sobolev et al., 2007). So, as it is following from Fig. 1.3b 100 Mn/Fe ratios for Schirmacher olivines are forming a mixing array of primary melts from peridotite to pyroxenite source and are different from MORB olivines. The further melting of pyroxenites at a depth of about 100 km resulted in the formation of melts enriched in Ni relatively to the liquids that could be produced by the melting of a peridotite source. The question on the reasons of the heterogeneity of the plume mantle remains unsolved. It could be related to processes accompanying material ascent from the core–mantle boundary, where fragments of the earlier subducted oceanic crust probably accumulate, or to the interaction of the ascending hot peridotitic mantle with the lower parts of continental blocks at depths of 200–300 km. In either case, eclogite melting could result in the appearance of pyroxenites and their subsequent involvement into the derivation of basaltic magmas. The estimated amount of pyroxenite component admixture in the melting peridotite source by Ni content in olivine from the basalt: $X_{Px}Ni = 10.5402 \times (NiO \times FeO/MgO) - 0.4368$ (Sobolev et al., 2007) is different for two studied samples (Fig. 1.3c). It is from 20 to 30% in average for sample 47139-7 and about 70% for sample 47225-7. It is noteworthy, that olivines from sample 47225-7 become more enriched with Ni in the course of differentiation which could be connected with stronger contamination in the process of crystallization. It is necessary to point out that the values of $^{143}Nd/^{144}Nd$ and $^{208}Pb/^{204}Pb$ isotope ratios correlate with this conclusion reflecting more radiogenic values for lead and lower εNd for sample 47225-7.

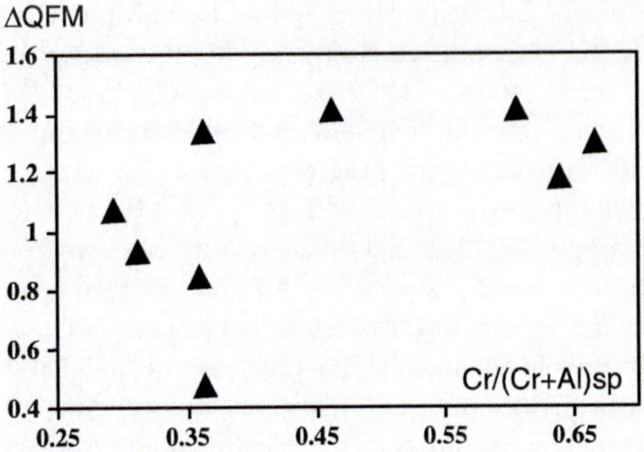

Fig. 1.4 Spinel Cr index *vs* ΔQFM plot for Schirmacher Oasis dolerites

Mostly olivine contains spinel inclusions which compositions are typical of mantle melts. Relative deviations from the QWM buffer (ΔQWM as – lg (fO_2)) are about 1–1.2 (Fig. 1.4).

Clinopyroxene composition provides a mean for pressure estimation (Nimis and Ulmer, 1998). Using this approach Schirmacher Oasis dyke samples yields pressure estimates at 1–3 kbar (calculated at mean temperature $T = 1{,}100°C$), which points out to a fairy shallow crystallization. Similar crystallization pressure interval was also calculated for western Dronning Maud Land magmas, which is evident from model pressure histograms (Fig. 1.5).

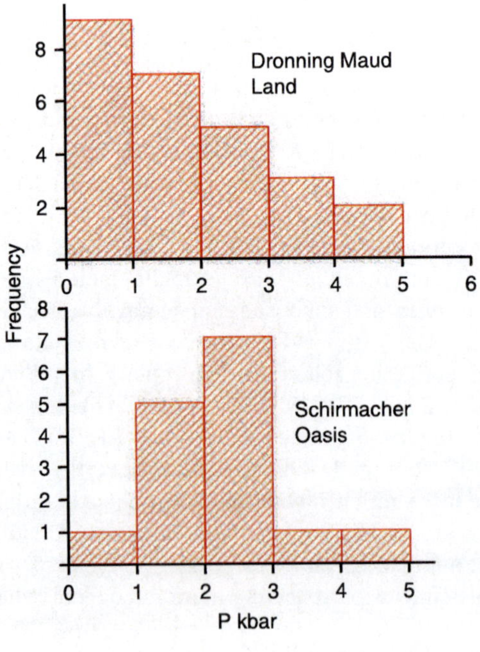

Fig. 1.5 Crystallization pressure histograms for mafic rocks from Schirmacher Oasis and Dronning Maud Land (Migdisova et al., 2004). Pressure estimation uncertainties does not exceed 1.7 kbar

The Geochemical Character and Evolution of Mantle Karoo–Maud Superplume in Antarctica

The contents of lithophile elements in the samples are given in Table 1.2 and illustrated in the spider diagram in Fig. 1.6a. The primitive mantle-normalized (Sun and McDonough, 1989) patterns indicate different degrees of lithophile element enrichment in the dolerites. There are distinct positive Pb and negative Ta and Nb anomalies. Two tendencies could be marked for magmas of Schirmacher Oasis. The first one is described by light negative Ta and Nb anomalies (samples: 47206-3, 47333-1, 47139-7) and the second tendency is the absence of such anomaly. Sample 47201-4 (altered dolerite), which was classified as a derivative of alkaline magmas, shows high contents of all elements, but retains all of the specific geochemical anomalies. All samples have high Ba contents and show high trace element ratios between a highly to moderately incompatible and a moderately incompatible element (La/Sm, La/Nb, La/Yb, Sm/Yb, Table 1.2). We compare trace element ratios on the Fig. 1.6b, where high La/Ce is indicator of incompatible element enrichment of melts generated during the melting from an enriched source, and high Gd/Yb reflects the melting of garnet-free source (Kamenetsky and Maas, 2002). One can see that Schirmacher magmas have high Gd/Yb ratios (1.5–3) and high La/Ce (0.4–0.5). These values are higher than for MORB, and different from the Macquarie Island glasses field by the lower La/Ce ratios. Figure 1.6b also shows the absence of significant input of material from depleted source into Schirmacher melts producing. At the same time, the mantle source composed by substances with 85% of old recycling MORB and 15% low continental crust (Stracke and Bourdon, 2009) is quite close to the Schirmacher dolerites and could be reason for shifting the Shirmacher melts composition from higher Gd/Yb and La/Ce ratios in the direction of lower values of these ratios (Fig. 1.6b).

The isotopic analysis of dolerites from the Schirmacher Oasis (Table 1.2) revealed the radiogenic compositions of Sr ($^{87}Sr/^{86}Sr$ of 0.7045–0.7047) and Pb ($^{208}Pb/^{204}Pb$ of 37.98–38.2 and $^{207}Pb/^{204}Pb$ of 15.45–15.52). The relatively high $^{206}Pb/^{204}Pb$ (17.9–18.2) and low $^{143}Nd/^{144}Nd$ values (0.51275–0.51255) also indicate the enrichment of the dolerite source compared with the depleted oceanic mantle (Hauri et al., 1994). Reverse correlation of the initial values of Sr isotope ratio with epsilon Nd signatures (Fig. 1.7) could reflect the effect of the lower crust contamination of the dolerite mantle source.

The lead isotopic ratios of basalts from the Schirmacher Oasis are different from those of the DML basalts, which have even more radiogenic $^{206}Pb/^{204}Pb$ ratios, up to 18.7–18.8. Nonetheless, in the lead isotope diagram, all these rocks plot along a common trend, which extends from the relatively depleted dolerites of the Schirmacher Oasis to the more enriched basalts of the western DML (Fig. 1.8) (Belyatsky et al., 2006) and the Transantarctic Mountains (Ferrar Province), which underwent extensive crustal contamination (Elliot et al., 1999).

This diagram also shows the isotopic compositions of ancient Proterozoic anorthosite dykes recalculated to the emplacement age of Mesozoic dolerites. It

Table 1.2 Trace element (ppm) concentrations and Sr, Nd, and Pb isotope data for rocks of Schirmacher Oasis

component	47133-1	47137-20	47139-7	47201-4	47206-3	47225-6	47225-7
Ba	304	342	283	4,216	129	313	258
Th	2.08	1.35	0.69	15.02	1.02	1.34	1.17
U	0.43	0.34	0.16	1.58	0.16	0.24	0.21
Nb	3.2	9.8	3.7	15.5	3.4	10.9	8.4
Ta	0.20	0.61	0.21	0.75	0.20	0.84	0.66
La	12.70	10.24	5.34	63.72	7.06	11.92	10.15
Ce	30.0	21.6	11.7	123.8	16.4	28.1	23.1
Pb	5.00	2.67	2.17	9.43	1.50	4.05	2.93
Pr	4.21	2.95	1.65	14.38	2.42	4.14	3.30
Sr	154	298	286	831	221	352	321
Nd	19.7	14.1	8.22	58.1	12.4	20.6	16.1
Sm	5.27	3.70	2.34	10.38	3.61	5.59	4.36
Zr	20	83	52	249	83	149	101
Hf	0.95	2.16	1.36	6.35	2.25	3.86	2.70
Eu	1.18	1.23	0.76	2.30	1.19	1.75	1.44
Gd	5.94	3.91	2.55	7.31	3.98	5.66	4.63
Dy	6.51	3.49	2.49	5.16	3.68	4.78	4.01
Y	36.5	17.8	12.8	26.5	18.1	22.6	18.8
Er	3.98	1.83	1.30	2.53	1.83	2.28	1.98
Yb	3.76	1.60	1.15	2.26	1.56	1.84	1.69
Ho	1.39	0.68	0.49	0.97	0.70	0.88	0.77
Lu	0.55	0.23	0.17	0.33	0.22	0.27	0.24
La/Nb	1.56	7.25	5.33	1.03	3.29	8.11	7.16
La/Sm	2.41	2.77	2.28	6.14	1.96	2.13	2.33
La/Yb	3.38	6.40	4.67	28.24	4.54	6.49	6.00
Sr/Nd	1.19	0.90	1.31	0.66	0.62	0.98	0.89
Gd/Yb	0.03	0.12	0.11	0.24	0.12	0.17	0.14
La/Ce	16.19	15.94	17.71	20.79	16.61	13.04	12.67
age (Ma)	170	170	170	170	170	170	170
[Nd], ppm	19.76	n.d.	7.920	55.08	11.88	19.76	17.32
$^{147}Sm/^{144}Nd$	0.16061	n.d.	0.16814	0.10985	0.17377	0.16212	0.16068
$^{143}Nd/^{144}Nd$	0.512536	n.d.	0.512649	0.511578	0.512720	0.512656	0.512585
err (2S)	4	n.d.	4	4	2	3	8
[Rb], ppm	33.32	n.d.	112.4	186.4	11.14	30.46	28.24
[Sr], ppm	161.3	n.d.	285.4	612.0	229.2	359.2	341.5
$^{87}Rb/^{86}Sr$	0.59815	n.d.	1.13929	0.88193	0.14066	0.24536	0.23932
$^{87}Sr/^{86}Sr$	0.711137	n.d.	0.707522	0.714170	0.704928	0.705283	0.705227
err (2S)	15	n.d.	19	17	15	21	14
eNd	–1.21	n.d.	0.83	–18.80	2.10	1.10	–0.25
$(^{87}Sr/^{86}Sr)t$	0.70969	n.d.	0.70477	0.71204	0.70459	0.70469	0.70465
$^{206}Pb/^{204}Pb$	n.d.	n.d.	18.149	18.197	17.969	18.062	17.925
$^{207}Pb/^{204}Pb$	n.d.	n.d.	15.494	15.535	15.508	15.511	15.494
$^{208}Pb/^{204}Pb$	n.d.	n.d.	37.934	39.621	38.176	38.208	38.137

Note: LILE concentration and isotope analysis were determined at Karpinsky Geology Institute (St. Petersburg) by ICP MS (detailed in Sushchevskaya et al., 2009). Err (2S) – corresponds to isotope ratios error at 95% confidence level and shows figures at the last decimal place. eNd $(^{87}Sr/^{86}Sr)t$ – initial isotope composition corresponding sample at the time of dyke emplacement (170 Ma).

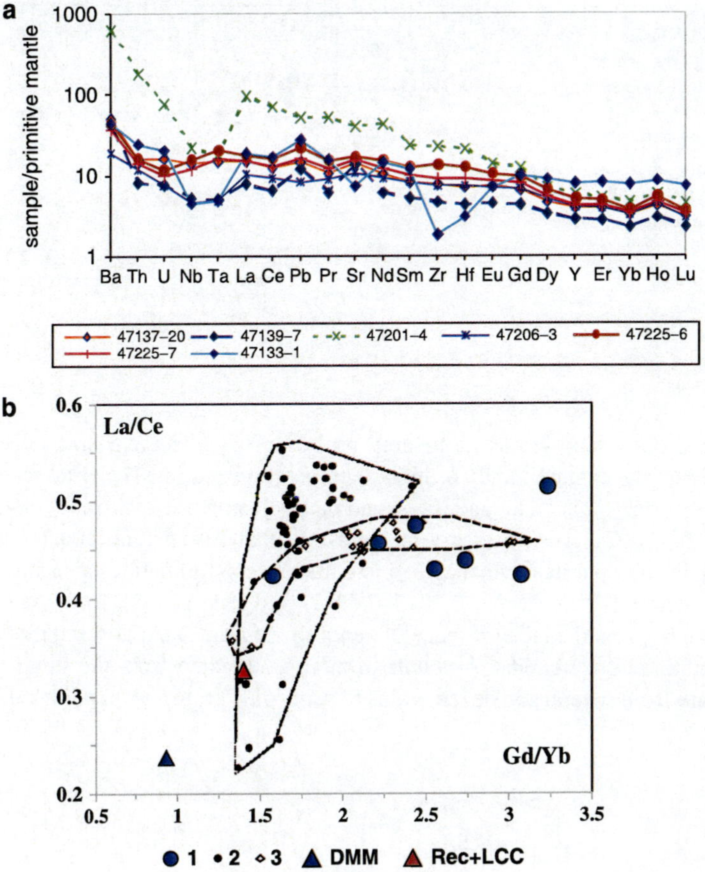

Fig. 1.6 Trace element composition of Schirmacher Oasis dolerites: (**a**) primitive mantle-normalized (Sun and McDonough, 1989) patterns of lithophile elements for the studied samples; (**b**) correlation between Gd/Yb and La/Ce ratios for olivine-bearing dolerites (1) and comparison with the enriched magmas from Macquarie Island ophiolite (2) (Kamenetsky and Maas, 2002) and Bouvet triple junction glasses (3) (Sushchevskaya et al., 2003). Triangles show depleted source (DMM) and enriched source (Rec+LCC) close to composition of 85% old recycling MORB and 15% low continental crust according to (Stracke and Bourdon, 2009)

is clearly seen that they plot in the field of the Mesozoic basalts of Antarctica. As was noted above, the ancient dykes often spatially associate with plume-related dykes and have similar strike and dip angles, at least in the Schirmacher Oasis. The spreading of Mesozoic plume magmas evidently occurred mainly along highly permeable weakened zones in the lithosphere, and the ancient dyke zones could serve as magma conduits.

Such a process of plume melt penetration along the zones of ancient dykes was also characteristic of the Karoo plume in the south-eastern part of the African continent at 180–173 Ma (Jourdan et al., 2004, 2006).

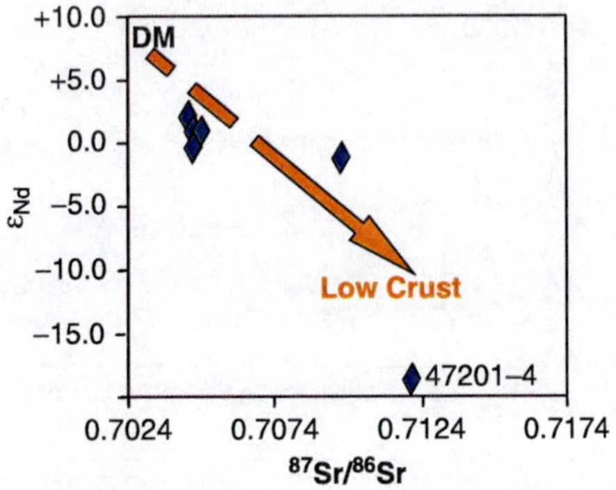

Fig. 1.7 Isotope characteristics of dolerites evidence mainly to the effect of lower crust contamination of primary melts. The most isotope enriched sample 47201-4 is a highly carbonatized and potassium enriched altered dolerite

The inherited old zircon grains discovered in the dolerites of the Schirmacher Oasis could reflect the contamination of higher temperature Jurassic tholeiitic magmas by the material of ancient andesites and/or enclosing metamorphic rocks (Belyatsky et al., 2007). It is necessary to point out, that isotope and geochemical characteristics of dolerites evidence to insignificant contamination of primary melts by crust material ($^{87}Sr/^{86}Sri = 0.7045–0.7047$, $\varepsilon Nd = +1: +2$). At the same time, there have been marked the presence of xenoliths of host magmatic and metamorphic rocks in the dyke bodies. Amounts of zircon separated from the samples under study vary from several and dozen grains to some milligrams. Mineralogically

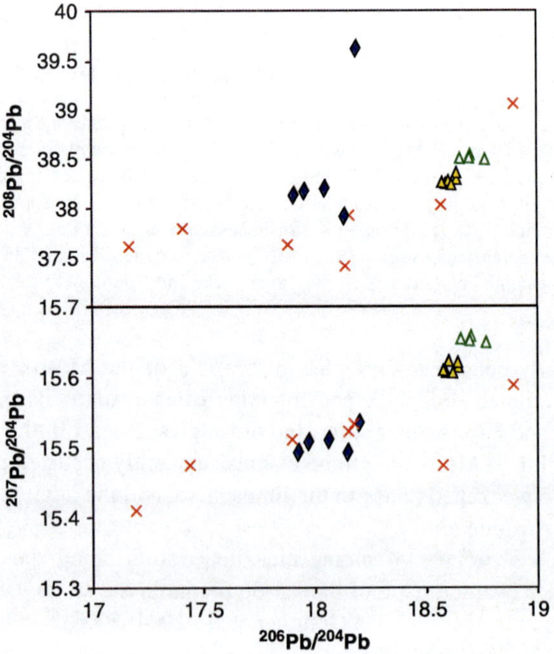

Fig. 1.8 Similarity of the Pb isotopic characteristics of the Mesozoic dolerites (*rhombs*) of Schirmacher Oasis and the spatially associated ancient dykes: Precambrian dykes of Schirmacher Oasis (our unpublished data) – *green triangles*, Ferrar Province basalts (Antonini et al., 1999) – *yellow triangles* and DML basalts (Belyatsky et al., 2002) – crosses

studied zircons are characterized by development of rounded (partly molten) prismatic, polyfaceted and isometrical forms. From one hand it evidences to the obvious capturing of zircons from host rocks and from the other hand, it points to polygenetic origin of zircon (magmatic and metamorphic as well).

Local U-Pb isotope analysis (SHRIMP II) has allowed to determine some discrete events in geological history of the region: 560–640, 690–770, 970–1,010, 1,715–1,820, about 2,500, 2,810 m.y. ago. Zircon composition (trace element and REE patterns) allows to mark out with considerable reliability primary magmatic zircons and metamorphic ones and, also, to evaluate by model mixing (Sambridge and Compston, 1994) relative share of different genetic rock types in the structure of the region: 2,810 Ma magmatic rocks – 4% of total crustal volume, 2,500 Ma mainly metamorphic rocks – 7%, 1,770 Ma magmatic rocks – 15%, 990 Ma magmatic rocks – 8%, 770 Ma magmatic and metamorphic rocks – 21%, 610 Ma magmatic and metamorphic rocks – 45%. It is necessary to point out, that together with well known stages of geological history (Pan-African, Grenvillian, and early Proterozoic) developed in this region, it was possible for the first time to identify Archaean crustal component, the presence of which had been supposed earlier on the base of Nd isotope composition. At the same time none of the single Jurassic zircon grains within more than two hundred studied zircons was identified up to now.

Prominent features of the Karoo–Maud plume magmatism in Antarctica are the presence of high-magnesium volcanics and the wide occurrence of dolerite dykes and sills and olivine-rich gabbroid intrusions (Leitchenkov et al., 2003). Many of the dolerites are chemically similar to slightly enriched tholeiites (Luttinen and Furnes, 2000), which suggests that the plume magma source was similar to the depleted oceanic mantle. The plagioclase K–Ar age of basaltic lavas from the Vestfjella Mountains is 180 Ma, which is very close to the ages of the basalts of the Kirwanveggen (Luttinen and Furnes, 2002) and mafic dykes and flows of south-eastern Africa (183 $\pm$ 1 Ma) (Duncan et al., 1997; Jourdan et al., 2005). The Sm–Nd mineral (pyroxene and plagioclase) and bulk-rock age of the dolerites from the Schirmacher Oasis is 171 $\pm$ 24 Ma, which coincides within the uncertainty with the age of the Jurassic magmatism of DML and southeastern Africa (Belyatsky et al., 2006). Thus, based on the geochemical, isotopic, and petrological similarity and simultaneous formation, the Mesozoic basic magmatism of the Schirmacher Oasis can be interpreted with a high degree of certainty as a derivative of the activity of the Karoo–Maud mantle plume (both in the western DML of Antarctica and in the Karoo province of south-eastern Africa).

Conclusions

The investigation of the Mesozoic (about 170 Ma) basaltic magmatism of the Schirmacher Oasis reveals the dyke olivine-bearing dolerites and gabbro-dolerites crystallized at 1–2 kbar (calculated at mean temperature T°C = 1,100). Olivine shows a wide compositional range between $Fo_{91.5}$ to Fo_{55}, which proves strong influence of accumulation processes during melt differentiation. Clinopyroxene crystallization followed precipitation of olivine and plagioclase. Clinopyroxene has wide *mg#* variation from 81 to 69 and low Cr/Al (0.01–0.14) and Na/Al (0.1–0.15) ratios.

Schirmacher Oasis magmas are similar in petrological sense to the previously studied rocks of western Dronning Maud Land, region of the Vestfjella Mountains (Luttinen and Furnes, 2000), which are interpreted as the manifestation of the Karoo–Maud plume in Antarctica.

The results of olivine study strongly confirm that the Schirmacher melt extraction model should include melting of heterogeneous mantle source. Our opinion – this source was a pyroxenite-bearing peridotite. The melting of pyroxenites at a depth of about 100 km resulted in the formation of melts enriched in Ni relatively to the liquids that could be produced by the melting of a peridotite source solely. The average estimations of NiO content in the most magnesian olivines exceed up to 0.4 wt% that is significantly higher than for MORB magnesian olivines, for which the typical NiO content is about 0.25 wt%.

Melting model by Sobolev et al. (2007) shows that share of pyroxenites within melting source could vary from 30 to 80%. High ratios of incompatible elements Gd/Yb (2–3.5) testify possible presence of garnet within a mantle source. Mixture of accumulated melts from peridotite and pyroxenite in different proportion forms various primary melts.

Pyroxenite component within mantle peridotite could be formed from old oceanic crust (recycling component) or as a part of low continental crust, which was involved into melting process during Karoo–Maud plume upwelling. Melting of a pyroxenite-bearing peridotite produces a lot of primary melts of variable composition. Thus, for the most alkaline studied samples which represent a lower degree of melting, the variation in trace element ratios shows that the source of enrichment was the subducted component with low Th and Nb anomaly. But for other samples we have only little anomalies or there are no any anomalies at all and this indicates the process of mixing melts from pyroxenites and peridotite matrix was operated on different depths and melting degree.

The isotope compositions of Schirmacher Oasis dolerites demonstrate clearly radiogenic signatures – $^{87}Sr/^{86}Sr$: 0.7045–0.7047, $^{208}Pb/^{204}Pb$: 37.98–38.2; $^{207}Pb/^{204}Pb$: 15.45–15.52 and low $^{143}Nd/^{144}Nd$ values 0.51275–0.51255; and reflect the crust contamination process, which took place during plume upwelling and emplacement into the upper crust. The magmas of the initial stage of plume activity (region of western DML) appeared to be the most contaminated. Thus the study of plume-related magmatism within the Antarctica proves the distribution of Karoo–Maud plume, which is determined by the splitting of Gondwana continent and formation of the Indian Ocean, to the east – from Queen Maud Land towards Schirmacher Oasis. The time and spatial position of the dykes indicate the eastward spreading of the plume material from DML to the Schirmacher Oasis was within at least 10 Ma.

Acknowledgements Dmitry Gladkochub and Rajesh K. Srivastava are thanked for their positive comments that helped significantly to improve the paper. This work was financially supported by the Russian Foundation for Basic Research, project no. 09-05-00256.

References

Antonini P, Piccirillo EM, Petrini R et al (1999) Enriched mantle – Dupal signature in the genesis of the Jurassic tholeiites from Prince Albert Mountains (Victoria Land, Antarctica). Contrib Mineral Petrol 136: 1–19

Belyatsky BV, Prasolov EM, Sushchevskaya NM et al (2002) Specific features of the isotopic composition of Jurassic magmas in the Dronning Maud Land, Antarctica. Dokl Earth Sci 386: 855–858

Belyatsky BV, Sushchevskaya NM, Leichenkov GL et al (2006) Magmatism of the Karoo–Maud superplume in the Schirmacher Oasis, East Antarctica. Dokl Earth Sci 406: 128–131

Belyatsky BV, Rodionov N, Savva EV, Leitchenkov GL (2007) Zircons from mafic dykes as a tool for understanding of composition and structure of continental crust on the example of Mesozoic olivine dolerite dykes, Schirmacher Oasis, Antarctica. Geophys Res Abstr 9: 10509

Cox KG (1992) Karoo igneous activity and the early stages of the break-up of Gondwanaland. In: Storey BC, Alabaster T, Pankhurst RJ (eds) Magmatism and the Causes of Continental Break-up, Geol Soc London Spec Publ., Bath, UK, 68: 137–148

Duncan RA, Hooper PR, Rehacek J et al (1997) The timing and duration of the Karoo igneous event, Southern Gondwana. J Geophys Res 102: 18127–18138

Elliot DH, Fleming TH, Kyle PR, Foland KA (1999) Long-distance transport of magmas in the Jurassic Ferrar large igneous province, Antarctica. Earth Planet Sci Lett 167: 89–104

Hauri EH, Whitehead JA, Hart SR (1994) Fluid dynamic and geochemical aspects of entrainment in mantle plumes. J Geophys Res 99: 24275–24300

Hinz K, Krause W (1982) The continental margin of Queen Maud Land. In: Antarctica seismic sequences, structural elements and geological developments. Geol Jahrb 23: 17–41

Hoch M, Rehkamper M, Tobschall HJ (2001) Sr, Nd, Pb and O isotopes of minettes from Schirmacher Oasis, East Antarctica: A case of mantle metasomatism involving subducted continental material. J Petrol 42: 1387–1400

Hoch M, Tobschall HJ (1998) Minettes from Schirmacher Oasis, East Antarctica – indicators of an enriched mantle source. Antarctic Sci 10: 476–486

Jourdan F, Feraud G, Bertrand H et al (2004) The Karoo Triple Junction questioned: Evidence from Jurassic and Proterozoic 40Ar/39Ar ages and geochemistry of the Giant Okavango Dyke Swarm (Botswana). Earth Planet Sci Lett 222: 989–1006

Jourdan F, Feraud G, Bertrand H et al (2005) Karoo large igneous province: Brevity, origin, and relation to mass extinction questioned by new 40Ar/39Ar age data. Geology 33: 745–748

Jourdan F, Feraud G, Bertrand H et al (2006) Basement control on dyke distribution in large igneous provinces: Case study of the Karoo Triple Junction. Earth Planet Sci Lett 241: 307–322

Kamenetsky VS, Maas R (2002) Mantle-melt evolution (dynamic source) in the origin of a single MORB suite: Perspective from magnesian glasses of Macquarie Island. J Petrol 45: 1909–1922

Laiba AA, Vorob'ev DM, Gonzhurov NA, Tolyunas YuV(2002) Preliminary results of geological studies of the Schirmacher Oasis during the 47th Russian Antarctic Expedition. In: Proceedings of Scientific Conference on the Study and Protection of the Environment in Antarctica, St. Petersburg, Russia, 2002, AANII, St. Petersburg 64–66 [in Russian]

Leitchenkov GL, Masolov VN (1997) Tectonic and magmatic history of the Eastern Weddell Sea region. In: Ricci CA (ed) The Antarctic Region: Geologic Evolution and Processes. Terra Antarctica Publication, Siena: 461–466

Leitchenkov GL, Miller H, Zatsepin EN (1996) Structure and mesozoic evolution of the eastern Weddell Sea, Antarctica: History of early Gondwana break-up. In: Storey BC, King EC, Livermore RA (eds) Weddell Sea Tectonics and Gondwana Break-up, Geol Soc London Spec Publ., Bath, UK, 108: 175–190

Leitchenkov GL, Sushchevskaya NM, Belyatsky BV (2003) Geodynamics of the Atlantic and Indian Sectors of the South Ocean. Dokl Earth Sci 391: 675–678

Luttinen AV, Furnes H (2000) Flood basalts of Vestfjella: Jurassic magmatism across an Archean–Proterozoic lithospheric boundary in Dronning Maud Land, Antarctica. J Petrol 41: 1271–1305

Migdisova NA, Sushchevskaya NM, Luttinen AV, Mikhalskii EM (2004) Variations in the composition of clinopyroxene from the basalts of various geodynamic settings of the Antarctic region. Petrology 12: 176–194

Nimis P, Ulmer P (1998) Clinopyroxene geobarometry of magmatic rocks Part 1: An expanded structural geobarometer for anhydrous and hydrous basic and ultrabasic systems. Contrib Mineral Petrol 133: 122–135

Sambridge MS, Compston W (1994) Mixture modelling of multi-component data sets with application to ion-probe zircon ages. Earth Planet Sci Lett 128: 373–390

Sobolev AV, Hofmann AW, Kuzmin DV et al (2007) The amount of recycled crust in sources of mantle-derived melts. Science 316: 412–417

Sobolev AV, Hofmann AW, Sobolev SV, Nikogosian IK (2005) An olivine-free mantle source of Hawaiian shield basalts. Nature 434: 590–597

Stracke A, Bourdon B (2009) The importance of melt extraction for tracing mantle heterogeneity. Geochim Cosmochim Acta 73: 218–238

Sun SS, McDonough WF (1989) Chemical and isotopic systematics of oceanic basalts: Implications for mantle composition and processes. In: Saunders AD, Norry MJ (eds) Magmatism in the Ocean Basins, Geol Soc London Spec Publ., Bath, UK, 42: 313–345

Sushchevskaya NM, Belyatsky BV, Leichenkov GL, Laiba AA (2009) Evolution of the Karoo–Maud Mantle plume in Antarctica and its influence on the magmatism of the early stages of Indian Ocean opening. Geochem Int 47: 1–17

Sushchevskaya NM, Migdisova NA, Belyatsky BV, Peyve AA (2003) Genesis of enriched tholeiitic magmas in the western segment of the Southwest Indian Ridge, South Atlantic Ocean. Geochem Int 41: 1–20

Vuori SK, Luttinen AV (2003) The Jurassic gabbroic intrusions of Utpostane and Muren: Insights into Karoo-related plutonism in Dronning Maud Land, Antarctica. Antarctic Sci 15: 283–301

White R, McKenzie D (1989) Magmatism at rift zonees: The generating of volcanic continental margins and flood basalts. J Geophys Res 94: 7685–7729

Chapter 2
The Late Archaean Uauá Mafic Dyke Swarm, São Francisco Craton, Brazil, and Implications for Palaeoproteozoic Extrusion Tectonics and Orogen Reconstruction

Elson P. Oliveira

Introduction

Cratons and Precambrian orogenic belts are key geological sites to help understand how the continental crust grows and how the continents assemble and break up. Mafic dykes play an important role in these tectonic studies because they may form dyke swarms that emplace parallel to a continental margin, or form radiating swarms branching out from a common magmatic centre where continental breakup may ultimately take place. As such they are useful geological records for supercontinent reconstruction (e.g. Ernst and Buchan, 1997; Wingate and Gidding, 2000; Pesonen et al., 2003; Gladkochub et al., 2010; French and Heaman, 2010) and breakup (e.g. Corrêa Gomes and Oliveira, 2000; Shellnutt et al., 2004; Mason and Brewer, 2004; Kullerud et al., 2006). Some dyke swarms remain undeformed in the craton/continent interior as the failed arm of a rift and in this circunstance they may preserve magma flow indicators that help infer the potential locus of magma propagation with associated basalts and layered mafic-ultramafic complexes (e.g. Gibson et al., 1987; Maurice et al., 2009; Klausen et al., 2010).

Mafic dykes are a major component of the São Francisco Craton in Brazil (Sial et al., 1987; Corrêa Gomes and Oliveira, 2000) but their ages and tectonic significance are poorly constrained. Of particular interest are the Palaeoproterozoic (or older) dyke swarms in the northeastern part of the craton, which are related to the evolution of the Archaean-Palaeoproterozoic Itabuna-Salvador-Curaçá orogen. This orogen is the result of Palaeoproterozoic collision of three Archaean blocks/microcontinents (Barbosa and Sabaté, 2004), one of which – the Serrinha block – contains two main mafic dyke swarms that show field relations relevant for the tectonic reconstruction of orogenic belts. Therefore, the aim of this paper is to present new field relations and geochronology for one of these mafic dyke swarms (the Uauá mafic dyke swarm) and discuss its relevance for tectonic reconstruction of Palaeoproterozoic orogenic belts.

E.P. Oliveira (✉)
Department of Geology, Institute of Geosciences, University of Campinas, Campinas, Brazil
e-mail: elson@ige.unicamp.br

R.K. Srivastava (ed.), *Dyke Swarms: Keys for Geodynamic Interpretation*,
DOI 10.1007/978-3-642-12496-9_2, © Springer-Verlag Berlin Heidelberg 2011

Regional Geology

The main geological features of the São Francisco Craton are outlined in Teixeira and Figueiredo (1991) and Teixeira et al. (2000). In general, the São Francisco Craton consists of Archaean to Palaeoproterozoic high-grade (migmatite, granulite) gneisses and granite-greenstone supracrustal terranes overlain by Meso- to Neoproterozoic platform-type cover (Fig. 2.1).

The high-grade terranes, mostly exposed in Bahia state (Fig. 2.1), are separated into the Neoarchaean Jequié migmatite-granulite complex (Alibert and Barbosa, 1992; Barbosa and Sabaté, 2004), the Mesoarchaean to Palaeoproterozoic Serrinha block (Mello et al., 2006; Rios et al., 2009; Oliveira et al., 2002, 2010a), and the Neoarchaean Itabuna-Salvador-Curaçá orogen (Delgado et al., 2003, Oliveira et al., 2004a, 2010a; Barbosa et al., 2008).

Tectonic models proposed for the Itabuna-Salvador-Curaçá orogen involve basement reworking (Silva et al., 1997; Oliveira et al., 2000, 2004a, b), accretion of continental margin and oceanic arcs, and plutonic complexes to Archaean blocks (Teixeira and Figueiredo, 1991; Delgado et al., 2003; Oliveira et al., 2004b, 2010a)

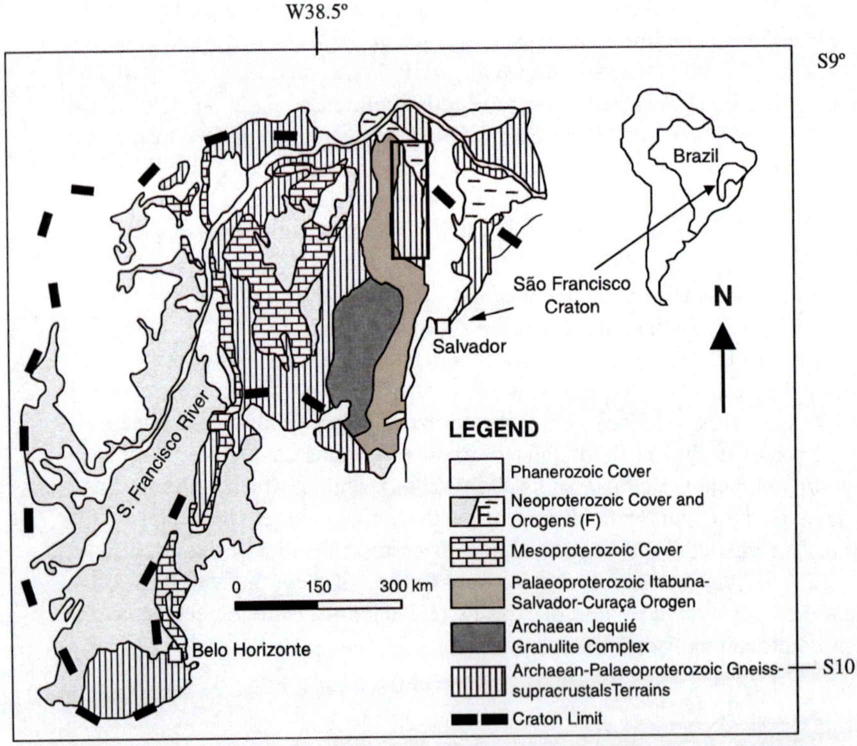

Fig. 2.1 The São Francisco Craton with the main tectonic units and location of the northern and southern segments of the Itabuna-Salvador-Curaçá orogen. *Box indicates* Fig. 2.2

and final continental collision between 2,084 and 2,039 Ma, i.e. the age interval for the peak of high-grade metamorphism (Silva et al., 1997; Mello et al., 2006; Oliveira et al., 2002, 2010a).

Barbosa and Sabaté (2004) proposed that four Archaean tectonic blocks, namely Gavião, Serrinha, Jequié and the Itabuna-Salvador-Curaçá orogen have collided during the Palaeoproterozoic orogeny.

Because the Uauá mafic dykes lie within the Serrinha block, geological information on this block is relevant.

Serrinha Block

The Serrinha block, or microcontinent (Fig. 2.2) comprises a basement complex of migmatites, banded gneisses, orthogneisses, mafic dykes and mafic-ultramafic complexes. The banded gneiss unit is the end product of deformation of migmatites

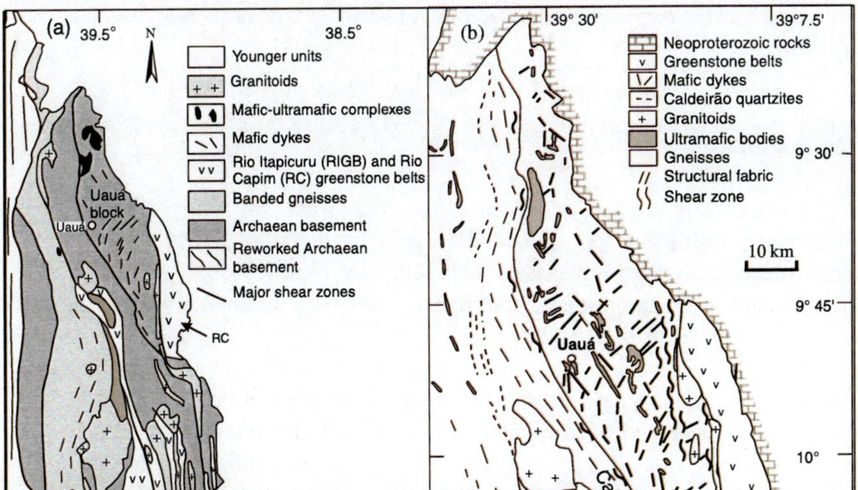

Fig. 2.2 Geological map of the Serrinha block (modified after Souza et al., 2003) with location of the Uauá block, Caldeirão belt, mafic dykes, and the Rio Capim (RC) and Rio Itapicuru greenstone belts (RIGB). Their corresponding towns indicate the Uauá and Santa Luz mafic dyke swarms

and mafic dykes; this mafic dyke swarm is called here the Santa Luz mafic dykes. The basement is overlain, or lies in tectonic contact with supracrustal sequences of the Rio Itapicuru and Rio Capim belts, and of the Caldeirão shear belt. Granites intrude all units. The Uauá, Jacurici and Retirolândia gneiss-migmatite complexes (Fig. 2.2) are Mesoarchaean domains (3,152–2,933 Ma, cf. Oliveira et al., 2010a), or minor blocks in the Serrinha block.

Rio Itapicuru greenstone belt (RIGB in Fig. 2.2a) is a low grade metamorphic supracrustal sequence approximately 180 km long and 30 km wide, divided by Kishida and Riccio (1980) into three lithostratigraphic units: (i) the basal mafic volcanic unit composed of massive and pillowed basaltic flows interlayered with chert, banded iron-formation, and carbonaceous shale; (ii) the intermediate to felsic volcanic unit with metadacites, metandesites and metapyroclastic rocks, and (iii) a metasedimentary pelitic-psammitic unit composed mainly of metapelites and minor chemical sedimentary rocks. The geochronological data indicate a Palaeoproterozoic evolution for this belt, between 2,170 and 2,080 Ma (Mello et al., 2006; Oliveira et al., 2010a).

The Rio Capim belt (RC in Fig. 2.2a) and Caldeirao belt lie in contact with the Uauá block and for this reason the three tectonic units will be described in detail in the next section.

The Uauá Block

The Uauá block (Fig. 2.2b) is bordered to the west by the Archaean-Palaeoproterozoic Caldeirão belt, and to the east it is in fault contact with the Palaeoproterozoic Rio Capim greenstone belt (Oliveira et al., 2010b), or unconformably overlain by Neoproterozoic continental shelf metasedimentary rocks of the Sergipano orogen (Bueno et al., 2009).

The basement of the Uauá block consists mostly of NW-trending banded gneisses of unknown age intruded by layered anorthosite, peridotite and diorite complexes, and tonalite-granodiorite bodies (Mascarenhas and Sá, 1982; Oliveira et al., 1999; Cordani et al., 1999). Most of these rocks have been metamorphosed under granulite facies conditions and later retrogressed to amphibolite grade. Mesoarchaean ages are widespread in the Uauá block. Paixão and Oliveira (1998) obtained a 3,161 $\pm$ 65 Ma whole-rock Pb-Pb isochron for anorthosites of the Lagoa da Vaca layered anorthosite complex and zircon Pb-evaporation age of 3,072 $\pm$ 20 Ma for orthogranulites, while Cordani et al. (1999) presented zircon U-Pb SHRIMP ages between 3.12 and 3.13 Ga for the Capim tonalite. Several other Archaean felsic igneous bodies occur in the Uauá block, some of which, i.e. the Uauá quarry enderbitic granulite and a gneissic granodiorite to the southeast of Uauá town had their zircon grains respectively dated at 2,933 $\pm$ 3 and 2,991 $\pm$ 22 Ma (Oliveira et al., 2002).

The Caldeirão belt (Fig. 2.2b) comprises a 10 km-wide sheared sequence of steeply dipping quartzites, sillimanite-cordierite-garnet gneiss, granodioritic orthogneisses, mafic rocks and migmatites, all metamorphosed under amphibolite facies conditions. SHRIMP U-Pb age dating indicates a 3,150 Ma for the orthogneisses (Oliveira et al., 2002). To the south, the shear belt dismembers into

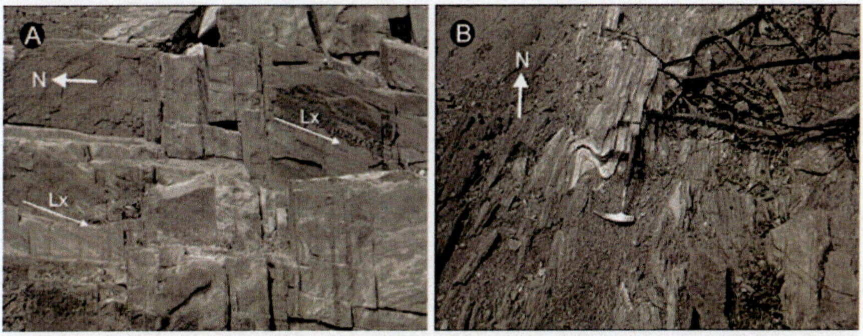

Fig. 2.3 Structural features in shear zone that limits the Uauá block and the Rio Capim greenstone belt. (**a**) Upright mylonite with low dipping stretching lineations (Lx). (**b**) Syn-milonitization asymmetrical fold indicating dextral kynematics

narrow sinistral strike-slip shear zones, one of which (the Main Shear Zone of Chauvet et al., 1997) continues for over 150 km across the Rio Itapicuru greenstone belt. The transition from this belt to the Uauá block is gradational and marked by refolding of older structures in the latter, granite and pegmatite intrusions, and development of shear zones. Syn-deformational titanite in mafic dykes (Oliveira et al., 2000) and detrital zircons in quartzites (Oliveira et al., 2002) constrain a maximum deposition age of 2,700 Ma for the metasedimentary rocks and the regional metamorphic age between 2,039 and 2,077 Ma.

The Rio Capim greenstone belt is a relatively small, 4-km wide, 20-km long, N-, to NW-trending belt of deformed and metamorphosed mafic to felsic volcanic rocks and associated pelitic rocks, intruded by a few plutons ranging in composition from gabbro/diorite to granite (Winge, 1981; Jardim de Sá et al., 1984; Oliveira et al., 2010b). From NW to SE the belt shows mineral assemblages indicative of increasing metamorphic grade from low amphibolite- to granulite facies (Jardim de Sá et al., 1984). Zircon U-Pb ages for the igneous rocks fall in the range 2,148–2,128 Ma (Oliveira et al., 2010b). The Rio Capim greenstone belt lies in contact with the Uauá block, to the west, along the 20–500 m-thick upright Galo do Ouro shear zone with N-S-trending, S-dipping sub-horizontal stretching lineations. S-C relations and syn-mylonitisation asymmetric folds indicate a dextral kynematics for this shear zone (Fig. 2.3). Sheared mafic dykes of the Uauá block close to the shear zone yielded a K-Ar age of ca. 2.0 Ga (Bastos Leal et al., 1994).

The Uauá Mafic Dykes

Dyke Types and Structural Features

Two mafic dyke swarms intrude the Archaean gneissic basement of the Uauá block, both bimodal, i.e. norite-tholeiite. One dyke swarm trends NW and is made up of metamorphosed tholeiite (now amphibolite) and norite-pyroxenite dykes, whereas

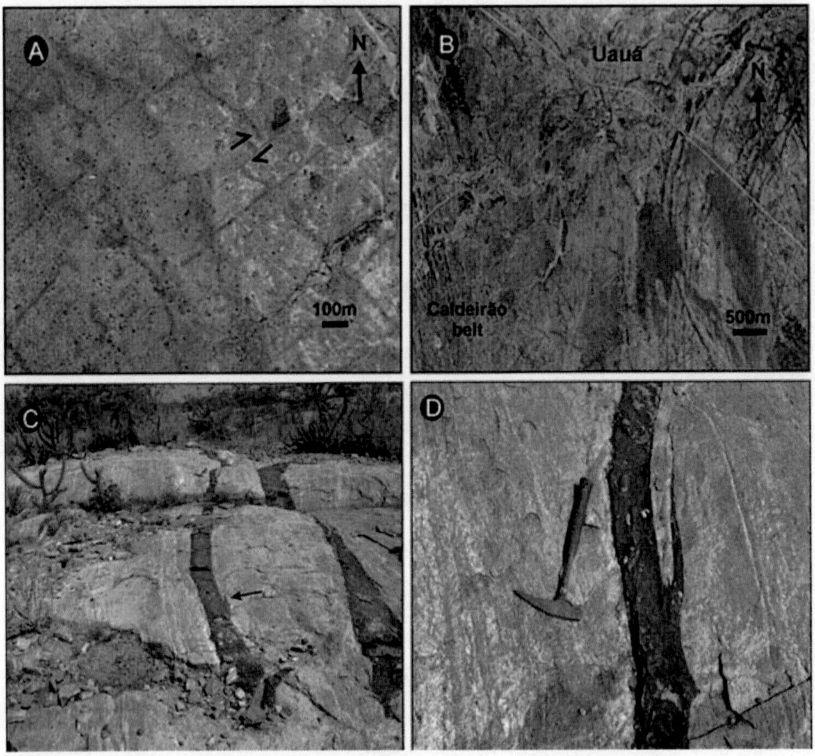

Fig. 2.4 Field aspects of the Uauá dykes. (**a**) Aerial photograph showing crosscutting relations between the younger NE-trending dykes and the older NW-trending dykes. (**b**) Aerial photograph showing the curvature of the younger NE-trending dykes as they approach the Caldeirão belt. (**c**) Disrupt dykes in the Caldeirão belt. (**d**) Branching of the main dyke in (**c**), indicating magma flow towards northeast

the other trends NE and is made up of non-metamorphic tholeiite and norite dykes. On the basis of crosscutting relations and drag folds (Fig. 2.4a) the NE-trending dykes are younger than the NW-trending dykes. In general, the dykes are less than a metre to 30 metres in width and several kilometres long. The norite dykes of the older dyke swarm are mostly coarse-grained cumulates, sometimes with well preserved plagioclase oikocrysts and bronzite-hypersthene chadacrysts, whereas those of the young swarm are fine-grained and show quench textures, such as radiating pyroxene needles. A few quench-textured norite dykes trend also to NW and in this situation distinction between younger and older norite dykes is more difficult. However, the quench-textured norites show Sm-Nd model ages younger than the cumulate norites but similar to the NE-trending tholeiite dykes, and this feature appears to be distinct characteristics of the two norite dykes.

The most interesting structural feature of the NE-trending dykes is the change in their direction. Apart from minor faults and occasional sheared margins, in the central part of the Uauá block the dykes are not significantly deformed and trend to northeast. However, towards the western boundary of the Uauá block, the dykes gradually change direction from NE-SW, through N-S to NW-SE until they become parallel to the Caldeirão shear belt and disappear (Fig. 2.4b). Some of these dykes can be traced within the transition to the Caldeirão belt where they are disrupted or boudined, and metamorphosed to amphibolites (Fig. 2.4c). In spite of deformation, in low-strain zones the dykes still preserve flow indicators, such as dyke branching (Fig. 2.4d), suggesting magma flow towards northeast.

Sm-Nd Geochronology

Earlier attempts to date the Uauá dykes by the U-Pb technique on zircon grains have not succeeded owing to zircon inheritance from the basement. For this reason, whole-rock Sm-Nd isotope data were obtained for both NE-trending tholeiite and norite dykes. The analyses were carried out at the Isotope Geochemistry Laboratory of the University of Kansas, USA, following the general procedures of Patchett and Ruiz (1987). Depleted mantle Nd model ages (T_{DM}) were calculated according to DePaolo (1988), and ISOPLOT (Ludwig, 1999) was used to calculate isochrons.

Four norite and six tholeiite dykes were analysed (Table 2.1), all of them with T_{DM} ages in the range 2.52–2.83 Ga. The older NW-trending dykes have T_{DM} > 3.0 Ga and were not included here. The Sm-Nd reference isochron obtained with the ten dyke samples yielded the age 2,589 ± 86 Ma (Fig. 2.5). Although this result can be revised in the future, as more robust age data are available (e.g. U-Pb in baddeleyite), the 2,589 Ma age is considered as the approximate emplacement timing of the NE-trending dykes.

Table 2.1 Sm-Nd isotope data for samples of the NE-trending Uauá mafic dykes

Sample	Dyke type	Sm (ppm)	Nd (ppm)	$^{147}Sm/^{144}Nd$	$^{143}Nd/^{144}Nd$	(± 1σ)	T_{DM} (Ga)
MP-07	Quench norite	4.823	21.501	0.13561	0.511727	14	2.59
MP-34	Quench norite	5.313	24.197	0.13274	0.511637	13	2.67
EO-119.1	Quench norite	4.022	16.162	0.15048	0.511957	12	2.62
EO-118.2	Tholeiite	1.887	5.952	0.19174	0.512684	33	2.52
MP-02	Tholeiite	1.821	5.587	0.19707	0.512766	14	–
UA96-3.2	Tholeiite	1.494	4.664	0.19367	0.512684	13	2.83
LR-34-F	Tholeiite	1.410	4.446	0.19174	0.512668	19	2.66
LR-38-F	Tholeiite	1.567	5.070	0.1868	0.512581	13	2.66
UA96-2.1	Tholeiite	1.881	6.020	0.1889	0.5126	14	2.79
EO-47.2	Quench norite NW-trending	4.382	17.610	0.15046	0.51197	16	2.61

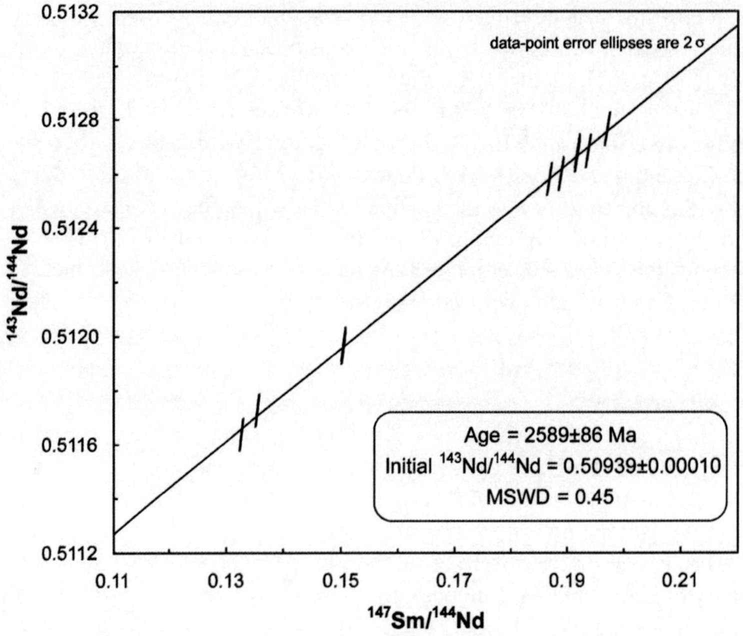

Fig. 2.5 Whole-rock Sm-Nd isochron for the young tholeiite-norite dykes

Discussion

The Uauá block is a fault-bound Mesoarchaean terrane sandwiched between two contrasting Palaeoproterozoic tectonic domains, namely the Caldeirão shear belt to the west and the Rio Capim greenstone belt to the east.

Two bimodal norite-tholeiite mafic dyke swarms occur in the Uauá block, the youngest of which trends northeast and is ca. 2,580 Ma old. This dyke swarm shows a curvature in dyke direction from NE, through NS to NW as the Caldeirão belt is approached. This feature is interpreted as large-scale drag folds, indicating a left-lateral displacement of the Uauá block in relation to the Caldeirão belt. On the eastern boundary of the Uauá block, kynematic indicators in shear zones show rigth-lateral displacement of the Uauá block relative to the Rio Capim belt. Overall, combination of the sinistral kinematics in the Caldeirão belt with the dextral kinematics in shear zones between the Uauá block and the Rio Capim belt indicates bulk movement of the Uauá block from present-day south to north.

Northwards extrusion of the Uauá block has occurred during the final oblique continent collision that formed the Itabuna-Salvador-Curaçá orogen. Time constraints for this collision is provided by SHRIMP U-Pb ages on metamorphic rims on zircon grains from quartzite of the Caldeirão belt and orthogneisses of the Itabuna-Salvador-Curaçá orogen (Silva et al., 1997; Oliveira et al., 2002, 2010a), showing high-grade metamorphic ages between 2,082 and 2,074 Ma. However, extrusion may have lasted longer on the basis of U-Pb age dating on metamorphic syn-deformation titanite grains in boudined mafic dyke of the Caldeirão belt, with the age 2,039 ± 2 Ma (Oliveira et al., 2000).

Terrane extrusion is a common feature in Phanerozoic colisional orogenic belts such as the Alpine-Himalayan belt (e.g. Tapponnier et al., 1982; Lacassin et al., 1997; Hollingsworth et al., 2008) and in Neoproterozoic belts (e.g. Goscombe et al., 2005; Van Schmus et al., 2008; Bueno et al., 2009), but its recognition in Archaean and Palaeoproterozoic orogens is poorly constrained. Indeed, in the Palaeoproterozoic Nagssugtoqidian orogen (West Greenland) Manatschal et al. (1998) describe deformation phases indicative of extrusion associated with exhumation of the belt, while in the Trans-Hudson orogen, which is considered as the Palaeoproterozoic analogue of modern accretionary orogens, Corrigan et al. (2009) just speculate that the Sask Craton extruded southward during impingement of the Superior Craton ca. 1.83 Ga. In contrast, the gradual change in direction (drag folds) of the NE-trending, Neoarchaean Uauá mafic dyke swarm during oblique continent collision is a first-order example of lateral extrusion in Palaeoproterozoic orogens.

How far has the Uauá block displaced from its original location is unknown but the suggested northeastwards magma flow direction deduced from dyke branching may be useful for regional correlations and metallogenesis. If the NE-trending Uauá dykes were the failed arm of an Archaean rift then the complement magma plumbing system, with basalts and the deep-seated mafic-ultramafic equivalents, should be located somewhere to the west-southwest (present-day coordinates) of the Uauá block. Figure 2.6 shows the younger Uauá mafic dyke swarm and potential correlative mafic-ultramafic bodies and greenstone belts in a pre-drift configuration of the São Francisco-Congo Craton.

The nearest mafic rocks that could fit the triple arm of a rift is the Santa Luz mafic dyke swarm which build up the mafic portion of the Serrinha block regional banded gneiss and trends mostly to north and northwest (Fig. 2.2). However, the recent zircon U-Pb SHRIMP age of 2,705 ± 5 Ma for this swarm (EP Oliveira, unpublished data) makes correlation with the NE-trending Uauá dykes unsustainable. Also, no ~2.58 Ga greenstone belt is known in this part of the São Francisco Craton or in the Congo Craton. In the former the Rio Capim and Rio Itapicuru greenstone belts are Palaeoproterozoic (2.17–2.08 Ga, cf. Oliveira et al., 2010a) whereas the Mundo Novo greenstone belt, to the west of the Itabuna-Salvador-Curaçá orogen is ca. 3.3 Ga old (Peucat et al., 2002). Mafic-ultramafic bodies are common in the Itabuna-Salvador-Curaçá (shown as black elipses in Fig. 2.6) and may correlate in age with the Uauá dykes. Indeed, the Caraiba and São José do Jacuípe noritic-gabbroic complexes are the only rocks dated so far and they are similar in age to the Uauá dykes (ca. 2,580 Ma, cf. Oliveira et al., 2010a). Farther south, several mafic-ultramafic complexes have been recognized, of which the Ni sulphide-bearing Mirabela gabbro and the Rio Piau layered anorthosite are potential correlatives to the Uauá dykes; however, their ages have not been constrained yet. In the Congo Craton the youngest greenstone belt (L in Fig. 2.6) was dated at 2,970–2,940 Ma (whole-rock Pb-Pb isochron and zircon Pb-Pb evaporation), whereas the Kinguele line of mafic-ultramafic complexes yield a 2,783 Ma Pb-Pb isocron (Feybesse et al.,

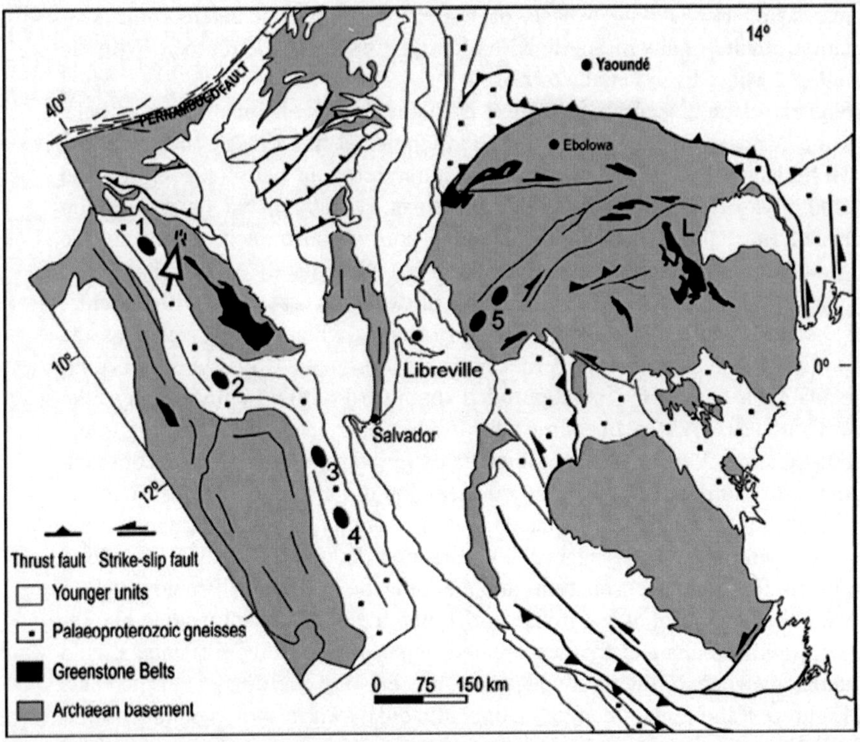

Fig. 2.6 Pre-drift geological correlation between Archaean-Palaeoproterozoic terranes of the São Francisco-Congo Craton showing the NE-trending Uauá mafic dykes in relation to the main greenstone belts and mafic-ultramafic bodies *(black elipsis)*. *Unfilled arrow* indicates the Uauá dykes and their probable magma-flow direction. Geology simplified from Barbosa (1994) and Feybesse et al. (1998). Mafic-ultramafic complexes: 1-Caraíba, 2-São José do Jacuípe, 3-Mirabela, 4-Rio Piau, 5-Kinguele

1998). Therefore, on the basis of these ages it appears that the São Francisco Craton is the most potential area to look for tectonic correlation with the Uauá dykes, an exercise that must be pursued in the future.

Acknowledgments EPO thanks WR. Van Schmus for laboratory facilities at Kansas University, USA. The Brazilian FAPESP (96/03821-8, 06/06222-1) and CNPq (300845/91-0, 302590/2008-0) provided research grants. The manuscript benefited from the positive comments of Peng Peng.

References

Alibert C, Barbosa J (1992) Ages U-Pb déterminés à la SHRIMP sur des zircons du complexe de Jequié, craton du São Francisco, Bahia, Brésil. Societé Géologique de France, éd., 14e RST, Toulouse, Abstract, p. 4

Barbosa JSF (1994) (coordinator). Geological Map of Bahia State, 1:1.,000,000, Superintendence of Geology and Mineral Resources. (in Portuguese)

Barbosa JSF, Peucat J-J, Martin H, Silva FCA, Moraes AM, Corrêa–Gomes LC, Sabaté P, Marinho MM, Fanning CM (2008) Petrogenesis of the late-orogenic Bravo granite and surrounding high-grade counry rocks in the Palaeoproterozoic orogen of Itabuna-Salvador-Curaçá block, Bahia, Brazil. Precamb Res 167: 35–52

Barbosa JSF, Sabaté P (2004) Archaean and Paleoproterozoic crust of the São Francisco Craton, Bahia, Brazil: Geodynamic features. Precamb Res 133: 1–27

Bastos Leal LR, Teixeira W, Piccirillo EM, Menezes Leal AB, Girardi VAV (1994) Geocronologia Rb/Sr e K/Ar do enxame de diques máficos de Uauá, Bahia (Brasil). Geochimica Brasiliensis 8: 99–114

Bueno JF, Oliveira EP, McNaughton N, Laux JH (2009) U–Pb dating of granites in the Neoproterozoic Sergipano Belt, NE-Brazil: Implications for the timing and duration of continental collision and extrusion tectonics in the Borborema Province. Gondwana Res 15: 86–97

Chauvet A, Silva FCA, Faure M, Guerrot C (1997) Structural evolution of the Paleoproterozoic Rio Itapicuru granite-greenstone belt (Bahia, Brazil): The role of synkinematic plutons in the regional tectonics. Precamb Res 84: 139–162

Cordani UG, Sato K, Nutman A (1999) Single zircon SHRIMP determination from Archean tonalitic rocks near Uauá, Bahia, Brazil. Proceedings II South American Symposium on Isotope Geology, 27–30

Corrêa Gomes LC, Oliveira EP (2000) Radiating 1.0 Ga mafic dyke swarms of Eastern Brazil and Western Africa: Evidence of post-assembly extension in the Rodinia Supercontinent? Gondwana Res 3: 325–332

Corrigan D, Pehrsson S, Wodicka N, De Kemp E (2009) The Palaeoproterozoic Trans-Hudson Orogen: A prototype of modern accretionary processes. In: Murphy JB, Keppie JD, Hynes AJ (eds) Ancient Orogens and Modern Analogues. Geol Soc Lond Spec Public 327: 457–479

Delgado IM, Souza JD, Silva LC, Silveira Filho NC, Santos RA, Pedreira AJ, Guimarães JT, Angelim LAA, Vasconcelos AM, Gomes IP, Lacerda Filho JV, Valente CR, Perrota MM, Heineck CA (2003) Geotectônica do Escudo Atlântico. In: Bizzi LA, Schobbenhaus C, Vidotti RM, Gonçalves JH (eds). Geologia, Tectônica e Recursos Minerais do Brasil. CPRM, 227–334

DePaolo DJ (1988) Neodymium Isotope Geochemistry: An Introduction. Springer, Berlin Heidelberg

Ernst RE, Buchan KL (1997) Giant radiating dyke swarms: Their use in identifying Pre-Mesozoic large igneous provinces and mantle plumes. In: Mahoney JJ, Coffin MF (eds) Large Igneous Provinces: Continental, Oceanic, and Planetary Flood Volcanism. AGU Geophys Monogr 100: 297–333

Feybesse JL, Johan V, Triboulet C, Guerrot C, Mayaga-Mikolo F, Bouchot V, Eko N'dong J (1998) The West Central African belt: A model of 2.5–2.0 Ga accretion and two-phase orogenic evolution. Precamb Res 87: 161–216

French JE, Heaman LM (2010) Precise U–Pb dating of Paleoproterozoic mafic dyk eswarms of the Dharwar craton, India: Implications for the existence of the Neoarchean supercraton Sclavia. Prec Res doi:10.1016/j.precamres.2010.05.003

Gibson IL, Sinha MN, Fahrig WF (1987) The geochemistry of the Mackenzie dyke swarm, Canada. In: Halls HC, Fahrig WF (eds). Mafic Dyke Swarms. Geol Assoc Can Spec Paper 34: 109–121

Gladkochub DP, Pisarevsky SA, Donskaya TV, Ernst RE, Wingate MTD, Söderlund U, Mazukabzov AM, Sklyarov EV, Hamilton MA, Haneeg JA (2010) Proterozoic mafic magmatism in Siberian craton: An overview and implications for paleocontinental reconstruction. Prec Res doi:10.1016/j.precamres.2010.02.023

Goscombe B, Gray D, Hand M (2005) Extrusional tectonics in the core of a transpressional Orogen; the Kaoko Belt, Namibia. J Petrol 46: 1203–1241

Hollingsworth J, Jackson J, Walker R, Nazari H (2008) Extrusion tectonics and subduction in the eastern South Caspian region since 10 Ma. Geology 36: 763–766

Jardim de Sá EFJ, Souza ZS, Fonseca VP, Legrand JM (1984) Relações entre "greenstone belts" e terrenos de alto grau: o caso da faixa Rio Capim, NE da Bahia. Proceedings XXXIII Congresso Brasileiro de Geologia, Rio de Janeiro, 2615–2629

Kishida A, Riccio L (1980) Chemostratigraphy of lava sequences from the Rio Itapicuru Greenstone Belt, Bahia, Brazil. Precamb Res 11: 161–178

Klausen MB, Söderlund U, Olsson JR, Ernst RE, Armoogam M, Mkhize SW, Petzer G (2010) Petrological discrimination among Precambrian dyke swarms: Eastern Kaapvaal craton (South Africa). Prec Res doi:10.1016/j.precamres.2010.01.013

Kullerud K, Skjerlie KP, Corfu F, Rosa JD (2006) The 2.40 Ga Ringvassøy mafic dykes, West Troms Basement Complex, Norway: The concluding act of early Palaeoproterozoic continental breakup. Precamb Res 150: 183–200

Lacassin R, Maluski H, Leloup PH, Tapponnier P, Hinthong C, Siribhakdi K, Chuaviroj S, Charoenravat A (1997) Tertiary diachronic extrusion and deformation of western Indochina: Structural and 40Ar/39Ar evidence from NW Thailand. J Geophys Res 102(B5): 10023–10037

Ludwig KR (1999) Isoplot/Ex – A geochronological toolkit for Microsoft Excel. Berkeley Geochronological Center Spec Public No. 1a

Manatschal G, Ulfbeck D, van Gool J (1998) Change from thrusting to syncollisional extension at a mid-crustal level: An example from the Palaeoproterozoic Nagssugtoqidian Orogen (West Greenland). Can J Earth Sci 35: 802–819

Mascarenhas JF, Sá JHS (1982) Geological and metallogenic patterns in the Archean and Early Proterozoic of Bahia State, Eastern Brazil. Rev Bras Geoc 12: 193–214

Mason AJ, Brewer TS (2004) Mafic dyke remnants in the Lewisian Complex of the outer Hebrides, NW scotland: A geochemical record of continental break-up and re-assembly. Precamb Res 133: 121–141

Maurice C, David J, O'Neil J, Francis D (2009) Age and tectonic implications of Paleoproterozoic mafic dyke swarms for the origin of 2.2 Ga enriched lithosphere beneath the Ungava Peninsula, Canada. Precamb Res 174: 163–180

Mello EF, Xavier RP, McNaughton NJ, Hagemann SG, Fletcher I, Snee L (2006) Age constraints on felsic intrusions, metamorphism and gold mineralization in the Paleoproterozoic Rio Itapicuru greenstone belt, NE Bahia State, Brazil. Mineral Deposita 40: 849–866

Oliveira EP, Carvalho MJ, McNaughton NJ (2004a) Evolução do segmento norte do Orógeno Itabuna-Salvador-Curaçá: Cronologia da acresção de arcos, colisão continental e escape de terrenos. Boletim Geologia USP – Série Científica 4, 41–53

Oliveira EP, Lafon J-M, Souza ZS (1999) Archaean-Proterozoic transition in the Uauá Block, NE São Francisco Craton, Brazil: U–Pb, Pb–Pb and Nd isotope constraints. First International Symposium on Tectonics of the Brazilian Geological Society, Lençóis-BA, 12–15 May. Abstract Volume, pp. 38–40

Oliveira EP, McNaughton NJ, Armstrong R (2010a) Mesoarchaean to Palaeoproterozoic Growth of the Northern Segment of the Itabuna-Salvador-Curaçá Orogen, São Francisco Craton, Brazil. In: Kusky T, Mingguo Z, Xiao W (eds) The Evolving Continents: Understanding Processes of Continental Growth. Geol Soc Lond Spec Pub 338: 263–286. doi: 10.1144/SP338.13

Oliveira EP, Mello EF, MacNaughton NJ (2002) Reconnaissance U-Pb geochronology of early Precambrian quartzites from the Caldeirão belt and their basement, NE São Francisco Craton, Bahia, Brazil: Implications for the early evolution of the Palaeoproterozoic Salvador-Curaçá Orogen. J South Amer Earth Sci 15: 284–298

Oliveira EP, Souza ZS, Corrêa Gomes LC (2000) U-Pb dating of deformed mafic dyke and host gneiss: Implications for understanding reworking processes on the western margin of the Archaean Uauá block, NE São Francisco Craton, Brazil. Rev Bras Geoc 30: 149–152

Oliveira EP, Souza ZS, McNaughton NJ, Lafon JM, Costa FG, Figueiredo AM (2010b) The Rio Capim volcanic-plutonic-sedimentary belt, São Francisco craton, Brazil: geological, geochemical and isotopic evidence for oceanic arc accretion during Palaeoproterozoic continental collision. Gondwana Res doi:10.1016/j.gr.2010.06.005

Oliveira EP, Windley BF, McNaughton NJ, Pimentel M, Fletcher IR (2004b) Contrasting copper and chromium metallogenic evolution of terranes in the Palaeoproterozoic Itabuna-Salvador-Curaçá Orogen, São Francisco Craton, Brazil: New zircon (SHRIMP) and Sm-Nd (model) ages and their significance for orogen-parallel escape tectonics. Precamb Res 128: 143–165

Paixão MAP, Oliveira EP (1998) The Lagoa da Vaca complex: An Archaean layered anorthosite body on the western edge of the Uauá Block, Bahia, Brazil. Rev Bras Geoc 28: 201–208

Patchett PJ, Ruiz J (1987) Nd isotopic ages of crust formation and metamorphism in the Precambrian of eastern and southern Mexico. Contrib Mineral Petrol 96: 523–528

Pesonen LJ, Elming SA, Mertanen S, Pisarevsky S, D'Agrella-Filho MS, Meert JG, Schmidt PW, Abrahamsen N, Bylund G (2003) Palaeomagnetic configuration of continents during the Proterozoic. Tectonophysics 375: 289–324

Peucat JJ, Mascarenhas JF, Barbosa JS, Souza FS, Marinho MM, Fanning CM, Leite CMM (2002) 3.3 Ga SHRIMP U-Pb zircon age of a felsic metavolcanic rock from the Mundo Novo greenstone belt in the São Francisco craton, Bahia (NE Brazil). J South Amer Earth Sci 15: 363–373

Rios DC, Davis DW, Conceição H, Davis WJ, Rosa MLS, Dickin AP (2009) Geologic evolution of the Serrinha nucleus granite–greenstone terrane (NE Bahia, Brazil) constrained by U–Pb single zircon geochronology. Precamb Res 170: 175–201

Shellnutt JG, Dostal J, Keppie JD (2004) Petrogenesis of the 723 Ma Coronation sills, Amundsen basin, Arctic Canada: Implications for the break-up of Rodinia. Precamb Res 129: 309–324

Sial AN, Oliveira EP, Choudhuri A (1987) Mafic dyke swarms of Brazil. In: Halls HC, Fahrig WF (eds) Mafic Dyke Swarms. Geol Assoc Can Spec Paper 34: 467–481

Silva LC, McNaughton NJ, Melo RC, Fletcher IR (1997) U–Pb SHRIMP ages in the Itabuna-Caríba TTG high-grade complex: The first window beyond the Paleoproterozoic overprinting of the eastern Jequié craton, NE Brazil. Proceedings 2nd International symposium on granites and associated mineralizations. Sociedade Brasileira de Geologia, Salvador-Bahia: pp. 282–283

Souza JD, Kosin M, Melo RC, Oliveira EP, Carvalho MJ, Leite CMM (2003) Guia de excursão – Geologia do segmento norte do orógeno Itabuna-Salvador-Curaçá. Revista Brasileira de Geociências 33(1-suplemento): 27–32

Tapponnier P, Peltzer G, Armijo R, Le Dain AY, Cobbold P (1982) Propagating extrusion tectonics in Asia: New insights from simple experiments with plasticine. Geology 10: 611–616

Teixeira W, Figueiredo MCH (1991) An outline of Early Proterozoic crustal evolution in the São Francisco craton, Brazil: A review. Precamb Res 53: 1–22

Teixeira W, Sabaté P, Barbosa J, Noce CM, Carneiro MA (2000) Archean and Paleoproterozoic Tectonic Evolution of the São Francisco Craton. In: Cordani UG, Milani EJ, Thomaz Filho A, Campos DA (eds) Tectonic Evolution of South America. 31st International Geological Congress, Rio de Janeiro: 101–137

Van Schmus WR, Oliveira EP, Silva Filho AF, Toteu SF, Penaye J, Guimarães IP (2008) Proterozoic Links Between the Borborema Province, NE Brazil, and the Central African Fold Belt. In: Pankhurst R, Brito Neves BB, Trouw R, de Wit M (eds) West Gondwana: Pre-Cenozoic Correlations Across the South Atlantic. Geol Soc Lond Spec Pub 294: 69–99

Wingate MTD, Gidding JW (2000) Age and palaeomagnetism of the Mundine Well dyke swarm, Western Australia: Implications for an Australia–Laurentia connection at 755 Ma. Precamb Res 100: 335–357

Winge M (1981) A Sequência Vulcano-sedimentar do Grupo Capim – Bahia: Caracterização geológica e modelo metalogenético. Master Dissertation, Departamento de Geociências, Universidade de Brasília: 120 pp

Chapter 3
An Integrated Study of Proterozoic Dykes, Dharwar Craton, Southern India

E.J. Piispa, A.V. Smirnov, L.J. Pesonen, M. Lingadevaru, K.S. Anantha Murthy, and T.C. Devaraju

Introduction

Prominent globally distributed mafic dyke swarms represent the feeder systems of large igneous provinces (LIPs) (e.g. Ernst and Buchan, 1997). While the geometry of a dyke swarm can be reliably determined by aerial/satellite imagery and aeromagnetic mapping, mafic dykes also often preserve pristine paleomagnetic and geochemical signatures (e.g. Halls, 2008). The integration of paleomagnetic and geochemical data with precise U-Pb geochronology has proven to be a powerful tool to study continental reconstructions, the morphology of the geomagnetic field and the source mantle properties (e.g. Bleeker, 2004; Buchan et al., 2007).

The Proterozoic mafic dyke swarms exposed in the Dharwar craton (Southern India) are of special interest because the craton has been a principal constituent of several ancient supercontinents (e.g. Rogers, 1996; Heaman, 2008). The paleomagnetism and geochemistry of Dharwar dykes have been extensively studied (e.g. Radhakrishna and Joseph, 1993; Dawson and Hargraves, 1994; Chatterjee and Bhattacharji, 2001; Halls et al., 2007; Radhakrishna et al., 2007; Devaraju et al., 2008; Jayananda et al., 2008; Pradhan et al., 2009) and a number of reliable age determinations have been reported (e.g. French, 2007; French et al., 2008; Halls et al., 2007). However, the number and emplacement age of the swarms is still a

E.J. Piispa (✉)
Department of Geological and Mining Engineering and Sciences, Michigan Technological University, Houghton, MI 49931, USA
e-mail: ejpiispa@mtu.edu

A.V. Smirnov
Department of Geological and Mining Engineering and Sciences, Michigan Technological University, Houghton, MI 49931, USA

L.J. Pesonen
Division of Geophysics and Astronomy, University of Helsinki, FI-00014 Helsinki, Finland

M. Lingadevaru and K.S. Anantha Murthy
Department of Applied Geology, Kuvempu University, Shankaraghatta 577451, Karnataka, India

T.C. Devaraju
Department of Studies in Geology, Karnatak University, Dharwad 580001, Karnataka, India

R.K. Srivastava (ed.), *Dyke Swarms: Keys for Geodynamic Interpretation*,
DOI 10.1007/978-3-642-12496-9_3,

subject of ongoing debate. In this communication, we report new results of our paleomagnetic and geochemical analyses of several Proterozoic Dharwar dykes and discuss their geodynamical implications.

Geology and Sampling

The basement of Dharwar craton consists of the 3.4–2.7 Ga Peninsular gneisses, the Sargur Group (>3.0 Ga) and Dharwar Supergroup (2.9–2.6 Ga) greenstone sequences, and 2.6–2.5 Ga granitoids (Friend and Nutman, 1991; Chadwick et al., 2000). The regional metamorphic grade changes gradually from amphibolite/greenschist facies in the north towards a high-grade granulite facies in the south (Radhakrishna and Joseph, 1993). The Archean basement is overlain by Paleoproterozoic sediments including the crescent-shape Cuddapah basin (<1.9 Ga) (Fig. 3.1).

Multiple cross-cutting mafic dyke swarms transect the Dharwar craton (Fig. 3.1), especially in its eastern part, trending in three dominant directions: NW-SE, E-W and NE-SW (e.g. Murthy, 1995). Their composition ranges from tholeiites or alkali-olivine basalts to metadolerites/metanorites (e.g. Rao and Pupper, 1996). Radiometric age data and observed cross-cutting relationships indicate that most of the Dharwar swarms were emplaced during the Proterozoic with the peak of emplacement activity in the Paleoproterozoic (e.g. French, 2007).

The oldest E-W trending swarm, representing the Bangalore LIP, was emplaced at ~2,366 Ma (e.g. Zachariah et al., 1995; French, 2007). Two slightly younger dyke swarms, the N-S trending Kandlamadugu swarm (2,213.6 ± 3.9 Ma) and NW-SE trending Somala dyke swarm (2,209.3 ± 2.8 Ma) have been identified south of Cuddapah (French, 2007). These events were followed by a proposed Mahhubnagar LIP (Ernst and Srivastava, 2008) at ~2,180 Ma (Pandey et al., 1997; French et al., 2004) represented by dykes trending from NW-SE to E-W with a ~60° fanning. We speculate that based on the fanning geometry and nearly coeval emplacement ages, the Somala and Kandlamadugu dykes could be a part of the Mahhubnagar LIP event. The next younger Southern Bastar–Cuddapah LIP event supposedly took place at ~1,890 Ma and is represented by E-W, ENE-WSW dykes in the Dharwar craton (Halls et al., 2007; Ernst and Srivastava, 2008).

The age of the NE-SW trending Karimnagar dyke swarm is poorly defined. While the dykes were correlated with the Gwalior traps based on paleomagnetism (e.g. Rao et al., 1990), this interpretation has been recently questioned (Halls et al., 2007) and the swarm has been suggested to be a part of the Bangalore LIP (Kumar et al., 2010). A 1,027 ± 13 Ma age was reported from an E-W trending great dyke of Penukonda, southwest of the Cuddapah basin (Pradhan et al., 2009). However, a much older age of 2,365.9 ± 1.5 Ma was determined from supposedly the same dyke by French (2007).

We collected 230 hand samples from 30 dykes (12 E-W, 5 NW-SE, 6 NE-SW and 5 N-S trending) (Fig. 3.1) and from their baked and unbaked host rocks when possible. The dykes vary in width from 0.5 to 100 m with the mode of 20–30 m. The samples were taken from quarry and roadside outcrops, river cuts, and small field

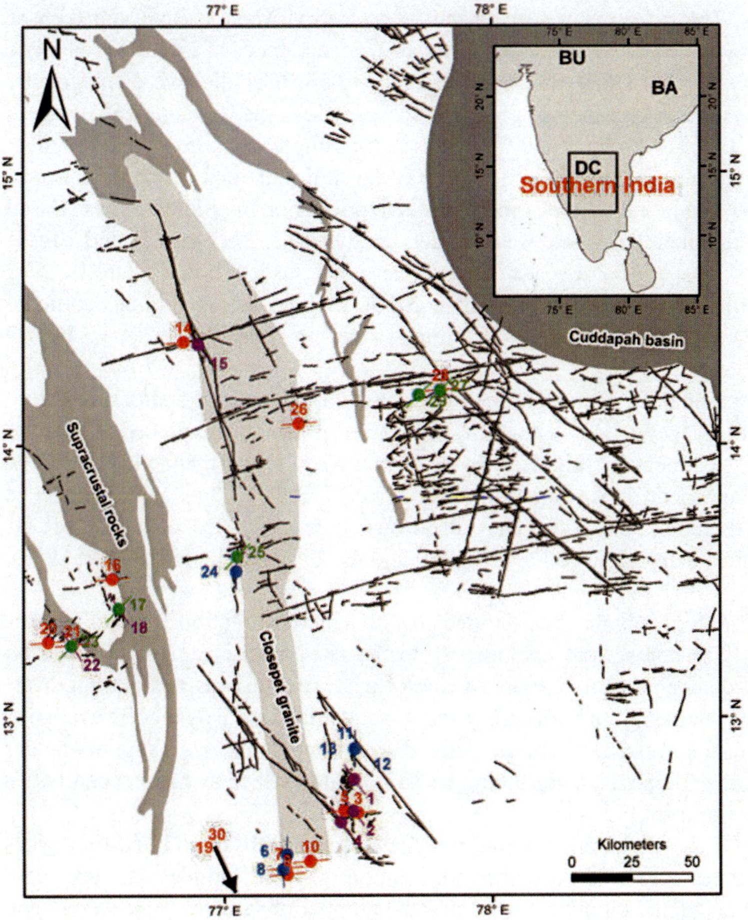

Fig. 3.1 Mafic dyke swarms of the Dharwar craton (after Halls et al., 2007). The numbered circles show our sampling sites. *Red*, *blue*, *green*, and *purple circles* correspond to the E-W, N-S, NE-SW, and NW-SE trending dykes, respectively. *Inset*: the *rectangle* shows the study area in the Indian peninsula and the letters denote the approximate centers of the Dharwar (DC), Bastar (BA), and Bundelkhand (BU) cratons

exposures. The paleomagnetic samples were oriented using both sun and magnetic compass. One sample per dyke from the chilled margin or from fresh interior (if chilled margin was not present) was also taken for geochemical analyses.

Results

Rock- and Paleomagnetism

The basic petrophysical properties: density, magnetic susceptibility, natural remanent magnetization (NRM), and Koenigsberg (Q) ratio were determined for all the samples. The dyke mean susceptibilities varied between 10 μSI and 0.25 SI with an average of 0.02 SI. Typical temperature dependences of low-field magnetic susceptibility showed Curie temperatures close to 585°C indicative of magnetite. Magnetic hysteresis parameters plot into the pseudosingle-domain (PSD) region of the Day plot (Day et al., 1977) close to the single domain – multi domain (SD-MD) mixture models by Dunlop (2002). Overall, our rock magnetic analyses suggest PSD magnetite as the principal magnetic carrier in studied dykes. Based on our rock magnetic analyses, we excluded ten dykes (1, 2, 11–13, 17 and 20–23) from further consideration because of their very low susceptibility values (<1,000 μSI) indicating strong alteration (Bleeker, 2010). We also excluded three additional dykes (4, 5, and 25) affected by lightning strikes as indicated by very large Q values (> 60).

Paleomagnetic directions were measured by detailed thermal and/or alternating field (AF) demagnetization. The demagnetization process was preceded by low-temperature treatment in liquid nitrogen to remove viscous magnetizations. The remanent magnetization was measured using a 2G DC SQUID magnetometer. The remanence directions and site mean directions were calculated using the principle component analysis (Kirschvink, 1980) and the Fisher statistics (Fisher, 1953), respectively.

We were able to obtain well-defined paleomagnetic directions from 17 dykes (Table 3.1). The cores were excluded from the site mean calculations based on randomized demagnetization behavior, and/or pervasive viscous or lightning overprints. When samples from both chilled margin and interior of a dyke were available, the former usually resulted in better defined directions (smaller maximum angular deviations). In general, the margin samples were characterized by higher coercivities than the interior ones.

A relatively large scatter of the within-site directions observed in several dykes (e.g. 6, 8 and 30) may be a result of minor rotations of some samples with respect to their in-situ positions. Finally we note that the number of samples for dykes 14 and 28 is insufficient for meaningful statistical analysis.

Most samples revealed two or three components of natural remanent magnetization (Table 3.1; Fig. 3.2). For the majority of dykes, the characteristic remanent magnetization (ChRM) was isolated within a narrow (10–25°C) unblocking temperature range above 550°C or within the high-coercivity part (>80 mT) of AF demagnetization spectra. The ChRM directions from E-W trending dykes plot into three groups (2–5 dykes per group) (Fig. 3.2a; Table 3.1). We calculated two group means for the steep up-component of Group 1 (Table 3.1), with and without the ChRM from the great dyke of Penukonda (dyke 28) which has contradicting age determinations. The means are not significantly different from each other. A significantly different ChRM direction is observed from dyke 9 and its baked host rock; however the direction is characterized by large uncertainties (Table 3.1). The ChRM directions for the N-S to NW-SE trending dykes are more scattered (Fig. 3.2b). Nevertheless, three of them are sufficiently close to calculate a mean direction (Table 3.1; group 4). The two NE-SW-trending dykes show similar directions of ChRM and intermediate coercivity components (Fig. 3.2c). A zigzag shape of the

Table 3.1 Paleomagnetic directions determined from the Dharwar dykes

Dyke #	Lat (°)	Long (°)	Trend	N/r	D (°)	I (°)	a95 (°)	k
3	12.65	77.50	E-W	17/1	142.8	–79.3	6.3	33.0
14	14.38	76.85	E-W	2/0	99.5	–69.1	69.1	15.3
16	13.51	76.58	E-W	16/2	76.5	–81.8	11.7	10.9
26	14.07	77.28	E-W	4/1	71.4	–66.6	11.9	61.1
28	14.20	77.81	E-W	4/0	98.2	–68.2	6.9	180.0
28[d, e]	14.20	77.81	–	3/0	24.2	–71.5	23.3	29.1
Group 1 mean with 28	**–**	**–**	**–**	**5**	**94.2**	**–74.1**	**9.4**	**67.8**
Group 1 mean w/o 28	**–**	**–**	**–**	**4**	**92.7**	**–75.6**	**12.2**	**57.7**
7	12.45	77.23	E-W	4/9	22.0	71.9	30.8	9.9
10	12.47	77.32	E-W	4/1	17.99	66.9	7.92	135.6
Group 2 mean	**–**	**–**	**–**	**2**	**19.8**	**69.4**	**11.4**	**481**
19	12.08	77.33	E-W	14/4	216.3	–4.7	11.6	12.7
19L[e]	12.08	77.33	E-W	17/2	355.9	–0.2	5.3	46.6
30	12.04	77.31	E-W	5/0	224.1	1.9	25.6	9.9
Group 3 mean	**–**	**–**	**–**	**2**	**220.2**	**–1.4**	**22.6**	**124**
8[e]	12.45	77.22	N-S	5/5	245.6	40.4	17.1	20.9
18[e]	13.40	76.61	NW-SE	6/2	82.0	–18.7	4.9	187.8
6	12.50	77.23	N-S	5/0	84.8	66.9	25.3	10.1
15	14.37	76.91	NW-SE	6/0	37.5	62.1	15.5	19.6
24	13.54	77.05	NNW-SSE	9/0	236.3	–47.5	13.7	15.0
Group 4 mean[f]	**–**	**–**	**–**	**3**	**57.8**	**60.1**	**23.1**	**29.6**
27[e]	14.20	77.81	NE-SW	12/2	68.3	–9.8	6.5	45.1
27[d, e]	14.20	77.81	–	2/0	50.8	–17.3	8.4	882.2
27 m	14.20	77.81	–	14/2	65.9	–10.9	6.5	38.9
29	14.18	77.73	NE-SW	4/0	66.5	–1.4	11.9	61.1
29[d, e]	14.18	77.73	–	2/0	65.0	14.0	35.6	51.2
Mean of 27 m and 29	**–**	**–**	**–**	**2**	**66.2**	**–6.1**	**21**	**144**
27i	14.20	77.81	NE-SW	4/0	303.3	37.6	26.8	12.7
29i	14.18	77.73	NE-SW	7/1	316.2	38.4	19.9	10.2
Mean	**–**	**–**	**–**	**2**	**309.7**	**38.1**	**22.2**	**128**
27b	14.20	77.81	NE-SW	3/0	56.5	–47.8	15.3	66.0
29c	14.18	77.73	NE-SW	2/0	25.0	68.5	6.3	1,595.2
9	12.42	77.23	E-W	6/1	300.4	18.3	21.6	10.6
9[d]	12.42	77.23	–	2/0	316.2	13.9	35.4	51.8
9m	**–**	**–**	**–**	**2**	**304.6**	**17.3**	**16.1**	**12.8**

Dyke #: the dyke number; Lat (Long): site latitude (longitude); N/r: the number of cores included/rejected for the dyke mean calculation; D (I): declination (inclination); a95: the radius of the 95% confidence circle; k: the precision parameter. Bold font shows between-site mean directions; m – the mean of a single dyke and baked contact; i – an intermediate coercivity component; L – a low coercivity component; b – probably baked by dyke 28; c – an anomalous ChRM direction observed from dyke 29.

[d]Baked contact.

[e]Not used to calculate the mean.

[f]The mean calculated using the antipodal direction from site 24.

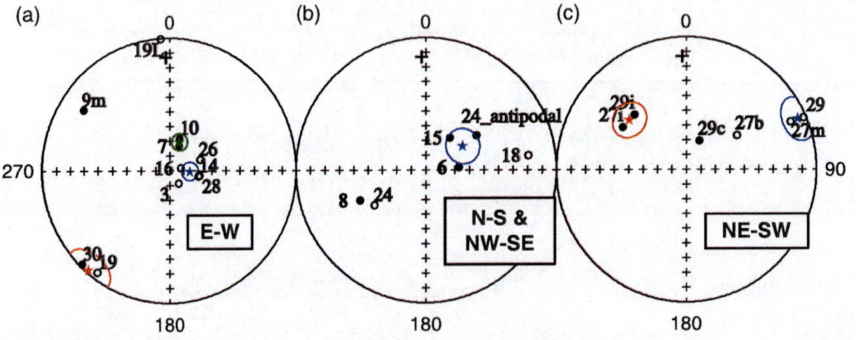

Fig. 3.2 Paleomagnetic directions from the Dharwar dykes. *Solid/open circles* show the *down/up* inclinations. The cross marks the present earth field (PEF) direction at the study area. (**a**) The E-W dykes. The *blue*, *green* and *red stars* show the mean ChRM directions for the dyke groups 1, 2 and 3, respectively (Table 3.1). (**b**) The N-S to NW-SE dykes. The star shows the between-site mean direction for the dyke group 4. (**c**) Two NE-SW dykes. The *blue* and *red stars* show the mean directions of ChRM and intermediate components, respectively. The other directions shown are described in the text and Table 3.1

Zijderveld plots for these dykes suggests a possibility of chemical alteration. We also note that one sample (29c) resulted in a strikingly different (steep down) ChRM direction. One sample from dyke 27 (27b), closest to dyke 28, yields a much steeper component probably reflecting a partial baking by the latter (Fig. 3.2c).

Geochemistry

Selected major and trace elements are given in Table 3.2. Loss on Ignition (LOI) is low for all samples consistent with a lack of alteration as suggested by review of thin sections. Samples in this study are sub-alkaline (Fig. 3.3). Their extrusive equivalents vary from basalt to andesite. When plotted in an AFM diagram (FeO^t – Na_2O+K_2O – MgO; after Irvine and Baragar, 1971) they follow a pattern of iron enrichment indicative of tholeiitic magmas.

Srivastava (2006) has demonstrated trace element differences between NW-SE dykes of different ages in Bastar craton (Fig. 3.4a). Dykes 27 and 29 of this study show the highest enrichment of trace elements similar to that found in the ~1,890 Ma dykes. The rest of the samples in this study have a trace element geochemical signature that falls intermediate between the two older age groups from the Bastar craton. Halls et al. (2007) reported fewer trace element data from the E-W trending Dharwar dykes (Fig. 3.4b). We note that the younger ~1,894 Ma dyke of Halls et al. (2007) closely follows the REE pattern of ~1,890 Ma dykes from the Bastar craton (Srivastava, 2006). This dyke did not yield a stable paleomagnetic direction. Halls et al. (2007) also reported other E-W dykes with poor paleomagnetic data or with a steep down direction. These dykes share similar major element characteristics with the ~1,890 Ma Bastar dykes (Fig. 3.3) and two of these dykes have trace element patterns similar to the ~1,890 Ma dykes of the Bastar craton.

Table 3.2 Relevant major and trace element geochemistry for selected Dharwar dykes. The samples were analyzed by Activation Labs. Ltd, Ancaster, Canada using lithium metaborate/tetraborate fusion followed by ICP for whole rock analyses and ICP/MS for trace elements (Actlabs package WRA+trace4litho). The detection limits for the trace elements in ppm: 20 for Ni, 2 for Rb and Sr, 4 for Zr, 0.1 ppm for Ce,Yb and Ta and 1 for Nb

Dyke # Dyke trend	29 NE-SW	27 NE-SW	15 NW-SE	24 N-S	28 E-W	16 E-W	14 E-W
wt%	–	–	–	–	–	–	–
SiO2	57.49	50.42	54.01	52.59	54.41	52.24	55.19
MgO	2.69	5.43	6.13	7.38	10.26	6.67	8.88
Na2O	3.23	2.59	2.88	1.90	1.62	2.15	2.33
K2O	2.31	0.83	1.02	0.31	0.59	0.58	1.25
TiO2	1.893	1.883	1.287	0.637	0.506	0.918	0.832
P2O5	0.27	0.26	0.12	0.08	0.08	0.14	0.16
LOI	0.56	0.65	0.32	0.37	0.66	0.30	0.17
Ppm	–	–	–	–	–	–	–
Ni	< 20	50	130	130	150	90	130
Rb	90	35	31	11	27	25	54
Sr	283	283	330	136	153	141	288
Zr	234	138	124	54	49	75	89
Nb	20	9	7	2	3	4	6
Ce	72.8	35.6	35.7	12.3	17.9	17.0	31.4
Yb	3.4	3.1	1.9	1.7	1.3	2.6	1.4
Ta	0.9	0.4	0.4	< 0.1	0.1	0.1	0.3

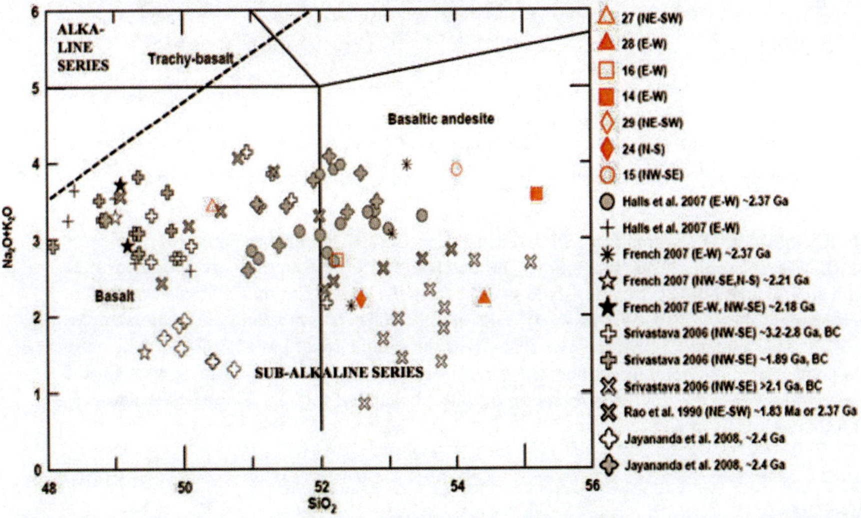

Fig. 3.3 A close-up of total alkali-silica (TAS) diagram (after Le Maitre, 2002). The *dotted line* divides subalkaline basalts from alkaline basalts (after Irvine and Baragar, 1971). The *color* symbols represent the dykes from this study. The other symbols represent the data from prior studies (see the legend). BC: the Bastar craton

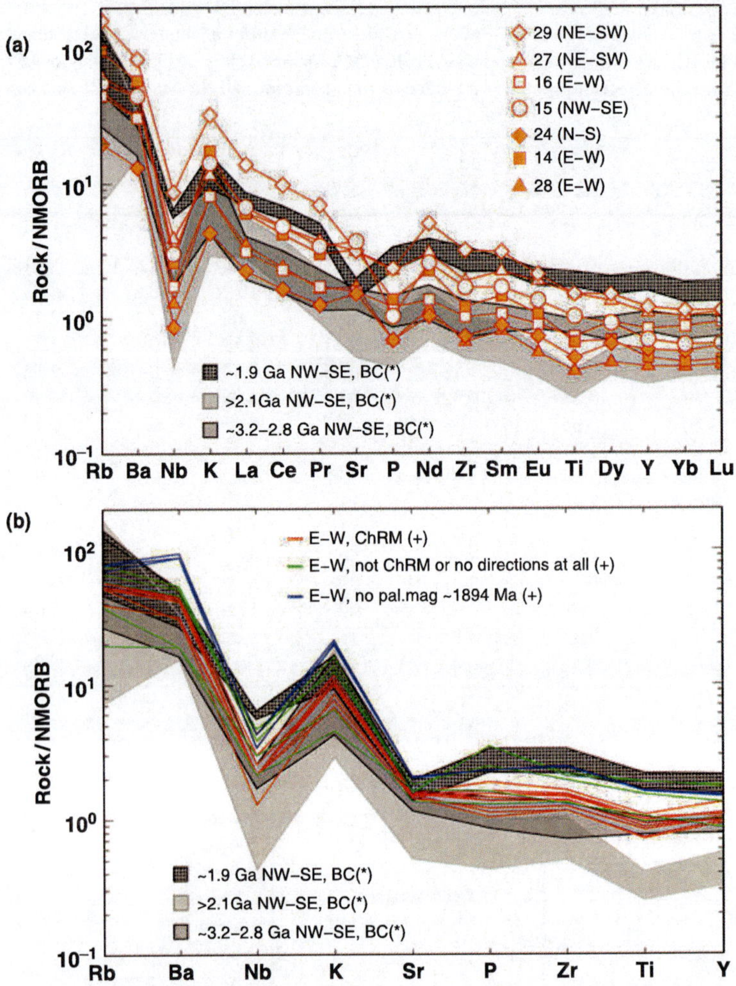

Fig. 3.4 Rock/NMORB (Normalized Mid Ocean Ridge Basalt) spider diagrams. (**a**) The *color* symbols show the samples from this study. The hatched and *grey* patterns correspond to three geochemically distinct NW-SE trending dyke swarms from the Bastar Craton (*: Srivastava, 2006). (**b**) Trace element data from E-W dykes (+: Halls et al., 2007). The *red lines* show the samples with a ~2,370 Ma steep up ChRM; the *blue lines* show the samples dated at ~1,894 Ma, which did not yield stable paleomagnetic directions; the *green lines* show the undated samples with ChRM directions opposite to the ~2,370 Ma direction or with no stable direction. The hatched and *grey* patterns are the same as in (A)

Discussion

Most of the magnetization components that we determined from the E-W trending dykes are consistent with the directions reported in prior works (e.g. Halls et al., 2007 and references therein). In particular, the steep up direction (Fig. 3.2a) was identified as the component A from the ~2,366 Ma E-W dykes and the steep down directions observed in dykes 7 and 10 as the component Ar by Halls et al. (2007). In addition, one of our E-W dykes (19), from the southern part of the study area (Fig. 3.1) revealed a shallow N direction denoted as the B-component by Halls et al. (2007). The B-component has been interpreted as an overprint acquired during the conversion from brown to black clouding (additional magnetite exsolution) during a later tectonic event. In accordance with the prior observations, we found abundant black-clouded feldspars in dyke 19. We note on passing that our petrographic analyses have confirmed the southward increase in plagioclase cloudiness reported by Halls et al. (2007) for the Dharwar craton.

The geochemical characteristics of the ~2,366 Ma E-W dykes with steep up directions (this work, French, 2007; Halls et al., 2007) do not significantly differ from one another (Fig. 3.3). The observed minor variation may reflect the differences in analytical techniques and/or temporal evolution of magma source over a protracted period of time. We note that the geochemistry of the ~3.2–2.8 Ga and >2.1 Ga NW-SE dyke swarms from the Bastar craton (Srivastava, 2006) is similar to that of the ~2,366 Ma E-W dykes (Figs. 3.3 and 3.4). Although no paleomagnetic data are available for the older Bastar craton dykes, this observation invites speculation about a similar/shared magma source for the oldest dyke swarms in the Dharwar and Bastar cratons. This in turn could indicate that these cratons were geographically contiguous already at ~2,366 Ma.

The steep down direction that we observed from the N-S to NW-SE dykes (Fig. 3.2b) is close to that observed from the E-W dykes 7 and 10. One possibility is that the latter dykes belong to the younger ~2,180 Ma Mahhubnagar LIP. However, in our opinion this possibility is not consistent with the observed dyke swarm geometries. Based on their trend and location, the dykes 7 and 10 are more likely to belong to the older Bangalore LIP in which case their paleomagnetic directions may represent the reversed A-component as discussed by Halls et al. (2007). If true, this would indicate a considerably slow plate movement between ~2,370 and ~2,200 Ma. We note however that the steep down-component was also observed in one N-trending ~800 Ma alkaline dyke (Dawson and Hargraves, 1994), so the possibility of an overprint cannot be ruled out. Additional paleomagnetic and rock magnetic investigations are needed to resolve this problem.

The shallow NE component observed from the NE-SW dykes 27 and 29 is similar to the directions reported from the same area (Kumar and Bhalla, 1983; Pradhan et al., 2009) and from the Karimnagar dykes (Subba Rao and Radhakrishna Murthy, 1985). The baked contact test for the dykes is inconclusive: while the baked host rocks give a similar direction to the dyke direction (Table 3.1), we could not obtain any consistent direction from the unbaked rocks. We note that the zigzag-shaped Zijderveld plots indicate the presence of a chemical remanent magnetization (CRM). The age of the CRM (the components 27i and 29i) should be limited between ~2,366 and ~1,030 Ma. The origin of this component remains unknown.

The cross-cutting relationship of our dyke 28 (supposedly the great dyke of Penukonda) with the dykes 27 and 29 is not clear from our field observations. However, Kumar and Bhalla (1983) suggested a younger age of the great dyke of

Penukonda based on its observed cutting through a NE-SW trending dyke; their interpretation is consistent with the ~1,027 Ma age reported by Pradhan et al. (2009). The younger age of this dyke is further supported by a steeper ChRM component from sample 27b that may indicate a partial baking by dyke 28 (such partial baking was also suggested by Kumar and Bhalla, 1983). On the other hand, the steep up ChRM component of dyke 28 is consistent with its much older (~2,370 Ma) age found by French (2007).

The geochemical characteristics of NE-SW trending Karimnagar dyke swarm (Rao et al., 1990) show a largely bimodal distribution overlapping with both the ~2,366 Ma (Halls et al., 2007; French, 2007) and ~1,890 Ma dyke swarms (Srivastava, 2006; Halls et al., 2007) (Fig. 3.3). The Karimnagar dykes also yield two different paleomagnetic directions: a NE shallow up and steep up (Rao et al., 1990; Subba Rao and Radhakrishna Murthy, 1985). These observations indicate that the Karimnagar swarm may actually represent two generations of dyke emplacement. Precise age determinations are crucial to test this hypothesis (Kumar et al., 2010).

Another possibility is that the NE shallow up direction seen in dykes 27 and 29 represents a later overprint due to the formation of the Gwalior traps and Cuddapah basin intrusions (related to the ~1,890 Ma LIP). However, such overprinting is not supported by the very low LOI values indicating negligible alteration. We feel that the apparent contradiction may be explained with the co-existence of three different nearly parallel E-W trending dyke swarms of different age (~2,370 Ma, ~1,890 and ~1,030 Ma). Such a possibility can be supported by visual inspection of satellite images and earlier studies indicating episodic magmatic intrusive activity along preferred directions (Chatterjee and Bhattacharji, 2001).

We note that the geochemical fingerprints of the ~1890 Ma dykes and the NE-SW dykes studied here are similar to those of the ~1890 Ma NW-SE trending BD2 dyke swarm from the Bastar craton (Srivastava, 2006; French et al., 2008; but see also Radhakrishna et al., 2007) (Figs. 3.3 and 3.4). Therefore both the BD2 dykes and the ~1,890 Ma E-W to NE-SW trending Dharwar dykes with shallow NE directions can be related to the formation of Cuddapah basin. A detailed paleomagnetic study of the Bastar craton dykes is needed to clarify this problem.

Conclusions

The results of our paleomagnetic and geochemical investigations are consistent with the results reported for the Dharwar mafic dykes by other authors. Our study hints that the Bastar and Dharwar cratons could have been amalgamated before ~2,370 Ma. In addition, an integrated analysis of the paleomagnetic and geochemical data performed here supports the existence of at least two different ~E-W-trending dyke swarms (~2,370 and ~1,890 Ma) in the Dharwar craton, both possibly reaching the Bastar craton. While geochemistry seems to differentiate these two dyke swarms fairly well, the primary origin of their magnetization is still debatable. Our results, currently based only on limited sampling, should therefore be considered preliminary. Additional paleomagnetic and geochemical analyses as well as precise age determinations are needed to decipher the complex geological history of the Dharwar craton. Our study, nevertheless, demonstrates the potential of combining paleomagnetism with geochemistry and accurate age data.

Acknowledgments The first author's (E.J.P.) field work in India in 2010 and geochemical analyses were supported by the Stan Dyl Endowed PhD Fellowship (MTU). We thank Ted Bornhorst for his invaluable help in acquiring and interpreting the geochemical data. We also thank Henry Halls for his constructive comments on this manuscript.

References

Bleeker W (2004) Taking the pulse of the Planet Earth: A proposal for a new multi-disciplinary flagship project in Canadian Solid Earth Sciences. Geosci Can 31: 179–190

Bleeker W (2010) The use of hand-held magnetic susceptibility meters in the field: an invaluable tool in regional studies of dyke swarms. In: Srivastava RK and Chalapathi Rao NV (eds) 6th International Dyke Conference – Abstracts, AS Prints and Stationers, Varanasi, India: p. 15

Buchan KL, Goutier J, Hamilton MA, Ernst RE, Matthews WA (2007) Paleomagnetism, U-Pb geochronology and geochemistry of Lac Esprit and other dyke swarms, James Bay area, Quebec, and implications for Paleoproterozoic deformation of the Superior province. Can J Earth Sci 44: 1–23

Chadwick B, Vasudev VN, Hegde GV (2000) The Dharwar craton, southern India, interpreted as the result of late Archean oblique convergence. Precamb Res 99: 91–111

Chatterjee N, Bhattacharji S (2001) Petrology, geochemistry and tectonic settings of the mafic dikes and sills associated with the evolution of the Proterozoic Cuddapah Basin of South India. Proc Indian Acad Sci Earth Planet Sci 110(4): 433–453

Dawson EM, Hargraves RB (1994) Paleomagnetism of Precambrian swarms in the Harohalli area, South of Bangalore, India. Precamb Res 69: 157–167

Day R, Fuller MD, Schmidt VA (1977) Hysteresis properties of titanomagnetites: Grain size and composition dependence. Phys Earth Planet Int 13: 260–267

Devaraju TC, Alapieti TT, Sudhakara TL, Kaukonen RJ (2008) Mafic Dyke Swarms of Volcanic Arc, Ocean Floor and N-MOR Basalt Affinity with Destructive Plate Margin Emplacement Features in the Northern Segment of Western Dharwar Craton. In: Srivastava RK, Sivaji C, Chalapathi Rao NV (eds) Indian Dykes: Geochemistry, Geophysics and Geochronology. Narosa Publishing House Pvt Ltd, New Delhi: 215–237

Dunlop DJ (2002) Theory and application of the day plot (Mrs/Ms versus Hcr/Hc) 1. Theoretical curves and tests using titanomagnetite data. J Geophys Res doi:10.1029/2001JB000486

Ernst RE, Buchan KL (1997) Giant radiating dyke swarms: Their use in identifying pre-Mesozoic large igneous provinces and mantle plumes. In: Mahoney J and Coffin M (eds) Large Igneous Provinces: Continental, Oceanic, and Planetary Volcanism, AGU Geophys Monogr Ser 100: 297–333

Ernst RE, Srivastava RK (2008) India's place in the Proterozoic world: Constraints from the Large Igneous Province (LIP) record. In: Srivastava RK, Sivaji C, Chalapathi Rao NV (eds) Indian Dykes: Geochemistry, Geophysics and Geochronology. Narosa Publishing House Pvt Ltd, New Delhi: 41–56

Fisher RA (1953) Dispersion on a sphere. Proc R Soc A 217: 295–305

French JE (2007) U-Pb dating of Paleoproterozoic mafic dyke swarms of the south Indian Shield: Implications for paleocontinental reconstructions and identifying ancient mantle plume events. PhD Thesis, University of Alberta, Edmonton, Alberta

French JE, Heaman LM, Chacko T, Rivard B (2004) Global mafic magmatism and continental break-up at 2.2 Ga: evidence from the Dharwar Craton, India. Geol Soc Am 36: 340 (Abstracts with programs)

French JE, Heaman LM, Chacko T, Srivastava RK (2008) 1891–1883 Ma Southern Bastar-Cuddapah mafic igneous events, India: A newly recognised large igneous province. Precamb Res 160: 308–322
Friend CRL, Nutman AP (1991) SHRIMP U–Pb geochronology of the Closepet granite and Peninsula gneisses, Karnataka, South India. J Geol Soc India 32: 357–368
Halls HC (2008) The Importance of Integrating paleomagnetic Studies of Proterozoic Dykes with U-Pb Geochronology and Geochemistry. In: Srivastava RK, Sivaji C, Chalapathi Rao NV (eds) Indian Dykes: Geochemistry, Geophysics and Geochronology. Narosa Publishing House Pvt Ltd, New Delhi: 1–22
Halls HC, Kumar A, Srinivasan R, Hamilton MA (2007) Paleomagnetism and U–Pb geochronology of easterly trending dykes in the Dharwar craton, India: Feldspar clouding, radiating dyke swarms and the position of India at 2.37 Ga. Precamb Res 155: 47–68
Heaman LM (2008) Precambrian large igneous provinces: An overview of geochronology,origin and impact on earth evolution. J Geol Soc India 72: 15–34
Irvine TN, Baragar WRA (1971) A guide to the chemical classification of the common volcanic rocks. Can J Earth Sci 8: 523–548
Jayananda M, Manesha N, Srivastava RK, Mahabaleswar B, Blais S (2008) Petrology and Geochemistry of Paleoproterozoic High-Magnesian Norite and Dolerite Dyke Swarms from the Halagur-Satnur areas, Eastern Dharwar Craton, Southern India. In: Srivastava RK, Sivaji C, Chalapathi Rao NV (eds) Indian Dykes: Geochemistry, Geophysics and Geochronology. Narosa Publishing House Pvt Ltd, New Delhi: 239–261
Kirschvink JL (1980) The least squares line and plane and the analysis of paleomagnetic data. Geophys J R Astron Soc 62: 699–718
Kumar A, Bhalla MS (1983) Paleomagnetics and igneous activity of the area adjoining the southwestern margin of the Cuddapah basin, India. Geophys J R Astron Soc 73: 27–37
Kumar A, Halls HC, Hamilton MA (2010) Paleomagnetism and U-Pb Geochronology of the Karimnagar Dykes Dharwar Craton, India. In: Srivastava RK and Chalapathi Rao NV (eds) 6th International Dyke Conference – Abstracts, AS Prints and Stationers, Varanasi, India: p. 6
Le Maitre RK (2002) Igneous Rocks; A classification and glossary of terms, 2nd edn. Cambridge University Press, Cambridge
Murthy NGK (1995) Proterozoic mafic dykes in southern peninsular India: A review. In: Devaraju TC (Ed) Mafic Dyke Swarms of Peninsular India, Geol Soc India Mem 33: 81–98
Pandey BK, Gupta JN, Sarma KJ, Sastry CA (1997) Sm-Nd, Pb-Pb and Rb-Sr geochronology and petrogenesis of the mafic dyke swarm of Mahbubnagar, South India: Implications for Paleoproterozoic crustal evolution of the Eastern Dharwar Craton. Precamb Res 84: 181–196
Pradhan VR, Meert JG, Pandit MK, Kamenov G, Gregory LC, Malone SJ (2009) India's changing place in global Proterozoic reconstructions: A review of geochronologic constraints and paleomagnetic poles from the Dharwar, Bundelkhand and Marwar cratons. J Geodyn doi:10.1016/j.jog.2009.11.008
Radhakrishna T, Joseph M (1993) Proterozoic paleomagnetism of the south Indian shield and tectonic constraints. Geol Soc India Mem 26: 321–336
Radhakrishna T, Krishnendu NR, Balasubramonian G (2007) Mafic dyke magmatism around the Cuddapah Basin: Age constraints, petrological characteristics and geochemical inference for a possible magma chamber on the southwestern margin of the basin. J Geol Soc India 70: 194–206
Rao VP, Pupper JH (1996) Geochemistry and petrogenesis and tectonic setting of Proterozoic mafic dykes swarms, East Dharwar Craton, India. J Geol Soc India 47: 165–174
Rao JM, Rao GVSP, Patil SK (1990) Geochemical and paleomagnetic studies on the middle Proterozoic Karimnagar mafic dyke swarm. In: Parker AJ, Rickwood PC, Tucker DH (eds) Mafic Dykes and Emplacement Mechanisms: pp. 373–382
Rogers JJW (1996) A history of continents in the past three billion years. J Geol 104: 91–107
Srivastava RK (2006) Precambrian mafic dyke swarms from the central Indian Bastar craton: Temporal evolution of the subcontinental mantle. In: Hanski E, Mertanen S, Ramo T and

Vuollo J (2006) Dyke Swarms – Time markers of Crustal Evolution, Taylor & Francis Group plc, London: 147–159
Subba Rao YV, Radhakrisha Murthy IV (1985) Paleomagnetism and ages of dolerite dykes in Karimnagar District, Andhra Pradesh, India. Geophys J R Astron Soc 82: 331–337
Zachariah JK, Hanson GN, Rajamani V (1995) Postcrystallization disturbance in the neodymium and lead isotope systems of metabasalts from the Ramagiri schist belt, southern India. Geochim Cosmochim Acta 59: 3189–3203

Chapter 4
Proterozoic Mafic Dykes from the Southern Margin of Cuddapah Basin, India: Part 1 – Geochemistry and Petrogenesis

M.R. Goutham, K.V. Subbarao, C.V.R.K. Prasad, J.N. Walsh, and V. Damodara Reddy

Introduction

The dyke swarms around the margin of the Cuddapah Basin are predominantly doleritic and gabbroic and few others are amphibolitic, peridotitic or syenitic. The petrochemical, geophysical and tectonic aspects of these dykes are discussed in detail by Murthy et al. (1987). The eastern margin of the basin is devoid of any dyke swarms and the basin itself is characterized by extensive arc-concentric dolerite-picrite sills, basaltic- andesitic and rhyolitic lava flows and tuffs of different episodes. At least three major generations of dyke emplacement were recognized between about 2,100 and 600 Ma with trends of NE, EW, WNW to NW, NNE to ENE and NS (Murthy et al., 1987). These are intrusives in to the Archaean basement and overlain by the sedimentary formations of the Cuddapah Basin. They are thus post-Archaean and pre-Cuddapah in age. Murthy (1995) considers the dense swarms of dykes along the margin of the Cuddapah Basin to have been formed in the distended crust by mantle plume activity around hot spots. Based on the similarity of the chemistry of the dykes and the rift volcanics Damodara Reddy et al. (1995) consider that the dykes might have intruded along a stress system that formed during early abortive attempts to break up the continental crust. A combine study of geochronology and palaeomagnetsm (Halls et al., 2007) from the Dharwar Craton

M.R. Goutham (✉)
Dr. K.S. Krishnan Geomagnetic Research Laboratory, Indian Institute of Geomagnetism, Allahabad 221 505, India
e-mail: gouthammr@rediffmail.com

K.V. Subbarao
University Centre for Earth and Space Sciences, University of Hyderabad, Hyderabad 500 046, India

C.V.R.K. Prasad
AMARARAMA, Vijayawada, India

J.N. Walsh
Department of Geology, Royal Holloway, University of London, Egham, TW 20 OEX, UK

V. Damodara Reddy
Department of Geology, Sri Venkateswara University, Tirupati, India

R.K. Srivastava (ed.), *Dyke Swarms: Keys for Geodynamic Interpretation*,
DOI 10.1007/978-3-642-12496-9_4, © Springer-Verlag Berlin Heidelberg 2011

suggests that a large radial dyke swarm was emplaced between 2,418 and 2,367 Ma which cuts across the structural grain of the craton. The pervasive mafic magmatism around 1.88 Ga in several Archaean cratons suggests a global scale mantle plume activity during this time (French et al., 2008). Based on the presence of ~1,890 Ma magmatic event in both the Bastar and Dharwar Cratons, it was hypothesized that a possible large radiating dyke swarm with the convergent point lying towards east of the Eastern Dharwar Craton (French et al., 2008; Ernst and Srivastava, 2008). Apart from sills, there is a possibility of a major layered intrusion of 1,890 Ma within the Cuddapah Basin (Radhakrishna et al., 2007). Various independent studies reveal that the magmatism and the sedimentation in the Cuddapah Basin range from ~1,900 to 600 Ma (Anand et al., 2003).

Ikramuddin and Stueber (1976); Pandey et al. (1997) and Radhakrishna et al. (1991), Radhakrishna and Joseph (1995), Radhakrishna (1999) and Devaraju (1995) have made detailed studies on the dykes of different regions in the southern India, which provides relevant information on various aspects of dyke magmatism and crustal evolutionary processes in the Precambrian. The present study attempts to bring out the chemical characteristics and the petrogenesis of the dykes.

Field Characteristics

The study area includes over 50 square km area in Chittoor district of Andhra Pradesh (covered by Survey of India toposheet Nos. 57O/3, 57O6, 57O/7 and 57O/11) at the south of the southern margin of the Cuddapah Basin. The dykes occur as intrusives in to the Archaean basement, trending mostly in E-W direction. The width of the dykes ranges from few centimeters to about 100 m with varying lengths. They are medium to coarse grained and are dark in colour suggesting their mafic nature. All the dykes are generally doleritic and tholeiitic in character and shows sharp contact with the country rock. Due to quarrying of some of the dykes for dimensional stone, fresh exposures are available for sampling.

Sampling and Analytical Technique

Twenty one samples from the eighteen dykes have been collected from five localities namely Chandragiri, Tirupati, Errasupalli, Surendranagar and Chittoor from the study area (Fig. 4.1). These were analyzed for major, trace and REE using ICP-MS technique at the Royal Holloway college, University of London, UK (after Thompson and Walsh, 1989). The data are presented in Table 4.1.

Results and Discussion

Major, trace and REE and elemental ratios for the dykes are presented in Table 4.1 along with CIPW norms. The normative mineralogy shows that five dykes (D1, D5, D9, D10 and D17) are quartz normative while all other dykes (D2, D3, D4, D4A, D6, D8, D12, D13, D13A and D16) are olivine normative. To bring out the

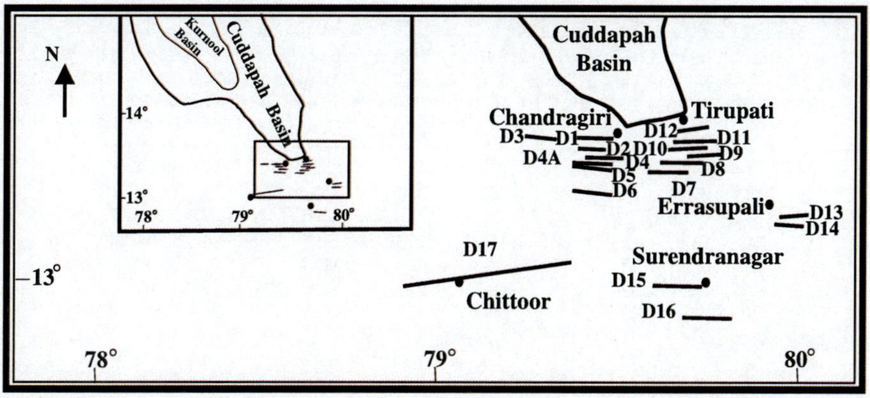

Fig. 4.1 Location map showing sampling sites of the dykes from the southern margin of the Cuddapah Basin. Numbers indicate the dyke numbers given in Table 4.1. Lengths of the dykes are not to scale

distribution of the normative minerals and also to characterize the rocks they are shown in the Q-Hy-Di-Ol-Ne diagram of Thompson, 1984 (Fig. 4.2). Most of the dykes fall in the olivine tholeiite field. The AFM diagram (Fig. 4.3) also shows that they are tholeiitic with a few less differentiated (analysis Nos. 9, 18, 20 and 22). Since few trace elements are considered immobile in the secondary geological processes, plots of the dykes are shown in YTC (Y= Y + Zr, T=$TiO_2 \times 100$, C = Cr) diagram of Davies et al., 1979 (Fig. 4.4). The classification of dykes as tholeiites is confirmed while a differentiation trend is exhibited from magnesian suite to more evolved tholeiites.

The Mg number [Mg# = molar 100 MgO/(MgO+FeO)] ranges from 64.1 to 33.8 (Table 4.1). Along the tholeiitic trend there is a gradual decrease of Mg number from the C corner towards the YT line (Fig. 4.4). This is probably an expression of the differentiation of the parental magma giving rise to the series of dykes. In the Chandragiri area (dyke Nos.D1 to D6, analysis Nos. 1–9) there are less differentiated to more differentiated dykes where as in the Tirupati area (dyke Nos.D8 to D12, analysis Nos. 10–15) all are well-differentiated dykes. Comparatively the dykes from Errasupalli (dyke Nos.D13 & D13A, analysis Nos. 16–18) Surendranagar (dyke No D16, analysis No 20) and Chittoor (dyke No. D17, analysis No.21–22) areas are all least differentiated.

Trace and REE Geochemistry

In order to make a proper comparison between the dykes under study and other well studied basic rocks of India and elsewhere, mantle normalized abundance patterns of selected Proterozoic dykes adjoining the Cuddapah Basin, upper and lower crusts, Chitradurga granite, parts of Deccan basalt province, Réunion and Hawaiian basalts (Fig. 4.5a–c) are considered. A close look at the literature of the Proterozoic dykes in and around the study area reveals an incomplete data set. In view of this, it was not

Table 4.1 Major, trace and REE analysis with elemental ratios and CIPW norm

Dyke no	D1	D2	D3	D4	D4A	D4A	D4A	Dyke No.	D5	D6	D8
Analysis no	1	2	3	4	5	6	7	Analysis No	8	9	10
Sample no	CGR5	BPT1	PLR3	MLP3	MLP - P1	MLP- P2	MLP - P3	Sample no	BRP1	MVP2	TML4
Group	IV	III	IV	II	I	I	I	Group	III	II	IV
SiO_2	50.5	49.91	49.92	48.13	45.27	44.25	41.3	SiO_2	50.77	48.64	49.29
TiO_2	1.4	1.17	0.99	3.29	2.74	2.43	2.99	TiO_2	1.13	1.01	1.71
Al_2O_3	12.5	13.42	14.29	13.17	12.24	13.45	14.76	Al_2O_3	13.93	9.8	12.7
Fe_2O_3	15.3	14.28	13.61	16.25	15.04	16.37	17.16	Fe_2O_3	14.3	12.52	16.93
MnO	0.2	0.19	0.19	0.2	0.19	0.2	0.21	MnO	0.2	0.18	0.22
MgO	5.2	5.96	6.28	4.58	6.59	6.89	6.7	MgO	5.15	11.65	4.82
CaO	9.9	8.9	9.08	6.69	6.49	5.47	4.73	CaO	9.86	8.58	8.66
Na_2O	2.2	2.21	2.3	3.71	1.38	1.84	2.12	Na_2O	2.44	1.89	1.96
K_2O	0.6	1.45	1.27	1.71	3.46	2.66	1.85	K_2O	0.64	0.4	1.3
P_2O_5	0.1	0.1	0.12	0.58	0.55	0.48	0.58	P_2O_5	0.13	0.1	0.21
LOI	0.89	1.5	1.2	0.5	3.08	3.48	4.29	LOI	0.89	1.8	1.09
Total	98.59	99.09	99.25	98.81	97.03	97.52	96.69	Total	99.44	96.57	98.89
Ba	151.0	150	267	586	1111	659	566	Ba	260	215	220
Rb	41.0	75	52	69	161	118	73	Rb	30	29	98
Sr	131.0	115	159	332	61	54	60	Sr	148	210	129
Cs	2.1	1.38	0.82	1.64	3.23	2.98	1.53	Cs	0.99	0.62	3.83
Zr	122.0	97	87	189	228	205	240	Zr	101	103	157
Hf	3.3	2.8	1.93	2.78	6.45	5.38	6.23	Hf	2.31	2.84	4.44
Nb	5.8	4.5	2.6	7.2	17.8	15.1	17.1	Nb	3.3	4.4	7.7
Ta	0.6	0.47	*0.2*	0.54	1.17	0.92	1.32	Ta	1.09	0.58	0.71
Y	37.5	32.7	25.4	19.5	58.7	46.3	56.4	Y	29.6	20	46.9
Cr	99.0	194	144	31	68	60	72	Cr	46	896	45
Ni	56.0	80	106	67	57	63	76	Ni	68	390	56
V	288.0	266	292	277	270	240	261	V	319	211	249
Cu	115.0	102	134	55	17	34	19	Cu	183	126	99
Zn	123.0	98	102	135	186	207	229	Zn	107	90	132
U	1.0	0.84	0.38	0.4	2.77	2.19	2.79	U	0.5	0.63	1.16
Th	4.5	3.69	1.84	1.84	4.11	3.55	3.92	Th	1.95	2.73	5.52
Li	23.0	18	15	35	56	68	85	Li	11	21	34
Sc	43.0	42	36	23	34	34	35	Sc	38	26	40
Tl	0.1	0.09	0.12	0.12	0.89	0.65	0.41	Tl	0.23	0.16	0.22
La	12.7	10.1	7.3	15.8	26.7	16.1	19.7	La	8.3	12.9	16
Ce	26.9	22.4	16.7	35.4	56.1	39.9	48.9	Ce	19.4	28	33.7
Pr	3.0	2.4	1.6	4.2	6.6	4.8	6.4	Pr	2.1	3.2	3.9
Nd	17.8	15.28	11.98	23.14	31.79	24.7	31.09	Nd	13.62	17.09	21.92
Sm	4.4	3.68	2.59	4.7	7.23	5.96	8.01	Sm	3.15	3.71	5.36
Eu	1.0	0.96	0.7	1.48	1.37	1.18	1.32	Eu	0.87	0.95	1.23
Gd	4.5	3.78	2.71	4.13	6.75	6.08	7.52	Gd	3.42	3.79	5.88
Dy	4.9	4.73	3.31	3	7.93	6.33	8.31	Dy	3.75	3.19	6.38
Ho	1.2	1.14	0.81	0.66	1.75	1.41	1.65	Ho	0.96	0.74	1.49
Er	3.1	2.97	2.17	1.59	4.79	3.87	4.29	Er	2.39	1.7	3.85
Yb	3.4	3.02	2.15	1.3	4.91	3.94	4.51	Yb	2.78	1.57	4.15
Lu	0.5	0.48	0.34	0.22	0.75	0.6	0.66	Lu	0.42	0.23	0.67
Mg #	37.9	43.1	45.5	33.9	44.4	43.2	41.2	Mg #	39.4	62.8	33.9
Th/Ta	7.2	7.9	9.2	3.4	3.5	3.9	3.0	Th/Ta	1.8	4.7	7.8
La/Yb	3.8	3.34	3.4	12.15	5.4	4.1	4.4	La/Yb	2.99	8.22	3.86
Nb/Th	1.3	1.22	1.4	3.9	4.3	4.3	4.4	Nb/Th	1.7	1.6	1.4
Nb/La	0.46	0.45	0.36	0.46	0.67	0.94	0.87	Nb/La	0.40	0.34	0.48
$[La/Sm]_N$	1.87	1.77	1.82	2.17	2.39	1.75	1.59	$[La/Sm]_N$	1.70	2.25	1.93
$[Gd/Yb]_N$	1.12	1.04	1.04	2.63	1.14	1.28	1.38	$[Gd/Yb]_N$	1.02	2.00	1.17

D9	D10	D10	D12	D12	Dyke No.	D13	D13A	D13A	D16	D17	D17
11	12	13	14	15	Analysis no	16	17	18	20	21	22
TML7	TML8	TML10	TML16	TML20	Sample no	PKP6	ERP - A	ERP - B	KCR1	35D3	35D4
IV	IV	III	III	III	Group	IV	II	II	II	II	II
49.21	49.56	50.74	48.96	49.52	SiO_2	49.55	49.37	48.28	53.42	53.12	53.41
1.66	1.3	1.27	1.94	1.37	TiO_2	1.15	1.09	0.99	1.07	1.06	1.14
12.66	13.59	12.72	13.46	13.26	Al_2O_3	13.46	9.96	9.57	9.61	9.53	10.39
16.27	14.96	15.45	16.39	15.77	Fe_2O_3	15.16	12.35	12.44	13.1	12.67	12.97
0.22	0.19	0.19	0.2	0.19	MnO	0.2	0.18	0.18	0.16	0.16	0.16
4.85	5.24	5.01	4.63	4.86	MgO	6.22	10.67	12.17	8.21	7.69	7.32
9.45	7.86	7.67	8.06	8.56	CaO	10.46	8.54	8.64	8.11	7.54	7.36
2.09	2.42	2.53	2.51	2.95	Na_2O	2.02	2.0	1.78	3.04	3.05	3.45
0.32	1.54	0.83	1.38	0.71	K_2O	0.47	1.01	1.12	1.29	1.38	1.41
0.21	0.24	0.16	0.23	0.16	P_2O_5	0.1	0.11	0.09	0.18	0.17	0.19
1.8	1.4	1.5	2.2	1.6	LOI	0.7	1.09	1.49	0.39	0.3	1.1
98.74	98.30	98.07	99.96	98.95	Total	99.49	96.37	96.75	98.58	96.67	98.9
119	351	325	496	231	Ba	132	239	195	605	530	582
16	65	33	73	28	Rb	28	46	48	41	43	39
172	214	151	185	201	Sr	110	223	214	425	442	470
0.5	4.31	1.05	3.07	2.57	Cs	2.92	1.31	1.73	1.42	1.28	1.24
156	118	123	131	114	Zr	91	103	96	107	103	113
4.12	2.91	3.7	3.77	2.83	Hf	2.46	2.99	2.54	2.87	2.7	2.67
7.3	4.8	5.3	7.3	4.2	Nb	3.9	4.8	3.7	6.7	6.1	6.3
0.72	0.52	0.62	0.79	0.5	Ta	0.5	0.3	0.2	0.4	0.66	0.68
46.8	29.8	39.8	40.9	33.4	Y	29.3	21.5	18.5	18.5	17.2	17.3
54	72	49	66	43	Cr	191	903	1059	468	403	415
60	67	61	66	55	Ni	86	353	396	217	202	185
243	235	335	459	355	V	265	220	207	210	199	203
93	60	145	152	177	Cu	104	143	136	177	171	176
114	116	122	123	110	Zn	109	95	88	97	96	103
1.16	0.38	0.75	0.69	0.67	U	0.57	0.7	0.58	0.86	0.85	0.81
5.51	2.58	3.07	2.79	2.5	Th	2.88	3.04	2.57	3.55	3.26	3.38
23	33	32	20	27	Li	16	25	27	13	12	13
40	34	38	34	35	Sc	43	26	26	21	20	19
0.08	0.31	0.29	0.32	0.2	Tl	0.08	0.2	0.38	0.24	0.18	0.12
15.5	17.1	13	15.1	10.3	La	8.6	13.9	12	17.3	15.9	16.5
33.3	34.2	27.3	34.3	23.3	Ce	18.7	29.6	26.2	37.1	34.8	35.1
3.9	3.8	3.4	4.0	2.6	Pr	1.9	3.3	2.9	4.3	3.9	3.9
21.44	22.03	20.01	21.77	15.29	Nd	13.77	18.92	16.84	21.95	20.81	20.41
5.21	4.22	4.6	5.44	3.5	Sm	3.23	4.11	3.53	4.3	3.91	3.88
1.2	1.25	1.12	1.42	1.01	Eu	0.89	1.15	0.92	1.1	1.0	0.99
5.54	4	4.74	5.21	3.81	Gd	3.44	3.99	3.29	3.81	3.33	3.4
6.45	3.91	5.2	6.02	4.04	Dy	4	3.56	2.89	3.04	2.56	2.57
1.45	0.97	1.24	1.33	1.03	Ho	0.94	0.79	0.69	0.62	0.61	0.59
3.88	2.49	3.36	3.52	2.62	Er	2.56	1.81	1.61	1.57	1.42	1.4
4.24	2.48	3.65	3.91	2.71	Yb	2.61	1.61	1.33	1.38	1.19	1.27
0.64	0.42	0.58	0.59	0.44	Lu	0.43	0.24	0.21	0.18	0.17	0.19
34.9	38.8	36.9	33.8	35.8	Mg #	42.5	61.2	64.1	53.5	52.7	50.9
7.7	5.0	5.0	3.5	5.0	Th/Ta	5.8	10.1	12.9	8.9	4.9	5.0
3.66	6.9	3.56	3.9	3.8	La/Yb	3.3	8.6	9.0	12.5	13.4	13.0
1.3	1.9	1.7	2.6	1.7	Nb/Th	1.4	1.6	1.4	1.9	1.9	1.9
0.47	0.28	0.41	0.48	0.41	Nb/La	0.45	0.35	0.31	0.39	0.38	0.38
1.92	2.62	1.83	1.79	1.90	$[La/Sm]_N$	1.72	2.19	2.20	2.60	2.63	2.75
1.08	1.33	1.07	1.1	1.16	$[Gd/Yb]_N$	1.09	2.05	2.05	2.28	2.31	2.21

Table 4.1 (continued)

Dyke no	D1	D2	D3	D4	D4A	D4A	D4A	Dyke No.	D5	D6	D8
Analysis no	1	2	3	4	5	6	7	Analysis No	8	9	10
Sample no	CGR5	BPT1	PLR3	MLP3	MLP - P1	MLP- P2	MLP - P3	Sample no	BRP1	MVP2	TML4
Group	IV	III	IV	II	I	I	I	Group	III	II	IV
CIPW norm								CIPW norm			
Q	1.86	–	–	–	–	–	–	Q	0.2	–	–
Or	3.29	8.59	7.47	10	21.23	16.29	11.53	Or	3.76	2.47	7.65
Ab	18.35	18.85	19.53	31.28	12.17	16.15	1.92	Ab	20.63	16.57	16.66
An	22.7	22.52	24.93	14.16	17.71	21.43	20.8	An	24.98	17.7	21.98
Ne	–	–	–	–	–	–	–	Ne	–	–	–
Di	21.56	17.8	16.25	12.8	10.13	2.93	–	Di	19.34	21.02	16.7
Hy	26.25	21.16	19.5	1.8	12.84	11.19	11.02	Hy	25.73	28.96	29.59
Ol	–	5.58	7.29	18.89	15.71	22.5	24.54	Ol	–	8.26	0.31
Mt	3.05	2.95	2.8	3.41	3.34	3.44	3.46	Mt	2.86	2.76	3.38
Ilm	2.61	2.24	1.88	6.24	5.42	4.81	6.01	Ilm	2.15	2	3.25
Ap	0.28	0.22	0.26	1.27	1.25	1.1	1.33	Ap	0.28	0.24	0.45
Mg#	37.9	43	45.5	33.9	44.3	43.2	41.2	Mg#	39.4	62.8	33.9

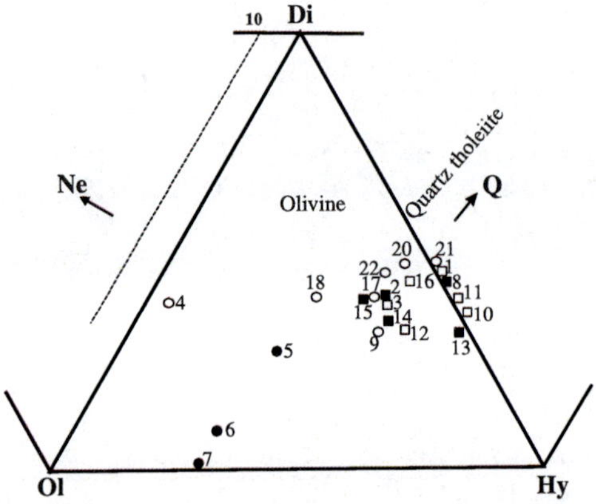

Fig. 4.2 Normative Q-Hy-Dy-Ol-Ne diagram (After Thompson, 1984). Numbers indicate the analysis number given in Table 4.1; Legend: *Solid circles* – Group I, *open circles* – Group II, *solid square* – Group III, *open square* – Group IV

found possible to make a proper comparison between the published analyses and the analytical data of the rocks under study. However, an attempt has been made with the available information and by constructing rock-mantle normalized spidergrams. In both the groups (published analysis – Chatterjee and Bhattacharji, 2001 and present analysis) concentrations of Rb and K are high and show depletion in Nb, P and Sr. At first approximation there appears to be some similarities. But it could be improper to infer any close link between the two groups mentioned above until a proper data set is available for the dykes studied by earlier workers.

Based on a critical study of the geochemistry of these rocks, four chemically distinct groups (I, II, III and IV) have been made. The chemical contrasts in all the four groups are shown through selected elemental ratios in Table 4.2 and also through chondrite normalized REE abundance patterns (Fig. 4.6 a–d). Of course, as in any classification there are overlaps of results.

Rock – mantle normalized abundance spidergrams for one representative sample from each group is included in the following section.

D9	D10	D10	D12	D12	Dyke No.	D13	D13A	D13A	D16	D17	D17
11	12	13	14	15	Analysis no	16	17	18	20	21	22
TML7	TML8	TML10	TML16	TML20	Sample no	PKP6	ERP - A	ERP - B	KCR1	35D3	35D4
IV	IV	III	III	III	Group	IV	II	II	II	II	II
					CIPW norm						
1.91	–	1.76	–	–	Q	–	–	–	–	0.15	–
1.88	9.17	4.94	8.17	4.23	Or	2.76	6.12	6.82	7.59	8.29	8.35
17.92	20.71	21.73	21.22	25.11	Ab	16.91	17.05	15.56	25.7	26.29	29.34
24.42	21.9	21.19	21.42	20.95	An	26.1	15.67	15.26	8.81	8.39	8.74
–	–	–	–	–	Ne	–	–	–	–	–	–
18.26	13.48	13.83	14.53	17.6	Di	20.82	22.34	23.16	25.1	23.71	22.19
28.74	23.41	30.53	20.6	18.69	Hy	24.9	24.61	18.33	24.57	27.71	21.44
–	5.1	–	6.48	7.2	Ol	3.14	8.55	15.87	2.77	–	4.28
3.15	3.08	3.14	3.27	3.18	Mt	2.95	2.77	2.8	2.96	2.93	2.96
3.2	2.51	2.45	3.71	2.63	Ilm	2.17	2.13	1.94	2.04	2.05	2.19
0.47	0.54	0.37	0.5	0.34	Ap	0.22	0.26	0.19	0.39	0.39	0.41
34.9	38.8	36.9	33.8	35.7	Mg#	42.5	61.2	64.1	53.5	52.7	50.9

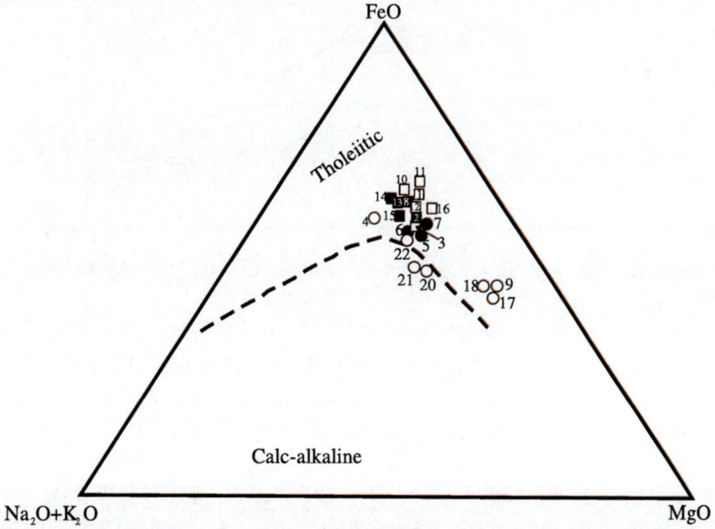

Fig. 4.3 AFM diagram showing Calc-Alkaline & tholeiitic trends (After Irvine and Baragar, 1971). Numbers indicate the analysis number given in Table 4.1. For legend see Fig. 4.2

Group I

Chemically sample MLP – P1 of dyke D4A with high TiO_2 content (2.74%) and Nb/La (0.67 as compared to the chondrite value of about 1.0) and Nb/Th (4.3 as compared to typical intraplate alkali basalts, 10–15) looks like post-collision al basalt (see Sun and McDonough, 1989), which has combined features of intraplate and subduction related basalts. Such features could also be imposed by strong crustal contamination. High Li (56–85 ppm), K_2O (2.66–3.46%) and Rb (118–161 ppm) may not be magmatic but may only indicate the role of secondary processes or involvement of sediment, because if this were a primary feature, one

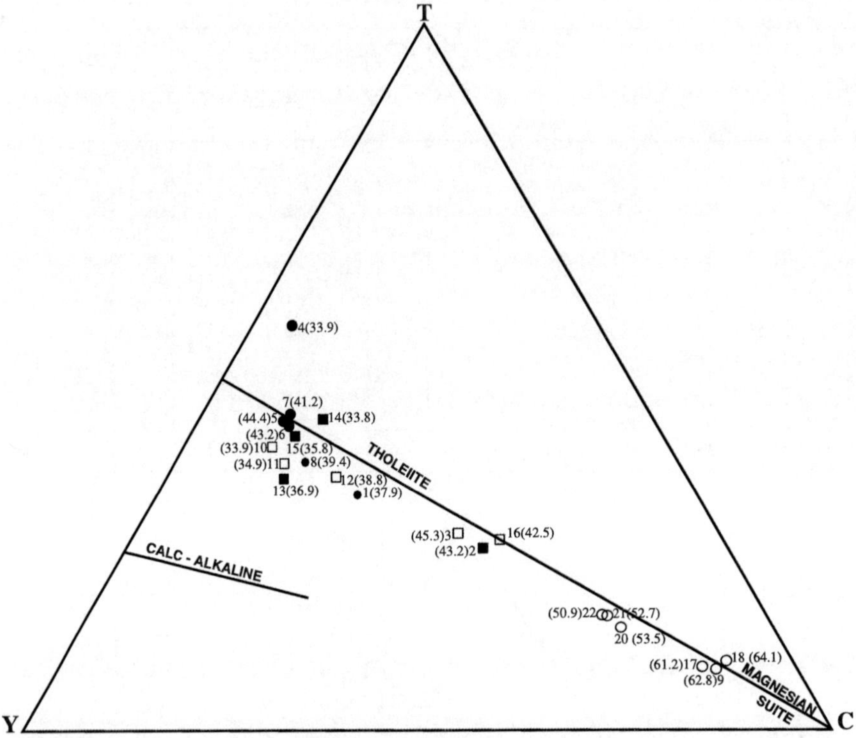

Fig. 4.4 YTC Diagram of Davies et al. (1979). Numbers indicate the analysis number given in Table 4.1. Numbers in parenthesis are Mg#. For legend see Fig. 4.2

would expect the presence of mica. On the other hand samples MLP-P2 and MLP-P3 with high Nb/La (about 0.9), low Nb/Th (4.3–4.4), and very low Th/U (about 1.5) again look like intraplate basalts (see Sun and McDonough, 1989). These could not simply reflect less crustal contamination. There are no typical established mantle-derived basalts with similar elemental ratios. However, Island Arc Basalts have low Th/U values similar to that of this dyke. As shown in Fig. 4.7a, the dyke patterns are nearly similar to those of upper crust in respect of Rb to P, but show higher concentrations of Ti, Gd to Yb with respect to upper and lower crusts and the Chitradurga granite.

On the rock-mantle normalized plots dyke D4A (sample MLP-P2 – Fig. 4.7) has higher amounts of Rb, Ba, U, Nb, Th, K than highly fractionated Deccan Bushe flows but shows nearly similar concentrations of P, Nd, Zr, Hf and Gd to Yb to Deccan Poladpur flows (less contaminated than Bushe flows – Fig. 4.7b). These dyke rocks have distinct negative Sr, Eu, Nb and Ta anomalies possibly reflecting sediment subduction, which can superimpose a crustal signature (Sun and McDonough, 1989). If we filter off this cover it is possible to determine the original magmatic features and explain high Nb/La (about 0.9) ratio that is close to chondritic value of 1.0. Thus the overall chemistry of these dykes looks like

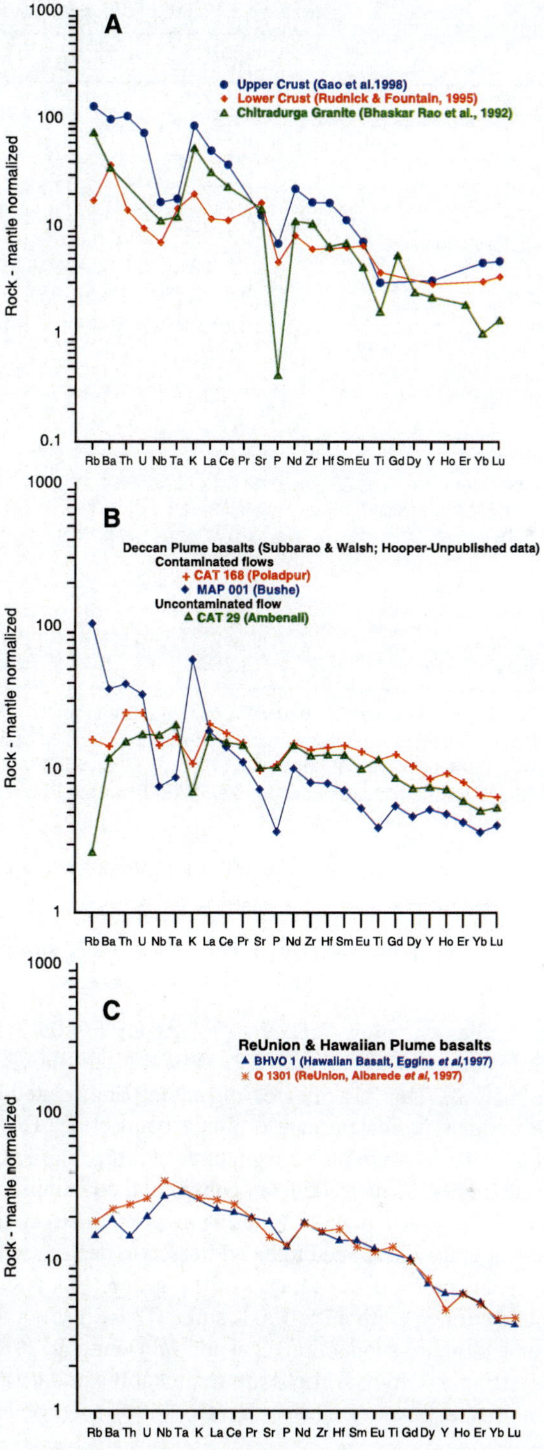

Fig. 4.5 Rock/Mantle normalized abundance patterns in (**a**) Average upper and lower crusts and Chitradurga granite (**b**) Deccan plume basalt and (**c**) Reunion and Hawaiian plume basalts

Table 4.2 Some elemental ratios of different rock formations

	Nb/La	Nb/Th	$[La/Sm]_N$	$[Gd/Yb]_N$
1. Cuddapah dykes				
Group I	0.67 - 0.94	4.3–4.4	1.59–2.39	1.14–1.38
Group II	0.31–0.46	1.4–3.9	2.17–2.75	2.0–2.63
Group III	0.45–0.48	1.22–2.6	1.7–1.9	1.02–1.16
Group IV	0.2 - 0.48	1.3–1.9	1.72–2.62	1.04–1.33
2. Deccan Formations				
Ambenali	1.0–1.15	8.98–10.92	–	–
Poladpur	0.74–0.77	4.62–4.98	–	–
Bushe	0.34–0.42	1.42–1.8	–	–
3. Réunion (Q 1301)(pure Plume end member)	1.39	12.09	–	–
4. Hawaiian Basalt (BHVO-1)	1.25	15.47	–	–
5. Chondrite	1.03	8.48	–	–
6. Primitive mantle	1.03	8.38	–	–
7. N –MORB	0.93	19.41	–	–
8. E – MORB	1.31	13.83	–	–
9. OIB	1.29	12	–	–
10. Avg. lower crust			–	–
11. Upper crust	0.34	1.34	–	–
12. Lower crust	0.62	4.16	–	–
13. Chitradurga granite	1.36		–	–

References: 1: Present work; 2: Deccan Formation – Lightfoot et al. (1990); Subbarao and Walsh, New Deccan standards, Unpublished data; Hooper – unpublished data; 3: Réunion – Albarede et al. (1997); 4: Hawaiian Basalt – Eggins et al. (1996); 5–9: Sun and McDonough (1989); 10, 11: Gao et al. (1998); 12: Rudnick and Fountain (1995), 13: Bhaskara Rao et al. (1992).

intraplate basalts with some crustal contamination or involvement of subduction modified lithospheric mantle (metasomatised).

Group II (dyke Nos. D4, D6, D13A, D16, D17; sample Nos. MLP3, MVP2, ERP-A, ERP-B, KCR-1, 35D3, 35D4 – see Table 4.1)

These samples with low TiO_2 (~1%), Al_2O_3 (~10%), high MgO (7–12%) and HREE depletion [$(Gd/Yb)_N \approx 2.2$, Figs. 4.8 and 4.6b], suggest large degree of mantle melting. This high degree of melting and related high temperatures of the mantle perhaps indicates mantle plume connection. However, they all have low Nb/La (≈ 0.4 relative to plume signatures of more than 1; see Table 4.2). The contribution of LREE seems to be from continental contamination/melting of subducted modified lithospheric mantle. Filtering away such effect and getting the real plume component is the basic need to be addressed in dealing with all these dykes.

The mantle normalized patterns of this group (Fig. 4.8b) display negative Nb, Ta and P, and strong positive Rb, Ba, K spike. These distinct features indicate sediment addition and/or arc subduction (Sun and McDonough, 1989).

Interestingly, Group II rocks are remarkably similar to the highly contaminated Deccan Bushe flows (Lightfoot et al., 1990) as revealed in the rock – mantle

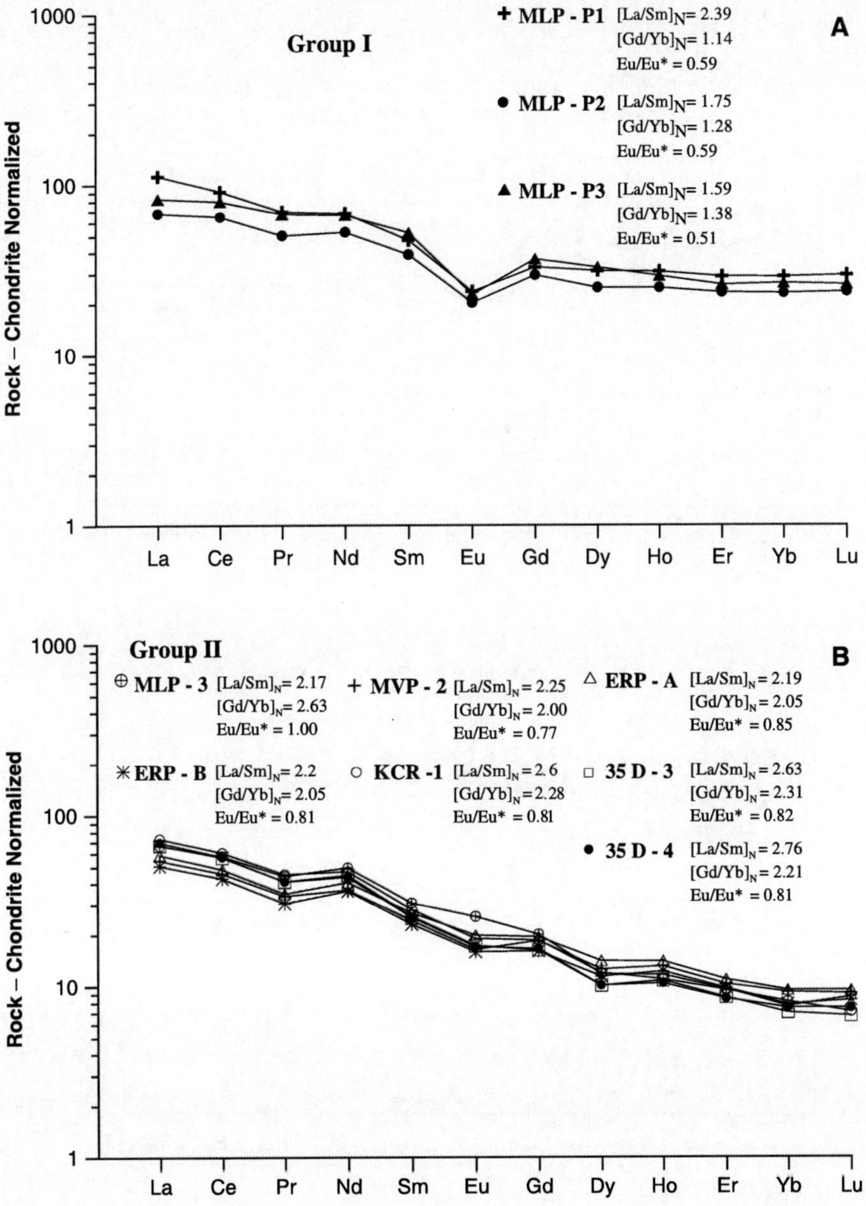

Fig. 4.6 Rock-chondrite normalized REE abundances of (**a**) Grouup I, (**b**) Group II, (**c**) Group III and (**d**) Group IV of Cuddapah dykes

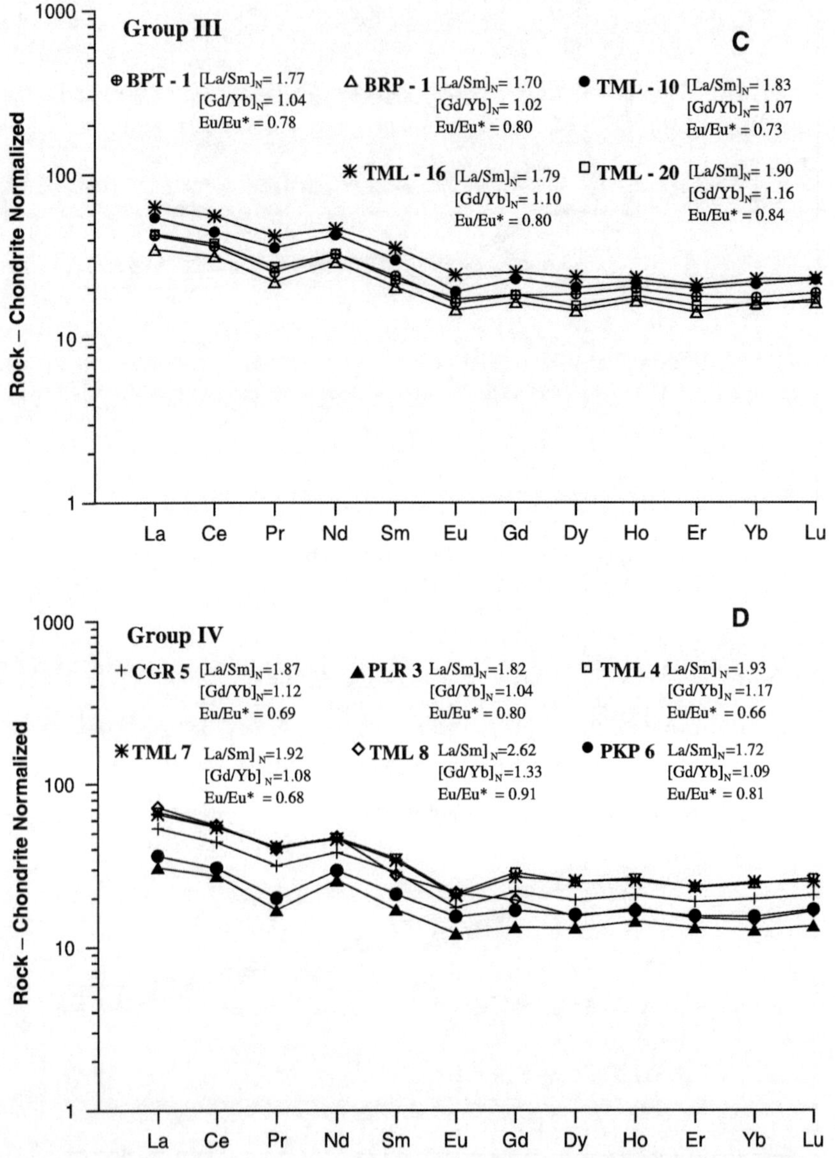

Fig. 4.6 (continued)

normalized patterns (Fig. 4.8b). The Deccan Bushe rocks have been explained in terms of contamination of Ambenali magma (least contaminated) with component similar to the Archaean granites underlying the Deccan.

Therefore, by analogy, Group II rocks may be considered to share similar recipe as Deccan Bushe and invoke the involvement of granitic contaminant. Further, the rock – mantle normalized pattern of Group II also lies between the upper and lower

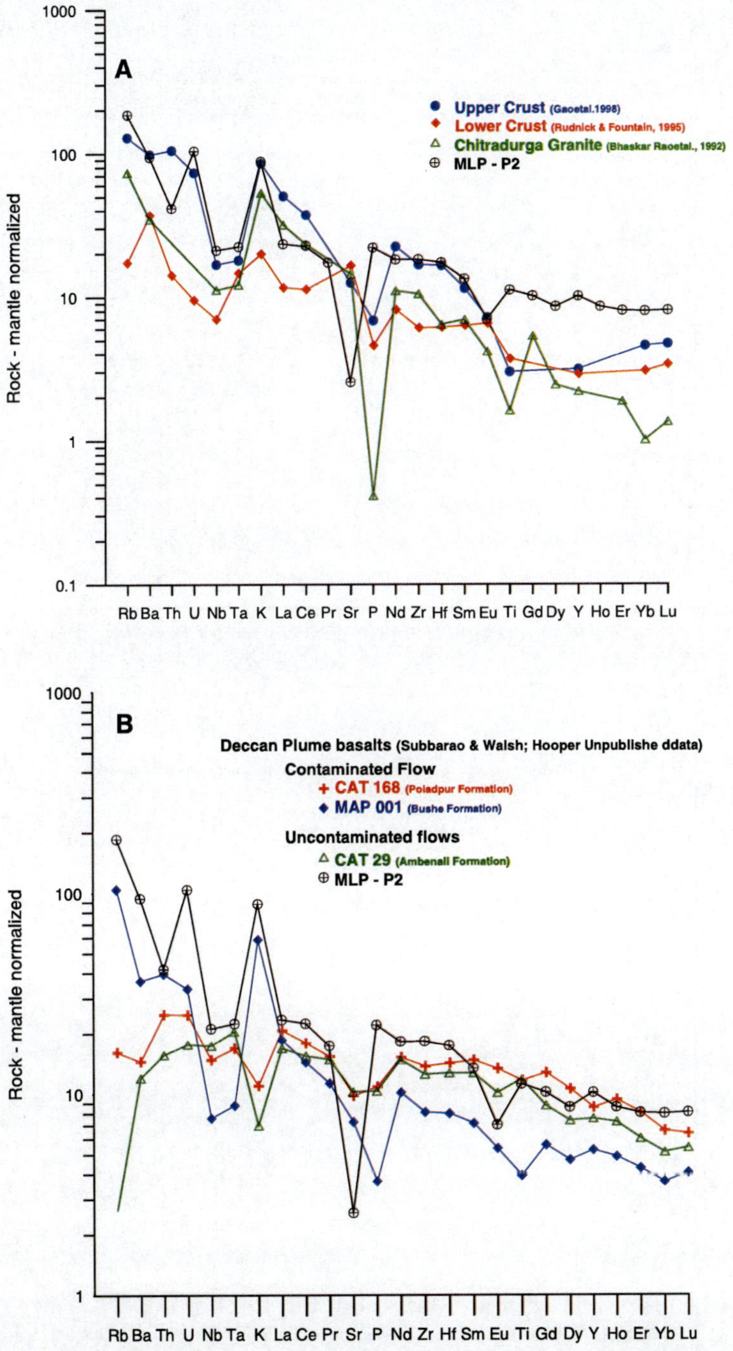

Fig. 4.7 A comparison of rock – mantle normalized pattern of representative sample No. MLP – P2 from Group I with average composition of upper and lower crusts and Chitradurga granite (**a**) and with representative contaminated and uncontaminated Deccan Plume Basalts (**b**)

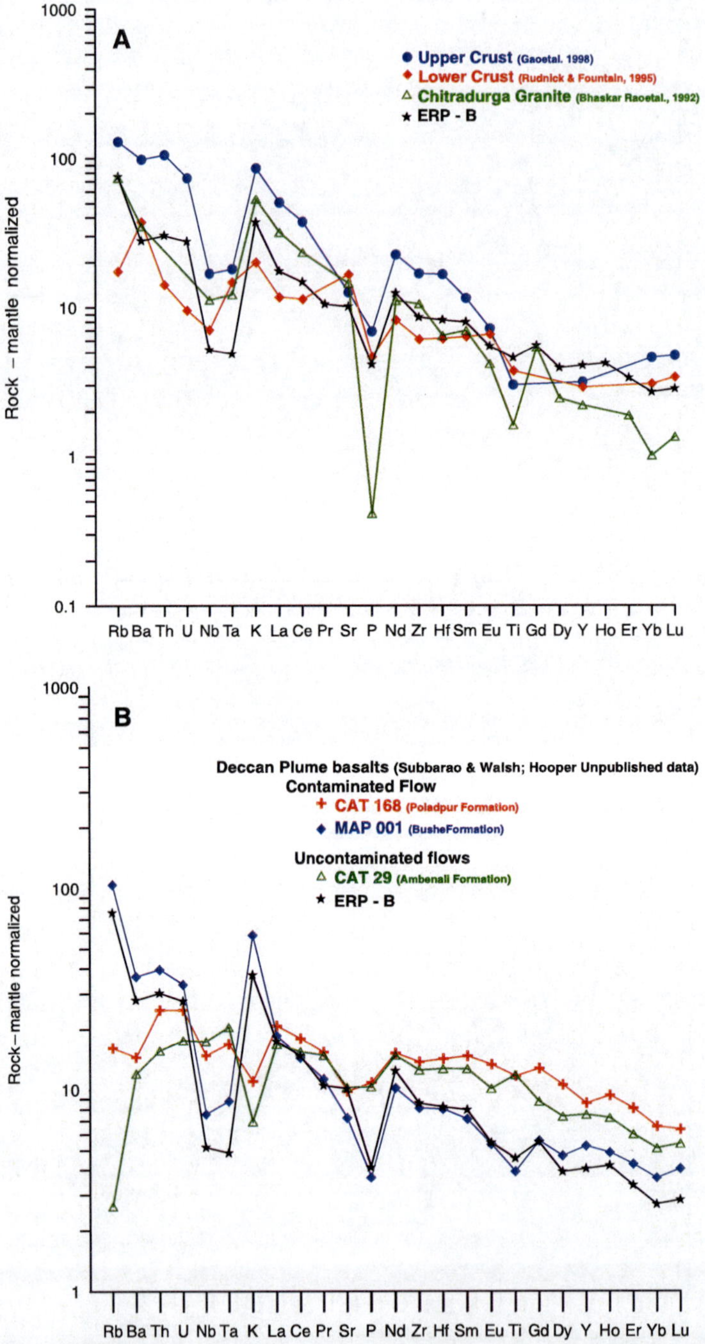

Fig. 4.8 A comparison of rock – mantle normalized pattern of representative sample No. MLP – P2 from Group II with average composition of upper and lower crusts and Chitradurga granite (**a**) and with representative contaminated and uncontaminated Deccan Plume Basalts (**b**)

crustal patterns and also partly matches with the Chitradurga granite (Fig. 4.8a). All these possibly lead to "subduction-related" environment for Group II dykes, either by assimilation of crustal (including sediment) component, or by the partial melting of and/or assimilation of, a subcontinental lithospheric mantle enriched by an earlier subduction process.

Group III (dyke Nos. D2, D5, D10, D12; sample Nos. BPT1, BRP1, TML10, TML16, TML20)

Chemically, the compositions of these dykes are nearly similar to Group II, except the LREE is less enriched [$(La/Sm)_N \approx 1.5$] and also with less HREE depletion $(Gd/Yb)_N \approx 1.1$ (Table 4.1, Fig. 4.6c). In general, these values also match fairly well with the Deccan Bushe flows (Table 4.2, Fig. 4.9b) and also lie in between those of the upper and lower crust as well as Chitradurga granite (mixed characters; Fig. 4.9a). All these features again indicate nearly similar genetic history like that of Group II (i.e. crustal contamination).

Group IV (dyke Nos. D1, D3, D8, D9, D10 and D13; sample Nos. CGR5, PLR3, TML4, TML7, TML8 and PKP6)

This Group is characterized by the presence of lower concentrations of Rb, Ba and K than the rest (Table 4.1). The rock – mantle normalized pattern of this Group match well with Deccan Poladpur flows (Fig. 4.10b), which are considered as having developed by contamination of Ambenali magma-type (least contaminated) with ancient granitic crust together with small amounts of gabbroic fractionation (Lightfoot et al., 1990). This can also be explained by mixing of the Ambenali and Bushe magmas (less contaminated than Bushe).

The chemical characteristics as well as elemental ratios of Group IV dykes indeed support lesser degree of upper crustal contamination than the Deccan Bushe (Table 4.2; Fig. 4.10b). There is a close similarity in the concentrations of Th, U, Nb, Ti, Sr, P, Nd, Zr between the Group IV rocks and the upper crustal rocks (Fig. 4.10a).

In essence, by uncovering/filtering all these strong effects of superimposed crustal signatures, it is possible to see through the original pristine plume component of the Cuddapah magmatic activity which would be similar to Deccan (Ambenali magma type), Reunion and the Hawaiian (Fig. 4.5b, c) basaltic magmatism.

How Do We Uncover/Filter Crustal Signatures?

A semi quantitative model is attempted to filter the superimposed crustal signatures over the basic magma in the study area. A detailed geochemical study indicates a mix of uncontaminated or least contaminated Ambenali like -magma ($\approx$ 80–90%) with about 5% of gabbro, about 10% of upper crustal and/or sediments as the possible recipe to explain the genesis of the dykes under study. This assumption is made because of the similarities between Bushe Formation of the Deccan and the rocks under study; and the Bushe magma can be produced with the proportions given above (Lightfoot et al., 1990). The enriched nature of K, Rb, Ba, Th, U and

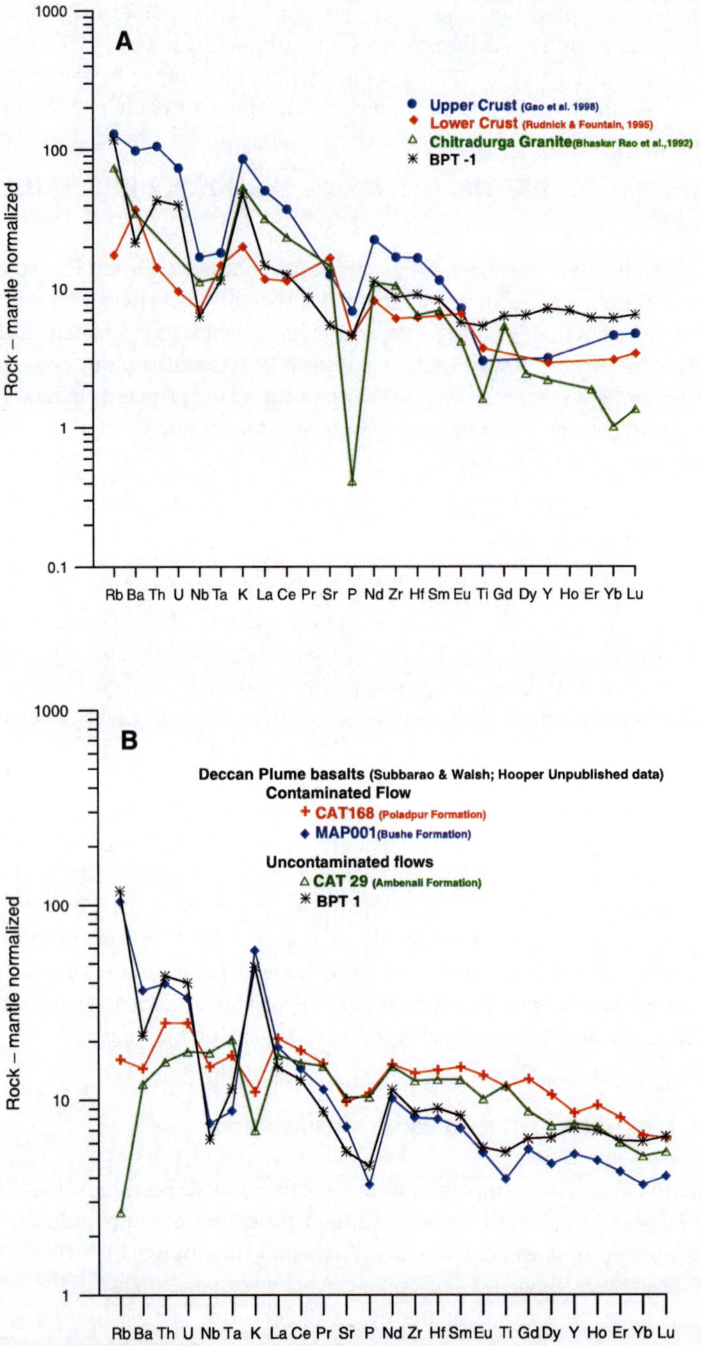

Fig. 4.9 A comparison of rock – mantle normalized pattern of representative sample No. MLP – P2 from Group III with average composition of upper and lower crusts and Chitradurga granite (**a**) and with representative contaminated and uncontaminated Deccan Plume Basalts (**b**)

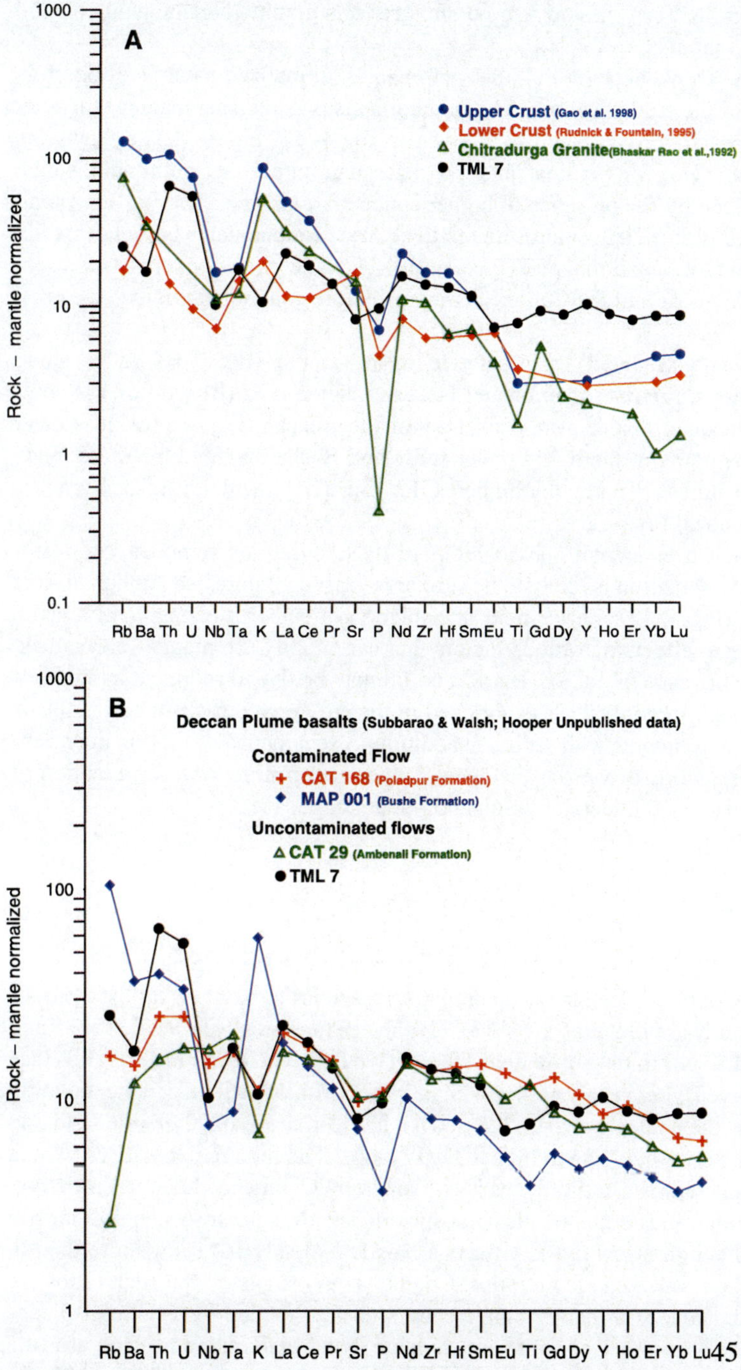

Fig. 4.10 A comparison of rock – mantle normalized pattern of representative sample No. MLP – P2 from Group IV with average composition of upper and lower crusts and Chitradurga granite (**a**) and with representative contaminated and uncontaminated Deccan Plume Basalts (**b**)

depletion of Sr, P, Ti of these dykes possibly reflects granitic contaminant coupled with sediment inputs.

Against the above background, the rock-mantle normalized patterns of possible components of the Cuddapah magma (contaminants plus original magma) together with a representative dyke sample is given in Fig. 4.11. Figure 4.11a shows the spidergram of average upper crust and Chitradurga granite from South India, which are characterized by the presence of higher concentrations of K, Rb, Ba, Th, U and depletion of Sr, P and Ti, in contrast to relatively least contaminated basic Formation of the Deccan (Ambenali magma) (Fig. 4.11b). The latter is very likely to represent the initial composition of the Cuddapah magma before contamination with the crust or sediment.

A highly contaminated representative dyke sample (ERP- B) under study together with a similarly contaminated Deccan magma type (Bushe flow) is compared with uncontaminated Ambenali flow of the Deccan (Fig. 4.11c). It is clear that the spidergrams of the dykes under study and Bushe magma type share many chemical signatures (in particular higher K, Rb, Ba, Th, U) with crustal and granitic materials as shown Fig. 4.11c.

If the enriched elemental compositions of these dykes are removed, the resultant magma composition is likely to be similar to least contaminated Ambenali -like magma type (Fig. 4.12). Enrichment or depletion of selected elements in a given system is due to effects of various factors that can modify the mantle- normalized spidergrams. Increase in Cs, Rb, Ba, K, and Pb may be due to sediment addition or arc environment. In general, Sr is enriched in the arc-derived rocks whereas this is depleted in environments with sediment addition. A schematic description of enrichments and depletions of various elements due to one or more, or a combination of factors was discussed in detail by Sun and McDonough, 1989.

Mantle Plume

Th/Ta and La/Yb ratios provide important constraints for the magma sources of the mafic dykes Condie (1997). In his diagram for Palaeo Proterozoic dykes the dykes of the present study are plotted (Fig. 4.13). Five dykes (analysis Nos. 9, 12, 17, 21 and 22 in Table 4.1) fall in the norite field. These have high Th/Ta and La/Yb ratios (see Table 4.1). Five Dykes (analysis Nos.4, 5, 6, 7, 8, 13, 14, 15, 16) falls in the tholeiitic field whereas the analysis nos. 1, 2, 3 and 11 fall just above the tholeiitic field and 18 fall above norite field due to higher Th/Ta ratios. The high ratio of Th/Ta (which means Ta depleted) for norite dykes is explained by Condie as due to their derivation from Archaean sub continental lithosphere. The tholeiite dykes exhibit mostly <5 Th/Ta, although a few dykes (analysis Nos.1, 2, 3, 11, 16) show more than 5, while they show variable La/Yb ratios. Condie (1997) suggests that such ratios are typically indicative of a mantle plume source to the tholeiite dykes. He also suggested that the occurrence of norite and tholeiite dykes in the same swarms, as is the case with the present dykes which are from the swarms at the south of the Cuddapah Basin, can be accounted for if plumes supply heat to partially melt the lithosphere to

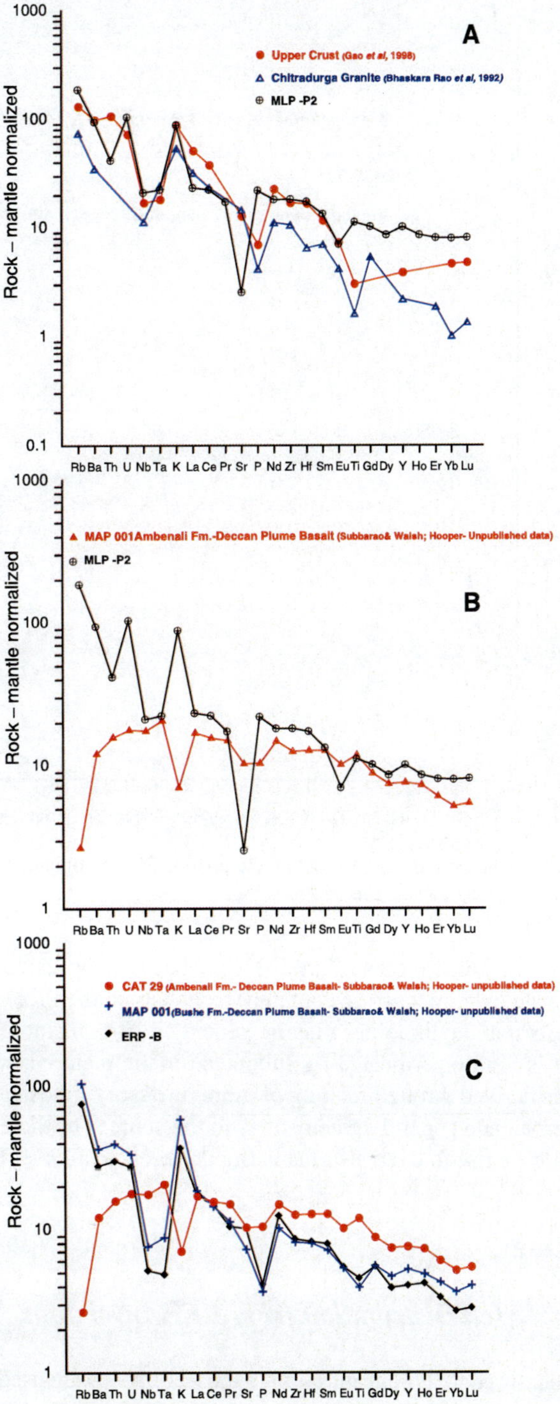

Fig. 4.11 Rock mantle normalized patterns of (**a**) Upper crust, Chitradurga Granite and representative Cuddapah sample (MLP-P2); (**b**) uncontaminated Deccan Plume Basalt (Ambenali) and representative Cuddapah sample (ERP-B); (**c**) Contaminated (Bushe), uncontaminated (Ambenali) Deccan Plume Basalt with representative Cuddapah sample (ERP-B)

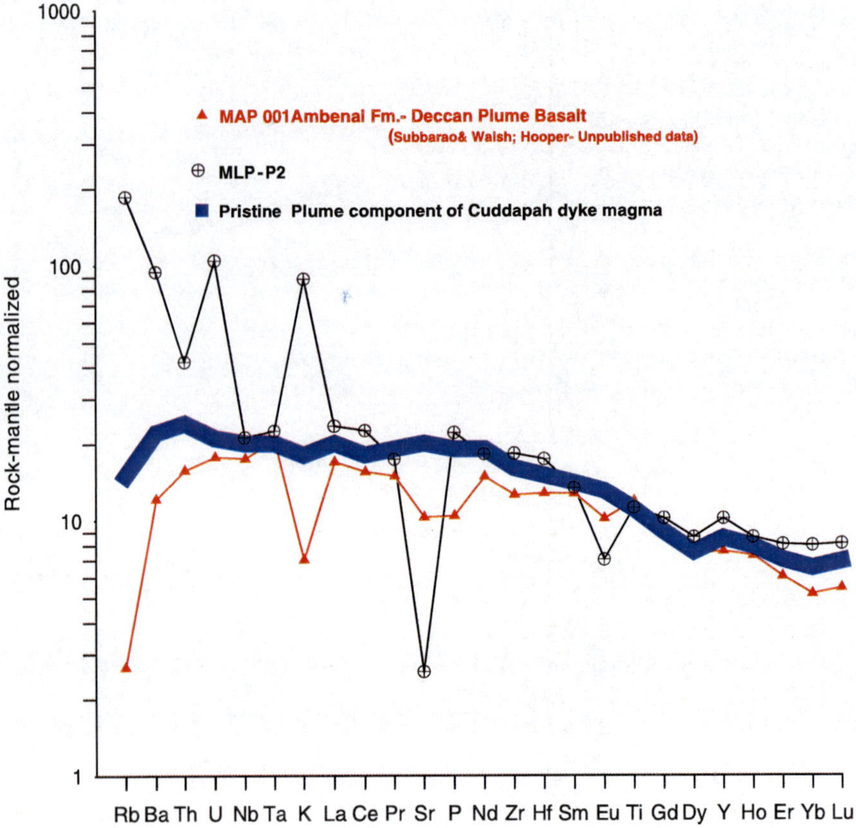

Fig. 4.12 Schematic diagram showing the pristine Plume component of Cuddapah magma after filtering the strong crustal effects

produce the norite magma. Tarney (1992) has also discussed the problem of tholeiite and norite dykes occurring contemporaneously in the Proterozoic dyke swarms and suggests that the tholeiites may be generated from material, which may be mantle, which is highly permeated by subducted mafic material, while the noritic magma may be derived through melting of more refractory lithosphere. Thus there seems to be a broad and general agreement as to the sources of tholeiite and norite magmas, which gave rise to dyke swarms in the Proterozoic all over the world.

Melting

Partial Melt Distribution from REE Inversions

The measured REE concentration in the basic rocks are used to infer the depth range and percentage of mantle melting and also temperature of mantle source by making

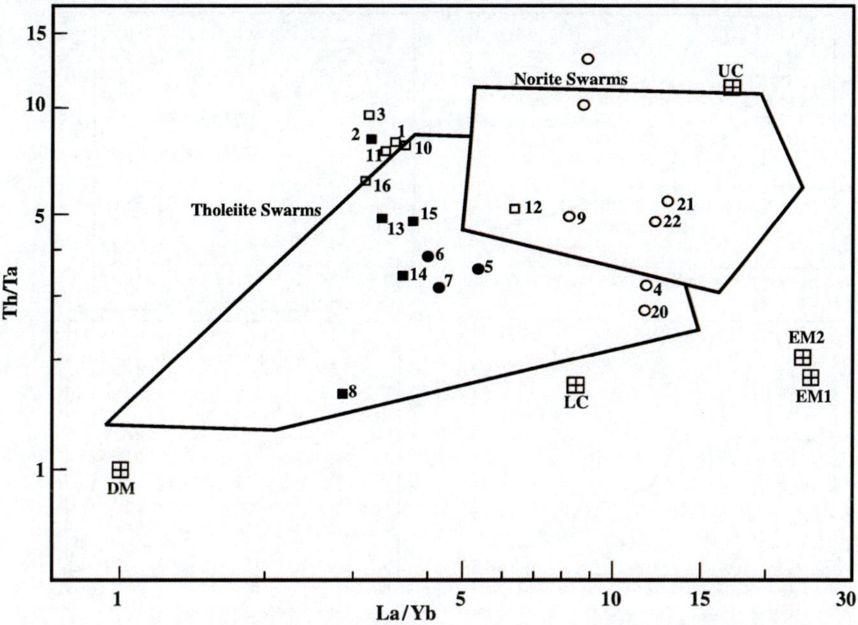

Fig. 4.13 Th/Ta and La/Yb diagram showing fields of distribution of Palaeoproterozoic norite and tholeiite dyke swarms (Condie, 1997). UC-upper crust, LC-lower crust, DM-depleted mantle, EM-enriched mantle. Numbers indicate analysis number given in Table 4.1. For legend see Fig. 4.2

use of the REE inversion technique of McKenzie and O'nions (1991) and White and McKenzie (1995). The temperature of mantle controls the depth of the base of the melting, while the thickness of the overlying lithospheric lid determines the top of the melting region. In the inversion studies it is required to know the initial REE concentration in the parent magma. Four Cuddapah dykes (MVP-2, ERP-B, KCR-1 and 35D4) having more than 6% of MgO are selected for inversion modeling. The inversion results of the dykes under study are presented in Fig. 4.14.

Top left graph (Fig. 4.14a) shows observed and calculated REE concentrations normalized to MORB source. Mean value for each element is shown by dots and the vertical bars indicate one standard deviation, estimated from the variance of the data. The dark line shows the best fit to the data obtained by fractional- melting inversions. Figure 4.14b and c show the observed (dots) and predicted (solid lines) concentrations of major and trace elements using the melt fraction inferred from the REE inversion fit. Figure 4.14d shows the melt fraction with depth requires to produce the fit observed in Fig. 4.14a. Uncorrected (dark heavy line) and corrected for fractionation of 14.5% (thin line) melt distributions are also included in this figure. The depth at which melting started is around 100 km with the upper limit around 70 km. The maximum melt fraction estimated from the inversion modeling is close about 20% (see Fig. 4.14d) the total melt thickness generated after fractionation correction is about 4.8 km. This upper depth estimated around 70 km possibly indicates

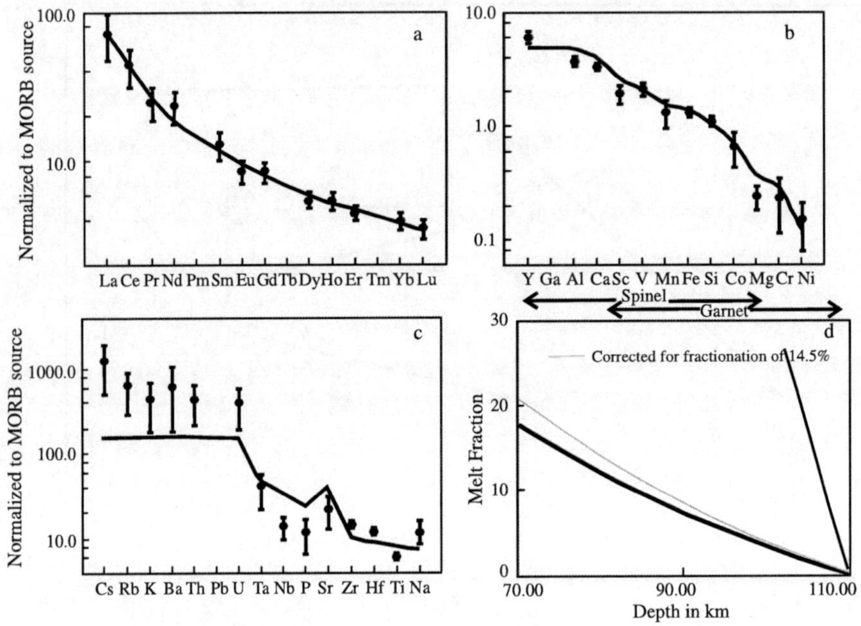

Fig. 4.14 REE inversion modeling results of Cuddapah dykes

that the Cuddapah Mantle plume lies beneath older and thicker South Indian lithosphere. Interestingly a similar depth analogy occurs in the case of younger plumes such as Hawaii and the Reunion where melting stopped around 75 km owing to thickness of the overlying plate (White and McKenzie, 1995).

The mantle source used in this modeling study is primitive mantle that was first depleted and later enriched. To perform such a complicated processes to match the data the amount of melting that generates the melt is poorly constrained, because any addition of metasomatic melt increases the melt fraction. Further it is not possible to constraint the potential temperatures when we are melting metasomatically enriched lithosphere, because its major element compositions may not be similar to MORB source. Primitive mantle has essentially some major element composition as the MORB source and that's why inversion techniques of White and Mckenzie (1995) works well for large active plumes. The Nb and Ti concentrates are also consistent with a subduction influence. Further hydrous phases like amphibole also lower the solidus temperature.

Despite the fact the mantle potential temperatures are not well constrained it is quite possible the mantle enrichment vis a vis the influx of metasomatic fluids into the South Indian lithosphere may perhaps be responsible to generate unique ultrapotassic rocks such as lamprophyres, lamproites and alkali basalts surrounding the Cuddapah Basin. A detailed inversion modeling studies coupled with Ar/Ar geochronological dating would be necessary on different alkali rich rocks to properly document the geochemical nature of the source region as well as the different mantle enrichment events in the geological past.

Summary and Conclusions

These studies so far give a good evidence of mantle plume source and also crustal contamination (mostly ancient granite and sometimes amphibolites) in the basic dykes under study. These rocks also demonstrate crystal fractionation with progressive change from intense crustal contamination (sediment and granitic material) to less contamination. This probably implies a depletion of fusible constituents within the crust or the isolation of magma chamber from the country rocks. By and large the chemistry of these dykes seems to be intrapalte basalts with crustal contamination or contribution of subduction-modified lithospheric mantle. Perhaps, these dykes might have formed in a subduction related environment either by assimilating crustal component or by partial melting of an enriched subcontinental lithospheric mantle. The lower concentrations of Nb, La and Ta are typically consistent with a subduction influence.

REE inversion studies indicate melting ($\approx$ 20%) of metasomatically enriched lithosphere. They also show that the depth of melting is around 100 km with an upper limit of 70 km, which possibly indicates that the mantle plume, by which these rocks have been formed, lies beneath the old thick south Indian lithosphere at a depth of 70 km.

Interestingly the mantle -normalized spidergrams for group II & III dyke rocks just identical to the Deccan Bushe flows, which have been proved to be contaminated. Further, such a close match between some of these dyke compositions of Proterozoic age and the Deccan Bushe flows of Cretaceous age possibly indicates the presence of similar magma sources, igneous processes as well as depth of melting. It seems very likely that both in the Proterozoic and in the Cretaceous times the Peninsular Indian continent experienced very similar tectono-volcanic activity.

Acknowledgements MRG, CVRKP and VDR are thankful to Department of Science & Technology, Govt. of India for providing the financial assistance to carry out this study. KVS is grateful to Dan McKenzie of Cambridge University for introducing him to REE inversion modeling. Authors are thankful to Nilanjan Chatterjee and Rajesh K. Srivastava for their critical and constructive comments and suggestions which considerably elevated the quality of the paper. We are also thankful to Viswanathan, IIT, Mumbai for his suggestions on the earlier draft of the paper.

References

Albarede F, Luais B, Fitton G, Semet M, Kaminski E, Upton BGJ, Bachelery P, Cheminee JL (1997) The geochemical regimes of piton de la fournaise volcano (ReUnion) during the last 530 000 years. J Petrol 38(2): 171–201

Anand M, Gibson SA, Subbarao KV, Kelley SP, Dickin AP (2003) Early-Proterozoic melt generation processes beneath the intra-cratonic Cuddapah Basin, Southern India. J Petrol 44(12): 2139–2171

Bhaskara Rao YJ, Sivaraman TV, Pantulu GVC, Gopalan K, Naqvi SM (1992) Rb-Sr ages of late Archaean metavolcanics and granites, Dharwar craton, South India and evidence for early Proterozoic thermotectonic event(s). Precamb Res 59: 145–170

Chatterjee N, Bhattacharji S (2001) Petrology, geochemistry on tectonic settings of the mafic dikes and sills associated with the evolution of the Proterozoic Cuddapah Basin of South India. Proc Indian Acad Sci (Earth Planet Sci) 110(4): 433–453

Condie KC (1997) Sources of Proterozoic mafic dyke swarms: Constraints from Th/Ta and La/Yb ratios. Precamb Res 81: 3–14

Damodara Reddy V, Anjanappa K, Prasad CVRK, Subba Reddy N (1995) Magnetic Properties and Chemistry of Precambrian dykes from Peninsular India. In: Devaraju TC (ed) Dyke swarms of Peninsular India. Geol Soc Ind 33: 99–110

Davies JF, Grant RWE, Whitehead RES (1979) Immobile trace elements and Archaean volcanic stratigraphy in the Timmins mining area, Ontario. Can J Earth Sci 16: 305–311

Devaraju TC (1995) Dyke swarms of Peninsular India (ed). Mem 33. Geol Soc Ind 451 p

Eggins SM, Woodhead JD, Kinsley LPJ, Mortimer GE, Sylvester P, McColloch MT, Hergt JM, Handler MR (1996) A simple method for the precise determination of ≥ 40 trace elements in geological samples by ICPMS using enriched isotope internal standardization. Chem Geol 134: 311–326

Ernst RE, Srivastava RK (2008) India's place in the Proterozoic world: Constraints from the large igneous province (LIP) record. In: Srivastava RK, Sivaji C, Chalapathi Rao NV (eds) Indian Dykes: Geochemistry, Geophysics and Geochronology. Narosa Publishing House Pvt. Ltd, New Delhi: 41–56

French JE, Heaman LM, Chako T, Srivastava RK (2008) 1891–1883 Ma Southern Bastar-Cuddapah mafic igneous events, India: A newly recognized large igneous province. Precamb Res 160: 308–322

Gao S, Luo TC, Zhang BR, Zhang HF, Han YW, Zhao ZD, Hu YK (1998) Chemical composition of the continental crust as revealed by studies in East China. Geochemic et Cosmochemica Acta 62(11): 1959–1975

Halls HC, Kumar A, Srinivasan R, Hamilton MA (2007) Paleomagnetism and U-Pb geochronology of easterly trending dykes in the Dharwar Craton, India: Feldspar clouding, radiating dyke swarms and the position of India at 2.37 Ga. Precamb Res 155: 47–68

Ikramuddin M, Stueber AM (1976) Rb –Sr ages of Precambrian dolerite and alkaline dykes, Southeast Mysore state, India. Lithos 9: 235–241

Irvine TN, Baragar WRA (1971) A Guide to the chemical classification of the common volcanic rocks. Can J Earth Sci 8: 523–548

Lightfoot PC, Hawkesworth CJ, Devey CW, Rogers NW, Van Calsteren PWC (1990) Source and differentiation of deccan trap lavas: Implications of geochemical and mineral chemical variations. J Petrol 31(5): 1165–1200

McKenzie D, O'nions RK (1991) Partial melt distributions from inversions of rare- earth element concentrations. J Petrol 32: 1021–1091

Murthy NGK (1995) Proterozoic mafic dykes in Southern peninsular India: A Review. In: Devaraju TC (ed) Dyke swarms of Peninsular India. Geol Soc Ind 33: 81–98

Murthy YGK, Babu Rao V, Guptasarma D, Rao JM, Rao MN, Bhattacharji S (1987) Tectonic, Petrochemical and geophysical studies of Mafic Dyke Swarms around the Proterozoic Cuddapah Basin, South India. In: Halls HC, Fahrig WF (eds) Mafic Dyke Swarms. Geol Assoc Can spec pap 34: 303–316

Pandey BK, Gupta JN, Sarma KJ, Sastry CA (1997) Sm-Nd, Pb-Pb and Rb-Sr geochronology and petrogenesis of the mafic dyke swarm of Mahaboobnagar, South India: Implications for Palaeoproterozoic crustal evolution of Eastern Dharwar Craton. Precamb Res 84: 181–196

Radhakrishna T (1999) Dyke swarms in South Indian Shield: Nature of mantle sources and Neoproterozoic super continent assembly. DST News letter, Deep Continental Studies in India 9(2): 16–20

Radhakrishna T, Gopakumar K, Murali AV, Mitchell JG (1991) Geochemistry and Petrogenesis of Proterozoic mafic dykes in North Kerala, Southwestern Indian Shield- Preliminary results. Precamb Res 49: 235–244

Radhakrishna T, Joseph M (1995) Proterozoic Palaeomagnetism of the mafic dyke swarms in the high-grade region of South India. Precamb Res 76: 31–46

Radhakrishna T, Krishnendu NR, Balasubramonian G (2007) Mafic dyke magmatism around the Cuddapah basin: Age constraints, petrological characteristics and geochemical inference for a possible magma chamber on the southwestern margin of the basin. J Geol Soc India 70: 194–206

Rudnick RL, Fountain DM (1995) Nature and composition of the continental crust: A lower crustal perspective. Rev Geophys 33: 267–309

Sun SS, McDonough WF (1989) Chemical and isotopic systematics of oceanic basalts: Implications for mantle composition and processes. In: Saunders AD, Norry MJ (eds) Magmatism in the Ocean Basins. Geol Soc Lond Spec Publ 42: 313–345

Tarney J (1992) Geochemistry and significance of mafic dyke swarms in the Proterozoic. In: Condie KC (ed) Proterozoic Crustal Evolution. Elsevier Publications, Amsterdam: 151–179

Thompson RN (1984) Dispatches from the basalt front. 1. Experiments. Proc Geol Ass 95: 249–262

Thompson M, Walsh JN (1989) Handbook of ICP spectrometry. Chapman & Hall Publications: London, England, 316 p

White RS, McKenzie D (1995) Mantle plumes and flood basalts. J Geophys Res 100: 17543–17585

Chapter 5
Proterozoic Mafic Dykes from the Southern Margin of Cuddapah Basin, India: Part 2 – Palaeomagnetism and Ar/Ar Geochronology

M.R. Goutham, K.V. Subbarao, C.V.R.K. Prasad, J.D.A. Piper, and D.P. Miggins

Introduction

Precambrian dyke swarms can define key data for constraining apparent polar wander paths (APWPs) when the palaeomagnetic data are related to field tests and linked to age dates (e.g. Halls et al., 1994; Buchan et al., 2000). They may then provide data points within long time intervals during which other suitable rock types are absent or poorly represented, although the intervals of emplacement recorded may be brief (e.g. Heaman, 1997). Within the basement terranes of south-central India the dyke swarms offer the main opportunity for defining the APWP between consolidation and cooling of the basement in Late Archaean times (e.g. Piper et al., 2003) and the deposition of supracrustal successions such as the Cuddapah Supergroup (Goutham et al., 2006). Basic dykes intruding the Archaean metamorphic basement in southern India have been investigated palaeomagnetically by a number of studies over several decades. Unfortunately the temporal control of the results is generally poor and their value for defining the kinematics of the Indian Shield during the Precambrian times is correspondingly limited. Recently attempts have been made to rectify this

M.R. Goutham (✉)
Dr. K.S. Krishnan Geomagnetic Research Laboratory, Indian Institute of Geomagnetism, Allahabad 221 505, India
e-mail: gouthammr@rediffmail.com

K.V. Subbarao
University Centre for Earth and Space Sciences, University of Hyderabad, Hyderabad 500 046, India

C.V.R.K. Prasad
AMARARAMA, Vijayawada, India

J.D.A. Piper
Geomagnetism Laboratory, Department of Earth and Ocean Sciences, University of Liverpool, Liverpool L69 7ZE, UK

D.P. Miggins
Denver Argon Geochronology Laboratory, USGS, Denver Federal Center, Denver, CO 80225, USA

R.K. Srivastava (ed.), *Dyke Swarms: Keys for Geodynamic Interpretation*,
DOI 10.1007/978-3-642-12496-9_5, © Springer-Verlag Berlin Heidelberg 2011

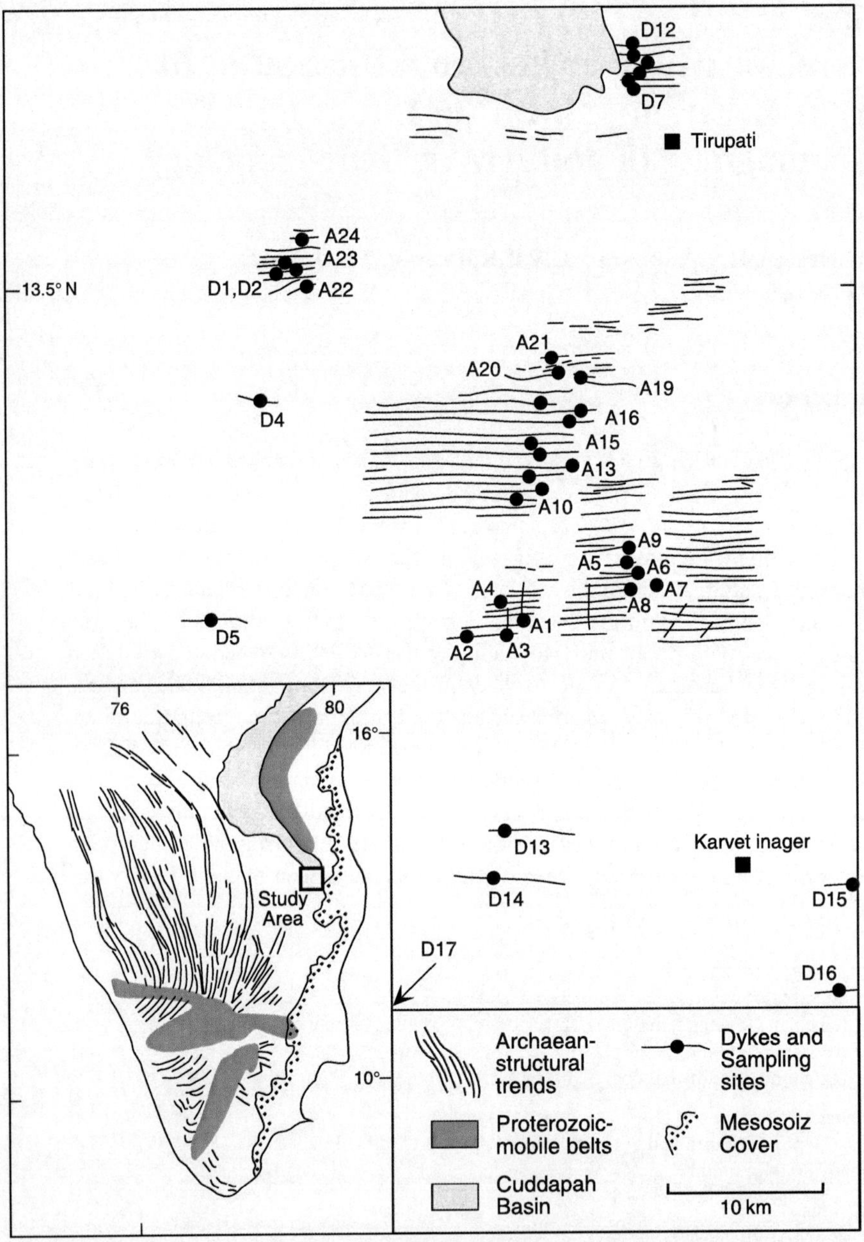

Fig. 5.1 Geological map of the region near the southern margin of the Cuddapah Basin showing the generalized distribution of dolerite dykes and the locations of sampling sites of this study. The *inset* map shows the regional location of the study area with major tectonic features of the Precambrian basement of southern India after Drury and Holt (1980). Note the contrast in trends between the dykes of this study and the dominant N-NNW trends observed in the shield further to the south west

deficiency by combining palaeomagnetic and age dating studies on specific dyke suites (Radhakrishna and Joseph, 1993, 1994, 1996; Pradhan et al., 2010).

Such integrated studies have not previously been extended to the dense swarms of dykes intruding the granite and gneissose basement in the vicinity of the Cuddapah Basin in southern India (Fig. 5.1 inset). With a view to determining palaeomagnetic poles constrained by radiometric studies we report here a study of dyke swarms emplaced into the shield near the southern margin of the Cuddapah Basin (Fig. 5.1) with the palaeomagnetic investigation supported by $^{40}Ar/^{39}Ar$ geochronology.

Geological Setting and Previous Study

The Cuddapah Supergroup of Peninsular India (see Fig. 5.1, inset) was deposited during Proterozoic times in a crescent shaped intracratonic basin covering an area of about 45,000 km in Andhra Pradesh (King, 1872; Krishnan, 1982; Nagaraja Rao et al., 1987; Chatterjee and Bhattacharji, 2001). The overlying Kurnool Group occurs mainly in two outliers and is considered to be stratigraphically equivalent to the higher Vindhyan formations of central India. According to Aswathanarayana (1964) Cuddapah sedimentation began around 1,080 Ma. However, based on Rb-Sr dating of sills near Pulivendla, yielding ages of 980 and 1,359 Ma, Crawford and Compston (1973) considered that sedimentation began appreciably earlier. Bhattacharji (1987) dated magmatism in the Cuddapah Basin ranging in age from 1,700 to 1,000 Ma, whilst Bhaskara Rao et al. (1995) reported a Rb-Sr age on a mafic sill near Pulivendla of 1,800 Ma and Zachariah et al. (1999) determined a Pb-Pb age of 1,800 Ma from the Vempalle Limestone overlying the Gulcheru Quartzite, the oldest Formation in the Cuddapah Basin. French et al. (2008) report a U-Pb age of 1,882 Ma for the Pulivendla sill and Ar-Ar data by Chatterjee and Bhattacharji (2001) indicate an age of $\sim$1,879 Ma. A recent Ar-Ar age of 1,900 Ma on sills intruding the lower Cuddapah formations (Anand et al., 2003) also indicates a greater antiquity for the beginning of Cuddapah sedimentation than suggested by the earlier work.

The cratonic basement to the Cuddapah Supergroup consists of coarse-grained granites and gneisses of predominantly equigranular texture and inferred Archaean age (Murty et al., 1987). Numerous basic dykes, both in swarms and as isolated dykes, intrude this basement (Suryanarayana and Anjanappa, 1975 and Fig. 5.1). Most dykes are sub-parallel with E-W strikes and are cut across by rare dykes with N-S strike (Fig. 5.1). Dyke widths range from a few centimeters to a hundred meters and show sharp chilled contacts with the country rock; textures, grain sizes and colours are typically doleritic. Since these dyke are not seen to penetrate the Cuddapah Supergroup they could all be older than 1,900 Ma in age. However, this evidence is not unambiguous because dyke swarms are also observed to penetrate up to the level of sedimentary unconformities without significantly penetrating into them.

Current constraints on the history of circum-Cuddapah dyke emplacement are weak. Anjanappa (1972) considered them to be emplaced at the end of the Archaean

Eon whilst subsequent study by Kumar and Bhalla (1983) of dykes SW of the Cuddapah Basin concluded that they comprise at least three episodes of igneous emplacement prior to deposition of the Cuddapah Supergroup; dykes with similar magnetization directions were found to have comparable strike directions. Tectonic, petrochemical and geophysical evidence from dykes within 20–50 km of the Cuddapah Basin margin were interpreted in terms of at least five stages of dyke emplacement between 2,100 and 600 Ma by Murty et al. (1987). The inferred oldest dyke groups were found to be all tholeiitic in composition and apparently older than 1,700 Ma in age, whilst the remainder were found to be either tholeiitic or alkaline and have possible ages of 1,700–600 Ma.

Methods of Study

A total of 182 oriented block samples were collected from 41 dykes in the vicinity of Tirupati at locations shown in Fig. 5.1 where trends are predominantly E-W. Sample orientations were by both Sun and magnetic compasses. Up to four separately-oriented blocks were collected from each dyke and cores drilled out in the laboratory to obtain a total of 492 specimens. Natural remanent magnetizations (NRMs) were measured by "Minispin" magnetometers. Approximately 50% of the specimens were then progressively demagnetized to evaluate stability of NRMs and isolate component structures using either a Schonstedt thermal demagnetizer or Molspin alternating field (a.f.) demagnetizer. Component structures were resolved from orthogonal projections of the remanence vector, Characteristic Remanent Magnetisations (ChRMs) were identified, and their equivalent directions calculated by principal component analysis.

Mineral separates and groundmass concentrates were separated from the dyke samples for $^{40}Ar/^{39}Ar$ geochronology using standard heavy liquid and magnetic techniques following crushing, grounding and sieving of selected samples. Prepared samples were irradiated in the USGS TRIGA reactor in Denver, Colorado to produce ^{39}Ar from ^{39}K by neutron bombardment. After irradiation, ^{40}Ar (radiogenic)/^{39}Ar (potassium) ratios of the samples and standards were determined and techniques described by Snee (2002) used to produce $^{40}Ar/^{39}Ar$ spectra.

Results

Palaeomagnetism

Sample susceptibilities (χ) range from 0.9 to 43.1 $\times$ 10 SI units and seventy 1% have Q_n ratios greater than 1. Although this would tend to reflect stable NRM behaviours, the sample is biased by some very high lightning-induced Isothermal Remanent Magnetizations (IRMs) and samples with this latter characteristic have been excluded from demagnetisation. Domain states of the ferromagnets were evaluated by hysteresis and susceptibility behaviours at low temperature (KLT spectra). Most dyke specimens show KLT peaks at –150°C observed where the magnetocrystalline anisotropy constant of magnetite changes from negative to positive (the Verwey Transition). This effect is observed in samples containing a ferromagnet dominated by MD magnetite but is suppressed in non-stoichiometric (partially oxidised) magnetite (Radhakrishnamurty, 1993); uniform curves characteristic of single domain (SD) or pseudo-single domain (PSD) grain structures were only rarely observed. The ratio RS is the susceptibility at liquid nitrogen temperature (–196°C) to that at room temperature and averages ~0.5 in these dolerite dykes, which is typical of domain structures in magnetite with MD:(SD + PSD) > 1. Hysteresis loops derived from representative specimens at room temperature (25°C) and at low temperature (–196°C) show two types of behavior. An increase in coercive force (H_c) at low temperature (–196°C) compared to coercivity at room temperature indicates the presence of MD and cation-deficient magnetite (maghemite); a decrease is a signature of mixed SD and MD grains states in magnetite (Radhakrishnamurty, 1993). Both behaviours were present in the dyke samples and provide a general indication that mixed domain states are present in these dykes with MD predominating.

NRM directions in 24 of the 41 dykes yielded significant within-site ChRM groupings. Directional behaviors were either complex or unstable at the remaining 17 dykes and included large NRMs dominated by lightning-induced IRMs; these are excluded from further consideration. Pilot demagnetization tests on specimens from each dyke (mostly in 50°C intervals during thermal treatment and 5 mT intervals during a.f. treatment) reveal low unblocking temperature/coercivity components of presumed secondary origin removed by 150–200°C during thermal treatment and by about 5 mT during a.f. treatment to recover the ChRM, typically by the Curie point of low-Ti magnetite. Remaining specimens were subjected to progressive thermal demagnetization at 50 or 100° intervals to 500, 580, 600, and 650°C, or over a range of a.f. fields above 5 mT. Behaviors of stable samples to this treatment are usually straightforward with the removal of one or two lower blocking temperature/coercivity components removed in initial stages to recover convergent ChRMs; typical orthogonal diagrams of vector behaviors are shown in Fig. 5.2.

ChRM directions identified within the group of 24 stable dykes indicate that several ages of emplacement are represented here. Directions in 10 dykes have southerly declinations and shallow inclinations and since this is approximately comparable to the late Phanerozoic palaeofield direction in this region, we refer to it as a "reversed" direction (RD) although its polarity with respect to the Proterozoic field is not defined. Four dykes with westerly declination and shallow negative inclination are defined as intermediate magnetization directions (ID) and 10 dykes show "normal" (i.e. northerly) magnetization directions (ND). Within the 10 ND dykes two subgroups are identified differing mainly in inclination (Table 5.1). The first group ND1) has northward declination and intermediate inclination whilst the second group (ND2) shows northwesterly declination and shallow inclination approximately antiparallel to the RD group. A single dyke (D17) located to the southwest

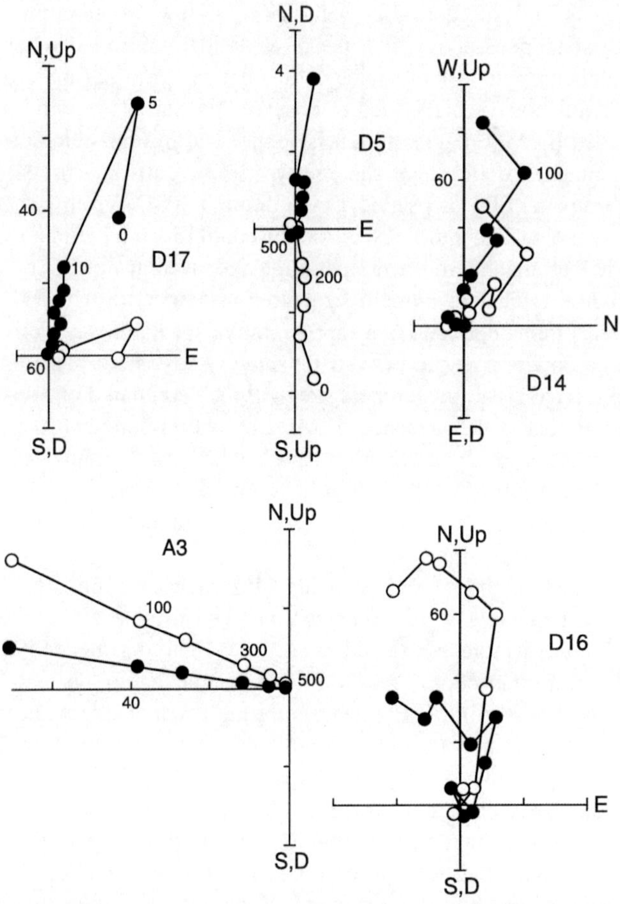

Fig. 5.2 Demagnetisation behaviours of dyke samples shown as orthogonal projections of the remanence vector. The figures refer to temperatures (°C) or a.f. fields (mT.) of treatment and the large figures refer to magnetization intensity in $\times 10^{-3}$ A.m/kg

of the region shown in Fig. 5.1 at 79° 3′E, 35° 11′N yields a contrasting site mean direction of D/I = 33/–29° (ND3) and could record a different dyke emplacement from the remainder of the collection, as suggested by an age determination from this unit (Sect. 4.2 and Table 5.3).

In summary, the results of thermal and a.f. demagnetization of dykes D14, A1, A3, A22 yield stable ChRMs with a mean (ID) direction of D/I = 279/–22°. This group includes the only two members of the sparse N-S trend sampled in this study (see Fig. 5.1). Dykes D1, D4, D15, A2, A4, A9, A11, A13, A18 and A20, all with E-W trends, yield reversed ChRM directions and a mean (RD) of D/I = 174/8°. Dykes D5, D6, D10, D12 and D16, again all with E-W trend, have normal (ND1) ChRM directions and yield a mean of D/I = 7/–61° and dykes A15, A19, A21

Table 5.1 Summary of Palaeomagnetic results from Tirupati Dykes, South Cuddapah Basin margin, Eastern India

		Site coordinates						Palaeomagnetic pole		
Dyke	n	°N	°E	D°	I°	α_{95}	K	°N	°E	Dp/dm
(i) Normal group NDI										
D5	5	13.48	79.28	7.6	−57.6	10.9	50.2			
D6	7	13.45	79.33	12.4	−52.8	9.2	44.0			
D10	3	13.63	79.40	346.8	−59.3	19.6	40.5			
D12	7	13.65	79.40	12.5	−65.6	9.0	46.2			
D16	4	13.36	79.48	17.3	−66.0	12.9	51.5			
Mean:	**5**	**(R = 4.96)**		**7.0**	**−60.7**	**7.7**	**100.0**	**34.4**	**253.1**	**9.0/11.7**
(ii) Normal group ND2										
A15	4	13.51	79.38	350.7	−16.5	2.7	1,159.7			
A19	4	13.55	79.38	351.0	−21.3	3.6	654.7			
A21	4	13.55	79.38	346.5	−21.3	2.6	1,205.5			
A24	4	13.58	79.30	347.5	−11.3	2.7	1,138.1			
Mean:	**4**	**(R = 3.99)**		**348.9**	**−17.6**	**5.9**	**240.1**	**64.9**	**286.0**	**3.2/6.2**
(iii) Normal dyke ND3										
D17	9	13.20	79.02	33.0	−29.4	8.4	38.5	46.4	209.6	–
(iv) Intermediate group (ID)										
D14	4	13.40	79.36	284.4	−23.9	15.8	34.8			
A1	8	13.46	79.36	274.6	−24.0	2.2	612.1			
A3	8	13.45	79.35	277.2	−22.9	5.0	122.9			
A22	8	13.56	79.30	279.0	−15.3	2.9	369.5			
Mean:	**4**	**(R = 3.99)**		**278.8**	**−21.6**	**6.4**	**204.7**	**5.6**	**336.4**	**5.6/6.8**
(v) Reversed group (RD)										
D1	7	13.57	79.30	176.1	−2.1	25.3	6.6			
D4	10	13.58	79.30	180.3	9.3	10.8	21.0			
D15	7	13.40	79.36	197.0	17.6	21.5	8.8			
A2	8	13.46	79.33	176.6	8.8	2.6	441.1			
A4	8	13.45	79.33	162.0	2.3	3.0	334.8			
A9	8	13.48	79.36	169.0	14.4	3.9	199.5			
A11	8	13.50	79.35	172.0	14.5	2.3	572.1			
A13	8	13.50	79.35	174.9	4.4	3.0	348.0			
A18	8	13.53	79.36	174.4	3.8	2.1	705.9			
A20	8	13.53	79.38	162.8	6.5	4.7	138.7			
Mean:	**10**	**(R = 9.82)**		**174.4**	**8.1**	**6.9**	**50.1**	**−71.6**	**97.3**	**3.5/6.9**

R is the length of the resultant vector derived from n samples yielding a mean direction with declination D and inclination I; k is the Fisher precision parameter (= (n–1)/(n–R)) and alpha 95 is the radius of the cone of 95% confidence about the mean direction. Dp and dm are the radii of the oval of 95% confidence about the derived pole position in the colatitude direction and at right angles to it respectively.

and A24 yield a mean direction (ND2) of D/I= 349/–18°; these dykes all lie to the south or southwest periphery of the present study area. Dyke mean magnetization directions are summarized in Fig. 5.3 and site and group mean results are listed in Table 5.1.

The Deccan Igneous Province emplaced near the Mesozoic-Cenozoic boundary at ~65 Ma is a major feature of the Phanerozoic palaeomagnetic imprint within the Indian subcontinent and has produced widespread overprinting in supracrustal rocks of central and northern India. However, the rocks of this study show no tendency for magnetizations to cluster either near the Deccan normal (D/I = 347/–48° to 339/–57°) or reversed (D/I = 157/+57°) field directions and are clearly uninfluenced by this overprint (Fig. 5.3). Unfortunately dyke margins with the country rock were often poorly exposed and otherwise uniformly in contact with coarse-grained granite gneisses that did not permit application of palaeomagnetic contact tests. However, the recognition of several distinct groups of magnetizations implies that no later blanket remagnetisation has influenced this segment of the Precambrian basement.

D17 yielding a direction remote from the groups RD, ID, ND1 and ND2 is likely to represent a single short cooling event; it will therefore have recorded a component of the ancient secular variation to yield a virtual geomagnetic pole. The remaining groups are compiled from a number of individual dykes and means are expected to record time averages of the palaeofield and yield palaeomagnetic poles (Table 5.1). Group ND2 is close to anti parallel to group RD. However, application of a reversal test (McFadden, 1990) shows that the angle between the two axes is 10.9° and marginally exceeds the critical value of 8.1°. Since the test is negative we evaluate pole positions from the ND2 and RD directions separately although it is possible that they are of similar age.

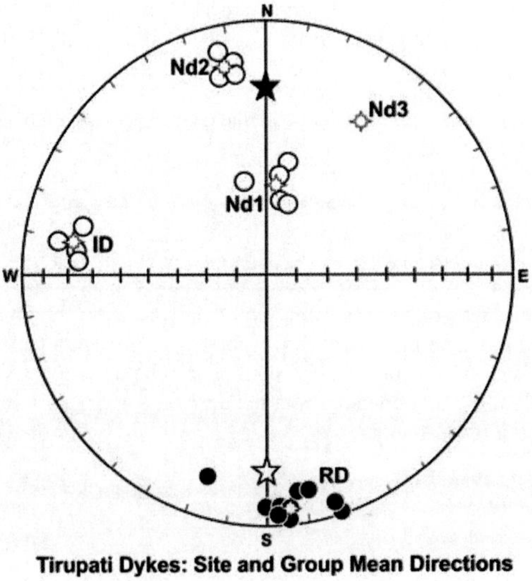

Fig. 5.3 Equal area projection showing site mean directions (*circles*) derived from dolerite dykes of this study. *Open symbols* are plots on the upper hemisphere and *closed symbols* are lower hemisphere plots. The *stars* are mean directions and ND1, ND2, ND3, ID and RD identify groups of directions discussed in the text. The large five pointed stars are the mean directions of the recent normal and reversed polarity dipole field in the study area

Geochronology and Chemistry

$^{40}Ar/^{39}Ar$ geochronology was undertaken on separates from three plagioclase crystals and a biotite crystal from dykes D4, D15 and D17, and on a groundmass concentrate from aphyric dyke D14. The results are summarized in Table 5.2 and argon release spectra shown in Fig. 5.4. Samples yield age estimates of 1,206 Ma for the single dyke D17 (the ND3 direction), 936 Ma for D14 (the ID direction) and estimates of 929, 821 for D4 and 813 for D15 (the RD direction). As noted in Table 5.2, a maximum age is given for each sample. All plagioclase step-heating spectra indicate the presence of excess argon, most likely derived from abundant fine-grained inclusions contained within the plagioclase and/or excess argon incorporated into the crystal before crystallization.

The age spectrum for samples from D4 and D15 show a classic "saddle-shaped" pattern, whereas sample from D17 shows a less pronounced saddle. These types of spectra are interpreted to indicate the presence of excess argon and the lower part of the saddle is generally considered to be a close approximation to the true or maximum age of the sample (see Fig. 5.4), but may also be younger. The sample from D14 has a low temperature age of 935 Ma and climbs to an age of 4,053 Ma most likely indicating excess argon because the older ages are geologically unreasonable. The biotite age is a maximum because only two heating steps were obtained due to the relatively small amount of sample analyzed. The biotite drops in apparent age from 1,368 to 929 Ma, indicating the possible presence of excess argon. Interpretation of such age spectra should be viewed with caution and supported by independent field relationships. Isochron ages were determined to compare apparent ages obtained from the age spectrum and composition of trapped (nonradiogenic) component(s) of the sample.

Chemical analyses from four dykes comprising D6 and D16 (ND1 group), D13 (RD group) and D17 (ND3) are listed in Table 5.3. They show high Fe and Mg contents and low alkali metal contents, although K_2O:(Na_2O + CaO) ratios are relatively high. These compositions identify dominant tholeiiitic within-plate characteristics according to the criteria of Pearce and Cann (1973).

Interpretation

Since qualifications to the nature of the $^{40}Ar-^{39}Ar$ results (see Table 5.2 footnote) render the link between these ages and emplacement/magnetization of the dykes of this study uncertain, we initially address their interpretation in the context of trends and palaeomagnetic results from other dyke swarms in the shield of south-central India. A previous study was conducted on the dykes in the vicinity of Tirupati by Anjanappa (1975) and although demagnetization and component analysis in that study were not clear, the derived directions of magnetization agree closely with the present investigation: 11 sites in E-W trending dykes yielded a mean of D/I = 350/–12° corresponding with our groups ND2 and RD, whilst 3 sites yielded a mean direction of D/I = 278/–22° correlating with group ID.

Table 5.2 $^{40}Ar/^{39}Ar$ results and incremental heating analyses of basic dykes bordering the Cuddapah Basin

Temp °C	$^{40}Ar_R$[a]	$^{39}Ar_K$	$^{40}Ar_R/^{39}Ar_K$[b]	$^{39}Ar/^{37}Ar$	$\%^{40}Ar_R$	$\%^{39}Ar$	Apparent age[c] ($\pm\sigma$)
D17 (ND3) Plagioclase							
650	4.217	0.0220	192.1	3.66	93.9	0.5	3,422.1 ± 2.8
800	9.413	0.2916	32.3	3.57	98.0	6.8	1,206.4 ± 1.4
900	13.638	0.4197	32.5	3.09	99.3	9.8	1,212.3 ± 1.7
1,000	40.519	0.7194	56.3	2.12	99.3	16.8	1,765.4 ± 1.8
1,100	71.665	0.8417	85.1	1.72	98.9	19.7	2,265.4 ± 2.0
1,200	149.382	1.1667	128.1	2.58	98.2	27.3	2,820.8 ± 2.2
1,300	114.012	0.6849	166.5	1.19	97.9	16.0	3,204.6 ± 2.3
1,450	29.278	0.1342	218.2	0.60	98.0	3.1	3,619.3 ± 3.4

[a]Mineral separates and a groundmass concentrate were separated from 4 rock samples. The samples were crushed, ground, and sieved to 60–120 mesh size (250–125 μm). Samples were passed through magnetic separator and heavy liquids and then handpicked to greater than 99% purity. All samples then were cleaned with reagent-grade acetone, alcohol, and deionized water and air-dried in an oven at 95°C. Between 3 and 34.5 mg of mineral and 99.2 mg of groundmass concentrate were wrapped in aluminum foil packages and encapsulated in silica vials along with neutron-fluence standards prior to irradiation. The standards for this experiment are hornblende MMhb-1 with percent K = 1.555, $^{40}Ar_R = 1.624 \times 10^{-9}$ mole/gm, and K-Ar age = 523.1 Ma (Samson and Alexander, 1987) and FCT sanidine with an internally calibrated age of 28.03 Ma as measured against MMhb-1. For irradiation, an aluminum canister was loaded with six silica vials, each containing samples and standards similar to that described by Snee et al. (1988). Standards were placed between every sample as well as at the top and bottom of each silica vial. Samples were irradiated in one irradiation package in the TRIGA reactor at the US Geological Survey in Denver, Colorado. The length of the irradiation was 120 h at 1 MW. The irradiation package was rotated at 1 rpm during irradiation. All samples and standards were analyzed in the Denver Argon Laboratory of the US Geological Survey using a Mass Analyser Products 215 rare-gas mass spectrometer on a Faraday-cup collector. Each sample was heated in a double-vacuum low-blank resistance furnace (similar to that described by Staudacher et al., 1978) for 20 min, in a series of 2–8 steps, to a maximum of 1,585°C, and analyzed using the standard stepwise heating technique described by Snee (1982). Each standard was degassed to release argon in a single step at 1,250°C for Mnhb-1 hornblende. For every argon measurement, five isotopes of argon (^{40}Ar, ^{39}Ar, ^{38}Ar, ^{37}Ar, and ^{36}Ar) are measured. Detection limit at the time of these experiments was 9.736×10^{-13} moles of argon. Standard techniques were employed to produce $^{40}Ar/^{39}Ar$ age spectra, apparent K/Ca diagrams, and isochron diagrams as described by Snee (2002).

Table 5.2 (continued)

Temp °C	$^{40}Ar_R$[a]	$^{39}Ar_K$	$^{40}Ar_R/^{39}Ar_K$[b]	$^{39}Ar/^{37}Ar$	$\%^{40}Ar_R$	$\%^{39}Ar$	Apparent age[c] ($\pm\sigma$)
D15 (RD), Plagioclase							
650	0.454	0.0073	61.9	0.69	74.5	12.7	1,851.8 ± 2.0
850	0.257	0.0095	27.0	0.62	81.8	16.4	1,041.5 ± 12.0
950	0.324	0.0121	26.8	0.39	95.9	20.9	1,037.0 ± 4.1
1,050	0.272	0.0139	19.7	0.31	96.0	23.9	813.0 ± 3.7
1,150	0.798	0.0059	136.0	0.24	97.2	10.1	2,880.4 ± 4.1
1,250	0.709	0.0034	206.2	0.25	97.5	5.9	3,502.5 ± 9.8
1,350	0.596	0.0036	167.5	0.31	96.5	6.1	3,186.3 ± 6.6
1,450	0.523	0.0023	229.6	0.21	95.1	3.9	3,669.9 ± 31.0
D14 (ID), groundmass concentrate							
650	4.405	0.177	22.9	2.54	92.4	3.4	935.6 ± 1.1
700	8.022	0.2254	35.6	3.91	99.0	4.4	1,302.0 ± 1.5
800	36.031	0.9680	37.2	5.09	99.4	18.8	1,343.9 ± 1.5
900	14.431	0.3631	39.7	3.22	98.9	7.1	1,407.1 ± 1.5
1,000	32.618	0.7094	46.0	1.14	99.1	13.8	1,554.1 ± 1.6
1,100	87.679	1.8251	48.0	0.26	99.5	35.5	1,600.2 ± 1.7
1,200	44.033	0.6368	69.1	0.21	99.2	12.4	2,014.8 ± 1.9
1,300	39.870	0.1402	284.4	0.04	98.6	2.7	4,052.7 ± 3.4
1,450	24.639	0.0945	260.7	0.04	98.5	1.8	3,912.9 ± 3.3

[b]Abundance of "Radiogenic ^{40}Ar" and "K-derived ^{39}Ar" is measured in volts and calculated to five decimal places. Voltage may be converted to moles using 1.160×10^{-12} moles argon per volt signal. "$^{40}Ar_R/^{39}Ar_K$" is calculated to three decimal places. All three are rounded to significant figures using analytical precision.

[c]"$^{40}Ar_R/^{39}Ar_K$" has been corrected for mass discrimination. Mass discrimination was determined by calculating the $^{40}Ar/^{36}Ar$ ratio of aliquots of atmospheric argon pipetted from a fixed pipette on the extraction line; the ratio during these experiment was between 296.6 and 299.1, which was corrected to 295.5 to account for mass discrimination. "$^{40}Ar_R/^{39}Ar_K$" was corrected for all interfering isotopes of argon including atmospheric argon. ^{37}Ar and ^{39}Ar, which are produced during irradiation, are radioactive and their abundances were corrected for radioactive decay. Abundances of interfering isotopes from K and Ca were calculated from reactor production ratios determined by irradiating and analyzing pure CaF_2 and K_2SO_4; the K_2SO_4 was degassed in a vacuum furnace prior to irradiation to release extraneous argon. Corrections for Cl-derived ^{36}Ar were determined using the method of Roddick (1983). Production ratios for this experiment were determined $(^{40}Ar/^{39}Ar)_K$, $(^{38}Ar/^{39}Ar)_K$, $(^{37}Ar/^{39}Ar)_K$, $(^{36}Ar/^{37}Ar)_{Ca}$, $(^{39}Ar/^{37}Ar)_{Ca}$, and $(^{38}Ar/^{37}Ar)_{Ca}$; measured values are available upon request.

Table 5.2 (continued)

Temp °C	$^{40}Ar_R$ [a]	$^{39}Ar_K$	$^{40}Ar_R/^{39}Ar_K$ [b]	$^{39}Ar/^{37}Ar$	$\%^{40}Ar_R$	$\%^{39}Ar$	Apparent age[c] ($\pm\sigma$)
D4 (RD), Plagioclase							
650	2.568	0.07031	36.5	1.36	95.6	3.3	1,289.4 ± 1.4
950	7.794	0.3864	20.2	0.61	96.4	17.9	821.4 ± 1.9
1,050	20.166	0.7584	26.6	1.45	99.5	35.1	1,019.5 ± 1.2
1,150	14.174	0.4422	31.9	0.81	99.2	20.6	1,168.9 ± 1.3
1,250	20.382	0.3398	60.0	0.57	99.0	15.7	1,800.9 ± 2.9
1,400	15.238	0.1601	95.2	0.27	99.3	7.4	2,369.9 ± 2.1
D4 (RD), Biotite							
850	0.661	0.0173	38.1	7.63	80.6	27.9	1,368.3 ± 5.5
1,300	1.013	0.0448	22.6	7.77	97.5	72.1	929.1 ± 1.1
Summary of $^{40}Ar/^{39}Ar$ age results for dyke samples							
Dyke	**Mineral:**	°N	°E	**Isochron age**	**Initial $^{40}Ar/^{36}Ar$**	Maximum age	
D17	Plagioclase	13.2	79.02	1,053 ± 90	4,170 ± 484	**1,206.4 ± 1.4**	
D15	Plagioclase	13.4	79.36	834.1 ± 99	814 ± 160	**813.0 ± 3.7**	
D14	Groundmass	13.4	79.36	926.0 ± 153	18,300 ± 2,630	**935.6 ± 1.1**	
D4	Plagioclase	13.6	79.3	1,020.0 ± 183	978 ± 2,200	**821.4 ± 1.9**	
D4	Biotite			–	–	**929.1 ± 1.1**	

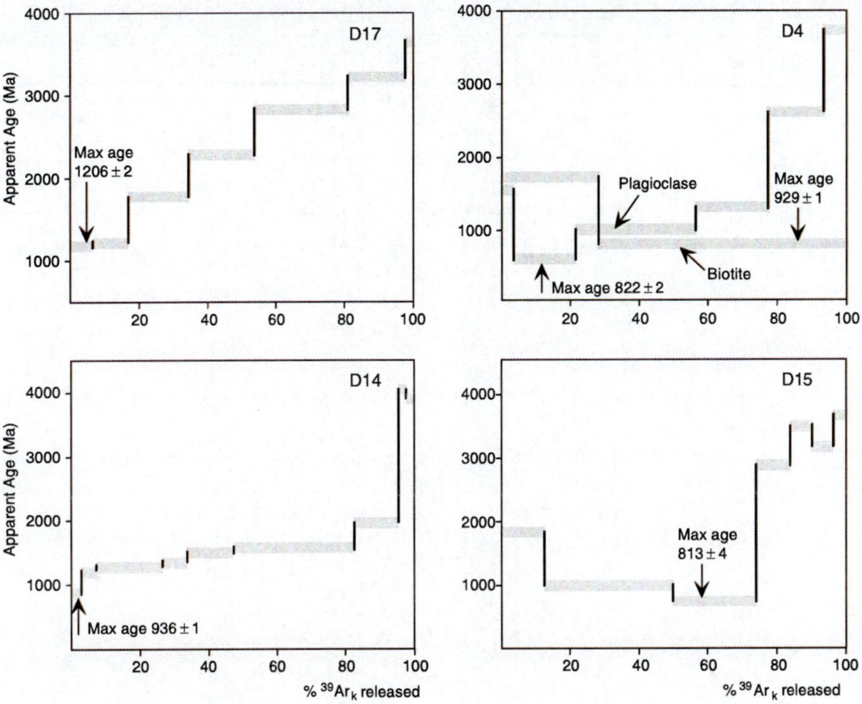

Fig. 5.4 40Ar/39Ar age specta for 4 samples in this study

All other studies on dyke swarms within the Indian Shield have been conducted on swarms to the west of the Cuddapah Basin (Poornachandra Rao, 2005) and further to the south west bordering the Charnockite Belt (Venkatesh et al., 1987; Radhakrishna and Joseph, 1993, 1994, 1996; Radhakrishna and Joseph, 1994). The latter dykes have dominant N-S to NNW-SSE trends in the west and E-W trends in the east (see Fig. 5.1 inset) with the former contrasting in trend with most of the dykes of this study. They are considered to record, at least in part, a major ~1,800 Ma Large Igneous Province (LIP, see Radhakrishna and Joseph, 1996; Ernst and Srivastava, 2008; French et al., 2008) but also include older Palaeoproterozoic dykes (Piper et al., 2003; Halls et al., 2007). With the exception of the Dharmapuri Dykes (D/I =0/+9°, K-Ar ages of 1,565–1,805) the palaeomagnetic results from all these dykes yield steeper inclinations and/or contrasting declinations to the directions shown in Fig. 5.3 (Radhakrishna and Joseph, 1994, 1996; Halls et al., 2007). A similar conclusion applies to results from the N-S to NW-SE trending dykes studied by Venkatesh et al. (1987) in Tamil Nadu; only the single ChRM direction from dyke ND3 is represented in the directional populations resolved by these authors. The dykes on the western side of the Cuddapah Basin studied by Poornachandra Rao (2005) comprise conjugate NW-SE and NE-SW sets contrasting with dyke trends of this study. A wide scatter of K-Ar and Ar^{39}–Ar^{40} age determinations

Table 5.3 Geochemical Data for dykes of this study

Dyke no	D6	D13	D16	D17
SiO_2	48.64	48.28	53.42	53.41
TiO_2	1.01	0.99	1.07	1.14
Al_2O_3	9.8	9.57	9.61	10.39
Fe_2O_3	12.52	12.44	13.1	12.97
MnO	0.18	0.18	0.16	0.16
MgO	11.65	12.17	8.21	7.32
CaO	8.58	8.64	8.11	7.36
Na_2O	1.89	1.78	3.04	3.45
K_2O	0.4	1.12	1.29	1.41
P_2O_5	0.1	0.09	0.18	0.19
LOI	1.8	1.49	0.39	1.1
Total (%):	96.57	96.75	98.58	98.9
	ppm	ppm	ppm	Ppm
Ba	215	195	605	582
Rb	29	48	41	39
Sr	210	214	425	470
Zr	103	96	107	113
Hf	2.84	2.54	2.87	2.67
Nb	4.4	3.7	6.7	6.3
Ta	0.58	0.2	0.4	0.68
Y	20	18.5	18.5	17.3
U	0.63	0.58	0.86	0.81
Th	2.73	2.57	3.55	3.38
La	12.9	12	17.3	16.5
Ce	28	26.2	37.1	35.1
Pr	3.2	2.9	4.3	3.9
Nd	17.09	16.84	21.95	20.41
Sm	3.71	3.53	4.3	3.88
Eu	0.95	0.92	1.1	0.99
Gd	3.79	3.29	3.81	3.4
Dy	3.19	2.89	3.04	2.57
Ho	0.74	0.69	0.62	0.59
Er	1.7	1.61	1.57	1.4
Yb	1.57	1.33	1.38	1.27
Lu	0.23	0.21	0.18	0.19
Th/Ta	4.7	12.9	8.9	5.0
La/Yb	8.22	9.0	12.5	13.0

between 1,884 and 644 Ma suggest that at least some of these dykes belong to the ~1,800 Ma LIP event. The somewhat variable ChRM directions reported from these dykes mostly have intermediate to steep inclinations and declinations contrasting with those shown in Fig. 5.3; the exceptions are five dyke mean directions close to ND1. Rao et al. (1990) also report results from NE-SW trending dykes in the vicinity of Karimnagar to the south west (18.5°N, 79°E) of the present study region and compare them with dykes intruding the northern part of the Cuddapah Basin with which they appear to coalesce. Directions of magnetization are uniformly NE negative similar to the isolated dyke ND3; our $^{40}Ar/^{39}Ar$ age determination of 1,206 Ma is therefore an estimate for the age of this swarm; this suggests a possible link with the extensive Mackenzie and Jotnian age dolerite magmatism elsewhere and with a possible ~1,200 Ma revision of the age of the Harohalli Dyke pole (Radhakrishna and Matthew, 1996, Pradhan et al., 2008, 2010). The remaining majority data from this study comprising the dense E-W dyke swarm at the southern margin of the Cuddapah Basin and yielding magnetization groups ID, ND2 and RD thus appear to define a discrete dyke swarm which is unrepresented in the high-grade terranes to the west and south.

The $^{40}Ar/^{39}Ar$ determinations support this by implying emplacement of ID and RD early in Neoproterozoic times, during which interval the region rotated through ~90°. However, this time frame is appreciably later than the commencement of Cuddapah sedimentation at ~1,900 Ma as implied by the evidence summarized in Sect. 2 and it is surprising to find no evidence for equivalent intrusive or extrusive igneous rocks within the basin. Two interpretations are possible; either (i) the dykes record late Mesoproterozoic-Neoproterozoic continental rifting contemporary with post-Grenville rifting events elsewhere (Windley, 1995) and did not migrate significantly upwards into the overlying supracrustal succession, or (ii) the dykes are of Palaeoproterozoic (or even late Archaean) age and have been overprinted by Late Mesoproterozoic rifting and the Eastern Ghats (Grenville age) orogeny.

The latter explanation might be expected to have produced a uniform regional signature rather than the range of discrete magnetizations determined here. Also the range of $^{40}Ar/^{39}Ar$ ages is embraced by widespread basic dyke injection in consolidating Proterozoic shields in India (Pradhan et al., 2010) and elsewhere including shields with a probable close temporal relationship to India in Neoproterozoic times (e.g. Radhakrishna and Joseph, 1996; Wingate and Giddings, 2000). Hence we interpret the poles in the context of the former explanation but qualify this interpretation with the view that an independent method such as U-Pb geochronology is required to give more confidence to the palaeopole ages.

Poles from India assigned to the interval ~1,200–800 Ma are summarized in Table 5.4. A placement of the Indian Shield adjacent to western North America in the "Rodinian" supercontinent reconstruction (Dalziel, 1991; Hoffman, 1991) is no longer tenable mostly specifically because it requires continental break-up hundreds of millions of years before the geological evidence defines this event near the dawn of the Cambrian (and well before the primary ~850–550 Ma sedimentary correlation), and because it widely separates Gondwana cratons with strong Late Archaean-early Proterozoic affinities (Piper, 2007). It now appears that the Siberian Shield occupied the position conventionally assigned to India (Piper, 2011). Hence we instead evaluate the proposition that India comprised a component of the proto-Gondwana wing in "Palaeopangaea" (Piper, 2010) during Meso-Neoproterozoic times. Owing to paucity of data the position of the Indian Shield within this assemblage is not yet tightly defined. It is also apparent that peripheral parts of Palaeopangaea, including the Gondwana group, underwent significant reorganization during Grenville orogenesis at ~1,100 Ma. These continental movements render the interpretation of any poles with only approximate ages assigned to these times (such as the data in Table 5.4) highly uncertain. Rotational operations of

Table 5.4 A Compilation of Palaeomagnetic poles from India provisionally assigned to the interval ~1,200–800 Ma

Rock unit	Code	Age (Ma)	N	E	A95	References
ND3 dyke	Id3+	–1,206	46	210[a]	8	This study
NDI dykes	Nd1+		34	253	11	This study
ND2 dykes	Nd2+		65	286	5	This study
Harohalli dykes	HD+	1,200[b]	27	79	9	7,544
Kaimur series	KS+	1,150	82	286	6	3,467
ID dykes	ID++	–936	6	336	7	This study
Majhgawan kimberlite	MK++	1,074	37	213	13	Gregory et al., 2006
Karimnagar dykes	KD++	c	32	197	9	Rao et al., 1990
Anatapur dykes, older	Ad1++	–	28	176	14	Pradhan et al., 2010
Anatapur dykes, younger	Ad2++	1,026	10	211	10	Pradhan et al., 2010
RD dykes	RD++	(929–813)	72	27	5	This study

The figures in the reference column are pole numbers in the Global Palaeomagnetic Database.
[a]Virtual geomagnetic Pole.
[b]Revised age suggested by Pradhan et al. (2008). Age assignments in brackets are indications from the Ar_{39}–Ar_{40} spectra but are not regarded as definitive.
[c]Possible similar age to poles MK and AD2 on basis of similar palaeomagnetic direction and same polarity.

India to Laurentia of anticlockwise rotation of 152° about an Euler Pole at 27.5°S, 152.0°N before ~1,100 Ma and 142° about an Euler pole at 48°S, 160.5°E in post-Grenville times are suggested (Piper, 2010).

The suggested ~1,200–800 Ma database of palaeomagnetic poles from India listed in Table 5.4 is tested in Fig. 5.5 with these rotations applied and the poles plotted as stars. Some key data points from other shields rotated into Palaeopangaea are also shown on this figure after Piper (2010). Prior to Grenville orogenesis and relative movements at ~1,100 Ma we observe fair agreement with the possible 5 oldest poles indicating that ND2 (and possibly the other ND Poles and HD) are a signature of widespread ~1,200–1,300 Ma basic dyke injection epitomized by the Mackenzie and Jotnian LIP events in the Laurentian and Fennoscandian Shields respectively.

In post-Grenville times the apparent polar wander (APW) is embraced by the Keweenawan polar track and the Grenville-Sveconorwegian loop (Fig. 5.5). On this path we find of poles ID, KD and RD according with APW during these times; RD plots within the Grenville Loop in a position appropriate to the Ar^{39}–Ar^{40} age results (this might also accommodate ND2). The KD and Vindhyan poles are consistent with the collective path at ~1,030 Ma. However although the Anatapur dykes identify an age shift (AD1→AD2) compatible with motion along the Kewenawan Track and their assigned age of ~1,020 Ma, their positions are shifted to the west and suggest that some adjustment to the reconstruction may be appropriate here.

The dyke swarms along the southern margin of the Cuddapah basin occur in parallel groups with individual members very close to each other and with only thin country rock screens in between. These dykes evidently include several ages

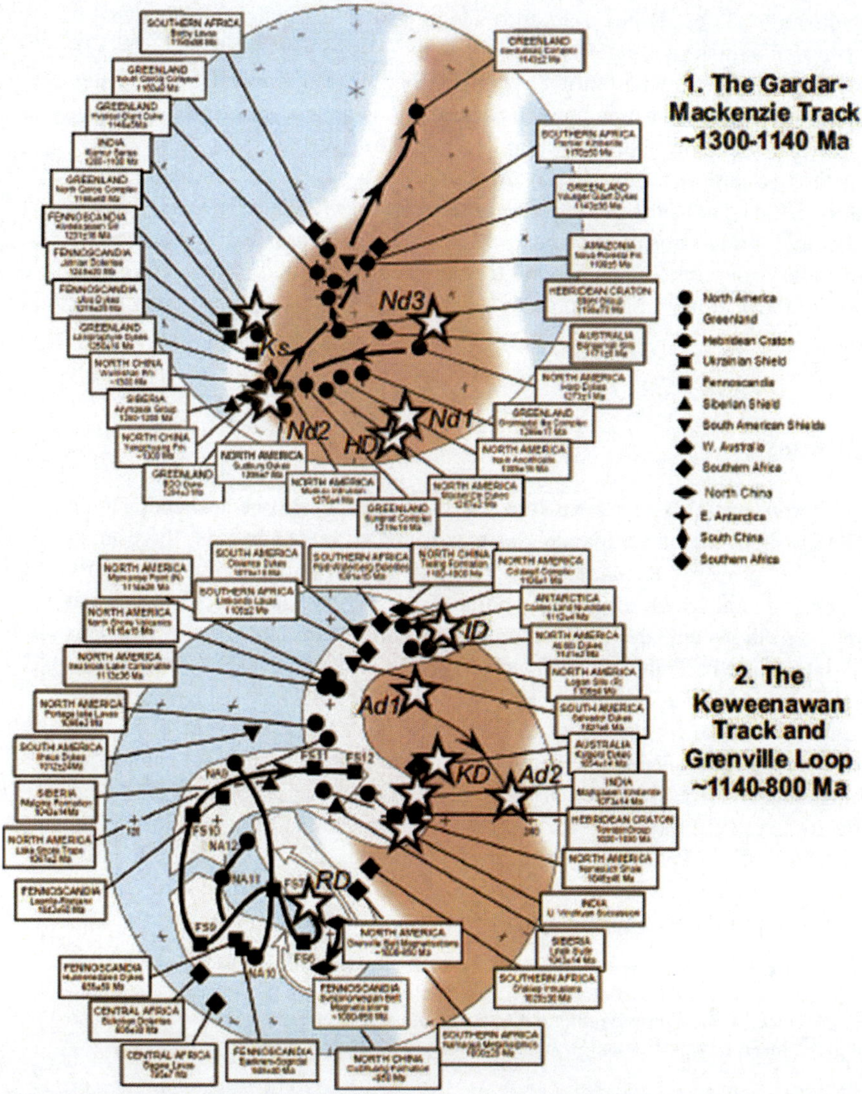

Fig. 5.5 The apparent polar wander path of Palaeopangaea (**a**) between ~1,300 and 1,040 Ma and (**b**) ~1,040 and 800 Ma after Piper (2010). The poles from India provisionally assigned to the interval ~1,200–800 Ma are plotted as stars and listed in Table 5.4. They are rotated into the Laurentian reference frame using operations given in the text for comparison with the master path. The continental area of Palaeopangaea is indicated by the *brown colour*

of magnetization although they were intruded along stress systems so close that it difficult to distinguish one parallel dyke from the other (Venkatesh et al., 1987). Joint pattern study and quartz fabric analyses of the intruded granite from the study area suggest that the dykes in this region represent tension fractures trending almost east-west although they are also temporally related to a compressive stress system acting in the N-S direction (Suryanarayana and Anjanappa, 1975). Thus these dykes could be considered to have been intruded in association with E-W tensional stresses that prevailed during abortive attempts to break-up the continental crust at some stage of Proterozoic times. This general tectonic setting would be compatible with similarity of the chemistry of these dykes to rift volcanics (Damodara Reddy et al., 1995).

Conclusions

The E-W Proterozoic dyke swarm into the basement of east-central India lying immediately south of the Cuddapah Basin is found to record at least three magnetization, and probably emplacement, events of which two were probably nearly contemporaneous although of opposite polarity. $^{40}Ar/^{39}Ar$ age results suggest that these magnetizations are late Mesoproterozoic to early Neoproterozoic in age. This is much later than the Palaeoproterozoic age of the commencement of sedimentation in the Cuddapah Basin and suggests that either the dykes have not invaded the overlying supracrustal succession or that they are much older than ages defined by $^{40}Ar/^{39}Ar$ age spectra. The former possibility is suggested by preservation of discrete magnetization events. An assessment of these and other poles from the Indian Shield assigned to the interval ~1,200–800 Ma suggests that they fall close to the integral APW path of Palaeopangaea during these times although some refinement of India's position within the proto-Gondwana continental grouping during these times would appear to be required.

Acknowledgements The authors are grateful to late C. Radhakrishnamurty for his help and suggestions in the early stage of this work. MRG and CVRKP are thankful to Department of Science & Technology, Govt. of India for providing the financial assistance to carry out this study. We are grateful to Joe Meert for a review and Rajesh K. Srivastava for careful editorial advice.

References

Anand M, Gibson SA, Subbarao KV, Kelley SP, Dickin AP (2003) Early-Proterozoic melt generation processes beneath the intra-cratonic Cuddapah Basin, Southern India. J Petrol 44: 2139–2171

Anjanappa K (1972) Application of Palaeomagnetism in estimating the age of the dyke swarms in Peninsular India. Curr Sci 41: 106–108

Anjanappa K (1975) Palaeomagnetism and age of dolerite dykes of the Tirupati area, Chittoor district, Andhra Pradesh, In: Recent Researches in Geology, Hindustan Publishers Co. 2: 162–169

Aswathanarayana A (1964) Age of the Cuddapahs, India. Nature 193: 470–471

Bhaskara Rao YJ, Pantulu GVC, Damodara Reddy V, Gopalan K (1995) Time of early sedimentation and volcanism in the Proterozoic Cuddapah Basin, South India. In: Devaraju TC (ed) Evidence form the Rb-Sr age of Pulivendla mafic sill, Dyke swarms of Peninsular India, Geol Soc Ind Bangalore 33: 329–338

Bhattacharji S (1987) Lineaments and igneous episodes in the evolution of intracratonic Proterozoic basins of India. In: Saha AK (ed) Geological evolution of Peninsular India – Petrological and Structural aspects, Recent Researches in Geology, Hindustan Publ. Co. 13: 1–16

Buchan KL, Mertanen S, Park RG, Pesonen LJ, Elming SA, Abrahamsen N, Bylund G (2000) Comparing the drift of Laurentia and Baltica in the Proterozoic: The importance of key palaeomagnetic poles. Tectonophysics 319: 167–198

Chatterjee N, Bhattacharji S (2001) Petrology, geochemistry and tectonic settings of the mafic dykes and sills associated with the evolution of the Proterozoic Cuddapah basin of south India. Proc Indian Acad Sci Earth and Plan Sci 110: 433–453

Crawford AR, Compston W (1973) The Age of the Cuddapah and Kurnool systems, South India. J Geol Soc Aust 19: 453–464

Dalziel IWD (1991) Pacific margins of Laurentia and east Antarctica-Australia as a conjugate pair: Evidence and implications for an Eocambrian supercontinent. Geology 19: 598–601

Damodara Reddy V, Anjanappa K, Prasad CVRK, Subba Reddy N (1995) Magnetic properties and chemistry of Precambrian dykes from Peninsular India. Geol Soc India Bangalore In: TC Devaraju (Ed.) Dyke swarms of Peninsular India 33: 99–110

Drury SA, Holt RW (1980) The tectonic framework of the South Indian craton: A reconnaissance involving LANDSAT imagery. Tectonophysics 65: T1–T15

Ernst RE, Srivastava RK (2008) India's place in the Proterozoic world: Constraints from the Large Igneous Province (LIP) record. In: Srivastava RK, Sivaji C, Chalapathi Rao NV (eds) Indian Dykes: Geochemistry, Geophysics and Geochronology. Narosa Publishing House Pvt. Ltd, New Delhi: 41–56

French JE, Heaman LM, Chacko T, Srivastava RK (2008) 1891–1883 Ma southern Bastar craton-Cuddapah mafic igneous events, India: A newly recognized large igneous province. Precam Res 160: 308–322

Goutham MR, Raghubabu K, Prasad CVRK, Subbarao KV, Damodara Reddy V (2006) A Neoproterozoic Geomagnetic field Reversal from India: Stratigraphic Significance and implication to the formation of Gondwana. J Geol Soc India 67: 221–233

Gregory LC, Meert JG, Pradham V, Pandit MK, Torsvik TH (2006) A palaeomagnetic and geochronolgic study of the Majhgawan kimberlite, India: Implications for the age of the Vindhyan Supergroup. Precam Res 149: 65–75

Halls HC, Kumar A, Srinivasan R, Hamilton MA (2007) Paleomagnetism and U- Pb geochronology of eastern trending dykes in the Dharwar craton, India: Feldspar clouding, radiating dyke swarms and the position of India at 2.37 Ga. Precam Res 155: 47–68

Halls HC, Palmer HC, Bates MP, Phinney WC (1994) Constraints on the nature of the Kapuskasing structural zone from the study of Proterozoic dyke swarms. Can J Earth Sci 31: 1182–1196

Heaman LM (1997) Global mafic magmatism at 2.5 Ga: Remanents of an ancient large igneous province? Geology 25: 299–302

Hoffman PF (1991) Did the breakout of Laurentia turn Gondwanaland inside out? Science 252: 1409–1412

King W (1872) The Kadapa and Kurnool formations of Madras Presidency. Memoir Geol Surv India 8: 320 p

Krishnan MS (1982) The Geology of India and Burma. CBS Publications, New Delhi

Kumar A, Bhalla MS (1983) Palaeomagnetics and igneous activity of the area adjoining the southwestern margin of the Cuddapah Basin, India. Geophys J Royal Astronom Soc 73: 27–37

McFadden PL (1990) A new fold test for palaeomagnetic studies. Geophys J Int 103: 163–169

Murty YGK, Babu Rao V, Guptasarma D, Rao JM, Rao MN, Bhattacharji S (1987) Tectonic, Petrochemical and Geophysical studies of Mafic Dyke Swarms around the Proterozoic

Cuddapah Basin, South India. In: Halls HC and Fahrig WF (eds) Mafic Dyke Swarms. Geol Assoc Canada Spec Paper 34: 303–316

Nagaraja Rao BK, Rajurkar ST, Ramalingaswamy G, Ravindrababu B (1987) Stratigraphy, Structure and Evolution of the Cuddapah Basin, Memoir 6, Geol Soc India 33–86

Pearce JA, Cann JR (1973) Tectonic setting of basic volcanic rocks determined using trace element analysis. Earth Planet Sci Lett 19: 290–300

Piper JDA (2007) The Neoproterozoic Supercontinent Palaeopangaea. Gondwana Res 12: 202–227

Piper JDA (2010) Palaeopangaea in Meso-Neoproterozoic times: The palaeomagnetic evidence and implications to continental integrity, supercontinent form and Eocambrian break-up. J Geodyn 50: 191–223

Piper JDA (2011) SWEAT and the end of SWEAT: The Laurentia-Siberia configuration during Meso-Neoproterozoic times, International Geology Review, DOI:10.1080/00206814.2010.527679

Piper JDA, Mallik SB, Bandyopadhyay G, Mondal S, Das AK (2003) Palaeomagnetic and rock magnetic study of a deeply-exposed continental crustal section: The charnockite belt of Southern India. Precam Res 121: 185–219

Poornachandra Rao GVS (2005) Orthogonal dykes around the Cuddapah Basin – palaeomagnetic study. J Indian Geophys Un 9: 1–11

Pradhan VR, Meert JG, Pandit MK, Kamenov G, Gregory LC, Malone SJ (2010) India's changing place in global Proterozoic reconstructions: A review of geochronologic constraints and palaeomagnetic poles from the Dharwar, Bundelkhand and Marwar cratons. J Geodyn 50: 224–242

Pradhan VR, Pandit MK, Meert JG (2008) A cautionary note of the age of the paleomagnetic pole obtained from the Harohalli dyke swarms, Dharwar craton, southern India. In: Srivastava RK, Sivaji Ch, Chalapati Rao NV (eds) Indian Dykes: Geochemistry, Geophysics, and Geochronology. Narosa Publishing Ltd, New Delhi, India: 339–352

Radhakrishna T, Joseph M (1993) Proterozoic Palaeomagnetism of South Indian shield and tectonic constraints. Mem Geol Soc India 25: 321–336

Radhakrishna T, Joseph M (1994) Proterozoic Palaeomagnetism of the mafic dyke swarms in the high-grade region of South India. Precam Res 76: 31–46

Radhakrishna T, Joseph M (1996) Late Precambrian (850–800 Ma) palaeomagnetic pole for the south Indian Shield from the Harohalli alkaline dykes: Geotectonic implications for Gondwana reconstructions. Precam Res 80: 77–87

Radhakrishnamurty C (1993) Magnetism and Basalts. Geol Soc Ind Bangalore 26: 208

Rao JM, Rao GVSP, Patil SK (1990) Geochemical and palaeomagnetic studies on the Middle Proterozoic Karimnagar mafic dyke swarm, India. In: Parker AJ, Rickwood PC, Tucker DH (eds) Mafic Dykes and Emplacement Mechanisms. Balkerma, Rotterdam: 373–382

Roddick JC (1983). High precision intercalibration of ^{40}Ar-^{39}Ar standards. Geochimica Cosmochimica Acta 47: 887–898

Samson DD, Alexander EC (1987) Calibration of interlaboratory ^{40}Ar/^{39}Ar dating standard, MMhb-1. Chem Geol (Isotope Geosci Sec) 66: 27–34

Snee LW (1982) Emplacement and cooling of the Pioneer Batholith, south-western Montana [Ph.D. thesis], Ohio State University, Columbus 320 pp

Snee LW (2002) Argon Thermochronology of mineral deposits – A review of analytical methods, formulations, and selected applications. US Geol Sur Bull B: 2194

Snee LW, Sutter JF, Kelly WC (1988) Thermochronology of economic mineral deposits: Dating the stages of mineralization at Panasqueira, Portugal, by high precision ^{40}Ar/^{39} Ar age-spectrum techniques on muscovite. Econ Geol 83: 335–354

Staudacher TH, Jessburger EK, Dorflinger D, Kiko J (1978) A refined ultrahigh – vacuum furnace for rare-gas analysis. J Phys E Sci Instrum 11: 781–784

Steiger RH, Jager E (1977) Subcommission on geochronology: Convention on the use of decay constants in geo- and cosmochronology. Earth Planet Sci Lett 36: 359–362

Suryanarayana KV, Anjanappa K (1975) Joint pattern and Dyke trends in the Tirupati area, Andhra Pradesh, India. Proc Ind Natl Sci Acad Part A 41: 86–402

Venkatesh AS, Poornachandra Rao GVS, Prasad NTV, Bhalla MS (1987) Palaeomagnetic and Geochemical studies on Dolerite dykes from Tamil Nadu, India. Precam Res 34: 291–310
Windley BF (1995) The Evolving Continents. Wiley, New York, NY
Wingate MTD, Giddings JW (2000) Age and palaeomagnetism of the Mundine Well dyke swarm, Western Australia: Implications for an Australia-Laurentia connection at 755 Ma. Precam Res 100: 335–357
Zachariah JK, Bhaskar Rao YJ, Srinivasan R, Gopalan K (1999) Pb, Sr, and Nd isotope systematics of Uranium mineralised stromatolitic dolomites from the Proterozoic Cuddapah Supergroup, South India: Constraints on age and Provenance. Chem Geol 162: 49–64

Chapter 6
Palaeoproterozoic Dykes and Event Stratigraphy of the South Indian and Western Australian Cratons: Implications for Assembly of an Archaean Supercontinent "SIWA" and Its Breaking

S. Mohanty

Introduction

The Archaean–Proterozoic boundary signifies accretion of ~80% of the earth's crust to form continental masses. Application of the concepts of continental drift and plate tectonics has led to identification of supercontinents in geological past. Palaeomagnetic investigations together with the worldwide orogenic events, and widespread continental rifting, anorogenic magmatism and emplacement of mafic dyke swarms provide testing tools for identification of supercontinental assembly and their break-up. The continental reconstructions for the older rocks are hampered by wide spread deformation, metamorphism and magmatism in many cratonic blocks. However, successful reconstruction of Neoproterozoic supercontinent Rodinia has invited suggestions for the existence of a Palaeoproterozoic to Mesoproterozoic supercontinent "Columbia" (Rogers and Santosh, 2002, 2009). The oldest protocontinental nuclei the Ur block (Australia, India, Madagascar, Zimbabwe and Kaapvaal) collided with the Nena (North America, Siberia, Baltica and Greenland) and the Atlantica (Northwest Africa and South America) during the global-scale collisional events between 2,100 and 1,500 Ma to form the oldest supercontinental assembly named as "Columbia" (Fig. 6.1). In this hypothesis the Indian and Western Australian cratons are assumed to be part of the Ur block. Although the assembly of India – Western Australia – East Antarctica during Neoproterozoic is fairly well understood (Fig. 6.2), the exact juxtaposition and the relationship between the litho-tectonic developments of these blocks during Palaeoproterozoic – Mesoproterozoic time are yet to be understood properly. On the basis of similar geochemical characters of mafic dyke swarms Srivastava et al. (2000), Srivastava and Singh (2003), Srivastava (2008) supported the supercontinent assembly of Columbia, and suggested juxtaposition of East Antarctica and Eastern-Central-Southern Indian cratons. In this paper an attempt is made to examine the

S. Mohanty (✉)
Department of Applied Geology, Indian School of Mines, Dhanbad 826004, India
e-mail: mohantysp@yahoo.com

R.K. Srivastava (ed.), *Dyke Swarms: Keys for Geodynamic Interpretation*,
DOI 10.1007/978-3-642-12496-9_6, © Springer-Verlag Berlin Heidelberg 2011

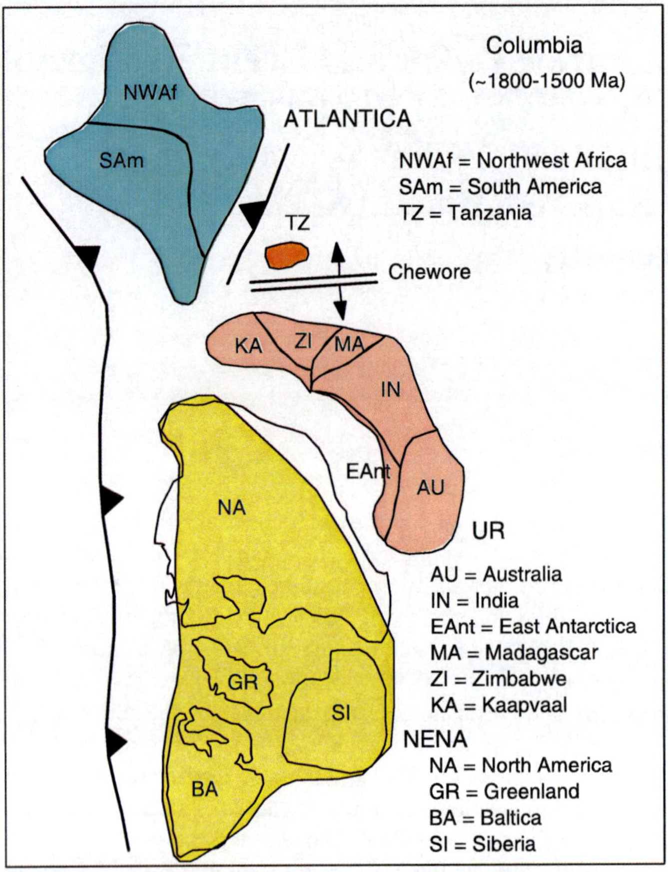

Fig. 6.1 Configuration of the supercontinent Columbia, showing the protocontinental nuclei NENA, UR and ATLANTICA (from Rogers and Santosh, 2002)

palaeomagnetic database and Palaeoproterozoic evolution of the Indian and Western Australian cratons to determine the possible movement patterns for the assembly of Columbia.

Geological Setup of the Indian Craton

The Indian craton is divisible into two tectonic blocks (the North Indian Block and the South Indian Block) separated by a prominent E-W trending mountain belt (the Satpura Mountains) in the central part of India (Fig. 6.2). Both the tectonic blocks comprise of Archaean gneisses and schists, unconformably overlained by the Proterozoic rocks (Naqvi and Rogers, 1987). The Archaean basement and its Proterozoic cover rocks constitute five protocontinental nuclei of India (the Western Indian Craton and the Bundelkhand craton of the North Indian Block; and the

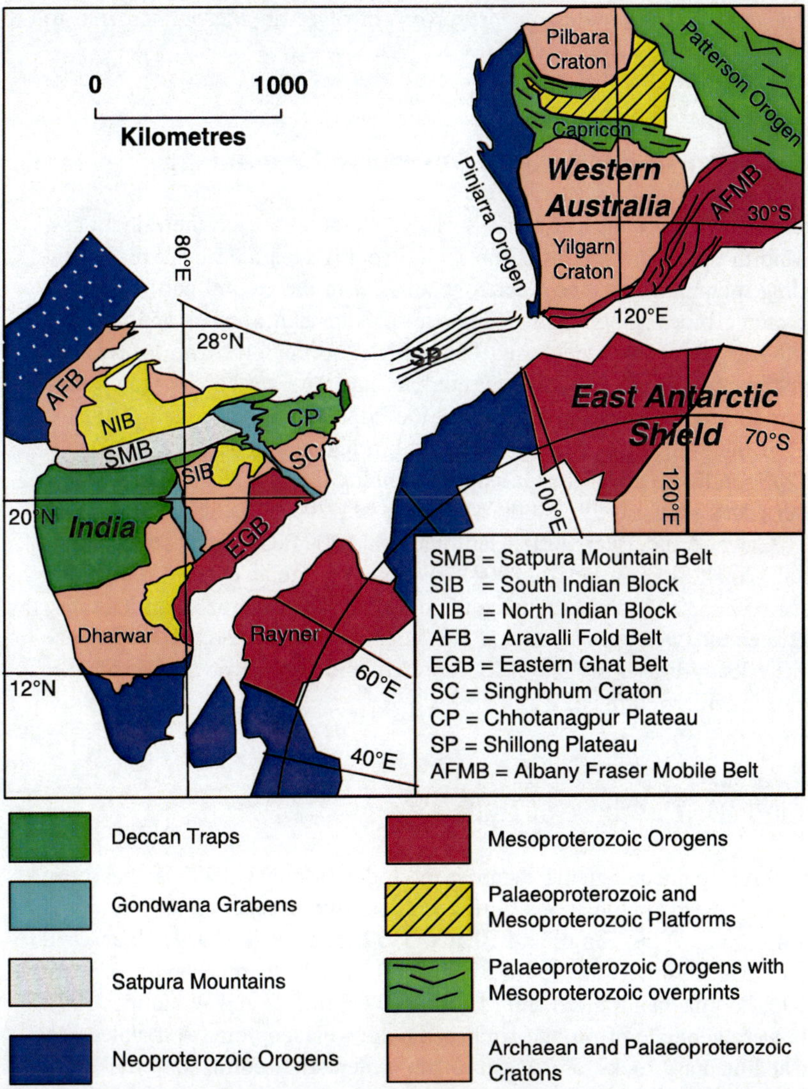

Fig. 6.2 A part of the supercontinent Rodinia, showing the juxtaposition of the Indian Shield, East Antarctic Shield and Western Australian Shield (after King, 1973; Powell et al., 1988, 2001)

Singhbhum, Bastar, Dharwar cratons of the South Indian Block). These nuclei were involved in multiple deformation and metamorphism. Undeformed platformal sediments constitute several basins (the Cuddapah, Chhattisgarh, Bhima and Pakhal basins of the South Indian block; and the Vindhyan and Marwar basins of the North Indian Block). Three ancient mountain belts (the Aravalli Mountain belt of Western India, the Satpura Mountain belt of central India, and the Eastern Ghats along the East Coast of India), and the southern peninsula (the Southern Granulite Terrane) have Archaean to Palaeoproterozoic precursors involved in Mesoproterozoic to Neoproterozoic overprints.

Geological Setup of the Western Australian Craton

The Western Australian craton comprises of two tectonic blocks (the Pilbara craton in the north and the Yilgarn craton in the south) separated by a prominent E-W trending mountain belt (the Capricorn orogen) in the central part (Fig. 6.2). The two cratonic blocks dominantly comprise of Archaean granites and gneisses interlayered with greenstone belts and banded iron formations. The Proterozoic metamorphic rocks occur in the Gascoyne complex and the Capricorn orogen, which represent the collisional phase amalgamating the Pilbara and the Yilgarn cratons at $\sim$1,800 $\pm$ 30 Ma. An older orogeny (Glenburgh orogeny at $\sim$1,980 $\pm$ 20 Ma) is present in the southern part of the Gascoyne Complex (Myers, 1995; Martin et al., 1998, 2000; Clark et al., 1999, 2000; Wingate et al., 2000; Collins, 2003; Evans et al., 2003; Cawood and Tyler, 2004; Halilovic et al., 2004; Occhipinti et al., 2004; Pirajno et al., 2004; Sheppard et al., 2004). Two more orogenic events (Mangaroon orogeny at 1,650 $\pm$ 30 Ma and Edmundian Orogeny of Neoproterozoic age) are also present between the two cratonic blocks. The Yilgarn craton is also bordered on the south by the Albany-Fraser orogen of Mesoproterozoic age and on the west by the Neoproterozoic Pinjarra orogen.

Palaeomagnetic Database of India

The Precambrian palaeomagnetic database of India (Mohanty, 2006a) has been compiled from several publications during the last five decades (Athavale et al., 1963; Verma et al., 1968; Singh and Bhalla, 1972; Klootwijk, 1974; Verma and Prasad, 1974; McElhinny et al., 1978; Bhalla et al., 1980; Poornachandra Rao and Bhalla, 1996; Poornachandra Rao and Mishra, 1997; Schmidt and Williams, 1999). These data were generated from the Archaean dykes; Palaeoproterozoic volcanics, dykes and sedimentary rocks; and Meso-to Neoproterozoic sedimentary rocks and volcanics. Analysis of the apparent polar wandering path for the Precambrian rocks of India demonstrates six "hairpin bends", suggesting complex behaviour of the Indian craton during the Precambrian time (Mohanty, 2005, 2006a). These bends are identified at 2,500 $\pm$ 100 Ma, 1,950 $\pm$ 50 Ma, 1,800 $\pm$ 50 Ma, $\sim$1,500 Ma, $\sim$1,200 and $\sim$850 Ma. The first event signifies the breaking of the Archaean craton and development of rift to passive margin sedimentation process. The second event was associated with the development of two major orogenic events (the Aravalli Orogeny of Western India and the Satpura Orogeny of Central India). The redistribution of the stress patterns during these orogenic events possibly gave rise to the development of the Cuddapah basin and the Gwalior basin. The 1,800 $\pm$ 50 Ma event marks the second deformation and granulite facies metamorphism of the Satpura

Mountain belt and deposition of the Lower Vindhyan rocks. The Mesoproterozoic events at ~1,500 and 1,200 Ma represent the development of the Delhi and Eastern Ghats orogenies. During this time the rocks of the Chhattisgarh, Kaimur and Rewa Groups were deposited. The Neoproterozoic event of 850 Ma marks the dispersal of the Rodinia, leading to the deposition of the Bhander Group and Malani Rhyolites.

The two major and most prominent hairpin bends in the apparent polar wandering path of India (1,950 ± 50 Ma and 1,800 ± 50 Ma events) represent the time period when the Glenburgh and Capricorn orogeny took place between the Yilgarn craton and the Pilbara craton. The tectono-stratigraphic evolutions of the Satpura Mountain belt separating the North Indian and South Indian Blocks, and of the Capricorn Orogen-Gascoyne Complex show identical patterns from Archaean to Palaeoproterozoic (Mohanty, 2010). Therefore, more detailed examination of the juxtaposition of these two blocks is required.

Significance of ~2,400 Ma Dykes of Dharwar and Yilgarn Cratons

The palaeomagnetic investigations of ~2,400 Ma dyke swarms of the Dharwar and the Yilgarn cratons (Halls et al., 2007) have shown that these cratons were located at ~75° and ~50°, respectively (Fig. 6.3a). These studies also suggest that the Dharwar craton of India and the Yilgarn craton were possibly close to each other at the Archaean–Proterozoic boundary. However, the model proposed by Halls et al. (2007) shows the Yilgarn craton at the position presently occupied by the Himalayas and the Tibetian plateau (Fig. 6.3a). As the palaeomagnetic data do not constrain the

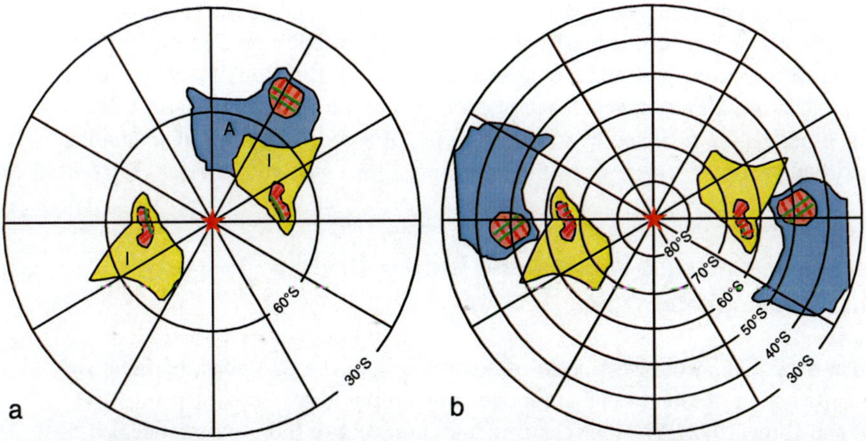

Fig. 6.3 Position of India and Western Australia at ~2,400 Ma (**a**) as proposed by Halls et al. (2007), and (**b**) the proposed model after rotating the Western Australia by 180° and shifting it to the east coast of India

Table 6.1 Compilation of published paleomagnetic data for Archaean–Palaeoproterozoic dykes of Yilgarn craton, Western Australia

Sl. No.	Dyke	N	I	D	λ_o	λ_p	ϕ_p	α_{95}	References
1.	Widgiemooltha	11	–67°	242°	–50°	8°	337°	8°	Evans, 1968
2.	YA	6	81°	61°	72°	–22°	134°	10°	Giddings, 1976
3.	YD	10	46°	258°	27°	–24°	45°	10°	Giddings, 1976
4.	YF	10	51°	345°	32°	25°	103°	11°	Giddings, 1976
5.	Ravensthorpe	4	83°	114°	76°	–38°	136°	13°	Giddings, 1976

longitude of a continental block, Halls et al. (2007) had used the parity of orientation of the Archaean dykes for their model and suggested that either craton can be placed at any other position along the same latitude. The Archaean palaeomagnetic data from the Yilgarn craton has a 2,410 ± 2 Ma key pole from the Widgiemooltha dyke (Evans, 1968); and another dyke from the Perth region (YA) is believed to be of same age (Giddings, 1976). These two dykes show opposite inclination and polarity (Table 6.1). Therefore for the southern hemispheral position of the Yilgarn craton the polarity data for YA dyke suite was taken and the Yilgarn craton was rotated by 180° to be placed adjacent to the east coast of India. It is found that the gap between the South Indian craton and the Yilgarn craton reduces substantially and a very good match is obtained between the coastlines of the two blocks (Fig. 6.3b). At this position the NW-SE trending basic dyke swarms of the Bastar craton match in orientation with the ~E-W trending dyke swarms of the Yilgarn craton (Fig. 6.4). The palaeomagnetic database for India (Mohanty, 2006a) was re-examined to assess the tectonic significance of the Archaean–Proterozoic dykes and sedimentary rocks up to 1,500 Ma (Table 6.2). It is observed that the dykes of ~2,400 Ma (Sl. No.10–15 in Table 6.2) show high inclination similar to the Yilgarn dykes. It may be noted that the north directions for the Yilgarn craton and the Dharwar craton at ~2,400 Ma determined from palaeomagnetic data also show matching orientation (Fig. 6.4). As three of the major key elements of continental reconstruction (i.e. palaeolatitude, palaeo-north directions and mafic dyke orientations; cf. Ernst and Buchan, 2001; Bleeker and Ernst, 2006) suggest for joint evolution of the South Indian and Yilgarn cratons, other geometrical and temporal elements of reconstruction were examined more critically.

Coastline of Australia Against the Indian Craton During Palaeoproterozoic

The idea of matching the west coast of Australia with the east coast of India for the reconstruction of Gondwanaland model was proposed by Crawford and Oliver (1969) and King (1973). The idea changed later in favour of India–Antarctica coastline fit (Crawford, 1974; Powell et al., 1988). The concept was re-examined by Harris (1993, 1995), but was discarded on the basis of apparent continuity of the tectonic trends of the Central Indian Tectonic Zone (CITZ) on to the Albany-Fraser

Table 6.2 Compilation of published paleomagnetic data for Archaean–Palaeoproterozoic dykes, volcanics and sedimentary rocks of South Indian Block

Sl. No.	Dyke	N	I	D	λ_o	λ_p	ϕ_p	α_{95}	References
1.	Newer Dolerite I	7	09	344	05	66	310	22	Verma and Prasad, 1974
2.	Newer Dolerite II	1	46	142	27	–28	125	13	Verma and Prasad, 1974
3.	Peninsular gneiss	9	43	285	25	18	010	12	Prasad et al., 1999
4.	Clospet granite	5	46	300	27	32	009	18	Prasad et al., 1999
5.	Mysore 1 NW	20	–48	329	–29	39	292	7	Hasnain and Qureshy, 1971
6.	Tiptur 1	6	–21	287	–11	–14	152	13	Bhalla et al., 1980
7.	Tiptur 2	2	33	334	18	–65	180	22	Bhalla et al., 1980
8.	Ananthapur 1	5	–34	320	–19	38	298	13	Anil Kumar and Bhalla, 1983
9.	Bastar and Malanjkhand	–	26	273	14	08	002	–	Basumallick et al., 2002
10.	Mysore 2 WNW	4	–75	141	–62	–33	237	10	Hasnain and Qureshy, 1971
11.	Harohalli dyke	16	–88	131	–86	–15	254	33	Prasad et al., 1999
12.	Harohalli A	7	–81	095	–72	14	060	7	Halls et al., 2007
13.	Sargur	6	79	302	69	22	057	–	Bhalla et al., 1980
14.	Pitepani volcanics	5	–41	343	–23	43	285	13	Poornachandra Rao and Mishra, 1997
15.	Bangarpet	–	–81	129	–72	24	063	–	Bhalla et al., 1980
16.	Newer Dolerite III	2	82	056	74	30	094	16	Verma and Prasad, 1974
17.	Holenarsipur	3	62	042	43	43	117	11	Bhalla et al., 1980
18.	Ananthapur 1	6	–69	057	–56	–07	047	7	Anil Kumar and Bhalla, 1983
19.	Ananthapur 2	6	–72	071	–51	03	047	5	Anil Kumar and Bhalla, 1983
20.	Karimnagar 1	5	–63	040	–44	17	230	13	Subba Rao and Murthy, 1980
21.	Mysore NW	9	–62	018	–43	32	242	7	Hasnain and Qureshy, 1971
22.	Mysore NW	5	–59	008	–40	38	248	10	Hasnain and Qureshy, 1971
23.	Mysore 3 EW	12	50	026	31	61	127	13	Hasnain and Qureshy, 1971
24.	Dharwar	–	44	011	26	74	116	–	Singh and Bhalla, 1972
25.	Mysore EW	12	–30	002	–16	60	253	9	Hasnain and Qureshy, 1971
26.	Sitagota volcanic	4	–28	257	–15	17	161	8	Poornachandra Rao and Mishra, 1997
27.	Tirupati NS	–	–22	278	–11	05	337	–	Anjanappa, 1972
28.	Karimnagar 2	–	12	068	06	23	168	9	Subba Rao and Murthy, 1980
29.	Ananthapur	–	–08	057	–04	31	182	–	Anil Kumar and Bhalla, 1983

Table 6.2 (continued)

Sl. No.	Dyke	N	I	D	λ_o	λ_p	ϕ_p	α_{95}	References
30.	Ananthapur	–	–07	064	–04	23	177	–	Anil Kumar and Bhalla, 1983
31.	Kolar EW	27	–18	037	–09	47	197	2	Damodara Reddy and Prasad, 1979
32.	Hyderabad EW	10	–16	032	–08	49	205	7	Anil Kumar and Rao, 1979
33.	Hyderabad EW	13	–03	044	–02	43	188	5	Verma et al., 1968
34.	Tirupati EW	4	–18	349	–09	65	285	6	Anjanappa, 1972
35.	Tirupati EW (Reverse)	7	09	171	04	–70	107	6	Anjanappa, 1972
36.	Kolar EW	14	–06	316	–03	43	331	4	Damodara Reddy and Prasad, 1979
37.	Kolar EW	19	–09	324	–04	50	324	3	Damodara Reddy and Prasad, 1979
38.	Harohalli B	7	02	000	00	79	257	9	Halls et al., 2007
39.	Cuddapah 1	19	–06	298	–03	26	338	4	Prasad et al., 1987
40.	Cuddapah 2	10	–12	294	–06	25	339	7	Prasad et al., 1987
41.	Cuddapah 3	18	–05	300	–01	26	333	9	Prasad et al., 1987
42.	Cuddapah 4	29	00	140	00	–55	324	7	Prasad et al., 1987

Mobile Belt (both the coastlines being on opposite sides of the line of continuity) and on the assumption that both the tectonic zones are of Grenvillian time. Recent tectonic studies in the CITZ indicate it to be a Palaeoproterozoic orogenic belt overprinted by the Grenvillian event (Mohanty, 2000, 2003, 2006b; Zhao et al., 2002, 2003). Mohanty (2010) has shown that the CITZ is a Palaeoproterozoic orogenic belt (~2,100 Ma) rejuvenated during the Satpura Orogeny correlatable with the Capricorn Orogeny (~1,800 Ma) on the northern side of the Yilgarn craton. This gives an apparent offset of ~1,200 km. Tectonic studies in the Western Australian craton have shown that the Yilgarn craton is bound on the south by the Albany-Fraser belt, which has been involved in a dextral displacement of ~500 km during Mesoproterozoic (~1,300 Ma). The western margin of the Yilgarn craton is affected by a sinistral strike slip movement along the Darling Range fault during Neoproterozoic time (~850 Ma). When we remove these Mesoproterozoic and Neoproterozoic strike slip movements of the Yilgarn craton from its reconstructed position during Gondwana/Rodinia, it comes exactly to the east of the position occupied by the Eastern Ghats. Geochronological studies in the Eastern Ghats (Rickers et al., 2001; Dobmeier and Simmat, 2002) suggest that it an orogenic belt which formed during the Grenvillian orogeny (~1,000 Ma) and finally thrusted over the Bastar craton at ~540 Ma (Collins and Pisarevsky, 2005). Therefore, the author proposes to fit the Yilgarn craton of the Western Australia with the South Indian craton at the position of the Eastern Ghats at ~2,400 Ma (Fig. 6.4).

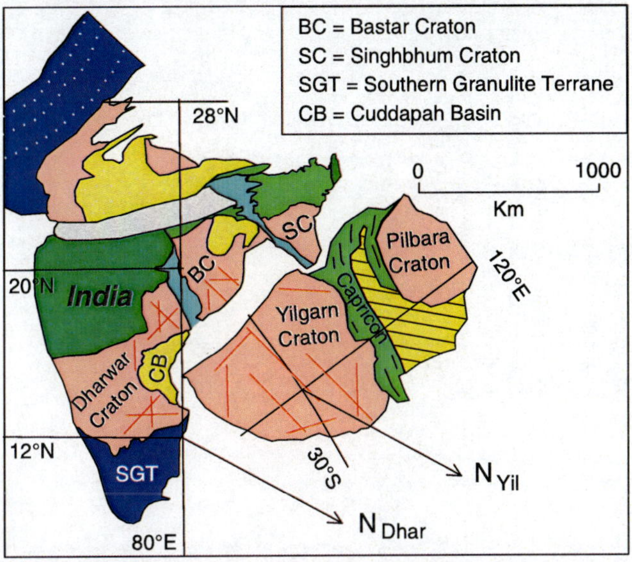

Fig. 6.4 Configuration of an Archaean–Palaeoproterozoic supercontinent "SIWA". Representative basic dykes in the South Indian (SI) and the Western Australian (WA) *blocks* are shown by *red colour*. *Arrows* marked as N_{Yil} and N_{Dhar} are north directions for Yilgarn craton and Dharwar craton at ~2,400 Ma, respectively. The other legends are same as in Fig. 6.2

Barcodes for Igneous and Tectonothermal Events for the South Indian and Yilgarn Cratons

The Large Igneous Province (LIP) records for continental reconstruction of the Indian Craton were analysed by Ernst and Srivastava (2008) and led to identification of events at 2,370, 2,180 and 1,890 Ma. In the present analysis igneous and tectonothermal events of the Dharwar craton and Bastar craton of South India and the Yilgarn craton of Western Australia were compiled from the published literature (Table 6.3). It is observed that all these cratons were developed prior to 3,400 Ma with dominantly TTG gneisses and their precursors forming the basement over which BIF bearing sediments were deposited with associated komatiitic basalts. These rocks were deformed and metamorphosed under high-grade metamorphic condition at ~3,300 Ma and subsequently intruded by granites of 2,900–3,100 Ma age. Gold-bearing granite-greenstone association of 2,600–2,900 Ma age are common to these three cratons. Granite intrusions and high-grade metamorphism took place in these cratons between the time period from 2,450 to 2,550 Ma, followed by intrusion of mafic dyke swarms at ~2,400 Ma. Thus the Dharwar – Bastar – Yilgarn cratons show identical evolutionary history till 2,400 Ma. Deposition of sedimentary rocks of Moogie metamorphic and their equivalents (siliciclastic sediments,

Table 6.3 Archaean–Palaeoproterozoic events stratigraphy of South Indian and Western Australian Cratons

Time (Ma)	Dharwar craton	Bastar craton	Yilgarn craton
1,600–1,700	Deformation of Cuddapah Supergroup	Keshkal granite 1,639 ± 40 (16)	Mangaroon orogeny Durlacher supersuite granites (1,620–1,670 Ma) Pooranoo metamorphics Coarse clastics and flood basalts
1,700–1,800	Cuddapah Supergroup Upper clastics, argillite, poorly bedded cherts and carbonates Tadpatri Formation 1,756 ± 29 (12)	–	Basic Dykes 2,000–1,600 Moorarie supersuite granites 1,780–1,810 Capricorn orogeny (1,780–1,830 Ma)
1,800–1,900	Cuddapah Supergroup Vempalle lava Mafic sill 1,885 ± 3 (11) Coarse clastics Mafic dyke swarms 1,817 ± 24; 1,879 ± 5 (10f)	Tholeiite dykes (BD2) 1,883 ± 2; 1,891 ± 1 (15)	Morrisssey metamorphics Narracoota volcanics 1,890 Ma Clastic and chemical sediments (BIF) Basic dykes 2,000–1,600
1,900–2,100	Dykes along Cuddapah margin 1,915 ± 28 (10e) Sm-Nd model age of lamproites 2,065–2,132	Bastar Granite 2,095 ± 118 (14)	Dalgarina supersuite intrusions ~1,975 Glenburgh orogeny (2,000–1,965 Ma) EGP Kimberlite 2,025 ± 10 (3)
2,100–2,400	Mahabubnagar dike 2,208 ± 220 (10c), 2,173 ± 64 (10d) Hyderabad dike 2,270 ± 213 (10c) E Dharwar dykes 2,365 ± 1 (10b)	High-Mg mafic dykes Paliam-Darba granite 2,308 ± 48 (13) Sambalpur granite 2,347 ± 16 (12)	Basic Dykes 2,000–1,600 Moogie metamorphics

Table 6.3 (continued)

Time (Ma)	Dharwar craton	Bastar craton	Yilgarn craton
2,400–2,600	East Dharwar basic dykes 2,420 ± 246 (10a) Granulite event 2,473 ± 15 (9) Granite Plutons 2,513 ± 5 (Clospet) (8) Younger gneissic complex 2,532 ± 3 (7a), 2,631 ± 6 (7b) Kolar granite 2,540 ± 4 (6)	Amphibolite dykes (BD1) Bijli Rhyolite 2,462 ± 25 (11) Dongarhgarh granite 2,465 ± 22 (10) Markampara granite 2,480 ± 3 (9) Malanjkhand granite 2,490 ± 8 (7) 2,478 ± 9 (8) Kawadgaon granite 2,497 ± 152 (6) Katekalyan migmatite 2,530 ± 89 (5) Leuco–granite 2,573 ± 139 (4)	Widgiemooltha dyke 2,420 ± 30 (2) Jimberlana dyke 2,411 ± 55 (1) Granitoid (TTG) intrusion Halfway gneiss ~2,540 Granitoid intrusion
2,600–2,900	Dharwar Supergroup/Kolar schist belt: (auriferous) Greywacke Carbonate, manganese-iron formation 2,639 ± 32 (5) Polymictic conglomerate BIF 2,720 ± 7 (4) Alternating orthoquartzite and pillowed basic volcanics and acidic volcanics 2,746 (3a, b) – 2,911 (3c) QPC	Abujhmar/Sonakhan/ Bailadila/Sakoli Group: (auriferous) BIF Mafic and felsic volcanics Coarse clastics	Granite greenstone successions: (auriferous) Komatiitic basalts, felsic volcanics, BIF, clastics, dolomite, turbidite (EGF 2,690–2,710; Murchison flood basalt –2; granite ~2,830; Marda greenstone belt 2,850 ± 25; Lake Johnston 2,903 ± 5)
2,900–3,100	Kuncha granite 3,080 ± 76 (2) TTG granite 3,127 ± 8 (6), TTG gneiss 3,140 (7)	Sukma granite 3,018 ± 61 (3)	Murgoo gneiss 2,950 Granites (~3,070)

Table 6.3 (continued)

Time (Ma)	Dharwar craton	Bastar craton	Yilgarn craton
3,100–~3,800	Amphibolite – granulite metamorphism	Amphibolite – granulite metamorphism	Amphibolite – granulite metamorphism 3,300–3,350
	Older gneissic complex (Sargur Supergroup): fuchsite quartzite with barite bands, BIF, manganiferous carbonates, graphitic schists and chromiferous layered ultramafic-mafic-anorthosite.	Bengpal/Amgaon Group:	D_3 deformation
		Deposition of sediments	Deposition of sediments: 3,350–3,400; coarse clastics, BIF, pelite
		Bengpal basalts	Murchison flood basalt –1
			Basic/ultramafic dykes
	Sargur basalts/komatiites		
	Banded tonalite-trondhjemite-granite gneiss (TTG): Gorur gneiss 3,358 ± 66 (2)	Sukma gneiss 3,509 ± 14 (2c)	Dugel gneiss 3,400
		Kapasi tonalite 3,561 ± 11 (2b),	D_2 deformation
	Detrital zircon ~3,580 (1)	Dali-Rajahara granite 3,582 ± 4 (2a)	
	–	Markampura TTG gneiss ~3,610 (1)	Meeberrie gneiss ~3,680
			D_1 deformation
			Manfred complex ~3,780

References for the table

Dharwar Craton: 1: U-Pb age (Nutman et al., 1992); 2: Rb-Sr isochron age (Beckinsale et al., 1980); 3a: Pb-Pb age Zachariah et al., 1995; 3b: Pb-Pb age (Balakrishnan et al., 1990); 3c: Sm-Nd age (Anil Kumar et al., 1996); 4: U-Pb age (Trendall et al., 1997); 5: Pb-Pb age (Russel et al., 1995); 6: Pb-Pb zircon age (Jayananda et al., 2000); 7: Pb-Pb zircon age (Krogstad et al., 1988, 1991); 8: U-Pb age (Friend and Nutman, 1991); 9: Sm-Nd age (Jayananda and Peucat, 1996); 10a: Rb-Sr age (Ikramuddin and Stuber, 1976); 10b: U-Pb age (Halls et al., 2007); 10c: Rb-Sr age (Murthy, 1995); 10d: Sm-Nd age (Pandey et al., 1997); 10f: Ar-Ar age (Chatterjee and Bhattacharji, 2001); 11: U-Pb IDTIMS age of baddeleyite (French et al., 2004, 2008); 12: Pb-Pb age of U minerals (Zachariah et al., 1999).

Bastar Craton: 1: Rb-Sr WR age (Sarkar and Gupta, 1990); 2a: U-Pb zircon age (Rajesh et al., 2009); 2b: U-Pb zircon age (Ghosh, 2004); 2c: U-Pb zircon age (Sarkar et al., 1993); 3–4: Pb-Pb WR age (Sarkar and Gupta, 1989); 5: Rb-Sr age (Bandyopadhyay et al., 1990); 6: Rb-Sr WR age (Singh and Chabria, 1999); 7: Re-Os age (Stein et al., 2004); 8: U-Pb zircon age (Panigrahy et al., 2002); 9: U-Pb zircon age (Sarkar et al., 1990); 10–11: Rb-Sr whole rock age (Krishnamurthy et al., 1988); 12: Rb-Sr WR age (Choudhary et al., 1996); 13: Rb-Sr age (Ramesh Babu et al., 1993); 14: Rb-Sr WR age (Sarkar et al., 1994); 15: U-Pb IDTIMS age of baddeleyite (French et al., 2008); 16: Rb-Sr WR age (Sarkar et al., 1990).

Western Australia: Myers (1995); Chen et al. (2003); Cawood and Tyler (2004); Collins (2003); Evans et al. (2003); Halilovic et al. (2004); Martin et al. (1998, 2000); Occhipinti et al. (2004); Pirajno et al. (2004); Sheppard et al. (2004); Wingate et al. (2000); 1: Sm-Nd Age (Fletcher et al., 1987); 2: Rb-Sr age (Evans, 1968); 3: Re-Os age (Graham et al., 2004).

BIF and glacial diamictites) in the Mt Bruce Supergroup took place from 2,400 to 2,100 Ma between the Yilgarn craton and the Pilbara craton. These were involved in Glenburgh orogeny between 2,000 and 1,965 Ma. During the same time basic dykes and kimberlite intruded the Yilgarn craton. In the Indian cratons intrusions of several granitic plutons ranging in age from 2,237 to 2,347 Ma took place in the Bastar craton and intrusions of several basic dykes took place along the margins of the Cuddapah basin. Thus, a common tectonic evolution of the south Indian and Yilgarn cratons, possibly forming a supercontinental block till ~2,100 Ma is established from the stratigraphic records. The ~2,100 Ma event comprising of intrusions of basic dykes and kimberlites were possibly related to the breaking of this supercontinent (discussed later).

A New Supercontinental Model: SIWA

The matching configuration of the western boundary of the Yilgarn craton and the South Indian craton along the position of the Eastern Ghats, together with the orientation of mafic dykes, magnetic north direction for the corresponding time period and identical tectonothermal events of the South Indian and Yilgarn cratons (Fig. 6.4) suggests the existence of a supercontinent at least up to ~2,400 Ma. This supercontinent is proposed to be named here as "SIWA" – an acronym for South India and Western Australia. It may be noted that the oldest rocks found in the Yilgarn craton are located in the Narryer gneissic terrane in the northwest. The oldest rocks reported from India are from the Bastar and Singhbhum cratons which are located close to the Narryer gneissic terrane in the configuration proposed here. The ~2,400 and ~1,900 Ma basic dyke events were related to the breaking process of "SIWA", possibly in several phases. Intrusions of several dyke swarms in the eastern margin of the South Indian craton were related to different phases of breaking between 2,400 and 1,900 Ma. The final breaking process gave rise to the development of the Cuddapah basin with deposition of sediments and associated volcanic flows and sills (1,885 Ma) along the margin of the South Indian block (Table 6.3). The age data of the igneous activity (mostly volcanic flows and sills) in the Cuddapah basin imply that the breaking and deposition of clastic sediments took place much before 1,885 Ma, possibly between 1,950 and 2,000 Ma. Deposition of clastic and chemical sediments and Narracoota volcanic (1,890 Ma) in the Morrisssey metamorphics on the northern margin of the Yilgarn craton also took place during the breaking process at ~2,000 Ma. The separation of the Yilgarn craton and its collision with the Pilbara craton took place at ~1,780–1,830 Ma, giving rise to the Capricorn orogeny. On the other hand the deposition of sediments in the Cuddapah basin continued, possibly giving rise to a passive margin set up. Subsequently, the northern margin of the Yilgarn craton was involved in another orogenic event (Mangaroon orogeny) between 1,620 and 1,670 Ma, and the sediments of the Cuddapah basin were deformed at ~1,650 Ma.

Conclusions

The paper reviews and integrates recent geological and geochronological data from India and the Western Australian craton and suggests:

1. existence of a pre-Rodinian cratonic nuclei "SIWA" before 2,400 Ma;
2. the break-up of SIWA at ~2,000 Ma giving rise to palao-Indian ocean;
3. the sedimentation of the Cuddapah Supergroup in a rift-valley/passive margin set-up;
4. the separation of the Yilgarn craton from the South Indian craton and its northward movement lead to the development of a collisional mountain along the Capricorn orogen at ~1,800 Ma.

At present high precision geochronological data from the Yilgarn and South Indian cratons are scarce to prepare a detailed barcode for detailed understanding of the events between 2,400 and 1,900 Ma. High precision data from the individual dykes would help in locating the exact point of plume head responsible for dyke formation and breaking of the supercontinent "SIWA".

Acknowledgements I am thankful to D.K. Paul, Ex-Director, ISM for making arrangement for my place in the team to take a geological traverse from Durgapur in Chhotanagpur Gneiss terrane through the Singhbhum craton to the Vishakhapatnam Sector of the Eastern Ghats during an International Conference held at Bhubaneswar, which helped me to understand the key problems of the geology of eastern India and the Eastern Ghats. Assistance from the Indian School of Mines helped to attend the International conference on Supercontinent Assembly held at Fremantle. Helpful suggestions from the reviewer John J.W. Rogers and the Editor Rajesh K. Srivastava are thankfully acknowledged.

References

Anil Kumar, Bhalla MS (1983) Palaeomagnetics and igneous activity of the area adjoining the south-western margin of the Cuddapah basin, India. Geoph J R Astrn Soc 73: 27–37

Anil Kumar, Bhaskar Rao YJ, Sivaraman TV, Gopalan K (1996) Sm-Nd ages of Archaean metavolcanics of the Dharwar craton, South India. Precamb Res 80: 205–216

Anil Kumar, Rao NTVP (1979) Palaeomagnetic studies of a Precambrian dyke near Hyderabad. Proceeding 3rd Workshop on "Status, Problems and Programmes in Indian Peninsular Shield, Hyderabad: 146–152

Anjanappa K (1972) Application of palaeomagnetism in estimating the age of the dolerite dykes of Peninsular India. Curr Sci 41: 106–108

Athavale RN, Radhakrishnamurthy C, Sahasrabudhe PW (1963) Palaeomagnetism of some Indian rocks. Geophys J Roy Astron Soc 7: 304–313

Balakrishnan S, Hanson GN, Rajamani V (1990) Pb and Nd isotope constraints on the origin of high Mg and tholeiite amphibolites, Kolar schist belt, southern India. Contrib Min Pet 107: 279–292

Bandyopadhyay BK, Bhoskar KG, Ramachandra HM, Roy A, Khadse VK, Mohan M, Rao KS, Ray Barman T, Bishui PK, Gupta SN (1990) Recent geochronological studies in parts of the Precambrian of central India. Geol Surv India Spec Publ 28: 199–210

Basumallick S, Mondal S, Das AK, Majumdar D, Goswami S (2002) Palaeomagnetic and rock magnetic study of Proterozoic rocks of Central India. Deep Continental Studies in India: Newsletter 12(1): 2–5

Beckinsale RD, Drury SA, Holt RW (1980) 3360 m.y. old gneisses from the South Indian Craton. Nature 283: 469–470

Bhalla MS, Hansraj A, Prasada Rao NTV (1980) Paleomagnetic studies of Precambrian dykes from Hunsur, Holenarsipur and Tiptur in Karnataka, India. Geoviews 5: 157–172

Bleeker W, Ernst RE (2006) Short-lived mantle generated magmatic events and their dyke swarms: The key unlocking Earth's paleogeographic record back to 2.6 Ga. In: Hanski E, Mertanen S, Rämö T, Vuollo J (eds) Dyke Swarms – Time Markers of Crustal Evolution. Taylor & Francis Group, London: 3–26

Cawood PA, Tyler IM (2004) Assembling and reactivating the Proterozoic Capricorn orogen: Lithotectonic elements, orogenies and significance. Precamb Res 128: 201–218

Chatterjee N, Bhattacharji S (2001) Petrology, geochemistry and tectonic settings of the mafic dikes and sills associated with the evolution of the Proterozoic Cuddapah Basin of south India. Proc Indian Acad Sci (Earth Planet Sci) 110: 433–453

Chen SF, Greenfield JE, Nelson DR, Riganti A, Wyche S (2003) Lithostratigraphy and tectonic evolution of contrasting greenstone successions in the central Yilgarn Craton, Western Australia. Precamb Res 127: 249–266

Choudhary AK, Naik A, Mukhopadhyay D, Gopalan K (1996) Rb-Sr dating of Sambalpur granodiorite, western Orissa. J Geol Soc India 47: 503–506

Clark DJ, Hensen BJ, Kinny PD (2000) Geochronological constraints for a two-stage history of the Albany-Fraser Orogen, Western Australia. Precamb Res 102: 155–183

Clark DJ, Kinny PD, Post NJ, Hensen BJ (1999) Relationships between magmatism, metamorphism and deformation in the Fraser Complex, Western Australia: Constraints from new SHRIMP U-Pb zircon geochronology. Aust J Earth Sci 46: 923–932

Collins AS (2003) Structure and Age of the Northern Leeuwin Complex, Western Australia: Constraints from Field Mapping and U-Pb Isotopic Analysis. Aust J Earth Sci 50:585–599

Collins AS, Pisarevsky SA (2005) Amalgamating eastern Gondwana: The evolution of the Circum-Indian Orogens. Earth Sci Rev 71: 229–270

Crawford AR (1974) Indo-Antarctica Gondwanaland, and the distribution of granulite belt. Tectonophysics 22: 141–157

Crawford AR, Oliver RL (1969) The Precambrian geochronology of Ceylon. Geol Soc Aust Spec Publ 2: 183–306

Damodara Reddy V, Prasad CVRK (1979) Palaeomagnetism of the dykes from Kolar Gold Mines, India. J Geol Soc India 20: 489–500

Dobmeier C, Simmat R (2002) Post-Grenvillean transpression in the Chilka Lake area, Eastern Ghats Belt – implications for the geological evolution of peninsular India. Precamb Res 113: 243–268

Ernst RE, Buchan KL (2001) Large mafic magmatic events through time and links to mantle-plume heads. In: Ernst RE, Buchan KL (eds) Mantle Plumes: Their Identification Through Time, Geol Soc America Spec Pap 352: 483–575

Ernst RE, Srivastava RK (2008) India's Place in the Proterozoic World: Constraints from the Large Igneous Province (LIP) Record. In: Srivastava RK, Ch S, Chalapathi Rao NV (eds) Indian Dykes: Geochemistry, Geophysics and Geochronology. Narosa Publishing House Pvt. Ltd, New Delhi, India: 41–56

Evans ME (1968) Magnetisation of dikes: A study of the paleomagnetism of the Widgiemooltha dike suite, Western Australia. J Geophys Res 73: 3261–3270

Evans DAD, Sircombe KN, Wingate MTD, Doyle M, McCarthy M, Pidgeon RT, Van Niekerk HS (2003) Revised geochronology of magmatism in the western Capricorn orogen at 1805–1785 Ma: Diachroneity of the Pilbara-Yilgarn collision. Aust J Earth Sci 50: 853–864

Fletcher IR, Libby WG, Rosman KJR (1987) Sm-Nd dating of the 2411 Ma Jimberlana Dyke, Yilgarn Block, Western Australia. Aust J Earth Sci 34: 523–525

French JE, Heaman LM, Chacko T, Rivard B (2004) Global mafic magmatism and continental breakup at 2.2 Ga: Evidence from the Dharwar craton, India. Geol Soc Am Abs Prog 36: 340

French JE, Heaman LM, Chacko T, Srivastava RK (2008) 1891–1883 Ma Southern Bastar-Cuddapah mafic igneous events, India: A newly recognized large igneous province. Precamb Res 160: 308–322

Friend CRL, Nutman AP (1991) SHRIMP U-Pb geochronology of the Closepet granite and Peninsular Gneiss, Karnataka, south India. J Geol Soc India 38: 357–368

Ghosh JG (2004) 3.56 Ga tonalite in the central part of the Bastar craton, India: Oldest Indian date. J Asian Earth Sci 23: 359–364

Giddings JW (1976) Precambrian paleomagnetism in Australia I: Basic dykes and volcanic from the Yilgarn Block. Tectonophysics 30: 91–108

Graham S, Lambert D, Shee S (2004) The petrogenesis of carbonatite, melnoite and kimberlite from the Eastern Goldfields Province, Yilgarn Craton. Lithos 76: 519–533

Halilovic J, Cawood PA, Jones JA, Pirajno F, Nemchin AA (2004) Provenance of the Earaheedy Basin: Implications for assembly of the Western Australian Craton. Precamb Res 128: 343–366

Halls HC, Kumar A, Srinivasan R, Hamilton MA (2007) Paleomagnetism and U-Pb geochronology of easterly trending dykes in the Dharwar craton, India: Feldspar clouding, radiating dyke swarms and the position of India at 2.37 Ga. Precamb Res 155: 47–68

Harris LB (1993) Correlations between the Central Indian Tectonic zone and the Albany Mobile Belt of Western Australia: Evidence for a continuous Proterozoic orogenic belt. In: Findlay RH, Unrug R, Banks MR, Veevers JJ (eds) Gondwana 8: assembly, evolution and dispersal. A.A Balkema, Rotterdam: 165–180

Harris LB (1995) Correlation between the Albany, Fraser and Darling Mobile Belts of Western Australia and Mirnyy to Windmill Islands in the East Antarctic Shield: Implications for Proterozoic Gondwanaland reconstructions. Mem Geol Soc India 34: 47–71

Hasnain I, Qureshy MN (1971) Paleomagnetism and geochemistry of some dikes in Mysore State, India. J Geophys Res 76: 4786–4795

Ikramuddin M, Stuber AM (1976) Rb-Sr ages of Precambrian dolerites and alkaline dykes, southeast Mysore State, India. Lithos 9: 235–241

Jayananda M, Moyen JF, Martin H, Peucat JJ, Auvray B, Mahabaleswar B (2000) Late Archaean (2550–2520 Ma) juvenile magmatism in the Eastern Dharwar craton, southern India: Constraints from geochronology, Nd-Sr isotopes and whole rock geochemistry. Precamb Res 99: 225–254

Jayananda M, Peucat JJ (1996) Geochronological frame work of Southern India Archaean and Proterozoic Terrains in Southern India. In: Santosh M, Yoshida M (eds) Archaean and Proterozoic terrains in southern India, Field Science Publishers, Osaka, Japan, Gondwana Res Group Memoir 3: 53–75

King LC (1973) An improved reconstruction of Gondwanaland. In: Tarling DH, Runcorn SK (eds) Implications of continental drift to the earth sciences. Academic Press, London: 851–865

Klootwijk CT (1974) Palaeomagnetic data from the Precambrian Gwalior Traps, Central India. Tectonophysics 21: 181–195

Krishnamurthy P, Chaki A, Pandey BK, Chimote JS (1988) Geochronology of the granite-rhyolite suites of the Dongargarh Supergroup, central India. Proceeding 4th National Symposium Mass Spectrometry, Bangalore, EPS-2: 1–5

Krogstad EJ, Hanson GN, Rajamani V (1988) U-Pb ages and Sr and Nd isotope data for gneisses near the Kolar schist belt: evidence for the juxtaposition of discrete Archaean terranes. In: Ashwal LD (ed) Workshop on the deep continental crust of South India, LPI Tech Report 88–06: 181–186

Krogstad EJ, Hanson GN, Rajamani V (1991) U-Pb ages of zircon and sphene for two gneissic terranes adjacent to Kolar schist belt, south India: Evidence for separate crustal evolution histories. J Geol 99: 801–816

Martin DMcB, Clendenin CW, Krapez B, McNaughton NJ (1998) Tectonic and geochronological constraints on late Archaean and Palaeoproterozoic stratigraphic correlation within and between the Kaapvaal and Pilbara Cratons. J Geol Soc Lond 155: 311–322

Martin DMcB, Powell CMcA, George AD (2000) Stratigraphic architecture and evolution of the early Paleoproterozoic McGrath Trough Western Australia. Precamb Res 99: 33–64

McElhinny MW, Cowley JA, Edwards DJ (1978) Palaeomagnetism of some rocks from Peninsular India and Kashmir. Tectonophysics 50: 41–54

Mohanty S (2000) Proterozoic granites of Madhya Pradesh and Bihar: A compilation. In: First Meeting of IGCP-426 "Proterozoic granites and crustal evolution", Jaipur, 226–264

Mohanty S (2003) Structural-Stratigraphic relations in Precambrian rocks of Sausar belt, central India. Gondwana Geol Soc Spec Publ 7: 109–117

Mohanty S (2005) Apparent Polar Wandering Path of the Proterozoic System of India: Implication for Tectonic Evolution. In: Mohanty S, Chakraborty PP, Dhar YR, Majumder T (eds) Proterozoic System of India: Evolution and Economic Potential, Dhanbad, Abstract Volume: 23–23

Mohanty S (2006a) Apparent polar wandering path of the Proterozoic system of India: Implication for tectonic evolution of the Indian subcontinent. Indian J Geol 78: 19–35 (Published January 2009)

Mohanty S (2006b) Structural evolution of the Sausar belt, central India: Implications for tectonic evolution of the Central Indian Tectonic Zone. Indian J Geol 78: 135–142 (Published January, 2009)

Mohanty S (2010) Tectonic evolution of the Satpura Mountain Belt: A critical evaluation and implication on supercontinent assembly. J Asian Earth Sci 39: 516–526. doi: 10.1016/j.jseaes.2010.04.025

Murthy NGK (1995) Proterozoic mafic dykes in Southern Peninsular India: A review. Mem Geol Soc India 33: 81–98

Myers JS (1995) The generation and assembly of an Archaean supercontinent: Evidence from the Yilgarn craton, Western Australia. Geol Soc Lond Spec Publ 95: 143–154

Naqvi SM, Rogers JJW (1987) Precambrian Geology of India. Oxford University Press, Oxford

Nutman AP, Chadwick B, Ramakrishna M, Viswanatha MN (1992) SHRIMP U-Pb ages of detrital zircons from Sargur supracrustal rocks in Western Karnataka, Southern India. J Geol Soc India 39: 367–374

Occhipinti SA, Sheppard S, Passchier C, Tyler IM, Nelson DR (2004) Palaeoproterozoic crustal accretion and collision in the southern Capricorn Orogen: The Glenburgh Orogeny. Precamb Res 128: 237–255

Pandey BK, Gupta JN, Sarma KJ, Sastry CA (1997) Sm-Nd, Pb-Pb and Rb-Sr geochronology and petrogenesis of the mafic dyke swarm of Mahbubnagar, South India: implications for Paleoproterozoic crustal evolution of the Eastern Dharwar Craton. Precamb Res 84: 181–196

Panigrahy MK, Misra KC, Bream B, Naik RK (2002) Genesis of the granitoid affiliated copper–molybdenum mineralization at Malanjkhand central India: facts and problems: extended abstract. In: Proceedings of the 11th Quadrennial IAGOD Symposium and Geocongress, Windhoek, Namibia (electronic)

Pirajno F, Jones JA, Hocking RM, Halilovic J (2004) Geology and tectonic evolution of the Palaeoproterozoic basins of the eastern Capricorn Orogen, Western Australia. Precamb Res 128: 315–342

Poornachandra Rao GVS, Bhalla MS (1996) Magnetostratigraphy of Vindhyan Supergroup. J Geol Soc India 47: 29–32

Poornachandra Rao GVS, Mishra DC (1997) A Proterozoic APWP of cratons from either side of the Narmada-Son lineament and the continuity of the Indian subcontinent. Geol Surv India Misc Publ 63: 129–139

Powell CMcA, Pisarevsky S, Wingate MTD (2001) A new shape for Rodinia. Gondwana Res 4: 736–737

Powell CMcA, Roots SR, Veevers JJ (1988) Pre-breakup continental extension in East Gondwanaland and the early opening of the eastern Indian ocean. Tectonophysics 155: 261–283

Prasad CVRK, Pulla Reddy V, Subba Rao KV, Radhakrisnamurthy C (1987) Palaeomagnetism and the crescent shape of the Cuddapah basin. Geol Soc India Mem 6: 331–347

Prasad JN, Satyanarayana KVV, Gawli PB (1999) Palaeomagnetic and low-field AMS studies of Proterozoic dykes and their basement rocks around Harohalli, South India. J Geol Soc India 54: 57–67

Rajesh HM, Mukhopadhyay J, Beukes NJ, Gutzmer J, Belyanin GA, Armstrong RA (2009) Evidence for an early Archaean granite from Bastar craton, India. J Geol Soc Lond 166: 193–196

Ramesh Babu PV, Pandey BK, Dhana Raju R (1993) Rb-Sr ages on the granite and pegmatitic minerals from Bastar-Koraput pegmatite belt, Madhya Pradesh and Orissa, India. J Geol Soc India 42: 33–38

Rickers K, Mezger K, Raith MM (2001) Evolution of the continental crust in the Proterozoic Eastern Ghats Belt, India and new constraints for Rodinia reconstruction: Implications from Sm-Nd, Rb-Sr and Pb-Pb isotopes. Precamb Res 112: 183–210

Rogers JJW, Santosh M (2002) Configuration of Columbia, a Mesoproterozoic supercontinent. Gondwana Res 5: 5–22

Rogers JJW, Santosh M (2009) Tectonics and surface effects of the supercontinent Columbia. Gondwana Res 15: 373–380

Russel J, Chadwick B, Krishna Rao B, Vasudev VN (1995) Whole-rock Pb/Pb isotopic ages of Late Archaean limestones, Karnataka, India. Precamb Res 78: 261–272

Sarkar G, Corfu F, Paul DK, McNaughton NJ, Gupta SN, Bishui PK (1993) Early Archaean crust in Bastar craton, central India – A geochemical and isotopic study. Precamb Res 62: 127–137

Sarkar G, Gupta SN (1989) Dating of Early Precambrian granite complex of Bastar district, Madhya Pradesh (Abstract). Rec Geol Surv India 122(2): 29–30

Sarkar G, Gupta SN (1990) Dating of Early Precambrian granite complex of Bastar district, Madhya Pradesh (Abstract). Rec Geol Surv India 123: 2

Sarkar G, Gupta SN, Bishui PK (1994) New Rb-Sr isotopic age and geochemistry of granite gneisses from southern Bastar: Implication for crustal evolution. Indian Minerals 48: 7–12

Sarkar G, Paul DK, McNaughton NJ, Misra VP (1990) A geochemical and Pb, Sr isotopic study of the evolution of granite-gneisses from the Bastar craton, central India. J Geol Soc India 35: 480–496

Schmidt PW, Williams GE (1999) Palaeomagnetism of the Mesoproterozoic Gangau Formation, Basal Vindhyan Supergroup, Madhya Pradesh. Mem Geol Soc India 44: 75–85

Sheppard S, Occhipinti SA, Tyler IM (2004) A 2005–1970 Andean-type batholith in the southern Gascoyne Complex, Western Australia. Precamb Res 128: 257–277

Singh J, Bhalla MS (1972) Preliminary palaeomagnetic studies on igneous rocks of U.P., Andhra Pradesh and Mysore. Curr Sci 41: 92–94

Singh Y, Chabria T (1999) Late Archaean-Early Proterozoic Rb-Sr isochron age of granite from Kawadgaon, Bastar district, Madhya Pradesh. J Geol Soc India 54: 405–409

Srivastava RK (2008) Global Intracratonic Boninite-Norite Magmatism during the Neoarchean–Paleoproterozoic: Evidence from the Central Indian Bastar Craton. Int Geol Rev 50: 61–74

Srivastava RK, Singh RK (2003) The Paleoproterozoic dolerite dyke swarm of the southern Bastar Craton, Central-East India: Evidence for the Columbia supercontinent. Mem Geol Soc India 52: 163–177

Srivastava RK, Singh RK, Verma R (2000) Juxtaposition of India and Antarctica during the Precambrian: Inferences from geochemistry of mafic dykes. Gondwana Res 3: 227–234

Stein HJ, Hannah JL, Zimmerman A, Markey RJ, Sarkar SC, Pal AB (2004) A 2.5 Ga porphyry Cu–Mo–Au deposit at Malanjkhand, Central India: Implications for late Archean continental assembly. Precam Res 134: 189–226

Subba Rao YV, Murthy IVR (1980) Magnetic and Palaeomagnetic investigations over a dolerite dyke near Hasanpur village in Karimnagar district, AP. Geophys Res Bull 18: 9–16

Trendall AF, De Laeter JR, Nelson DR, Mukhopadhyay D (1997) A precise zircon U-Pb age for the base of the BIF of the Mulaingiri Formation (Bababudan Group, Dharwar Supergroup) of the Karnataka Craton. J Geol Soc India 50: 161–170

Verma RK, Prasad SN (1974) Palaeomagnetic study and chemistry of Newer Dolerites from Singhbhum, Bihar, India. Can J Earth Sci 11: 1043–1054

Verma RK, Pullaiah G, Hasnain I (1968) Palaeomagnetic study of a dyke near Hyderabad, India. Bull Natl Geophys Res Inst 6: 79–86

Wingate MTD, Campbell IH, Harris LB (2000) SHRIMP baddeleyite age for the Fraser Dyke Swarm, southeast Yilgarn Craton, Western Australia. Aust J Earth Sci 47: 309–313

Zachariah JK, Bhaskar Rao YJ, Srinivasan R, Gopalan K (1999) Pb, Sr and Nd isotope systematics of uranium mineralized stromatolitic dolomites from the Proterozoic Cuddapah Supergroup, south India: Constraints on age and provenance. Chem Geol 162: 49–64

Zachariah JK, Hanson GN, Rajamani V (1995) Post-crystallisation disturbances in the Nd and Pb isotope systematics of metabasalts from the Ramagiri Schist Belt, South India. Geochimica Cosmochimica Acta 59: 3189–3203

Zhao G, Cawood PA, Wilde SA, Sun M (2002) Review of global 2.1–1.8 Ga orogens: Implications for a pre-Rodinia Supercontinent. Earth Sci Rev 59: 125–162

Zhao GC, Sun M, Wilde SA, Li SZ (2003) Assembly, accretion and breakup of the Paleo-Mesoproterozoic Columbia Supercontinent: Records in the North China Craton. Gondwana Res 6: 417–434

Chapter 7
Petrology and Mineral Chemistry of Picrite Sill from Peddakudala-Velpula Area, in Southwestern Part of the Proterozoic Cuddapah Basin, Andhra Pradesh, India

V.V. Sesha Sai

Introduction

Mafic-ultramafic magmatism constitutes an integral part of evolution of Dharwar craton, having taken place repeatedly at various stages of its development (Srivastava et al., 2008; Ernst and Srivastava, 2008; Jayananda et al., 2008 and Srivastava and Ahmed, 2009). Basaltic sills and lava flows (~1.88 Ga) are commonly intercalated with sediments in the lower part of the sedimentary sequence in the intra-cratonic, mid-Proterozoic Cuddapah basin of southern India (Chatterjee and Bhattcharji, 2001). The Proterozoic Cuddapah basin of Peninsular India covering an area of around 44,000 km^2 in the state of Andhra Pradesh, is one of the well known Proterozoic sedimentary basins in the world (Nagaraja Rao and Ramalingaswamy, 1976). Contemporaneous with the sedimentation, Cuddapah basin has witnessed number of phases of igneous activity during the course of its evolution Six phases of igneous activity are recorded in Cuddapah basin (Nagaraja Rao et al., 1987). Sub-aerial eruption of basic lava flows after the deposition of Vempalli Formation marks the first phase of igneous activity within the basin. The volcanic activity represented by fine grained basic rocks and tuffs within the Tadpatri Formation represents the second phase of igneous activity and intrusion of sills of picritic and doleritic composition into the Vempalle and Tadpatri succession marks the third phase of igneous activity. The barium and iron oxide rich volcanic activity which was contemporaneous with the Pullampet/Cumbum Formation represents the fourth phase, while intrusion of basic dykes and alkaline rocks (riebeckite syenites of Giddaluru-Racherla-Idamakallu-Dhupadu and Chelima Lamproite) into the rocks of Nallamalai Fold Belt represents the fifth phase of igneous activity. The sixth phase of igneous activity is represented by intrusion of granitic rocks into the Nallamallai Group in the northeastern fringes of the Cuddapah basin. However, later work (Sesha Sai, 2005) revealed that the Nakerikallu and Ipuru granitoids are traversed by dolerite dykes. A number of NNW-SSE to NW-SE trending sills of picrite,

V.V. Sesha Sai (✉)
Petrology Division, Geological Survey of India, Shillong 793003, India
e-mail: seshubb@yahoo.co.in

R.K. Srivastava (ed.), *Dyke Swarms: Keys for Geodynamic Interpretation*,
DOI 10.1007/978-3-642-12496-9_7,

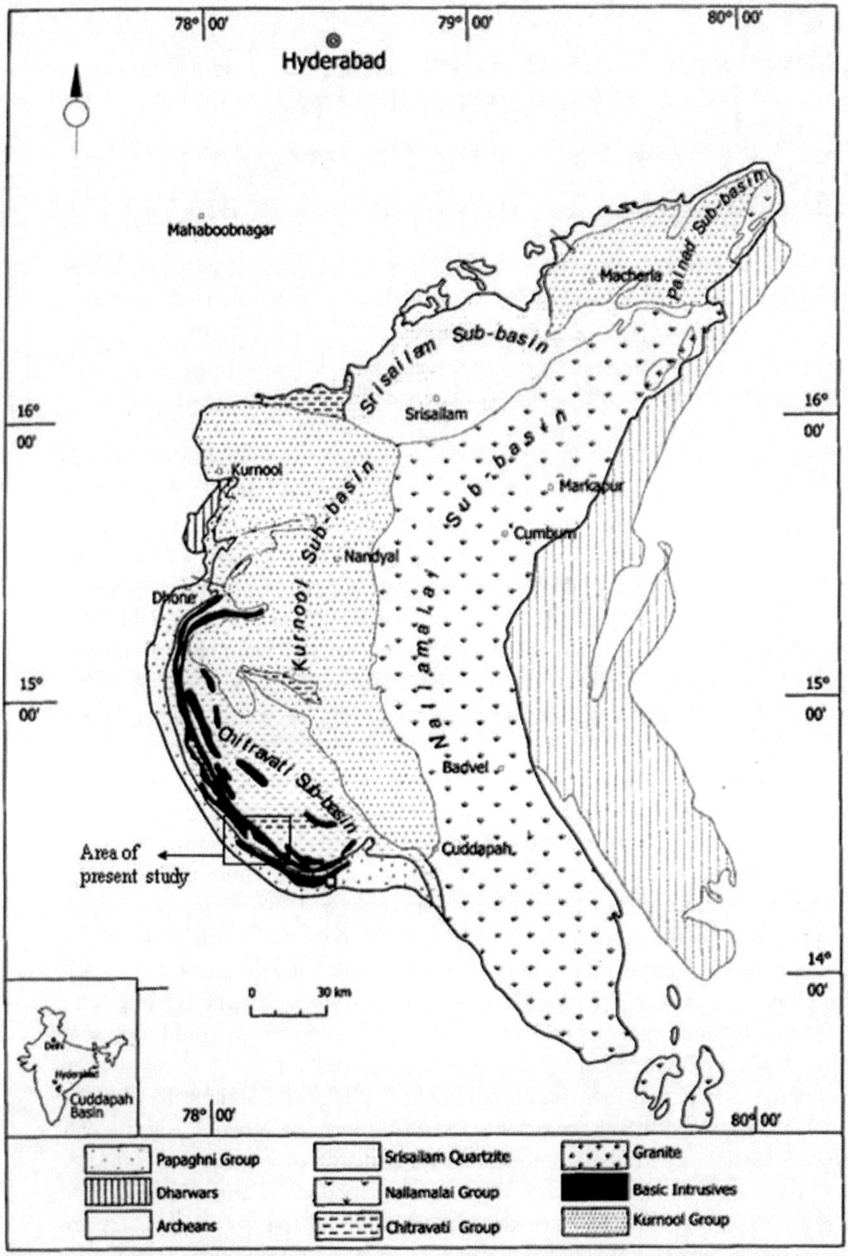

Fig. 7.1 Geological map of Cuddapah basin showing the spatial distribution of igneous rocks (after Nagaraja Rao et al., GSI 1987). *Inset* map showing the position of Cuddapah basin in India

dolerite and olivine gabbro compositions are emplaced parallel to the south western margin of Cuddapah basin (Fig. 7.1). The present paper pertains to petrology and mineral chemistry of a major NW-SE trending picrite body in southwestern part of Cuddapah basin.

Igneous Activity in Tadpatri Formation in Pulivendla-Muddanuru Section

A number of mafic-ultramafic sills exhibiting sharp contact with the shales and dolomite of Tadpatri Formation are noticed to around Pulivendla, Dondlavagu, Tonduru-Mallela section and Velupucherla area in south western part of Cuddapah basin. A NNW-SSE to NW SE trending composite sill made of two components i.e. picrite and dolerite has emplaced the shales and dolomites of Tadpatri Formation and is exposed over a stretch of 12 km from Peddakudala in northwest to Velpula in southeast. Field relationship indicates that dolerite cuts the picrite sill (Sesha Sai, 2006). Dolerite is medium grained, mesocratic, and massive showing spheroidal weathering and petrographically composed of laths of plagioclase, subhedral augite (altered at places to amphibole), accessory ilmenite and pyrite. Development of

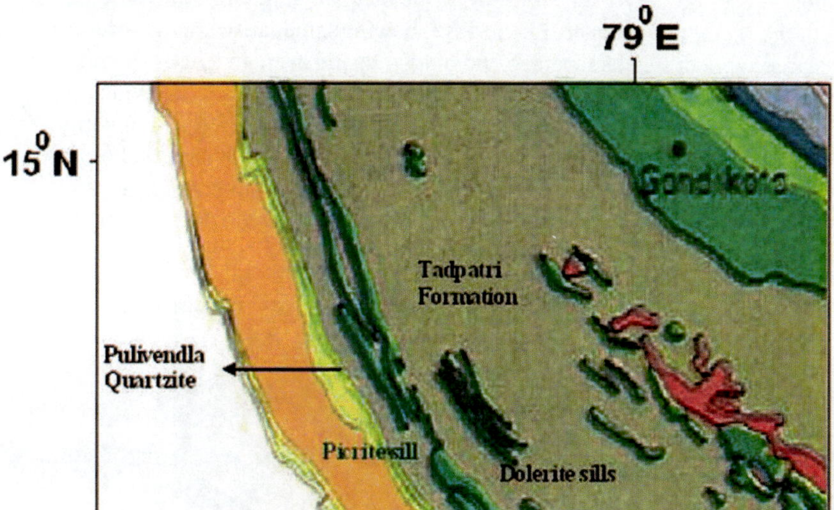

Fig. 7.2 Geological map of South western part of Cuddapah basin showing disposition of basic flows over Vempalle Formation and picrite and dolerite sills in Tadpatri formation (after, GSI, 1981)

sub-ophitic and ophitic textures is common in dolerites. To north of Peddakudala picrite is noticed as minor patches within the mesocratic dolerite. Towards south east the picrite becomes prominent and near Velpula it attains maximum thickness and represents the cumulus phase. On surface, picrite is melanocratic shows typical pitted weathering and vary in width from 15 to 20 m near Peddakudala to more than 100 m near Velpula. In the middle part of Tadpatri Formation as noticed in Tonduru-Mallela-Muddanuru section three major igneous rocks are noticed viz. dolerite sills, felsic tuffs and gabbro sills (porphyritic at places). These sills are emplaced along NW-SE striking and northeasterly dipping (12°) reddish ferruginous sub-arkosic quartz arenites and greyish green splintery shales interbedded with felsic tuffs and limestone bands. An olivine gabbro is noticed traversing the tuff and quartzite is noticed to west of Velupucherla in upper part of Tadpatri Formation. The mafic-ultramafic sills of Tadpatri Formation are emplaced parallel to the south western margin of Cuddapah basin (Fig. 7.2).

Petrology and Mineral Chemistry of Picrite Sill

Petrographic studies indicate that the picrite shows cumulus texture and is modally composed of olivine (~ 38%), clinopyroxene (~ 25%), plagioclase (~ 20%), phlogophite (~ 8%) orthopyroxene (~ 5%), while magnetite, ilmenite and chrome-spinel form the accessory opaques. Plagioclase showing lamellar twinning occur as relatively large subhedral laths (average width 0.5 mm and average length 1.5 mm), while olivine occur as equant grains that are perfectly euhedral at places. Olivine and plagioclase represent the cumulus phase, while phlogopite is noticed as randomly oriented flakes. Co-precipitation of cumulus plagioclase and olivine resulted in development of conspicuous poikilitic texture in the rock, wherein euhedral to sub-hedral grains of olivine chadacrysts are enclosed in the host plagioclase okiocrysts

Fig. 7.3 Poikilitic texture in picrite; *olivine grains* (Ol) enclosed in plagioclase

Fig. 7.4 Phlogopite (Ph) showing typical pleochroism in shades of orangish *red colours*

(Fig. 7.3). Phlogopite is pleochroic in shades of orangish red colours (Fig. 7.4) and most of the phlogopite flakes range in length from 0.3 to 0.4 mm. At places it is noticed that phlogopite partially transects olivine. Exsolution lamellae of orthopyroxene are noticed in clinopyroxene, while ilmenite lamellae are commonly noticed along the cleavage planes in phlogopite. Olivine is altered to serpentine along the cracks (Figs. 7.5 and 7.6).

EPMA analyses were carried out at Central Petrological Laboratory, GSI, Kolkata by CAMECA Sx 100. Analyses conditions: Accelerating voltage: 15 kV, current: 12 nA. Beam size: 1 μ. All natural standards have been used except for Mn and Ti for which synthetic standards have ben used. Mineral chemistry studies carried out with the aid of EPMA (Table 7.1) indicate that olivine is forestrite (with 0.14% NiO), orthopyroxene is enstatite and plagioclase is labradorite in composition. Orthopyroxene (Cr_2O_3 – 0.38% and NiO – 0.10%) and phlogopite (Cr_2O_3 – 0.89 and NiO – 0.12%) analysed significant chromium

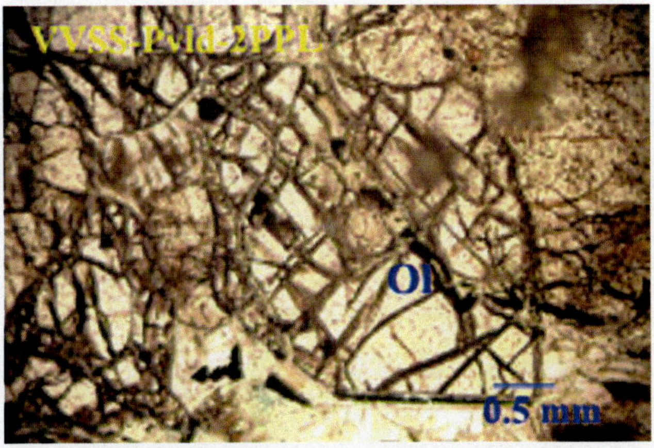

Fig. 7.5 Euhedral olivine (Ol) altered to serpentine along cracks

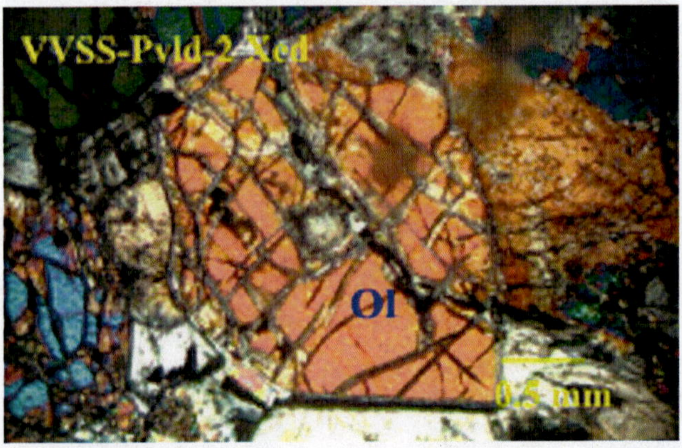

Fig. 7.6 Same as Fig. 7.5 under crossed nicols

Table 7.1 Mineral chemistry of the mineral phases in Picrite from Peddakudala-Velpula area in south western part of Cuddapah basin, Andhra Pradesh

Oxide (%)	Olivine	Opx	Phlogopite	Cpx	Plagioclase	Magnetite
SiO_2	38.65	55.55	39.00	55.73	49.4	0.00
TiO_2	0.00	0.38	2.61	0.39	0.00	5.64
Al_2O_3	0.01	1.01	13.91	2.60	31.3	5.58
MnO	0.00	0.27	0.01	0.41	0.00	0.32
MgO	40.15	29.45	20.73	19.60	0.10	2.43
FeO	20.62	10.76	8.22	7.55	0.88	56.93
CaO	0.13	1.95	0.01	11.23	14.72	0.00
K_2O	0.00	0.00	6.94	0.01	0.17	0.01
Na_2O	0.00	0.06	1.26	0.75	3.20	0.04
Cr_2O_3	0.00	0.38	0.89	0.00	0.00	23.13
NiO	0.14	0.10	0.12	0.11	0.00	0.19
Total	99.95	99.99	93.71	98.40	99.64	94.27
Cations						
Si	0.998	1.958	5.678	2.022	9.092	0.000
Al	0.000	0.042	2.386	0.111	6.784	2.158
Ti	0.000	0.010	0.285	0.011	0.000	1.393
Fe	0.445	0.362	1.001	0.229	0.132	15.628
Mn	0.005	0.008	0.001	0.013	0.000	0.089
Mg	1.546	1.539	4.449	1.060	0.027	1.190
Ca	0.004	0.074	0.001	0.437	2.879	0.002
K	0.000	0.000	1.290	0.001	0.040	0.006
Na	0.000	0.004	0.356	0.053	1.142	0.025
Cr	0.000	0.007	0.103	0.000	0.000	6.003
Ni	0.003	0.002	0.014	0.003	0.000	0.050
Species	Forestrite	Enstatite	Phlogopite	Augite	Labradorite	Magnetite

and nickel contents alongwith magnetite (Cr_2O_3 – 23.13% and NiO – 0.19%) and chrome-spinel (Cr_2O_3 – 36.61 and NiO – 0.19%). The notable Cr_2O_3 and NiO contents in silicate and oxides phases analysed in picrite is also substantiated by trace elements analyses of picrite carried out by ICP-MS where in Cr value ranging from 2,157 to 2,162 ppm, Ni values ranging from 981 to 982 ppm were recorded along with Co values ranging from 130 ppm to 134 ppm and V values ranging from 151 to 153 ppm (Sesha Sai, 1994).

Trace and REE Geochemistry: Trace element analyses of the cumulus picrite indicate Cr content of 2,157–2,898 ppm and Ni contents of 919–982 ppm (Table 7.2). Forestrite which is the most abundant mineral with ~ 38% by mode and with 0.14% NiO has controlled the Ni partitioning while, enstatite (0.38% Cr_2O_3),

Table 7.2 Trace element and rare earth element analyses of Picrite sill in Tadpatri formation from Peddakudala-Pulivendla-Velpula area in south western part of Cuddapah basin

Element	Sesha Sai (1994) Velpula-Pulivendla area, SW part of Cuddapah basin		Anand et al. (2003) Pulivendla area, SW part of Cuddapah basin	
Ba	41	42	97	106
Co	130	134	106.3	108.2
Cr	2,157	2,162	2,857	2,898
Cu	62	63	48.9	59.5
Nb	1.5	1.8	1.77	2.37
Ni	981	982	919	919
Rb	9.6	12	14.3	19.2
Sc	29	30	22.8	28.3
Sr	48	49	66.4	67.9
V	151	153	117	137
Y	8.6	10.6	8.47	10.63
Zr	25	27	32.8	44
La	2.69	2.71	6.40	7.93
Ce	4.56	4.60	15.4	19.1
Pr	1.28	1.31	1.4	1.8
Nd	2.04	2.14	5.83	6.91
Sm	0.63	0.83	1.3	1.53
Eu	0.14	0.13	0.42	0.53
Gd	0.36	0.39	1.64	2.07
Tb	–	–	0.27	0.32
Dy	0.76	0.78	1.75	2.08
Ho	–	–	0.36	0.44
Er	0.94	0.96	1.06	1.37
Tm	–	–	0.17	0.18
Yb	0.26	0.47	1.11	1.22
Lu	0.15	0.17	0.16	0.17
Rb/Sr	0.2	0.24	0.21	0.28
Zr/Y	2.90	2.54	3.8724	4.1392
Sm/Yb	2.42	1.76	1.17	1.25
La/Sm	4.26	3.26	4.92	5.18
Sm/Nd	0.30	0.38	0.22	0.22

magnetite (23.13% Cr_2O_3) and chrome-spinel (Cr_2O_3 – 36.61%) controlled the partitioning of Cr. Relatively enriched HFSE contents i.e. Zr (25–44 ppm) and Y (8.6–10.63 ppm), Zr/Y ratios from 2.54 to 4.13, La/Sm ratios (3.26–5.18) and Sm/Yb ratios (1.17–2.42) indicate a Mg rich primary mantle source. Rb/Sr ratios (0.21–0.28) and relatively Sm/Nd ratios (0.22–0.38) however indicate partial crustal contamination. Relatively dense picritic magmas may be trapped at or near the base of the crust where they become contaminated with lower crustal wall rocks and differentiate to form layered sills with gabbroic upper parts and olivine + pyroxene cumulate lower parts (Arndt and Goldstein, 1989).

Discussions

Paleo-Proterozoic dyke swarms of 1,769.1 $\pm$ 2.5 Ma are reported in North China craton (Halls et al., 2002), Fennoscandia and Greenland (Raimo Lahtinen et al., 2008), Sao Francisco craton, Brazil (Wilson Teixeira and Mario Cesar Heredia Figueiredo, 1991), North Atlantic and Baltic cratons (Bridgwater et al., 1995), Uruguay (Maurizio Mazzucchelli et al., 1995), Proterozoic sedimentary strata in the Baraga Basin in northern Michigan (Xin Ding et al., 2010) and wide spread Precambrian mafic-ultramafic magmatism has been recorded in Yilgarn and Pilbara cratons, Hamersley Basin, and Sylvania Inlier of Australia (Aus. Geo News, 2009). Paleoproterozoic Large Igneous Province has been identified in the southern Bastar craton and nearby Cuddapah basin from the adjacent Dharwar craton, India and this record of ~1.9 Ga mafic/ultramafic magmatism associated with concomitant intracontinental rifting and basin development preserved along much of the south-eastern margin of the south Indian shield is a widespread geologic phenomenon on Earth (French et al., 2008). The ultramafic picrite sill and the associated NW-SE to NNW-SSE trending dolerite sills and olivine gabbro bodies occurring in Tadpatri Formation in Peddakudala-Pulivendla-Velpula section and Tonduru-Mallela-Velupucherla section in SW part of Cuddapah basin indicate the surface expression and corroborate with the observation made during the geophysical studies by Tewari and Rao (1987), who indicated the presence of a high density igneous body beneath the SW part of Cuddapah basin. A positive gravity anomaly at the center of the concentric sills indicates a large lopolithic cupola that probably originated through mantle upwelling during the early part of the basin's history (Chatterjee and Bhattcharji, 2001). ^{40}Ar–^{39}Ar laser fusion determination on the phlogopite mica from the mafic-ultramafic sill complex of Tadpatri Formation, constrain an age of initial phase of extrusion in the basin at 1.9 Ga (Anand et al., 2003). Lineament reactivation in NW-SE direction in response to mantle perturbation intensified middle Proterozoic times leading to emplacement of several cross cutting dykes in SW part of Cuddapah basin (Chatterjee and Bhattcharji, 2001). Mantle derived xenoliths of garnet-lherzolite composition were reported from this picrite body (Das Gupta, 1986). Basalt-eclogite transformation or olivine spinel transformation are the source of the basic magma emplaced in lower Cuddapahs (Bhattacharji and Singh, 1984) and these processes involve thermal boundary layer of the mantle.

Conclusions

Emplacement of the forestrite rich picrite sill along the NW-SE to NNW-SSE trending deep seated fractures within the lower Cuddapah Tadpatri sediments indicates a major phase of mantle derived Early Proterozoic magmatism contemporaneous with sedimentation within the stable intracratonic Cuddapah basin in southern Peninsular India. Conspicuous poikilitic relationship between cumulus plagioclase and olivine in picrite indicate co-precipitation of both the phases. Supercooling results in early, precumulus crystallisation of metastable, non-equilibrium plagioclase, followed by co-precipitation of cumulus plagioclase and olivine (Tegner and Wilson, 1994). Forward modeling of trace and major elements and inverse modeling of REE suggest that the primary melts that gave rise to emplacement of mafic-ultramafic sill Complex in Tadpatri Formation was generated due to partial melting of lherzolite mantle source (Anand et al., 2003). Paleo-Proterozoic mafic-ultramafic magmatism in and around the southwestern margin of Cuddapah basin reflects the tectono-magmatic processes that have played a significant role in evolution of the Cuddapah basin and the adjacent granite-greenstone terrene to its west. Available Rb–Sr age of 1,817 ± 24 Ma (Bhasker Rao et al.,1995) and ^{40}Ar–^{39}Ar laser fusion age of 1.9 Ga for the phlogopite mica from mafic-ultramafic sill complex of Tadpatri Formation (Anand et al., 2003), indicate that the deposition of sedimentary strata in the lower part of Cuddapah Super Group has commenced during Paleo-Proterozoic times.

Acknowledgements Sudipta Lahiri, Deputy Director General, and Jayapaul D, Director, Petrology Division of GSI, NER, Shillong are thanked for the support and encouragement. Basav Kumar Chatopadyaya, Shyamal Kumar Sengupta, Geologists (Sr.) and Sandip Nandy, Geologist (Jr.) of Central Petrological Laboratory, GSI, Kolkata, are thankfully acknowledged for EPMA analyses. A. Suryaprakasa Rao, Osmania University is thanked for his valuable encouragement, while Abhijit Roy and S. Rama Murthy of GSI, NER, are thanked for their help in processing some of the figures in the paper. J. P. Shrivastava is thankfully acknowledged for critically reviewing the paper and rendering valuable suggestions in finalisation of MS.

References

Anand MA, Gibson SA, Subba Rao KV, Kelley SP, Dicken AP (2003) Early Proterozoic melt generation process beneath the intracratonic Cuddapah basin, Southern India. J Petrol 44: 2139–2171

Arndt NT, Goldstein SL (1989) An open boundary between lower continental crust and mantle: Its role in crust formation and crustal recycling. Tectonophysics 161: 201–212

Aus. Geo News (December, 2009) Issue No. 96. Revealing Archean mafic-ultramafic magmatism and mineral prospectivity across Australia. pp. 2

Bhasker Rao YJ, Pantulu GVC, Damodar Reddy V, Gopalan K (1995) Time of early sedimentation and volcanism in the Proterozoic Cuddapah basin, South India: Evidence from Rb-Sr age of Pulivendla mafic sill. Geol Soc India Mem 33: 329–338

Bhattacharji JC, Singh RN (1984) Thermo mechanical structure of the southern part of Indian sheild and its relevance to Precambrian basin evolution. Tectonophysic 105: 1103–1120

Bridgwater D, Mengel F, Fryer B, Wagner P, Søren CH (1995) Early Proterozoic mafic dykes in the North Atlantic and Baltic cratons: Field setting and chemistry of distinctive dyke swarms. Geol Soc Lond Spl Publ 95: 193–210

Das Gupta PK (1886) Picrite borne mantle xenoliths from Cuddapah basin, South India. Indian J Earth Sci 13: 333–338

Ding X, Li C, Edward M, Ripley DR, Sandra K (2010) The Eagle and East Eagle sulfide ore-bearing mafic-ultramafic intrusions in the Midcontinent Rift System, upper Michigan: Geochronology and petrologic evolution. Geochem Geophys Geosyst 11: q03003, 22 pp

Ernst RE, Srivastava RK (2008) India's place in the Proterozoic world: Constraints from the Large Igneous Province (LIP) record. In: Srivastava RK, Sivaji C, Chalapathi Rao NV (eds) Indian Dykes: Geochemistry, Geophysics and Geochronology. Narosa Publishing House Pvt. Ltd, New Delhi: 41–56

French JE, Heaman LM, Chacko T, Srivastava RK (2008) 1891–1883 Ma Southern Bastar-Cuddapah mafic igneous events, India: A newly recognised large igneous province. Precamb Res 160: 308–322

Geological Survey of India (1981) Explanatory brochure on Geological and Mineral Map of Cuddapah Basin pp 1–21

Halls HC, Li J, Davis D, Hou G, Zhang B, Qian X (2002) A precisely dated Proterozoic palaeomagnetic pole from the North China craton, and its relevance to palaeocontinental reconstruction. Geophys J Int 143: 185–203

Jayananda M, Mahesha N, Srivastava RK, Mahabaleshwar B, Bkais S (2008) Petrology and geochemistry of Paleo Proterozoic high magnesium norite and dolerite dyke swarms from Halagur-Satnur areas, Eastern Dharwar craton, Southern India. In: Srivastava RK, Sivaji C, Chalapathi Rao NV (eds) Indian Dykes: Geochemistry, Geophysics and Geochronology. Narosa Publishing House Pvt. Ltd, New Delhi: pp. 239–260

Lahtinen R, Adam AG, Victor AM (2008) Paleo Proterozoic evolution of Fennoscandia and Greenland. Geol Surv Greenland Episodes 31: 20–28

Mazzucchelli M, Rivalenti G, Piccirillo EM, Girardi VAV, Civetta L, Petrini R (1995) Petrology of the Proterozoic mafic dyke swarms of Uruguay and constraints on their mantle source composition. Precamb Res 74: 177–194

Nagaraja Rao BK, Ramalingaswamy G (1976) Some new thoughts on the stratigraphy of the Cuddapah Super Group. Seminar on Kaladgi-Badami-Bhima-Cuddapah Super Group, Mysore, India: pp. 17–20

Nagaraja Rao BK, Ramalingaswamy G, Rajurkar ST (1987) Stratigraphy, structure and evolution of the Cuddapah basin. Purana basins of Peninsular India. Mem. 6 Geol Soc India

Nilanjan C, Somdev B (2001) Petrology, geochemistry and tectonic setting of the mafic dykes and sills associated the evolution of the Proterozoic Cuddapah basin, South India. Indian Acad Sci (Earth Planet Sci) 110: 433–453

Sesha Sai VV (1994) Geological and geochemical studies of basic igneous rocks associated with Vempalle Formation around Pulivendla area, Cuddapah basin. Unpub. Ph.D. theses, Osmania University, Hyderabad. pp. 1–69

Sesha Sai VV (2005) Petrographic and Petrochemical characterisation of Proterozoic granites in Northeastern fringes of Cuddapah basin, Andhra Pradesh. National Seminar on Proterozoic Systems of India. Evolution and Economic Potential. ISM Dhanbad.

Sesha Sai VV (2006) Igneous activity in Proterozoic Cuddapah basin. Vol. on Res. papers in DST sponsored field workshop in Cuddapah Basinconducted by GSITI, Hyderabad: 53–62

Srivastava RK, Ahmed T (2009) Precambrian mafic magmatism in the Indian shield: retrospect and prospect. J Geol Soc India 73: 7–11

Srivastava RK, Sivaji C, Rao NVC (2008) India's place in the Proterozoic world: Constraints from the Large Igneous Province (LIP) record. In: Srivastava RK, Sivaji C, Chalapathi Rao NV (eds) Indian Dykes: Geochemistry, Geophysics and Geochronology. Narosa Publishing House Pvt. Ltd, New Delhi: pp. 41–56

Tegner G, Wilson JR (1994) Textures in a pokilitic olivine gabbro cumulate: Evidence for Supercooling. Mineral Petrol 54: 161–173

Teixeira W, Cesar M, Figueiredo H (1991) An outline of Early Proterozoic crustal evolution in the São Francisco craton, Brazil: A review. Precamb Res 53: 1–22

Tewari HC, Rao VK (1987) A high velocity intrusive body in the upper crust in the south western part of Cuddapah basin as delineated by Deep Seismic Sounding and gravity modeling. Purana Basins of Peninsular India. Mem 6 Geol Soc India

Chapter 8
Mantle-Derived Carbonate Fluid Alteration and Gold Mineralization in Southern Granulite Terrain, Wynad, India

K.L. Pruseth, V. Ravikant, S. Varghese, and R. Krishnamurthi

Introduction

Large-scale carbonate deposition and replacement in shear zones from various terrains, that are part of regional carbonate alteration, usually host significant gold mineralization (e.g. Kerrich, 1990; Kerrich et al., 1987; Groves et al., 1989; Barnicoat et al., 1991; Goldfarb et al., 2001; Groves et al., 2003). Though the carbonate permeations are generally perceived as evidence for localized, high CO_2 fluxes into the crust, the source(s) of these fluids is controversial. Evidence for regional carbonate alteration of the crust by fluids from mantle-derived magmas (alkalic basalt, syenite and carbonatite) was presented for the Attur shear zone of Tamil Nadu (Fig. 8.1a, map from Santosh and Sajeev, 2006) by Wickham et al. (1994). We relate the occurrences of the auriferous veins of Wynad (Fig. 8.1b) (Binu Lal et al., 2003) to similar carbonate alteration zones, along the regional Moyar shear zone, where gold-hosted quartz-pyrite hydrothermal veins are well known, but the source rocks for the gold remains unclear.

Here, unlike the Archaean greenstone-hosted gold deposits normally found in India, we report newly identified quartz-carbonate (both ferroan calcite and ankerite) dykes of Wynad that have a direct continuity with the auriferous quartz veins. Distinct mantle value of initial Sr isotopic composition from these dykes, strongly suggests that this orogenic gold was carried and precipitated by mantle-derived carbonate-rich hydrothermal fluids. This mantle component probably was initiated due to structural and thermal events, synchronous with or post-dating orogeny (Goldfarb et al., 2001), specifically the Pan-African orogeny along the Moyar shear zone.

K.L. Pruseth (✉)
Department of Geology & Geophysics, Indian Institute of Technology, Kharagpur-721302, India
e-mail: klpruseth@gmail.com

V. Ravikant
Indian Institute of Science Education and Research-Kolkata, Mohanpur 721452, India

S. Varghese and R. Krishnamurthi
Department of Earth Sciences, Indian Institute of Technology, Roorkee 247667, India

R.K. Srivastava (ed.), *Dyke Swarms: Keys for Geodynamic Interpretation*,
DOI 10.1007/978-3-642-12496-9_8, © Springer-Verlag Berlin Heidelberg 2011

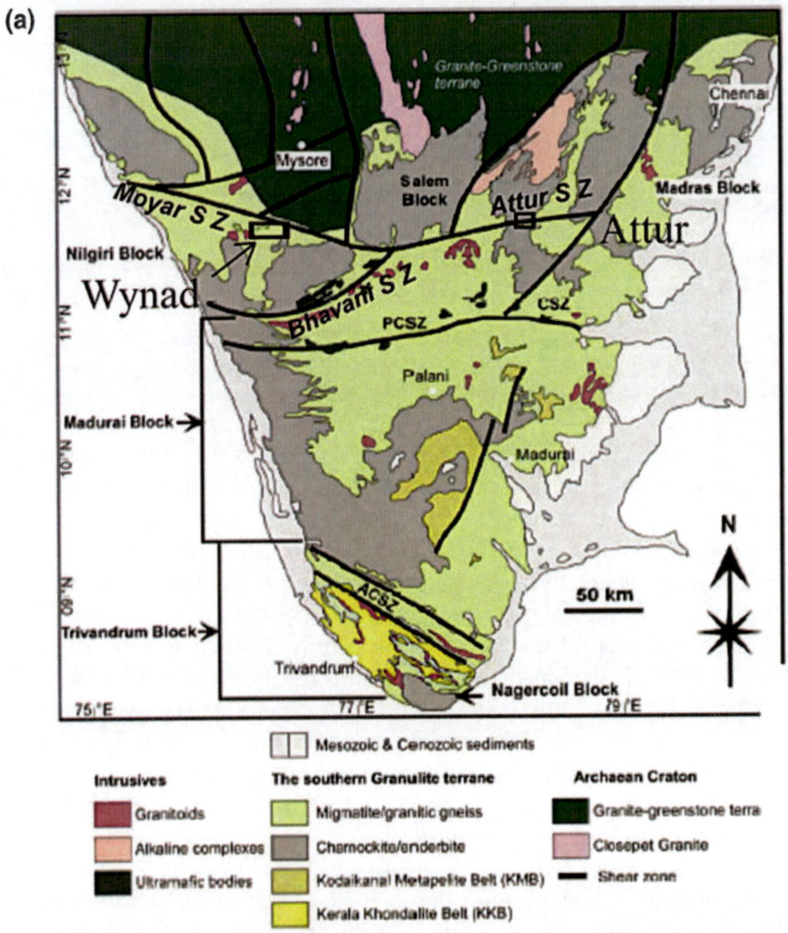

Fig. 8.1 (**a**) Geological map of the Southern Granulite Terrain (from Santosh and Sajeev, 2006) showing location of the Wynad gold field, south of the Moyar shear zone; location of the Attur shear zone, Tamil Nadu is also shown. (**b**) Sketch geological map of the Wynad gold field (after Binu Lal et al., 2003)

Geologic Setting

The Wynad Gold Field (WGF) occurs within the northernmost part of the Southern Granulite Terrain, north of the ~2.5 Ga enderbitic Nilgiri massif and immediately south of the Moyar shear zone (Fig. 8.1a). The WGF comprises a network of auriferous hydrothermal quartz veins (individual veins of length >100 m, average thickness of ~1.5 m) within Precambrian migmatitic tonalitic gneisses and granulites (Nair and Suresh, 1994; Binu Lal et al., 2003., Malathi and Srikantappa, 2005). Whereas the enderbitic remnants are part of the Late Archaean Nilgiri

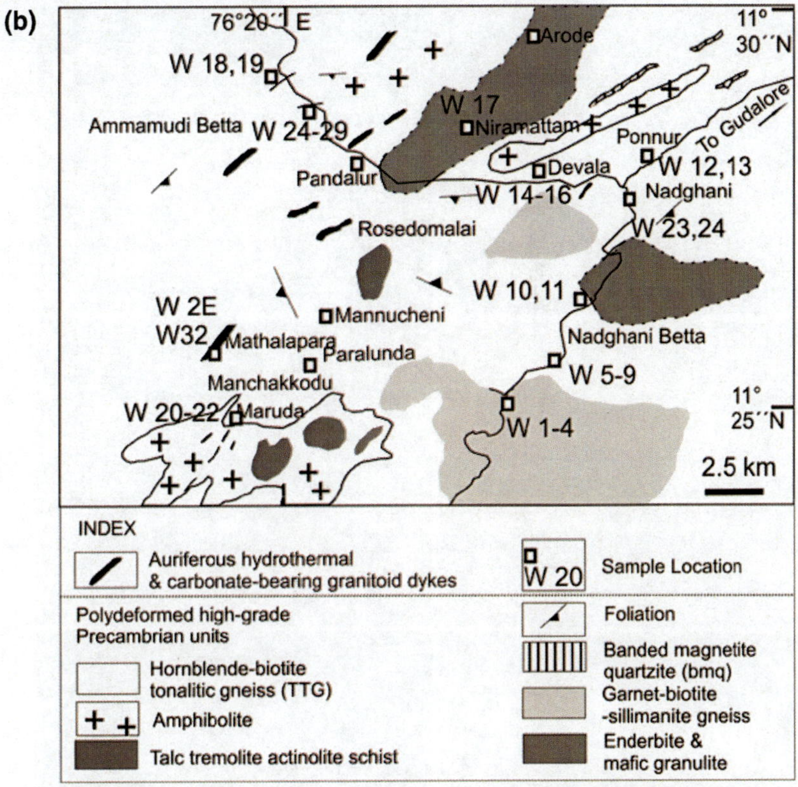

Fig. 8.1 (continued)

granulites (Ghosh et al., 2004), lenticular banded magnetite-quartzite, amphibolite (retrogressed mafic granulite) and minor charnockite are interbanded with the polydeformed migmatitic tonalitic gneisses; a few late, undeformed intrusives of pegmatites have been observed.

Gold-bearing quartz veins, with minor pyrite, muscovite and calcite, form a network that cross-cuts amphibolite and hornblende-biotite quartzo-feldspathic (tonalitic) gneisses to the south and southwest of Wynad (around Maruda, Devala and Pandalur, shown on Fig. 8.1b). Furthermore, these high-grade metamorphic, polydeformed rocks are intruded by unusual pinkish to grey quartz-carbonate dykes and pods, and the auriferous veins can be clearly traced to these dykes (Fig. 8.2a, b).

The most important observation is that there appear to be stages in the hydrothermal evolution of these quartz veins and from field evidence they are seen to emanate from distinct, hitherto unrecognized, quartz-carbonate dykes and pods. These stages recording different mineral assemblages from source rock to barren milky white quartz veins are given below according to their probable sequence of formation:

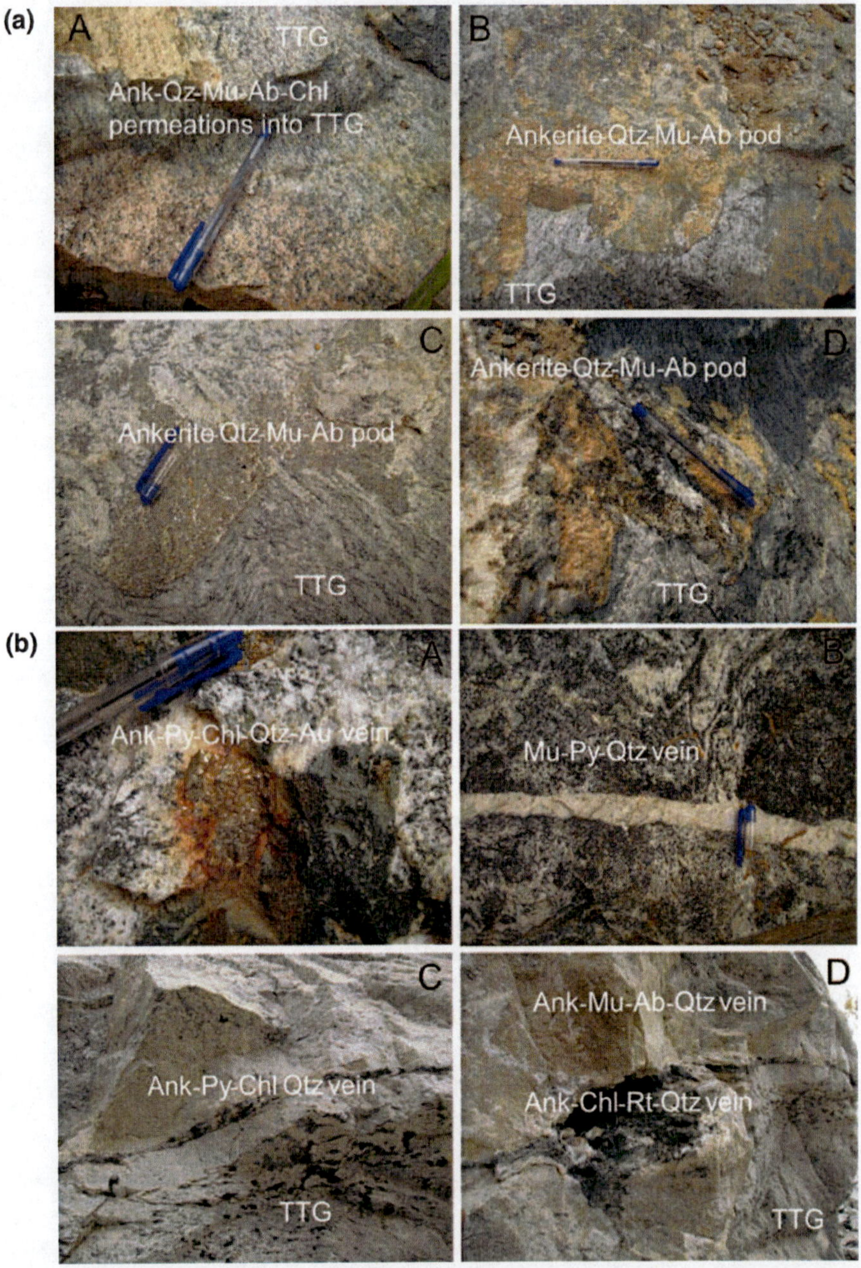

Fig. 8.2 (**a**) A to D. Field relations of the quartz-carbonate muscovite, albite, chlorite dykes intruding the tonalitic gneisses (TTG) from around Pandalur and Maruda, Wynad (carbonate alteration). (**b**) A to D. Field relations of auriferous hydrothermal veins clearly showing emplacement along dilatant and brittle fractures as observed around Maruda

Quartz-carbonate dykes: quartz + pyrite ± pyrrhotite + muscovite + albite + chlorite + carbonate + hematite ± gold ± zircon (Fig. 8.2a, A & D)
Quartz-carbonate dykes (sub-volcanic textured): with megacrysts of muscovite and ankerite and ferroan calcite (Fig. 8.2a, B & C)
Hydrothermal vein (primary auriferous quartz veins): quartz + pyrite ± pyrrhotite + chalcopyrite + gold + muscovite ± chlorite ± carbonate (Fig. 8.2b, A, B & C)
Hydrothermal vein: secondary veins with vugs that contain quartz, chlorite, rutile and scalenohedral pink calcite (Fig. 8.2b, D)
Milky white quartz veins: with traces of muscovite.

Analytical Methods

Pyrite separated from chips of the Wynad hydrothermal veins were washed in deionized water, dried and crushed manually to coarse grains in an agate mortar. Clean crystals of pure pyrite were hand picked from this crushed mixture immersed in ethanol and dried. Water-soluble components of fluid inclusions were released by powdering the sulfides in an agate mortar in ~5–10 ml of ultrapure water (crush-leach method). Similar methods have been successfully used by Bottrell et al. (1988), Banks et al. (1992), Christensen et al. (1995) and Channer et al. (1999). The fluid was rinsed off twice through a grade 292 Sartorius filter paper and collected; fine mineral powder was collected from the dried filter paper.

Trace elements were analyzed on a Perkin-Elmer Sciex ELAN DRC-e ICP-MS at the Institute Instrumentation Centre, IIT, Roorkee (following Ravikant, 2010). About 30 mg of the pyrite fluid and residue, and powdered samples of seven white calcites was digested in Savillex vials using conc. HCl + HNO_3 (3:1) mixture followed by conc. HCl, and diluted to 100 ml in ultrapure 2% HNO_3. Ten ppb of Rh and In were added as internal standards. The dried fluid leachate was dissolved in 10 ml ultrapure water and used for analysis. Basalt rock reference standards United States Geological Survey (USGS) BCR 2 and Geological Survey of Japan (GSJ) JB 2 were used to externally calibrate element concentrations. Analytical precisions were <6% for HFSE and transitional elements, <2% for Ba, Sr and Rb, <5% for LREE and <7% for MREE and HREE. Trace element concentrations of selected samples are given in Tables 8.1 and 8.2.

For Rb-Sr geochronology, the USGS and GSJ rock reference samples AGV 2, BCR 2, W 2 and JB 2, JGb 1 and JA2, respectively, were used to check the accuracy of the procedure. Whole rock, muscovite separates, carbonate (calcite) from the quartz-carbonate dykes and pyrite residues from the auriferous hydrothermal veins were digested using conventional conc. HF + HNO_3 and conc. HCl and conc. HNO_3 + HCl method, respectively. Rubidium and Sr from all samples and fluid leachates were separated using conventional cation exchange chromatography, on primary quartz glass columns packed with 4 ml BioRad resin AG 50×50 200–400 mesh, using 2N HCl as eluent. Strontium (in 1 M HNO_3) and Rb (in MilliQ water) were loaded on degassed single W filaments with 1 μl TaF_5 activator.

Table 8.1 Trace element concentrations (in μg/g) and diagnostic ratios in selected fluids from Wynad pyrite (pyfl), chondrite normalizing values of McDonough et al. (1991)

Element	W21 Cpyfl	W21 Dpyfl	W21 Epyfl	W22 ACpyfl	W22 BCpyfl	W23 ACpyfl	W23 Bpyfl	W24 Apyfl	W26 Apyfl	W27 Apyfl	W27 Bpyfl	W28 Apyfl	W28 Bpyfl
Co	28	219	70	31	8.5	14	12	5.5	3.3	5.7	2.5	0.92	0.40
Cr	6.5	7.2	7.9	8.4	7.6	7.5	6.9	9.9	3.3	6.9	5.5	6.4	4.2
Ni	77	139	72	105	44	38	47	96	5.0	20	32	9.4	3.9
Rb	287	341	361	348	189	263	127	38	16	100	75	41	15
Ba	16	23	13	12	8.4	6.3	7.8	7.5	1.5	5.8	4.3	3.1	1.0
Th	0.9	0.4	0.6	0.4	0.8	0.6	1.2	0.2	0.1	0.7	0.8	0.9	0.9
U	84	130	106	97	103	69	89	62	30	47	42	45	13
Nb	0.31	0.22	0.25	0.22	0.31	0.22	0.13	0.14	0	0.13	0.11	0.13	0
Ta	3.2	3.6	3.9	1.5	4	1.2	1.7	0.56	0	0.93	0.32	0.7	0.8
Sr	87	130	71	71	60	29	58	48	1.7	25	22	17	3.0
Hf	0.13	0.14	3	0.13	0.1	0.03	0.03	0.02	0.01	0.01	0.03	0.03	0.04
Zr	0.13	0.14	0.15	0.13	0.14	0.11	0.15	0.03	0.02	0.16	0.12	0.11	0.04
Y	0.32	2.9	0.34	0.25	0.37	0.36	0.14	0.17	0.12	0.22	0.15	0.24	0.14
La	2.3	32	3.1	1.8	2.3	2.1	1.3	0.54	1.3	1.4	1.3	1.5	1.4
Ce	1.6	14	2.0	1.2	1.7	1.5	0.91	0.32	0.85	1.0	1.0	1.2	1.0
Pr	1.3	10	2.0	1.1	1.6	1.4	0.8	0.33	0.65	0.92	0.91	1.1	0.83
Nd	1.1	8.1	1.4	0.96	1.4	1.4	0.77	0.26	0.48	0.75	0.70	0.85	0.67
Sm	1.2	4.6	1.3	1.0	1.2	1.0	0.59	0.38	0.33	0.66	0.59	0.56	0.45

Table 8.1 (continued)

Element	W21 Cpyfl	W21 Dpyfl	W21 Epyfl	W22 ACpyfl	W22 BCpyfl	W23 ACpyfl	W23 Bpyfl	W24 Apyfl	W26 Apyfl	W27 Apyfl	W27 Bpyfl	W28 Apyfl	W28 Bpyfl
Eu	1.4	2.7	1.0	0.91	0.81	0.87	0.80	0.43	0.35	0.60	0.48	0.73	0.37
Gd	1.0	3.6	0.92	0.68	0.77	0.74	0.57	0.26	0.30	0.48	0.46	0.65	0.34
Tb	0.62	4.5	0.83	0.46	0.72	0.60	0.33	0.10	0.25	0.36	0.44	0.58	0.31
Dy	0.57	3.9	0.56	0.40	0.58	0.50	0.27	0.13	0.23	0.36	0.37	0.42	0.28
Ho	0.64	4.42	0.63	0.52	0.60	0.70	0.29	0.13	0.16	0.38	0.35	0.47	0.33
Er	0.65	4.0	0.53	0.45	0.52	0.63	0.24	0.11	0.19	0.28	0.34	0.44	0.22
Tm	0.71	3.30	0.60	0.50	0.59	0.70	0.32	0.14	0.18	0.34	0.32	0.42	0.30
Yb	1.0	4.3	0.85	0.83	0.96	1.1	0.44	0.21	0.35	0.64	0.63	0.78	0.41
Lu	0.71	2.57	0.60	0.50	0.59	0.70	0.32	0.14	0.18	0.34	0.32	0.42	0.30
Σ REE	15	102	16	11.3	14.4	14	7.9	3.3	5.7	8.4	8.2	10.1	7.1
$(La/Sm)_{CN}$	1.9	6.9	2.5	1.7	1.9	2.0	2.2	1.2	3.8	2.1	2.2	2.7	3.0
$(La/Yb)_{CN}$	2.1	7.5	3.7	2.1	2.4	1.8	3.0	2.3	3.5	2.2	2.0	1.9	3.3
$(Dy/Yb)_{CN}$	0.53	0.92	0.66	0.48	0.60	0.44	0.62	0.60	0.65	0.56	0.58	0.53	0.69

Table 8.2 Trace element concentrations (in μg/g) and diagnostic ratios in selected calcite from auriferous hydrothermal veins of Wyand, chondrite normalizing values of McDonough et al. (1991)

Element	W32B cc	W32C cc	W32F cc	W32G cc	W32H cc	W32I cc	W32J cc
Co	15	11	15	4.7	1.8	6.9	4.9
Cr	1.4	2.3	1.6	2.1	1.4	1.7	1.9
Sc	13	5.5	17	3.4	0	5.4	5.5
Ni	4.5	3.1	2.5	1.9	2.5	2.7	3.2
Rb	1.5	2.4	1.9	6.1	3.6	2.1	2.4
Ba	5	6.3	5.6	18	5.6	7.9	4.2
Th	0.80	0.25	0.23	0.22	0.13	0.21	0.22
U	0.27	0.06	0.10	0.10	0.05	0.04	0.05
Nb	3.8	1.9	1.8	2.2	1.9	1.3	2.0
Ta	16	5.2	6	7.2	4.5	5.2	7.1
Sr	8,165	6,162	8,547	5,522	6,469	7,389	7,230
Hf	4.1	2.1	4.2	4.7	1.1	0.96	0.78
Zr	65	26	48	58	14	7.7	8.7
Y	1,268	684	1,489	150	156	638	519
La	41	13	37	18	8.9	17	20
Ce	24	9.2	24	11	5.9	11	13
Pr	20	7.7	21	9.3	4.9	9.3	11
Nd	16	6.6	18	7.4	3.9	7.8	8.8
Sm	12	5.2	15	4.0	2.6	6.0	6.1
Eu	41	11	32	5.8	6.3	14	14
Gd	29	9.7	27	4.4	4.4	11	11
Tb	16	7.8	21	3.0	2.5	8.0	7.2
Dy	16	8.1	21	2.6	2.4	8.6	6.5
Ho	17	9.2	21	2.5	2.4	9.1	6.9
Er	15	8.3	18	2.3	2.2	8.6	6.0
Tm	16	9.5	20	2.3	2.2	8.9	6.5
Yb	25	14	31	3.3	3.7	15	10
Lu	14	7.7	17	1.7	1.9	7.8	5.8
Σ REE	303	128	323	78	55	142	1,323
$(La/Sm)_{CN}$	3.4	2.5	2.4	4.4	3.4	2.8	3.2
$(La/Yb)_{CN}$	1.6	0.95	1.2	5.4	2.4	1.1	1.9
$(Dy/Yb)_{CN}$	0.61	0.58	0.67	0.77	0.65	0.58	0.63

Isotopic analyses of Rb and Sr were carried out on a Thermo Finnigan Triton T1 thermal ionization mass spectrometer, in static multicollector mode, at the National Facility for Geochronology and Isotope Geology at the Indian Institute of Technology, Roorkee. Mass fractionation corrections in Sr isotopic analyses were done with $^{86}Sr/^{88}Sr = 0.1194$. The precision of $^{87}Sr/^{86}Sr$ (0.03%) ratios are the external precisions based on replicate determinations of standards; precision on the $^{87}Rb/^{86}Sr$ ratio is 1% by IDTIMS. During the period of this study, the mean $^{87}Sr/^{86}Sr$ ratio was 0.71026 ± 0.00008 (2σ, $n = 21$) for the NBS 987 Sr standard. The total procedural blank was < 0.7 ng for Sr and ~0.1 ng for Rb. The Rb-Sr isotopic data (Table 8.3) were regressed using the ISOPLOT-Ex program of Ludwig (2001) and ages were calculated with 2σ errors. Rb and Sr concentrations were calculated offline by a double spike correction program.

Table 8.3 Rb-Sr isotopic data for carbonate–bearing dykes and pyrite residues of auriferous hydrothermal veins of Wynad

Sample number and location	Rb (μg/g)	Sr (μg/g)	$^{87}Rb/^{86}Sr^{a}$	$^{87}Sr/^{86}Sr \pm 2\sigma^{b}$	$^{87}Sr/^{86}Sr \pm 2\sigma^{b}$
Dykes					
W32E whole rock	8.9	446	0.057	0.70377 ± 21	–
W32E Fe-calcite	5.3	584	0.026	0.70301 ± 21	–
W32E musc 1	62	142	1.28	0.71099 ± 21	–
W32E musc 2	47	56	2.41	0.71838 ± 21	–
W2E whole rock	10	445	0.067	0.70324 ± 20	–
W2E Fe-calcite	6.4	533	0.0349	0.70300 ± 20	–
W2E musc	53	171	0.906	0.70824 ± 20	–
Pyrite residues					Pyrite fluids (only IC)
W21A py	3.3	7.4	1.30	0.70847 ± 20	–
W21C py	7.0	14	1.41	0.70858 ± 20	–
W21D py	2.3	7.8	0.859	0.70854 ± 20	–
W21E py	2.6	9.5	0.793	0.70878 ± 20	–
W22A py	3.5	16	0.679	0.70824 ± 20	0.70897 ± 20
W23A py	3.5	11	0.951	0.70826 ± 20	0.70894 ± 20
W23B py	3.2	11	0.806	0.70829 ± 20	0.70910 ± 20
W24A py	3.6	7.5	0.980	0.71175 ± 20	0.70886 ± 20
W26A py	4.2	15	0.813	0.70821 ± 20	0.70787 ± 20
W27B py	3.5	12	0.836	0.70823 ± 20	0.70841 ± 20
W28B py	2.5	9.5	0.757	0.70852 ± 20	0.70858 ± 20
W29A py	1.9	7.3	0.768	0.70864 ± 20	0.70992 ± 20

[a]Error of 1% on this ratio.
[b]Errors on least significant digits.

Mineral chemical analyses of selected phases were carried out on a Cameca SX100 electron probe microanalyzer, at the Institute Instrumentation Centre, IIT Roorkee, with an accelerating voltage of 15 kV and an electron beam current of 20 nA focused to <1 μ spot size. Counting times were 20 s on peak and 10 s on background; synthetic and natural standards were used for calibration of elements.

Results

Petrology

Gold-bearing quartz-pyrite veins, emplaced within faults and shears, represent late-stage hydrothermal veins originating from ~0.5 m thick, undeformed, pink quartz-carbonate dykes. Both hypidiomorphic and subvolcanic textures with carbonate and muscovite megacrysts set within a fine-grained groundmass are seen that have an assemblage comprising quartz + pyrite + phengite (muscovite-celadonite) + chlorite ($X_{Fe\ 0.79–0.84}$) + ankerite ($Ca_{0.47–0.58}$ $[MgFe]_{0.42–0.49}$) + ferroan calcite ($Ca_{0.88–0.93}$

$Fe_{0.06–0.09}$) + albite ($Ab_{0–10}$) + gold + hematite + zircon. The sub-volcanic textured hydrothermal vein comprises quartz + pyrite ($Fe_{0.89–1.0}$ S_2) + muscovite + chlorite + ankerite ($Ca_{0.50–0.51}$ $[MgFe]_{0.46–0.49}$) + ferroan calcite ($Ca_{0.83–.0}$ $Fe_{0.0–0.29}$) + albite + gold + hematite + zircon, whereas the primary hydrothermal veins which are being mined for gold have additional pyrrhotite and chalcopyrite ($Cu_{0.91–0.96}$ $Fe_{0.99–1.1}$ S_2). Secondary hydrothermal veins contain vugs and additional rutile.

Gold is exclusively hosted by pyrites in the auriferous quartz veins which are essentially arsenian (up to 1,700 ppm As). Binu Lal et al. (2003) have provided excellent photomicrographs of native gold occurring within pyrite. Powder X-ray diffraction patterns of these pyrites contain excess peaks indicating minor distortion of their structure. Compared to the ideal pyrite composition these pyrites have relatively low Fe (Fe $_{0.82–0.90}$ S_2 to $Fe_{0.89–1.0}$ S_2) and random spots analyzed by electron probe showed Au concentrations varying from 0.11 to 0.17 wt%. Brighter white patches within a background of grey visible in back scattered electron images have near-ideal pyrite compositions ($Fe_{0.92–1.0}$ S_2 to $Fe_{0.96–1.0}$ S_2). Grey pyrites show consistently low totals suggesting presence of volatiles but are particularly rich in Au and As. Energy dispersive spectroscopy revealed presence of C, O, Ca, Cl and K within grey pyrite suggesting inclusions of CO_2, calcite and KCl crystals. Similar densely populated gas-rich fluid inclusions are also seen in quartz from the auriferous hydrothermal veins (Fig. 8.3).

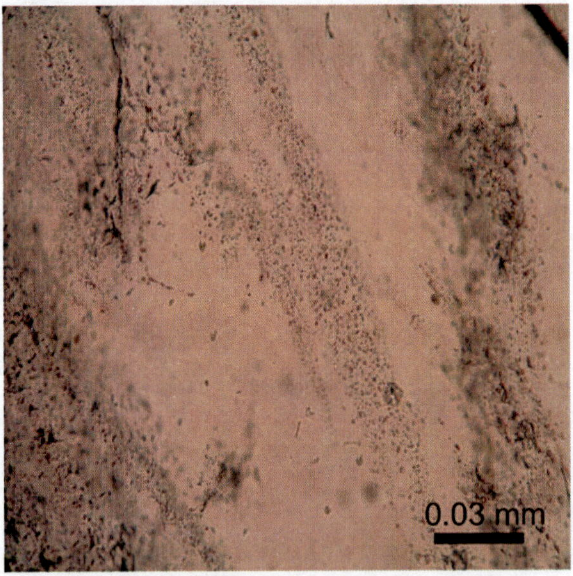

Fig. 8.3 CO_2 and H_2O rich gas-rich fluid inclusions in quartz from auriferous hydrothermal veins of Wynad

Geochemistry of Fluids

As the pyrites were considered to be diagnostic of gold mineralization (being transported as sulfide complexes and precipitated at low temperatures along with pyrite, Binu Lal et al., 2003), we extracted the water-soluble component of the fluid inclusions by crushing selected pyrite samples dipped in MilliQ water. The results of such analyses are shown in Table 8.1 and the chondrite- (after Sun and McDonough, 1989) and primitive upper mantle (PUM)-(after McDonough et al., 1991) normalized patterns are shown in Figs. 8.4a and b. The most diagnostic

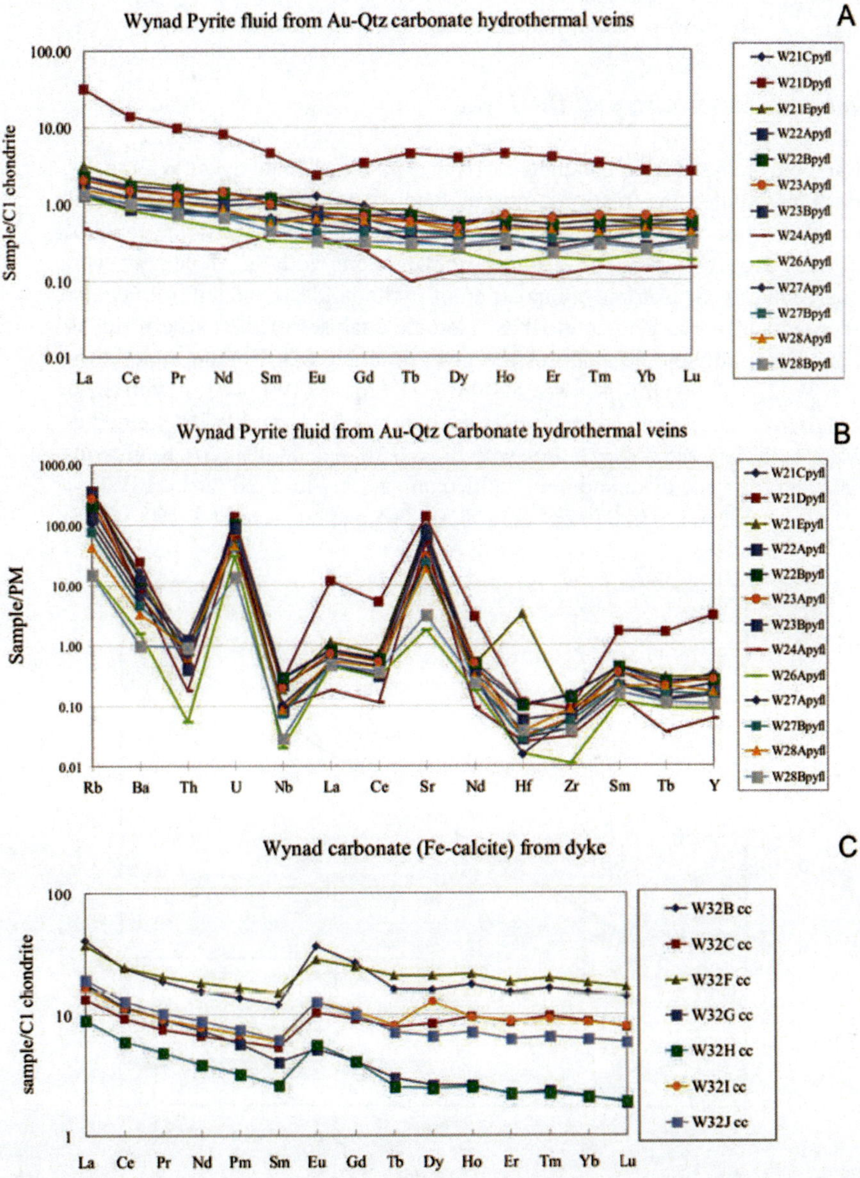

Fig. 8.4 (**a**) Chondrite (C1 of Sun and McDonough, 1989) normalized REE plot of fluid extracted from the pyrite. (**b**) Primitive mantle normalized (McDonough et al., 1991) spidergram of fluid extracted from the pyrite and ferroan calcite (**c**)

feature of these fluids is that they are in general depleted in trace elements with respect to the PUM except for Rb, Ba, Sr and U and have a near flat chondrite normalized REE pattern, suggesting juvenile source(s) to these fluids. Ferroan calcite from the auriferous hydrothermal veins were also analyzed for their trace and REE contents (Table 8.2) and are shown on Fig. 8.4c which show subequal LREE/HREE and positive Eu anomalies, characteristic of derivation from plagioclase breakdown.

Sr Isotope Geochemistry of the Dykes

Rb-Sr isotopic analyses were conducted from a quartz-carbonate dyke (W32) using the phases muscovite (fine fraction), muscovite (coarse flakes), whitish calcite, and whole rock fraction and a typical hydrothermal vein that cross-cuts the tonalitic gneiss (W2E) from which ankerite and muscovite along with its whole rock were analyzed. Similarly, Sr isotopic composition of pyrite fluids, separated from twelve hydrothermal auriferous veins and Rb-Sr isotopic analyses of the pyrite residues, were conducted; all isotopic analyses are given in Table 8.3. Isotopic ratios from the dyke, W32, yielded an Rb-Sr age of 443 ± 18 (2σ), MSWD $= 1.6$, with $Sr_i = 0.7031 \pm 0.0003$ (2σ) whereas the auriferous hydrothermal vein, W2E, yielded an age of 420 ± 35 (2σ), MSWD $= 0.038$, with $Sr_i = 0.7028 \pm 0.0003$ (2σ). Regressing all seven data from the dyke and vein yielded an age of 446 ± 26 (2σ), MSWD $=$ 2 with $Sr_i = 0.7029 \pm 0.0004$ (2σ) (Fig. 8.5). The age-corrected (at 450 Ma) Sr

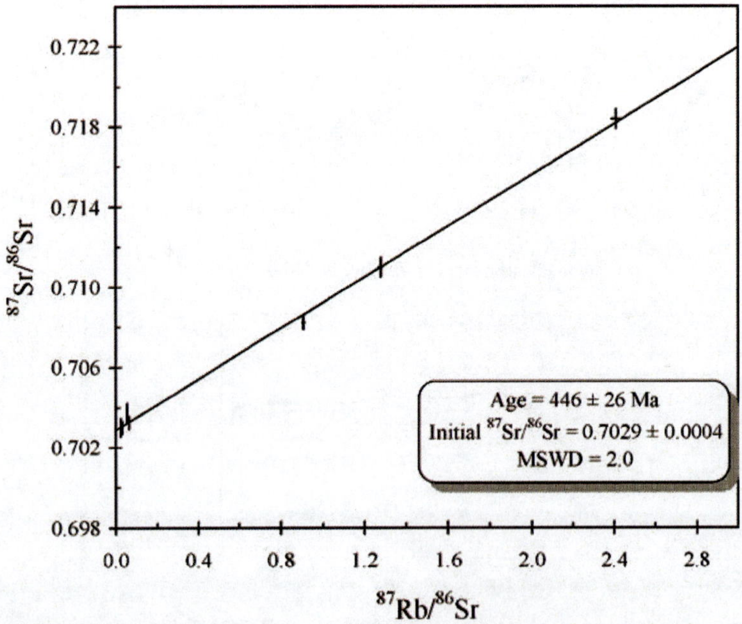

Fig. 8.5 Rb-Sr isochron diagram regressing phases from both quartz-carbonate dyke (sample W32) and auriferous hydrothermal vein (sample W2E) of SW Wynad

initial ratios of the pyrite residues also are tightly constrained to between 0.70216 and 0.70389; regressing the pyrite residues along with the seven-points from the quartz-carbonate dykes yields an isochron age of 445 ± 35 (2σ), MSWD = 7.6 with an $Sr_i = 0.7031 \pm 0.0005$.

Discussion

Generally gold mineralization in southern India is ascribed to magmatic intrusives, hydrothermal exhalatives, metamorphic outgassing in greenstone belts or to a combination of all of them. Our observations indicate that the auriferous hydrothermal veins of the Wynad Gold Field, one of the very few occurrences in the Southern Granulite Terrain, are undeformed rocks intrusive into polydeformed high-grade metamorphic rocks. Rather than by any crustal scale shear zone, their emplacement was controlled by late-stage brittle fracture-fault systems that are clearly younger than any tectonothermal event. This makes these dykes potentially amenable to geochemical study.

Mantle-derived (cryptic) intrusions (alkali basalt, syenite and carbonatite) of Neoproterozoic age (~800 Ma) acted as source(s) to the extensive carbonate influx in the Attur shear zone as deciphered by Wickham et al. (1994). Although regional carbonate alteration zone features of the Attur area have petrologic similarities with those observed in the Wynad area and both these occurrences lie proximal to the E-W trending Moyar-Attur crustal-scale shear zone, these alteration zones do not seem to be coeval. From the Attur area, whole rock, muscovite and carbonate isochrons yield dates of 439 and 494 Ma with high initial $^{87}Sr/^{86}Sr$ ratios of 0.7095 and 0.7074, respectively. Although these dates are similar to those obtained in this study, from the Wynad quartz-carbonate dykes (~450 Ma), the initial $^{87}Sr/^{86}Sr$ ratios of the Wynad samples are very low (0.7030) and distinctly resemble mantle ratios in contrast to the higher initial ratios of the Attur samples as reported by Wickham et al. (1994), who interpret these ages as that of retrogression and hydration postdating granulite-facies metamorphism in the Southern Granulite Terrain.

As discussed by Barker (1989), if the carbon of a carbonatite is juvenile being derived from identifiable or suspected magmas, then carbonate-rich rocks, can be considered as having been derived from cryptic carbonatite intrusions. Because no relative LREE enrichment was observed in our pyrite fluid analyses (Fig. 8.3a) and in the absence of any reported carbonatite intrusions around Wynad, we discount the link to carbonatite for gold mineralization in these quartz-carbonate veins. Furthermore, the carbon of the carbonate fraction from auriferous hydrothermal veins in the Wynad area was shown, using stable isotope ratios of carbonate inclusions in quartz, by Santosh et al. (1995) to have been derived from the mantle. This is also clearly supported by our low initial Sr isotopic ratios from the quartz-carbonate dykes. As carbonate influx occurred definitely postdating the ~500–550 Ma Pan-African tectonothermal event (deformation ceased by ~490 Ma, e.g. Meiβner et al., 2002) in the Wynad area, this would then constrain the age of the mantle/lower crust-derived fluid influx to be late Pan-African. In addition, we have shown that the fluids trapped within the pyrites (ubiquitously associated with gold) possess a near-flat chondrite normalized REE pattern, again strongly suggesting that the fluids were directly derived from the mantle/lower crust, and not from any carbonatite or lamprophyre melt.

The predominance of quartz and carbonate with a low sulfide-content in the auriferous veins, emplaced along brittle-ductile fractures, and the association of chlorite-carbonate-sericite alteration of the host rock suggest the orogenic nature of the gold mineralization (Goldfarb et al., 2001). Further, the age of the mineralization (~450 Ma) links its genesis to the structural and thermal events of the Pan-African orogenic cycle in the Southern Granulite Terrain. Thus the distinct mantle/lower crust signatures of these dykes further corroborates that the gold mineralization is of orogenic type and occurred due to mantle/lower crust-derived CO_2–rich fluids (carbonate flux).

Conclusions

1. In the Wynad Gold Field of the Southern Granulite Terrain, auriferous quartz-pyrite veins, emplaced within faults and shears, represent late-stage hydrothermal veins originating from ~0.5 m thick, undeformed, pink quartz-carbonate dykes. These quartz-carbonate dykes have been dated at ~450 Ma, with very low initial $^{87}S/^{86}Sr$ isotopic ratios (0.7029) and have nearly flat chondrite normalized REE values.
2. In conjunction with earlier isotopic analyses of carbon from carbonates of auriferous veins that indicates its mantle/lower crustal derivation, the Sr isotopic data and nearly flat chondrite normalized REE pattern, strongly suggest a mantle-derived carbonate fluid that transported the gold. The gold mineralization is orogenic type related to post-Pan-African orogeny processes.

Acknowledgements We are thankful to Mihir Deb, who reviewed the manuscript, for his valuable suggestions that significantly helped improve the readability of the manuscript.

References

Banks DA, Wayne D, Miller M, Scrivner R (1992) Composition and origin of ore-fluids in the SW-England orefield. Pan American Conference on Research in Fluid Inclusions (PACROFI) 4: 12

Barker DS (1989) Field relations of carbonatites. In: Bell K (ed) Carbonatites-Genesis and Evolution. Unwin Hyman, London: 38–69

Barnicoat AC, Fare RJ, Groves DI, McNaughton NJ (1991) Synmetamorphic lode-gold deposits in high-grade settings. Geology 19: 921–924

Binu Lal SS, Sawaki T, Wada H, Santosh M (2003) Ore fluids associated with the Wynad gold mineralization, southern India: Evidence from fluid inclusion microthermometry and gas analysis. J Asian Earth Sci 22: 171–187

Bottrell SH, Yardley BWD, Buckley F (1988) A modified crush-leach method for the analysis of fluid inclusion electrolytes. Bull Mineral 111: 279–290

Channer DM, De R, Bray CJ, Spooner ETC (1999) Integrated cation-anion/volatile fluid inclusion analysis by gas and ion chromatography: Methodoloy and examples. Chem Geol 154: 59–82

Christensen JN, Halliday AN, Leigh KE, Randell RN, Kesler SE (1995) Direct dating of sulfides by Rb-Sr: A critical test using the Polaris Mississippi Valley-type Zn-Pb deposit. Geochim et Cosmochim Acta 59(24): 5191–5197

Ghosh JG, DeWit M, Zartman RE (2004) Age and tectonic evolution of Neoproterozoic ductile shear zones in the Southern Granulite Terrain of India, with implications for Gondwana studies. Tectonics 23 doi: 10.1029/2002TC001444

Goldfarb RJ, Groves DI, Gardoll D (2001) Orogenic gold and geologic time: A global synthesis. Ore Geol Rev 18: 1–75

Groves DI, Barley ME, Ho SE (1989) Nature, genesis, and tectonic setting of mesothermal gold mineralization in the Yilgarn Block, Western Austarlia. In: Keays RR, Ramsay WRH, Groves DI (eds) The Geology of Gold Deposits. Econ Geol Monogr 6: 71–85

Groves DI, Goldfarb RJ, Robert F, Craig JRH (2003) Gold deposits in metamorphic belts: Overview of current understyanding, outstanding problems, future research and exploration significance. Econ Geol 98: 1–29

Kerrich R (1990) Carbon-isotope systematics of Archean Au-Ag vein deposits in the Superior Province. Can J Earth Sci 27: 40–56

Kerrich R, Fryer BJ, King RW, Willmore LM, van Hees E (1987) Crustal outgassing and LILE enrichment in major lithospheric structures, Archean Abitibi greenstone belt: Evidence on the source reservoir from strontium and carbon isotope tracers. Contrib Mineral Petrol 97: 156–168

Ludwig KR (2001) User Manual for Isoplot/Ex version 2.47. A geochronological tool kit for Microsoft Excel, Berkeley Geochronology Center Special Publication 2: 19

Malathi MN, Srikantappa C (2005) Composition and evolution of fluids and timing of gold mineralization in the Malappuram-Gudalur and Bhavani shear zone, Nilambur, Kerala. J Geol Soc India 65: 689–702

McDonough WF, Sun S-s, Ringwood AE, Jagoutz E, Hofmann AW (1991) K, Rb, and Cs in the earth and moon and the evolution of the earth's mantle. Geochim Cosmochim Acta 56: 1001–1012

Meiβner B, Deters P, Srikantappa C, Kohler H (2002) Geochronological evolution of the Moyar, Bhavani and Palghat shear zones of southern India: Implications for east Gondwana correlations. Precamb Res 114: 149–175

Nair RS, Suresh CM (1994) Gold mineralization in Kappil prospect, Malappuram district, Kerala. J Geol Soc Ind 43: 573–583

Ravikant V (2010) Palaeoproterozoic (~1.9 Ga) extension and rifting along the eastern margin of the Eastern Dharwar Craton, SE India: New Sm-Nd isochron age constraints from anorogenic mafic magmatism in the Neoarchean Nellore greenstone belt. J Asian Earth Sci 37: 67–81

Santosh M, Nadeau S, Javoy M (1995) Stable isotope evidence for the involvement of mantle-derived fluids in Wynad gold mineralization, South India. J Geol 103: 718–728

Santosh M, Sajeev K (2006) Anticlockwise evolution of ultrahigh temperature granulites within continental collisional zones in southern India. Lithos 92: 447–464

Sun S-s, McDonough WF (1989) Chemical and isotopic systematics of oceanic basalts: Implications for mantle composition and processes. In Saunders AD, Norry MJ (eds) Magmatism in ocean basins. Geol Soc Lond Spl Publ No 42: 313–345

Wickham SM, Janardhan AS, Stern RJ (1994) Regional alteration of the crust by mantle-derived magmatic fluids, Tamil Nadu, South India. J Geol 102(4): 379–398

Chapter 9
Mafic Dykes of Rewa Basin, Central India: Implications on Magma Dispersal and Petrogenesis

Trisha Lala, A.K. Choudhary, S.K. Patil, and D.K. Paul

Introduction

Like most Large Igneous Provinces (LIP), mafic dykes are an integral component of the Deccan Volcanic Province. It has been estimated that more than 2,000,000 km^3 of lava erupted over a continental area of 500,000 km^2 in a short span of 0.5 million years (White and McKenzie, 1989) (Fig. 9.1). In the Deccan Volcanic Province (DVP) mafic dykes are abundant in the Western Ghats and Konkan coastal belt in the west and in the Son-Narmada-Tapi river basins but rare in the eastern part (Karkare and Srivastava, 1990). Recent studies of these dykes by Ray et al. (2007), Sheth et al. (2004), Chandrasekharam et al. (1999), Bhattacharji et al. (1996), have brought out their petrological and geochemical details.

The Narmada-Son lineament in central India is a prominent linear feature across India extending to Madagascar. It is considered to be a continental wrench system (Chakraborty et al., 2003) formed in the Precambrian, or a mid-continental rift system (Beane et al., 1986; Shanker, 1991). Mafic dykes have been considered to be the feeder to the mafic flows in parts of the DVP. Sedimentary basins of Gondwana age formed along the Son-Narmada lineament (Chakraborty et al., op.cit.) as extensional or as strike-slip basins. Mafic dykes occur in these basins along well-defined faults. Rajmahal – Bengal Basin–Sylhet basalts (~115 Ma, Baksi et al. 1987; ~118 Ma, Kent et al., 2002) occur towards the eastern end of the lineament in eastern India. There are several mafic and ultramafic dykes in the vicinity of the Rajmahal basalt occurrences in the coal-bearing Raniganj and Jharia Gondwana basins. One such prominent dyke, the Salma dyke, a ferrogabbro, was dated at 64.4 ± 0.3 Ma (Kent

D.K. Paul (✉)
Department of Geology, Presidency College, Kolkata 700073, India
e-mail: dalimpaul@yahoo.co.in

T. Lala
Department of Geology, Presidency College, Kolkata 700073, India

A.K. Choudhary
Indian Institute of Technology, Roorkee 247667, Uttaranchal, India

S.K. Patil
Dr. K.S. Krishnan Geomagnetic Research Laboratory, Allahabad 221505, India

R.K. Srivastava (ed.), *Dyke Swarms: Keys for Geodynamic Interpretation*,
DOI 10.1007/978-3-642-12496-9_9,

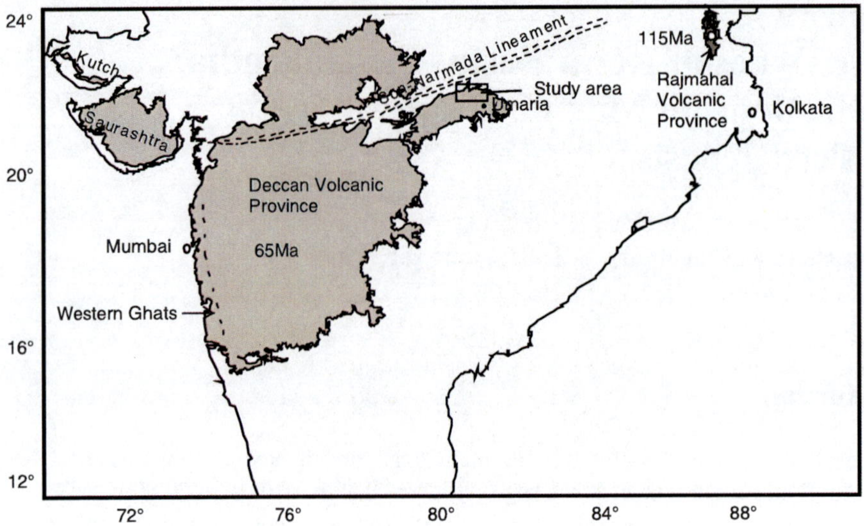

Fig. 9.1 Map showing Deccan Volcanic Province, DVP (65 Ma) and Rajmahal volcanic rocks (115 Ma) in eastern India. The present study area is in the eastern end of DVP

et al., 1997). Though geographically the Salma dyke is closer to the Rajmahal basalt occurrence, the age of 64.4 Ma is very close to that of eruption of Deccan volcanics.

A possibility exists that the magma which was emplaced as the Salma dyke belongs to the Deccan Province but moved eastward at sub-crustal depths. In this scenario similar dykes would be expected in the area between the western Deccan Province and the Rajmahal basalt occurrences.

Detailed study of the mafic dykes of the eastern Deccan Volcanic Province (DVP) is, however, rare in contrast to the dykes of western DVP. We have taken up a petrological and geochemical study of the mafic dykes around Shahdol, Umaria and Chirimiri in central India. These have been collectively discussed as Rewa dykes in this paper. The objective is to explore (1) any petrological linkage between the dykes of the main DVP with those of eastern India, (2) compositional variation between the mafic dykes within the Deccan basalts and the older (Carboniferous) Gondwana basin, and (3) evolutionary changes of magma migration from west to east.

Mode of Occurrence

The Deccan Volcanic Province (DVP) is intruded by a complex dyke-sill network comprising at least three regional scale dyke swarms (Auden, 1949, Deshmukh and Sehgal, 1998; Srivastava et al., 1996; Karkare and Srivastava, 1990; Bondre et al., 2006). The northernmost of these is the ENE-WSW trending Narmada-Satpura-Tapi swarm, which apparently continues westward into the Saurashtra Peninsula. The NNW-SSE trending Konkan or west coast dyke swarm represents the southern arm. The middle arm is much shorter in comparison to other arms and is exposed NE of Mumbai, named the Western Ghats swarm. The area forming the eastern-most fringe area of the Narmada-Satpura-Tapi swarm of mafic dykes that intrude the Gondwana sediments around Shahdol, Manendragarh, Chirimiri and Umaria in the Southern Rewa basin of central India (Fig. 9.2) (Paul et al., 2008) has been studied to assess compositional variation. The orientation and dimensions of twenty-nine dykes have been measured from the published geological maps of the Geological Survey of India (2005). Altogether twelve dykes have been selected for petrological and geochemical studies from an area of about 8,000 km^2. North of Shahdol, one mafic dyke, trending NW-SE is exposed along the Son River (Fig. 9.2) intruding the Gondwana (Permo-Triassic) sediments. It has developed a chilled margin characterized by fine grain size along the contact zone. The exposed dyke is 120 m long with a width of 6 m. Smaller NE-SW trending dykes cut across the main NW-SE dyke here and show thin ramifications into the country rock of Gondwana sandstone. Another large dyke trending NE-SW, is exposed along a nala bed in Akhrar, north of Shahdol. The mafic dykes in Manendragarh also intrude the Gondwana sediments and trend NE-SW (Fig. 9.2). These dykes vary from 200 m to 17 km in length and 4–0 m in width. These dykes also trend NE-SW and have a shorter length varying from 500 m to 2.5 km. However, dykes intruding the Gondwana sediments at Chirimiri show a dominant E-W trending pattern.

Age of the Rewa Dykes

The dykes in the area cut across Gondwana sediments in Shahdol and Chirimiri while those around Umaria cut the Deccan volcanics. Kent et al. (1997, 2002) described the mafic dykes from southwest of Rajmahal Volcanic Province (from Kalidaspur, Raniganj; Meghatari, Koderma) and showed that these dykes were of Cretaceous age, close to the age of the Rajmahal basalts (i.e. around 115 Ma). But the prominent Salma dyke, a ferrodolerite, was of Palaeogene age (64.4 Ma) (Kent et al., 1997).

The dykes of Rewa have not been dated isotopically. As the Umaria dykes cut across the Deccan lavas, these are younger than the flows. Palaeomagnetic determinations have been made on the Shahdol and Chirimiri dykes during the present study. The paleomagnetic results on the dykes are given in Table 9.1. Two dykes from Shahdol RB 5 gave normal polarity (348/–7.25) and RB-13 also show normal polarity (338/–21). Amongst the Chirimiri dykes, RB-61, RB-75 show reverse polarity (180/65; 159/38) and RB-64 shows normal polarity (308/–60). Overall mean polarity for these five dykes was obtained by inverting the reverse polarity. Mean ChRm direction for these five dykes has been computed as $D = 339.53°$, $I = -39.18°$, $\alpha_{95} = 27.26°$ and $k = 8.83$. Palaeomagnetic pole at 40.40°N and 286.62°E has been computed by averaging the Virtual Geomagnetic Pole (Fig. 9.3). This is close to the mean Deccan Super Pole (Vandamme et al., 1991). The palaeomagnetic data, therefore, indicates that the dykes of the eastern DVP are Deccan-related.

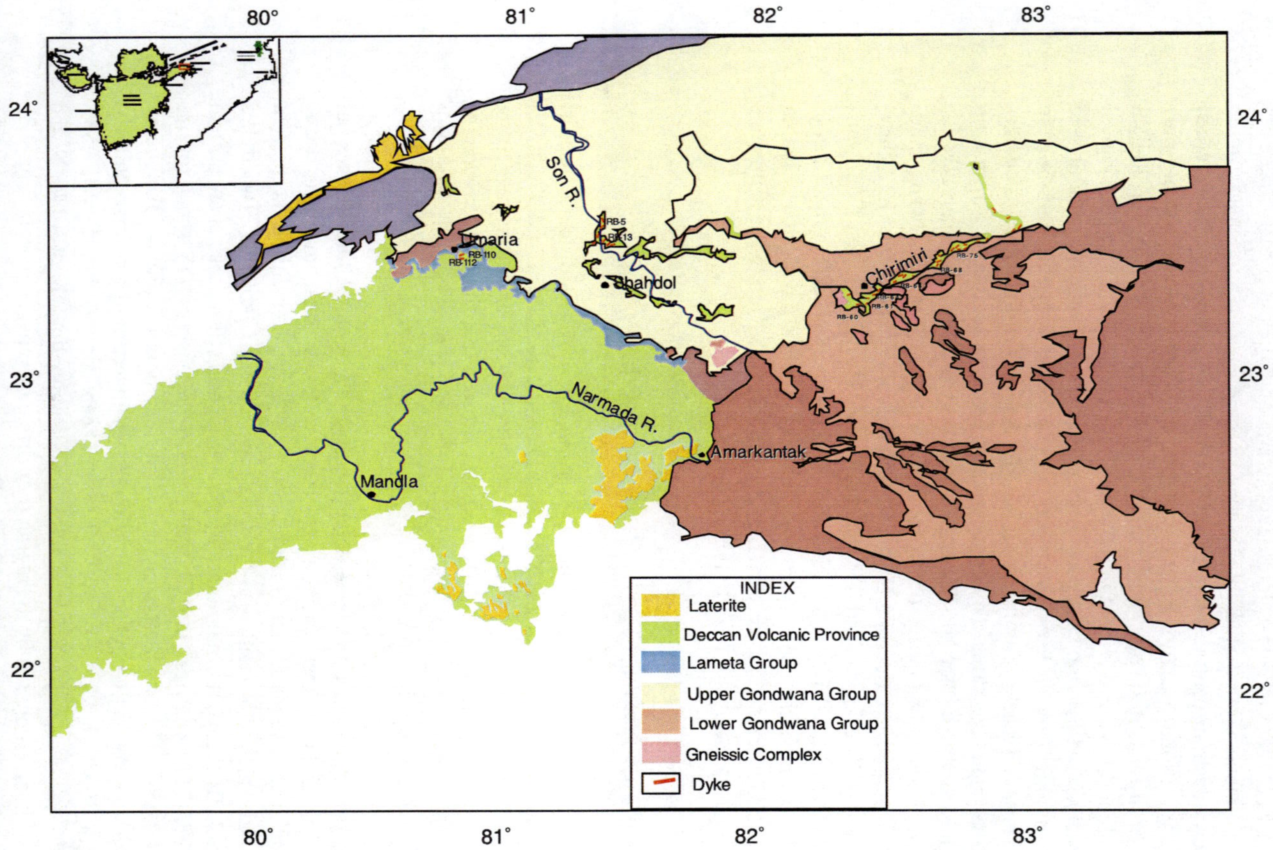

Fig. 9.2 Geological map of the area around Shahdol and Chirimiri based on published Quadrangle Geological maps of Geological Survey of India (2005). The dyke samples are indicated in *bold numbers*, omitting prefix RB, for clarity. The *inset* map shows the extent of Deccan Volcanic Province

Table 9.1 Paleomagnetic results of Rewa dykes

Sample ID	Height	D	I	A_{95}	K	n	VGP Lat	VGP Long	dp	dm	Polarity
RB-5	397	348	−7.25	–	–	1	60.52	286.17	2.53	5.03	N
RB-13	330	338	−21	11.6	33.9	6	49.45	295.59	6.42	12.2	N
RB-61	418	180	65	17	7.46	12	−19.91	82.11	22.14	27.44	R
RB-75	428	159	38	9.6	21.35	12	−41.12	108.4	6.71	11.35	R
RB-64	506	308	−60	6.59	44.35	12	9.82	299.34	7.53	9.96	N

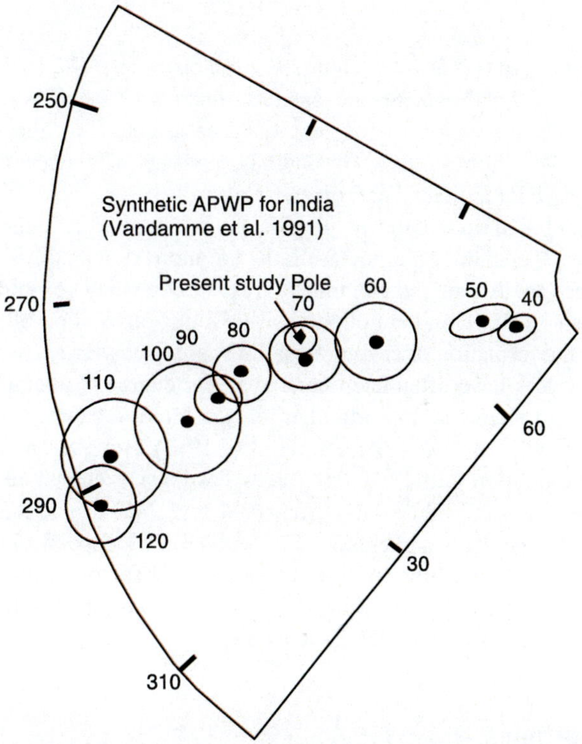

Fig. 9.3 Plot of present study VGP on synthetic APWP for India during the last 120 Ma (APWP from Vandamme et al. 1991)

Petrography

The mafic dykes consist primarily of plagioclase, pyroxene, glass and opaque minerals. Olivine is rare and occurs as anhedral grains (Fo_{29-54}) varying from ~2 to 0.5 mm in size. Phenocrysts of plagioclase feldspar and clinopyroxene sometimes occur in clusters forming glomeroporphyritic texture. Plagioclase also occurs as finer laths forming aggregates. The modal compositions of some of the dyke samples are given in Table 9.2. Phenocryst components (constituting about 39–59 volume %)

Table 9.2 Modal composition of dykes, Rewa Basin

	RB 5	RB 13B	RB 60	RB 66	RB 110	RB 112	RB-68A	RB-64
Plagioclase (P)	28.4	25.9	28.1	39.7	43	29.6	15.7	28.1
Plagioclase (Gm)	40.1	31	36.2	27.9	24.5	27.5	49.3	38.9
Clinopyroxene (P)	15.3	12.7	15.7	13.5	15.6	13.3	12.9	19.8
Clinopyroxene (Gm)	12.1	12.6	6.7	10.7	9.5	10.9	6.9	6.4
Olivine	–	8.4	8.7	–	–	16.2	10.6	–
Opaque	4.1	9.4	4.6	8.2	7.4	2.5	4.6	6.8

P: Phenocryst and Gm: Groundmass

consist of plagioclase feldspar (15.7–43 volume %), clinopyroxene (12.7–19.8 volume %), olivine (8.4–16.2 volume %) and opaque minerals (2.5–9.4 volume %). Magnetite and ilmenite, disseminated in the groundmass, constitute the opaque minerals. Grains are 4–0.4 mm in size. Groundmass consists of glass, microlites of plagioclase feldspar, clinopyroxene and opaque minerals.

Irregular and scattered sericitisation of plagioclase is observed in most of the samples. In particular, the smaller plagioclase laths are more sericitised. The phenocrysts of plagioclase are less altered and mostly fresh, though may be cloudy. The sericitisation has been localised and is not pervasive. But devitrification of glass is pervasive leading to the formation of cryptocrystalline aggregates.

Loss on ignition values have often been used to assess degree of alteration and geochemical mobility of elements. In particular, loss-on-ignition values have been used to assess the reliability of ^{39}Ar–^{40}Ar ages (Baksi, 1987). However, no definite correlation of loss-on-ignition with ^{39}Ar–^{40}Ar ages has been documented. In the samples of our study, the LOI values vary from 0.94 to 1.24% (Table 9.3). Two samples, RB 110 and RB 64, have higher LOI of 2.21 and 2.79% respectively. These two samples are not petrographically different from the other samples, nor do they differ in trace element abundances (see Table 9.2). Although RB 5 has high Ba, Rb and Th compared to other dykes, the LOI value is low.

Mineral Composition

Major element compositions of minerals of the dykes were determined by wavelength dispersive method with a CAMECA SX100 electron microprobe using natural mineral and synthetic standards. Accelerating voltage of 20 kV and beam current of 20 nA was used.

Compositions of representative minerals of four dykes are given in Table 9.4. Sample RB 13a is from the core and RB 13d is from the border zone (Fig. 9.2) of the Son River dyke. The composition of the plagioclase in RB 13a varies from An_{42} to An_{79}. In a zoned plagioclase phenocryst, 250 μm long, the core has composition $An_{71.6}$ to $_{79.9}$, but the rim has composition of An_{42} to An_{59} indicating a prolonged crystallisation. In contrast, in the medium sized 250 μm long plagioclase in RB 13d the composition varies from $An_{62.5}$ to $An_{64.8}$. The anorthite content in the finer

Table 9.3 Major oxide (wt%), trace elements (ppm) and normative composition of Rewa dykes

Sample I.D	RB-64	RB-60	RB304	RB-68a	RB-66	RB112	RB110	RB-75	RB61	RB-5	RB13b	RB13d	Avg	JB-2	JB-2 certified value[a]
SiO_2	49.2	48.6	50.9	48.4	50.5	49.1	49.4	47.3	49.6	49.2	48.2	48.3	49.1	53.21	53.2
TiO_2	2.79	2.89	3.10	3.01	3.23	3.80	3.05	2.58	2.50	2.78	1.93	1.85	2.79	1.19	1.19
Al_2O_3	14.8	13.0	14.4	12.2	14.0	12.4	14.0	12.6	14.8	15.1	15.2	15.9	14.03	14.65	14.67
Fe_2O_3	14.3	14.8	14.3	15.9	14.9	16.1	14.8	16.4	14.3	14.0	13.4	13.0	14.66	14.32	14.34
MnO	0.17	0.18	0.17	0.19	0.17	0.2	0.17	0.2	0.17	0.16	0.16	0.16	0.18	0.19	0.2
MgO	5.40	7.20	4.23	7.32	4.43	4.65	4.96	8.57	5.28	5.35	6.97	7.05	5.95	4.61	4.66
CaO	9.86	10.5	7.95	10.1	9.53	9.43	9.63	10.1	9.19	8.13	9.19	9.23	9.40	9.86	9.89
Na_2O	2.93	2.41	1.97	2.19	2.53	2.52	2.15	2.30	1.85	2.75	2.72	2.69	2.42	2.01	2.03
K_2O	0.69	0.42	0.64	0.57	0.77	0.57	0.65	0.51	0.37	0.82	0.37	0.36	0.56	0.41	0.42
P_2O_5	0.36	0.31	0.35	0.34	0.47	0.39	0.37	0.32	0.25	0.34	0.20	0.20	0.33	0.09	0.1
LOI	2.79	0.97	0.94	1.24	1.15	0.94	2.21	1.05	1.01	1.25	1.24	1.10			
Total	**103.29**	**101.28**	**98.95**	**101.46**	**101.68**	**100.1**	**101.39**	**101.93**	**99.32**	**99.88**	**99.58**	**99.84**	**99.37**	**100.54**	**100.7**
Norm															
Q	0.42	0.58	11.14	1.25	5.8	5.07	5.83	0.001	7.68	2.67	0.0006	2.14	–	–	–
Or	4.08	2.47	3.78	3.36	4.55	3.36	3.84	3.01	2.18	4.84	2.18	2.12	–	–	–
Plag	49.9	43.74	45.16	40.24	45.9	42.1	44.75	41.95	46.57	49.62	51.11	52.94	–	–	–
Di	17.65	21.69	7.17	21.5	16.34	20.92	15.52	20.65	10.75	10.17	13.23	11.67	–	–	–
Hy	16.95	20.04	18.72	21.47	15.22	13.95	16.97	15.66	20.22	20.17	20.38	24.1	–	–	–
Ol	–	–	–	–	–	–	–	7.77	–	–	2.3	–	–	–	–
Mt	4.08	4.24	4.08	4.55	4.27	4.6	4.21	4.69	4.08	4.00	3.82	3.72	–	–	–
Il	5.32	5.51	5.89	5.7	6.08	7.22	5.89	4.97	4.75	5.32	3.61	3.61	–	–	–
Ap	0.78	0.67	0.76	0.74	1.02	0.84	0.8	0.69	0.54	0.71	0.43	0.43	–	–	–

Table 9.3 (continued)

Sample I.D	RB-64	RB-60	RB304	RB-68a	RB-66	RB112	RB110	RB-75	RB61	RB-5	RB13b	RB13d	Avg	JB-2	JB-2 certified value[a]
Trace elements ppm															
Sc	33.3	31.0	28.9	37.9	31.9	39.4	39.0	33.5	29.9	35.3	38.0	39.8	34.8	55.373	54.4
V	383	314	376	395	328	497	498	412	363	427	404	410	400.7	568.483	578
Cr	47.9	54.4	35.8	58.8	29.1	31.7	30.8	47.6	42.3	42.9	57.3	62.1	45.1	27.783	27.4
Co	65.6	61.3	46.3	69.3	63.0	63.9	56.9	66.8	53.7	74.4	79.2	80.1	65.0	39.587	39.8
Ni	13.9	54.0	16.1	19.3	12.3	25.4	25.8	17.4	27.9	23.2	34.7	36.5	25.6	14.197	14.2
Cu	277	223	278	272	356	327	353	252	238	300	211	216	275	229.381	227
Zn	156	116	88	193	159	97	134	152	73	150	119	122	130	110.299	110
Ga	21.5	19.0	22.6	20.6	25.1	26.6	27.3	22.6	21.5	25.6	22.1	22.3	23.1	17.068	17
Rb	15.2	6.4	18.7	12.9	15.4	10.4	15.5	7.71	7.88	25.5	9.83	9.87	13.0	6.192	6.2
Hf	1.34	1.18	4.88	1.35	1.94	5.34	5.16	1.23	3.25	5.41	3.37	3.44	3.16	1.449	1.42
Pb	3.23	3.45	4.575	3.20	4.35	5.07	6.43	3.06	3.451	4.79	4.66	4.52	4.23	5.432	5.4
Sr	277	195	194	186	246	226	238	235	209	210	223	224	222	178.581	178
Y	36.4	35.0	40.9	36.9	49.7	45.9	44.7	34.6	30.8	47.0	31.6	32.3	38.8	25.007	24
Zr	44.5	38.5	214.4	43.6	64.5	232.8	222.8	39.7	145.7	247.8	149.1	150.8	132.8	51.952	51.4
Nb	9.35	7.09	16.9	8.81	12.4	18.4	16.1	7.18	12.3	20.8	12.2	12.38	12.82	0.774	0.8
Cs	2.05	0.90	0.55	0.44	0.78	0.13	0.19	1.07	0.83	1.55	0.32	0.28	0.76	0.911	0.9
Ba	148	73.5	127	153	173	133	177	74.6	94.5	292	83.8	77.3	134	207.082	208
Th	–	0.86	2.85	1.03	1.57	1.37	1.45	0.76	1.204	3.01	1.41	1.49	1.55	0.333	0.33

[a] JB-2 Reference: Govindraju K (1994).

Table 9.4 Representative mineral composition of Rewa dykes

	Plagioclase					Pyroxene							Olivine		Glass			Magnetite		Illmenite
Sample	RB-13A		RB-13D			RB-13A	RB-13D	RB-5					RB-13D		RB-5			RB-13D	RB-5	RB-5
Position		Rim	Core	Rim	Rim		Core	Rim		Core		Rim	Core	Rim						
Comment					80 μ	80 μ	50 μ													
SiO_2	53.82	57.81	49.89	52.96	51.68	47.83	50.03	51.12	50.31	50.65	51.33	50.68	35.09	31.8	68.38	68.48	72.26	0.05	0.11	0.04
TiO_2	0.06	0.03	0.04	0.03	0.07	2.02	1.03	1.1	1.23	1.15	0.96	1.3	0	0.14	0.04	0.05	0.02	25.26	23.3	50.72
Al_2O_3	27.69	25.61	30.62	28.4	28.66	4.15	2	1.78	1.81	1.91	1.71	1.94	0	0.01	14.12	15.46	15.7	1.74	0.86	2.06
CaO	12.01	8.95	14.95	12.49	13.46	18.53	13.94	14.69	14.49	14.25	11.22	18.13	0.37	0.44	1.07	0.07	1.46	0	0.42	0.12
Cr_2O_3	0	0	0.04	0	0.05	0.02	0.01	0.14	0.01	0.08	0	0.25	0	0	0.01	0	0	0.04	0.02	0.00
FeO	0.91	0.71	0.5	0.37	0.69	12.25	16.8	18.09	17.41	13.62	17.11	14.21	37.72	53.7	0.39	0.15	0.33	68.54	67.42	47.56
MnO	0	0	0	0.02	0	0.19	0.4	0.32	0.29	0.37	0.38	0.28	0.42	0.81	0.03	0.03	0	0.47	1.41	0.35
MgO	0.43	0.03	0.17	0.14	0.15	13.15	14.47	12.99	13.68	15.22	16.43	12.78	25.31	12.04	0.07	0.01	0.04	0.35	0.04	0.30
Na_2O	4.51	6.3	3.19	4.52	3.84	0.2	0.23	0.21	0.17	0.21	0.07	0.24	0.02	0	1.34	1.01	4.57	0.01	0	0.00
K_2O	0.2	0.36	0.08	0.12	0.15	0	0.02	0.02	0.02	0	0	0.05	0	0.03	11.55	13.27	4.57	0.01	0	0.00
BaO	0	0	0	0	0	0.08	0.02	0.02	0	0	0	0	0	0	0	0.06	0.43	0.4	0.4	0.80
P_2O_5	0	0.02	0.02	0	0	0.09	0.04	0.03	0.04	0	0.02	0.1	0.04	0.11	1.15	0	0.01	0.02	0	0.00
Total	**99.624**	**99.83**	**99.51**	**99.05**	**98.75**	**98.51**	**98.98**	**100.5**	**99.47**	**97.48**	**99.23**	**99.96**	**98.97**	**99.07**	**98.15**	**98.59**	**99.39**	**96.88**	**93.98**	**101.95**
Oxygens	8	8	8	8		6	6	6	6	6	6	6	4	4				4	4	6
Si	2.436	2.595	2.286	2.418	2.377	1.801	1.881	1.903	1.888	1.904	1.906	1.887	0.997	0.991	3.026	2.998	3.039	0.002	0.003	0.00
Ti	0.002	0.001	0.001	0.001	0.002	0.057	0.029	0.031	0.035	0.033	0.027	0.036	0	0.003	0.001	0.002	0.001	0.676	0.457	0.63
Al	1.477	1.355	1.654	1.528	1.554	0.184	0.089	0.078	0.08	0.085	0.075	0.085	0	0	0.737	0.798	0.778	0.073	0.026	0.04
Cr	0	0	0.001	0	0.002	0.001	0	0.004	0	0.002	0	0.007	0	0	0	0	0	0.001	0	0.00
Fe3	0.039	0.021	0.021	0.021	0.021	0.082	0.082	0.081	0.082	0.082	0.081	0.081	0.009	0.01	0.256	0.253	0.243	0.305	1.054	0.67
Fe2	0.034	0.027	0.019	0.014	0.027	0.386	0.528	0.563	0.547	0.428	0.531	0.442	0.897	1.4	0.014	0.005	0.012	2.041	0.415	0.66
Mn	0	0	0	0.001	0	0.006	0.013	0.01	0.009	0.012	0.012	0.009	0.01	0.021	0.001	0.001	0	0.014	0.031	0.01
Mg	0.029	0.002	0.012	0.01	0.01	0.738	0.811	0.721	0.765	0.853	0.909	0.709	1.072	0.559	0.005	0.001	0.003	0.019	0.002	0.01
Ca	0.582	0.431	0.734	0.611	0.663	0.747	0.562	0.586	0.583	0.574	0.446	0.723	0.011	0.015	0.051	0.003	0.066	0	0.012	0.00
Na	0.396	0.548	0.283	0.4	0.342	0.015	0.017	0.015	0.012	0.015	0.005	0.017	0.001	0	0.115	0.086	0.373	0.001	0	0.00
K	0.012	0.021	0.005	0.007	0.009	0	0.001	0.001	0.001	0	0	0.002	0	0.001	0.653	0.742	0.245	0	0	0.00
Sum	**5.008**	**5**	**5.018**	**5.01**	**5.008**	**4.016**	**4.013**	**3.993**	**4.002**	**3.987**	**3.992**	**4**	**2.998**	**3.001**	**4.86**	**4.889**	**4.759**	**3.133**	**2**	**2.01**

grains in the groundmass is comparable. One plagioclase grain enclosed entirely within a clinopyroxene has anorthite content of $An_{64.1}$. In another dyke sample RB 68a (Table 9.4) the plagioclase shows a wide variation of anorthite content ($An_{42.3}$ to $An_{66.5}$). This also reflects a prolonged crystallisation history.

The clinopyroxene grains in RB13a have composition ranging from $Wo_{38.3}Fs_{23.9}En_{37.8}$ to $Wo_{34.5}Fs_{41}En_{24.5}$. It is clear that the fine clinopyroxenes have higher Fe with decreasing Mg. In the sample RB13d collected from the border zone, the clinopyroxenes show significant variation in CaO and MgO from core to rim ($Wo_{38.8}Fs_{19.1}En_{42.1}$ to $Wo_{26.9}Fs_{57.1}En_{16}$). The clinopyroxenes from RB 5 show variation from $Wo_{32.3}Fs_{33.6}En_{32.1}$ to $Wo_{24.8}Fs_{29.2}En_{45.9}$. One large grain, 250 μm long, of clinopyroxene has been analysed from rim to rim. Variation of some constituents across the grain is shown in Table 9.4.

The opaque minerals that occur in the groundmass mostly as anhedral grains include titaniferous magnetite. TiO_2 content varies from 23.9 to 29.7 wt% and total iron as FeO varies from 63.9 to 68.5 wt%. The few ilmenites analysed show rather uniform composition: Fe as FeO around 47% and TiO_2 around 5% (Table 9.4).

Geochemistry

Samples for geochemical analyses were selected on the basis of petrographic freshness. Small chips of fresh rocks were cleaned with water and crushed in a Jaw crusher and Pulveriser. Between crushing and pulverising, an aliquot of the second sample was processed to remove any memory of the first sample. Major and trace element (including rare earth elements) abundances of twelve mafic dykes are presented in Table 9.3. The major element analyses were carried out using X-ray fluorescence technique with a Philips–PRO PW 2,440 fully automatic, microprocessor controlled X-ray spectrometer equipped with a 4 KW X-ray generator. International rock standards of US Geological Survey and Geological Survey of Japan were used to prepare calibration curves. Total iron was measured as Fe_2O_3. The trace elements were determined using a Perkin Elmer SCIEX, Model 6,100 ELAN DRCII ICP-Mass Spectrometer with a Meinhard nebulizer for sample introduction (see Balaram and Gnaneshwar Rao, 2003 for details). The major and trace element analyses were carried out at the National Geophysical Research Institute at Hyderabad. The major and trace element compositions of the International Standard, JB2, were determined during the present analysis. The results on JB2, given in Table 9.3, indicate the accuracy of the present data.

Of the twelve samples analysed, ten are quartz-normative, and the other two are olivine-normative. All the samples plot in the basalt field in a total alkali-silica diagram (Le Bas et al., 1986) and are of sub-alkaline variety (Irvine and Baragar, 1971). Two samples from the Son River bed (Table 9.3) have the lowest TiO_2 contents of 1.85 and 1.93% and P_2O_5 (0.2%). The dyke samples of Sangamner (Bondre et al., 2006) and Pachmari (Sheth et al., 2009) are very close to our samples in the TAS diagram (Fig. 9.4). The concentration of incompatible trace elements is variable (e.g. Zr from 38.5 to 247.8 ppm, Y from 31.6 to 49.7 ppm, Nb from 7 to 20.8 ppm). When P_2O_5 is plotted against Zr, the Chirimiri dykes form a cluster different from

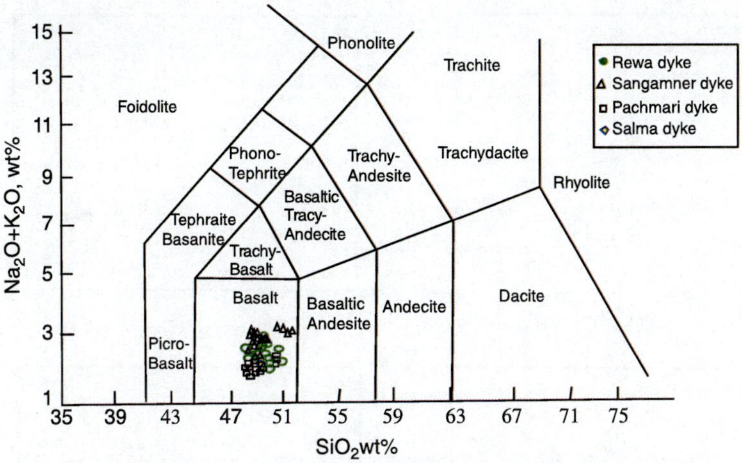

Fig. 9.4 Total alkali-silica diagram (Le Bas et al., 1986) showing Rewa dykes (*present study*), Pachmari dykes (Sheth et al., 2009), Sangamner dykes (Bondre et al., 2006) and Salma dykes from eastern India (Kent et al., 1997). All the dyke samples fall in an overlapping field

the others and follow a different tend (Fig. 9.5a). This clustering is a reflection of lower Zr contents in the Chirimiri dykes (Table 9.3). In a plot of the incompatible trace element ratios, Nb/Zr vs Zr/Y, also the Chirimiri dykes fall in a different group (Fig. 9.5b) separating these from the Shahdol dykes. Significantly, in a P_2O_5 vs TiO_2 plot, the dykes including the Chirimiri samples show a positive correlation (Fig. 9.5c) indicating fractional crystallisation. This is corroborated by the variation of Cr with MgO (Fig. 9.5d) and of Nb/Y against Zr (not shown). Large variations of Zr and TiO_2 contents in Deccan basalts, as found in the dykes of this study, were interpreted to indicate very extensive fractionation (Keays and Lightfoot, 2009). The composition of the Salma dyke from near the Rajmahal volcanic field (Kent et al., 1997) is also similar to the mafic dykes of the Deccan Volcanic Province, although some elements e.g. Zr, Y are similar to the Rajmahal basalts (see Patil and Arora, 2008). Among the samples, the samples RB 13b, RB 13d (Son River) and RB 60, RB 75 (from Chirimiri in the north eastern part) have higher MgO and lower CaO than the others. Also ΣREE contents are comparatively lower in the Son River dykes (86.44 ppm in RB 13b and 87.87 ppm in RB 13d) (Table 9.5). Among all the other Rewa dykes of this study, ΣREE varies from 83.83 to 141.69 ppm.

The REE's show moderate fractionation (Fig. 9.6) with $(La/Yb)_n$ varying from 2.1 to 3.6 on a chondrite-normalised plot, using the chondrite abundances of Evensen et al. (1978). There is a variation in the REE abundances, but the chondrite-normalised plots are very similar. Two samples, RB 61 and RB 75, have lower REE abundances than the others. There is an overlap in the chondrite-normalised fields of the mafic dykes and that of the Deccan flows from the neighbouring areas of Shahdol (Paul et al., 2008). Primitive mantle-normalised (Sun and McDonough, 1989) plots of selected incompatible trace elements (Fig. 9.7) indicate a prominent positive Pb peak between Ce and Sr in all the samples, and a negative Zr-Hf trough, between Nd and Sm. All the elements plotted in Fig. 9.7 show enrichment of

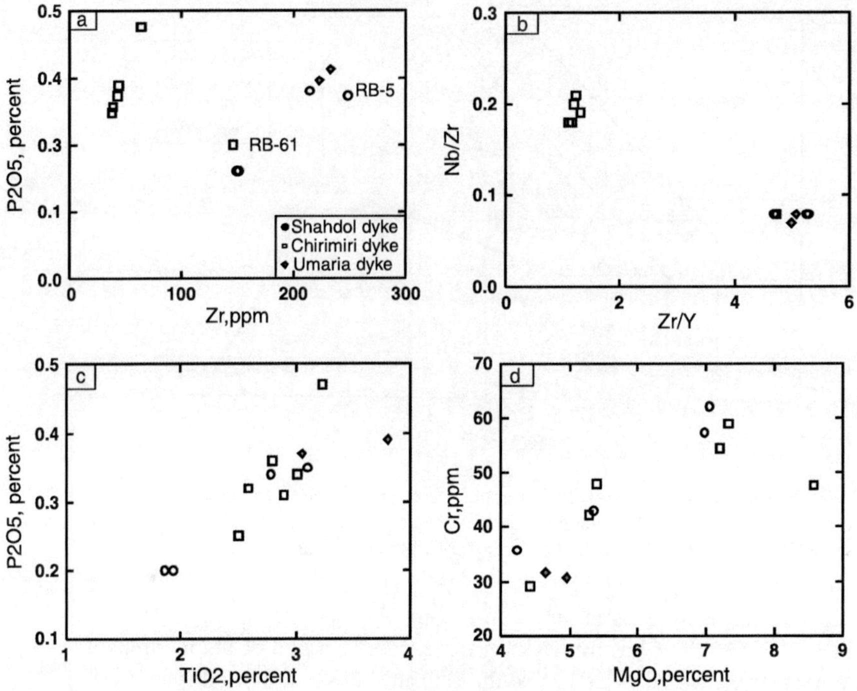

Fig. 9.5 Interelemental variation in Rewa dykes. (**a**) Variation of P_2O_5, percent vs Zr, ppm, showing two groups; Chirimiri dykes (*square*) and the Shahdol dykes, reflect contrasting Zr abundances. A positive correlation exists for the two groups separately. (**b**) Nb/Zr vs Zr/Y for all the dykes. (**c**) P_2O_5, percent vs TiO_2, percent, illustrating a strong positive correlation. (**d**) Plot of MgO, percent vs Cr, ppm showing positive correlation as a result of fractional crystallisation. Sample RB 75 is away from the main trend. All the symbols are as in Fig. 9.5a

10–40 times over the primitive mantle abundances. We have compared the primitive mantle-normalised average trace element abundances of the Rewa dykes with those of Sangamner (Bondre et al., 2006), Pachmari (Sheth et al., 2009) and Salma dyke (Kent et al.1997) in Fig. 9.8, taking data from Table 9.6. There is a striking similarity in the distribution patterns (Fig. 9.7).

Among all the samples ΣREE shows a significant positive correlation with SiO_2, TiO_2, K_2O, P_2O_5, Ba, Rb, Sr, Nb, Ga, V, Hf and Sc but negative correlation with MgO, Co. This would suggest fractionation of plagioclase and clinopyroxene (Sen, 1995) in the samples.

Sr-Nd Isotope Geochemistry

Analytical Methods

About 75–100 mg of representative rock powders were digested with triple distilled concentrated HNO_3 and HF in Teflon vials. About 50 mg of ^{87}Rb, ^{84}Sr and mixed

Table 9.5 Rare earth elements and Sr-Nd isotopic data of Rewa dykes

Sample I.D	RB64	RB60	RB304	RB-68a	RB-66	RB112	RB110	RB-75	RB61	RB-5	RB-13b	13d	JB2	JB-2 Certified value[a]
La	11.2	9.20	17.2	10.5	16.3	14.5	16.0	8.12	9.91	18.8	10.4	10.7	2.36	2.37
Ce	29.8	25.5	41.4	28.7	43.3	40.8	42.8	23.2	25.55	46.5	26.5	27.2	6.87	6.77
Pr	3.62	3.23	4.65	3.56	5.24	4.88	4.9	2.99	3.06	5.28	3.15	3.2	0.95	0.96
Nd	22.7	21.3	27.4	22.6	32.9	29.9	28.5	19.8	18.5	30.6	18.8	19.3	6.66	6.7
Sm	5.74	5.79	6.9	5.68	8.00	7.96	7.2	5.40	5.10	7.75	5.08	5.02	2.28	2.25
Eu	2.05	2.05	2.07	2.01	2.74	2.47	2.28	2.03	1.64	2.53	2.09	1.81	0.88	0.86
Gd	6.98	7.08	8.89	6.96	9.59	9.54	8.68	6.75	6.44	9.66	6.36	6.47	3.14	3.28
Tb	1.11	1.19	1.36	1.18	1.57	1.52	1.41	1.10	1.02	1.52	1.04	1.04	0.62	0.62
Dy	6.12	6.14	6.80	6.34	8.38	7.63	7.16	6.05	5.12	7.80	5.41	5.27	3.70	3.66
Ho	1.20	1.17	1.37	1.23	1.66	1.50	1.45	1.16	1.02	1.57	1.05	1.09	0.82	0.81
Er	3.74	3.50	4.10	3.80	5.08	4.46	4.54	3.52	3.11	4.64	3.17	3.27	2.67	2.63
Tm	0.60	0.57	0.64	0.62	0.83	0.67	0.72	0.55	0.48	0.72	0.49	0.51	0.44	0.45
Yb	3.11	2.68	3.42	3.10	4.20	3.65	3.92	2.77	2.58	3.78	2.53	2.61	2.48	2.51
Lu	0.46	0.39	0.51	0.45	0.61	0.53	0.59	0.40	0.37	0.56	0.37	0.38	0.38	0.39
REE	98.47	89.72	126.72	96.67	140.40	130.01	130.05	83.83	83.96	141.69	86.38	87.80	–	–
87Sr/86Sr (t)	–	–	–	0.706213	0.709416	0.705326	0.704625	0.707651	–	0.705679	0.709434	0.714403	–	–
143Nd/144Nd (t)	–	–	–	0.511957	0.512669	0.512802	0.512629	0.512674	–	0.511677	–	0.512675	–	–
εNd(t)	–	–	–		2.25	4.83	1.46	2.34	–		–	2.35	–	–

[a]JB-2: Govindraju K (1994).

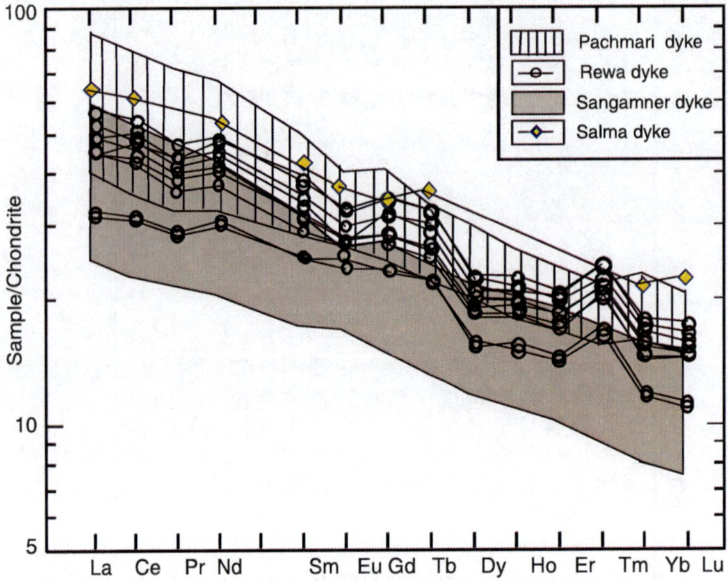

Fig. 9.6 Chondrite-normalised rare earth element abundances in mafic dykes (Rewa data from Table 9.3), Pachmari (Sheth et al., 2009), Sangamner (Bondre et al., 2006) and Salma (Kent et al., 1997) dykes. The chondritic abundance values are from Evensen et al. (1978)

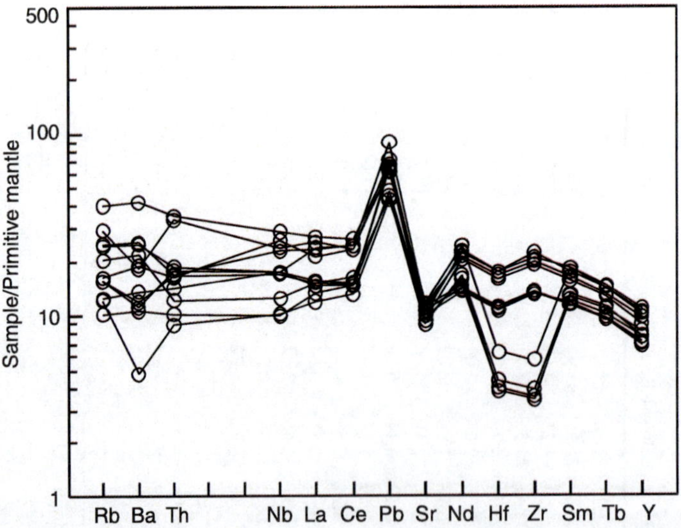

Fig. 9.7 Primitive mantle-normalised (Sun and McDonough, 1989) element patterns for Rewa dyke samples. Note the prominent Pb peaks and Zr-Hf troughs

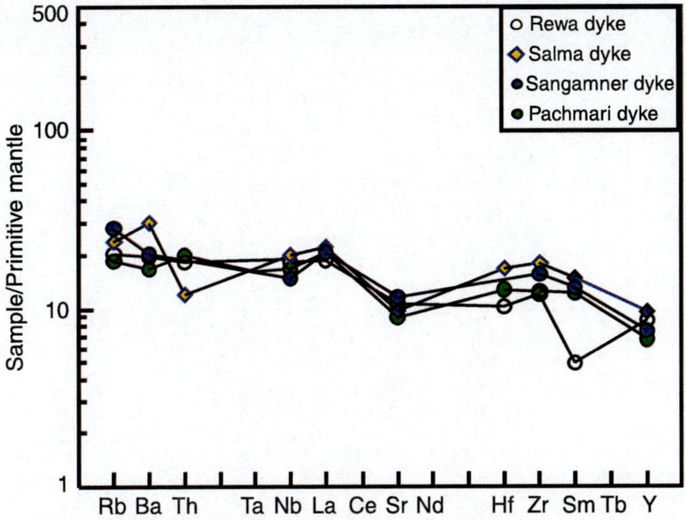

Fig. 9.8 Primitive mantle – normalised element (Sun and McDonough, 1989) patterns comparing the average abundance values of Sangamner, Pachmari and Salma dykes. The data is included in Table 9.6

^{150}Nd-^{149}Sm spikes with concentrations of 3.056, 8.4559, 0.78252 and 0.66519 μg/g respectively were added to the rock solutions.

Rb, Sr and REE were separated using HCl acid eluents with Bio-Rad AG 50W-X8 200–400 mesh H^+ form cation exchange resin. Sm and Nd were separated from the REE fraction by using LN Spec (Part No C-50) resin from Eichrom. Nd and Sm were eluted with 0.25 N HCl and 0.75 N HCl respectively. Isotopic analyses were carried out using a fully automatic Thermo Fisher TRITON multi-collector Thermal Ionisation mass spectrometer at the Indian Institute of Technology, Roorkee.

Rb, Sr, Sm and Nd concentrations in sample were calculated off-line after correcting for fractionation according to the double spike correction method. The isotopic compositions ^{87}Sr/^{86}Sr and ^{143}Nd/^{144}Nd were also re-calculated with the corrections for spike contribution. Reproducibility of the concentration measurements is monitored periodically by measuring the concentration of these elements in USGS rock standards. The total procedure blank in the laboratory at the present time is less than 8 ng of Sr and less than 1 ng for Nd.

The ^{87}Sr/^{86}Sr ratios were normalized to ^{86}Sr/^{88}Sr = 0.1194. Mean value for NIST Sr standard SRM-987 was ^{87}Sr/^{86}Sr = 0.710248 ± 20 (2σ). The ^{143}Nd/^{144}Nd ratios were normalized to ^{146}Nd/^{144}Nd = 0·7219. Average value for the Japanese Nd standard JNdi-1 was 0·512108 ± 16(2σ).

Table 9.6 Compositional variation of mafic dykes from Deccan Volcanic Province

		Rewa (present study)		Sangamner		Pachmari	
	Salma	Range	Avg (n = 12)	Range	Avg (n = 27)	Range	Avg (n = 18)
SiO_2	49.55	47.3–50.9	49.05	49.02–52.31	50.05	48.40–50.64	49.25
TiO_2	2.91	1.85–3.80	2.79	1.29–3.27	2.52	2.10–3.27	72.5
Al_2O_3	12.2	12.2–15.9	14.03	11.93–14.57	13.13	11.80–12.87	12.35
Fe_2O_3(t)	16.44	13.0–16.1	14.66	11.16–17.32	15.02	15.12–17.26	15.06
MnO	0.22	0.16–0.20	0.18	0.16–0.24	0.21	0.17–0.28	0.21
MgO	5.57	4.32–8.57	5.95	4.76–8.97	5.84	4.46–7.15	6.28
CaO	9.95	7.95–10.47	9.40	9.18–11.27	9.98	9.58–11.44	10.31
Na_2O	2.36	1.85–2.93	2.42	1.83–2.69	2.45	1.47–2.14	1.95
K_2O	0.57	0.36–0.82	0.56	0.20–0.99	0.41	0.16–0.78	0.46
P_2O_5	0.29	0.2–0.39	0.33	0.14–0.40	0.25	0.18–0.42	0.26
Trace elements, ppm							
Sc	43	28.9–39.8	34.8	26–38	31.7	31.89–38.96	34.81
V	414	314–498	401	246–489	338	354.3–512.8	419.13
Cr	84	29.1–58.8	45.1	41–557	144.7	237.30–684.3	480.53
Co	–	46.3–80.1	65.0	44–54	43.6	44.33–57.92	49
Ni	65	16.1–54.0	25.6	41–228	81.3	37.42–88.37	64.15
Rb	15	6.4–25.5	13.0	10–35	18.15	2.06–24.65	11.75
Sr	205	186–277	222	203–287	242.8	161.90–241.20	189.03
Sm	6.53	5.02–8.00	2.15	3.51–7.47	5.72	4.31–7.61	5.42
Y	43	30.8–47.0	38.8	17–46	34.2	25.49–38.69	30.24

Table 9.6 (continued)

	Salma	Rewa (present study)		Sangamner		Pachmari	
		Range	Avg (n = 12)	Range	Avg (n = 27)	Range	Avg (n = 18)
Zr	199	38.5–247.8	133	104–237	171.8	114.10–227.40	138.3
Nb	14.2	7.09–20.8	12.8	7–17	10.5	7.32–20.80	11.78
Ba	211	74.0–292	134	62–225	141.9	70.14–310.7	115.32
La	15.2	8.12–18.8	12.7	9.20–21.6	15.8	9.42–24.18	14.37
Ce	37.6	23.2–46.5	33.4	22.1–50.7	37.26	21.67–52.55	32.37
Yb	3.72	2.53–4.20	3.19	1.54–3.91	2.83	2.54–3.86	3.03
Lu	0.58	0.37–0.61	0.47	0.22–0.48	0.405	0.34–0.52	0.45
Hf	5.13	1.23–5.34	3.16	–	–	2.98–5.76	3.89
Ta	0.89	0.53–4.28	1.73	–	–	0.42–1.22	0.67
Th	2.02	0.76–3.01	1.55	–	–	3.19–1.29	1.68
Isotopic abundance							
$(^{87}Sr/^{86}Sr)i$	0.70482	0.704625–0.714403		0.70555–0.70877		0.70516–0.71589	
$(^{143}Nd/^{144}Nd)i$	0.512667	0.511677–0.512802		0.511923–0.512784		0.512594–0.512689	
εNdi	1.00	1.46–4.83		–13.4 to 3.00		0.8–110.9	

Reference: Salma: Kent et al. (1997); Sangamner: Bondre et al. (2006); Pachmari: Sheth et al. (2009).

Isotopic Results

The Nd-Sr isotopic compositions of seven dyke samples analyzed in the present study are given in Table 9.5. All isotopic ratios have been time-corrected for 65 Ma, the age of Deccan volcanic eruption.

The measured $^{87}Sr/^{86}Sr$ ratios of the dykes range from 0.704667 to 0.71447. The corresponding $(^{87}Sr/^{86}Sr)i$ values vary from 0.704625 to 0.714403. The Sr isotopes are distinctly more radiogenic, excepting sample RB 110 that has a low $(^{87}Sr/^{86}Sr)_i$ ratio of 0.704625, compared to normal mantle-derived rocks. Measured $^{143}Nd/^{144}Nd$ in all the seven samples vary from 0.511677 to 0.512871. For analytical uncertainty, the initial $^{143}Nd/^{144}Nd$ of two samples. RB 5 and RB 68 a, could not be calculated. The initial $^{143}Nd/^{144}Nd$ values of the remaining five samples, RB 66, RB 75, RB 13d, RB 110, and RB 112, vary from 0.512629 to 0.512802 (εNd = + 1.46–+ 4.83) (Table 9.5; Fig. 9.9). Considering the Sr and Nd isotopic

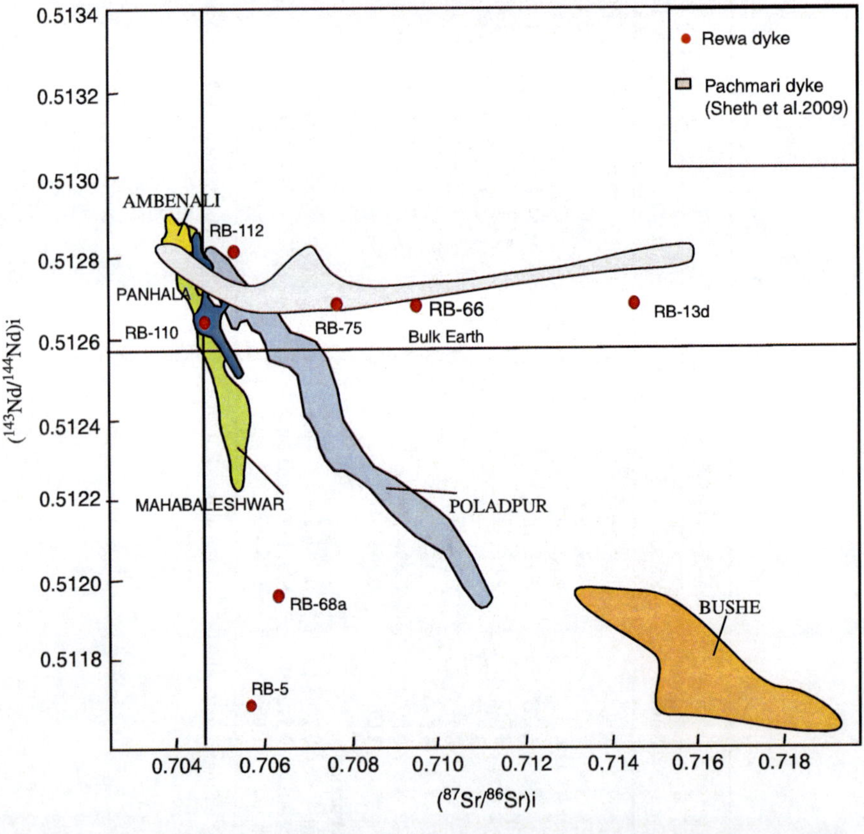

Fig. 9.9 Sr-Nd isotopic data for Rewa dykes (this study) compared with the standard south western Deccan Formations (Lightfoot and Hawkesworth, 1988). Also included is the field for Desur sub unit (after Cox and Hawkesworth, 1985). For sample numbers RB 5 and RB 68a, the measured $^{143}Nd/^{144}Nd$ are plotted. Sr and Nd isotopic compositions obtained on standards are indicated in the text. Also plotted are Sangamner (Bondre et al., 2006) and Pachmari (Sheth et al., 2009) dykes

composition, the samples can be grouped in two groups: one having a nearly constant Nd isotopic abundance but variable Sr isotopic composition, and the other having nearly constant $^{87}Sr/^{86}Sr$ at 0.706 but lower $^{143}Nd/^{144}Nd$ values (Fig. 9.9). Petrographical characters of the samples do not indicate pervasive alteration. Taken all the samples together, LOI (Table 9.2) has a positive correlation with Sr, Rb and Ba but has no relationship with the Sr and Nd isotopic composition. For example, RB 5 has highest Rb of 25.5 ppm, Ba (292 ppm) and Th (3.01 ppm) (Table 9.2) but the initial $^{87}Sr/^{86}Sr$ ratio, 0.705679, is among the lowest. In the case of dyke RB 13 d, the initial $^{87}Sr/^{86}Sr$ ratio of 0.7144403 and the concentations of the elements Rb, Ba and Th are low. We therefore conclude that the isotopic composition has not been affected by alteration.

There is no relation between $^{87}Sr/^{86}Sr$ ratios and SiO_2 or 1/Sr ratios ruling out simple crustal contamination. The Sr, Nd isotopic characters in the south western Deccan Formations have been generally interpreted as simple mixing between an Ambenali-type magma with a lithospheric mantle end member (Sheth et al., 2009). A more complicated two-stage mixing has been proposed for these Formations by Peng et al. (1994) to explain the Nd-Pb isotopic arrays. We explain the observed variable Sr, Nd isotopic composition of five Rewa dyke samples (RB 110, RB 112, RB 75, RB 66, RB 13d) by binary mixing between an Ambenali-type magma and a lithospheric end member with enriched $^{87}Sr/^{86}Sr$ ($\sim$0.716) but almost constant $^{143}Nd/^{144}Nd$ around 0.5127.

Discussion

Mafic dykes constitute an important component of Large Igneous Provinces. Many researchers have taken up geochemical as well as isotopic studies to understand the relationship between the dykes and the flows. In an effort to establish linkage with established Deccan stratigraphy of the Western Ghats, dykes of the Western and Central Deccan Volcanic Province (DVP) have received attention. Details of the distribution of mafic dykes and their chemical characters for the northeastern part of the DVP are not available (Vanderkluysen et al., 2004). Such studies on the dykes from other regions such as the eastern and northeastern DVP are essential to develop a regional database. The present study is one such attempt on the mafic dykes of the Rewa Basin in the eastern most part of DVP. We examined regional compositional change of these dykes and the dispersal of Deccan magma on a wider context.

Regional Composition

The Son River dykes are more primitive with higher MgO and lower TiO_2 contents than other samples. Sen and Cohen (1994) established the contemporaneity of the Chala-Delakhari sill of Pachmari region and Deccan volcanism of the Western Ghats based on two ^{40}Ar–^{39}Ar ages of 66.1 ± 0.3 and 65.1 ± 0.3 Ma. We have compared the composition of the Pachmari and Sangamner mafic dykes with the Rewa dykes of the present study (Table 9.6). In the total alkali – silica diagram (Fig. 9.4) these dykes plot in the basalt field as do the Rewa dykes. The primitive mantle-normalised values of some trace elements of these dykes are plotted in Fig. 9.8. The Salma dyke of the Gondwana coalfield of the Rajmahal Province is also shown in this figure for a comparison. The normalised patterns are strikingly similar, except slightly enriched Ba and depleted Th in the Salma dyke. This similarity would suggest similar petrogenetic processes for the dykes.

The chondrite-normalised rare earth element patterns of Rewa, Pachmari and Sangamner along with the Salma dykes (Fig. 9.6) are also very similar, $(La/Yb)_n$ ratios of all the dykes vary within 2.1 and 3.6. Total REE contents are, however, variable. There is overlap in the chondrite-normalised fields with the Sangamner dykes having higher total REE, and Pachmari dykes having lower total REE than the Rewa dykes. The Salma dyke REE pattern is also similar but with a higher REE contents than the Rewa and Pachmari dykes. The Salma dyke has REE contents similar to the enriched Sangamner dyke samples. Sen and Cohen (1994) concluded that the Bushe, Poladpur and Ambenali Formations of the Western Ghats are chemically distinct from the northeastern lavas, the latter having low Ba, Rb and K.

The Sr and Nd isotopic data of the present study (Table 9.5) generally fall within the Sr-Nd isotopic range for the Deccan basalts. The range of $^{87}Sr/^{86}Sr$ in the Rewa samples (0.704625–0.714403) is within the reported values for Bushe Formation where pronounced crustal contamination has been established (Keays and Lightfoot, 2009, and many references therein). The Rewa and Pachmari dykes (Sheth et al., 2009) show a similar pattern with a flat mixing curve in a $^{87}Sr/^{86}Sr$ – $^{143}Nd/^{144}Nd$ plot (Fig. 9.9). The mafic *basaltic flows* of the Rewa basin also display a broadly similar Sr-Nd isotopic relationship (our unpublished data). The εNd values of both Rewa and Pachmari dykes are positive. Barring two samples, RB 68a and RB 5, the Rewa dykes have εNd values from + 1.46 to + 4.83. Mixing between two end members, an Ambenali-type magma and another with high $^{87}Sr/^{86}Sr$ would be required to derive the Sr-Nd array of these samples (Fig. 9.9) (*cf.* Yan et al., 2008). But to obtain the isotopic composition of RB 68a and RB 5 (Fig. 9.9) mixing between an Ambenali-type end member with another component with low Sm/Nd or $^{143}Nd/^{144}Nd$ around 0.5117 is required.

Dispersal and Migration

It is well known that the Deccan volcanic rocks attain the highest thickness, 2 km, in the Western Ghats, and gradually thin down towards east. In Amarkantak (Fig. 9.2) the thickness of the Deccan flows is estimated at 170 m (Geological Survey of India, 1995). Towards east, in and around Shahdol, the thickness of Deccan volcanic rocks is about 30 m overlying the Gondwana sediments. The frequency and number of mafic dykes in the Deccan rocks are also high in the west than in the east. The dominant dyke occurrences have been shown in Fig. 9.2. Whether the mafic dykes are the feeder to the Deccan flows is a subject of intense discussion. Recently Sheth et al. (2009) have discussed the difficulty of dyke-sill-flow relation from geochemical characters for the Pachmari dykes. Occurrence of volcanic vents with emanating dykes and feeding the flows has also been discussed (Misra, 2001, 2008). In the eastern DVP, the relation between the flows and the dykes has not been established from field relations. It is not known if separate vents of volcanic eruption exist in this part. However from the gradual diminishing thickness of the lava sequences, it appears that the flows migrated eastward from the western part (West, 1959). The reported age of 64.4 Ma for the Salma dyke (Kent et al., 1997) near the 115 Ma Rajmahal Volcanic Province in eastern India suggests contemporaneity of the former with the Deccan volcanics. Besides the Salma dyke, there are other Tertiary dykes in the vicinity of Rajmahal. In our earlier discussion we have shown the similarity of the geochemical characters of Salma with the Rewa dykes. Kent et al. (1997) suggested that the Tertiary dykes of eastern India are chemically and isotopically similar to the ferro-basaltic lavas from the upper Deccan Formations of Western Ghats. It is known that in the Raniganj (Gondwana) basins of eastern India, there are ultramafic alkaline dykes of kimberlitic (Group II) affinity which are about 115 Ma old (Kent et al., 1998), very similar to the age of the Rajmahal volcanic rocks (Baksi et al.,

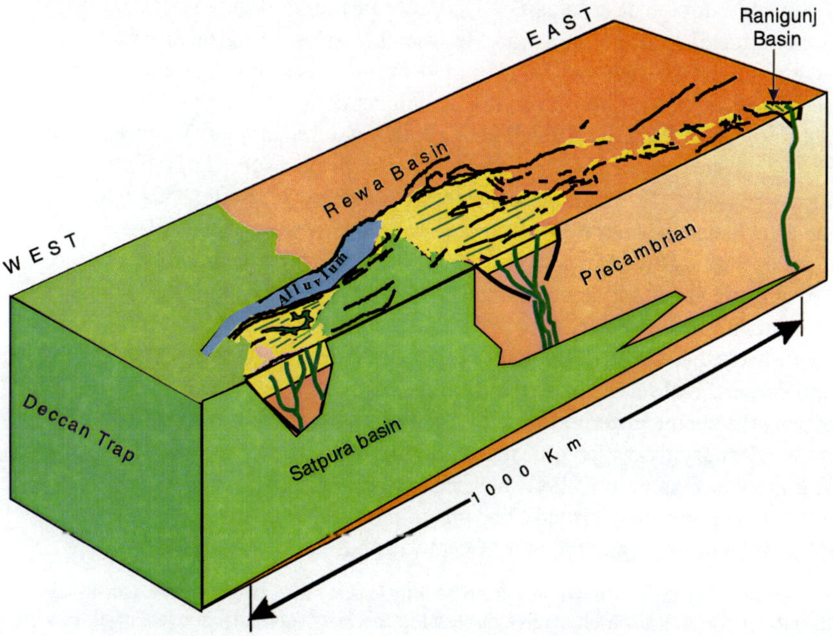

Fig. 9.10 Schematic diagram showing possible subsurface migration of Deccan magma from western India eastward and emplacement as dykes in the western periphery of DVP and the Gondwana basins of Rewa, Raniganj and other basins of eastern India along intrabasinal faults as pathways. Chemical similarity and $^{39}Ar-^{40}Ar$ age of 64 Ma for the Salma dykes seem to support the migration of magma

1987). Example of long distance migration of basaltic lava flow could be cited from the Pomona flow of the Columbia River basalt province that has been traced for almost 600 km from its source (Hooper, 1997), and the chemical and mineralogical homogeneity across long distances were interpreted to indicate *exceptionally* high flow rates.

Combining the field occurrences and the geochemical attributes it is suggested that Deccan magma migrated eastward subsurface from the western part (see also Paul, 2005) and was emplaced as dykes using the intrabasinal faults of the Gondwana basins as pathways (Fig. 9.10) over a distance of more than 500 km. The diminishing thickness of the Deccan volcanics as also the rare occurrence of mafic dykes in the eastern part could follow from the supply of magma. Migration of mafic magma over distances of 2,000 km for the Mackenzie dykes (Ernst and Baragar, 1992) or for a distance ranging from 150 to 350 km for the Pachmari dykes (Sheth et al.,2009) or north eastern Deccan lavas for over 900 km (Peng et al., 1998) also has been suggested.

Conclusions

The mafic dykes from the eastern part of Deccan Volcanic Province are of uniform chemical and mineralogical composition. Between the dykes two groups at Shahdol and the Chirimiri in the east can be identified on the basis of MgO content and some trace elements, particularly Zr. In major element composition the Rewa dykes are similar to those of Pachmari (Sheth et al., 2009) and Sangamner (Bondre et al., 2006). Sr and Nd isotopic composition of the dykes indicate a spread of (εNd)i from + 1.46 to + 4.83 and $(^{87}Sr/^{86}Sr)_i$ varies from 0.704625 to 0.714403. This range is within the established Deccan flow sequences from the Western Ghats. The Sr-Nd isotopic compositions of the Rewa dykes are similar to those of Pachmari, further to the west of our study area. Mixing of normal crustal source with a mantle component cannot explain the spread in the isotopic data. Mixing between an Ambenali – type magma with two other components with high $^{87}Sr/^{86}Sr$ and low Sm/Nd would be required to derive the observed Sr-Nd array (Fig. 9.9).

Comparing the chemical composition and isotopic ages, the mafic dykes of Rewa basin and the Tertiary dykes from near the Rajmahal Volcanic Province, it is suggested that Deccan magma travelled subsurface over long distance, at least of the order of 500 km, along the Narmada lineament and was emplaced as dykes along intrabasinal faults in Rewa and the other Gondwana basins of eastern India.

Acknowledgements We thank V Balaram and NN Murthy for help with the geochemical analyses and C Chakraborty for discussion on magma dispersal in the early stages. Comments of official reviewer, Ray W Kent, and the editor, Rajesh Srivastava, greatly improved the paper. We appreciate the comments of N C Ghose on an earlier draft. H N Bhattacharya extended various facilities at Presidency College, Kolkata. The Department of Science and Technology, Government of India provided financial support in the form a Research Fellowship to T Lala and a grant to D K Paul. DKP is grateful to the Indian National Science Academy for a Senior Scientist position during the course of this research.

References

Auden JB (1949) Dyke in western India – A discussion of their relationships with the Deccan Traps. Trans Nat Inst Sci India 3: 123–157

Baksi AK (1987) Critical evaluation of the age of the Deccan traps, India: Implications for flood basalt volcanism and faunal extinctions. Geology 15: 147–150

Baksi AK, Barman TR, Paul DK, Farrar E (1987) Widespread early cretaceous flood basalt volcanism in eastern India: Geochemical data from Rajmahal-Bengal-Sylhet Traps. Chem Geol 63: 133–141

Balaram V, Gnaneshwar R (2003) Rapid determination of REEs and other trace elements in geological samples by microwave digestion and ICP-MS. At Spectr 24(6): 206–212

Beane JE, Turner CA, Hooper PR, Subbarao KV, Walsh JN (1986) Stratigraphy, composition and form of the Deccan basalts, Western Ghats, India. Bull Volcanol 48: 61–83

Bhattacharji S, Chatterjee N, Wampler JM, Nayak PN, Deshmukh SS (1996) Indian intraplate and continental margins, rifting, lithospheric extension and mantle upwelling in Deccan Flow Basalt volcanism near the K/T boundary: Evidence from mafic dike swarms. J Geol 104: 379–398

Bondre NR, Hart WK, Sheth HC (2006) geology and geochemistry of the sangamner mafic dyke swarm, western Deccan volcanic orovince, India: Implications for regional stratigraphy. J Geol 114: 155–170

Chakraborty C, Mandal N, Ghosh SK (2003) Kinematics of the Gondwana basins of Peninsular India. Tectonophysics 377: 299–324

Chandrasekharam D, Mahoney JJ, Sheth HC, Duncan RA (1999) Elements and Nd-Sr-Pb isotope geochemistry of flows and dikes from the Tapi rift, Deccan flood basalt province. India J Volcanol Geothermal Res 93: 111–123

Cox KG, Hawkesworth CJ (1985) Geochemical stratigraphy of the Deccan Traps at Mahabaleswar, Western Ghats, India with implications for open system magmatic processes. J Petrol 26: 355–377

Deshmukh SS, Sehgal MN (1998) Mafic dyke swarms in the Deccan volcanic province of Madhya Pradesh and Maharashtra. In: Subbarao KV (ed) Deccan Flood basalts. Geol Soc India Mem 10: 323–340

Ernst RE, Baragar WRA (1992) Evidence from magnetic fabric for the flow pattern of magma in the Mackenzie giant radiating dyke swarm. Nature 356: 511–513

Evensen NM, Hamilton PJ, O'nions RK (1978) Rare earth abundances in chondritic meteorites. Geochim Cosmochim Acta 42: 1199–1212

Geological Survey of India (1995) Project CRUMANSONATA, Geoscientific studies of the Son-Narmada Lineament Zone, Geol Surv India Spec Publ10371 p

Geological Survey of India (2005) Geological map of Umaria and Shahdol

Govindraju K (1994) Compilation of working values and sample descriptions for 383 Geostandards Geostandards. Newsletter 18: 1–158

Hooper PR (1997) The Columbia River flood basalt province: Current status. In: Mahoney JJ, Coffin MF (eds) Large Igneous Provinces: Continental, Oceanic and Planetary volcanism. Geophysical Monograph, American Geophysical Union 100: 1–27

Irvine TN, Baragar WRA (1971) A guide to the chemical classification of the common rocks. Can J Earth Sci 8: 523–548

Karkare SG, Srivastava RK (1990) Regional dyke swarms related to the Deccan Trap Alkaline Province, India. In: Parker AJ, Rickwood PC, Tucker DH (eds) A.A. Balkema, Rotterdam: 335–347

Keays RR, Lightfoot PC (2009) Crustal sulphur is required to form magmatic Ni-Cu sulphide deposits: Evidence from chalcophile element signatures of Siberian and Deccan Trap basalts. Miner Deposita doi: 10.1007/S00126–009–0271–1

Kent RW, Kelley SP, Pringle MS (1998) Mineralogy and Ar/Ar geochronology of orangeites (Group II kimberlites) from the Damodar Valley, Eastern India. Mineral Mag 62: 313–323

Kent RW, Malcolm S, Pringle R, Dietmar M, Saunders AD, Ghose NC (2002) 40Ar/39Ar geochronology of the Rajmahal basalts, India, India, and their relationship to the Kerguelen plateau. J Petrol 43: 1141–1153

Kent RW, Saunders AD, Kempton PD, Ghose NC (1997) Rajmahal Basalts, Eastern India: Mantle sources and melt distribution at a volcanic rift margin. In: Mahoney JJ, Coffin MF (eds) Large Igneous Provinces: Continental, Oceanic and Planetary volcanism. Geophysical Monograph, American Geophysical Union 100: 145–182

Le Bas MJ, Le Maitre RW, Streckeisen AL, Zanettin P (1986) A chemical classification of volcanic rocks based on the total alkali-silica diagram. J Petrol 27: 745–750

Lightfoot P, Hawkesworth CJ (1988) Origin of Deccan Trap lavas: Evidence from combined trace element and Sr-, Nd- and Pb-isotope studies. Earth Planet Sci Lett 91: 89–104

Misra KS (2001) Lava channel/tube network in western Maharashtra: Study of their morphology, formation and transportation of Deccan lava. Geol Surv India Spec Publ 64: 209–221

Misra KS (2008) Dyke swarms and Dykes within the Deccan Volcanic Province, India. In: Srivastava RK, Sivaji Ch, Chalapathi Rao NV (eds) Indian Dykes: Geochemistry, Geophysics and Geochronology. Narosa Publishing House, New Delhi: 57–72

Patil SK, Arora BR (2008) Palaeomagnetic and Rock magnetic studies on the intrusives from Raniganj basin, Damodar valley: Linkage to the Rajmahal Volcanics. In: Srivastava RK, Sivaji Ch, Chalapathi Rao NV (eds) Indian Dykes: Geochemistry, Geophysics and Geochronology. Narosa Publishing House, New Delhi: 511–526

Paul DK (2005) Petrology and geochemistry of the Salma dike, Raniganj coalfield (lower Gondwana), eastern India: Linkage with Rajmahal or Deccan volcanic activity? J Asian Earth Sci 25: 903–913

Paul DK, Lala T, Chaudhary AK (2008) Petrology and geochemistry of mafic flows and dykes of eastern Deccan Volcanic province, India. 33rd International Geological Congress, Oslo, Norway

Peng ZX, Mahoney JJ, Hooper PR, Harris C, Beane JE (1994) A role for lower continental crust in flood basalt genesis? Isotopic and incompatible element study of the lower six formations of the western Deccan Traps. Geochim Cosmochim Acta 58: 267–288

Peng ZX, Mahoney JJ, Hooper PR, Macdougall JD, Krishnamurthy P (1998) Basalts of the north-eastern Deccan Traps, India: Isotopic and elemental geochemistry and relation to south western stratigraphy. J Geophys Res 103: 29843–29865

Ray R, Sheth HC, Mallik J (2007) Structure and emplacement of the Nandurbar-Dhule mafic dyke swarm, Deccan Traps and the tectonomagmatic evolution of flood basalts. Bull Volcanol 69(5): 537–551

Sen G (1995) A simple petrologic model for the generation of Deccan Trap magmas. Int Geol Rev 37: 825–850

Sen G, Cohen H (1994) Deccan intrusion, crustal extension, doming and size of the Deccan-Reunion plume head. In: Subbarao KV (ed) Volcanism. Wiley Eastern, New Delhi: 201–216

Shanker R (1991) Thermal and crustal structure of "SONATA". A zone of mid-continental rifting in Indian shield. J Geol Soc India 37: 211–220

Sheth HC, Mahoney JJ, Chandrasekharam D (2004) Geochemical stratigraphy of Deccan flood basalts of the Bijasan Ghat section Satpura Range. India J Asian Earth Sci 23: 127–139

Sheth HC, Ray JS, Ranjini R, Vanderkluysen L, Mahoney JJ, Kumar A, Shukla AD, Das P, Adhikary S, Jana B (2009) Geology and geochemistry of Pachmari dykes and sills, Satpura-Gondwana basin, Central India: Problems of dyke-sill-flow correlations in the Deccan traps. Contrib Mineral Petrol doi. 10.1007/S 004010–009–0387–4

Srivastava RK, Singh SB, Karkare SG, Chandra R (1996) Geochemistry of Deccan volcanics and dykes from the Nisarpur and Manpur sections, Madhya Pradesh. Gond Geol Mag 2: 135–144

Sun SS, McDonough WF (1989) Chemical and isotopic systematics of oceanic basalt: implications for mantle composition and processes. In: Saunders. A.D., Norry, M.J (eds) Magmatism in the ocean basins Geol Soc London, Special paper 42, Blackwell Scientific publisher, 313–345

Vandamme D, Coutillot V, Besse J, Montigny R (1991) Paleomagnetism and age determination of the Deccan Trap (India): Results of a Nagpur-Bombay traverse and review of earlier work. Rev Geophy 29: 159–190

Vanderkluysen L, Mahoney JJ, Hooper PR(2004) Implications for the emplacement of the Deccan Traps (India) from isotopic and elemental signatures of dikes. EOS: Trans A Geophys Union 85(47) Fall Meet Suppl Abstarct V51B–0561

West WD (1959) The source of the Deccan trap flows. J Geol Soc India 1: 44–51

White RS, McKenzie DP (1989) Magmatism at rift zones: The generation of volcanic continental margins and flood basalts. J Geophys Res 94: 7685–7729

Yan J, Chen YF, Xu XS (2008) Geochemistry of Cretaceous mafic rocks from the Lower Yangtze region, eastern China: Characteristics and evolution of the lithospheric mantle. J Asian Earth Sci 33(3–4): 177–193

Chapter 10
Tectonic Significance and Age of Doleritic Sill Near Bandhalimal in the Singhora Protobasin of Chhattisgarh Basin, Central India

D.K. Sinha, S.K. Jain, and Kusum P. Naganath

Introduction

The Proterozoic basins of India, in general, are devoid of igneous rock associations except in the Vindhyan, Chhattisgarh and Cuddapah Basins (Kale, 1991). The proportion of igneous rocks in the aggregate sequence from all the basins is not more than 3% of the total volume. These rocks occur as concordant and discordant bodies in the sedimentary sequence. Their importance is well established for indirect inference for geological age of lithological sequence, in which they have been emplaced. The sedimentary basins in particular require such igneous rocks, which can be utilized to unfold their geological history. The Chhattisgarh basin is known a little for the igneous rocks except a few dykes namely (a) dolerite traversing Gunderdehi (shale) Formation, Raipur Group of Chhattisgarh Basin, near Dhanagar and Mahadeopali of Raigarh district (Tripathi and Murti, 1981) (b) dolerite dyke intruding limestone of Raipur Group near Raipur, again in the upper sequence of the basin (Subba Rao et al., 2007) (c) tuffs (porcellanites) from basal Rehatikhol Formation (Sinha, 1997; Chakraborti, 1997; Bickford et al., 2009; and Das et al., 2009) and (d) recently identified ignimbrite and tuffaceous units towards the upper part of the Chhattisgarh succession [(Sukhda tuff and Sapos tuff) (Patranbis-Deb et al., 2007)]. These rocks are significant for age deduction of Chhattisgarh Basin.

Several workers (Kruzer et al., 1977; Murti, 1987, 1996; Moitra, 1990; Banerjee and Majumdar, 1999; Das et al., 2001) have suggested a late mid-Proterozoic to Neoproterozoic age for the rocks of the Chattisgarh Supergroup but it has been questioned in the light of latest work done by Patranabis-Deb et al., 2007; Das et al., 2009; and Bickford et al., 2009. The age of Chhattisgarh Basin is being considered >1,000 Ma on the basis of U-Pb ages of magmatic zircons reported from Sukhda

D.K. Sinha (✉)
Atomic Mineral Division, Department of Atomic Energy, Hyderabad 500016, India
e-mail: dksjai@gmail.com

S.K. Jain
Atomic Mineral Division, Department of Atomic Energy, Nagpur 440001, India

K.P. Naganath
Atomic Mineral Division, Department of Atomic Energy, Banglore 560072, India

R.K. Srivastava (ed.), *Dyke Swarms: Keys for Geodynamic Interpretation*,
DOI 10.1007/978-3-642-12496-9_10, © Springer-Verlag Berlin Heidelberg 2011

Tuff overlying Saradih Limestone of the Raipur Group (upper Chhattisgarh). On this evidence (Patranabis-Deb et al., 2007, 2008) have suggested pre-Neoproterzoic age for Upper part of Chhattisgarh Basin. Das et al., 2009 have dated monazite and zircon grains from tuff (porcellanites) unit occurring over Rehatikhol Formation and under Saraipalli Formation and proposed an age of ~1,500 Ma as the depositional/crystallization age of tuff. Bickford et al. (2009) have dated zircon from Singhora tuff and obtained an age of 1,405 ± 9 Ma. Thus the depositional age of rocks of Chhattisgarh Basin can be bracketed between 1,500 and 1,000 Ma period.

Present paper reports for the first time, Rb-Sr whole rock radiometric age of a basic sill intruded in to basal Rehatikhol Formation. The emplacement of dolerite is seen as a part of volcanic and tuff related activity which has further provided thermal gradient to Rehatikhol Formation. This has indirectly helped in the development of hydrothermal sulphides (Pb, Cu, Ag, Fe) and oxides (U and Fe) besides alterations in the fracture zones near Juba-Banjhapalli (Jain et al., 1998). The lead systematics collected from galena occurring with the sulphide association has also been presented to strengthen authors view point on age of Rehatikhol Formation. Such studies on these aspects have been attempted for the first time, from this part of the basin and enhance our knowledge of basins geological history. The petrography, geochemistry and tectonic environment of emplacement of dolerite have also been described to correlate it with the existence of volcanism during Singhora period of Chhattisgarh Basin and its bearing on age inference.

Regional Geology

The Chhattisgarh Basin is situated within the central Indian craton and surrounded by Sambalpur granitoids in the east (2,380 ± 44 Ma; Choudhary et al., 1996), Bilaspur-Raigarh-Surguja belt of Sausar Group in the north (1,541 ± 26 to 1,100 ± 20 Ma; Pandey et al., 1995), Nandgaon volcano-plutonics in the southwest (2,462 ± 25 to 2,039 ± 79 Ma; Pandy et al., 1995) and Khairagarh volcano-sedimentary rocks in the west (2,120 ± 35 Ma; Sinha, 1996). Metasediments and granitoids of Bastar craton (3,610 ± 336 to 2,110 ± 41 Ma; Pandey et al., 1995) lie to the south of the basin. Sonakham volcano-sediments and volcano-plutonics (2,347 ± 16 Ma; Pandey et al., 1995) trending NNW–SSE have divided Chhattisgarh Basin into Hirri subbasin in the west and Baradwar subbasin in the east having Singhora and Barapahar protobasins (Fig. 10.1). Das et al. (1992) suggested an intracratonic sag setting for Chhattisgarh Basin and classified it into three groups. The lowermost Singhora Group developed in an embryonic basin to the east, consists of four formations of arenite (Rehatikhol), argillite±carbonate (Saraipali), arenite (Bhalucona), and argillite±carbonate sequence (Chhuipali Formation). Sinha and Hansoti (1995) have proposed modifications to the stratigraphy of Singhora Group in which they have mentioned presence of porcellanites in the Rehatikhol, Saraipali and Chhuipali Formations and sill and dyke of basic compositions (Table 10.1). Singhora Group covers an area of over 1,000 km^2 with about 400 m thickness of sedimentary column. The Chandarpur Group, unconformably overlying the Singhora Group,

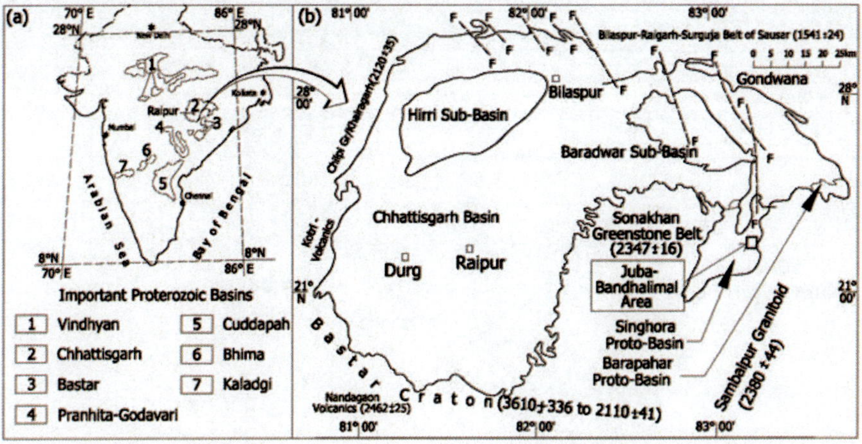

Fig. 10.1 (**a**) Location of Chhattisgarh Basin in India and (**b**) Juba-Bandhalimal area in Singhora Protobasin, Chhattisgarh Basin

Table 10.1 Stratigraphic succession according to Das et al. (1992) and Sinha and Hansoti (1995)

CHHATTISGARH	SUPERGROUP	Raipur Group		Dolerite Dykes, Shale and Limestone
		Unconformity		
		Chandarpur Group		Sandstone, shale
		Unconformity		
		SINGHORA GROUP	Chhuipali Fm.	Shale, limestone, porcellanite.
			Bhalukona Fm.	Glauconitic sandstone
			Saraipali Fm.	Shale, limestone, porcellanite.
			Rehatikhol Fm.	Conglomerate, sandstone, shale, porcellanite and doleritic sill.

Unconformity

Sambalpur Granitoid of Archaen to Lower Proterozoic Age

comprises mostly an arenitic litho-assemblage. It is divisible into three formations and grades conformably into Raipur Group at the top. Raipur Group consisting of argillite-carbonate suite is subdivided into six formations (Fig. 10.2). Recent work by Das et al., 2009 on Sighora Basin records structural overprinting which is formation specific and can be observed as meso-scale doubly plunging antiforms and synforms. They have observed that the area is cross cut by several dykes most of which are confined within the granite basement. However, the sill and dyke of Singhora Basin are not mentioned by them.

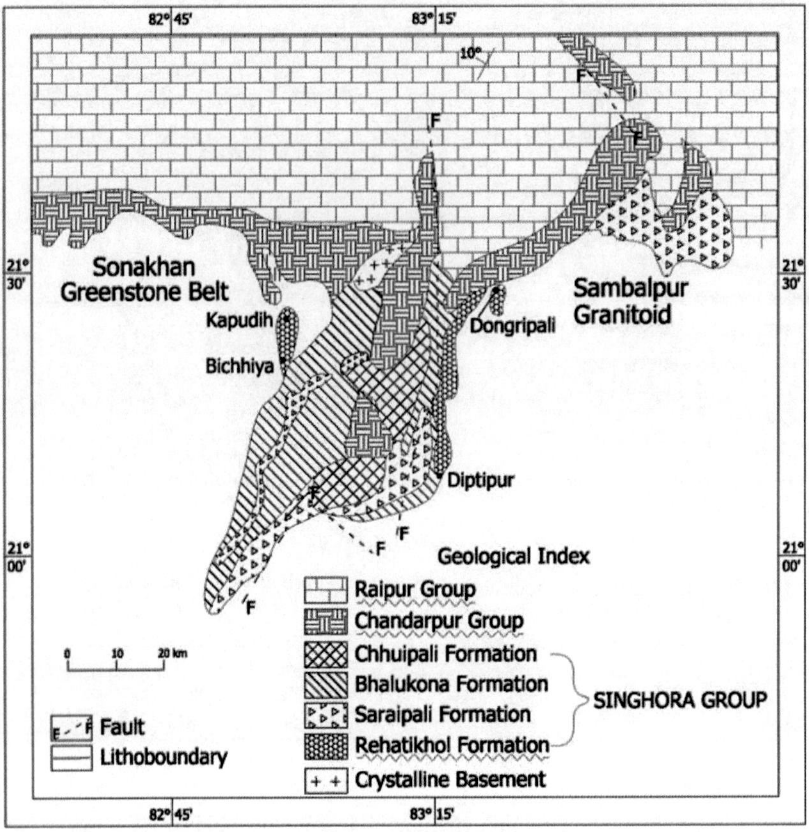

Fig. 10.2 Geological map of Singhora-proto basin, Chhattisgarh state (modified after Das et al., 1992)

Local Geology

The Rehatikhol Formation extends along the eastern and western margins of Singhora proto-basin (Fig. 10.2). In the eastern margin, it is exposed from Dongripali to Diptipur. The general strike of the formation is NE–SW with dips (5°–10°) towards NW. Similarly, it is exposed intermittently from Kapudih to Palasapali along western margin of Singhora Basin (Sinha et al., 1998). The rocks of Rehatikhol Formation in the western margin show a general NE–SW trend with gentle (5°–10°) southeasterly dip corresponding to basinal structure. The Rehatikhol Formation in the eastern margin commences with basal conglomerate having mostly quartz, clasts and minor lithic fragments of granite, schist and volcanics set in a clasts framework (Figs. 10.3 and 10.4). The conglomerate directly

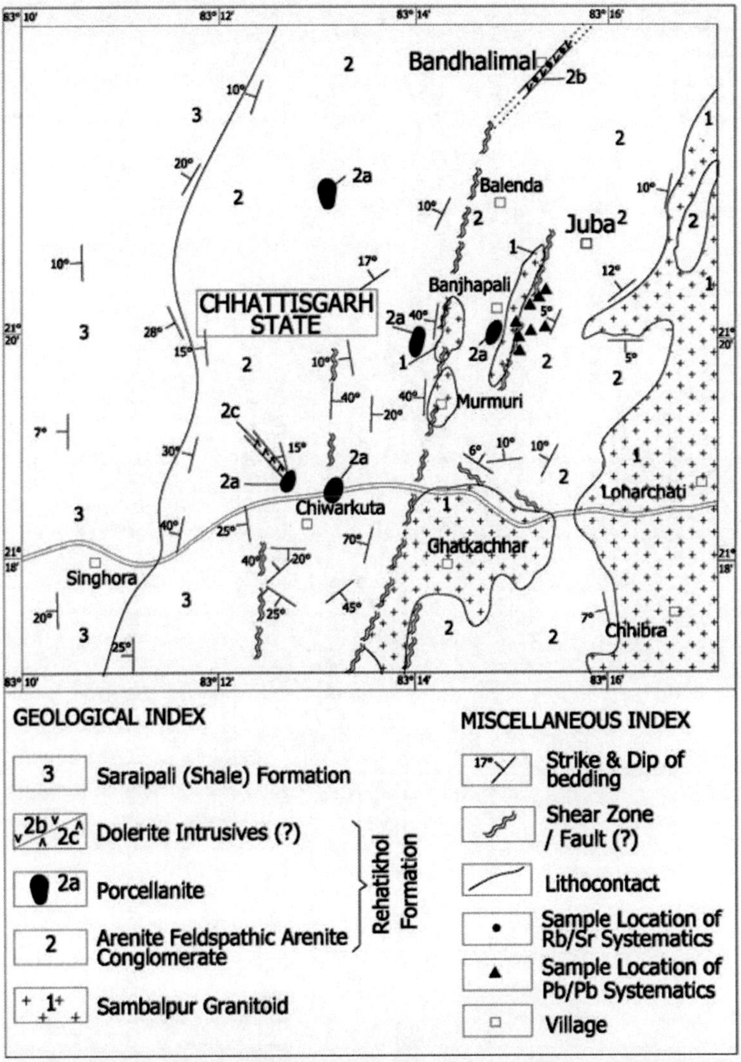

Fig. 10.3 Geological map of Bandhalimal–Juba-Chhibra area, Singhora protobasin, Chhattisgarh basin, India (modified after Das et al., 1992)

overlies Sambalpur Granitoid and is succeeded by greenish conglomerate/plebbly feldspathic arenite and very coarse grained greenish feldspathic to sub-feldspathic arenite. A thin pebbly horizon in the upper part marks the closing of first cycle of sedimentation. The second cycle of sedimentation commences with the deposition of cross bedded, coarse to medium grained arenite with layers of pebbly horizons.

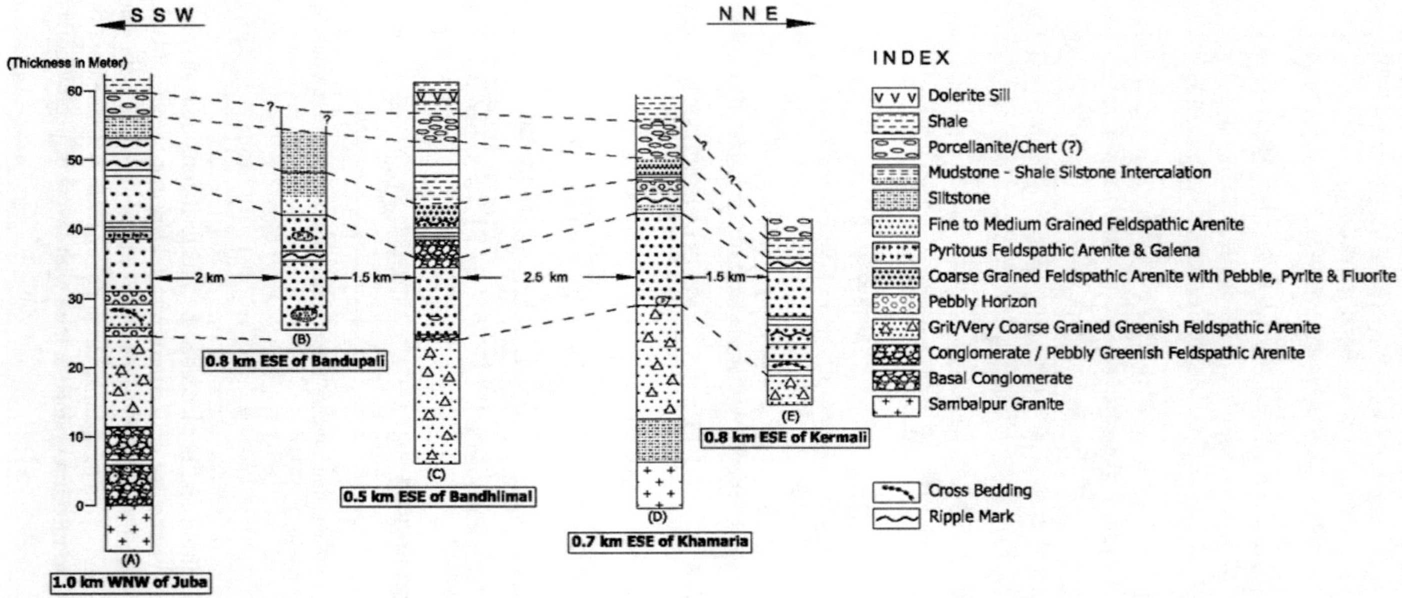

Fig. 10.4 Stratigraphic profiles in different sections of Juba-Bandhalimal-Kermeli section (T.S. 64O/7) showing dolerite sill

Grain size decreases upward and the arenite grades into siltstone-shale units and is overlain by porcellanitic units. This sequence is also associated with pyritous feldspathic arenite which is a part of the medium to coarse grained feldspathic arenite unit. Pyrite occurs as diagenetic mineral (δ^{34} values +31.58‰; n = 5) surrounding the clastic grains (Sinha et al., 2001). Shaly partings in the arenitic units are greenish and purple coloured, and sometimes calcareous. Black shale partings with pyrite layers are also exposed at Singhbahal. Presence of enormous sulphides and alternating sequence of arenite and argillites suggests that they were laid under shallow water oscillatory conditions (Sinha and Hansoti, 1995; Chakraborti, 1997 and Sinha et al., 1998). The Rehatikhol Formation has also been interpreted as conglomeratic or coarse arenaceous alluvial fan, braid-plain and delta deposits and overlying Saraipalli Formations as a storm-infested inner to distal shelf deposits (Chakraborty et al., 2009).

A doleritic sill (sometimes vesicular) is exposed near Bandhalimal village (21°22′20″: 83°15′15″) for ~2 km in NE–SW direction with an average thickness of 2 m within the shale unit of Rehatikhol Formation. Chilled margin in sill and the induration in the shale are very well developed (Fig. 10.5). Further trace of sill is not exposed. However extension of such activity cannot be ruled out. Sill has unique feature of vesicles arranged in a circular pattern which requires explanation (Fig. 10.6). Stratigraphic column recorded at Bandhlimal location (Fig. 10.4) depicts its relation with porcellanite within the Rehatikhol Formation. Since sill (with vesicles) is occupying at the top of sequence, it is missing in other sections. The importance of this basic sill has already been stated for inference of probable age of Singhora sedimentation.

Fig. 10.5 Field photographs of doleritic sill along with chilled margin

Fig. 10.6 Unusual feature (?) of vesicles arranged in *circular shape*

Petrography

The sill is aphanitic and melanocratic with several cavities and vugs filled with altered material. Randomly oriented plagioclase laths, and phenocrysts (andesine-labradorite), and olive green augite are chief constituents (Fig. 10.7). Calcite, skeletal ilmenite (Fig. 10.8) biotite (Fig. 10.9), chlorite (penninite), sericite (Fig. 10.10) and arenite fragments (Fig. 10.9) are the minor minerals. These minerals are accompanied by quartz (Fig. 10.9), clinozoisite, leucoxene, pyrite, chalcopyrite, and

Fig. 10.7 Intersertal texture in dolerite. *Triangular shaped area* (TA) filled with chlorite. X Nicol, TL

Fig. 10.8 Skeleton grains of ilmenite (sk-il) in dolerite. 1 Nicol, TL

Fig. 10.9 Partly assimilated fragments of arenite in dolerite. Biotite (Bio), plagioclase (Plg), altered potash feldspar (A-Kf) and quartz (Qz). X Nicol, TL

Fig. 10.10 Sericitization of plagiocalse (Sz Plg) with albite rims (Al-R). X Nicol, TL

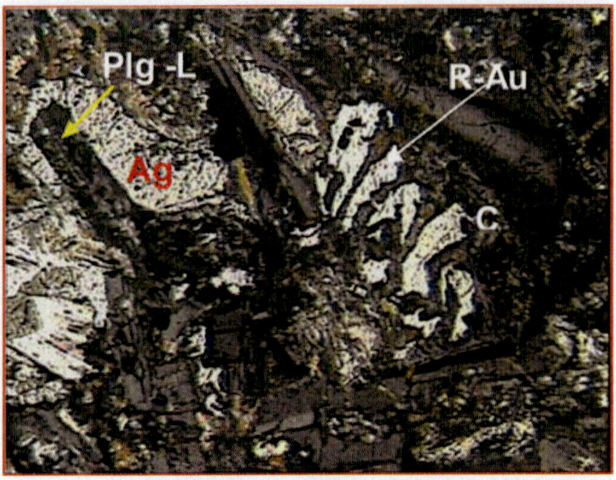

Fig. 10.11 Chloritization (C) of augite of dolerite. R-Au (Relict augite). Plagioclse laths (Plg-L) are enclosed in augite. X Nicol, TL

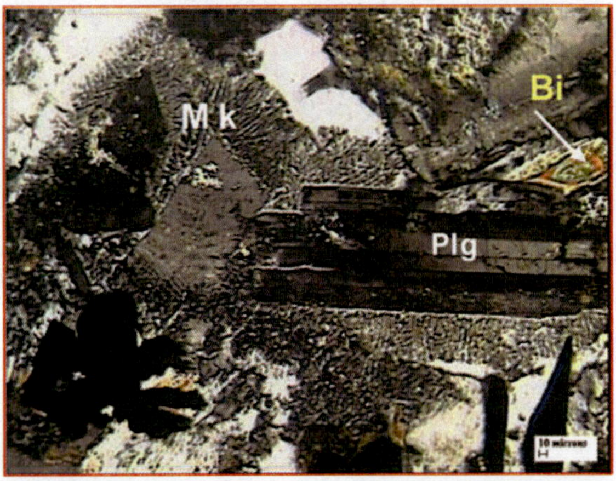

Fig. 10.12 Formation of uralite (Ur) from augite (Ag). X Nicol, TL

Fig. 10.13 Altered dolerite of Singhora Basin. Limonite (L.N) coating, Ilmenite (Il). X Nicol, TL

apatite as accessories. Plagioclase laths are partially to wholly enclosed within plates of augite forming ophitic to subophitic texture (Fig. 10.11). Plagioclase laths are also randomly oriented forming hexagonal to triangular shaped area, filled with chlorite, thereby forming intersertal texture (Fig. 10.7). On the basis of above mineralogical constituents and textural attributes the rock has been classified as dolerite. Two types of alteration in mineralogy of dolerite has been noticed namely alteration of ferro-magnesian minerals during the emplacement (Fig. 10.12) and deuteric alteration (Fig. 10.13). The former is common in the peripheral portion of dykes, causing the encrustation of goethite. Deuteric alteration is seen in relatively undeformed and dark green colored portion of dolerite body. The overall pattern of alteration is visible in the Fig. 10.14.

Fig. 10.14 Myrmekite (Mk) formation around plagioclase (Plg). X Nicol, TL

Geochemistry and Tectonic Environment

Major oxides and trace elements for 14 rock samples have been obtained by XRF for the chemical classification of dolerite and inference on the tectonic set-up (Tables 10.2 and 10.3). The range and average of 14 chemical analyses from these samples indicate that the concentration of SiO_2 ranges 39.4–52.8 wt%, av. 49.6. TiO_2 shows a variation from 0.36 to 3.83 wt%, av. 2.08 wt%, Al_2O_3 varies from 12.42 to 20.11 wt%, av 16.36 wt%, total iron as FeO ranges from 12.42 to 26.84 wt%, av. 15.53 wt%, MnO is low and varies from 0.12 to 1.73 wt%, av 0.40 wt%, MgO varies from 1.56 to 6.55 wt%, av. 4.91 wt%. CaO 4.55–9.97 wt%, av. 7.66 wt%, Na_2O varies from 0.6 to 2.92 wt%, av. 1.71 wt% and K_2O ranges from 0.11 to 2.14, av. 1.26. Mg number varies from 5.86 to 28.19, av. 20.39 reflect the relative stage of formation of basic rock. The Mg number (28.19) of these rocks suggests

Table 10.2 Major minor and trace elemental data of doleritic sill near Bandhalimal village, Chhattisgarh Basin (Oxides in % and trace in ppm)

S. No.	SiO_2	TiO_2	Al_2O_3	FeO^t	MgO	MnO	CaO	Na_2O	K_2O	P_2O_5	FeO^t/MgO	Ni	Co	Nb	Zr	Y	Ba	Rb	Sr	K/Sr	Y/Nb	Rb/Sr	K/Rb	A.I.	Mg No.
1.	51.34	2.29	16.64	13.11	4.84	0.14	8.85	1.93	0.63	0.2	2.709	54	38	11	264	26	570	32	474	11.03	2.36	0.07	163.41	1.81	22.18
2.	51.24	2.39	16.72	13.11	4.92	0.13	9.06	1.27	0.66	0.2	2.665	65	33	13	244	28	396	26	423	12.95	2.15	0.06	210.69	1.38	22.47
3.	50.52	2.28	16.73	12.65	5.21	0.13	8.73	2.16	0.76	0.19	2.428	57	34	13	217	27	460	34	444	14.21	2.08	0.08	185.53	2.28	24.13
4.	49.93	2.46	16.58	13.27	4.64	0.17	9.97	1.45	0.58	0.14	2.860	56	26	11	218	30	430	31	452	10.65	2.73	0.07	155.29	1.72	21.26
5.	50.45	2.13	16.79	13.07	6.07	0.13	8.14	1.81	0.85	0.23	2.153	62	39	12	268	25	530	38	433	16.29	2.08	0.09	185.66	2.10	26.40
6.	50.36	2.09	16.79	12.82	6.55	0.12	8.03	1.6	1.17	0.18	1.957	69	35	7	199	27	419	41	428	22.69	3.86	0.10	236.85	2.21	28.29
7.	49.02	2.14	16.66	13.36	6.45	0.14	8.47	1.47	1.19	0.19	2.071	75	38	12	235	28	519	41	450	21.95	2.33	0.09	240.90	2.60	27.16
8.	50.73	2.04	16.35	12.42	6.31	0.41	7.9	1.55	1.62	0.17	1.968	64	35	12	207	25	472	54	409	32.88	2.08	0.13	249.00	2.41	28.18
9.	52.5	0.43	20.11	20.1	1.62	0.5	6.79	2.92	0.11	0.11	12.407	62	33	9	206	27	411	49	423	2.16	3.00	0.12	18.63	1.88	5.86
10.	51.1	3.87	14.13	16.7	3.34	0.31	4.55	1.96	2.14	0.31	5.000	73	30	11	238	30	490	52	456	38.95	2.73	0.11	341.58	2.98	13.38
11.	39.4	0.36	12.6	26.84	4.6	1.73	7.7	1.09	0.16	0.22	5.835	55	38	6	190	23	381	27	401	3.31	3.83	0.07	49.19	−2.04	11.69
12.	49.45	3.83	15.44	16.68	7.2	0.29	5.78	0.6	2.12	0.29	2.317	58	28	12	237	31	480	50	447	39.36	2.58	0.11	351.92	2.48	25.00
13.	46	0.41	19.35	19.6	1.56	1.2	8.14	1.96	2.14	0.31	12.564	73	30	11	238	30	490	52	456	38.95	2.73	0.11	341.58	8.04	5.79
14.	52.8	2.33	14.08	13.64	5.47	0.14	5.12	2.1	3.56	0.23	2.494	61	39	10	215	29	530	42	499	59.21	2.90	0.08	703.52	3.40	23.65
Avg	49.63	2.08	16.36	15.53	4.91	0.40	7.66	1.71	1.26	0.21	4.24	63.14	34.00	10.71	226.86	27.57	469.86	40.64	442.50	23.19	2.68	0.09	245.27	2.37	20.39

Table 10.3 Major minor and trace elemental data of doleritic sill near Bandhalimal village, Chhattisgarh Basin (Oxides in % and trace in ppm)

Samples details	SiO_2	TiO_2	Al_2O_3	FeO^t	MgO	MnO	CaO	Na_2O	K_2O	P_2O_5	Ni	Co	Nb	Zr	Y	Ba	Rb	Sr	K/Sr	Y/Nb	Rb/Sr	Rb/Sr	K/Rb	A.I.	Mg No.
Unaltered basic	48.14	1.96	16.37	16.04	5.29	0.49	8.12	1.52	1.12	0.22	4	63	34	11	225	28	464	39	438.88	20.93	2.78	0.09	218.36	2.42	21.21
Altered basic	51.62	2.23	16.34	14.85	4.42	0.27	7.05	1.96	1.45	0.20	5	63	35	11	229	28	478	43	447.33	26.20	2.54	0.10	281.14	2.31	19.29

the evolved nature of magma and the low value of Mg number (5.86) suggest crustal contamination.

The major and trace elements of concentration of dyke resemble like continental tholeiite (Condie, 1997) which is further supported by various other chemical diagrams such as alkaline index {A.I. = $Na_2O + K_2O/(SiO_2–43) \times 0.17$} against Al_2O_3 content as per Middlemost (1975) (Fig. 10.15); SiO_2 vs. FeO^t/MgO diagram (Miyashiro, 1974) (Fig. 10.16); and total alkali vs. silica diagram (Cox et al., 1979) (Fig. 10.17). These figures are indicative of parent magma which was of tholeiitic composition. In the TiO_2-K_2O-P_2O_5 triangular diagram of Pearce et al. (1975) (Fig. 10.18), most of the samples plot in the continental field. Thus continental setting of sill is interpreted.

Alteration processes have facilitated varying degree of modification of the primary composition of sills, which is recognized by comparing the mineralogy and chemical composition of the altered samples with the samples displaying least alteration. Altered samples contain high Na_2O (up to 2.92 wt%), K_2O (up to 3.56 wt%), silica (up to 52.8 wt%), and FeO (up to 14.9 wt%) as compared to the unaltered dolerite, which is having relatively low Na_2O (2.12 wt%), K_2O (2.21 wt%) and FeO (13.5 wt%).

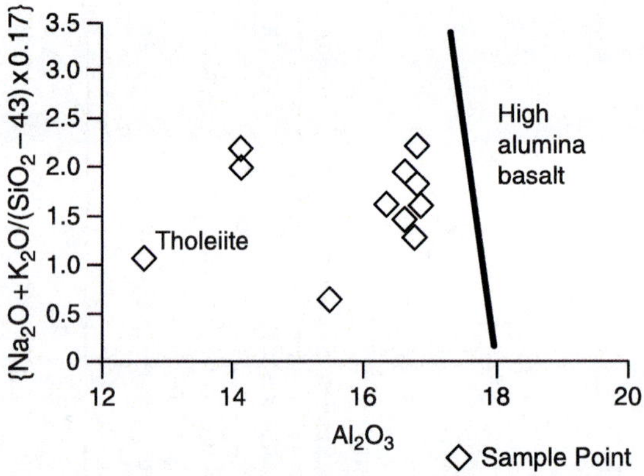

Fig. 10.15 Alkaline index (A.I.) = {$Na_2O + K_2O/(SiO_2–43) \times 0.17$} against Al_2O_3 content Middlemost (1975)

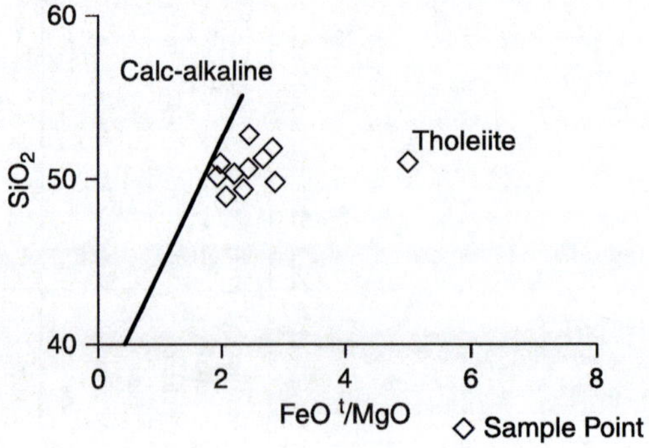

Fig. 10.16 SiO_2 vs. FeO^t/MgO diagram (after Miyashiro, 1974)

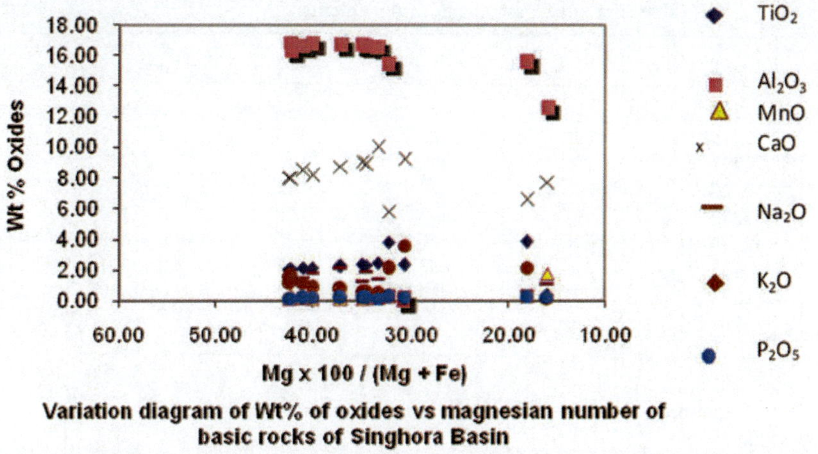

Fig. 10.17 Variation diagram of wt% of oxides vs magnesian number of doleriteic sill

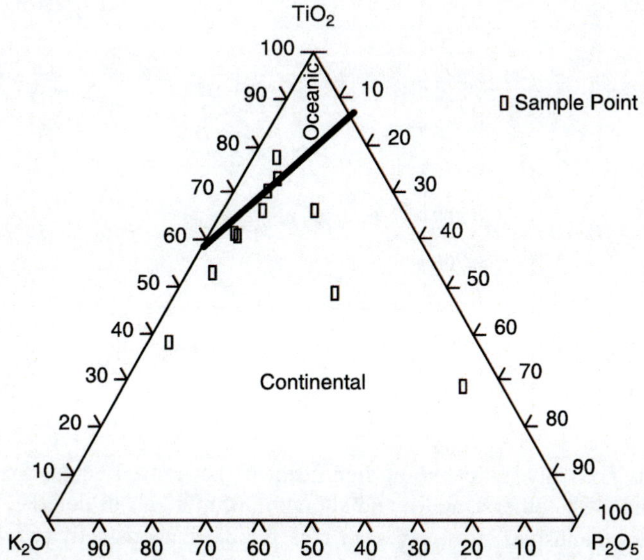

Fig. 10.18 TiO_2–K_2O–P_2O_5 triangular diagram of Pearce et al. (1975)

Radiometric Age

Rb-Sr whole rock estimation carried out at Geochronology Lab. AMD, Hyderabad, on 8 representative rock samples of dolerite, provide 1,100 Ma reference line (Table 10.4 and Fig. 10.19). Six samples are clustered around the reference line and may be considered for age purposes. The two deleted samples lie well above

Table 10.4 Rb/Sr systematics from dolerite sill

S.No.	Sample no.	Rb (ppm)	Sr (ppm)	$^{87}Rb/^{86}Sr$	$^{87}Sr/^{86}Sr$
1	2219*	17	450	0.11	0.71583
2	2218*	19	423	0.13	0.72291
3	2221	23	449	0.15	0.71256
4	2220	22	430	0.15	0.71166
5	2222	29	413	0.20	0.71189
6	2224	40	446	0.26	0.71448
7	2223	41	441	0.27	0.71321
8	2225	42	414	0.29	0.71437

Age = 1,121±742 Ma, Initial ratio = 0.7095 ± 0.002. *Samples ignored

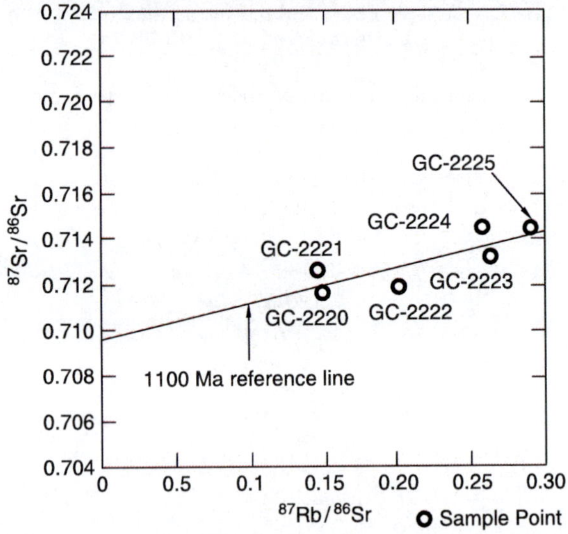

Fig. 10.19 Rb-Sr age from doleritic sill within Rehatikhol formation of Singhora Chhattisgarh Basin

this line, which is probably indicative of significant higher crustal contamination in them, compared to the other samples and alteration of minerals of dolerite. The Rb-Sr data is not conclusive, but it gives an idea for the igneous activity which occurred during geological history of Singhora Group.

In addition to above galena along with pyrite have also been observed in the fracture zones of Rehatikhol Formation near west of Juba (3 km south of Bandhalimal). The orientation of these fracture zones is sub-parallel to the sill. Hydrothermal activity in such fracture zone is considered coeval with the intrusion of sill (Sinha et al., 1998; Sinha and Jain, 2008). Pb isotopic ratios ($^{206}Pb/^{204}Pb$ = 19–20, $^{207}Pb/^{204}Pb$ = 16.4–16.7 and $^{208}Pb/^{204}Pb$ = 38.2–38.8) studied for eight galena samples indicate model ages varying from 660 to 1,100 Ma with "μ_2" ($^{238}U/^{204}Pb$) values of 13.2–13.6 for second stage (Table 10.5), based on two stage Pb growth model (Stacey and Kramers, 1975). This observation is consistent with the Rb/Sr age of

Table 10.5 Pb/Pb systematics from parallel fracture zone

S.No.	Sample No.	$^{206}Pb/^{204}Pb$	$^{207}Pb/^{204}Pb$	$^{208}Pb/^{204}Pb$	Age (Ma)	μ	K
1	DKS/JB/07	19.437	16.499	38.366	978	13.6	3.5
2	DKS/JB/09	19.232	16.511	38.474	1,105	13.7	3.6
3	DKS/JB/46	19.122	16.427	38.325	1,058	13.4	3.6
4	DKS/JB/15	19.574	16.454	38.667	845	13.3	3.5
5	DKS/JB/18	19.946	16.469	38.256	659	13.2	3.2
6	DKS/JB/23	19.413	16.502	38.454	995	13.6	3.5
7	DKS/JB/31	19.360	16.469	38.442	982	13.4	3.6
8	DKS/JB/336	20.035	16.629	38.808	825	13.9	3.4

sill as mentioned above. Considering two different aspects of age determination, it is proposed that the rocks of Rehatikhol Formation are older than 1,100 Ma.

Discussions and Conclusion

It is already mentioned that basic rock in the form of dolerite dykes traversing Gunderdehi (shale) Formation of the uppermost Raipur Group, near Dhangar and Mahadeopali of Raigarh district of Madhya Pradesh are known (Tripathi and Murti, 1981). Recently Subba Rao et al. (2007) have observed dolerite dyke near Raipur from two bore hole sections which is not exposed to surface and created doubt (Sinha, 2007). In addition to these two doleritic bodies, Sinha and Hansoti (1995) and Sinha et al. (1998) have also reported basic sill and porcellanite in the basal Rehatikhol Formation and showed presence of tuffaceous volcanic components in it. Subsequently tuff (porcellanites) have been studied in detail by Das et al., 2009, and for dating purpose by Bickford et al., 2009. Das et al (op. cit.) have reported the composition of tuff (porcellanites) varying from rhyodacite to andsite. According to them the tuff is extended over 9 km in length with maximunm thickness of 22 m and stratigraphycally positioned between Rehatikhol and Saraipali Formations. This observation however contradicts the view of Sinha et al., 1998 where presence of porcellanite is reported well within the Rehatikhol Formation and even from upper shaly formations namely Saraipalli and Chhuipalli. The basic sill reported by Sinha and Hansoti (1995) is exposed near Bandhalimal village, and extends for ~2 km in NE–SW direction with an average thickness of 2 m. It shows Chilled margin and the indurations in the shale, suggesting considerable temperature difference between host and emplaced rocks. Reddish coloured encrustations along border portion of rocks are due to greater concentration of iron bearing minerals and their alteration products. Relatively high content of K_2O (1.45 wt%, n=6) and Na_2O (1.95 wt%, n=6) in altered basic shows the presence of secondary albite (Fig. 10.10), the assimilation of host arenaceous rocks in intrusive dolerite (Fig. 10.9), formation of myrmekite (Fig. 10.13) and sericitization of feldspar. These altered features are also correlated with elevated content of SiO_2 (up to 51.6 wt% as against 48.1 wt%) and Ba (478 ppm) as against (463 ppm). Features are further accompanied by depletion of CaO (7.0 wt% as against 8.12 wt%) and MgO (4.4 wt% as against 5.2 wt%). The enrichment of Si, Na, K, Ba and depletion of Ca and Mg are attributed the reaction involved during assimilation of Na and K rich middle Proterozoic arenaceous rocks, when the sill was invading the sediments. Depletion of Ca and Mg in altered rock is due to the sericitization of plagioclase, formation of uralite from augite (Figs. 10.10 and 10.12) and chloritization of augite and biotite (Fig. 10.11). The above observations further suggest the composition of magma was extensively modified due to crustal contamination and alteration. The low Mg number (5.79–28.29) of these rocks suggest the evolved nature of magma. Extremely low value of Mg number (5.79) point towards the extensive crustal contamination as seen by subsequent myrmekite development.

The observations and interpretations made above generate two ideas for dolerite related geological history. Dolerite has embibed grains of quartz and other rock fragments from Rehatikhol Formation; therefore it is contaminated by crustal material. The doleritic magma was of evolved nature and probably had source, related with volcanism (porcellanitic) during the period of Chhattisgarh Basin development. Chhattisgarh Basin has experienced at least two phases of volcanism whose products namely tuff (porcellanite) and ignimbrites are concordentlly deposited in the Rehatikhol – Saraipali – Chhuipali Formations (Sinha, 1997; Chakraborti, 1997, and Das et al., 2009) and Saradih Limestone- Sarnadih sandstone formations (Basu et al., 2008; Patranabis-Deb et al., 2007, 2008). The composition of volcaniclastics varies from rhyolite to andesite but not to the basaltic range. Recent age data from these two stratigraphically separated horizons place them in age difference of 500 Ma (Rehatikhol ~1,500 Ma and Sukhda tuff ~1,000 Ma). Though Mukherjee and Ray (2008 and 2010) consider younger activity as related with basal formation of Chhattisgarh i.e. Lohardih Formation. The tectonic environment for tuffs (porcellanite) of Rehatikhol Formation has been interpreted by Das et al. (2009), where he has put them in the field of syncollisional and volcanic arc granite. Present study suggests that the parent magma of Bandhlimal sill was of tholeiitic composition. The tholeiitic composition of basalt is, therefore correlated with volcanic island environment where only minor amounts of alkali basalts and their derivatives erupted during terminal volcanism (Condie, 1997). Though in Rehatikhol Formation two doleritic bodies have been identified but their relation could not be established because of their limited exposures and spacial distance from each other. However inter relationship with each other cannot be ruled out. The Bandhalimal sill has NE–SW trend while Chiwarkuta dyke is emplaced along NW–SE (Fig. 10.3).

Considering Bandhlimal sill as a part of dyke activity of Singhora Group, it can be compared with the youngest age of important global intrusions. Major dyke swarms appear to have been episodically related with important ages of intrusions at 2,500, 2,390–2,370, 2,150–2,100, 1,900–1,850, 1,270–1,250, 1,150–1,100 Ma (Condie, 1997). Structural studies indicate that dykes are intruded at right angles to the minimum compression direction therefore NW–SE direction is considered to be minimum compression direction during pre-sill and post-Singhora period. This NW–SE minimum compression direction is perpendicular to basin margin which was in NE–SW. Several faults parallel to basin margin and Bandhlimal sill have been formed after deposition of Singhora Group of rocks. The NE–SW tectonism which was continuing in post-Singhora and pre-Bandhlimal sill period had been active even during the basin development and its filling. Evidences like the conglomerate and alluvial fan delta deposits of Rehatikhol and storm infested Saraipali shale were dominated by wave and storm systems and strong upliftment processes parallel to basin margin in NE–SW. The Bandhlimal sill has thus occupied the fracture parallel to NE–SW (basin margin) direction.

Rehatikhol Formation has been accepted as the oldest Formation of Chhattisgarh Supergroup (Das et al., 1992; Mukherjee and Ray, 2010). Radiometric age from Chaporadih Formation of Chandarpur Group, which is much younger to Rehatikhol Formation has yielded 700–750 Ma by K-Ar dating of authigenic glauconite (Kruzer et al., 1977). The palaeomagnetic studies for pole position of the younger Gunderdehi Formation by Verma et al. (1977) also give 1,250–1,300 Ma for Chandarpur Group. These dates are suggestive of older depositional period for Rehatikhol Formation. Therefore considering earlier data Singhora Group may be placed before 1,300 Ma chronologically. Recently Das et al. (2009) have proposed ∼1,500 Ma as depositional/crystallization age for the Singhora tuff, considering it's most frequent and exclusive presence within the tuffaceous unit. The age data has also been confirmed by Bickford et al. (2009) by zircon dating and assigning ∼1,400 Ma. Further Basu et al. (2008) have established ∼1,000 Ma age for the eruption/ emplacement of the Sukhda tuff. A concomitant thermal event in the basin at ∼1,000 Ma with related hydrothermal activity is also proposed by Das et al., 2009. Signatures of this thermal event are shown in the morphological expression of the monazite and zircon grains observed by them. Patranabis-deb et al (2007, 2008) supported above view. Bandhlimal sill and hydrothermal mineralizing phase in the fractures of Rehatikhol Formation by the present study proves ∼1,100 Ma thermal activity in the Singhora Basin. The sulphides and oxides have been precipitated by the ∼1,100 Ma old activity support the idea of Das et al., 2009. Data presented in this paper suggest time of initiation of Chhattisgarh Basin (>1,100 Ma) with the help of petrography, geochemistry and radiometric age of doleritc sill. This age estimation corroborates to the age deduced from the Conophyton species reported from Rehatikhol Formation (Sinha et al., 2002).

Therefore, on the basis of above evidences it is concluded that rocks of Singhora Group's are older than 1,100 Ma. Its stratigraphic position below Chandrapur Group is therefore justified. A thermal event of ∼1,100 Ma is marked by doleritic sill (may be a part of dyke swarm activity as known in the other parts of country/world) and related hydrothermal activity in the basin. These features have confirmed the classification of Singhora Basin under meso-Proterozoics.

Acknowledgements Authors are thankful to the Anjan Chaki Director, AMD, Hyderabad for granting the permission to publish this research paper in the Dyke Conference. Discussions held with P. S. Parihar Additional Director, AMD, Hyderabad and analytical support from scientists of XRF Lab, AMD, Nagpur, and Geochronology Lab AMD, Hyderabad, have provided significant inputs for this communication. Authors are thankful to all of them. The paper was reviewed by M. E. Bickford, Department of Earth Sciences, Syracuse University Syracuse, New York, USA, whose suggestions made authors to incorporate lead systematics from this area and enhance the scope of paper for meaningful interpretations. Authors are grateful to him for his constructive review.

References

Banerjee DM, Majumdar A (1999) On the late Neoproterozoic–early Cambrian transition events in parts of east Gondwanaland. Gondwana Res 2: 199–211

Basu A, Patranabis-Deb S, Schieber J, Dhang PC (2008) Stratigraphic position of the ~1,000 Ma Sukhda Tuff (Chhattisgarh supergroup, India) and the 500 Ma question. Precamb Res 167: 383–388

Bickford ME, Basu A, Patranabis-Deb S, Dhang P (2009) Depositional history of the mesoProterozoic Chhattisgarh Basin, Central India: Constraints from new shrimp zircon ages. Geol Soc of America Abs with Programs 41: 541

Chakraborti S (1997) Elucidation of the sedimentary history of the Singhora group of rocks, Chhattisgarh supergroup, Madhya Pradesh. Geol Surv Ind Rec 130: 184–187

Chakraborty PP, Sarkar A, Das K, Das P (2009) Alluvial fan to storm-dominated shelf transition in the Mesoproterozoic Singhora group, Chattisgarh supergroup, Central India. Precamb Res 170: 88–106

Choudhary AK, Naik A, Mukhopadhyay D, Gopalan K (1996) Rb-Sr dating of Sambalpur Granodiorite, Western Orissa. J Geol Soc Ind 47: 503–506

Condie KC (1997) Plate tectonic and crustal evolution. Pergamon Press, Elmsford, NY: 288

Cox KG, Bell JD, Pankhrst RJ (1979) The interpretation of igneous rocks. Allen and Unwin, London

Das N, Dutta DR, Das DP (2001) Proterozoic cover sediments of southeastern Chhattisgarh state and adjoiming part of orissa. Geol Surv India Spec Publ 55: 237–262

Das DP, Kundu A, Das N, Dutta DR, Kumaran K, Ramamurthy S, Thangavelu C, Rajaya V (1992) Lithostratigraphy and sedimentation of Chhattisgarh Basin. Indian Mineral 46: 271–288

Das K, Yokoyama K, Chakraborty PP, Sarkar A (2009) Basal tuffs and contemporaneity of the Chattisgarh and Khariar Basins based on new dates and geochemistry. Jour Geol 117: 88–102

Jain SK, Sinha DK, Verma SC, Singh R (1998) Ore mineragraphy of uraniferous polymetallic sulphides at Juba, Chhattisgarh Basin, Raipur district, Central India. Expl Res Atom Min 11: 25–36

Kale VS (1991) Constraints on the evolution of Purana Basins of Peninsular India. Jour Geol Soc India 38: 231–252

Kruzer H, Harse W, Kursten M, Schnitzer WA, Murthy KS, Srivastava NK (1977) K/Ar dates of two glauconites from the Chandarpur series (Chhattisgarh/India): On the Stratigraphic status of the Late Precambrian basins in Central India. Geol Jb B 28: 23–36

Middlemost EAK (1975) The basalt clan. Earth Planet Sci Lett 11: 337–364

Miyashiro A (1974) Volcanic rock series in Island area and active continental margin. American Jour Sci 274: 321–355

Moitra AK (1990) Chronologic implications of the stromatolites, microbiota and trace fossils of the Chattisgarh basin, Madhya Pradesh: Workshop on Precambrian of central India. Geol Surv India Spec Publ 28: 384–399

Mukherjee A, Ray RK (2008) SHRIMP ages of Zircon in the uppermost Tuff in Chattisgarh Basin in Central India Require ~500-Ma adjustment in Indian Proterozoic Stratigraphy: A discussion. Jour Geol 116: 537–539

Mukherjee A, Ray RK (2010) An Alternate View on the Stratigraphic Position of the ~1-Ga Sukhda Tuff vis-à-vis Chronostratigraphy of the Precambrians of the Central Indian Craton. Jour Geol 118: 325–332

Murti KS (1987) Stratigraphy and sedimentation in Chattisgarh Basin. *In* Purana Basins of Peninsular India, Geol Soc India Mem 6: 239–260

Murti KS (1996) Geology, sedimentation and economic mineral potential of the southcentral part of Chhattisgarh Basin. Geol Surv India Mem 125: 139

Pandey BK, Chabria T, Gupta JN (1995) Geochronological characterstic of the Proterozoic terrains of Peninsular India: Relevance to the first order target selection for uranium exploration. Expl Res Atom Min 8: 182–213

Patranabis-Deb S, Bickford ME, Hill B, Chaudhuri AK, Basu A (2007) SHRIMP ages of zircon in the uppermost tuff in Chattisgarh basin in central India require ∼500-Ma adjustment in Indian Proterozoic stratigraphy. Jour Geol 115: 407–415

Patranabis-Deb S, Bickford ME, Hill B, Chaudhuri AK, Basu A (2008) SHRIMP ages of zircon in the uppermost Tuff in Chattisgarh Basin in central India require ∼500 Ma adjustment in Indian Proterozoic stratigraphy: A reply. Jour Geol 116: 540–542

Pearce TH, Gorman BE, Birkett TC (1975) The TiO_2–K_2O–P_2O_5 diagram: A method of discriminating between oceanic and non-oceanic basalts. Earth Plan Sci Let 24: 419–426

Sinha DK (1996) Chrono-stratigraphy of the Khairagarh Basin, Central India: A reappraisal. Gond Geol Mag 11: 1–15

Sinha DK (1997) Porcellanite from the Singhora Group – An example of valcaniclastic sedimentation in the Chhattisgarh Supergroup Central India. *In* abstract volume of XIV Conv Ind Ass Sediment Madras: 9

Sinha DK (2007) A new find of younger dolerite dykes with continental flood basalt affinity from the Meso-Neoproterozoic Chhattisgarh Basin, Bastar Craton, Central India. Comments Jour Geol Soc India 70: 371–373

Sinha DK, Hansoti SK (1995) U and polymetallic sulphides bearing Juba member of Singhora Protobasin of Chhatisgarh supergroup, Raipur district, Madhya Pradesh. A new horizon. Jour Atom Min Sci 3: 1–10

Sinha DK, Jain SK (2008) uraniferous rhyolitic vein in the basement fractures of Singhora protobasin near Juba village, Raipur district, Chhattisgarh. Expl Res Atom Min 18: 119–124

Sinha DK, Jain SK, Choudhury DK, Veera Bhasker D (2002) Conophyton from Rehatikhol Formation, Chhattisgarh Basin, Central India. In: Guhey R, Thakur P, Sridhar Rao R, (eds) Proceedings volume on Recent Researches on the Proterozoic Basins of India, Pub by Govt Sci College Raipur: 27–34

Sinha DK, Raju KA, Veera Bhaskar D, Asha K (2001) Sulphur Isotopic Characteristics of pyrite and galena from the Singhora Group, Chhattisgarh supergroup, India: Genetic implications. Jour Geol Soc India 57: 171–177

Sinha DK, Tiwary A, Verma SC, Singh R (1998) Geology, sedimentary environment and geochemistry of uraniferous arenite of Singhora group, Chhattisgarh Basin, Raipur district, Central India. In: Tiwari RN, (ed) Proc Nat Symp Recent Res Sed Basin, Hindustan Pub Co Dehra-Dun: 150–171

Stacey JS, Kramers JD (1975) Approximation of terrestrial lead isotope evolution by a two- stage model. Earth Planet Sci Lett 26: 207–221

Subba Rao DV, Khan MWY, Sridhar DN, Naga Raju K (2007) A new find of younger dolerite dykes with continental flood basalt affinity from the Meso-Neoproterozoic Chhattisgarh Basin, Bastar Craton, Central India. Jour Geol Soc India 69: 80–84

Tripathi C, Murti KS (1981) Search for source rock of alluvial diamonds in the Mahanadi valley. Symp. on Vindhyans of Central India, Geol Surv India Misc Publ 50: 205–210

Verma RK, Samadder AK, Murti K (1977) Palaeomagnetic study of Gunderdehi shale of Raipur Group from Chhattisgarh basin in India. Indian Jour Earth Sci 4: 183–191

Chapter 11
Petrogenesis and Tectonic Setting of the Proterozoic Mafic Magmatic Rocks of the Central Indian Tectonic Zone, Betul Area: Geochemical Constraints

M.K. Mishra, S.J. Devi, T. Kaulina, K.C. Dass, S. Kumar, and T. Ahmad

Introduction

Studies world over have shown that Archean–Proterozoic was period for major crustal generation through mantle melting and its evolution. However, there is controversy regarding the rate of crustal growth through time. Major events of crustal addition is recorded during the Archean (2.7–3.2 Ga), the mid-Proterozoic (1.8–2.2 Ga) and in the late-Proterozoic (1.2–1.5) (Hawkesworth and Kemp, 2006). The highly variable composition of early Proterozoic crust compared to more magnesian rich basic volcanism suggests a drastic change in the growth of the continental crust towards the end of the Archaean period (Naqvi, 1981; Taylor and Mclennan, 1985; Pharaoh et al., 1987). These changes may be related to lowering of the mantle temperature as reflected in the absence of Komatiitic magmatism after late-Archean-early Proterozoic (Bickle, 1990; Ahmad and Rajamani, 1991). It is therefore important to study the magmatic rocks of the Archean and Proterozoic period to understand the crustal and related sub-continental mantle evolution in any Precambrian terrain.

In Central Indian Tectonic Zone, the basement Tirodi gneiss complex is overlain by the Betul supra-crustal rocks (Fig. 11.1). The basement rocks have been intruded by Padhar mafic-ultramafic Complex (PMUM) represented by pyroxenite and gabbroic rocks, north of the Betul town supra-crustal and numerous mafic dykes (originally doleritic now amphibolitic) are exposed all through the terrain. It is not clear whether these mafic-ultramafic rocks of Padhar have any relationship with the

T. Ahmad (✉)
Department of Geology, Centre for Advance Studies, University of Delhi, Delhi 110007, India
e-mail: tahmad001@yahoo.co.in

M.K. Mishra and S.J. Devi
Department of Geology, Centre for Advance Studies, University of Delhi, Delhi 110007, India

T. Kaulina
Geological Institute of the Kola Science Centre, Apatity, Russia

K.C. Dass and S. Kumar
Department of Geology, Centre for Advance Studies, University of Delhi, Delhi 110007, India

R.K. Srivastava (ed.), *Dyke Swarms: Keys for Geodynamic Interpretation*,
DOI 10.1007/978-3-642-12496-9_11, © Springer-Verlag Berlin Heidelberg 2011

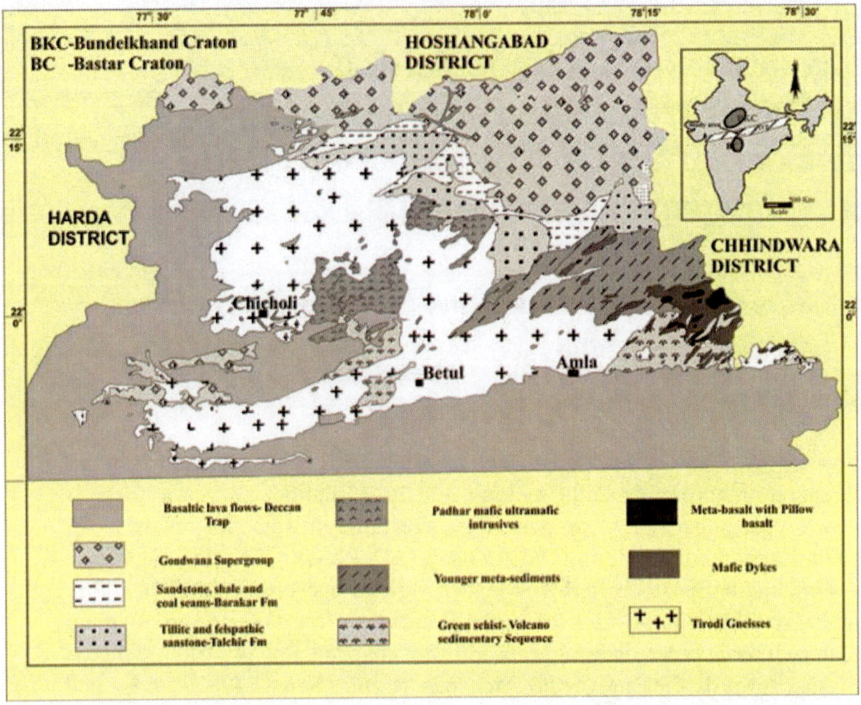

Fig. 11.1 Geological map of Betul area showing location of studied samples (modified after DRM-GSI complied by Joshi et al., and published by GSI Kolkata, 2002, 1st publ. and Roy and Chakraborti, 2008). *Inset* map of India showing outline of Central Indian Tectonic Zone (CITZ)

Fig. 11.2 Well preserved pillow structures in basalt of Betul supracrustal belt on Amla-Bordehi road

dykes and basic volcanic rocks of the Betul supra-crustals. The latter have excellent preservation of pillow lavas (Fig. 11.2).

In this contribution we discuss the genesis of the mafic-ultramafic intrusives and extrusives in relation to the basement gneiss. It will be interesting to see how much the basement rocks have influenced the mafic rocks in their genesis. Here we present comprehensive set of geochemical data on the Betul Supra-crustal Belt – an active continental margin (Roy and Prasad, 2003; Alam et al., 2009), which are important lithotectonic units within the CITZ. The present study aims at putting constraints on – source characteristic, petrogenesis, tectonic setting of their emplacement and lithospheric evolution in this part of the CITZ.

Geological Setting

The Precambrain Crust of India is represented by two crustal blocks namely the Bundelkhand in the north and the Bastar in the south, and the terrain between them is the CITZ (Fig. 11.1, and inset). An independent evolution of the two crustal blocks is suggested (Mondal et al., 2002), which is reflected by the drastic differences in their geological and geophysical characteristics (Qureshy and Iqbaluddin, 1992; Verma and Banerjee, 1992; Jagadeesh and Rai, 2008). The predominantly Proterozoic (<2.4 Ga) CITZ (Radhakrishna and Ramakrishna, 1988; Radhakrishna, 1989; Acharyya and Roy, 2000), is bounded by Son Narmada North Fault (SNNF) in the north and Central Indian Shear (CIS) in the south (Radhakrishna, 1989; Acharyya and Roy, 2000). The eastern and western limits of the CITZ are not well constrained (Ahmad et al., 2009).

The various tectonic activities since the Archaean in the CITZ, led to the growth and assembly of different lithotectonic units within the zone. The back arc/continental rift represented by Mahakoshal belt, the active continental margin represented by the Betul belt and the suture zone which is represented by the Ramakona-Katangi granulite belt with two more important lineaments called the Son-Narmada south Fault (SNSF) and Gavilgarh-Tan Shear Zone (GTSZ) (Yedekar et al., 1990; Jain et al., 1991).

The active continental margin (Roy and Prasad, 2003) represented by the Betul belt is a complex mosaic of number of tectonic blocks joined together. In such an environment addition of juvenile continental crustal material is a major phenomena that modifies the geochemical signature of the rocks of the belt. This E–W trending Betul belt is overlain by Gondwana sediments and Deccan Trap on all three sides in the north, west and south (Fig. 11.1). The Tirodi gneisses form the basement in the study area with occurrence of supra-crustal rocks such as quartzites, metapelites, bimodal volcanics of acid and basic compositions, meta-exhalites, calc silicates and BIF (Roy and Chakraborti, 2008). Large intrusive bodies of mafic-ultramafics rocks mainly represented by gabbro and pyroxenite are exposed in the western and north western part of Betul belt. It covers an area of about 160 km^2 with gabbroic rocks displaying variations from finer-grained micro gabbro to very coarse-grained gabbroic rocks.

Petrography

Pyroxenites of Padhar area are fresh, coarse grained inequigranular, composed of ortho-pyroxene and clino-pyroxene ± Opaque minerals ± olivine. Alteration of pyroxene to hornblende is common in these pyroxenites. The gabbroic rocks show inequigranular, hypidiomorphic texture. Plagioclase and pyroxene represent major minerals constituent while olivine and boitite are present as accessory minerals. Opaque minerals (mainly iron oxides) are present in ground mass and also in the form of small inclusion in both plagioclase and pyroxene as accessory mineral. They show ophitic to sub-ophitic texture. Mafic dykes of Padhar reveal the assemblage that is typical of dolerites and amphibolites i.e. containing pyroxenes ± amphibole ± feldspar ± biotite ± magnetite ± quartz. Feldspars occur both as phenocrysts and groundmass. Augite is the dominant pyroxene which occurs as euhedral to subhedral phenocrysts and also as fine grained groundmass, which are often partially replaced by amphiboles along the margin. Most of the samples have ophitic to sub-ophitic textures. Betul pillow basalt show hypo-crystalline texture. They are composed of very fine grained inequigranular plagioclase + pyroxene +glass ± hornblende. Sphene and opaques are present as accessory mineral.

Geochemical Variation

Major and trace elements analysis of the studied samples were carried out at Department of Geology, Center of advance studies, University of Delhi, Indian Institute of Technology Roorkee (IIT, Roorkee) and Wadia Institute of Himalayan Geology, Dehradun. Major oxides for most of the samples were analyzed by using the XRF at Department of Geology, University of Delhi, while the REEs were analyzed at IIT, Roorkee using ICP-MS. In Wadia Institute of Himalayan Geology, Dehradun we used WD-XRF (Siemens SRS 3,000) for major and trace element for a few basement rocks, following the procedure of (Saini et al., 1998) and ICP-MS (Perkin Elmer, ELAN DRC-e) for REE. Precision and accuracy for ICP-MS technique is given in (Ahmad et al., 2005). Sm-Nd isotopic analysis were carried out at Geological Institute of Kola science centre, Apatity, Russia. The major and trace element including REE are presented in Table 11.1.

PMUM, dykes, pillow basalts have a restricted range of SiO_2 ($44.84 \leq SiO_2 \leq 52.48$ wt%) with an average 49.09 wt%. Most of the PMUM are pyroxenites with SiO2 > 45 wt%, very few samples have SiO_2 values < 45 wt% (*cf.* 3 out of 15 analyses: Roy et al., 2004). A range of variation is observed in MgO and Fe_2O_3 contents The Padhar pyroxenites have higher range of MgO ($13.14 \leq MgO \leq 17.7$ wt%) averaging 15.54 wt% and Fe_2O_3 ($11.57 \leq Fe_2O_3 \leq 12.95$ wt%) averaging 12.74 wt%. In Padhar gabbro MgO exhibit large variation ($5.17 \leq MgO \leq 13.24$ wt%) with an average of 8.15 wt%, Fe_2O_3 ($6.71 \leq Fe_2O_3 \leq 11.33$ wt%) averaging 8.71 wt%.

Table 11.1 Major and trace elements data of mafic magmatic rocks of the Betul area. Basement Tirodi Gneiss data are shown for comparison

	Pillow lava (n = 2)		Betul Dykes (n = 10)		Padhar pyroxinite (n = 10)		Padhar gabbro (n = 19)		Tirodi gneisses (n = 17)	
	a	b	a	b	a	b	a	b	a	b
SiO_2	51.33	0.71	49.63	1.92	50.85	1.09	47.65	1.36	72.61	2.29
TiO_2	0.71	0.07	0.79	0.10	0.54	0.13	0.63	0.27	0.30	0.08
Al_2O_3	13.11	1.75	14.02	0.40	4.28	1.60	17.04	2.19	13.08	1.31
Fe_2O_3	8.59	1.29	11.01	1.07	12.62	0.40	8.71	1.59	3.21	0.97
MnO	0.24	0.12	0.18	0.01	0.21	0.01	0.14	0.02	0.05	0.02
MgO	9.99	2.55	9.10	2.50	15.54	1.48	8.15	1.87	1.01	0.40
CaO	12.75	2.40	11.90	0.93	13.20	0.73	12.44	0.95	1.45	0.66
Na_2O	0.44	0.08	0.36	0.17	0.12	0.07	0.26	0.11	3.39	0.60
K_2O	2.79	1.74	1.73	0.64	0.55	0.24	1.85	0.30	3.87	1.57
P_2O_5	0.17	0.01	0.13	0.04	0.06	0.12	0.11	0.05	0.08	0.09
LOI	0.51	0.08	0.90	0.32	1.97	1.10	3.47	2.25	0.48	0.20
Total	100.61	10.81	99.74	8.10	99.93	6.95	100.44	10.97	99.53	8.19
Mg#	72.82	2.10	65.20	3.86	74.06	1.63	68.17	5.37	37.29	8.34
Trace elements and REE (in ppm)										
Rb	8	1	15	6	31	25	11	6	127	108
Ba	70	29	55	36	65	21	281	122	1431	618
Th	7	1	1	1	1	0.2	0.4	0.4	29	31
U	0.2	0.1	0.4	0.2	1	1	1	0.4	4	3
Nb	5	1	3	1	2	1	3	1	15	6
Sr	362	51	232	45	93	134	596	190	230	237
Zr	95	30	54	9	25	32	66	20	382	96
Y	20	7	17	1	11	3	8	3	47	28
Sc	27	10	35	3	33	7	24	6	6	2
V	131	11	188	18	119	18	121	55	26	11

Table 11.1 (continued)

	Pillow lava (n = 2)		Betul Dykes (n = 10)		Padhar pyroxinite (n = 10)		Padhar gabbro (n = 19)		Tirodi gneisses (n = 17)	
	a	b	a	b	a	b	a	b	a	b
Zn	739	925	90	14	102	38	237	135	60	35
Co	50	21	46	4	50	10	43	5	13	6
Cr	1,097	197	443	150	766	56	319	213	193	97
Ga	15	3	16	1	26	10	17	4	22	5
Ni	271	53	108	30	230	83	130	44	38	29
Cu	61	18	101	32	35	21	238	98	17	9
La	49.86	5.67	21.36	8.68	6.24	2.72	8.31	2.09	51.53	11.68
Ce	90.10	9.78	41.77	15.99	13.84	4.71	15.20	3.47	100.45	21.19
Pr	10.03	1.33	4.98	1.79	1.69	0.45	1.70	0.39	10.90	2.26
Nd	35.40	4.09	18.29	5.84	9.52	2.61	9.59	2.25	43.71	10.97
Sm	6.40	0.61	3.90	1.05	2.75	0.72	2.77	0.66	8.33	2.64
Eu	1.12	0.09	0.73	0.11	0.93	0.23	1.06	0.13	2.28	0.67
Gd	6.03	0.51	4.17	1.06	2.60	0.72	3.02	0.61	8.30	2.36
Tb	0.84	0.07	0.71	0.18	0.41	0.11	0.40	0.12	1.15	0.42
Dy	4.87	0.64	4.68	1.21	2.11	0.60	1.82	0.64	6.39	2.58
Ho	1.05	0.22	1.05	0.31	0.39	0.11	0.32	0.12	1.26	0.55
Er	3.44	1.00	3.66	1.26	1.16	0.29	1.03	0.35	3.23	1.59
Tm	0.56	0.22	0.63	0.25	0.17	0.04	0.14	0.05	0.51	0.25
Yb	3.54	1.69	4.16	1.80	1.05	0.26	0.91	0.32	3.53	1.74
Lu	0.56	0.32	0.69	0.32	0.15	0.03	0.12	0.04	0.57	0.29

a = mean, b = Std deviation, n = Number of samples.

In pillow lavas and dykes MgO variation shows (6.38≤ MgO ≤ 11.79 wt%) with an average of 8.68 wt%, Fe_2O_3 vary from 7.67 to11.14 wt% with an average of 10.30 wt%. A range of variation is also observed in the Al_2O_3 content. Padhar gabbroic rocks are have higher value of Al_2O_3 (13.50≤ Al_2O_3 ≤ 21.38 wt%) with an average of 17.4 wt% compared to Padhar pyroxenites (2.70≤ Al_2O_3≤ 8.28 wt%). Al_2O_3 content in dykes and pillow lavas vary from 11.87 to 14.41 wt%. CaO ranges from 9.90–14.45 wt% with an average of 12.58 wt%. TiO_2 show large variations from 0.34 to 1.27 wt% with an average of 0.65wt%. P_2O_5 ranges between 0.01 and 0.39 wt% with an average of 0.10 wt%. The K_2O abundances varies between K_2O (0.36 ≤ K_2O ≤ 4.02 wt%) with an average 1.55 wt%, on the other hand Na_2O show much lower abundance Na_2O (0.06 ≤ Na_2O ≤ 0.59 wt%) with an average 0.26 wt%. Many of the pillow lavas, dykes and pyroxenites show quite high amount of K_2O from > 1.5 upto 4 wt%, these samples also show high abundance of Ba, opening the possibility of K-metasomatism during their genesis. Padhar pyroxenite, gabbro, Betul dykes and pillow lavas have relatively high Mg# (54.7≤ Mg# ≤ 76.1) averaging 69.1.

In binary plot (Fig. 11.3a–d) Mg# (Mg number) versus major elements for the Betul mafic magmatic rocks (Pyroxenites, gabbro, Betul dykes, and pillow) exhibit overall negative relationship with the Mg# indicating their near primary characteristics. Pyroxenites depict different trend (Fig. 11.3a–c) which probably indicating their different petrogenetic history compared to other mafic rocks from this area.

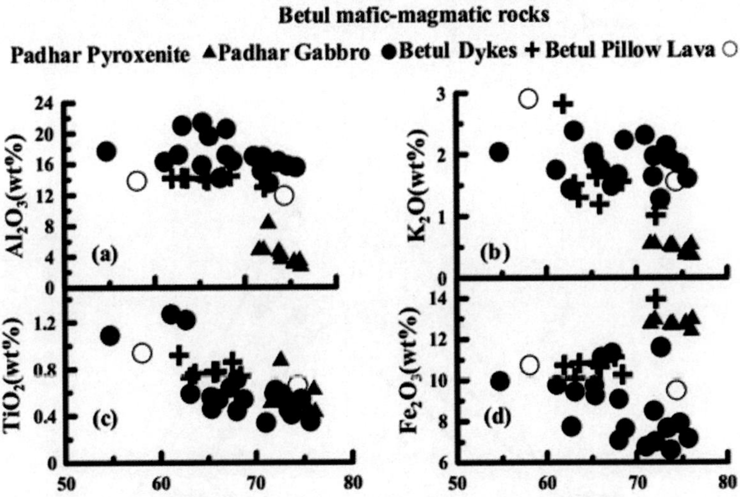

Fig. 11.3 Harker variation diagram (**a**) Mg# vs. Al_2O_3 wt% (**b**) Mg# vs.K_2O wt%, (**c**) Mg# vs. TiO_2 wt% (**d**) Mg# vs. Fe_2O_3wt% are binary plots for mafic magmatic rocks of the Betul supracrustal belt

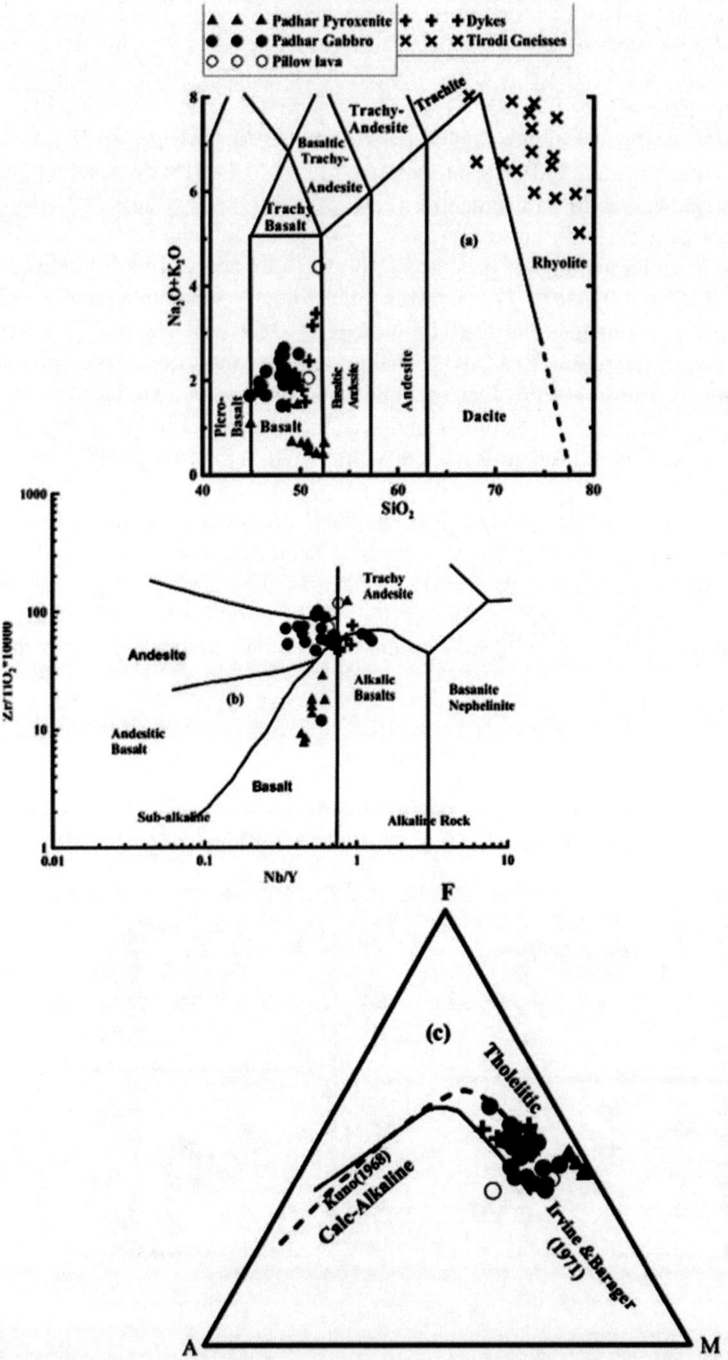

Fig. 11.4 (**a**) SiO_2 vs. $Na_2O + K_2O$ (TAS diagram: Le Bas et al., 1986), (**b**) Nb/Y vs. Zr/TiO_2*10,000 (Winchester and Floyd, 1977) and (**c**) A-F-M plot (trends from Irivine and Barager, 1971, Kuno, 1968)

Magma Types

In the major element classification of $Na_2O + K_2O$ wt% vs. SiO_2 wt% plot (TAS diagram: Le Bas et al., 1986) Padhar pyroxenite and gabbro, Betul mafic dyke and pillow basalt samples fall within category of basalt to basaltic andesite, whereas the basement Tirodi gneissic samples plot dominantly in dacite and rhyolite field (Fig. 11.4a). In trace elements ratio – ratio diagram of Nb/Y versus $Zr/TiO_2{*}10{,}000$ plot (Winchester and Floyd, 1977: Fig. 11.4b) all studied mafic samples plot in sub-alkaline series that range from basalt – andesitic basalt. On AFM diagram (A= $Na_2O + K_2O$, F = FeO^t, M = MgO: Irvine and Barager, 1971) (Fig. 11.4c), while the mafic magmatic rocks show tholeiitic affinity, the basement Betul gneissic rocks depict calc-alkaline trend.

Petrogenesis

Padhar mafic-ultramafic complex (PMUM).

Chondrite normalised REE (Fig. 11.5a, b) and Primitive Mantle normalised spider diagram patterns for Padhar pyroxenite and gabbro complexes [Fig. 11.5a (a–b) and Fig. 11.5b (a–b)] are very similar. There is considerable overlap in the REE patterns. The average ΣLREE for the Padhar gabbro and pyroxenite samples are 11.8

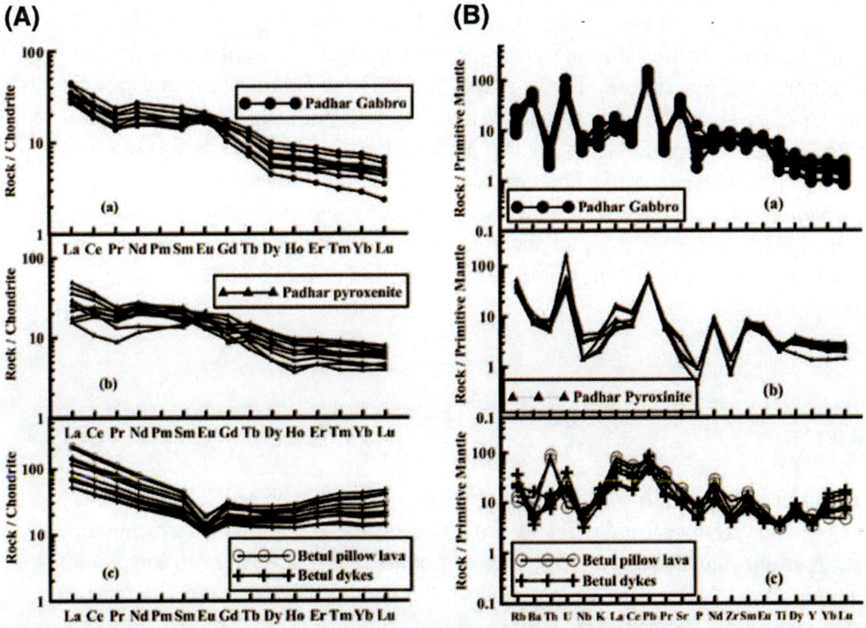

Fig. 11.5 (**A**) (a–c) C1 normalized rare earth element (REE) patterns and (**B**) (a–c) Primitive mantle normalized multi-element patterns for the mafic magmatic rocks of the Betul supra-crustal belt. Normalizing values from Sun and Mcdonough (1989)

ppm and 8.99 ppm and ΣHREE = 3.13 ppm and 2.74 ppm respectively. Padhar gabbro are more depleted in HREE than the pyroxenite as depicted by average Σ $(La/Yb)_N$ ratio of 6.53 and 4.27 for Padhar gabbro and pyroxenite respectively. There is not much difference in their LREE enrichment as Σ $(La/Sm)_N$ for gabbro is 1.93 and 1.46 for Pyroxenite. Enriched LREE characteristics probably indicate that they are derived from enriched mantle source which is further confirmed in spider diagram showing enrichment of LILE and depletion of HFSE. The variation in the trace elements abundances and fractionated LREE pattern of PMUM also indicate that all the samples of Padhar mafic-ultramafic complex were generated from similar sources but by varying degrees of partial melting. Primitive Mantle normalised spider diagram of Padhar mafic-ultramafic rocks show negative anomalies for Nb, Ti and P, while Ba, U, Pb and Sr show positive anomalies. Padhar sample shows negative Nb, smaller negative P and negative Ti anomalies, these features are commonly observed in the continental flood basalts as well as in arc-related rocks (Tarney et al., 1981; Weaver and Tarney, 1983; Thompson et al., 1983). Continental crust has a distinct negative Nb anomaly (Weaver and Tarney, 1985). The observed strong negative anomalies for P, Ti and positive anomalies for U, Pb in primitive mantle normalized spider diagram also suggests role of crustal contamination in the genesis of these mafic magmatic rocks.

Dykes and Pillow lava

Chondrite normalised REE [Fig. 11.5a (c)] and Primitive mantle normalised multi – element pattern [Fig. 11.5b (c)] of Betul dykes and pillow lavas show similarity in their pattern, both having highly enriched LILE, depleted HFSE and HREE. Which are again supporting their derivation from enriched source like those for the Padhar gabbro and pyroxenite. The average ΣLREE for Betul dykes and pillow lavas are 25.25 ppm and 56.26 ppm and ΣHREE = 4.85 ppm and 6.58 ppm respectively. For these average $\Sigma(La/Sm)_N$ are 3.53 and 5.02 and average $\Sigma(La/Yb)_N$ are 3.68 and 10.10 respectively. The characteristic feature of these rocks are negative europium (–Eu) anomaly, which probably indicate fractionation of plagioclases. Spider diagram of these rocks show negative anomaly of Nb, P, Ti, Zr, Y while Th, U, Pb show positive anomaly suggesting contamination of these mafic rocks by the crustal material.

Discussion

We have used ratio – ratio plot of incompatible element such as Ce/Yb vs Zr/Y (Fig. 11.6a) and Y/Nb vs Zr/Y (Fig. 11.6b) to access the possibility of crustal contamination of our mafic rocks. In Fig. 11.6 symbol # 1and 2 are Archean lower and middle crust ratio (Weaver and Tarney, 1984), # 3 is a mixture of 1 and 2 in a ratio of 3:1. Figure 11.6a show Padhar gabbro, dykes, one samples of pillow lava and average of Tirodi gneisses follow the probable crust mantle mixing line indicating that the parental melts of the Padhar gabbro and Betul dykes were contaminated by the crustal components such as # 3 and possibly by Tirodi gneiss (Fig. 11.6a).

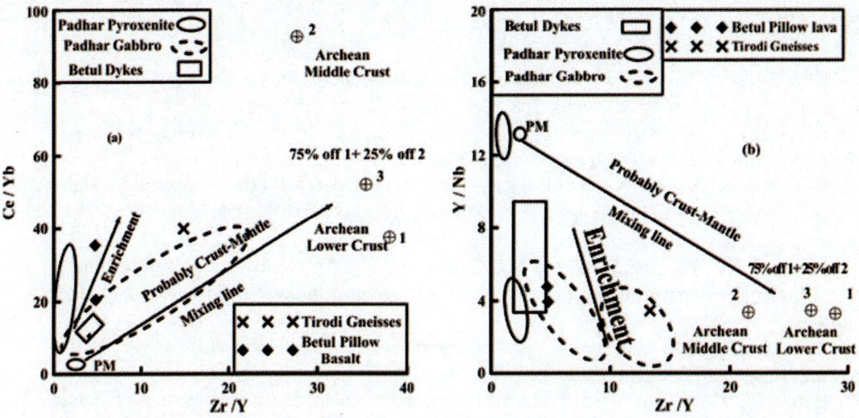

Fig. 11.6 Binary ratio-ratio plot of (**a**) Zr/Y vs. Ce/Yb, (**b**) Zr/Y vs.Y/Nb outlines for the mafic magmatic rocks of the Betul supra-crustal belt. # 1 and # 2 are Archean lower and middle crust ratios (Weaver and Tarney, 1984), # 3 is a mixture of 1 and 2 in the ratio of 3:1

The pyroxenite on the other hand follows the enrichment trend, probably indicating enrichment of the sources by mantle metasomatism (Ahmad and Tarney, 1993). In contrast Y/Nb vs Zr/Y plot (Fig. 11.6b) indicate enrichment of sources for the dykes and pyroxenites, discounting much possibility of crustal contamination in their genesis; however, the gabbros appear to have interacted by the Tirodi gneisses locally as indicated by Figs. 11.6a, b.

Limited Sm-Nd isotopic data for the studies samples indicate that the protolith of the PMUM were probably derived at ca. 2,770 Ma, much before that of the protoliths for the basic volcanics at about 1,951, 2,054 and 2,320 Ma (Alam et al., 2009). Sm-Nd data on the basement Tirodi Gneisses indicate extraction of their protoliths from about 2,960–2,413 Ma (unpublished data). Thus the basement gneiss and the PMUM rocks appear to have acted together as basement for the younger mafic magmatic dykes and flows. These basement rocks appear to have influenced the evolution of the younger volcanics and dykes through limited contamination.

The negative Nb anomaly is very commonly observed in many Phanerozoic continental rift volcanics (Weaver and Tarney, 1983., Thompson et al., 1983; Ahmad and Tarney, 1994; Ahmad and Rajamani, 1991; Ahmad and Bhat, 1987). The observed geochemical characteristics, associated continentally derived sediments and typical granite-gneiss basement probably indicate a rift tectonic environment for our study area.

Acknowledgment We thank DST, New Delhi for the grant of funds under ILTP (TA and TK) and SERC projects to TA. We also thank the University of Delhi for all the facilities and partial funding under the Minor Project given to TA. We thank colleagues of WIHG geochemistry laboratory for the analytical support. Nishchal, Kabita, Meraj, Naushad, Kumar Batuk and Deepak Joshi are especially thanked for the help during the different field works. T.C. Devaraju is thanked for the constructive comments. Rajesh Srivastava is thanked for his editorial comments and for inviting us to contribute this paper to this special volume.

References

Acharyya SK, Roy A (2000) Tectonothermal history of the Central Indian Tectonic Zone and reactivation of major faults/shear zones. Jour Geol Soc India 55: 239–256

Ahmad T, Bhat MI (1987) Geochemistry and petrogenesis of the Mandi–Darla volcanics, Northwestern Himalayas. Precamb Res 37: 231–256

Ahmad T, Harris NBW, Islam R, Khanna PP, Sachan HK, Mukherjee BK (2005) Contrasting mafic magmatism in the Shyok and Indus suture zone: Geochemical constraints. Himalayan Geol 26: 33–40

Ahmad T, Longjam KC, Fouzdar B, Bickle MJ, Chapman HJ (2009) Petrogenesis and tectonic setting of bimodal volcanism in the Sakoli mobile belt, Central Indian Shield. Island Arc 18: 155–174

Ahmad T, Rajamani V (1991) Geochemistry and petrogenesis of the basal Aravalli volcanics near Nathdwara, Rajasthan, India. Precamb Res 49: 185–204

Ahmad T, Tarney J (1993) North Indian Proterozoic volcanics, products of lithosphere extension: Geochemical studies bearing on the lithospheric derivation rather than crustal contamination. In: Casshyap SM, (ed) Rifted basins and aulacogens, geological and geophysical approach. Gyanodaya Prakashan Nainital, India: 130–147

Ahmad T, Tarney J (1994) Geochemistry and Petrogenesis of late Archean Aravalli volcanics, basement enclaves and granitoids, Rajasthan. Precamb Res 65: 1–23

Alam M, Naushad M, Wanjari N, Ahmad T (2009) Geochemical characterizations of mafic magmatic rocks of the Central Indian shield: Implication for Precambrain crustal evolution. In: Ahmad T, Hirsch F, Charusiri P, (eds) Journal of the Virtual Explorer: 32, paper 8, doi:10.3809/jvirtex.2009.00246

Bickle MJ (1990) Mantle evolution. In: Hall RP, Hughes DJ, (eds) Early Precambraon basic Magmatism. Blackie, Glasgow (and Chapman and Hall, New York): 111–135

Hawkesworth CJ, Kemp AIS (2006) Using hafnium and oxygen isotopes in zircons to unravel the record of crustal evolution. Chem Geol 226: 144–162

Irvine TN, Barager WR (1971) A guide to chemical classification of common volcanic rocks. Can Jour Earth Sci 8: 523–548

Jagadeesh S, Rai SS (2008) Thickness, composition, and evolution of the Indian Precambrain crust inferred from broadband seismological measurements. Precamb Res 162: 4–15

Jain SC, Yedekar DB, Nair KKK (1991) Central Indian Shear zone: A major Precambrian crustal boundary. Jour Geol Soc India 37: 521–548

Kuno H (1968) Differentiation of basaltic magmas. In: Hess HH, Poldervaart AA (eds) Basalts: The Poldervaart Treatise on Rocks of Basaltic Composition, 2. New York, Interscience: 623–688

Le Bas MJ, Le Maitre RW, Streckeiisen A, Zanettin B (1986) A chemical classification of volcanic rocks based on the total alkali-silica diagram. Jour Petrol 27: 745–750

Mondal MEA, Goswami JN, Deomurari MP, Sharma KK (2002) Ion microprobe 207Pb/206Pb ages of zircon from the Bundelkhand massif, northern India: Implication for crustal evolution of the Bundelkhand-Aravalli protocontinent. Precamb Res 117: 85–100

Naqvi SM (1981) The oldest supracrustal of the Dharwar Craton, India. Jour Geol Soc India 22: 458–469

Pharaoh TC, Beckinsale RD, Rickard D, (eds) (1987) Geochemistry and mineralization of Proterozoic volcanic suites. Geol Soc London Spec Publ no 33: 552

Qureshy MN, Iqbaluddin (1992) A review of the geophysical constraints in modeling the Gondwana crust in India. Tectonophysics 212: 141–151

Radhakrishna BP (1989) Suspect Tectono-Stratigraphic Terrain elements in the Indian Subcontinent. Jour Geol Soc India 34: 1–24

Radhakrishna BP, Ramakrishna M (1988) Archaean-Proterozoic boundary in India. Jour Geol Soc India 32: 263–278

Roy A, Chakraborti K (2008) Precambrain Mafic-Ultramafic Magmatism in Central Indian suture zone. Jour Geol Soc India 72: 123–140

Roy A, Chore SA, Viswakarma LL, Chakraborty K (2004) Geology and petrochemistry of Padhar mafic-ultramafic complex from Betul Belt: A study on arc type magmatism in Central India Tectonic zone. Geol Surv India Spec Publ no 84: 297–318

Roy A, Prasad HM (2003) Tectonothermal events in Central Indian Tectonic Zone (CITZ) and its implications in Rodinian crustal assembly. Jour Asian Earth Sci 22: 115–129

Saini NK, Mukherjee PK, Rathi MS, Khanna PP, Purohit KK (1998) A new geochemical reference sample of Granite (DG- H) from Dalhousie, Himachal Himalaya. Jour Geol Soc India 52: 603–606

Sun SS, McDonough WF (1989) In: Saunders AD, Norry MJ (eds) Chemical and isotopic systematics of oceanic basalts: Implication for mantle composition and processes Magmatism in Ocean Basin. Geol Soc Spec Publ 42: 313–345

Tarney J, Saunders AD, Mattey DP, Wood DA, Marsh NG (1981) Geochemical aspects of back arc spreading in the Scotia sea and western Pacific. Phillos Trans R Soc London A 297: 79–202

Taylor SR, Mclennan SM (1985) The continental crust: Its composition and evolution. Blackwell Scientific Publications, Oxford

Thompson RN, Morrison MA, Dickin AP, Hendry GL (1983) Continental flood basalts: Arachmids rule OK? In: Hawkerwarth CJ, Narry MJ, (eds) Continental basalt and mental xenoliths. Shiva publications, Nantwich: 155–185

Verma RK, Banerjee P (1992) Nature of continental crust along the Narmada-Son lineament inferred from gravity and deep seismic sounding data. Tectonophysics 202: 357–397

Weaver BL, Tarney J (1983) Chemistry of the sub continental mantle inferences from Archean and Proterozoic dykes and continental flood basalts. In: Hawkerwarth CJ, Narry MJ, (eds) Continental basalt and mental xenoliths. Shiva publications, Nantwich: 209–229

Weaver BL, Tarney J (1984) Empirical approach to estimating the composition of the continental crust. Nature 310: 575–577

Weaver BL, Tarney J (1985) Major and trace element composition of the continental lithosphere. In: Pollack HN, Rama Murthy Z, (eds) Structure and evolution of the continental lithosphere. Pergamon, Oxford: 39–68

Winchester JA, Floyd PA (1977) Geochemical discrimination of different magma series and their differentiation products, using immobile elements. Chem Geol 20: 325–344

Yedekar DB, Jain SC, Nair KKK, Dutta KK (1990) The Central Indian collision suture. Jour Geol Surv India Spec Publ no Precambrian of Central India 28: 1–37

Chapter 12
Petrology, Geochemistry and Petrogenesis of Early Precambrian Mafic Dyke Swarm from Dondi-Bhanupratappur-Keshkal Area, Central Bastar Craton, India

Gulab C. Gautam and Rajesh K. Srivastava

Introduction

Large dyke swarms emplaced within Precambrian and Phanerozoic cratons help to recognize Large Igneous Provinces (LIPs) and are interpreted as evidence for continental break-up and the remnants of mantle-plume related magmatism (Coffin and Eldholm, 1994, 2005; Ernst and Buchan, 1997, 2003; Ernst et al., 2007; Bryan and Ernst, 2008). Many giant dyke swarms are recognized in the Archaean cratons of the Indian Shield (Murthy, 1987, 1995; Halls et al., 2007; French et al., 2008; Ernst and Srivastava, 2008; Srivastava et al., 2008; French and Heaman, 2010). Bastar craton has also experienced many episodes of mafic magmatism, particularly mafic dykes, during Precambrian (Crookshank, 1963; Ramakrishnan, 1990; Ramachandra et al., 1995; Subba Rao et al., 2004; Srivastava, 2006a, b). Different Precambrian mafic dyke swarms exposed in the southern Bastar craton have distinct petrological and geochemical characteristics. These include Meso-Neoarchaean (~3.0–2.7 Ga) sub-alkaline mafic dykes (BD1), Neoarchean-Paleoproterozoic (~2.5–2.4 Ga) boninite-norite dykes (BN), and Paleoproterozoic (~1.9 Ga) sub-alkaline mafic dykes (BD2) (Srivastava and Gautam, 2008, 2009). Mafic dyke swarms of similar characteristics are also encountered from the northern Bastar craton (Subba Rao et al., 2004; Chalapathi Rao and Srivastava, 2009). On the basis of precise U-Pb geochronology French et al. (2008) have recognized remnant of a Paleoproterozoic Large Igneous Province – comprising the southern Bastar craton and Cuddapah basin from the adjacent Dharwar craton.

Numerous mafic dykes are also well exposed in the central part of Bastar craton but very limited petrological and geochemical information is available (Ramachandra et al., 1995). The present paper is aimed to provide new petrological and geochemical data (major oxides and few trace elements) on mafic dykes exposed

R.K. Srivastava (✉)
Department of Geology, Centre of Advanced Study, Banaras Hindu University, Varanasi 221 005, India
e-mail: rajeshgeolbhu@yahoo.com

G.C. Gautam
Department of Geology, Centre of Advanced Study, Banaras Hindu University, Varanasi 221 005, India

R.K. Srivastava (ed.), *Dyke Swarms: Keys for Geodynamic Interpretation*,
DOI 10.1007/978-3-642-12496-9_12,

in and around Dondi, Bhanupratappur and Keshkal areas, which fall in central part of Bastar craton. This is an attempt to fill in the gaps for better understanding of Precambrian mafic magmatism in the Bastar craton.

Geological Setting

The rectangular shape Bastar Craton is bounded by three prominent rifts on three sides viz. NW–SE trending Godavari rift, NW–SE trending Mahanadi rift, and ENE–WSW trending Narmada-Son rift and fourth side by a NE–SW trending Eastern Ghats Mobile Belt (Fig. 12.1; after Naqvi and Rogers, 1987). It comprises a vast tract of granitoids with inliers of supracrustal rocks of the Dongargarh, Sakoli, Sukma, Bengpal, and Bailadila Series and these are overlain by many unmetamorphosed Proterozoic sedimentary basins (Crookshank, 1963; Ramakrishnan, 1990; Ramachndra et al., 2001; Chaudhuri et al., 2002). The supracrustal rocks include different types of metamorphic rocks, a variety of mafic rocks and unmetamorphosed late Proterozoic sedimentary rocks. These sedimentary rocks are deposited in intracratonic basins which are developed in an intracratonic rift setting (Chaudhuri et al., 2002; Patranabis-Deb et al., 2007).

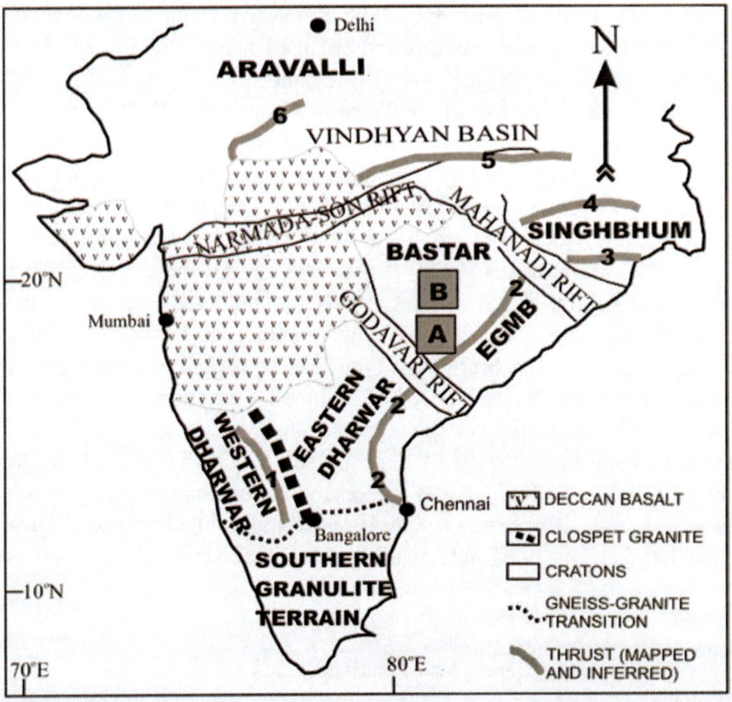

Fig. 12.1 Major cratons and structural features of India (after Naqvi and Rogers, 1987). Major structural features are: 1. Small thrusts in western Dharwar craton; 2. Eastern Ghats front; 3. Sukinda; 4. Singhbhum; 5. Son Valley; and 6. Great Boundary fault. EGMB: Eastern Ghat Mobile Belt. (**A**): Southern Bastar; (**B**): Study area

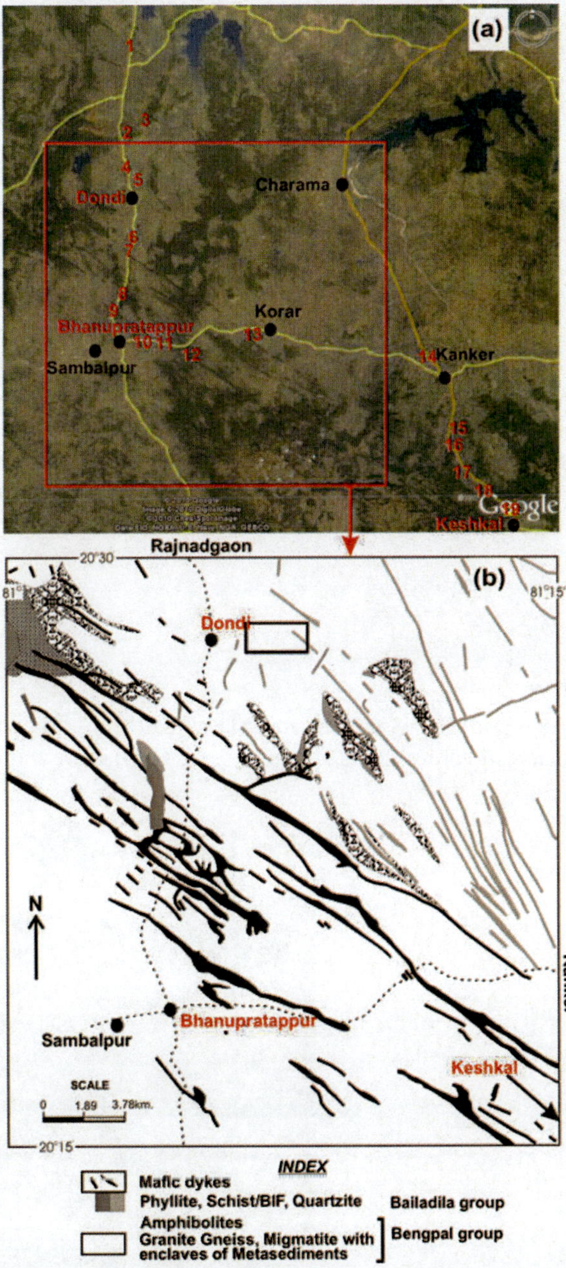

Fig. 12.2 (**a**) Google Earth map of the study area. Numbers indicate location of geochemically analyzed samples. (**b**) Simplified map of the mafic dyke swarm of a part of the Dondi-Bhanupratappur-Keshkal area, central Bastar craton (modified after Ramachandra et al., 1995)

Ramachandra et al. (1995) described the Bhanupratappur-Keskal mafic dyke swarm. Numerous long mafic dykes are well exposed in the study area (see Google Map; Fig. 12.2a). Similar to mafic dykes of southern Bastar craton (Srivastava and Gautam, 2008), mafic dykes of present study, which are exposed around Dondi, Bhanupratappur and Keshkal, also intrude Bengpal and Baildila Group of rocks (Fig. 12.2b). There are many geological evidences available to support emplacement of this mafic dyke swarm in an intracratonic setting (Crookshank, 1963; Ramakrishnan, 1990; Ramachandra et al., 1995). These mafic dykes trend in NW–SE. They vary in length from a few tens of meters to 35 km, whereas width varies from less than 50–200 m (Fig. 12.3). A sharp contact is noticed between dyke and country rock at many places (Fig. 12.3a). Most of the mafic dykes show amphibolite grade of metamorphism but many mafic dykes, particularly exposed around Keshkal, have preserved original igneous texture and mineralogy. Many metadolerites dykes are coarse-grained (Fig. 12.3b). No radiometric age data is available for these dykes and therefore difficult to deduce any age of emplacement of these dykes. But on the basis of field relationship observed by Ramachandra et al. (1995) and during the present study, it is suggested that probably these dykes are emplaced during Paleoproterozoic.

Petrography

Thin section study of mafic dykes of present area classified them into two types: (i) metabasites and (ii) metadolerites. Metabasites (Fig. 12.4a, b) show amphibolite

Fig. 12.3 Field photographs. (**a**) A metabasite dyke exposed near Bhanupratappur showing sharp contact with granite gneiss. (**b**) Coarse-grained metadolerite dyke encountered near Keshkal

Fig. 12.4 Photomicrographs of the studied mafic dykes. (**a**) Medium-grained metabasite showing granoblastic texture. (**b**) Coarse-grained metabasite. Two sets of oblique cleavages in hornblende and cloudy nature of feldspars are clearly visible. (**c**) Coarse-grained metadolerite. Fresh clinopyroxenes and cloudy feldspars are seen. (**d**) Metadolerite with ophitic texture. Width of each photograph is equivalent to 4 mm

facies metamorphic textures and mineral assemblages. They are medium- to coarse-grained with granoblastic textures and contain hornblende, plagioclase, chlorite, quartz, epidote, apatite, sphene, and iron oxides. Two sets oblique cleavages are very well visible in many thin sections (Fig. 12.4b). Metadolerites (Fig. 12.4c, d) also show metamorphic effect but they still have igneous textures. Many samples have fresh clinopyroxene grains (Fig. 12.4c). These samples show typical ophitic or sub-ophitic textures and contain plagioclase, clinopyroxenes, hornblende, apatite, and iron-oxides. Labradorite and oligoclase are common plagioclase present in dolerites. In some samples all pyroxenes are altered to hornblende, yet they show ophitic texture (Fig. 12.3d). Samples collected from Keshkal are mostly metadolerites. Cloudy and sericitised plagioclase feldspar is common in both types. An important observation is that these dykes show uniform petrographic characteristics all along the strike. Ramachandra et al. (1995) also reported ultramafic, transitional gabbro-norite, gabbroic anorthosite, gabbro and quartz gabbro. But in the present study no such rocks are noticed. Probably these types are identified on the basis of mineral compositions observed at places in the thin sections.

Geochemistry

During the field work only fresh and least altered samples were collected for the laboratory work. Most samples were collected from the middle of the dykes to avoid any contamination. Nineteen samples were selected for their whole rock major and few trace elements analyses (Table 12.1). These samples were analysed at Activation Laboratories, Ancaster, Ontario, Canada. ICP-OES (Model: Thermo-Jarret Ash ENVIRO II) was used to analyze major elements, whereas ICP-MS (Model: Perkin Elmer Sciex ELAN 6000) was used to determine trace element concentrations. The precision is approximately 5 and 5–10% for the major oxides and trace elements respectively, when reported at 100x detection limit. Several international geochemical reference material samples (STM1, MAG1, BIR1, DNC1, SY3, GBW, W2, etc.) were run along with these samples in order to assess the accuracy of data.

Geochemically studied samples show uniform compositions; a limited variation in major and trace element composition is observed. SINCLAS computer program (Verma et al., 2002) is used for CIPW norm calculations. Most samples are quartz normative and few show olivine normative. This suggests their compositional variation from quartz thoeliite to olivine tholeiite. For further classification total alkali and silica (TAS) diagram is used. On the TAS classification (Fig. 12.5a; Le Maitre, 2002) studied rocks are classified as basalt or basaltic andesite. Sub-alkaline tholeiitic nature of these samples is also observed on this plot (Fig. 12.5a; Irvine and Baragar, 1971). SINCLAS computer programme (Verma et al., 2002) also confirms sub-alkaline basalt and basaltic andesite nature of the samples of present study. One of the samples is classified as boninite. This is because this sample has high silica, high magnesium and low titanium composition and shows primitive composition (Mg#$\sim$70). Mg number (Mg#) of other samples varies between 43.89 and 74.21. Jensen's cation classification scheme (Fig. 12.5b; Jensen, 1976) is also used for the studied rocks; they show variation from high-iron tholeiite to high-magnesium tholeiite compositions. Few samples show basaltic komatiitic composition. These observations suggest that mafic dyke samples of central Bastar craton are sub-alkaline tholeiitic in nature and have variation from quartz tholeiite to olivine tholeiite with close affinity to high magnesium melt.

Harker-type variation plots (Figs. 12.6 and 12.7) are prepared to observe crystallization behavior of studied samples. Most of the studied samples are affected by post-magmatic processes like hydrothermal alteration and metamorphism, therefore it is expected that a few elements may be mobile during these secondary processes. Elements thought to be mobile during the post-magmatic secondary processes like hydrothermal alteration and metamorphism are Sr, K (alkalis), Rb, Ba, etc. (Pearce and Cann, 1973; Seewald and Seyfried, 1990; Verma, 1992; Condie and Sinha, 1996). These elements belong to large-ion lithophile element (LILE) group. On the other hand, high-field strength elements (HFSE) such as Ti, Zr, Y, Nb, P etc. are thought to be relatively immobile under hydrothermal alteration and mid-amphibolite facies metamorphic conditions (Pearce and Cann, 1973; Winchester

Table 12.1 Whole rock major oxides (wt%), trace elements (ppm) and CIPW Norms of the mafic dykes from Dondi, Bhanupratappur and Keshkal areas, Bastar craton

	1	2	3	4	5	6	7	8	9	10	11	12	13	14	15	16	17	18	19
Major oxides (wt%)																			
SiO_2	49.30	49.33	49.65	49.30	49.66	49.95	52.02	49.25	49.49	50.48	50.06	51.50	47.55	48.80	49.67	51.85	51.78	51.32	51.57
TiO_2	0.66	0.99	0.33	0.33	0.98	1.08	0.58	1.02	0.86	0.49	1.10	0.52	1.59	0.93	0.62	0.43	0.61	0.57	0.56
Al_2O_3	16.52	14.99	15.45	15.15	14.03	13.91	13.17	14.38	15.26	14.35	13.51	12.59	13.10	14.17	14.81	13.49	11.48	11.23	11.38
Fe_2O_3	10.57	13.48	13.04	10.12	14.11	15.21	11.50	15.96	13.01	11.25	14.71	11.34	15.70	14.96	11.81	9.91	13.21	13.28	10.56
MnO	0.15	0.18	0.18	0.16	0.20	0.21	0.17	0.18	0.18	0.17	0.19	0.17	0.21	0.20	0.18	0.16	0.17	0.18	0.15
MgO	6.97	6.64	6.63	8.45	6.32	5.88	7.40	5.34	5.74	8.01	5.51	8.79	7.00	6.64	7.10	8.99	9.58	10.51	12.08
CaO	12.60	10.62	10.82	10.64	10.67	10.44	9.66	9.50	10.45	11.40	9.73	9.98	10.32	9.60	11.10	10.93	8.74	8.42	7.44
Na_2O	2.14	2.15	2.16	2.50	1.97	2.21	2.15	2.21	2.24	1.72	2.09	1.57	1.85	2.36	2.23	1.63	1.69	1.59	2.05
K_2O	0.36	0.44	0.24	0.66	0.33	0.39	0.97	0.94	0.52	0.44	0.74	0.81	0.73	1.02	0.44	0.40	1.08	1.21	1.07
P_2O_5	0.06	0.12	0.11	0.08	0.10	0.12	0.08	0.16	0.11	0.07	0.16	0.08	0.17	0.13	0.08	0.07	0.11	0.12	0.12
LOI	1.4	1.12	1.03	2.34	0.93	1.21	2.26	1.07	1.93	1.20	1.47	1.95	1.49	1.05	0.91	1.13	0.99	1.12	2.13
Total	100.70	100.10	100.30	99.73	99.29	100.60	99.95	100.01	99.79	99.59	99.27	99.30	99.72	99.88	98.94	99	99.44	99.57	99.13
Mg#	60.65	53.52	54.30	66.13	51.15	47.47	61.82	43.89	50.77	62.47	46.68	66.10	51.03	50.92	58.43	69.53	64.60	66.57	74.21
Trace elements (ppm)																			
Cr	420	200	200	280	90	110	250	90	80	130	80	450	160	150	90	340	560	700	1,010
Ni	100	130	120	170	90	90	130	110	90	150	90	130	110	110	110	150	160	200	390
Rb	10	13	7	27	28	17	62	49	36	34	40	46	55	75	19	16	73	78	44
Ba	73	91	82	143	77	129	226	325	142	108	235	186	135	172	137	121	274	327	385
Sr	127	115	110	158	99	134	131	173	144	112	156	121	238	165	129	116	174	187	262
Nb	2	4	4	3	3	4	3	4	4	4	5	3	12	4	2	2	3	4	5
Zr	38	59	60	44	48	74	55	75	65	40	87	48	95	65	39	39	56	56	78
Y	14	21	22	10	22	24	16	26	22	12	28	14	21	23	15	12	14	11	13

Table 12.1 (continued)

	1	2	3	4	5	6	7	8	9	10	11	12	13	14	15	16	17	18	19
CIPW norm																			
Q	–	0.34	1.04	–	2.53	1.86	4.00	0.52	1.38	1.97	3.45	4.81	–	–	–	4.97	3.52	2.14	0.62
Or	2.16	2.66	1.44	4.04	2.01	2.35	5.92	5.69	3.17	2.67	4.53	4.96	4.45	6.18	2.68	2.44	6.55	7.34	6.58
Ab	18.40	18.60	18.62	21.92	17.15	19.06	18.79	19.16	19.59	14.94	18.32	13.78	16.15	20.48	19.45	14.21	14.68	13.82	18.04
An	34.95	30.63	32.35	29.18	29.29	27.40	24.20	27.20	31.06	30.93	26.20	25.83	26.10	25.69	29.99	29.16	21.09	20.47	19.43
Di	23.05	18.54	17.83	20.32	20.32	20.49	20.23	16.77	18.07	21.79	18.94	20.28	21.20	18.47	21.54	21.21	18.45	17.56	14.52
Hy	13.44	23.97	23.61	11.72	23.33	23.02	21.87	24.68	21.81	24.02	22.65	25.52	22.65	16.76	21.76	23.86	30.06	33.08	36.00
Ol	4.20	–	–	9.67	–	–	–	–	–	–	–	–	2.35	6.91	0.48	–	–	–	–
Mt	2.38	3.05	2.94	2.32	3.21	3.43	3.66	3.62	2.97	2.55	3.37	3.62	3.58	3.39	2.69	3.15	4.18	4.20	3.38
Il	1.28	1.93	1.922	0.65	1.92	2.10	1.13	1.99	1.69	0.96	2.17	1.02	3.11	1.82	1.22	0.85	1.20	1.11	1.11
Ap	0.14	0.29	0.26	0.19	0.24	0.28	0.19	0.38	0.26	0.17	0.39	0.19	0.41	0.31	0.19	0.17	0.26	0.29	0.31
TAS Class[n]	SAB	SAB	SAB	SAB	SAB	SAB	BA	SAB	SAB	SAB	SAB	BA	SAB	SAB	SAB	BON	SAB	SAB	SAB

SAB: sub-alkaline basalt; BA: basaltic andesite; BON: boninite; 1 to 5, 10 and 14 are metabasites and rest are metadolerites.

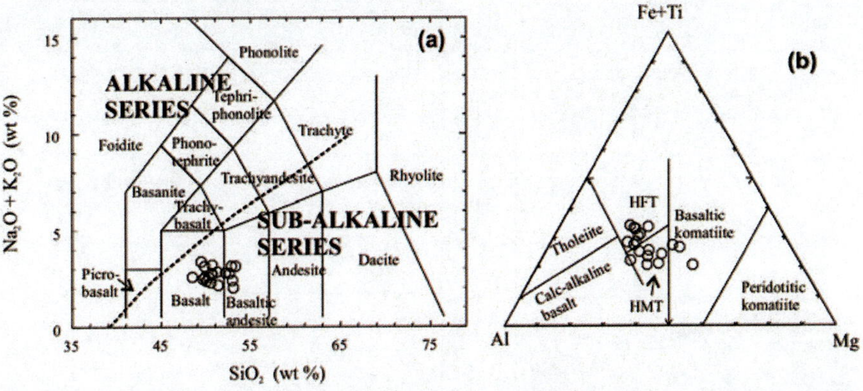

Fig. 12.5 (**a**) Total-alkali silica (TAS) classification diagram (after Le Maitre, 2002; *dotted line* divides sub-alkaline basalts from alkaline basalts after Irvine and Baragar, 1971). (**b**) Jensen's cation plot (Jensen, 1976). HFT: high-iron tholeiite; HMT: high-magnesium tholeiite

and Floyd, 1976; Floyd and Winchester, 1978; Rollinson, 1993; Jochum and Verma, 1996).

Figure 12.6 presents variations of SiO_2, Cr and few LILE, plotted against MgO composition. It is difficult to observe any crystallization trend in most of the plots. But, in general, Cr, Ba and Rb show positive correlation with MgO. This suggests clinopyroxene and olivine fractionation during the crystallization. No such trend is observed in other plots. In contrast, HFSE shows good crystallization trends (Fig. 12.7); all plotted oxides or trace elements increase with differentiation (i.e. decreasing MgO). Few samples, those mostly belonging to metadolerite group plot slightly different from the other samples. Due to limited chemical data it is difficult to explain the reason for this. There is a possibility that metadolerite samples probably have different genetic history than metabasite samples. This can be observed from MgO versus SiO_2 (Fig. 12.6a) and MgO versus Alkalis (Fig. 12.6b) plots. Two different trends can be inferred from these plots but to corroborate this conclusion further geochemical data, particularly rare-earth elements and isotopic composition, are required.

Discussion

Petrographically mafic dykes exposed in and around Dondi, Bhanupratappur and Keshkal areas are classified as metabasites (metamorphosed under mid amphibolites facies P-T conditions) and metadolerites (altered but still have igneous texture and mineralogy). Metabasites are mainly exposed around Dondi and Bhanupratappur, whereas metadolerites are encountered around Keshkal. It is a possibility that intensity of metamorphism gradually decreases from NW (Dondi) to SE (Keshkal). Geochemically these rocks are sub-alkaline tholeiites (quartz to olivine). Zr/Y ratios

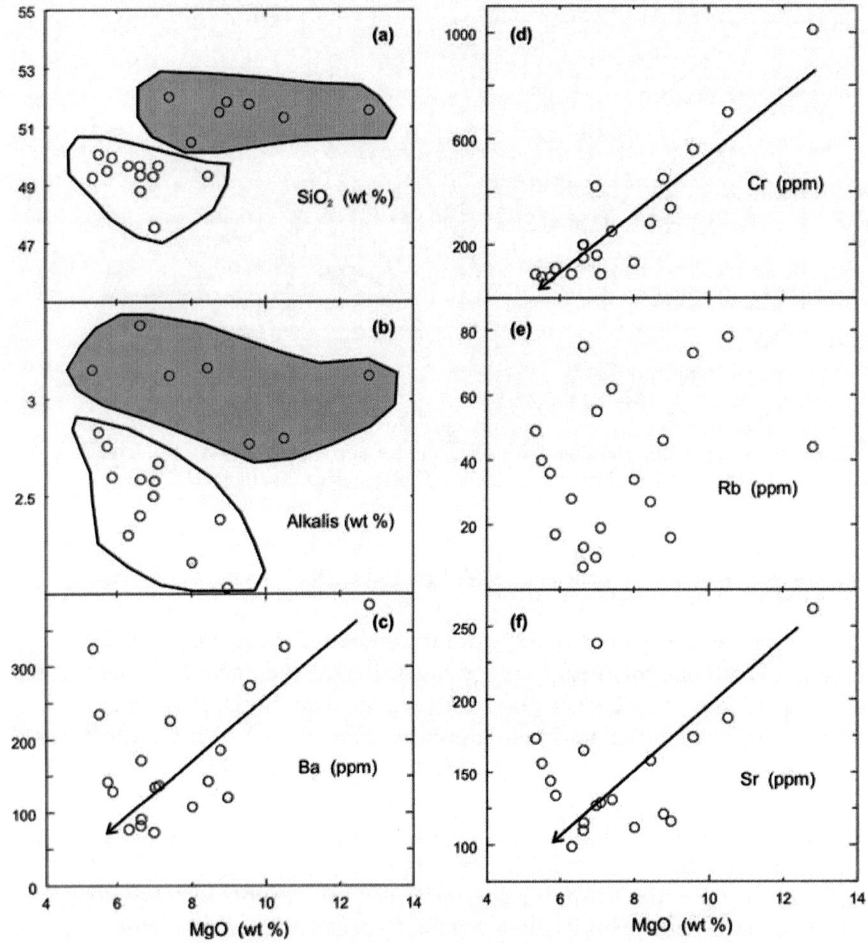

Fig. 12.6 Harker-type variation diagram plotted between MgO versus and SiO_2, Alkalis, Cr, LILE. *Dark shaded area* includes metadolerite samples whereas metabasites are in *un-shaded area*

(mostly below 4.5) also classify these mafic dyke samples as tholeiites and clearly discriminate from calc-alkaline nature (Barrett and MacLean, 1994). Two samples have slightly higher Zr/Y ratios (5.09 and 6.0 respectively) and show transitional nature. High Mg# observed in the many studied rocks suggests their derivation from high-Mg thoeiitic magma. Although major oxides and LILE do not show any clear crystallization trend but HFSE shows good crystallization trends. This suggests co-genetic nature of many of the studied mafic dyke samples. Some samples do not follow crystallization trends. This might be because of many reasons like post-magmatic alterations (including metamorphism), crustal contamination, or they have some different genetic history as observed in MgO versus SiO_2 and

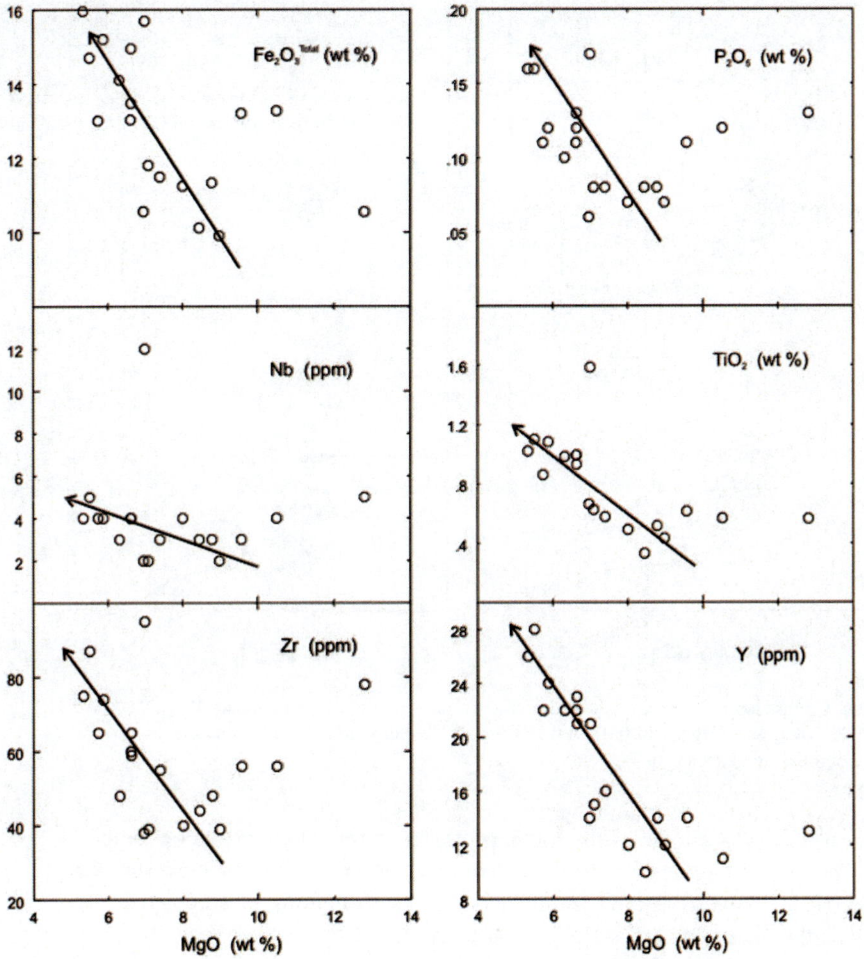

Fig. 12.7 Harker-type variation diagram plotted between MgO versus selected HFSEs

MgO versus alkalis plots (Fig. 12.6). With a limited geochemical data it is difficult to assess these assumptions.

Figure 12.8 is prepared for comparison of present geochemical data with the well-studied southern Bastar mafic dyke swarms. Srivastava and co-workers have identified three distinct mafic dyke swarms in the southern Bastar craton (Srivastava et al., 1996; Srivastava and Singh, 2004; Srivastava, 2006a, b; Srivastava and Gautam, 2007, 2008). These include two sets of sub-alkaline mafic dyke swarms, recognized as BD1 (~2.7 Ga) and BD2 (1.88 Ga) and a high-Mg boninite-norite (BN; ~2.4–2.5 Ga) mafic dyke swarm. From the comparison (Fig. 12.8) it is observed that mafic dykes of present study have close geochemical similarities with

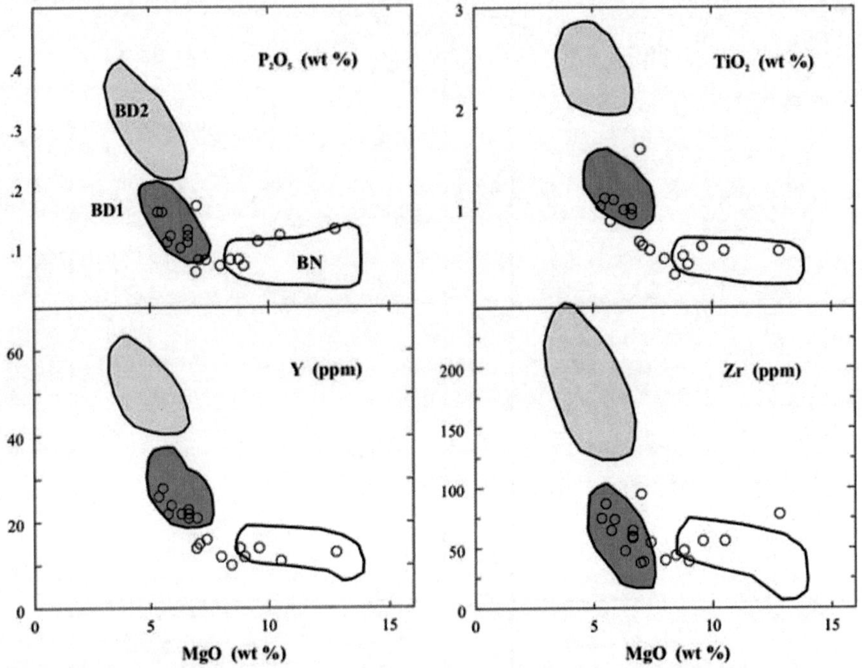

Fig. 12.8 Comparison of present data with mafic dykes from the southern Bastar area (Srivastava and Singh, 2004; Srivastava, 2006b). BD1, BD2 and BN belong to different mafic dyke swarms from the southern Bastar craton

BD1 swarm and BN swarm of the southern Bastar craton. On the basis of compatible and incompatible trace elements modeling Srivastava (2006a, b) suggested that BD1 mafic rocks were derived from ~15–20% batch melting of a depleted lherzolite mantle source, whereas BN swarm is produced by ~20–25% melting of a refractory mantle source. No sample of this study plot within the BD2 swarm field. It is interesting to observe close geochemical similarities between metadolerites of present study and BN dykes of southern Bastar. This is important because no boninite is as yet reported from the central Bastar craton. One sample (Table 12.1; S.No. 16) of present study is classified as boninite on the basis of its chemistry but further geochemical data, particularly rare-earth elements and isotopic data, is needed for any definite result.

Compatible-incompatible trace elements modelling have been successfully used to understand conditions for generation of different mafic magmas in the upper mantle and also to decipher their differentiation processes (Rajamani et al., 1985; Condie et al., 1987; Knoper and Condie, 1988). One of such models, based on Ni-Zr (after Rajamani et al., 1985), is used for the present mafic rocks (Fig. 12.9). This model indicates that mafic dykes exposed in the central Bastar craton were probably derived from a high-magnesium mafic melt generated through ~25% batch melting of a depleted mantle source followed by 30–40% fractional crystallization.

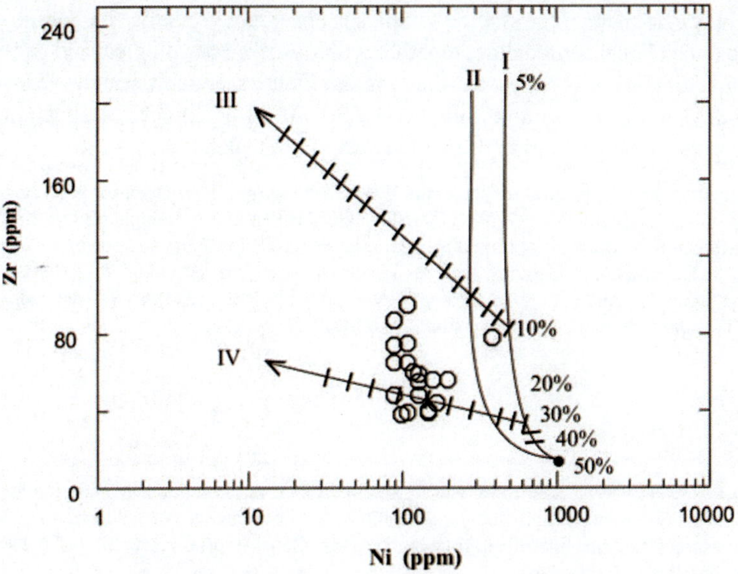

Fig. 12.9 Trace elements modelling based on Ni vs. Zr (after Rajamani et al., 1985). I and II: calculated batch melting curves at 1,850°C, 50 kb and 1,575°C, 25 kb respectively, and are marked with percentage of melting. III and IV: olivine fractionation trends at one atmosphere with ticks, marking increments of 5% olivine fraction from previous tick. Values for the mantle source (7.8 ppm Zr and 2,000 ppm Ni) are taken from Taylor and McLennan (1981) (source mode: 55% ol, 25% opx and 20% cpx; melting mode: 20% ol, 25% opx, 55% cpx)

The assembly and break-up of supercontinents throughout earth history is well documented (Rogers and Santosh, 2003; Santosh, 2010). Stabilization ages for the individual cratons play an important role in paleocontinental reconstructions at different times and mafic magmatism plays an important role in correlating the different cratonic blocks to facilitate reconstruction of many supercontinents. On the basis of available field, petrological, geochemical and geochronological data, the distinct mafic dyke swarms from the southern Bastar craton were inferred to be emplaced during the Neoarchean–Paleoproterozoic, Srivastava and Gautam (2008) suggested that Bastar craton was part of expanded Ur and Arctica supercontinents. Further study of mafic dykes of present study, particularly HFSE geochemistry and geochronology, would play role in paleocontinental reconstructions.

Conclusions

Petrological and geochemical data on the NW–SE trending metabasite and metadolerite mafic dykes exposed around Dondi, Bhanupratappur and Keshkal clearly classify them as sub-alkaline high-iron to high-magnesium tholeiites. Co-genetic relationship between many of the studied samples is observed. From the compatible-incompatible trace elements modeling, together with observed petrological and geochemical characteristics, it is suggested that a mafic melt generated through ~25% melting of a depleted mantle source followed by 30–40% fractional crystallization might be involved in the generation of mafic dykes of present study.

Acknowledgements RKS is thankful to University Grant Commission, New Delhi for providing a research grant for the present work {37–56/2009 (SR)}. GCG is thankful to Council of Scientific and Industrial Research, New Delhi for providing Senior Research Fellowship. Authors are thankful to Head of the Department of Geology, Banaras Hindu University for providing all necessary facilities during this work. Authors are grateful to N. V. Chalapathi Rao and H. M. Ramachnadra for their valuable comments on the earlier version of the MS.

References

Barrett TJ, MacLean WH (1994) Chemostratigraphy and hydrothermal alteration in exploration for VHMS deposits in greenstones and younger volcanic rocks. In: Lentz DR, (ed) Alteration and alteration processes associated with ore-forming systems. Geol Assoc Canada, Short Course Notes, Newfoundland 11: 433–467

Bryan SE, Ernst RE (2008) Revised definition of Large Igneous Provinces (LIPs). Earth Sci Rev 86: 175–202

Chalapathi Rao NV, Srivastava RK (2009) A new fi nd of boninite dyke from the Palaeoproterozoic Dongargarh super group: Inference for a fossil subduction zone in the Archaean of the Bastar craton, Central India. N Jb Miner Abh 186/3: 271–282

Chaudhuri AK, Saha D, Deb DK, Deb SP, Mukherjee MK, Ghosh G (2002) The Purana basins of southern cratonic provinces of India- A case for Mesoproterozoic fossil rifts. Gondwana Res 5: 23–33

Coffin MF, Eldholm O (1994) Large igneous provinces: Crustal structure, dimensions, and external consequences. Rev Geophysics 32: 1–36

Coffin MF, Eldholm O (2005) Large igneous provinces. In: Selley RC, Cocks R, Plimer IR, (eds) Encyclopedia of geology. Elsevier, Oxford: 315–323

Condie KC, Bobrow DJ, Card KD (1987) Geochemistry of Precambrian mafic dykes form the Southern superior province. In: Halls HC, Fahriig WF, (eds) Mafic dyke swarms, Geol Assoc Canada Spl Pap 34: 95–108.

Condie KC, Sinha AK (1996) Rare earth and other trace element mobility during mylonitization: A comparison of the Brevard and Hope Valley shear zones in the Appalachian Mountains, USA. Jour Met Geol 14: 213–226

Crookshank H (1963) Geology of Southern Bastar and Jeypore from the Bailadila range to Eastern Ghats. Mem Geol Surv India 87: 150

Ernst RE, Buchan KL (1997) Giant radiating dyke swarms: Their use in identifying pre-Mesozoic large igneous provinces and mantle plumes. In: Mahoney JJ, Coffin MF, (eds) Large igneous provinces: Continental, oceanic and planetary flood volcanism. Geophy Mono Series 100: 297–333

Ernst RE, Buchan KL (2003) Recognizing mantle plumes in the geological record. Annual Rev Earth Planet Sci 31: 469–523

Ernst RE, Buchan KL, Campbell IH (2007) Frontiers in large igneous province research. Lithos 79: 271–297

Ernst RE, Srivastava RK (2008) India's place in the Proterozoic world: Constraints from the Large Igneous Province (LIP) record. In: Srivastava RK, Sivaji C, Chalapathi Rao NV, (eds) Indian dyke: Geochemistry, geophysics and geochronology. Narosa Publishing House Pvt Ltd, New Delhi: 413–445

Floyd PA, Winchester JA (1978) Identification and discrimination of altered and metamorphosed volcanic rocks using immobile elements. Chem Geol 21: 291–306

French JE, Heaman LM (2010) Precise U–Pb dating of Paleoproterozoic mafic dyke swarms of the Dharwar craton, India: Implications for the existence of the Neoarchean supercraton Sclavia. Precamb Res, doi:10.1016/j.precamres.2010.05.003

French JE, Heaman LM, Chacko T, Srivastava RK (2008) 1891–1883 Ma Southern Bastar Cuddapah mafic igneous events, India: A newly recognized large igneous province. Precamb Res 160: 308–322

Halls HC, Kumar A, Srinivasan R, Hamilton MA (2007) Paleomagnetism and U-Pb geochronology of eastern trending dykes in the Dharwar craton, India: Feldspar clouding, radiating dyke swarms and the position of India at 2.37 Ga. Precamb Res 155: 47–68

Irvine TN, Baragar WRA (1971) A guide to chemical classification of the common volcanic rocks. Canadian Jour Earth Sci 8: 523–548

Jensen LS (1976) A new cation plot for classifying sub-alkaline volcanic rocks. Ontario Division Mines Misc Paper 66

Jochum KP, Verma SP (1996) Extreme enrichment of Sb, Tl, and other trace elements in altered MORB. Chem Geol 130: 289–299

Knoper MW, Condie KC (1988) Geochemistry and petrogenesis of Proterozoic amphibolites West Central Colarado. USA Chem Geol 67: 209–225

Le Maitre RW (2002) Igneous rocks: A classification and glossary of terms, 2nd edn. Cambridge University Press, Cambridge

Murthy NGK (1987) Mafic dyke swarms of the Indian shield. In: Halls HC, Fahriig WF, (eds) Mafic dyke swarms, Geol Assoc Canada Spl Pap 34: 393–400

Murthy NGK (1995) Proterozoic mafic dykes in southern peninsular India: A review. Geol Soc India Mem 33: 81–98

Naqvi SM, Rogers JJW (1987) Precambrian Geology of India. Oxford Monographs on Geology and Geophysics No. 6, Oxford University Press, New York

Patranabis-Deb S, Bickford ME, Hill B, Chaudhuri AK, Basu A (2007) SHRIMP age of zircon in the uppermost tuff in Chhattisgarh Basin in Central India reqire ~500- Ma adjustment in Indian Proterozoic stratigraphy. Jour Geol 115: 407–415

Pearce JA, Cann JR (1973) Tectonic setting of basic volcanic rocks investigated using trace element analysis. Earth Planet Sci Lett 19: 290–300

Rajamani V, Shivakumar K, Hanson GN, Shirey SB (1985) Geochemistry and petrogenesis of amphibolite, Kolar Schist belt, South India; evidence for komatiitic magma derived by low percentage of melting of the mantle. Jour Pet 26: 92–123

Ramachandra HM, Mishra VP, Deshmukh SS (1995) Mafic Dykes in Bastar Precambrian: Study of the Bhanupratappur-Keshkal Mafic Dyke Swarms. Geol Soc India Mem 33: 183–208

Ramachnadra HM, Roy A, Mishra VP, Dutta NK (2001) A critical review of the tectonothermal evolution of the Bastar craton. Geol Surv India Spec Publ 55: 161–180

Ramakrishnan M (1990) Crustal development in Suthern Bastar Central Indian Craton. Geol Surv India Spl Pub 28: 44–66

Rogers JJW, Santosh M (2003) Supercontinents in Earth history. Gondwana Res 6: 357–368

Rollinson HR (1993) Using geochemical data: Evaluation, presentation, interpretation. Longman, Essex

Santosh M (2010) A synopsis of recent conceptual models on supercontinent tectonics in relation to mantle dynamics, life evolution and surface environment. Jour Geodynam 50: 116–133

Seewald JS, Seyfried WE (1990) The effect of temperature on metal mobility in sub-seafloor hydrothermal systems: Constraints from basalt alteration experiments. Earth Planet Sci Lett 101: 388–403

Srivastava RK (2006a) Precambrian mafic dyke swarms from the Central Indian Bastar craton: Temporal evolution of the subcontinental mantle. In: Hanski E, Mertanen S, Ramo T, Vuollo J, (eds) Dyke Swarms – time markers of crustal evolution. Taylor & Francis Group, London: 147–159

Srivastava RK (2006b) Geochemistry and petrogenesis of Neoarchaean high-Mg low-Ti mafic igneous rocks in an intracratonic setting, central India craton: Evidence for boninite magmatism. Geochem Jour 40: 15–31

Srivastava RK, Gautam GC (2007) Geochemistry of distinct mafic intrusive rocks from Darba-Kukanar and Kerlapal-Sukma-Mokhpal areas, southern Bastar craton: Further data on the early Precambrian mafic magmatism of central India. Jour Geol Soc India 69: 1176–1188

Srivastava RK, Gautam GC (2008) Precambrian mafic dyke swarms from the southern Bastar central India craton: Present and future perspectives. In: Srivastava RK, Sivaji C, Chalapathi Rao NV, (eds) Indian Dyke: Geochemistry, geophysics and geochronology. Narosa Publishing House Pvt Ltd, New Delhi: 367–376

Srivastava RK, Gautam GC (2009) Precambrian mafic magmatism in the Bastar craton, central India. Jour Geol Soc India 73: 52–72

Srivastava RK, Hall RP, Verma R, Singh RK (1996) Contrasting Precambrian Mafic Dykes of the Bastar craton, Central India: Petrological and geochemical characteristics. Jour Geol Soc India 48: 537–546

Srivastava RK, Singh RK (2004) Trace element geochemistry and genesis of the Precambrian sub-alkaline mafic dykes from central India craton: Evidence for mantle metasomatism. Jour Asian Earth Sci 23: 373–389

Srivastava RK, Sivaji C, Chalapathi Rao NV (2008) Indian dyke through space and time: Retrospet and prospect. In: Srivastava RK, Sivaji C, Chalapathi Rao NV, (eds) Indian Dyke: Geochemistry, geophysics and geochronology. Narosa Publishing House Pvt Ltd, New Delhi: 1–18

Subba Rao DV, Balaram V, Charan SN, Sridhar DN, Naqvi SM (2004) Proterozoic mafic dykes in and around Chattisgarh basin, central India: Inferences of compositional variation in mantle sources. DST-DCS News Lett 14: 1–5

Taylor SR, McLennan SM (1981) The composition and evolution of the continental crust: Rare earth element evidence from sedimentary rocks. Phil Trans R Soc London A 300: 381–399

Verma SP (1992) Seawater alteration effects on REE, K, Rb, Cs, Sr, U, Th, Pb, and Sr-Nd-Pb isotope systematic of mid-ocean ridge basalts. Geochem Jour 26: 159–177

Verma SP, Torres-Alvarado IS, Sitelo-Rodriguez ZT (2002) SINCLAS: Standard igneous norm and volcanic rock classificstion system. Comp Geosci 28: 711–715

Winchester JA, Floyd PA (1976) Geochemical magma type discrimination; application to altered and metamorphosed basic igneous rock. Earth Planet Sci Lett 28: 459–469

Chapter 13
Fluid Induced Metamorphism in a Suite of Mafic Dykes from PalaeoProterozoic Mahakoshal Group, Central India

Kasturi Chakraborty, Sanjoy Sanyal, and Pulak Sengupta

Introduction

Fluids that infiltrate rocks during metamorphism augment mass-transfer across contrasting lithologies, and as a consequence, develop metasomatic selvages along the fluid path (Ague, 2003; Abart, 1995). Detail petrology including modeling of metasomatic reactions that shaped mineral assemblages in selvages are, therefore, expected to provide crucial information regarding P–T-fluid regimes of metamorphic belts (reviewed in Philpotts and Ague, 2009). Palaeoproterozoic Mahakoshal Group of Central Indian Tectonic Zone (CITZ) exposes an interbanded sequence of marble-metapelites-quartzite that were intersected by a suite of mafic dykes near the town Jabbalpur, M.P. India. The entire rock assemblages underwent green schist to amphibolite facies metamorphism in presence of pervasive fluid flow during c.1.8 Ga tectonothermal event (reviewed in Roy and Devrajan, 2000; Roy and Prasad, 2003; Chakraborty, 2009). In this communication we present field relations, textures and compositions of minerals in selvages that were developed along margins of mafic dykes occurring in marble and metapelite. Our study have demonstrated that the selvages were formed due to extensive material transport between mafic dykes and their host rocks during pervasive fluid-flow under regional amphibolite facies metamorphism. Mineral assemblages in the selvages further demonstrate that regional metamorphism and the attendant fluid flow occurred along gentle geothermal gradient (high T-low P) that is consistent with the appearance of andalusite in the adjoining metapelite.

K. Chakraborty (✉)
Geological Survey of India, Southern Region, Hyderabad, Andhra Pradesh, India
e-mail: kasturi.c8@gmail.com

S. Sanyal and P. Sengupta
Department of Geological Sciences, Jadavpur University, Kolkata, India

R.K. Srivastava (ed.), *Dyke Swarms: Keys for Geodynamic Interpretation*,
DOI 10.1007/978-3-642-12496-9_13, © Springer-Verlag Berlin Heidelberg 2011

General Geology

The Palaeoproterozoic Mahakoshal Group occupies a part of the northern sector of the Central Indian Tectonic Zone (CITZ) and occurs in detached outcrops for a strike length of around 500 km from Narsinghpur in the west to Palamau in the east. The supracrustals have been subjected to at least three phases of deformation, of which the first two are most pronounced, amphibolite facies metamorphism and intrusion of granites and syenites in addition to dykes paralleling the regional trend (Roy and Devrajan, 2000). Based on regional field relations and available petrographic and geophysical data, a back arc riftogenic origin has been proposed for the Mahakoshal supracrustals (Roy and Prasad, 2003).

A small segment of the Mahakoshal Group of rocks has been exposed near the Son-Narmada South Fault (SNSF) close to Jabalpur. The outcrops are dominated by low to medium grade pelitic and psammopelitic schists, dolomitic marble, banded chert, calc silicate rich rocks and minor BIF. The dolomitic marble can broadly be divided from north to south into talc-zone, talc-tremolite zone and diopside zone. The pelites can similarly be subdivided into a northern garnet zone followed southward by staurolite- and andalusite-zones respectively. The ensemble has been intruded by mafic dykes paralleling the regional trend. Three episodes of deformation, viz. D_{1-3} are discernible in the area (Chakraborty, 2009). The regional foliation has developed parallel to axial plane of D_1 folds. Metamorphism initiated with D_1 and reached its climax in post-D_1-pre-D_2 to early syn-D_2 stage (Chakraborty, 2009). Mafic dykes, emplaced within staurolite and andalusite zones of metapelites and talc-tremolite zone of dolomitic marble, are broadly concordant to the regional foliation and along with regional foliation were subsequently folded during younger deformation (Fig. 13.1).

Fig. 13.1 (**a**) Mafic dykes intrusive in banded dolomitic marble. The dyke is discordant to primary banding. The *arrow points* to undigested marble raft preserved within mafic dyke. (**b**) Folded dyke in metapelite, regional foliation being axial planar to this fold

Field Relations of the Dykes Near Jabalpur

Dykes present within the dolomitic marble are restricted to the talc-tremolite zone. The host marble is largely constituted of medium grained recrystallized dolomite ± calcite interbanded with talc-rich or tremolite-rich bands. In the talc-rich bands, talc constitutes 60–80% of the rock with minor dolomite ± quartz ± tremolite. The talc-tremolite rich bands are dominated by tremolite (>40 vol %) and contain variable proportion of dolomite, quartz and rarely talc. The calc-silicate bands are parallel to the regional foliation.

The dykes are approximately <1–5 m wide and laterally continuous up to a maximum distance of 25 m. These mafic rocks are medium grained, greenish black in colour and are well foliated along their margins while the cores are almost massive in appearance. The foliated margin is comprised of talc and chlorite. Although the marginal foliation is concordant with the regional foliation, the mafic rocks themselves are distinctly discordant to the compositional banding preserved in the adjoining marble (Fig. 13.1a). Some undigested rafts of marble are also found to occur within these mafic rocks. Such features are unequivocally suggestive of their intrusive nature syn- or prior to the D_1 deformation that was responsible for the formation of the regional foliation.

Most of the mafic dykes within the metapelites occur in the staurolite-zone. However, although the former unit hosts a large number of mafic bodies, isolated intrusives (less than a meter thick) occur within all other pelitic units as well. The dykes have been folded during D_1, the regional foliation being axial planar to these folds (Fig. 13.1b). The cores of all these dykes are massive (similar in character to the ones described before) while the rims are more foliated due to preferred orientation of chlorite in these parts.

Petrography and Mineral Chemistry

Analytical Procedures

Mineral compositions were determined with a CAMECA MICROBEAM Electron Probe Micro-analyzer (Cameca SX51) at the Mineralogical and Petrological Institute, University of Bonn. Operating conditions were 15 kV accelerating voltage, 1–3 μm. beam diameter and 15 nA specimen current. Natural and synthetic standards were used and PAP correction procedure applied. Whole rock analyses of major and minor elements were carried out with a Philips X-Ray Fluorescence at the Mineralogical and Petrological Institute, University of Bonn, Germany.

The Massive Core of the Dykes

The mineral assemblage constituting the mafic dykes is hornblende + plagioclase + quartz + ilmenite ± chlorite. Remnant orthopyroxene was detected within the hornblendes. The hornblende form coarse randomly oriented blades and constitute about 65% of the rock (Fig. 13.2a). They bear numerous inclusions of quartz and ilmenite.

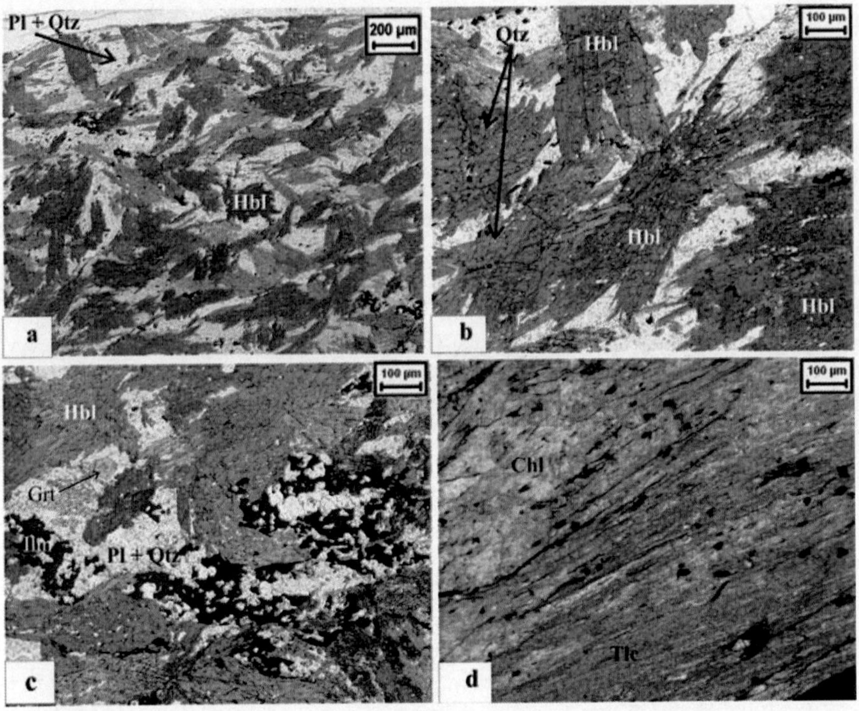

Fig. 13.2 (**a**) Coarse randomly oriented hornblende within fine grained recrystallized quartzofeldspathic matrix in mafic dykes (*Plane Polarized Light*). (**b**) Randomly oriented hornblende bearing inclusions of quartz within fine grained recrystallized quartzofeldspathic matrix in mafic dykes (*Plane Polarized Light*). (**c**) Sporadically occurring small euhedral garnets in mafic dykes intruded within metapelites (*Plane Polarized Light*). (**d**) Foliated talc-chlorite rock, foliation defined by strong preferred orientation of both talc and chlorite (*Plane Polarized Light*)

Some of these hornblende grains have been deformed by D_2 deformation. The amphiboles are set in a matrix of recrystallized polygonal aggregate of quartz and plagioclase (Fig. 13.2b). Garnet, which is found only in mafic dyke with in metapelites, is volumetrically small (<10 vol %) and occurs sporadically as small euhedral grains scattered within the quartzofeldspathic matrix (Fig. 13.2c). A few flakes of chlorite are scattered in the matrix. From the randomly oriented and occasionally kinked hornblendes that characterize the cores of all the dykes, it has been concluded that the hornblende-plagioclase-quartz equilibrium assemblage is post D_1 and pre-D_2 that has subsequently been deformed and kinked during D_2.

Representative microprobe analyses data for hornblende have been presented in Table 13.1. Amphiboles in the mafic dykes are tschermakite to magnesiohornblende (Leake et al., 1997). $MgFe_{-1}$ is an important substitution (Fig. 13.4a). In addition, the amphiboles also show edenite (Fig. 13.4b), tschermakite (Fig. 13.4c) and pargasite substitutions (Fig. 13.4d). In most plots, two distinct arrays are formed by amphiboles from dykes in metapelites and marble. This can be attributed generally to higher X_{Mg} for the hornblendes occurring within dykes in marble

Table 13.1 Microprobe analyses data for amphiboles from mafic dykes within metapelites and marbles

Rock type	Amphibolite dyke within metapelite				Amphibolite dyke within metapelite			Amhibolite dyke within marble				Amhibolite dyke within marble				
Sample No.	L 81 (2)n-a	L 81 (2)n-b	L 81 (2)n-c	L 81 (2)-a	L 81 (2)-b	L 81 (2)-c	L 81 (2)-d	M31 (1)-a	M31 (1)-b	M31 (1)-c	M31 (1)-d	M31 (1)-e	M31 (1)-f	M31 (1)-g	M31 (1)-h	M31 (1)-k
Point No.	4	8	12	2	4	7	13	2	4	8	9	10	11	12	14	16
Phase	Amph	Amph	Amph	Amph	Amph	Amph	Amph	Amph	Amph	Amph	Amph	Amph	Amph	Amph	Amph	Amph
SiO_2	42.45	42.64	42.70	41.63	41.33	42.68	42.79	43.63	42.47	44.70	42.82	43.80	43.71	43.14	42.35	42.00
TiO_2	0.47	0.37	0.33	0.34	0.39	0.36	0.48	0.39	0.46	0.43	0.40	0.40	0.42	0.46	0.49	0.45
Al_2O_3	14.53	14.73	15.85	13.67	13.52	14.36	14.03	11.64	13.35	12.36	14.52	13.00	12.66	13.47	13.41	12.68
Cr_2O_3	0.04	0.02	0.16	0.14	0.01	0.03	0.01	0.07	0.02	0.02	0.07	0.00	0.00	0.01	0.24	0.13
FeO	17.35	17.19	15.77	16.43	16.71	16.66	17.86	15.51	16.22	15.06	15.85	16.04	16.63	15.87	16.30	16.14
MnO	0.73	0.59	0.54	0.70	0.64	0.61	0.76	0.41	0.46	0.41	0.24	0.41	0.37	0.46	0.41	0.32
MgO	8.88	9.01	8.51	9.00	9.38	9.01	8.90	10.90	9.75	11.04	9.84	10.44	10.56	9.97	9.69	10.08
CaO	11.27	11.08	11.06	10.53	10.89	10.79	10.71	11.57	11.53	11.67	11.44	11.45	11.18	11.38	11.42	11.50
Na_2O	1.19	1.22	1.08	1.01	1.04	1.06	1.03	1.04	1.19	1.12	1.25	1.24	1.22	1.15	1.21	1.20
K_2O	0.26	0.26	0.26	0.28	0.23	0.23	0.27	0.34	0.32	0.25	0.26	0.23	0.23	0.33	0.31	0.40
Total	97.18	97.10	96.25	93.73	94.14	95.79	96.84	95.48	95.76	97.06	96.69	97.00	96.97	96.25	95.83	94.90
^{T}Si	6.27	6.29	6.33	6.34	6.27	6.36	6.33	6.48	6.33	6.51	6.31	6.42	6.42	6.38	6.32	6.32
^{T}Al	1.73	1.71	1.67	1.66	1.73	1.64	1.67	1.52	1.67	1.49	1.69	1.58	1.58	1.62	1.68	1.68
^{C}Al	0.80	0.85	1.09	0.80	0.69	0.88	0.77	0.52	0.68	0.64	0.83	0.67	0.61	0.73	0.68	0.57
^{C}Ti	0.05	0.04	0.04	0.04	0.04	0.04	0.05	0.04	0.05	0.05	0.04	0.04	0.05	0.05	0.06	0.05
$^{C}Fe_3$	0.76	0.75	0.52	0.82	0.96	0.73	0.88	0.95	0.84	0.80	0.71	0.80	0.88	0.76	0.80	0.93
^{C}Cr	0.00	0.00	0.02	0.02	0.00	0.00	0.00	0.01	0.00	0.00	0.01	0.00	0.00	0.00	0.03	0.02
^{C}Mg	1.95	1.98	1.88	2.04	2.12	2.00	1.96	2.41	2.17	2.40	2.16	2.28	2.31	2.20	2.15	2.26
$^{C}Fe_{2+}$	1.38	1.37	1.43	1.27	1.16	1.35	1.33	0.98	1.19	1.04	1.24	1.16	1.16	1.20	1.23	1.10
^{C}Mn	0.04	0.01	0.02	0.01	0.03	0.00	0.01	0.05	0.06	0.05	0.01	0.04	0.00	0.05	0.05	0.04

Table 13.1 (continued)

Rock type	Amphibolite dyke within metapelite				Amphibolite dyke within metapelite			Amhibolite dyke within marble				Amhibolite dyke within marble				
Sample No.	L 81 (2)n-a	L 81 (2)n-b	L 81 (2)n-c	L 81 (2)-a	L 81 (2)-b	L 81 (2)-c	L 81 (2)-d	M31 (1)-a	M31 (1)-b	M31 (1)-c	M31 (1)-d	M31 (1)-e	M31 (1)-f	M31 (1)-g	M31 (1)-h	M31 (1)-k
Point No.	4	8	12	2	4	7	13	2	4	8	9	10	11	12	14	16
Phase	Amph	Amph	Amph	Amph	Amph	Amph	Amph	Amph	Amph	Amph	Amph	Amph	Amph	Amph	Amph	Amph
^{B}Mg	0.00	0.00	0.00	0.00	0.00	0.00	0.00	0.00	0.00	0.00	0.00	0.00	0.00	0.00	0.00	0.00
$^{B}Fe_{2+}$	0.00	0.00	0.00	0.00	0.00	0.00	0.00	0.00	0.00	0.00	0.00	0.00	0.00	0.00	0.00	0.00
^{B}Mn	0.05	0.06	0.05	0.08	0.05	0.08	0.09	0.00	0.00	0.00	0.02	0.01	0.05	0.01	0.00	0.00
^{B}Ca	1.78	1.75	1.76	1.72	1.77	1.72	1.70	1.84	1.84	1.82	1.81	1.80	1.76	1.80	1.83	1.85
^{B}Na	0.17	0.19	0.19	0.20	0.18	0.20	0.21	0.16	0.16	0.18	0.17	0.19	0.19	0.19	0.17	0.15
^{A}Ca	0.00	0.00	0.00	0.00	0.00	0.00	0.00	0.00	0.00	0.00	0.00	0.00	0.00	0.00	0.00	0.00
^{A}Na	0.17	0.16	0.11	0.10	0.13	0.11	0.08	0.14	0.18	0.14	0.18	0.16	0.15	0.14	0.18	0.20
^{A}K	0.05	0.05	0.05	0.05	0.04	0.04	0.05	0.06	0.06	0.05	0.05	0.04	0.04	0.06	0.06	0.08
Total	15.22	15.21	15.16	15.15	15.18	15.15	15.13	15.16	15.23	15.16	15.23	15.21	15.20	15.20	15.24	15.25

Normalized on the basis of 23 oxygen

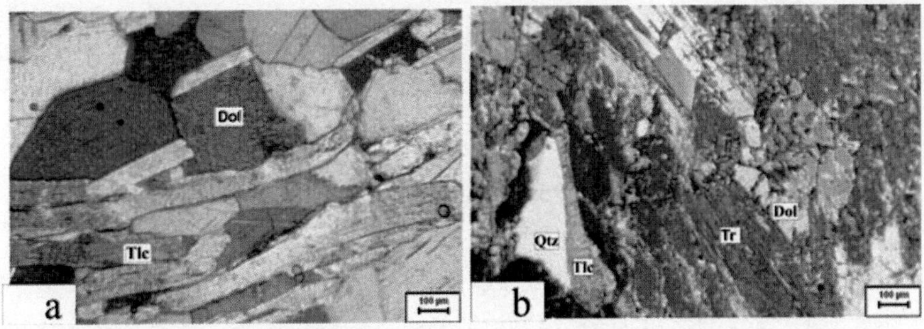

Fig. 13.3 (**a**) Strong preferred orientation of talc within dolomite (Crossed nicols). (**b**) Tremolite developing at the cost of dolomite, talc and quartz (Crossed nicols)

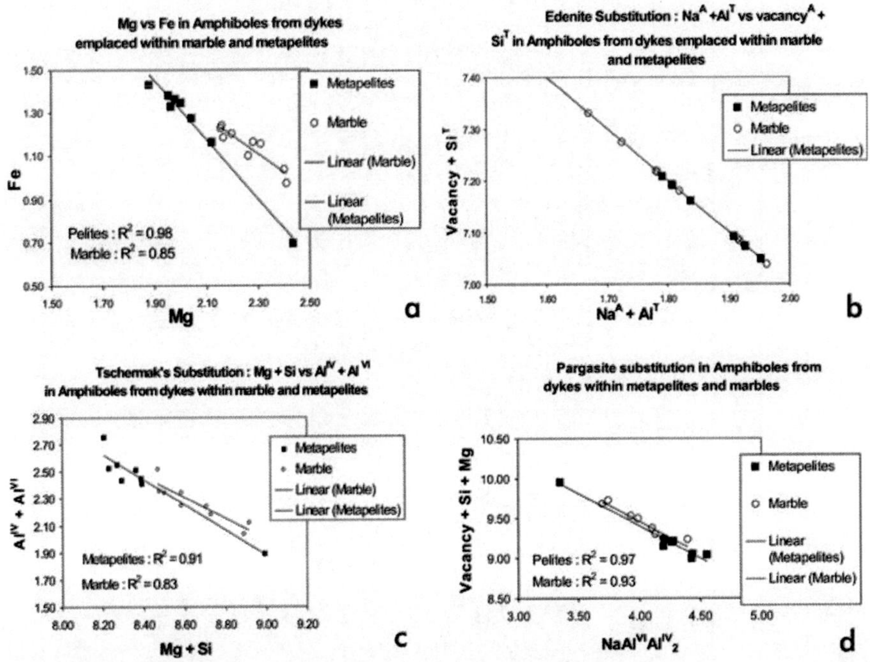

Fig. 13.4 (**a**) $MgFe_{-1}$ substitution in hornblende from dykes within metapelites and marble. (**b**) Edenite substitution in hornblende from dykes within metapelites and marble. (**c**) Tschermakite substitution in hornblende from dykes within metapelites and marble. (**d**) Pargasite substitution in hornblende from dykes within metapelites and marble

(1.74–2.47) compared to those occurring in metapelites (1.31–1.82) and to a lesser extent to the higher proportion of tetrahedral Al in the former. Plagioclase is anorthitic to bytownitic with X_{An} varying from 0.87 to 0.94 (Table 13.2). Garnet has 55% almandine and about 20% spessertine components. Grossular and pyrope contents vary between 10 and 11%. Andradite is less than 5% (Table 13.2). Orthopyroxenes have X_{Mg} in the range of 0.54–0.57 while CaO varies around 0.52–0.80 weight percent and MnO from 1.79 to 1.96 weight percent (Table 13.3). Maximum recorded

Table 13.2 Microprobe analyses data for plagioclase and garnet from mafic dykes within metapelites and marble

Rock type	Amphibolite dyke within metapelite					Amhibolite dyke within marble				Amphibolite dyke within metapelite					
Sample No.	L 81 (2)n-a	L 81 (2)n-b	L 81 (2)-a	L 81 (2)-b	L 81 (2)-d	M31-a	M31-b	M31-c	M31-d	M31-e	L 81 (2)n-a	L 81 (2)n-b	L 81 (2)n-c	L 81 (2)-a	L 81 (2)-c
Point No.	p	11	3	10	12	5	6	7	13	15	5	6	10	1	9
Phase	Pl	Pl	Pl	Pl	Pl	Pl	Pl	Pl	Pl	Pl	Grt	Grt	Grt	Grt	Grt
SiO_2	44.89	45.15	46.43	45.47	46.87	43.99	46.01	46.42	45.91	45.07	37.31	37.21	37.12	36.85	37.16
TiO_2	0.00	0.00	0.00	0.02	0.04	0.00	0.00	0.02	0.01	0.00	0.07	0.04	0.04	0.04	0.09
Al_2O_3	33.18	32.78	33.52	33.31	34.35	34.01	34.43	32.91	35.81	34.27	20.29	20.29	20.27	20.09	20.48
Cr_2O_3	0.01	0.02	0.00	0.00	0.02	0.00	0.02	0.00	0.00	0.01	0.00	0.05	0.06	0.09	0.04
FeO	0.63	0.02	0.13	0.01	0.10	0.19	0.13	0.14	0.24	0.04	26.61	27.06	26.82	26.05	26.12
MnO	0.00	0.01	0.01	0.04	0.00	0.02	0.02	0.02	0.04	0.00	8.57	8.77	8.82	9.21	9.36
MgO	0.48	0.00	0.00	0.02	0.00	0.00	0.00	0.10	0.00	0.00	3.02	2.90	3.00	3.06	3.03
CaO	18.92	19.87	17.99	19.28	19.02	19.00	19.16	17.78	18.44	18.86	4.39	3.76	3.84	4.00	3.96
Na_2O	0.84	0.73	1.19	0.77	0.69	1.10	0.92	1.20	1.16	1.35	0.00	0.00	0.00	0.01	0.02
K_2O	0.04	0.00	0.02	0.00	0.01	0.03	0.00	0.35	0.02	0.02	0.03	0.00	0.00	0.01	0.02
Total	99.00	98.58	99.29	98.93	101.10	98.35	100.70	98.95	101.63	99.62	100.30	100.10	99.97	99.42	100.27
Si	2.12	2.09	2.14	2.10	2.11	2.11	2.09	2.14	2.12	2.10	3.00	3.00	3.00	2.99	2.99
Ti	0.00	0.00	0.00	0.00	0.00	0.00	0.00	0.00	0.00	0.00	0.00	0.00	0.00	0.00	0.01
Al	1.81	1.87	1.82	1.87	1.87	1.85	1.88	1.84	1.86	1.87	1.92	1.93	1.92	1.92	1.93
Cr	0.00	0.00	0.00	0.00	0.00	0.00	0.00	0.00	0.00	0.00	0.00	0.00	0.00	0.01	0.00
Fe	0.02	0.00	0.01	0.00	0.00	0.01	0.01	0.01	0.01	0.00	1.79	1.82	1.81	1.77	1.76
Mn	0.00	0.00	0.00	0.00	0.00	0.00	0.00	0.00	0.00	0.00	0.58	0.60	0.60	0.63	0.64
Mg	0.03	0.00	0.00	0.00	0.00	0.00	0.00	0.01	0.00	0.00	0.36	0.35	0.36	0.37	0.36
Ca	0.94	0.99	0.92	0.95	0.94	0.94	0.95	0.88	0.91	0.93	0.38	0.32	0.33	0.35	0.34
Na	0.08	0.07	0.11	0.07	0.06	0.10	0.08	0.11	0.10	0.12	0.00	0.00	0.00	0.00	0.00
K	0.00	0.00	0.00	0.00	0.00	0.00	0.00	0.02	0.00	0.00	0.00	0.00	0.00	0.00	0.00
# Cation	5.01	5.01	5.00	5.00	4.99	5.01	5.01	5.00	5.00	5.03	9.96	9.96	9.96	9.97	9.98
O[a]	8.00	8.00	8.00	8.00	8.00	8.00	8.00	8.00	8.00	8.00	12	12	12	12	12

Table 13.2 (continued)

Rock type	Amphibolite dyke within metapelite					Amhibolite dyke within marble				Amphibolite dyke within metapelite					
Sample No.	L 81 (2)n-a	L 81 (2)n-b	L 81 (2)-a	L 81 (2)-b	L 81 (2)-d	M31-a	M31-b	M31-c	M31-d	M31-e	L 81 (2)n-a	L 81 (2)n-b	L 81 (2)n-c	L 81 (2)-a	L 81 (2)-c
Point No.	p	11	3	10	12	5	6	7	13	15	5	6	10	1	9
Phase	Pl	Pl	Pl	Pl	Pl	Pl	Pl	Pl	Pl	Pl	Grt	Grt	Grt	Grt	Grt
Xan	0.92	0.94	0.90	0.93	0.94	0.90	0.92	0.87	0.90	0.88					
Xab	0.07	0.06	0.10	0.07	0.06	0.09	0.08	0.11	0.10	0.11					
Xor	0.00	0.00	0.00	0.00	0.00	0.00	0.00	0.02	0.00	0.00					
X Fe_{+3}											0.03	0.02	0.02	0.03	0.02
XMg											0.12	0.11	0.12	0.12	0.12
Xfe_{+2}											0.55	0.57	0.56	0.54	0.54
X Mn											0.19	0.19	0.19	0.20	0.21
Xca											0.12	0.10	0.11	0.11	0.11

[a]Number of oxygens per formula unit.

Table 13.3 Representative microprobe analyses data for orthopyroxene and ilmenite from mafic dykes within metapelites and marble

Rock type	Amphibolite dyke within metapelite					Amphibolite dyke within metapelite			Amhibolite dyke within marble
Sample No.	L 81 (2)n-b	L 81 (2)-a	L 81 (2)-c	L 81 (2)-d	L 81 (2)-m	L 81 (2)n-a	L 81 (2)-a	L 81 (2)-c	M31(1)-d
Point No.	3	5	6	8	14	7	16	17	3
Phase	Opx	Opx	Opx	Opx	Opx	Ilm	Ilm	Ilm	Ilm
SiO_2	56.11	54.84	54.91	55.41	53.00	0.04	0.06	0.07	0.02
TiO_2	0.00	0.05	0.04	0.04	0.01	50.75	50.65	50.25	49.14
Al_2O_3	0.61	0.80	0.64	2.15	0.96	0.02	0.01	0.02	0.04
Cr_2O_3	0.00	0.02	0.01	0.00	0.08	0.10	0.01	0.04	0.05
FeO	23.70	23.88	23.99	21.98	23.09	45.55	45.77	45.95	48.06
MnO	1.96	1.96	1.79	1.95	1.90	2.05	2.22	2.20	1.31
MgO	17.94	17.99	19.06	18.99	18.99	0.64	0.49	0.61	0.64
CaO	0.55	0.59	0.58	0.52	0.80	0.02	0.09	0.15	0.01
Na_2O	0.02	0.04	0.03	0.03	0.06	0.00	0.00	0.01	0.06
K_2O	0.00	0.01	0.00	0.02	0.01	0.00	0.00	0.00	0.00
Total	100.91	100.18	101.05	101.09	98.90	99.17	99.32	99.31	99.32
Si	2.07	2.03	2.06	2.03	2.05	0.00	0.00	0.00	0.00
Ti	0.00	0.00	0.00	0.00	0.00	1.95	1.95	1.94	1.91
Al	1.00	0.04	0.03	0.10	0.04	0.00	0.00	0.00	0.00
Cr	0.00	0.00	0.00	0.00	0.00	0.00	0.00	0.00	0.00
Fe	0.74	0.77	0.71	0.71	0.74	1.95	1.96	1.97	2.07
Mn	0.06	0.06	0.06	0.06	0.06	0.09	0.10	0.10	0.06
Mg	1.00	1.03	1.04	1.01	0.99	0.05	0.04	0.05	0.05
Ca	0.02	0.02	0.02	0.02	0.03	0.00	0.01	0.01	0.00
Na	0.00	0.00	0.00	0.00	0.00	0.00	0.00	0.00	0.00
K	0.00	0.00	0.00	0.00	0.00	0.00	0.00	0.00	0.00
# Cation	3.92	3.95	3.92	3.92	3.93	4.04	4.05	4.06	4.09
O[a]	6.00	6.00	6.00	6.00	6.00	6.00	6.00	6.00	6.00
X(Mg)	0.55	0.55	0.57	0.56	0.54				

[a]Number of oxygens per formula unit.

Al_2O_3 is around 2.15 weight percent (Table 13.3). Ilmenites have 3–5 mol % pyrophanite (Table 13.3).

Talc Chlorite Bearing Metasomatic Margin

This is a fine grained foliated rock with chlorite rich lenses (modal chlorite 70–90% within the lenses) interwoven with talc rich bands (70–85% talc). Quartz and carbonate occur in minor amount (Fig. 13.2d). Minor actinolitic amphiboles (<10 vol %) are present in some samples. The chlorite and talc grains have a strong preferred orientation parallel to regional foliation (Fig. 13.2d). Polygonal aggregates of dolomite are present as small remnant lenses within talc (Fig. 13.2d). Quartz grains are small, lensoidal, aligned parallel to foliation and have strong undulose

Table 13.4 Representative microprobe analyses data for dolomite from dolomitic and calcite marble adjoining amphibolite dykes

Rock type				
Sample No.	TcDol	TcDol	TcDol	M69
Point No.	2	3	11	2
Phase	Dol	Dol	Cal	Cal
SiO_2	0.00	0.00	0.02	0.01
TiO_2	0.00	0.00	0.02	0.00
Al_2O_3	0.00	0.00	0.02	0.00
Cr_2O_3	0.01	0.00	0.00	0.00
FeO	0.08	0.13	0.07	0.03
MnO	0.00	0.00	0.00	0.07
MgO	22.98	22.88	0.03	0.45
CaO	34.96	32.75	56.94	57.49
Na_2O	0.01	0.00	0.00	0.02
K_2O	0.00	0.00	0.01	0.02
Total	58.05	55.78	57.10	58.10
Si	0.00	0.00	0.00	0.00
Ti	0.00	0.00	0.00	0.00
Al	0.00	0.00	0.00	0.00
Cr	0.00	0.00	0.00	0.00
Fe	0.00	0.00	0.00	0.00
Mn	0.00	0.00	0.00	0.00
Mg	0.95	0.98	0.00	0.02
Ca	1.04	1.01	1.99	1.98
Na	0.00	0.00	0.00	0.00
K	0.00	0.00	0.00	0.00
# Cation	2.00	2.00	2.00	2.00
O[a]	6.00	6.00	6.00	6.00
X(Mg)	0.48	0.49	0.00	0.01

[a]Number of oxygens per formula unit.

extinction. Remnant lenses of talc in chlorite-rich bands indicate that talc was possibly a reactant in chlorite formation. The timing of formation of this assemblage has been deduced as syn-D_1 since the preferred orientation of talc and chlorite in these metasomatised dyke-margins is largely parallel to the regional foliation.

Representative compositional data of the chlorites have been presented in Table 13.5. All the minerals are extremely magnesian. Chlorite from the talc chlorite schist is extremely rich in Mg ($X_{Mg} = 0.84$) and straddle the boundary of clinochlore and sheridanite in the Si vs Fe_{Total} diagram (Fig. 13.5). Talc is extremely Mg rich ($X_{Mg} = 0.94$, Table 13.5). The dolomite lenses have an X_{Mg} of ~0.48.

Table 13.5 Representative microprobe analyses data for chlorite and talc from Talc-chlorite rock

Rock type	Talc-Chl rock					
Sample No.	L25-a	L25-b	L25-m	L25-r	L25-m	L25-r
Point No.	2	3	6	8	4	5
Phase	Chl	Chl	Chl	Chl	Talc	Talc
SiO_2	28.74	28.30	28.95	29.12	61.30	62.40
TiO_2	0.06	0.06	0.07	0.05	0.02	0.04
Al_2O_3	21.27	22.32	21.99	21.21	0.79	0.37
Cr_2O_3	0.11	0.05	0.05	0.09	0.00	0.00
FeO	9.23	9.76	9.49	9.34	3.19	3.13
MnO	0.01	0.01	0.03	0.00	0.00	0.04
MgO	27.60	26.68	27.69	27.83	28.42	29.60
CaO	0.04	0.01	0.01	0.00	0.05	0.02
Na_2O	0.01	0.01	0.00	0.02	0.01	0.01
K_2O	0.00	0.02	0.01	0.00	0.06	0.02
F	0.00	0.00	0.00	0.00	nd	nd
Cl	0.00	0.00	0.00	0.00	nd	nd
Total	87.14	87.28	88.33	87.72	93.86	95.64
Si	5.58	5.49	5.54	5.61	7.97	7.97
Ti	0.01	0.01	0.01	0.01	0.00	0.00
Al	4.87	5.11	4.96	4.82	0.12	0.06
Cr	0.02	0.01	0.01	0.01	0.00	0.00
Fe	1.50	1.58	1.52	1.51	0.35	0.33
Mn	0.00	0.00	0.01	0.00	0.00	0.00
Mg	7.98	7.72	7.90	7.99	5.51	5.63
Ca	0.01	0.00	0.00	0.00	0.01	0.00
Na	0.00	0.00	0.00	0.01	0.00	0.00
K	0.00	0.01	0.00	0.00	0.01	0.00
F	0.00	0.00	0.00	0.00	–	–
Cl	0.00	0.00	0.00	0.00	–	–
Cation #	14.38	14.43	14.41	14.34	13.97	14.01
O[a]	28.00	28.00	28.00	28.00	22.00	22.00
X(Mg)	0.84	0.83	0.84	0.84	0.94	0.94

[a]Number of oxygens per formula unit.

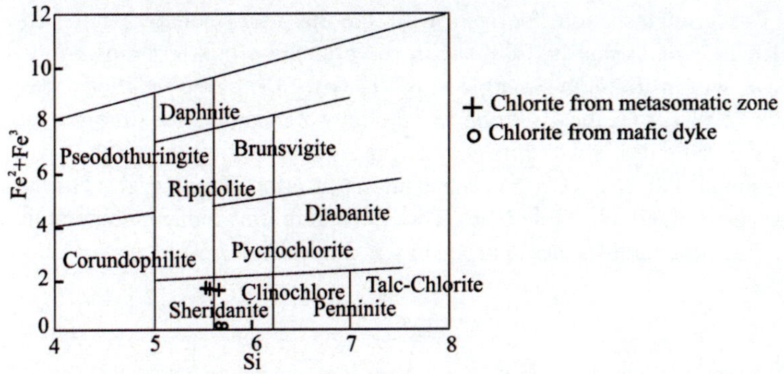

Fig. 13.5 Chlorite from the talc chlorite schist straddling boundary of clinochlore and sheridanite in the Si vs. Fe_{Total} diagram

Host Rock

Textural features demonstrate that talc (X_{Mg} ~0.94, Table 13.5) was formed after dolomite (X_{Mg} ~0.48, Table 13.4) and quartz/aqueous silica whereas formation of tremolite (X_{Mg} ~1.00) involved both talc and dolomite (Fig. 13.3a, b).

Physical Conditions of Metamorphism

Hornblende-Plagioclase thermometry (Blundy and Holland, 1990) for equilibrium assemblages suggests temperature around 400°C. A temperature range of 400–500°C for hornblende-plagioclase stability is also apparent from numerically

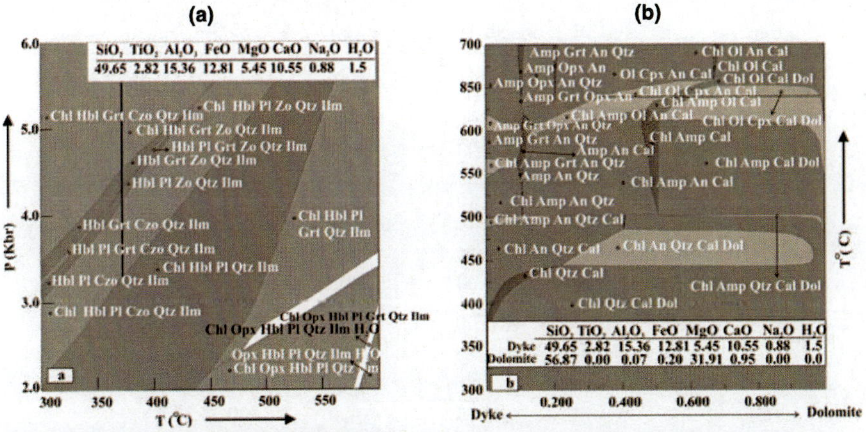

Fig. 13.6 (**a**) Numerically constrained pseudosections showing mineral assemblages that should result from metamorphism of mafic dykes in a given range of P–T conditions in presence of small quantities of aqueous fluid. (**b**) Numerically constrained pseudosections showing the possible assemblages that can be produced from cationic exchange between mafic dyke and adjoining talc dominant bands of dolomitic marble under anhydrous conditions in a given range of temperature

constrained P–T pseudosection (derived with the help of the computer program PERPLEX-07 of Connolly, 2005) with representative bulk composition of dykes intrusive within dolomitic marble (Fig. 13.6a). Garnet-biotite thermometry (Holdaway, 2000) from the Grt-zone of adjoining metapelites, a stratigraphic equivalent of the mafic dykes within talc-tremolite zone of dolomitic marble, gives temperature around 480 ± 22°C (the compositions of corresponding garnets and biotites have been presented in Table 13.6). Thus the maximum ambient temperature of the mafic dykes in course of metamorphism was around 450±50°C.

Table 13.6 Representative microprobe analyses data for garnet and biotite from garnet zone of metapelites

Rock type	Grt-mica schist		Grt-mica schist	
Sample No.	L9-1	L9-3	L9-1	L9-3
Point No.	R1	R8	1b	3b
Phase	Grt	Grt	Bio	Bio
SiO_2	37.04	36.25	35.20	34.42
TiO_2	0.00	0.00	1.37	1.42
Al_2O_3	21.12	20.61	22.24	17.55
Cr_2O_3	0.00	0.04	0.00	0.00
FeO	34.47	34.18	17.46	19.56
MnO	4.40	4.18	0.05	0.06
MgO	1.76	1.78	9.63	9.46
CaO	1.76	1.65	0.16	0.24
Na_2O	0.00	0.02	0.54	0.09
K_2O	0.01	0.00	6.70	7.17
Total	100.56	98.71	92.35	89.97
Si	2.99	2.99	2.67	2.75
Ti	0.00	0.00	0.08	0.09
Al	2.21	2.00	1.99	1.66
Cr	0.00	0.00	0.00	0.00
Fe	2.33	2.36	1.11	1.31
Mn	0.30	0.29	0.00	0.00
Mg	0.21	0.22	1.09	1.13
Ca	0.15	0.15	0.01	0.02
Na	0.00	0.00	0.08	0.01
K	0.00	0.00	0.55	0.73
O[a]	12.00	12.00	11.00	11.00
# Cation	8.00	8.01	7.58	7.70
X Fe	0.78	0.78	0.50	0.54
XMg	0.07	0.07	0.50	0.46
XMn	0.10	0.10	0.00	0.00
X Ca	0.05	0.05		

[a]Number of oxygen per formula unit.

Table 13.7 Bulk compositions of the end-member dyke, dolomitic marble and talc rich bands in marble used for construction of pseudosection

Rock type	Basic Dyke	Talc zone of marble	Dolomitic Marble
Sample No.	L 81 (2)-c	M-23-q	
SiO_2	49.65	56.87	56.87
TiO_2	2.82	0.00	0.00
Al_2O_3	15.36	0.07	0.07
Cr_2O_3	0.03	0.00	0.00
FeO	12.81	0.20	0.20
MnO	0.54	0.00	0.00
MgO	5.45	31.91	31.91
CaO	10.55	0.95	0.95
Na_2O	0.88	0.00	0.00
K2O	0.16	0.00	0.00
Total	98.27	90.00	88.78

Pressure from tetrahedral Al within amphibole (Anderson and Smith, 1995) varies from 3 to 4.5 (± 0.6) kbar in the temperature range of interest. Stability of andalusite in the adjoining metapelites conform a maximum pressure of ~ 4 kbar. Owing to high anorthite content of plagioclase ($X_{An} \geq 0.87$), barometer that involve the assemblage garnet-amphibole-plagioclase could not be used.

Numerically constrained pseudosections with representative bulk rock composition (Table 13.2) suggests that the assemblage talc + chlorite ± quartz ± dolomite ± calcite can neither be derived through metamorphism of the dykes in presence of a hydrous fluid (Fig. 13.6a) nor through temperature induced cationic exchange with adjoining marble under anhydrous condition (Fig. 13.6b, Table 13.7). The assemblage can however be stabilized under a restricted compositional range of binary H_2O–CO_2 fluid present within the host rock in course of prograde metamorphism. A series of T–X_{Bulk} pseudosections at 4 kbar with anhydrous dyke as one end member and talc-rich bands in dolomitic marble containing ~10 weight percent of fluid as other shows that the assemblage is stabilized only when X_{CO2} in the talcose bands is less than 0.6, with maximum stability in the range of X_{CO2} ~ 0.3–0.5 (Fig. 13.7a–d). With higher concentration of CO_2 in the fluid, orthopyroxene and plagioclase are likely to be stabilized (Fig. 13.7d).

Paragenesis of Talc-Chlorite Rock

It is evident from the foregoing sections that the low variance metasomatic assemblage observed at the contact of the mafic dykes and dolomitic marble could have developed under a restricted range of P–T–X_{fluid}–X_{bulk} conditions. Assuming Al_2O_3 to be largely immobile in course of metamorphism- a feature commonly observed in many metamorphic rocks, it can be presumed that the assemblage essentially represents metasomatised dyke and not metasomatized talc-rich calc silicate

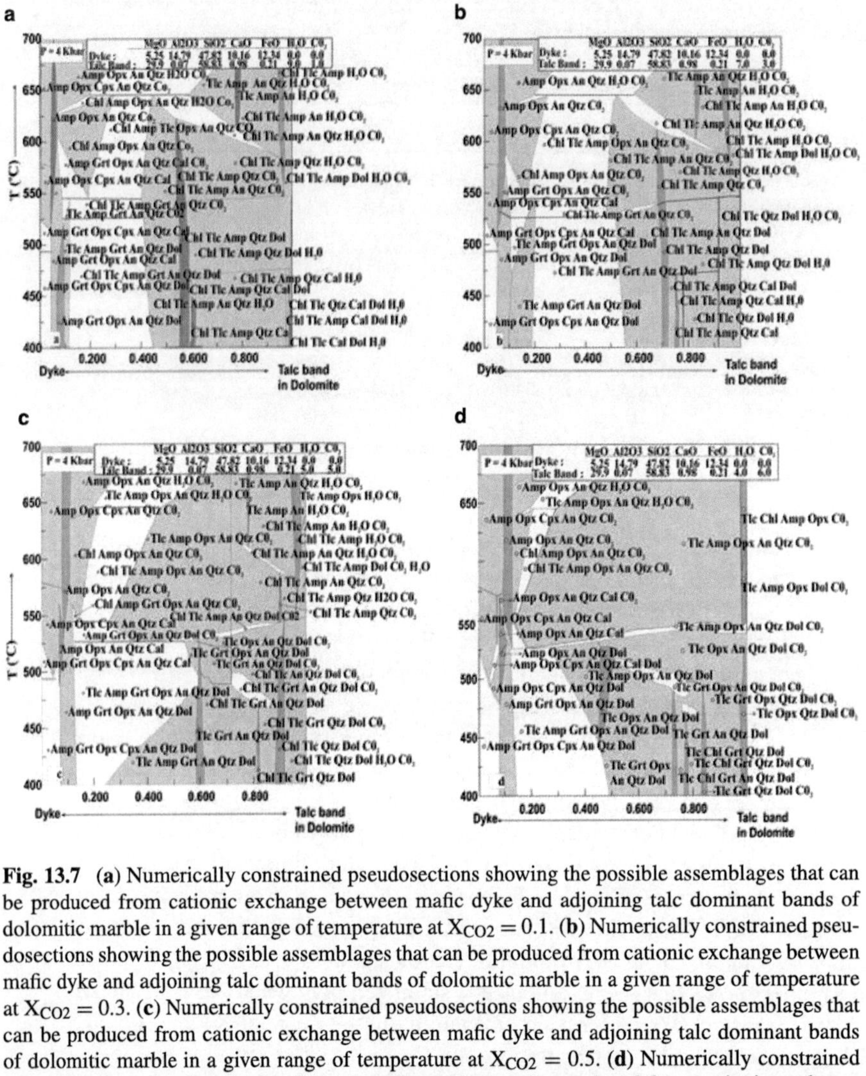

Fig. 13.7 (**a**) Numerically constrained pseudosections showing the possible assemblages that can be produced from cationic exchange between mafic dyke and adjoining talc dominant bands of dolomitic marble in a given range of temperature at $X_{CO2} = 0.1$. (**b**) Numerically constrained pseudosections showing the possible assemblages that can be produced from cationic exchange between mafic dyke and adjoining talc dominant bands of dolomitic marble in a given range of temperature at $X_{CO2} = 0.3$. (**c**) Numerically constrained pseudosections showing the possible assemblages that can be produced from cationic exchange between mafic dyke and adjoining talc dominant bands of dolomitic marble in a given range of temperature at $X_{CO2} = 0.5$. (**d**) Numerically constrained pseudosections showing the possible assemblages that can be produced from cationic exchange between mafic dyke and adjoining talc dominant bands of dolomitic marble in a given range of temperature at $X_{CO2} = 0.6$

bands present within dolomitic marble. Petrological features, however, suggest mass exchange (other than Al) between dolomitic marble and the mafic dyke. In the following sections an attempt will be made to explore the salient minerals reactions that were operative in the two contrasting lithologies. Owing to magnesian nature of the dolomite, talc and tremolite the chemical reactions will be construed in system $CaO–MgO–Al_2O_3–SiO_2–CO_2–H_2O$ (CMASV).

Extreme talc enrichment with almost total loss of CaO could have occurred in dolomitic marble by the reaction:

$$3\ \text{Dol} + 4\ \text{Qtz} / \text{SiO}_{2(aq)} + \text{H}_2\text{O} \rightarrow 3\ \text{Ca}^{+2} + \text{Tlc} + 4\ \text{CO}_2 \tag{1}$$

This requires permeation of aqueous silica and fluxing of Ca^{+2} from a system which is otherwise largely devoid of quartz. The parallelism of talcose bands to regional foliation planes and the strong preferred orientation of talc within these bands indicate that inhomogeneous permeation of silicic fluids occurred during D_1 deformation. The Ca^{+2} produced was fluxed out by the fluid (as is evident from presence of coarse calcite veins in the vicinity). But once the CaO was preferentially removed from the ambience, the talc continued to be stable until it breaks down to yield forsterite (Fig. 13.8a). In talc-chlorite rock that represents metasomatized mafic dyke rock, talc could form after magmatic pyroxenes by the reactions:

$$\text{Di} + \text{CO}_2 + \text{H}_2\text{O} \rightarrow \text{Tlc} + \text{CaCO}_3 + \text{Qtz} \tag{2a}$$

$$\text{Di} + \text{Dol} + \text{CO}_2 + \text{H}_2\text{O} \rightarrow \text{Tlc} + \text{CaCO}_3 \tag{2b}$$

$$2\text{Di} + \text{Dol} + \text{CO}_2 + \text{H}_2\text{O} \rightarrow \text{Tlc} + 3\ \text{Ca}^{+2} \tag{2c}$$

Presence of remnant talc lenses within chlorite attest to involvement of talc in chlorite formation. The later could have developed in the reaction:

$$\begin{aligned} 6.2\ \text{Pl} + 29\ \text{Ta} + 0.93\ \text{Ca}^{+2} + 13\ \text{CO}_2 &\rightarrow 10\ \text{Chl} + 6.5\ \text{Dol} \\ &+17.6\ \text{Qtz} + 84.3\ \text{Mg}^{+2} + 0.62\ \text{Na}^{+} + 9\ \text{H}_2\text{O} \end{aligned} \tag{3}$$

Reaction (3) operated in open system and involves considerable movement of different ions. The reaction has been balanced keeping in mind the volume proportions of chlorite, dolomite and quartz in the chlorite-rich bands. The reaction consumes plagioclase from mafic dyke and talc newly developed by reaction (1) and/or (2). The later also provides the Ca^{+2} required as a reactant. The fluid involved in the reaction has $X_{CO2} > 0.555$ which is within the limits of fluid composition discerned in the previous section. Consumption of plagioclase and talc in a different proportion could also have produced calcite instead of dolomite with chlorite by the reaction in system CMASV:

$$4\text{An} + \text{Tlc} + 17\text{Dol} + 15\ \text{H}_2\text{O} \rightarrow 4\ \text{Chl} + 21\text{Cal} + 13\ \text{CO}_2 \tag{4}$$

But petrographic study, coupled with Alizarin staining, shows the associated carbonate to be dominantly dolomite and therefore reaction (3) would have been more probable as compared to reaction (4). The Mg^{+2} released in the process was consumed in additional talc formation through magnesium metasomatism in the reaction:

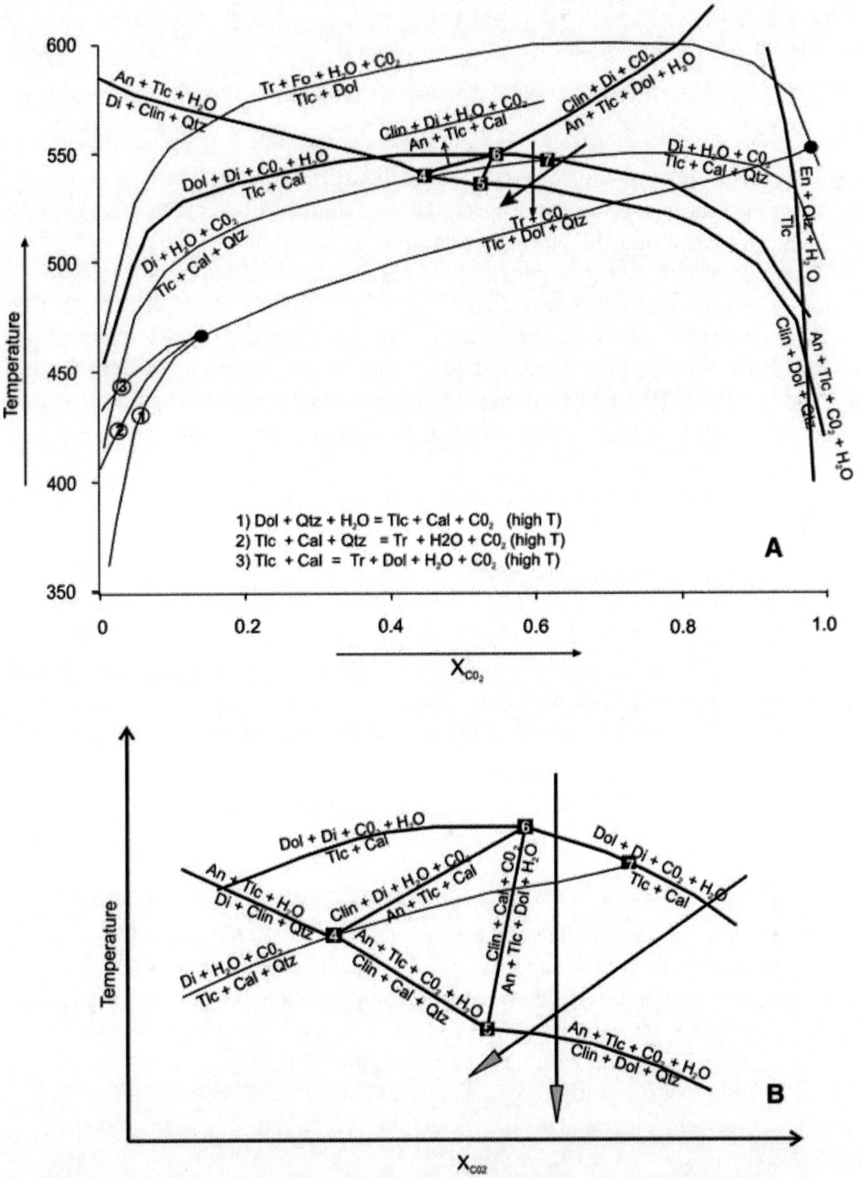

Fig. 13.8 (**a**) A portion of the T–X_{CO2} diagram constructed in pure Mg and Ca end member system with the help of the program THERMOCALC-326 (Powell and Holland, 1988). (**b**) An elaboration of the four numbered invariants (*filled squares* in Fig. 13.8a) and the univariants emanating from them. The two *arrow heads* represent the two possible paths for the reaction sequence– the *vertical arrow* shows the sequence occurring with fall in temperature in a fluid of constant composition while the inclined *arrow* is a path that should have been followed by the same reaction sequence should there been an increase in P_{H_2O} alongside drop in temperature

$$Dol + 3\,Mg^{+2}\,(aq) + 8H_4SiO_4.2H_2O\,(aq) = 2\,Tlc + 3Ca^{+2}\,(aq) + 6CO_2 + 30\,H_2O \quad (5)$$

Thus while Ca^{+2} produced in reaction (1) and (2) is taken up by reaction (3), the Mg^{+2} released in the later gets consumed in a cyclic talc-forming process in reaction (5) and no large scale fluid supported ion transport in and out of the system is required.

Figure 13.8a represent a portion of the T–X_{CO2} diagram constructed in pure Mg and Ca end member system with the help of the program THERMOCALC (v.3.26 Powell and Holland, 1988) and Fig. 13.8b is an elaboration of the four numbered invariants in Fig. 13.8a and the univariants emanating from them. It is evident from the diagrams that the sequence of talc formation followed by chlorite results from a mere drop in temperature at around $X_{CO2} \leq 0.6$ with or without any increase in P_{H2O} in the system (Fig. 13.8b).

Thus binary fluid permeation within the mafic dykes accompanied by cationic exchange (restricted to outer few centimeters, at best half a meter) between dykes and talc-rich magnesio-silicate layers in adjoining dolomitic marble led to the development of the metasomatic assemblage talc + chlorite ± quartz ± dolomite ± calcite. The core of the dyke retained pristine magmatic mineralogy during D_1 that underwent change only in post-D_1-pre-D_2 phase under elevated temperature and gave rise to the assemblage hornblende + plagioclase + quartz.

Estimation of Fluid Flux

A volumetric estimation of integrated fluid flux into the system in course of metasomatism can be made from calculations involving reaction progress variable ($\xi = \Delta n_x/\nu_x$, where Δn_x is the number of moles of the species x produced or consumed in the reaction and υ_x is the coefficient of the species x in the reaction; Rumble, 1982; Ferry and Burt, 1982; Spear, 1993) coupled with composition of the final equilibrium fluid, which, in its turn is a function of composition of the initial pore fluid (n_x^i being the initial number of moles of species x i.e. CO_2 for mixed H_2O–CO_2 equilibrium), composition of the fluid produced during reaction (n_x^p being the number of moles of species x produced by the reaction) and composition of the fluid infiltrating the system from an external reservoir (n_x^a the number of moles of species x added to the rock by infiltration from an external source). Substituting $n_{H_2O}^p = \upsilon_{H_2O}.\,\xi$ and $n_{CO_2}^p = \upsilon_{CO_2}.\,\xi$ in the relation:

$$X_{CO_2}^f = n_{CO_2}^f / (n_{CO_2}^f + n_{H_2O}^f) = (n_{CO_2}^i + n^p\,CO_2 + n^a\,CO_2) / (n_{CO_2}^i + n^p\,CO_2 + n_{CO_2}^a + n_{H_2O}^i + n_{H_2O}^p + n_{H_2O}^a) \text{ (R1, Spear, 1993)}$$

and putting n_{total}^i = (porosity) × (volume of rock)/(molar volume of fluid),

$n^i_{CO_2} = X^i_{CO_2} \cdot n^i_{total}$ and $n^i_{total} = (n^i_{CO_2} + n^i_{H_2O})$, a value of ξ can be derived in terms of mole fraction of fluids :

$$\xi = -[n^i_{total}(X^f_{CO_2} - X^i_{CO_2}) + n^a_{total}(X^f_{CO_2} - X^a_{CO_2})] / [X^f_{CO_2} \cdot \upsilon_{H_2O} - X^f_{H_2O} \cdot \upsilon_{CO_2}] \text{ (R2, Spear, 1993)}$$

For the present case, porosity per unit volume of the rock (i.e. per 100 cc) has been assumed to be 1% since higher porosity is unlikely for a rock undergoing deformation and metamorphic recrystallisation under greenschist facies conditions. The composition of the initial pore fluid is unknown. But since 1% porosity accounts for only very small amounts of fluid, the pore fluid composition will not be a significant factor in determining the final fluid composition. Therefore, the initial pore fluid composition has been assumed to be $X^i_{CO_2} = 0.9$. Also, for the sake of simplicity, it has been assumed that the infiltrating fluid is pure H_2O i.e. $X^a_{CO_2} = 0$. In effect, this leads to a conservative estimate of the fluid flux because any amount of $X^a_{CO_2}$ in the infiltrating fluid will only increase the amount of fluid required to achieve a given amount of reaction. The composition of the fluid produced during reaction i.e. $X^f_{CO_2}$ can be noted directly from Fig. 13.8. The number of moles of products required for calculation of ξ is obtained by dividing the modal volume of the product phases by their respective molar volumes (assuming that the volume of the solids remains practically constant under changed pressure-temperature conditions). υ_{H_2O} and υ_{CO_2} are noted directly from the reaction relations concerned.

With such inputs, n^a_{total} can be determined directly from the relation R2 and since molar volume of H_2O under the given P–T conditions is known from standard tables, the volume of infiltrating fluid is obtained as:

$$V^a_{fluid} = n^a_{total} \cdot \text{ molar volume of fluid at P-T condition of interest (Spear,1993)}$$

Once the volume of fluid infiltrated is known, the fluid/rock ratio can be calculated as:

$$F/R = \text{volume of infiltrated fluid/volume of rock considered} = V^a_{fluid} / 100.$$

The calculations show that for 10–80% talc production from dyke–dolomite interaction through reaction (2c), the volume of fluid that needs to interact with the rock is ~23–30% the volume of rock (Table 13.8). Similarly, a significant volume of fluid (nearly 25–42% the volume of rock) is required for chlorite production from talc in reaction (3). The corresponding fluid to rock volumetric ratio is 0.25–0.42 for 10–70% talc consumption in chlorite-poor and chlorite-rich layers and lenses respectively, Table 13.8).

Table 13.8 Fluid -rock ratio determined from modal abundances of the phases produced during metamorphism

Talc formation from Diopside (Reation 2c)							
vol rock (cc)	vol pore sp (cc)	$X(CO_2)$ initial	$X(CO_2)$ final	$X(CO_2)$ influx	talc (%) produced	n (talc)	F/R
100	1	0.90	0.6	0	10	0.151	0.23
100	1	0.90	0.6	0	50	0.755	0.27
100	1	0.90	0.6	0	80	1.209	0.30
Chlorite formation through Talc consumption (Reaction 3)							
vol rock (cc)	vol pore sp (cc)	$X(CO_2)$ initial	$X(CO_2)$ final	$X(CO_2)$ infl	talc (%) consumed	n (talc)	F/R
100	1	0.90	0.6	0	10	0.15	0.25
100	1	0.90	0.6	0	70	1.05	0.42

Concluding Remarks

This study demonstrates that a suite of mafic dykes that were emplaced within a sedimentary sequence (sandstone-shale-limestone) of Palaeoproterozoic Mahakoshal Group underwent superposed deformation and metamorphism. Fluid–rock interaction in a partially open system triggered substantial mass and fluid transfer across the contact between mafic dyke and enclosing dolomitic marble. This metasomatic process led to development of talc-chlorite rock at the marginal part of the mafic dyke. Quantitative geothermobarometry and numerically computed phase diagram indicate temperature of c.480 ± 22 °C during metasomatic alteration operative broadly within 10–15 km of crustal depth (c.4.5 ± 0.6 kbar). The mineralogical transformation of mafic dykes that was augmented by metasomatic alteration considerably changed the physical properties of rocks and allowing concentration of strain at the marginal part that in turn developed a pervasive foliation that is missing at the central part of the dyke.

Acknowledgement KC acknowledges C.S.I.R New Delhi for a doctoral research fellowship. PS acknowledges the fellowship of Alexander von Humboldt Foundation, Germany that enables us to generate analytical data at Mineralogy Petrology Institute, University of Bonn. We express our gratitude to M. M. Raith the then Director of Mineralogy Petrology Institute, University of Bonn for extending the laboratory facility and many stimulating discussions on several aspects of the present communication. Rajesh Sharma is thanked for his constructive commebnts on the earlier version of MS which helped lot in improving the MS. Field work for this study was sponsored by the CAS, Department of Geological Sciences, Jadavpur University.

References

Abart R (1995) Phase equilibrium and stable isotope constraints on the formation of metasomatic garnet-vesuvianite veins (SW Adamello, N. Italy). Contrib Mineral Petrol 122: 116–133

Ague JJ (2003) Fluid infiltration and transport of major, minor and trace elements during regional metamorphism of carbonate rocks, Wepawaung Schist, Connecticut, USA. American Jour Sci 303: 753–816

Anderson JL, Smith DR (1995) The effects of temperature and f_{O2} on the Al-in-hornblende barometer. Am Mineral 80: 519–559

Blundy JD, Holland TJB (1990) Calcic amphibole equilibria and a new amphibole plagioclase geothermometer. Contrib Mineral Petrol 104: 208–224

Chakraborty K (2009) Petrology of a suite of Metapelites and Metacarbonate rocks in and around Jabalpur. PhD. thesis, Jadavpur University

Connolly JAD (2005) Computation of phase equilibria by linear programming: A tool for geodynamic modeling and its application to subduction zone decarbonation. Earth Planet Sci Lett 236: 524–541

Ferry JM, Burt DM (1982) Characterization of mineral fluid composition through mineral equilibria. In: Reviews in mineralogy, mineralogical society of America, Chantilly, U.S.A, Ferry JM, (ed), vol 10. 207–262

Holdaway MJ (2000) Application of new experimental and garnet Margules data to the Garnet-Biotite geothermometer. Am Mineral 85: 881–892

Leake BE, Woolley AR, Arps CES, Birch WD, Gilbert MC, Grice JD, Hawthorne FC, Kato A, Kisch HJ, Krivovichev VG, Linthout K, Laird J, Mandarino JA, Maresch WV, Nickel EH, Rock NMS, Schumacher JC, Smith DC, Stephenson NCN, Ungaretti L, Whittaker EJW, Youzhi G (1997) Nomenclature of amphiboles: Report of the subcommittee on amphiboles of the international mineralogical association, commission on new minerals and mineral names. Am Mineral 82: 1019–1037

Philpotts PJ, Ague JJ (2009) Igneous and metamorphic petrology. Cambridge University Press, London

Powell R, Holland TJB (1988) An internally consistent thermodynamic dataset with uncertainties and corrections: 3: application methods, worked examples and a computer program. Jour MetGeol 6: 173–204

Roy A, Devrajan MK (2000) A reappraisal of the stratigraphy and tectonics of the Palaeo-Proterozoic Mahakoshal supracrustal belt, Central India. *In* Precambrian crust of Central and Eastern India. GSI Special Publ 57: 226–340

Roy A, Prasad HM (2003) Tectonothermal events in Central Indian Tectonic Zone (CITZ) and its implication in Rhodinian crustal assembly. Jour Asian Earth Sci 22: 115–129

Rumble DIII (1982) Stable isotope fractionation during metamorphic devoltilization reactions. Rev Mineral 10: 327–353

Spear FS (1993) Metamorphic phase equilibria and pressure-temperature-time paths, monograph 1. Mineralogical Society of America, Washington DC

Chapter 14
Petrology and Geochemistry of Metamorphosed Basic Intrusives from Chilka Lake Granulites, Eastern Ghats Belt, India: Implications for Rodinia Breakup

Sankar Bose, Kaushik Das, Supriya Chakraborty, and Hiroyuki Miura

Introduction

Mafic granulites represent important lithological member in most of the high-grade terrains, but their proper characterization is often attempted with caution. These rocks either represent metamorphosed basic sediment undergoing high pressure (HP) to ultra high pressure (UHP) metamorphic history (Boven et al., 1999; Galan and Marcos, 2000; Bhowmik and Roy, 2003; Garrido et al., 2006; Gross et al., 2009; Das et al., 2009), or as intrusives undergoing subsequent granulite-facies modification (Hoatson et al., 2005). While the former furnishes crucial information on deep-rooted crustal dynamics during the mountain building processes (Zhang et al., 2006; Liu et al., 2009), the latter provides information on mantle-derived thermal pulses in the lower crust (Hoatson et al., 2005). It is well-known that if the intrusive mafic rocks subsequently undergo granulite-facies metamorphism, all the primary magmatic characters get erased. However, field, petrographic and geochemical signatures sometimes preserve traces of the pre-metamorphic history of such rocks. In high-grade metamorphic terrains, post-emplacement granulite metamorphism substantially modifies major element chemistry owing to various degrees of fluid- and/or melt-induced metamorphic and metasomatic processes. Immobile trace element and REE chemistries, on the other hand, remain more robust and are likely to preserve signatures of pre-metamorphic history.

S. Bose (✉)
Department of Geology, Presidency College, Kolkata 700073, India
e-mail: sankar.bose@gmail.com

K. Das
Department of Geology, Bengal Engineering and Science University, Howrah 711103, India

S. Chakraborty
Department of Geology, Presidency College, Kolkata 700073, India

H. Miura
Department of Natural History Sciences, Hokkaido University, Sapporo 0600810, Japan

R.K. Srivastava (ed.), *Dyke Swarms: Keys for Geodynamic Interpretation*,
DOI 10.1007/978-3-642-12496-9_14,

Eastern Ghats Belt (EGB) of India exposes deep crustal section in 1,000 km long edifice along the eastern coastline of India. A litho-ensemble of diverse chemistry, this has long been recognized as a mobile belt that sutured cratonic blocks of India and east Antarctica during the assembly of the supercontinent Rodinia. With constant input from petrological, structural and isotopic data from two terrains in recent times, this correlation appears to be compelling. Despite the presence of mafic granulites in different locales in EGB, few petrological data are available from this rock (Bhui et al., 2007). So far, petrological data are available mostly from the large mafic-ultramafic intrusive complex from southern part of EGB, presently metamorphosed to granulite facies (Leelanandam, 1990, 1997; Leelanandam and Kumar, 2007). Excepting these, other studies are mainly focused on deciphering the metamorphic overprint on the mafic granulites (Dasgupta et al., 1993; Das et al., 2009). The pre-metamorphic history of most of the mafic granulites in the EGB is thus poorly known. However, those of the Chilka Lake area of northern part of the EGB are chosen for the present study as these could provide petrotectonic inputs which may help to understand the evolution of EGB in a holistic manner.

Geological Background

The study area is located north of the Chilka Lake anorthosite massif (Fig. 14.1) which belongs to the Eastern Ghats Province within the Proterozoic Eastern Ghats Belt (Dobmeier and Raith, 2003). Granulite facies rocks of diverse chemistry from this area preserve multiple stages of deformation and metamorphism (Bhattacharya et al., 1994; Sen et al., 1995; Dobmeier and Raith, 2000). The deformation history is complex with fold interference of four generations (D_1–D_4). A major part of the deformation history is interpreted to be dominated by transpressive regime when the nature of deformation was chiefly shear-induced (Dobmeier and Simmat, 2002). Metamorphic evolution has been construed from aluminous granulites and charnockites (Sen et al., 1995) which suggests multistage decompression-followed by-cooling retrogressive path ensued from UHT (1,100°C) high-*P* (> 10 kbar) peak. An event of UHT thermal metamorphism (>1,000°C) has also been characterized from Fe-Al-rich granulites and calc-silicate granulites along the contact of the anorthosite massif (Raith et al., 2007; Sengupta et al., 2008). Available geochronological data suggest that the earliest metamorphic signature dated back to 1,100–940 Ma followed by a sustained thermal regime during 800–500 Ma (Simmat and Raith, 2008; Upadhyay et al., 2009). The thermal metamorphism was operative during the emplacement of anorthosite, until recently, which was dated as c. 792 Ma (Krause et al., 2001; Dobmeier and Simmat, 2002). A recent TIMS U-Th-Pb age from the same anorthosite body unravels a much older emplacement age of c. 983 Ma (Chatterjee et al., 2008).

Two-pyroxene bearing mafic granulites form an integral part of a mixed rock suite that remain petrologically uncharacterized so far. The present endeavor therefore is to attempt a proper characterization of these rocks and their possible role in the tectonic evolution of the crustal growth history of the northern part of EGB, or

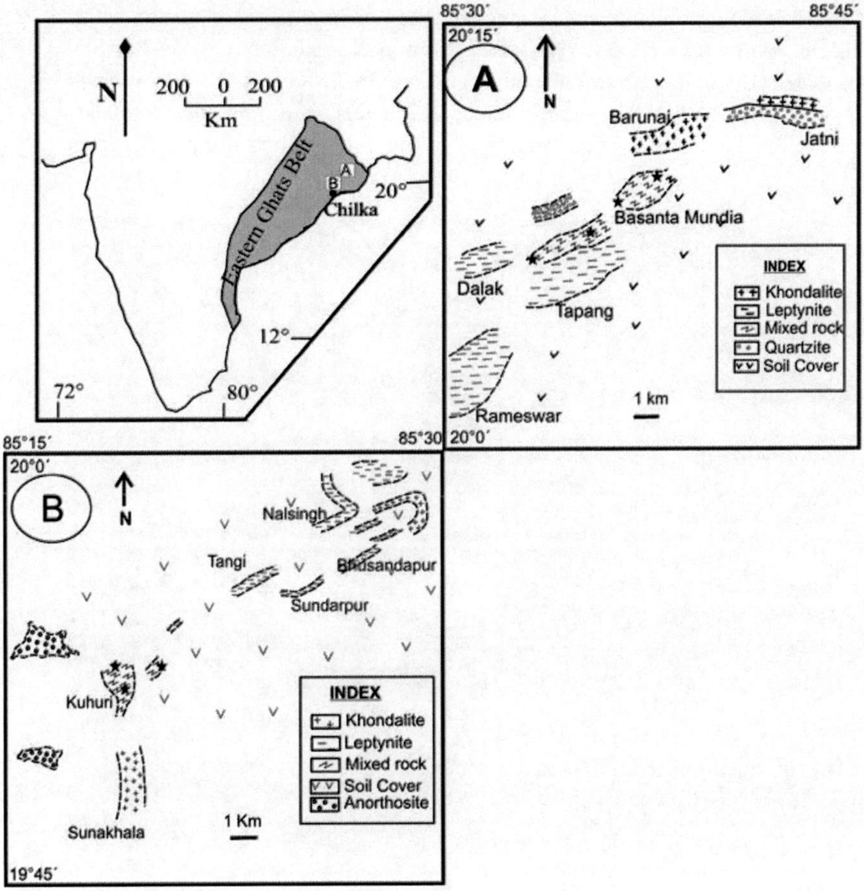

Fig. 14.1 Lithological map of the study area within the broad framework of EGB. Boxes A and B in the *inset* figure represent two *blocks* shown in the enlarged maps. Sample locations are shown by *star* fills (⋆)

the Chilka domain after Dobmeier and Raith (2003). This eventually leads to the evolutionary history of the supercontinent Rodinia of which EGB was an integral part. Mineral abbreviations used in this paper are followed after Kretz (1983).

Field Relations and Sample Description

Detailed mapping over an area of 400 sq. km north-west of the Chilka Lake anorthosite massif has identified several types of high-grade granulites. Major part of the area is composed of metapelitic rocks having the mineral assemblage Grt-Sil-Ilm-Qtz-Kfs-Pl (khondalite). Garnetiferous quartzofeldspathic gneiss (leptynite) and quartzite also occur as distinct rock units (Fig. 14.1). Mafic granulite

(Opx-Cpx-Pl-Ilm-Mag) occurs as a part of the "mixed suite of rocks" with charnockitic gneiss (Opx-Pl-Kfs-Qtz-Ilm), aluminous granulite (Grt-Spl-Spr-Sil-Crd-Qtz-Pl-Kfs-Bt) and quartzofeldspathic gneiss (Grt-Pl-Kfs-Qtz-Ilm) in various combinations (Fig. 14.2a). The quartzofeldspathic gneiss is migmatitic in appearance and occurs to host the other members of the suite, which usually occur as

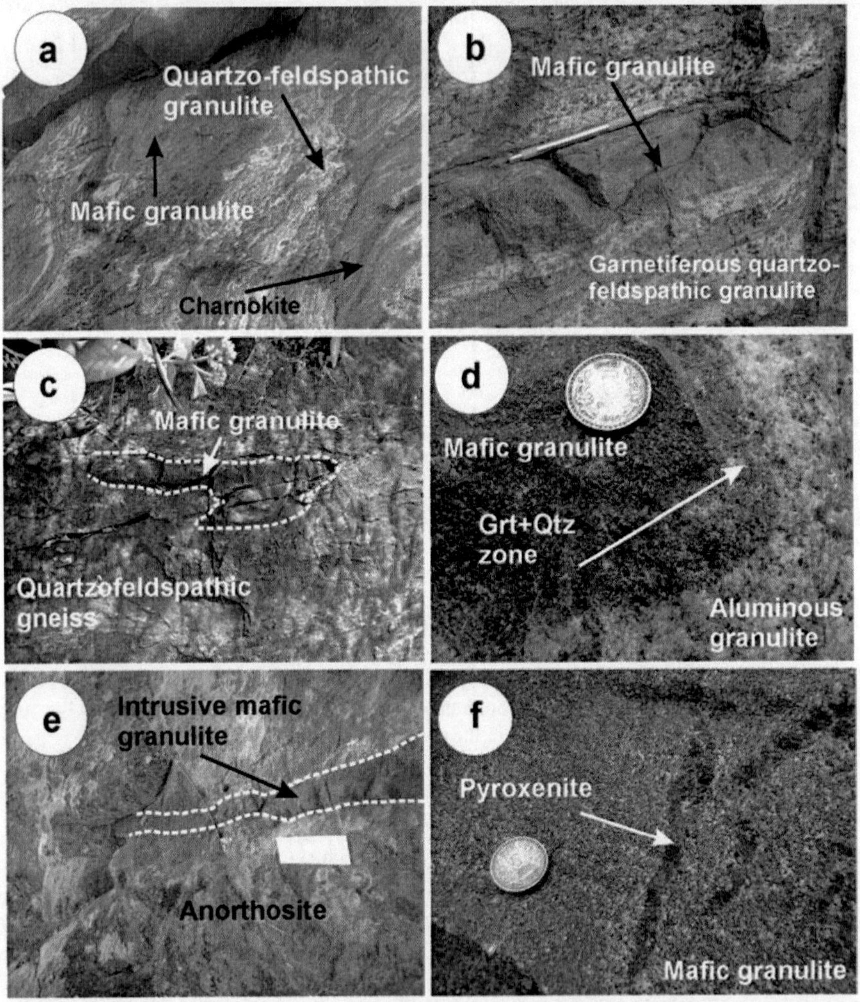

Fig. 14.2 Field features of the studied mafic granulites. (**a**) Occurrence of mafic granulite layers within the mixed suite of rocks consisting of quartzofeldspathic granulite, charnockite and aluminous granulite (not shown in this figure). (**b**) Mafic granulites layers paralleling the transposed S_2/S_3 foliation within the garnetiferous quartzofeldspathic gneiss. (**c**) Tight isoclinal folding of the mafic granulites layers by D_2–D_3 deformation. (**d**) The contact of mafic granulites with aluminous granulite within the mixed suite of rocks. Note the presence of garnet + quartz-bearing zone. (**e**) Mafic granulite layer intruding the anorthosite. (**f**) Freshly exposed surface of mafic granulite showing thin layers of ultramafic granulite (pyroxenite)

centimemer to meter-thick bands and lenses along the pervasive penetrative S_2 gneissic foliation (Fig. 14.2b) developed during D_2 deformation. S_2-foliation forming assemblages in metapelitic granulite contain porphyroblastic garnet, which includes internal folded foliation marked by biotite-sillimanite-quartz grains. This is the earliest foliation S_1 found in this region. The S_2 foliation shows tight isoclinal folding during D_3 and the S_3 foliation almost parallels the S_2 fabric at limbs resulting transposed S_2/S_3 fabric except the regional F_3 fold closures. However, these rocks are folded by subsequent deformation (D_2/D_3-D_4; Fig. 14.2c). All these features suggest that the mafic intrusion is pre- to syn-D_2 in broad time frame. The contact between the mafic granulite and associated rocks is sharp, but when the mafic granulite occurs adjacent to the aluminous granulite, the contact is marked by mm to cm-thick garnet-bearing zone (Fig. 14.2d). This zone either represents a metamorphosed chilled margin, or a post-emplacement metasomatic zone. Further south, the mafic granulite layers are found to intrude the anorthosite massif (Fig. 14.2e). The contact, however, is devoid of any mineral zone. The rock itself appears massive in freshly cut quarry faces, but shows crude foliation by planar arrangement of flat pyroxene and plagioclase grains in weathered surface. Discrete layers of pyroxenite are locally present within the mafic granulite (Fig. 14.2f).

The rocks sampled for the present study were collected from numerous quarry faces in between Kurarmal and Kuhuri area (Fig. 14.2a). While field relations were studied from different exposures, fresh samples from the quarry faces were used for petrographic and geochemical study. Apart from the mafic granulites, one sample from the Chilka Lake anorthosite massif was collected from the Sunakhala area (GPS 19°50.34´N, 85°18.53´E).

Petrography

The mafic granulites show substantial mineralogical variation. All the rocks contain mineral association orthopyroxene, clinopyroxene, plagioclase, ilmenite, magnetite and quartz, where modal variation in pyroxene grains is conspicuous. The rocks are coarse- to medium-grained without any preferred alignment. All the samples exhibit a granoblastic fabric consisting of orthopyroxene, clinopyroxene, plagioclase, ilmenite and magnetite. Quartz is present small amount. Texture shows strong imprints of metamorphism and deformation and relic igneous textures are hardly discernible. However, some porphyroblastic orthopyroxene grains contain lamallae and patches of exsolved clinopyroxene and vice versa (Fig. 14.3a). Such exsolution textures in pyroxene are commonly developed during subsolidus cooling. On the other hand, these could well be developed during metamorphic cooling as similar pyroxene exsolution textures are documented from UHT metamorphosed rocks (Harley, 1987). In the present case, it is difficult to ascertain which one was actually responsible for such textures. Although most of the igneous characters were lost during metamorphism, it was still possible for the rocks to retain such delicate intragrain fossil textures (cf. Frost and Chacko, 1989). Orthopyroxene and clinopyroxene grains form granoblastic aggregate in most of the samples. Variable amount

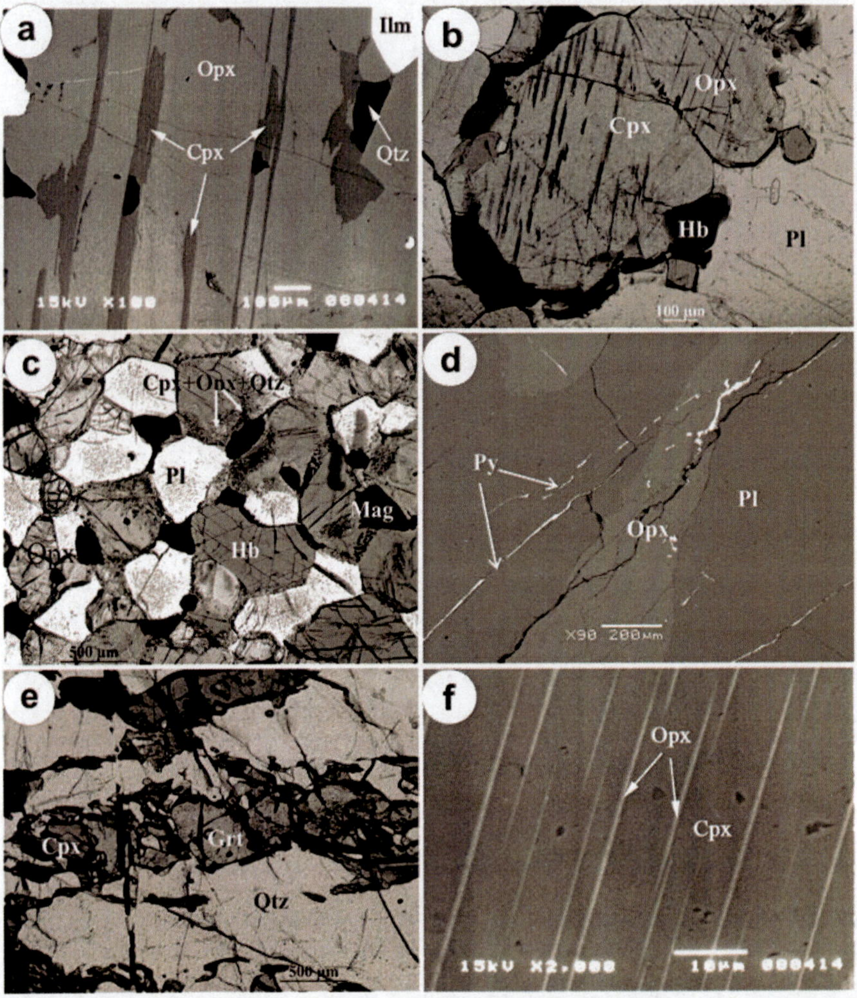

Fig. 14.3 Photomicrographs and BSE images showing textural features of mafic granulites and ultramafic granulite. (**a**) BSE image showing porphyroblastic orthopyroxene (Opx) contains exsolved lamellae and patches of clinopyroxene (Cpx). (**b**) Hornblende (Hb) replaces orthopyroxene (Opx) and clinopyroxene (Cpx) grains along the grain boundaries. (**c**) Prismatic hornblende (Hb) contains tiny intergrowth of granular pyroxene and quartz (Opx + Cpx + Qtz) at the grain periphery. (**d**) Pyrite (Py) veins cut through orthopyroxene (Opx) and plagioclase (Pl) grains (e) Discrete grains of garnet (Grt) and layers of quartz (Qtz) around clinopyroxene (Cpx) at the contact of mafic granulite and quartzofeldapathic gneiss. (**f**) BSE image showing megacrystic clinopyroxene (Cpx) in ultramafic granulite contains exsolved lamellae of orthopyroxene (Opx)

of hornblende and biotite are present in all the samples. These grains, particularly the hornblende grains, replace clinopyroxene and/or orthopyroxene along the fractures and margins (Fig. 14.3b). Both pyroxene and plagioclase grains show evidence of deformation by recrystallization and grain refinement. Some of the hornblende grains show tiny intergrowth of granular pyroxene and quartz along the margin (Fig. 14.3c). Veining by tiny sulfide minerals like pyrite and pyrrhotite is common within pyroxene and feldspar grains (Fig. 14.3d). This must have taken place due to fluid-induced processes during later stage of evolution.

Garnet appears only as discrete layers along with quartz at the contact of mafic granulite and the aluminous granulite (Fig. 14.3e). Profuse amount of quartz is present at this contact zone. Primary quartz is usually absent in the mafic granulites, but microscale quartz veins are prominent near such contact zones.

Layers and lenses of ultramafic granulite are locally present within the mafic granulite. These are essentially composed of clinopyroxene and orthopyroxene with subordinate amount of plagioclase and Fe-Ti oxides (ilmenite and magnetite). The overall texture is granoblastic with pyroxene megacrysts. Both the varieties of pyroxene show exsolution texture (Fig. 14.3f).

Mineral Chemistry

Chemical composition of representative samples were analyzed using JEOL-JXA 8,600 Electron Probe Micro Analyzer at the USIC, IIT Roorkee, India and JEOL-JSM 5,300 Electron Micro Probe at the Yokohama National University, Japan. Both the machines were operated at 15 KV acceleration voltage using 15 nA specimen current. 1–2 μm probe diameter for the spot size was used for point analyses. Natural standards were used and raw data were corrected by ZAF program. Same grains were analyzed in both the instruments to check the accuracy of the results, which remains satisfactory. Five samples of the mafic granulites were analyzed in detail and the analytical data are presented in Tables 14.1 and 14.2. Out of these, samples Ch-2B and Ch-KHQ6 preserve granulite-facies assemblage with minor presence of hornblende or biotite grains. Sample Ch-KHQ11 is retrogressed granulite with abundant hornblende grains. The sample Ch-BT6 represents the garnet-quartz-bearing contact zone of the mafic granulite shown in the Fig. 14.2d. Additionally, one sample of the ultramafic granulite (sample Ch-BT13) was also analyzed and the representative data are presented in Tables 14.1 and 14.2.

All the phases show extreme variation in phase chemical characters as given in Tables 14.1 and 14.2. Clinopyroxene grains show X_{Mg} values from 0.32 to 0.33 (sample Ch-BT6) to 0.75–0.81 (sample Ch-KHQ11). All the grains contain variable amount of Al, but Na content is negligible. TiO_2 and Cr_2O_3 contents are also low (<0.5 wt%). In sample Ch-2B, it shows minor Al-zoning from core (1.35 wt% Al_2O_3) to rim (1.96 wt% Al_2O_3) without any perceptible change in X_{Mg} value. In sample EG-KHQ6 however, the Al_2O_3 zoning is normal (2.8 wt% at core and 2.4 wt% at rim) with X_{Mg} values 0.64–0.66. Maximum alumina is present in clinopyroxene grains of the sample Ch-KHQ11, where it reaches up to 5.2 wt% with $X_{Mg} = 0.81$. The garnetiferous variety sample Ch-BT6 is least magnesian with clinopyroxene is Al-poor (1.8–1.4 wt% Al_2O_3) and iron-rich (X_{Mg} = 0.35–0.32). Orthopyroxene grains are comparatively less aluminous and less

Table 14.1 Representative mineral composition of orthopyroxene and clinopyroxene grains from studied granulites

Sample	Ch-BT6				Ch-2B		Ch-KHQ11	Ch-BT13		Ch-2B		Ch-KHQ11	Ch-BT6		Ch-BT13
Phase	Cpx									Opx					
Analysis No	3	6	24	69	24	31	13	2	5	9	12	16	5	44	6
SiO_2	48.65	48.74	48.41	49.04	50.85	51.12	51.41	52.83	53.76	48.14	48.67	53.50	46.34	46.44	54.78
TiO_2	–	0.27	–	–	–	–	0.87	0.20	0.19	–	0.40	–	–	–	0.07
Al_2O_3	1.62	1.57	1.39	1.79	1.96	1.35	5.22	1.64	1.52	0.54	1.31	2.50	0.88	0.67	1.28
Cr_2O_3	–	–	–	–	–	0.41	–	0.35	0.33	–	–	0.43	–	–	1.10
FeO	23.52	22.97	24.62	25.17	15.28	16.18	6.45	4.48	4.11	35.90	34.10	16.69	43.26	45.53	13.75
MnO	–	–	–	–	–	–	–	0.10	0.08	0.50	0.61	–	0.67	–	0.18
MgO	6.38	6.80	6.44	6.25	11.03	10.92	13.32	15.66	16.20	14.32	13.88	25.97	7.63	6.30	28.28
CaO	19.22	18.91	18.54	17.51	20.45	19.71	22.39	22.65	23.21	0.57	0.91	0.52	1.20	0.69	0.57
Na_2O	–	–	–	–	–	–	–	0.36	0.40	–	–	–	–	–	–
Total	99.38	99.27	99.40	99.77	99.56	99.69	99.67	98.27	99.84	99.97	99.86	99.61	99.98	99.63	99.15
O-basis															12
Si	3.89	3.89	3.89	3.91	3.90	3.93	3.79	3.93	3.94	3.86	3.87	3.89	3.87	3.92	3.85
Ti	–	0.02	–	–	–	–	0.05	0.01	0.01	–	0.02	–	–	–	0.00
Al	0.15	0.15	0.13	0.17	0.18	0.12	0.45	0.14	0.13	0.05	0.12	0.22	0.09	0.07	0.12
Cr	–	–	–	–	–	0.03	–	0.02	0.02	–	–	0.01	–	–	0.01
Fe^{2+}	1.58	1.53	1.65	1.68	0.98	1.04	0.40	0.28	0.25	2.41	2.27	1.01	3.02	3.21	0.83
Mn	–	–	–	–	–	–	–	–	–	0.03	0.04	–	0.05	–	0.01
Mg	0.76	0.81	0.77	0.74	1.26	1.25	1.46	1.74	1.77	1.71	1.64	2.81	0.95	0.79	3.04
Ca	1.65	1.62	1.60	1.50	1.68	1.62	1.77	0.90	0.91	0.05	0.08	0.04	0.11	0.06	0.02
Na	–	–	–	–	–	–	–	0.05	0.05	–	–	–	–	–	–
Total	8.03	8.02	8.04	8.00	8.01	7.99	7.92	7.08	7.08	8.11	8.05	7.98	8.09	8.05	7.87
X_{Mg}	0.33	0.35	0.32	0.31	0.56	0.55	0.79	0.86	0.87	0.42	0.42	0.74	0.24	0.20	0.78

Table 14.2 Representative mineral composition of garnet, plagioclase, ilmenite, magnetite and hornblende grains from studied granulites

Sample	Ch-2B	Ch-BT13		Ch-BT6				Ch-KHQ11		Ch-KHQ6	Ch-KHQ11
Phase	Pl			Gt				Hb		Ilm	
Analysis No	22	6	19	39	52	54	56	4	19	6	21
SiO_2	53.97	49.43	52.65	35.66	35.95	35.93	36.51	44.34	44.27	0.43	0.30
TiO_2	–	0.01	0.02	–	–	–	–	1.75	1.32	44.98	49.09
Al_2O_3	28.90	30.86	28.85	20.14	20.80	20.42	20.53	14.19	13.96	0.34	–
Cr_2O_3	–	0.01	–	–	–	–	–	–	–	–	–
Fe_2O_3[a]	–	–	–	–	–	–	–	–	–	15.26	9.30
FeO	0.46	0.08	0.10	34.54	33.53	34.38	33.73	9.92	9.24	38.04	37.41
MnO	0.37	–	–	0.97	0.98	1.35	1.07	0.44	–	–	1.25
MgO	–	–	0.02	1.79	1.63	1.06	1.06	14.38	15.39	1.42	3.26
CaO	10.33	14.73	12.48	6.63	6.92	6.86	6.95	12.31	12.63	0.30	–
K_2O	0.25	0.05	0.25	–	–	–	–	1.34	1.43	–	–
Na_2O	5.41	3.06	4.26	–	–	–	–	1.05	1.32	–	–
Total	99.68	98.26	98.63	99.73	99.81	100.00	99.85	99.72	99.55	100.77	100.60
O-basis	12							20		3	
Si	3.68	3.44	3.62	2.92	2.93	2.94	2.97	5.50	5.49	0.01	0.01
Ti	–	–	–	–	–	–	–	0.16	0.12	0.84	0.91
Al	2.32	2.52	2.34	1.95	2.00	1.97	1.97	2.08	2.04	0.01	–
Cr	–	–	–	–	–	–	–	–	–	–	–
Fe^{3+a}										0.29	0.17

Table 14.2 (continued)

Sample	Ch-2B	Ch-BT13		Ch-BT6				Ch-KHQ11		Ch-KHQ6	Ch-KHQ11
Phase	Pl			Gt				Hb		Ilm	
Analysis No	22	6	19	39	52	54	56	4	19	6	21
Fe^{2+}	0.03	–	0.02	2.37	2.28	2.35	2.30	1.03	0.96	0.79	0.77
Mn	0.02	–	–	0.07	0.07	0.09	0.07	0.05	–	–	0.03
Mg	–	–	–	0.22	0.20	0.13	0.13	2.66	2.84	0.05	0.12
Ca	0.75	1.10	0.92	0.58	0.60	0.60	0.61	1.63	1.68	0.01	–
K	0.02	0.00	0.02	–	–	–	–	0.21	0.23	–	–
Na	0.72	0.42	0.56	–	–	–	–	0.25	0.32	–	–
Total	7.53	7.50	7.48	8.11	8.07	8.08	8.04	13.57	13.68	2.00	2.00
X_{Mg}	–	–	–	0.08	0.08	0.05	0.05	0.72	0.75	–	–
X_{Ab}	0.48	0.28	0.37	–	–	–	–	–	–	–	–
X_{An}	0.51	0.72	0.61	–	–	–	–	–	–	–	–
X_{Or}	0.01	0.00	0.00	–	–	–	–	–	–	–	–
X_{Prp}	–	–	–	0.07	0.06	0.04	0.04	–	–	–	–
X_{Alm}	–	–	–	0.73	0.72	0.74	0.74	–	–	–	–
X_{Sps}	–	–	–	0.07	0.06	0.04	0.04	–	–	–	–
X_{Grs}	–	–	–	0.18	0.19	0.19	0.20	–	–	–	–

[a]Recalculated by stoichiometric balancing.

magnesian. Orthopyroxene coexisting with clinopyroxene in sample Ch-2B shows minor alumina zoning from core (1.3 wt%) to rim (0.5 wt%) with constant X_{Mg} value of 0.42. X_{Mg} value is very high (0.74) for orthopyroxene coexisting with clinopyroxene in the sample Ch-KHQ11. However, Opx grains in sample Ch-BT6 is extremely Fe-rich (X_{Mg} = 0.20–0.24) with low amount of alumina (<1 wt%). All the grains contain small amount of CaO (0.4–1.5 wt%) as Ca-Tschermak component. Plagioclase grains too show extreme variation in composition. In sample Ch-2B, plagioclase contains 47–51 mol% anorthite and 48–50 mole% albite components. Orthoclase component is either negligible or small (up to 2 mol%). Plagioclase in sample Ch-KHQ11 is almost pure anorthite. Ilmenite contains small amount of MgO (up to 3.3 wt%) as geikielite component. Magnetite is almost pure in most of the samples, but it contains 1.3–1.5 wt% Cr_2O_3 in sample Ch-KHQ11. Exsolved spinel is hercynite-spinel solid solution (X_{Mg} = 0.69–0.63) with small amount of Cr_2O_3 (up to 2 wt%) as chromite component. Garnet grains in the sample Ch-BT6 are almandine-grossular solid solution ($lm_{74\text{-}71}Grs_{20\text{-}18}$) with low amount of spessartine and pyrope components (4–6 mol%). Hornblende grains are magnesian (X_{Mg} = 0.75–0.72) and contain up to 1.9 wt% TiO_2 in sample Ch-KHQ11. Biotite (X_{Mg} = 0.67) contains up to 4.2 wt% TiO_2.

Both the varieties of pyroxene are magnesian in the ultramafic granulite (sample Ch-BT13; Table 14.1). Megacrystic orthopyroxene (X_{Mg}= 0.78) and clinopyroxene (X_{Mg}=0.86–0.87) are low in alumina (1.2–1.4 wt% for Opx and 1.6–1.9 wt% for Cpx). Exsolved clinopyroxene has similar composition to that of megacrystic variety. All the clinopyroxene grains are low-Na (<0.5 Na_2O) variety. Plagioclase contain 61–72 mol% anorthite component with very low orthoclase (<1 mol%).

Geothermobarometry

Since the studied rock represents crystallized mafic magma intruded in the crust, it is important to know the depth of emplacement as well as the pressure-temperature conditions of subsequent metamorphism. The latter is difficult to constrain because (1) for pressure estimation, no reliable barometric model could be applied for the present assemblages. Garnet is present in the contact zone, but no plagioclase is present in the same microdomain (2) For temperature estimation using the coexisting pyroxene assemblage, it is difficult to differentiate whether the estimated temperature represents subsolidus or metamorphic temperature.

Depth of Emplacement

Estimation of the depth of emplacement for the mafic magma can put significance control on the crust–mantle interaction. This can be done by using experimentally constrained phase diagram in the basaltic system (Irving, 1974) provided the bulk composition matches and there is no significant alteration after emplacement. As discussed by Bhui et al. (2007), this approach has limited applicability for metamorphosed mafic rocks. Phase assemblages in the contact aureole can also be used as garnet is commonly developed in such zones. This is again problematic for the present study as plagioclase does not coexist with garnet + orthopyroxene + quartz in the contact zone (Fig. 14.3e). A more realistic approach would be construction of pseudosection using internally consistent thermodynamic dataset for specific bulk composition (Holland and Powell, 1998; Connolly and Petrini, 2002). We have used the program PERPLEX (Connolly, 2005) to constrain the stability fields of stable subsolidus assemblages using the updated thermodynamic data of Holland and Powell (1998) as input parameters. While the solid solution models for phases like garnet, orthopyroxene, clinopyroxene are taken from Holland and Powell (1998), one for the plagioclase is taken from Newton et al. (1980). Quartz, ilmenite and magnetite are taken as pure phases. We have used the sample Ch-4B for pseudosection modelling which does not show much effect of hydration. The bulk composition is taken from XRF data of the respective samples and the appropriate chemical system was chosen as Na_2O–CaO–FeO–MgO–Al_2O_3–SiO_2–TiO_2–O_2 (NCFMASTO) to accommodate all the stable phases present in the rock. The obtained pseudosection (Fig. 14.4) shows the stability fields of different assemblages including the one found in the sample. The stability field of the subsolidus assemblage Opx + Cpx + Mt + Pl + Ilm + Qtz lies below c. 9.5 kbar pressure in the temperature <1,200°C. This implies that the upper limit for the depth of emplacement of the mafic magma was c. 32 km. One problematic aspect of this approach is the consideration of quartz in the primary assemblage. Experimental studies show that presence of primary quartz shifts the subsolidus assemblage towards lower pressure (Jaques and Green, 1980), while quartz-absent basic magma may crystallize at greater depths. In this study, there is no unequivocal clue that the quartz grains are primary. On the contrary, there are enough evidences of post-emplacement quartz veining from the surrounding rocks. Therefore, the estimated depth of 32 km should be considered as minimum value for the crystallization of the mafic magma.

Temperature Estimates

Coexisting pyroxene grains showing exsolution textures yield temperatures in the range 976–801°C using the model of Kretz (1982). Textural features suggest that this temperature values represent subsolidus cooling after the emplacement of the mafic magma. On the other hand, coexisting orthopyroxene and clinopyroxene granoblasts give a range of temperatures in different samples. All these pyroxene grains show evidence of metamorphism and deformation. The calculated temperatures range 946–755°C following the model of Kretz (1982). In single pair of grains, core-core data gives maximum temperature of 946°C, while the rim-rim data give 762°C. Coexisting clinopyroxene-ilmenite gives temperature of 844°C, while orthopyroxene-ilmenite pairs give temperature of 719°C after the model of Bishop (1980).

In the garnet-present domain, coexisting garnet-orthopyroxene gives 913°C for core composition, while the rim compositions yield a temperature of 649°C using the

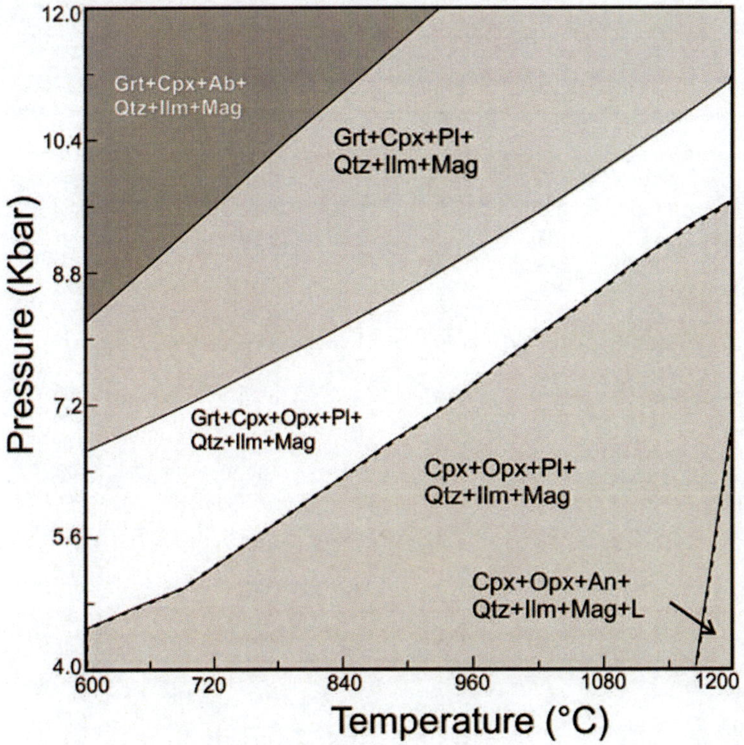

Fig. 14.4 Calculated pseudosection in the system Na_2O–CaO–FeO–MgO–Al_2O_3–SiO_2–TiO_2–O_2 (NCFMASTO) system for subsolidus assemblages using the bulk composition of the sample Ch-4B. The procedure for construction has been discussed in detail in the text. The field bounded by the *stippled line* contains the assemblage Opx + Cpx + Pl + Mag + Ilm + Qtz which should crystallize at pressure <9.5 kbar

method of Lee and Ganguly (1988). In the same domain, the garnet-clinopyroxene pair yields much restricted temperature range (841–758°C) as deduced from the thermometric model of Ellis and Green (1985).

Geochemistry

Seven samples of mafic granulite were used for bulk chemical analysis. All the samples were analyzed by XRF (major elements) using fused bead in Panalytical MagixPro machine at Hokkaido University and ICP-MS (trace elements and REE) at the Activation Laboratory, Canada. However, REE analysis was not done on sample Ch-4B (Table 14.3). Given the fact that all the rocks underwent deformation and granulite facies metamorphism subsequent to emplacement within the lower crust, the major element characteristics are unlikely to preserve the pristine magmatic history. Immobile trace element and REE characters however, are more reliable to trace back the early magmatic history and possible tectonic setting.

Table 14.3 Major, trace and rare earth element composition of the studied granulites

Sample No.	Ch-39D	Ch-3B	Ch-41/1C	Ch-DQ2	Ch-KHQ11	Ch-SQ2	Ch-4B	Ch-BT13
in wt %								
SiO_2	50.45	54.94	46.87	52.81	40.75	52.42	52.81	53.00
TiO_2	1.55	1.45	2.26	1.89	1.46	1.43	1.46	0.32
Al_2O_3	14.06	14.32	12.05	13.70	17.60	13.50	13.70	4.01
Fe_2O_3(T)	13.81	13.19	17.46	16.71	14.58	15.01	13.97	13.38
MnO	0.18	0.17	0.27	0.23	0.12	0.24	0.21	0.21
MgO	7.20	6.45	6.65	5.75	10.99	6.75	5.94	24.74
CaO	8.01	8.90	11.48	8.46	11.48	9.84	9.37	3.72
Na_2O	1.74	0.50	1.67	0.49	0.80	0.60	2.93	0.35
K_2O	1.99	0.06	0.50	<0.01	1.01	0.07	0.56	0.45
P_2O_5	0.24	0.14	0.23	0.23	0.17	0.13	0.14	0.04
LOI	0.53	0.09	0.12	0.05	0.61	0.13	0.28	0.08
Total	99.75	100.20	99.56	100.30	99.59	100.10	101.08	100.20
Mg#	51	49	43	41	60	47	46	77
in ppm								
Sc	38	47	56	53	38	50	48	28
Be	3	2	3	3	2	2	n.a.	n.a.
V	190	356	489	357	280	345	344	191
Cr	540	80	140	70	600	220	74	2,726
Co	43	51	66	50	56	47	64	95
Ni	50	310	<20	<20	240	<20	40	577
Cu	160	180	60	100	40	20	93	4
Zn	210	160	90	170	80	70	114	85
Ga	20	19	19	22	36	18	15	6
Ge	1.70	2.10	2.10	2.10	2.70	1.8	n.a.	n.a.
As	<5	11	<5	<5	<5	<5	n.a.	n.a.
Rb	100	1	4	<1	8	<1	10	39
Sr	108	91	130	114	106	124	146	14
Y	73.10	23.30	51.10	43.20	30.60	35.20	45.54	9
Zr	198	87	116	188	129	111	99	30
Nb	15	6.80	17.10	9.30	6.30	7.50	18.68	4
Mo	<2	<2	<2	<2	<2	<2	n.a.	n.a.
Ag	1.20	1.10	<0.5	1.20	1.10	<0.5	n.a.	n.a.
In	6.10	<0.1	0.10	0.10	<0.1	<0.1	n.a.	n.a.
Sn	4	2	6	5	6	5	n.a.	n.a.
Sb	1.20	2.30	1.70	2.20	1.90	1.50	n.a.	n.a.
Cs	1	<0.1	<0.1	<0.1	0.10	<0.1	n.a.	n.a.
Ba	138	20	48	37	70	61	179	99
La	41.90	6.55	14.60	14.50	17	9.18	20	4
Ce	108	16.20	42.50	34.40	35.60	23	44.75	<5
Pr	17.70	2.85	6.39	4.92	4.53	3.47	5.25	n.a.
Nd	65.80	11.80	27	21.70	18.50	15.90	24.57	n.a.
Sm	15.50	3.06	7.11	5.79	4.63	4.36	6.73	n.a.
Eu	1.98	1.05	1.54	1.38	1.55	1.16	1.32	n.a.
Gd	16.20	4.05	7.94	6.67	5.12	5.16	7.03	n.a.
Tb	2.61	0.68	1.44	1.24	0.88	0.96	1.24	n.a.
Dy	14.80	4.21	8.87	7.69	5.29	6.24	8.49	n.a.

Table 14.3 (continued)

Sample No.	Ch-39D	Ch-3B	Ch-41/1C	Ch-DQ2	Ch-KHQ11	Ch-SQ2	Ch-4B	Ch-BT13
Ho	2.74	0.86	1.81	1.55	1.07	1.28	1.84	n.a.
Er	7.35	2.56	5.23	4.56	3.08	3.77	5.17	n.a.
Tm	0.99	0.38	0.76	0.66	0.44	0.55	0.72	n.a.
Yb	5.87	2.42	4.67	4.19	2.60	3.37	4.23	n.a.
Lu	0.81	0.37	0.67	0.63	0.36	0.49	0.57	n.a.
Hf	5.30	2.40	3.30	5	3.40	3.30	n.a.	n.a.
Ta	0.94	0.57	1.30	0.71	0.26	0.58	n.a.	n.a.
W	<0.5	0.80	<0.5	0.6	1.20	1.80	n.a.	n.a.
Tl	0.58	<0.05	<0.05	<0.05	<0.05	<0.05	n.a.	n.a.
Pb	12	<5	<5	10	<5	<5	8	<5
Bi	<0.1	0.10	<0.1	<0.1	1.40	<0.1	n.a.	n.a.
Th	5.74	1.26	0.53	4.31	5.35	0.91	0.65	<0.1
U	0.88	0.77	0.51	1.57	3.88	0.96	n.a.	n.a.

n.a = not analyzed

Major Elements

All the samples show silica content characteristic of basic rock. Two of the samples namely Ch-KHQ11 and Ch-41/1C have relatively low silica content (40.75 and 46.87 wt% respectively) due to relative predominance of amphibole. There is a minor variation of Mg# in different samples (41–60), which is reflected also in the mineral chemical data. Alumina content varies from 12 to 18 wt%. In the TAS diagram, most of the samples are plotted in the fields of basic and intermediate. However, some are plotted beyond the prescribed fields presumably due to post-emplacement modification. This is also why we didn't attempt any plotting for major element variation.

Trace Elements

Immobile trace element distribution is less susceptible to post-magmatic changes and we used several of these element plots to understand the geochemical affinity of the basic magma and possible tectonic setting. Zr/Nb ratio are commonly used to distinguished between MORB (Zr/Nb$\geq$30) and OIB (Zr/Nb<10). Most of the samples having low Zr/Nb values ranging from 5.2 to 14.8. Samples Ch-DQ2 and Ch-KHQ11 have higher values of 20.2 and 20.4 respectively. Plots of Ti-Zr-Y reveal MORB-type character (Fig. 14.5a). The plot of Ti-Zr-Sr (Fig. 14.5b) also shows MORB character. However, the latter plot would change with the mobility of Sr under metamorphic regime which automatically shifts the plot positions towards the MORB field. Chondrite- and primitive mantle-normalized spider plots show enrichment of both LILE and HFSE (Fig. 14.5c, d). On the other hand, MORB-normalized plot shows enrichment of LILE more pronounced than HFSE (Fig. 14.5e). All of

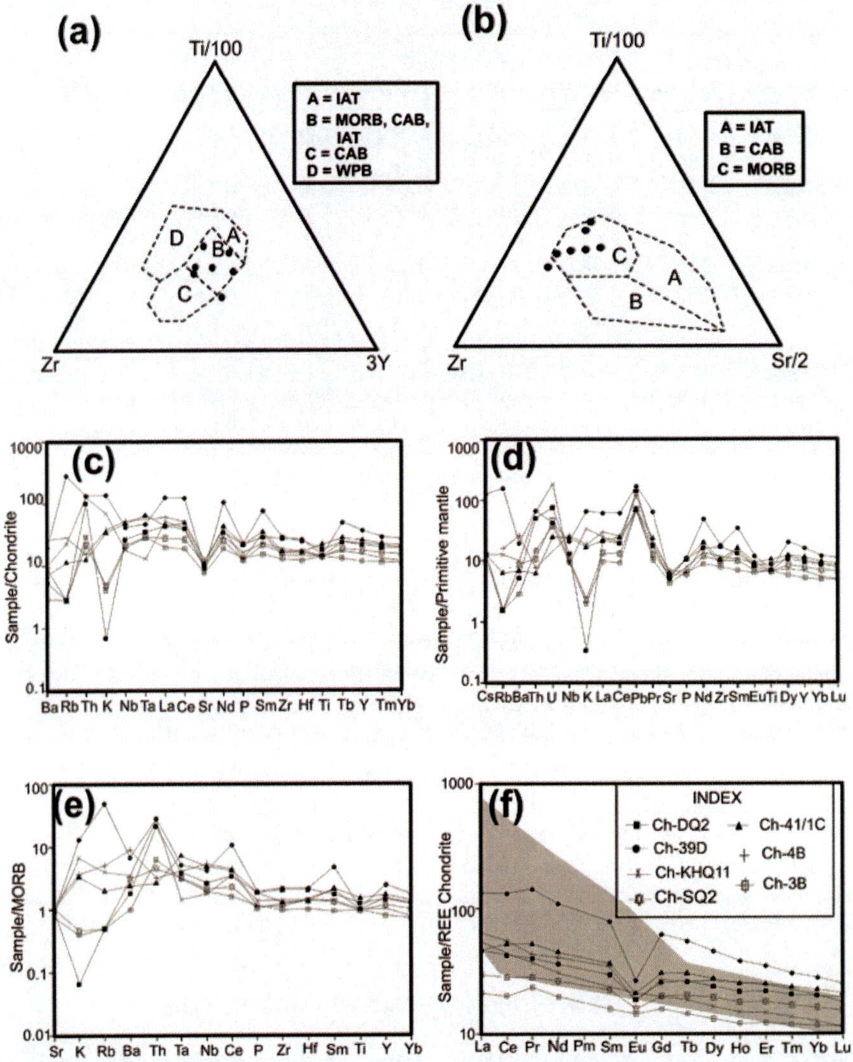

Fig. 14.5 Plots using trace and rare earth elements from the bulk chemical data. (**a**) Ti-Zr-Y discrimination diagram of the samples (after Pearce and Cann, 1973). Abbreviations used are IAT = Island arc tholeiites, MORB = Mid Oceanic Ridge basalt, CAB = Calc-alkali basalt, WPB = Within plate basalt. All the samples are plotted within the MORB/IAT/CAB field. (**b**) Ti-Zr-Sr discrimination diagram of the samples (after Pearce and Cann, 1973). (**c**) Chondrite-normalized, (**d**) Primitive mantle-normalized and (**e**) MORB-normalized trace element plots of the studied samples. (**f**) Chondrite-normalized plot of the rare earth elements of the samples. The shaded field demarcates the chondrite-normalized REE distribution of various suites of basaltic rocks of the East African Rift system (data taken from Kampunzu and Mohr, 1991)

these patterns however, are typical of melts generated from relatively undepleted mantle in intraplate settings (Winter, 2001). The scenario is slightly different from N-MORB type magma which generates from depleted mantle. Possibly, this melt has a substantial amount of P-MORB characteristics, thus showing a mixed enriched character.

Rare Earth Elements

Chondrite-normalized REE plot shows overall enrichment of LREE relative to HREE (Fig. 14.5f). Such LREE-enriched patterns are characteristic of the P-type MORB and ocean island basalt (OIB) developed within intraplate settings. On the other hand, N-type MORB is characterized by LREE-depletion (Wilson, 1989). Similar LREE-enriched pattern is also characteristic of continental rift setting (Cullers and Graf, 1984; Wilson, 1989; Kampunzu and Mohr, 1991). Almost all the samples show weak negative Eu anomaly. A very strong negative Eu anomaly is shown by the sample Ch-39D. LREE features such as $(La/Sm)_{MN}$ ratios [Normalization values are from Sun and McDonough (1989) and McDonough and Sun (1995)] can be used as an indicator for the extent of crustal contamination (Hoatson et al., 2005). Hoatson et al. (2005) showed that $(La/Sm)_{MN}$ ratio increases with progressive crustal contamination that reaches up to a value of 5.0. This ratio is low to moderate for the present samples (varying from 2.38 to 1.33), which implies lack of significant contamination by crustal processes.

Discussion

The studied mafic granulites, occurring within the mixed rock suite, show evidence for magmatic emplacement within the lower crust. In that sense these rocks are different than those formed by dehydration melting of hornblende-bearing protolith at deep crustal level. The emplacement of mafic magma occurred pre- to synkinematically with D_2 deformation as revealed from the structural data. The minimum depth of emplacement could be 32 km which is estimated from pseudosection analysis. Subsequent to the emplacement and subsolidus cooling, the rocks underwent high-grade metamorphism and deformation (D_2/D_3–M_2) when temperature reached at least up to 910°C. During this granulite facies metamorphic regime, the rocks underwent high-temperature recrystallization and subsequent grain coarsening that collectively impose the granoblastic texture. Pressure conditions during the metamorphism could not be determined in absence of any diagnostic mineral assemblage. Figure 14.4 vindicates the fact that the assemblage Opx + Cpx + Pl + Ilm + Mag + Qtz has extended stability field covering the entire gamut of granulite facies condition. Presence of variable amount of hornblende and biotite replacing the pyroxene grains suggests localized hydration of the granulites (M_3?). A second granulite facies metamorphism is inferred from the localized break down of the hornblende grains into granular intergrowth of orthopyroxene + clinopyroxene +

quartz. This can be regarded as the reworking of the retrogressed granulites by a separate tectonothermal event (D_4–M_4).

Bulk rock major element characteristics of the mafic granulites do not reveal the pre-metamorphic evolutionary history. Granulite metamorphism and associated deformation must have redistributed most of the LIL as well as mobile trace elements. Rare earth element distribution, on the other hand, is quite robust and likely to preserve the pristine characters. Chondrite-normalized REE plot suggests LREE-enriched pattern which is characteristic of P-type MORB, developed in intraplate setting. The pattern is similar to what is found in some of the continental rift-related basalt. However, alkaline igneous rocks, typical of rift setting, are characteristically absent in the vicinity. At the same time, enrichment of LILE and CO_2-rich volatiles within the fluid inclusions of the mafic granulites and associated rocks (Bose, unpub. data) suggest that the magma was derived from an enriched mantle source (Bailey, 1983).

Absence of geochronological data for the mafic granulites does not allow us to bracket the timing(s) of the emplacement of the mafic magma and subsequent metamorphism. However, sporadic age data from associated metasedimentary rocks and anorthosite can play a handy role and at least provide some framework for constraining these time events. The timing for the earliest metamorphism (M_1) has been loosely constrained as c. 1.25–1.10 Ga from the monazite U-Pb-Th ages of the metapelitic granulites of the Chilka Domain (Simmat and Raith, 2008). Some of high Mg-Al granulites (similar to those present in the mixed rock suite) preserve imprints of a thermal event at c. 950 Ma in monazite grains (Dobmeier and Simmat, 2002). Petrological significance of these two events, nevertheless, is poorly understood. The exhaustive study of monazite U-Th-Pb system, on the other hand, reveals a very strong thermal overprint at c. 800–770 Ma (Simmat and Raith, 2008). This is also confirmed by recent U-Pb zircon data (Upadhyay et al., 2009). Based on the petrological features present in the high Mg-Al granulites, we can correlate this event with the D_2–M_2 event. The mafic magma, thus emplaced just prior or close to c. 800 Ma. The upper limit for the timing of the magmatic emplacement is provided by the anorthosite, which is intruded by the mafic granulite. Zircon U-Pb TIMS data suggest that the anorthosite was emplaced at c. 983 Ma (Chatterjee et al., 2008), which is much older than the previously proposed data of c. 792 Ma by Krause et al. (2001). If we take the TIMS data as reliable, the mafic magmatism must have taken place in the time frame of c. 983–800 Ma. This has got an important bearing on the evolution of the Chilka Domain vis-à-vis the EGB.

The EGB exposed an eroded deep crustal section that witnessed multiple orogenic events throughout the Proterozoic time (see Karmakar et al., 2009). The type Eastern Ghats Province (Dobmeier and Raith, 2003) within the EGB records recurrent orogenic activities in the time frame of 1,250–500 Ma. The dominant tectonothermal event in this province occurred at c. 1,000–900 Ma (Shaw et al., 1997; Mezger and Cosca, 1999; Simmat and Raith, 2008; Upadhyay et al., 2009) and this part of the evolutionary history is correlated with the assembly of the supercontinent Rodinia. It is proposed that the EGB and the Rayner Complex evolved as a mobile belt suturing cratonic blocks of India and east Antarctica (Kelly et al., 2002;

Harley, 2003). Rodinia eventually broke up at the time period 825–725 Ma and this period witnessed emplacement of numerous dyke swarms in different fragments of the erstwhile Rodinia fragments (Li et al., 2008). Although, there is no direct proof of dyke activities from the field features of the mafic granulites, the coincidence of events tempts us to propose that the mafic magmatism in the Chilka domain was possibly linked with fragmentation of Rodinia.

Acknowledgements SB, KD and SC acknowledge Department of Science and Technology (DST), Government of India for providing the research grant (ESS/251/16/2005). We thank Rakoto Thierry for his help during preparation of rock powders and EPMA analyses. Tamal K. Ghosh is thanked for his help and expertise during EPMA work at IIT Roorkee. We are also thankful to Henry Halls for his comments during the 6th International Dyke conference 2010 where this work was presented. Insightful comments and suggestions from Saibal Gupta helped to revise and improve the quality of the manuscript.

References

Bailey DK (1983) The chemical and thermal evolution of rifts. Tectonophys 94: 585–597

Bhattacharya S, Sen SK, Acharyya A (1994) The structural setting of the Chilka Lake granulites-migmatites-anorthosite suite with emphasis on the time relation of charnockite. Precamb Res 66: 393–409

Bhowmik SK, Roy A (2003) Garnetiferous metabasites from the Sausar mobile belt: Petrology, *P-T* path and implications for the tectonothermal evolution of the Central Indian Tecnotic zone. Jour Petrol 44: 387– 420

Bhui AK, Sengupta P, Sengupta P (2007) Phase relations in mafic dykes and their host rocks from Kondapalle, Andhra Pradesh, India: Implications for the time–depth trajectory of the Palaeoproterozoic (late Archaean?) granulites from southern Eastern Ghats Belt. Precamb Res 156: 153–174

Bishop FC (1980) The distribution of Fe^{2+} and Mg between coexisting ilmenite and pyroxene with application to geothermometry. American Jour Sci 280: 46–77

Boven A, Theunissen K, Sklyarov E, Klerkx J, Melnikov A, Mruma A, Punzalan L (1999) Timing of exhumation of a high-pressure mafic granulite terrane of the Paleoproterozoic Ubende belt (West Tanzania). Precamb Res 93: 119–137

Chatterjee N, Crowley JL, Mukherjee A, Das S (2008) Geochronology of the 983-Ma Chilka Lake AnorthositeEasternGhats BeltIndia: implications for Pre-Gondwana tectonics. Jour Geol 116: 105–118

Connolly JAD (2005) Computation of phase equilibria by linear programming: A tool for geodynamic modeling and its application to subduction zone decarbonation. Earth Planet Sci Lett 236: 524–541

Connolly JAD, Petrini K (2002) An automated strategy for calculation of phase diagram sections and retrieval of rock properties as a function of physical conditions. Jour Met Geol 20: 697–708

Cullers RL, Graf JL (1984) Rare earth elements in igneous rocks of the continental crust; predominantly basic and ultrabasic rocks. In: Henderson P, (ed) Rare Earth geochemistry. Elsevier, Amsterdam: 510

Das S, Nasipuri P, Bhattacharya A, Swaminathan S (2009) The thrust-contact between the Eastern Ghats belt and the adjoining Bastar craton (Eastern India): Evidence from mafic granulites and tectonic implications. Precamb Res 162: 70–85

Dasgupta S, Sengupta P, Mondal A, Fukuoka M (1993) Mineral chemistry and reaction textures in metabasites from the Eastern Ghats belt, India and their implications. Mineral Mag 57: 113–120

Dobmeier C, Raith MM (2000) On the origin of "arrested" charnockitization in the Chilka Lake area, Eastern Ghats belt, India: A reappraisal. Geol Mag 137: 27–37

Dobmeier CJ, Raith MM (2003) Crustal architecture and evolution of the Eastern Ghats Belt and adjacent regions of India. In: Yoshida M, Windley BF, Dasgupta S, (eds) Proterozoic East Gondwana: Supercontinent assembly and breakup, Spec Publ 206, Geological Society of London, London. 145–168

Dobmeier C, Simmat R (2002) Post-Grenvillean transpression in the Chilka Lake area, Eastern Ghats belt-implication for the geological evolution of peninsular India. Precamb Res 113: 243–268

Ellis DJ, Green DH (1985) Garnet-forming reactions in mafic granulites from EnderbyLand, Antarctica: Implications for geothermometry and geobarometry. Jour Petrol 26: 633–662

Frost BR, Chacko T (1989) The granulite uncertainty principle: Limitations on thermobarometry in granulites. Jour Geol 97: 435–450

Galan G, Marcos A (2000) The metamorphic evolution of the high pressure mafic granulites of the Bacariza Formation (Cabo Ortegal Complex, Hercynian belt, NW Spain). Lithos 54: 139–171

Garrido CJ, Bodinier JL, Burg JP, Zeilinger G, Hussain SS, Dawood H, Chaudhry MN, Gervilla F (2006) Petrogenesis of mafic garnet granulite in the lower crust of the Kohistan Paleo-arc Complex (Northern Pakistan): Implications for intra-crustal differentiation of island arcs and generation of continental crust. Jour Petrol 47: 1873–1914

Gross AOMS, Droop GTR, Porcher CC, Fernandes LAD (2009) Petrology and thermobarometry of mafic granulites and migmatites from the Chafalote Metamorphic Suite: New insights into the Neoproterozoic *P–T* evolution of the Uruguayan-Sul-Rio-Grandense shield. Precamb Res 170: 157–174

Harley SL (1987) A pyroxene-bearing metaironstone and other pyroxene–granulites from Tonagh Island, Enderby Land, Antarctica: Further evidence for very high temperature (>980°C) Archaean regional metamorphism in the Napier Complex. Jour Metamorph Geol 5: 341–356

Harley SL (2003) The Archean to Cambrian crustal development of East Antarctica: Metamorphic characteristics and tectonic implications. In: Yoshida M, Windley BF, Dasgupta S, (eds) Proterozoic East Gondwana: Supercontinent assembly and breakup, Spec Publ 206, Geological Society of London, London. 203–230

Hoatson DM, Sun S, Claouélong JC (2005) Proterozoic mafic–ultramafic intrusions in the Arunta Region, central Australia Part 1: Geological setting and mineral potential. Precamb Res 142: 93–133

Holland TJB, Powell R (1998) An internally consistent thermodynamic data set for phases of petrological interest. Jour Met Geol 16: 309–343

Irving AJ (1974) Geochemical and high pressure experimental studies of garnet pyroxenite and pyroxene granulite xenoliths from the Delegate basaltic pipes, Australia. Jour Petrol 15: 1–40

Jaques AL, Green DH (1980) Anhydrous melting of peridotite at 0–15 Kb pressure and the genesis of tholeiitic basalts. Contrib Mineral Petrol 73: 287–310

Kampunzu AB, Mohr P (1991) Magmatic evolution and petrogenesis in the East African Rift system. In: Kampunzu AB, Lubala RT, (eds) Magmatism in extensional settings, the Phanerozoic African plate. Springer, Berlin: 85–136

Karmakar S, Bose S, Das K, Dasgupta S (2009) Proterozoic Eastern Ghats belt, India: A witness of multiple orogenies and its lineage with ancient supercontinents. Jour Virtual Explor, doi:10.3809/jv rtex.2009.00254

Kelly NM, Clarke GL, Fanning CM (2002) A two-stage evolution of the Neoproterozoic Rayner Structural Episode: New U-Pb sensitive high resolution ion microprobe constraints from the Oygarden group, Kemp Land, East Antarctica. Precamb Res 116: 307–330

Krause O, Dobmeier C, Raith MM, Mezger K (2001) Age of emplacement of massif-type anorthosites in the Eastern Ghats belt, India: Constraints from U-Pb zircon dating and structural studies. Precamb Res 109: 25–38

Kretz R (1982) Transfer and exchange equilibria in a portion of the pyroxene quadrilateral as deduced from natural and experimental data. Geochim Cosmochim Acta 46: 411–422

Kretz R (1983) Symbols for rock forming minerals. Am Miner 68: 277–279

Lee HY, Ganguly J (1988) Equilibrium compositions of coexisting garnet and orthopyroxene: Experimental determinations in the system FeO-MgO-Al_2O_3-SiO_2 and applications. Jour Petrol 29: 93–113

Leelanandam C (1990) The anorthosite complex and Proterozoic mobile belt of Peninsular India: A review. In: Naqvi SM, (ed) Precambrian continental crust and its economic resources. Elsevier: Amsterdam 409–436

Leelanandam C (1997) The Kondapalli layered complex, Andhra Pradesh, India: A synoptic overview. Gondwana Res 1: 95–114

Leelanandam C, Kumar V (2007) Petrogenesis and tectonic setting of the chromitites and chromite-bearing ultramafic cumulates of the Kondapalli layered complex, Eastern Ghats belt India: evidences from the texturalmineral chemical and wholerock geochemical studies. In: Leelanandam C, Rama Prasada Rao IB, Sivaji C, Santosh M, (eds) Indian continental crust and upper mantle, Mem Int Assoc Gondwana Res (IAGR) Kochi, Japan, 10. 89–107

Li ZX, Bogdanova SV, Collins AS, Davidson A, de Waele B, Ernst RE, Fitzsimons ICW, Fuck RA, Gladkochub DP, Jacobs J, Karlstrom KE, Lu S, Natapov LM, Pease V, Pisarevsky SA, Thrane K, Vernikovsky V (2008) Assembly, configuration, and break-up history of Rodinia: A synthesis. Precamb Res 160: 179–210

Liu X, Hu J, Zhao Y, Lou Y, Wei C, Liu X (2009) Late Neoproterozoic/ Cambrian high-pressure mafic granulites from the Grove mountains, East Antarctica: *P–T–t* path, collisional orogeny and implications for assembly of East Gondwana. Precamb Res 74: 81–199

McDonough WF, Sun S (1995) The composition of the Earth. Chem Geol 120: 223–254

Mezger K, Cosca MA (1999) The Thermal history of the Eastern Ghats belt (India), as revealed by U-Pb and ^{40}Ar-^{34}Ar dating of metamorphic and magmatic minerals: Implications for the SWEAT correlation. Precamb Res 94: 251–271

Newton RC, Charlu TV, Kleppa OJ (1980) Thermochemistry of the highest structural states plagioclases. Geochim Cosmochim Acta 44: 933–941

Pearce JA, Cann JR (1973) Tectonic setting of basic volcanic rocks determined using trace element analysis. Earth Planet Sci Lett 19: 290–300

Raith MM, Dobmeier C, Mouri H (2007) Origin and evolution of Fe-Al granulites in the thermal aureole of the Chilka Lake anorthosite, Eastern Ghats Province, India. Proc Geol Assoc 118: 87–100

Sen SK, Bhattacharya S, Acharya A (1995) A multi-stage pressure-temperature record in the Chilka Lake granulites: The epitome of the metamorphic evolution of Eastern Ghats, India. Jour Met Geol 14: 287–298

Sengupta P, Dasgupta S, Dutta NR, Raith MM (2008) Petrology across a calcareous rock–anorthosite interface from the Chilka Lake complex, Orissa: Implications for Neo-Proterozoic crustal evolution of the northern Eastern Ghats Belt. Precamb Res 162: 40–58

Shaw RK, Arima M, Kagami H, Fanning CM, Shiraishi K, Motoyoshi Y (1997) Proterozoic events in the Eastern Ghats Granulite belt, India: Evidence from Rb-Sr, Sm-Nd systematics, and SHRIMP dating. Jour Geol 105: 645–656

Simmat R, Raith MM (2008) U–Th–Pb monazite geochronometry of the Eastern Ghats belt, India: Timing and spatial disposition of poly-metamorphism. Precamb Res 162: 16–39

Sun SS, McDonough WF (1989) Chemical and isotopic systematics of oceanic basalts: Implications for mantle composition and processes. In: Sunders AD, Norry MJ, (eds) Magmatism in ocean basins, Spec Publ 42, Geological Society of London, London. 313–345

Upadhyay D, Gerdes A, Raith MM (2009) Unraveling sedimentary provenance and tectonothermal history of high to ultra-high temperature metapelites using zircon and monazite chemistry: A case study from the Eastern Ghats belt, India. Jour Geol 117: 665–683

Wilson M (1989) Igneous petrogenesis: A global tectonic approach. Unwin Hyman, London: 271

Winter JD (2001) An introduction to igneous and metamorphic petrology. Prentice Hall, New Jersey, NJ: 268

Zhang J, Zhao G, Sun M, Wilde SA, Li S, Liu S (2006) High-pressure mafic granulites in the Trans-North China Orogen: Tectonic significance and age. Gondwana Res 9: 349–362

Chapter 15
Mineralogical Control on Rheological Inversion of a Suite of Deformed Mafic Dykes from Parts of the Chottanagpur Granite Gneiss Complex of Eastern India

Sayan Ray, Sanjoy Sanyal, and Pulak Sengupta

Introduction

Infiltration driven processes during regional metamorphism cause significant change in chemistry and mineralogy of rocks, which under certain situations inflict a large change in seismic and rheological properties of the altered rocks (reviewed in Connolly, 2005).

In Jasidihi-Deoghar area of the state of Jharkhand, India (Fig. 15.1), a suit of mafic dykes were emplaced into and poly-deformed along with felsic country rock (now felsic gneiss) during Meso-to Neoproterozoic tectonic activities of the areas (reviewed in Sengupta, 1997, Ghosh and Sengupta, 1999). Owing to their thickness and higher competency contrast with the host felsic gneiss, the mafic dykes commonly stretched out and fragmented with the boudin neck being filled up with pegmatite (Fig. 15.2f). Field structures demonstrate that during the course of superposed deformation, the competency of the mafic dyke with respect to the host felsic gneiss and nodal pegmatite decreased and eventually reversed during progressive deformation (Sengupta, 1997, Ghosh and Sengupta, 1999). This work explores the petrological control of this rheological inversion and proposes a viable mechanism for the development of fish-head boudin.

Geological Backdrop

The areas in and around Jashidihi-Deoghar expose swarm of mafic dykes within the sea of composite felsic gneisses including augen gneiss and migmatite with cm to decimeter thick leucosomes with biotite–rich selvages (Fig. 15.2c reviewed in Ghosh and Sengupta, 1999). The entire rock ensemble was subjected to polyphase

S. Ray (✉)
Department of Geological Sciences, Jadavpur University, Kolkata 700032, India
e-mail: ry.syn05@gmail.com

S. Sanyal and P. Sengupta
Department of Geological Sciences, Jadavpur University, Kolkata 700032, India

R.K. Srivastava (ed.), *Dyke Swarms: Keys for Geodynamic Interpretation*,
DOI 10.1007/978-3-642-12496-9_15,

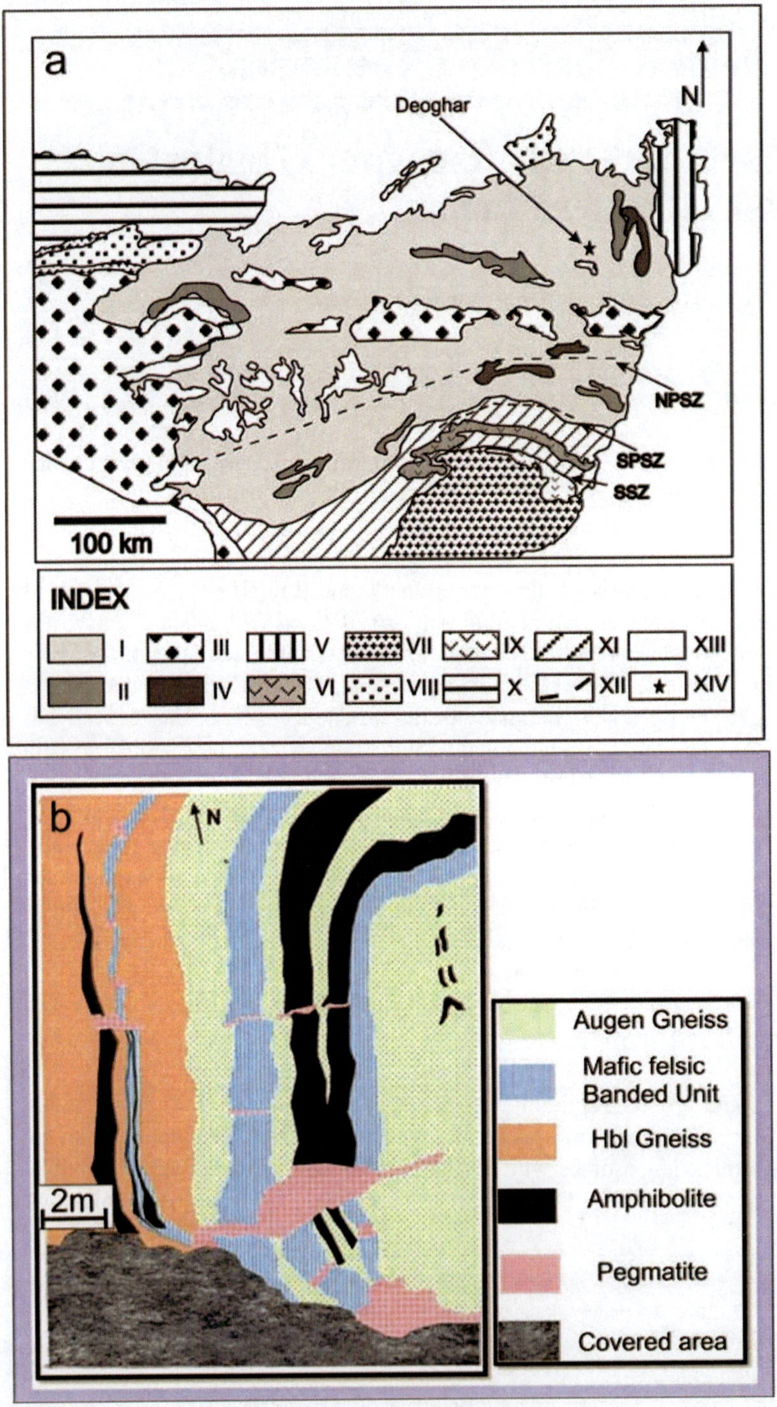

Fig. 15.1 (continued)

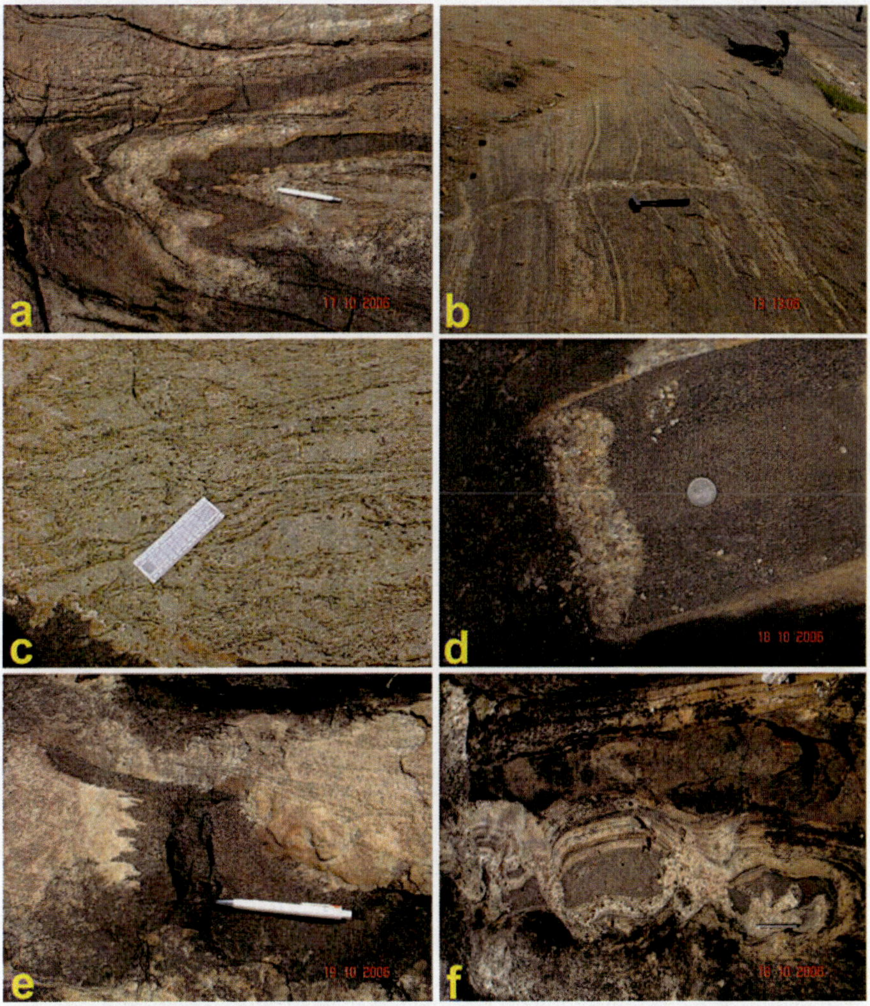

Fig. 15.2 (**a**) First generation fold (F_1) defined by interlayered fabric migmatite and Amphibolite dyke. (**b**) Foliation parallel as well as crosscutting pegmatite veins in fabric migmatite rock. (**c**) Development of nebular leucosomal patches with dark biotite selvege in fabric migmatitic rock (**d**) Development of fish head boudin with drawnout corners in Amphibolite. Coarse pegmatite concentration at the boudin neck. (**e**) Amphibolite band sends series of Flame fold and *Arrow head* folds within host Fabric migmatitic gneiss. (**f**) Composite boudin structure of Amphibolite-fabric migmatite interlayered unit within pegmatoidal leucosome of pointed cusp in amphibolite and rounded lobe in pegmatite

Fig. 15.1 (**a**) Regional map of part of East Indian Craton showing the location of the area (Deoghar – Jashidihi) (after Achariyya, 2003) *I:* CGGC, *II:* Meta sediments in CGGC, *III:* Gondwana rocks, *IV:* Granulite domains in CGGC, *V:* Rajmahal trap, *VI:* Dalma Volcanics, *VII:* Singhbhum and other granites, *VIII:* Mahakoshal group of rocks, *IX:* Dhanjori volcanics, *X:* Vindhyan supar group rocks, *XI:* Unclassified metamorphites, *XII:* Major lineaments and shear zones, *XIII*: Quaternary, *XIV*: Study area, (**b**) Outcrop scale map in parts of the study area showing lithological disposition deformation during which the thin and competent mafic dykes were repeatedly folded and produced different generations of boudins (Ghosh and Sengupta, 1999). A number of syntectonic pegmatite veins of different thickness (Fig. 15.2b) traversed studied areas. Small pods of pegmatite are also present in the zone of separation of mafic boudins (boudin neck). The rocks of the studied area represent part of the Meso-to Neoproterozoic Chotanagpur Granite Gneiss Complex (after here CGGC). Detail structural analyses of the studied area revealed three sets of folds and at least two generations of boudinage (Sengupta, 1997, Ghosh and Sengupta, 1999). The first fold (F_1), which is defined by mafic dykes, is isoclinal and has a roughly E-W axial planar foliation defined by oriented grains of amphibole and biotite (S_1). S_1 represents the most pervasive foliation of the region. Structural features demonstrate that bulk of the leucosomes in felsic gneisses were formed during the F_1 folding. The second folds (F_2), which are the dominant folds of the area, are isoclinal to tight with NNE striking axial plane. Interference of F_1- F_2 produced hook- and crescent-shaped outcrops. The third folds (F_3) are open to close and have subvertical E-W trending axial planes. Boudinaged layers of mafic dykes, pegmatite veins and leucosomes in felsic gneiss are the form surfaces of F_2 and F_3 folds. Limbs of both F_1–F_2 folds that are defined by mafic dykes show extensive development of boudins of varying size and shapes (Ghosh and Sengupta, 1999). Variation of the shapes of the boudins has been attributed to change in competence contrast between the mafic rocks and their felsic gneiss (Sengupta, 1997; Ghosh and Sengupta, 1999).

Field Features Demonstrating Changing Rheology of Mafic Dykes

During superposed deformation mafic dykes amidst felsic gneiss produced buckle folding that is represented by M, S and Z–shaped disharmonic folds at the hinge and limbs of larger folds (Fig. 15.2a). That the mafic dykes behaved as competent unit relative to the enclosing felsic gneiss during the superposed folding events is manifested by (1) profuse development of boudins and pinch-and-swell structure and (2) Class IB and IC types of fold geometry in mafic dykes where as class threefolds in the felsic gneiss (Ghosh and Sengupta, 1999). Syn-tectonic pegmatite veins in which the grain size is at least an order of magnitude higher than the mafic dykes and felsic gneiss, show buckle folds in the host of the latter rocks and occur at the neck region of mafic boudins. Figure 15.2f shows a composite boudin of mafic dykes and pegmatite whose boundary shows cuspate-lobate folds. In this structure, pegmatites occur at the cores of lobate folds suggesting that its competency was higher than the mafic dyke. These features indicate that pegmatite veins had higher competency relative to the other lithologies (Ramberg; 1955, Naha and Sen, 1965; Ramsay and Huber, 1987; Ghosh and Sengupta, 1999).

A number of features demonstrate that the difference in competency between mafic dykes and their enclosing rocks was reduced and the competency contrast eventually reversed during progressive deformation. The best example in support of reversal of competence between mafic boudin and their enclosing rocks is the formation of fish-head boudin with drawn-out corners around nodal pegmatite (Fig. 15.2d). Shape of these boudins has been interpreted as post boudinage flow of the boudin rind compared to its core and the adjoining lithologies (Ghosh and Sengupta, 1999). The reversal of competence is also manifested by formation of flame fold or arrowhead folds (Fig. 15.2e) on parasitic S and Z folds of mafic dykes with sharp outer arc that points towards the felsic gneiss (Kranck, 1953; Ghosh, 1993, Sengupta, 1997, Ghosh and Sengupta, 1999). These features suggest that the mafic dykes that initially behaved as competent unit became incompetent relative to the enclosing felsic gneiss.

Evolution of Mineral Assemblages in Mafic Dykes

Mutual Relationship Among the Minerals

Detail description of textural relations in mafic dykes is beyond the purview of this study. In the following sections, textures that are only related to rheological inversion have been described. In core of the mafic dykes clinopyroxene, orthopyroxene is rarely present. Plagioclase and ilmenite constitute the earliest assemblages. These minerals show a granoblastic fabric with polygonal grain boundaries (Fig. 15.3a, b). The polygonal aggregates of the early formed anhydrous minerals are extensively replaced with secondary amphibole (Fig. 15.3a-c). Secondary amphibole showing clinopyroxene pseudomorph is a common feature. Secondary amphibole is commonly associated with granules of titanite. Towards the contact with felsic gneiss, mafic dykes show development of biotite in profusion (Fig. 15.3e, f). Commonly the mafic boudins develop cm thick selvages rich in biotite. Biotite replaced amphibole and defines a distinct foliation parallel to S_1 (Fig. 15.3d). Biotite is the dominant mineral in the drawn out corners of the fish-head boudins, in flame folds and in the cuspate portion of the cuspate-lobate folds (Fig. 15.2f). Biotite is practically absent in core of the mafic dykes. A few grains of garnet are scattered in amphibole and biotite rich domains.

Mineral Chemistry

Composition of minerals was analyzed with EPMA (SX51) at the Department of Mineralogy and Petrology, University of Bonn, Germany. The detail of the analytical procedures is described in Sengupta et al. (1999). Representative mineral analyses of a fish-head mafic boudin are presented in Table 15.1. In terms of X_{Mg} (Mg/Mg+Fe) the ferromagnesian minerals can be arranged as Clinopyroxene (0.7)> Amphibole (0.59–0.66)> biotite (0.46)> ilmenite. Amphibole that develops on clinopyroxene is actinolite (Al_2O_3 = 2.42 wt%, X_{Mg} = 0.66). This is distinctly magnesian and less aluminous than the amphiboles that develops on plagioclase (Al_2O_3 = 10.51Wt%, X_{Mg} = 0.59, Table 15.1). Coexistence of amphiboles with

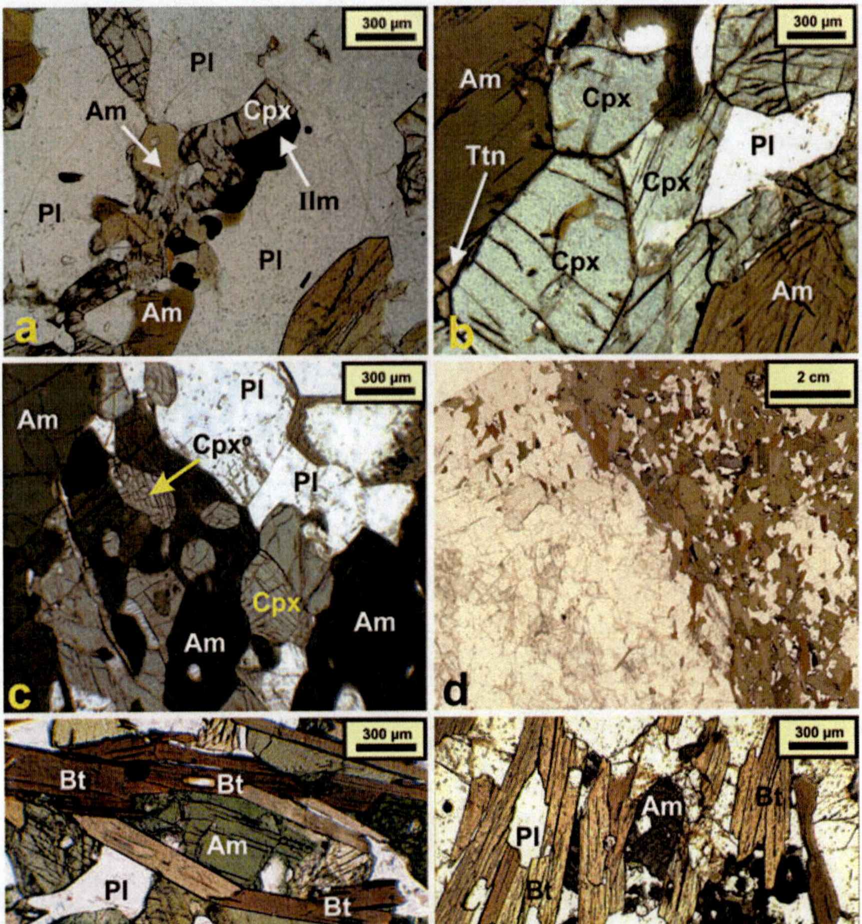

Fig. 15.3 (**a**) Recrystallized clinopyroxene grains partially replaced by amphibole near ilmenite and plagioclase. (**b**) Polygonal clinopyroxene (Cpx) and plagioclase (Pl) exhibit granoblastic texture. Secondary amphibole (Am) with clinopyroxene pseudomorph is also present. (**c**) Floating islands of recrystallised clinopyroxene (Cpx) within a replacive mass (**d**) Concentration of biotite at the boundary between amphibolites boudin and felsic gneiss. (**e**) and (**f**) Biotite (Bt) that replaced amphibole defines a distinct foliation at the fringe zone of boudin

contrast composition can't be explained. Composition of substrate mineralogy might have exerted a control on the composition of the secondary amphibole. Compositions of plagioclase, which doesn't show any distinct compositional zoning, straddle the boundary between andesine and labradorite (X_{An}= 0.53–0.54). Titanite has 2.86 wt% Al_2O_3. Whereas ilmenite has ~4 wt% MnO (Table 15.1).

Table 15.1 Representative compositions of minerals from a mafic boudin

Phase	Clinopyroxene	Amphibole	Amphibole	Plagioclase	Plagioclase	Biotite	Ilmenite	Titanite
				rim	core			
SiO_2	51.88	43.88	43.58	53.61	54.23	35.43	00.01	30.70
TiO_2	00.23	02.19	02.18	00.01	00.00	01.15	52.03	35.84
Al_2O_3	01.61	10.50	10.51	28.30	28.62	17.11	00.02	02.86
Cr_2O_3	00.00	00.01	00.10	00.02	00.00	00.01	00.00	00.00
FeO	09.93	13.52	13.96	00.19	00.06	20.06	40.97	02.62
MgO	12.72	11.69	11.25	00.03	00.00	09.49	00.09	01.26
MnO	00.33	00.13	00.13	00.00	00.00	00.20	03.97	00.10
CaO	21.96	11.66	11.44	11.45	11.53	00.14	00.01	25.45
Na_2O	00.39	01.30	01.49	05.51	05.51	00.00	00.01	00.00
K_2O	00.00	01.36	01.37	00.21	00.17	10.24	00.00	00.00
Total	99.05	96.23	96.00	99.32	100.12	93.83	97.12	98.83
Oxygen basis	6	23	23	8	8	22	3	5
Si	1.97	6.59	6.583	2.447	2.452	5.543	0.000	1.006
Ti	0.01	0.25	0.247	0.000	0.000	0.135	1.011	0.883
Al	0.07	1.86	1.872	1.523	1.526	3.155	0.001	0.110
Cr	0.00	0.00	0.012	0.001	0.000	0.001	0.000	0.000
Fe	0.31	1.70	1.763	0.007	0.002	2.625	0.885	0.072
Mg	0.72	2.62	2.532	0.002	0.000	2.213	0.004	0.061
Mn	0.01	0.02	0.016	0.000	0.000	0.027	0.087	0.003
Ca	0.89	1.88	1.852	0.560	0.559	0.023	0.000	0.894
Na	0.03	0.38	0.437	0.488	0.483	0.001	0.000	0.000
K	0.00	0.26	0.264	0.012	0.010	2.044	0.000	0.000
Cation sum	4.007	15.55	15.578	5.040	5.031	15.767	1.989	3.030
X_{Mg}	0.70	0.61	0.59	–	–	0.46	0.00	–
X_{An}	–	–	–	0.53	0.54	–	–	–

Mineral Reactions

Textural relations that are described above can be explained by two reactions-one produced amphibole from clinopyroxene and the other produced biotite from amphibole. In order to obtain balanced chemical reactions that explain formation of amphibole and biotite in the mafic dyke rocks a matrix operation technique called Singular Value Decomposition (SVD) was adopted utilizing the measured chemical compositions of the minerals (Fishes 1989, Lang et al., 2004). The computer program C-Space (Torres–Roldan et al., 2000) was used for this purpose and the required compositional matrix is presented in Appendix-1. The detail of the textural modeling technique has been described in Sengupta et al. (2009). The out put reactions are given below:

Amphibole Forming Reaction

$$12.70\ Plagioclase + 44.21\ Clinopyroxene + 8.91\ Ilmenite + 3.11\ K^{+} + 12.22\ H2O = 29.30\ Quartz + 12.42\ Amphibole + 1.00\ Mn^{2+} + 12.60\ Ca^{2+} + 9.69\ Titanite + 2.85\ Na^{+} \quad (1)$$

Biotite-Forming Reaction

$$10.22\ Plagioclase + 7.62\ Amphibole + 12.45\ Ilmenite + 16.47\ K^{+} + 1.46\ H2O = 9.09Bt + 10.02\ Quartz + 1.00\ Mn + 5.03\ Ca^{2+} + 14.83\ Titanite + 7.90\ Na^{+} \quad (2)$$

Reaction 1 explains the formation of amphibole + titanite from pyroxene + plagioclase + ilmenite (Fig. 15.3a, b) whereas; reaction (2) explains the growth of biotite from amphibole. Corroboration of reaction textures by the modeled chemical reactions justifies the view that SVD technique is a powerful tool in modeling reaction textures in multicomponent natural system (Lang et al., 2004, Sengupta et al., 2009). Several important features emerge from the balanced chemical reactions, some of which are presented below:

i. The assemblage pyroxene + plagioclase + ilmenite which presumably represent the magmatic assemblage of the mafic boudin were converted to amphibolite due to in filtration of H_2O and the process also released significant amount of Ca, Na and silica. A part of the Ca was locked in titanite and the rest along with Na were moved out of the system. Reaction 1 also supports the view that Al and Ti behaved as immobile components during amphibolite facies metamorphism.
ii. With continuing H_2O infiltration and K-metasomatism secondary amphibole, a product of reaction 1 and residual plagioclase, was converted to biotite + titanite releasing Ca and Na (reaction 2)
iii. Reactions 1 and 2 have large positive and negative volume change for the solid phases ($\Delta V = 181$ j/bar and –80.22j/bar respectively). Molar volume data are computed from Holland and Powell, 1998; 6 kbar and 700°C. The ΔV values remain virtually unchanged over a large range of P-T values). This follows that the amphibole forming reaction (reaction 1) can't progress significantly unless deformation creates pathways for fluid. Syntectonic growth of amphibole, therefore, was necessary for near complete transformation of the magmatic assemblage of mafic boudin to amphibole facies ensemble. Reaction 2, which has a negative volume change for solid phases, can create its own porosity and permeability (reaction enhanced permeability) and hence, should convert all the amphiboles to biotite. Restriction of biotite in the outer margin of the boudin, therefore, requires either impediment of fluid flow or lack of supply of K and

H_2O. In view of large ΔV of the reaction and syntectonic growth of biotite, it is presumed that supply of either or both of K and H_2O were responsible for restriction of biotite at the marginal part of the boudin.

Although the extant petrological data can't pinpoint the source of the metasomatic fluid, occurrence of syn-to post tectonic pegmatite including the pods at the boudin neck can be a viable source of H_2O and K. Crystallization of pegmatite presumably cut off the supply of the latter components and prevented the complete conversion of secondary amphibole to biotite.

Physical Conditions of Metasomatic Reactions

Barring a few garnet-bearing samples quantitative P-T estimation for the amphibolite facies assemblages of the mafic dykes is a difficult proposition. Nevertheless, to put some constrains on the physical conditions during which the magmatic mineralogy of the mafic rocks were converted to amphibole-and biotite-bearing assemblages, observed mineralogy of one mafic boudin were numerically simulated using the bulk composition of the boudin (XRF analyses done in the Mineralogy Petrology Department, University of Bonn Germany. Detail of the analytical procedure is given in Bhui et al., 2007). Computer program PERPLEX 07 (Connolly, 2005) was used for the numerical modeling along with the thermodynamic data of Holland and Powell, 1998, updated in 2004). The solution models used for the purpose were encapsulated in the file *solution.dat* of PERPLEX 07. It is evident from the computed phase diagram (Fig. 15.4) that the amphibole-bearing assemblage of the mafic boudin was formed within a narrow temperature interval of 720–740°C at pressure at or below 6 kbar. The inferred P-T values are in good agreement with the quantitative P-T estimates from the garnet-bearing assemblage (7±1 kbar, 700 ± 50°C, our unpublished data). That biotite developed at lower temperature is supported by garnet-biotite thermometry from the immediate felsic gneiss (550 ± 50°C, our unpublished data). The foregoing analyses indicate that poly phase deformation and infiltration driven metamorphism of mafic dykes occurred at mid crust (20–25 km crustal depth) at an elevated temperature (~ 740°C peak temperature) that corresponds to amphibolite facies metamorphism. The estimated temperature is in good agreement with partial melting of the protoliths of the felsic gneiss and formation of stromatic migmatite (Ashworth, 1985; Rushmer, 1991, Pattison et al., 2003).

Discussions

The foregoing analyses demonstrate that the mafic dykes that intruded the protolith(s) of the felsic gneiss of the studied area underwent polyphase deformation including at lest three sets of superposed folding and several generations of boudinage. Several generations of pegmatite veins accompanied repeated folding and boudinage. Felsic melts were also collected at the boudin necks. During the

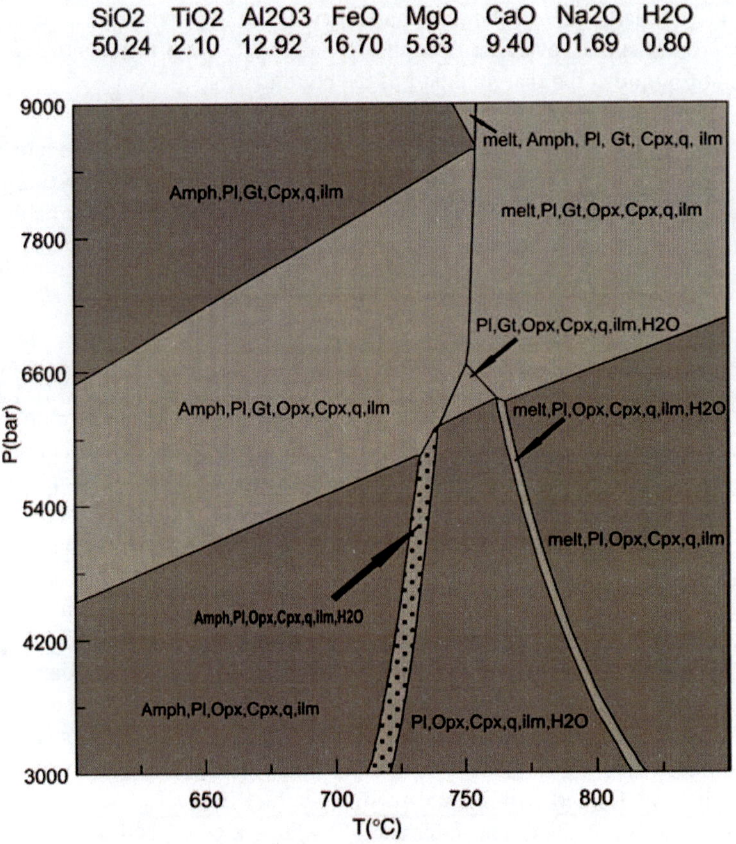

Fig. 15.4 P-T pseudosection of Amphibolite in the system $CaO–Na_2O–FeO–MgO–Al_2O_3–TiO_2–SiO_2–H_2O$ computed through PERPLEX 07 (Connolly, 2005). The H_2O content was so chosen that melt does not coexist with free H_2O. The stability field of the assemblage present in boudin core is stippled. The solution models of the different phases used in the diagram are as follows: Amph (Dale et al., 2005), Pl (Newton et al. 1980), Cpx (Holland and Powell, 1996), Opx (Holland and Powell, 1996), Chl (extended from Holland et al., 1998), Grt (Holland and Powell, 1998), Melt (White et al., 2001) All the solution models are encrypted in the file "solut_08.dat" and included in PERPLEX 07

superposed deformation, magmatic assemblage of the mafic dykes were recrystallized and transformed in to amphibolite-and biotite-bearing assemblages due to infiltration driven metamorphism. Integrating all the structural and petrological features a model is proposed for the genesis of fish-head boudin:

Stage 1: During this stage mafic rocks were subjected to layer parallel extension due to the drag of less viscous (less competent) protolith of the felsic gneiss (Fig. 15.5a).

Stage 2: Relatively more viscous mafic dykes eventually torned apart due to layer parallel drag inflicted by flow of less viscous felsic rock.

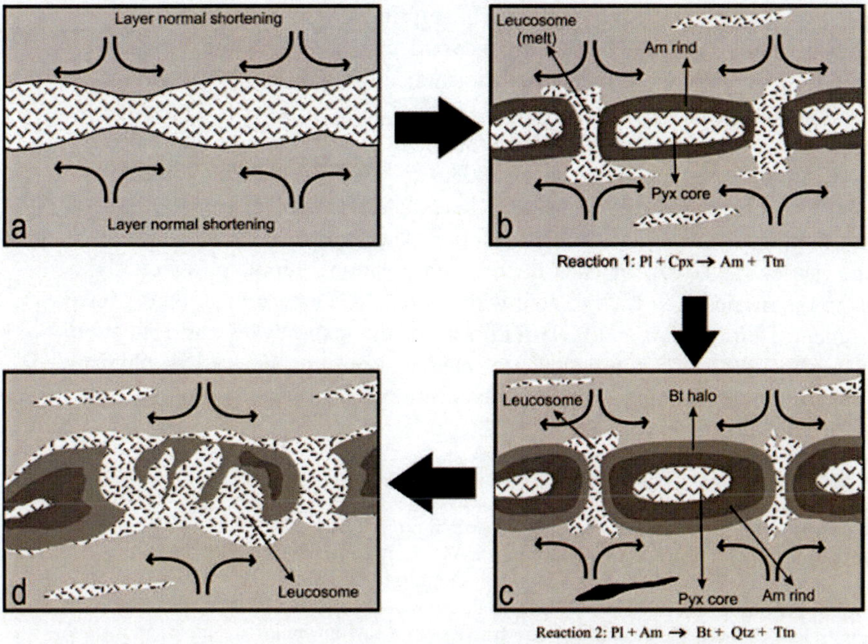

Fig. 15.5 Schematic diagram showing different stages of development of Fish-head boudin in Amphibolite (**a**) Development of pinch and swell structure in mafic dyke rock within Felsic migmatite due to layer normal compression (stage 1 in text); (**b**) Effect of continued compression and accompanying metamorphism led to formation of barrel-shaped boudin in mafic dyke rock with pyroxene core and amphibole rind while felsic migmatite host show generation of local melt pools segregated along boudin necks (stage 2 in text); (**c**) Crysllisation of local melt pool (now pegmatoidal leucosomes) near boudin neck releases aqueous fluid and K^+ that form biotite halo over amphibole; (**d**) Continued replacement of amphibole by biotite in a compressional field led to development of Fish-head boudin as well as cuspate and lobate folds in amphibolite boudins (stage 3 in text)

Recrystallization of magmatic pyroxene-bearing assemblage to amphibole-bearing assemblage (reaction 1) occurred during this stage. Syntectonic pegmatite veins got emplaced at high angle to the compression direction. Felsic rocks at the neck region of the mafic boudin that represented low-pressure domains got melted (Fig. 15.5b).

Stage 3: Deformation continued and crystallization of melt at the boudin neck converted amphibole to biotite (reaction 2) and produced biotite-rich rinds around the boudin. The biotite-rich rinds had lower viscosity compared to nodal pegmatite and the amphibole-bearing core of the mafic boudin. Flow of the less viscous rind around the relatively competent pegmatite pods produced fish-head boudin (Fig. 15.5d).

Similar mechanism can also envisage for the development of flame-folds and trapezoidal boudins described from the studied area (Ghosh and Sengupta, 1999).

Mafic dyke swarms is considered as a fingerprint of continental rifting (Ernst et al., 2001). Chatterjee et al., 2008a, 2010) have demonstrated that the felsic gneiss of the CGGC, which is the basement of the studied dyke rocks, underwent metamorphism at several times with 1.7–1.6 Ga being the earliest event. Extension of this Palaeoproterozoic crust during Mesoproterozoic time is indicated by intrusion of 1.55 Ga anorthosite (Chatterjee et al., 2008a) and ~1.5 Ga syanite (Roy Barman et al. 1994). Subsequently, the rocks of CGGC underwent polyphase deformation and metamorphism during c.1.3–1.1, 0.95–0.99 Ga and 0.75–0.80 Ga (reviewed in Chatterjee et al., 2010). In view of these observations, formation of the dyke swarms in the studied area is correlated with the c.1.5 Ga extension of their gneissic basement. Deformation and metamorphism of the mafic dykes and formation of fish-head boudins as a consequence of chemical reaction induced rheological inversion presumably took place during one or more of the Meso-to Neoproterozoic tectono-thermal events that affected the CGGC.

Acknowledgements SR, SS and PS acknowledge the financial support from the Center of Advanced Study, Department of Geological Sciences, Jadavpur University, Kolkata-700032. SS and PS acknowledge DST, India for funding a research project. They also thank DFG and AvH Foundation Germany for sponsoring research stay in Germany during 2005 and 2006. We express our gratitude to C. Ballhaus, director, Mineralogy Petrology Department, University of Bonn, Germany for providing analytical facility during the stay of SS and PS. We thank R. Goswami for his assistance in the field. We express our gratitude to S.K. Bhowmik for a thorough review which has improved the clarity of the manuscript. We thank Rajesh K. Srivastava for inviting us to contribute in this volume.

Appendix: Compositional Matrix for Textural Modeling

Phases#	Si	Ti	Al	Fe	Mg	Mn	Ca	Na	K	O	H
Bt	5.54	0.13	3.15	2.62	2.21	0.03	0.02	0.00	2.04	23.00	2.00
Pl	2.45	0.00	1.52	0.01	0.00	0.00	0.56	0.49	0.01	8.00	0.00
Qtz	1.00	0.00	0.00	0.00	0.00	0.00	0.00	0.00	0.00	2.00	0.00
Am	6.59	0.25	1.86	1.70	2.62	0.02	1.88	0.38	0.26	24.00	2.00
Cpx	1.99	0.01	0.07	0.30	0.70	0.01	0.87	0.03	0.00	6.00	0.00
Ilm	0.00	1.01	0.00	0.89	0.00	0.09	0.00	0.00	0.00	3.00	0.00
Mn	0.00	0.00	0.00	0.00	0.00	1.00	0.00	0.00	0.00	0.00	0.00
K	0.00	0.00	0.00	0.00	0.00	0.00	0.00	0.00	1.00	0.00	0.00
Ca	0.00	0.00	0.00	0.00	0.00	0.00	1.00	0.00	0.00	0.00	0.00
H_2O	0.00	0.00	0.00	0.00	0.00	0.00	0.00	0.00	0.00	0.00	2.00
Titanite	1.01	0.91	0.08	0.02	0.00	0.00	1.00	0.00	0.00	5.00	0.00
Na	0.00	0.00	0.00	0.00	0.00	0.00	0.00	1.00	0.00	0.00	0.00

References

Achariyya SK (2003) The nature of mesoproterozoic central Indian tectonic zone with exhumed and reworked older granulite. Gondwana Res 6(2): 197–214

Ashworth JR (1985) Introduction. In: Ashworth JR, (ed) Migmatites. Blackie, Glasgow and London: 1–35

Bhui UK, Sengupta P, Sengupta P (2007) Phase relations in mafic dykes and their host rocks from Kondapalle, Andhra Pradesh, India: Implications for the time-depth trajectory of the Paleoproterozoic (late Archean?) granulites from southern Eastern Ghats Belt. Precambrian Research 156: 153–174

Chatterjee N, Banerjee M, Bhattacharya M, Maji AK (2010) Monazite chronology, metamorphism-anatexis and tectonic relevance of the mid-Neoproterozoic Eastern Indian Tectonic Zone. Precamb Res 179: 99–120

Chatterjee N, Crowley JL, Mukherjee A, Das S (2008a) Geochronology of the 983 Ma Chilka Lake Anorthosite, Eastern Ghats Belt, India: Implications for pre-Gondwana tectonics. Jour Geol 116(2): 105–118

Connolly JAD (2005) Computation of phase equilibria by linear programming: A tool for geodynamic modeling and its application to subduction zone decarbonation. Earth Planet Sci Lett 236: 524–541

Dale J, Powell R, White RW, Elmer FL, Holland TJB (2005) A thermodynamic model for Ca-Na clinoamphiboles in Na2O-CaO-FeO-MgO-Al2O3-SiO_2-H_2O-O for petrological calculations. J Met Geol 23: 771–791

Ernst RE, Grosfils EB, M'Ege D (2001) Giant dike swarms: Earth, Venus, and Mars. Annu. Rev Earth Planet Sci 29: 489–534

Fisher G (1989) Matrix analysis of metamorphic mineral assemblages and reactions. Contrib Mineral Petrol 102: 69–77

Ghosh SK (1993) Structural Geology: Fundamentals and Modern Developments. Pergamon Press, London

Ghosh SK, Sengupta S (1999) Boudinage and composite boudinage in superposed deformations and syntectonic migmatisation. J Struct Geol 21: 97–110

Holland T, Baker J, Powell R (1998) Mixing properties and activity-composition relationships of chlorites in the system MgO-FeO-Al_2O_3-SiO_2-H_2O. Eur J Mineral 10: 395–406

Holland T, Powell R (1996) Thermodynamics of order-disorder in minerals. 2. Symmetric formalism applied to solid solutions. Am Mineral 81: 1425–1437

Holland TJB, Powell R (1998) An internally consistent thermodynamic data set for phases of petrologicalinterest. Jour Metamorph Geol 16: 309–343

Kranck EH (1953) Interpretation of gneiss structures with special references to Baffin Island. Proc Geol Assoc Canada 6: 59–68

Lang HM, Wachter AJ, Peterson VL, Ryan JG (2004) Coexisting clinopyroxene/spinel and amphibole/spinel symplectites in metatroctolites from the Buck Creek ultramafic body, North Carolina Blue Ridge. Am Mineral 89: 20–30

Naha K, Sen S (1965) Tectonic control in the formation of agmatites and pegmatites around Gumla, Bihar, eastern India. Neues Jahrbuch fur Mineralogie Abhandlungen 103: 116–125

Newton RC, Charlu TV, Kleppa OJ (1980) Thermochemistry of the high structural state plagioclases. Geochem Cosmochim Acta 44: 933–941

Pattison DRM, Chacko T, Farquhar J, McFarlane CRM (2003) Temperatures of granulite facies metamorphism: constraints from experimental phase equilibria and thermobarometry corrected for retrograde exchange. J Petrol 44: 867–900

Ramberg H (1955) Natural and experimental boudinage and pinch-and-swell structures. J Geol 63: 512–526

Ramsay JG, Huber MI (1987) The techniques of modern structural geology, 2: Folds and fractures. Academic Press, London

Ray Barman T, Bishui PK, Mukhopadhyay K, Ray JN (1994) Rb-Sr geochronology of the high-grade rocks from Purulia, West Bengal and Jamua-Dumka sector, Bihar. Ind Mineral 48: 45–60

Rushmer T (1991) Partial melting of two amphibolites: Contrasting experimental results under fluid absent conditions. Contrib Mineral Petrol 107: 41–59

Sengupta S (1997) Contrasting fabrics in deformed dykes and host rocks: Natural examples and a simplified model. In: Sengupta S, (ed) Evolution of geological structures in micro-to macro-scales. Chapman and Hall, London

Sengupta P, Bhui UK, Braun I, Dutta U, Mukhopadhyay D (2009) Chemical substitutions, paragenetic relations, and physical conditions of formation of hogbomite in the Sittampundi layered anorthosite complex, South India. Am Mineral 94(11-12): 1520–1534

Sengupta P, Sen J, Dasgupta S, Raith M, Bhui UK, Ehl J (1999) Ultra-high Temperature Metamorphism of Metapelitic Granulites from Kondapalle, Eastern Ghats Belt: Implications for the Indo-Antarctic Correlation. J Petrol 40: 1065–1087

Torres-Roldan RL, Garcia-Casco A, Garcia-Sanchez P (2000) CSpace: An integrated workplace for the graphical and algebraic analysis of phase assemblages on 32-bit wintel platforms. Comp Geosci 26: 779–793

White RW, Powell R, Holland TJ (2001) Calculation of partial melting equilibria in the system Na_2O-CaO-K_2O-FeO-MgO-Al_2O_3-SiO_2-H_2O (NCKFMASH). J Met Geol 19: 139–153

Chapter 16
Petrology and Geochemistry of Acid Dyke Rocks with Reference to their Bearing on Rare Metal and Rare Earth Mineralisation: Studies from Malani Igneous Suite, Northern Peninsular India

G. Vallinayagam

Introduction

The Neoproterozoic Malani Igneous Suite rocks are exposed predominantly in the Trans Aravalli Block (TAB) in western Rajasthan. They also occur in Tosham areas of Haryana, Nagar Parkar and Kirana Hills of Pakistan. The rocks of MIS are essentially of acidic with minor basic component and occur as volcanic, plutonic and dyke phases. They are characterized as A-type, anorogenic, HHP (High Heat Production) and derived by low degree partial melting of the source rocks from crust related to hot spot activity and controlled by lineaments (Pareek, 1981; Kochhar, 1984; Vallinayagam, 1988; Eby and Kochhar, 1990; Bhushan and Chandrasekaran, 2002; Kochhar, 2009; Singh and Vallinayagam, 2009). Acid dyke phase is exposed at Siwana, Dhiran, Nakora areas of Barmer District, Sankra of Jaisalmer, at Jalor and Jhunjhunu in Rajasthan and also in Tosham areas of Bhiwani District of Haryana (Fig. 16.1). The present paper provides comprehensive account of their identification and descriptions with special emphasize of bearing on rare metal, rare earth and radioactive mineral resources based on petrological and geochemical investigations.

Geological Setting

The rocks of dyke phase include viz. microgranite, rhyolite, trachyandesite, trachydacite in Siwana; microgranite, rhyolite in Jalor; rhyolite in Jhunjhunu; quartz-feldspar porphyry in Tosham hill. They intrude the rocks of acid volcanoplutonic phases. The dyke rocks are pink and grey in colours, fine grained, hard, massive and compact. They are trending mainly NE–SW direction in Siwana; in Jalor they are NNW–SSE and in Sankara NW–SE. Width ranges from few cm to few meters (average 2 m) and the length ranges from few meters to few kilometers. The acid

G. Vallinayagam (✉)
Department of Geology, Kurukshetra University, Kurukshetra 136 119, India
e-mail: gvallinayagam@rediffmail.com

R.K. Srivastava (ed.), *Dyke Swarms: Keys for Geodynamic Interpretation*,
DOI 10.1007/978-3-642-12496-9_16, © Springer-Verlag Berlin Heidelberg 2011

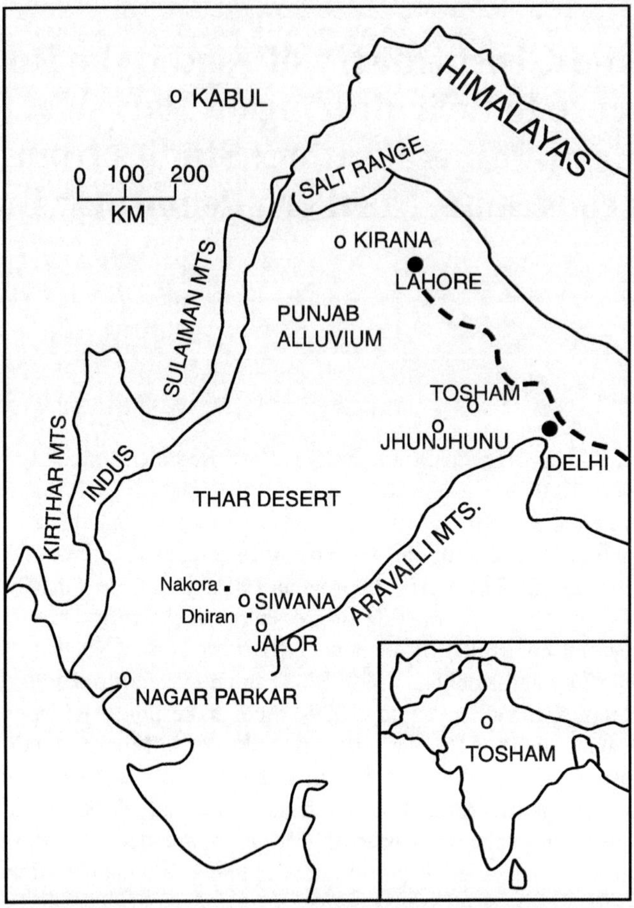

Fig. 16.1 Occurrence of the rocks of Malani Igneous Suite, Indian Subcontinent (modified after Vallinayagam, 2006)

dykes of MIS occur as radial/arc pattern, ring dykes and are observable prominently even from a distance on the outcrop due to arcuate shaped weathering pattern. Sharp contact between dyke and host rocks (granite and rhyolite) is noticed. The minerals in the dyke rocks are fine grained along the contact and become coarser towards the centre of the dyke. Few places mafic and felsic minerals observed as layers within them. Other striking features include the swelling and pinching nature and presence of host rock xenoliths within them. Siwana ring dyke is phyric (phenocryst of pink orthoclase) and aphyric also. They are pink, grey, green in colours with iron oxides/fluoride encrustation due to post magmatic alteration processes. Jalor rhyolite dykes are pink, yellowish pink, brownish blue in colours, phyric (alkali feldspar, quartz phenocrysts) and nonphyric (with flow structure). Jhunjhunu dyke rocks are pink in colour, fine grained and phyric. Tosham dyke rock contains large phenocysts (∼7 cm diameter) of quartz and alkali feldspar which are embedded in a fine grained, black groundmass.

Petrography

In Siwana, dyke rocks are hypersolvus, fine to medium grained and consist of quartz (modal mineralogy 24 wt%), alkali feldspar (perthite and orthoclase – 39%), arfvedsonite (10%), aegirine (4%), riebeckite (3%) as prominant minerals (Fig. 16.2 a, b). The alkali feldspars are occuring as phenocrysts, subhedral and altered into sericite and kaolin. Phenocryst/microcrystalline and subhedral varieties of quartz crystals occur in the groundmass. Granophyric and aphyric textures are observed. The alkali mafic minerals (op.cit) are occurring as meso and microphenocrysts. Riebeckite occur as acicular crystals, blue colour and are intergrown with arfvedsonite (Fig. 16.2a, b). Reddish brown colour, fine grained aenigmatite occur along with arfvedsonite. The accessory minerals are hematite, ilmenite, zircon, monazite and rutile. They occur in the quartzofelspathic groundmass. The silver grey coloured, anhedral ilmenite crystals are intergrown with arfvedsonite and at places it is altered to leucoxene. Anhedral, light yellow colour monazite is observed in groundmass. The dykes rocks in Jalor, Jhunjhunu, Tosham areas are subsolvus-hypersolvus and consist of quartz, orthoclase, albite, biotite and muscovite. In Tosham dykes show spherulitic to microspherulitic textures. Quartz crystals occur as embayed, corroded/shattered forms. Potash/soda feldspars are invariably altered to kaolin and sericite respectively. Biotite is deep brown colour and pleochroic (α – light brown to β – dark brown). The groundmass contains abundance of red, brown colour iron oxides, associated with quartzofeldspathic minerals.

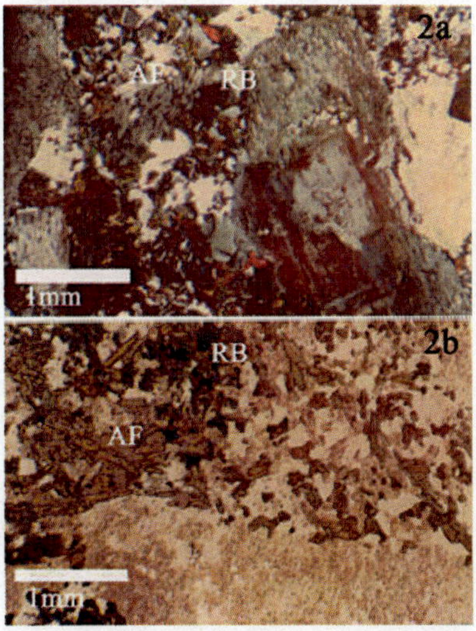

Fig. 16.2 (**a**) Perthite, quartz, arfvedsonite (AF) phenocrysts (**b**). Perthite phenocryst along with arfvedsonite and reiebeckite (RB) in microgranite dyke of Siwana ring Complex

Geochemistry

Siwana acid dyke rocks show high content of SiO_2 (69–74 wt%), Al_2O_3 (7–11%), Na_2O+K_2O (5–9%), total iron (5–9%), Rb (308–426 ppm); very high content of Nb (112–657 ppm), Ta (7–54 ppm), Zr (2,167–11,545 ppm), Hf (82–718 ppm), U (6–36 ppm), Th (27–162 ppm), Y (203–1,853 ppm), Ga (8–61 ppm), La (82–718 ppm), Ce (268–2,026 ppm) and HPU (Heat Production Unit) 4–21 μWm^{-3} and low content of MgO (0.2%), CaO (0.6%), Co (1 ppm), Cr (2 ppm), Ba (47–122 ppm), Sr (11–74 ppm) (Narayan Das et al., 1978; Jain et al., 1996; Vallinayagam, 1988, 1999, 2004). Acid Dykes of Jalor area are fairly enriched in SiO_2 (68–74%), total alkalies (5–7%), iron (3–5%), Rb, La, Ce, Nd, Sm, Zr, Ga, Y, Nb, and depleted in Ba, Sr, and Ti (Kochhar et al., 1995). Jhunjhunu acid dykes are high in silica, alkalies, Nb, Rb, U, Th, Sn, Y, total REE, HPU and low in Ca, Mg, Sr, Hf, Ta, Sc, Co, Ni (Sharma, 1994). Tosham dyke rock shows high silica (71%), total alkalies (9%), aluminum (14%), and low in iron (3.31%), CaO, MgO, TiO_2 (Kochhar, 1983; Khorana and Dhir, 2004). Sn (30 ppm), Li (3 ppm), total REE (3,000 ppm) contents for Tosham dyke are mentioned by Khorana and Dhir (2004), Vallinayagam (2006).

Siwana acid dyke rocks are peralkaline, acmite in CIPW normative whereas the dyke rocks of Jalor, Tosham, Jhunjhunu are peraluminous – minor amount of peralkaline and corundum/acmite in CIPW norm. In AFM diagram (Fig. 16.3), they are clustering toward the alkali-iron enrichment and magnesia deficiency edges. DI (Differentiation Index) vs. wt% of oxides diagram, silica, alumin, alkalies and iron show scatter of data whereas the phosphorous, titanium magnesia, calcium define a linear trend. This indicate nonorogenic tectonic environment in their emplacement. In Log (Y+Nb) vs Log Rb and Log Y vs Log Nb diagrams they also plot in "Within

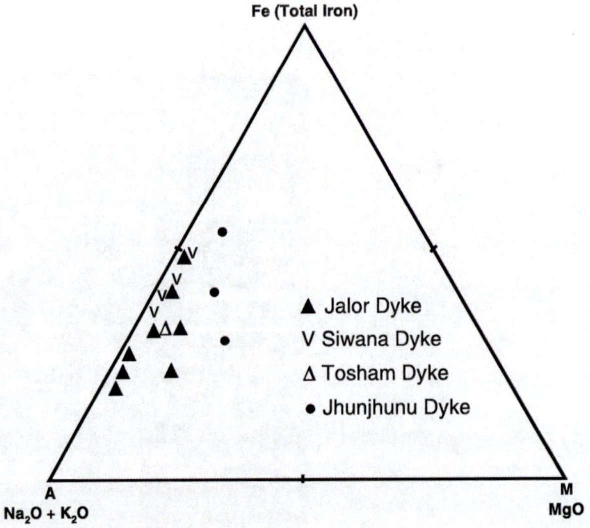

Fig. 16.3 AFM diagram for MIS dyke rocks

Plate Granite" field of Pearce et al. (1984). Primordial mantle normalized trace element diagram shows peaks of U, Th, Nb, La, Ce, Hf, Zr and trough of Ba, Sr, Ti, for them. Chondrite normalised REE diagram depicts them as high REE contents (4,500 ppm), enriched LREE with marked Eu negative anomalies profile. The dyke rocks show low in Ba/Rb (0.1), K/Rb (81), TiO_2/Ta (0.01) and displays negative correlation in Rb-Sr plot. These characters attest them as a product of progressive differentiation of the parent magma. In Rb-Ba-Sr ternary diagram (1981) they plot in the field of strongly differentiated granites. In K/Rb ratio – Rb/Sr ratio plot, they occupy in the mineralized granite fields.

Mineralisation Potential

Post orogenic dyke activities occur commonly at the end of large scale acid magmatism. Generally these dykes are characterized by a wide compositional milieu but volumetrically minor. They are grouped into three types based on the field relationship viz. dykes related to (i) continental breakup (ii) batholiths (iii) large igneous province. Dykes of MIS are related to the last above mentioned type. Anorogenic, A-type MIS magmatism marks the stabilization/cratonization of the continental crust of the northern part of the Indian Shield during Neoproterozic. Anorogeny is attributed to hot-spot activity and is due to the thermal process in the asthenosphere. Burke and Deway (1973) observed that anorogenic granites were emplaced above the mantle plume prior to development of intercontinental rifts. MIS acid magmatic phases are accompanied by post magmatic hydrothermal activities where the rate of transfer of heat and subsolidus fluid are predominant. Thus ore forming elements are concentrated in the deep crustal fluid and resulted the metallogenesis process in these dyke rocks. The combined tectonomagamtism are responsible for the heat dissipation, collection and concentration of Mo, W, Sn Nb, La, U, Th. Detailed field cum geochemical studies on dyke rocks of MIS identified the location of metal bearing dyke rocks which are interlinked with ore fluid migration and concentration. Currently available combined petrological and geochemical data depict that the acid dyke rocks of Siwana, Jhunjhunu and Tosham areas have potential for rare metals (Zr, Hf, Nb, Ta, Be, Li, Sn), rare earth (La, Ce, Y) and U, Th mineral resources.

Acknowledgements Author expresses his gratitude to R.K. Srivastava for encouragement to submit this manuscript; to V. Balaram, Hyderabad for geochemical analytical support, to DST, UGC, CSIR (Government of India) for financing the research activities on MIS for last two decades; to S.K. Bhushan, Ex.DDG (GSI, Govt. of India) for his fruitful review of the MS and to Chairman, Department of Geology, Kurukshetra University for providing research facilities.

Reference

Bhushan SK, Chandrasekaran V (2002) Geology and geochemistry of the magmatic rocks of the Malani Igneous Suite and Tertiary Alkaline Province of Western Rajasthan. Mem Geol Surv India 126: 1–129

Biste M (1981) Application of various geochemical proximity indicators to the tin favourability of South Sardinian granites. J Geochem Explor 15: 295

Burke KC, Deway JF (1973) Plume generated triple junction: Key indicators in applying plate tectonics to old rocks. J Geol 81: 406–433

Eby GN, Kochhar N (1990) Geochemistry and petrogenesis of the Malani igneous suite, North Peninsular India. J Geol Soc India 36: 109–130

Jain RB, Miglani TS, Kumar S, Swarnkar BM, Singh R (1996) Rare metal and rare earth rich peralkaline, agpaiitic granitoid dykes of Siwana ring complex, district Barmer, Rajasthan. Curr Sci 70: 854–858

Khorana RK, Dhir NK (2004) Post-Delhi acid magmatism in Tosham area and its bearing on tin-tungsten mineralization. Indian Miner 58: 89–98

Kochhar N (1983) Tusham ring complex, Bhiwani, India. Proc Indian Natl Sci Acad 49: 459–490

Kochhar N (1984) Malani Igneous Suite: Hot-spot magmatism and cratonization of the Northern part of the Indian shield. J Geol Soc India 25: 155–161

Kochhar N (2009) The Malani Supercontinent: Middle East connection during Late Proterozoic. In: Shrivastava KL, (ed) Economic mineralization. Scientific Publisher (India), Jodhpur: 15–25

Kochhar N, Dhar S, Sharma R (1995) Geochemistry and tectonic significance of acid and basic dyke associated with Jalor magmatism, west Rajasthan. Mem Geol Soc India No. 33: 375–389

Narayan Das GR, Begchi AK, Chaube DN, Sharma CV, Naveneetham KV (1978) Rare metal content, geology and tectonic setting of the alkaline complexes across the Trans-Aravalli region, Rajasthan. In: Verma PK, Verma VK, (eds) Recent researches in geology, vol 7. Hindustan Publishing Corporation, Delhi: 201–217

Pareek HS (1981) Petrochemistry and petrogenesis of Malani igneous suite, India. Geol Soc Am Bull 92: 206–273

Pearce JA, Harris NBW, Tindle AG (1984) Trace element discrimination diagrams for the tectonic interpretation of granitic rocks. J Petrol 25: 956–983

Sharma R (1994) High heat production (HHP) granites of Jhunjhunu area, Rajasthan, India. Bull Indian Geol Assoc 27: 55–61

Singh AK, Vallinayagam G (2009) Radioactive element distribution and rare-metal mineralization in anorogenic acid volcano-plutonic rocks of the Neoproterozoic Malani Felsic Province, Western Peninsular India. J Geol Soc India 73: 837–853

Vallinayagam G (1988) Geology and geochemistry of alkali granites and the associated acid volcanics around Mokalsar, district Barmer, W. Rajasthan, India and their bearing on the rift tectonics. Ph.D. Thesis, Punjab University, Chandigarh

Vallinayagam G (1999) Nb, Zr, REE rich acid dyke rocks from the Piplun area, Siwana Ring Complex, Western Rajasthan, India. In: Paliwal BS, (ed) Geological evolution of Northwestern India. Scientific Publishers, Jodhpur: 94–102

Vallinayagam G (2004) A report on rare metals and rare earths in the Siwana ring complex, Rajasthan. J Appl Geochem 2: 387–391

Vallinayagam G (2006) Geochemistry of rare metal and rare earth ore deposits: A study of the Malani igneous suite Trans-Aravalli block of NW Peninsular India. Indian J Geochem 21: 439–446

Chapter 17
Petrogenesis of Basaltic and Doleritic Dykes from Kawant, Chhotaudepur Province, Deccan Traps

K.R. Hari and Vikas Swarnkar

Introduction

Deccan Traps which cover about 10^6 sq. kms of Indian sub-continent is one of the largest flood basalt province of the world. These flows emplaced in a short time span at 65 Ma (Courtillot et al., 1988; Hofmann et al., 2000) are predominantly tholeiitic basalts and its derivatives. Few occurrences of alkali-olivine basalts, rhyolites, carbonatites, nephelinites, lamprophyres, granites etc. have been reported from different parts of western India and also along the Narmada-Son rift valley (eg:- Beane and Hooper, 1988; Greenough et al, 1988; Hari, 1988; Karkare and Srivastava, 1990; Gwalani et al., 1993; Srivastava, 1994).

Kawant, a part of the Chhotaudepur Province of Deccan Traps is situated in the lower Narmada valley (Fig. 17.1). In the present area, dyke swarms of syenite, tinguaite, lamprophyres, carbonatites etc are seen traversing through Deccan basalts (Sukeshwala and Avasia, 1972; Karkare and Srivastava, 1990; Viladkar and Avasia, 1992). Krishnamurthy et al (2000) reported picrite dyke of "ankaramite" type with high Mg olivine ($Fo_{60–88}$) from Kawant area. In the present paper, we present a detailed mineralogical and geochemical account of the E-W & SE-NW trending basaltic and doleritic dykes of Kawant area to assess various petrogenetic aspects.

Mineral Chemistry and Petrology

All the five basaltic dykes have olivine, clinopyroxene and minor amounts of feldspar as phenocryst embedded in a fine grained matrix of clinopyroxene, feldspar, opaques and glass (Fig. 17.2a, b). Two generation of olivine grains are noticeable in almost all sections. Serpentinization and iddingsitization are perceptible along the cracks and rims of olivine grains. Zoning in the phenocrystal grains are also

K.R. Hari (✉)
Government V.Y.T.P.G. Autonomous College, Durg, Chhattisgarh 491001, India
e-mail: krharigeology@gmail.com

V. Swarnkar
Government V.Y.T.P.G. Autonomous College, Durg, Chhattisgarh 491001, India

R.K. Srivastava (ed.), *Dyke Swarms: Keys for Geodynamic Interpretation*,
DOI 10.1007/978-3-642-12496-9_17, © Springer-Verlag Berlin Heidelberg 2011

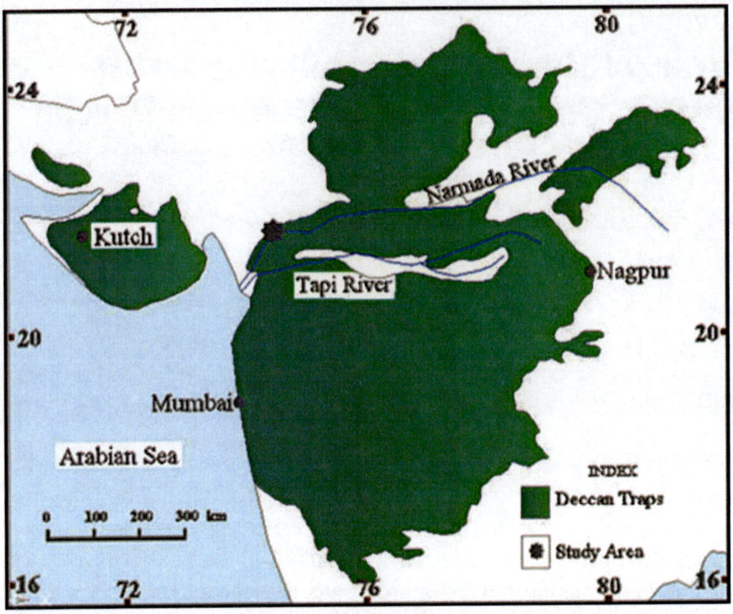

Fig. 17.1 Map of Deccan Traps showing the location of the study area

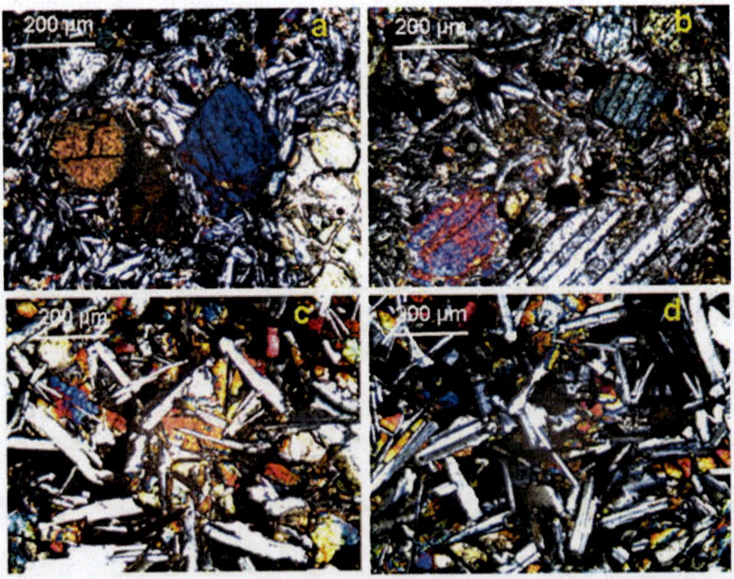

Fig. 17.2 Photomicrographs of basalt (**a** and **b**) and dolerite (**c** and **d**). Porphyritic texture with the phenocrysts of olivine, pyroxene and feldspars are noticeable in basalt. Ophitic and sub-ophitic textures are seen in dolerite

noticeable. Chemical compositions of pyroxene, olivine, feldspar and opaques of D-21 sample were analyzed using a CAMECA SX 100 electron microprobe at the Institute of Mineralogy and Mineral Resources, Technical University of Claustal, Germany by Dr. N.V. Chalapathi Rao. A beam current of 20 nA acceleration voltage of 20 kV and a beam diameter of 1 micron was used. Both natural as well as synthetic standards were employed and the obtained data is presented in Tables 17.1, 17.2, 17.3 and 17.4. The Fo% of olivine varies from 79 to 85. Minor compositional variations within the individual pyroxene grains are noticeable. Pyroxene grains are either diopsidic or augitic composition and not much chemical variations are noticeable between the phenocryst and groundmass grains. However considerable variations are noticeable in the feldspar grains. Some grains represent alkali feldspars in the mesostasis, whereas the other phenocrystal grains are plagioclase with varying An contents (An_{26}–An_{69}). Titanomagmatite is the main opaque phase identified in these rocks.

Dolerite dykes exhibit typical ophitic and subophitic texture (Fig. 17.2c, d) there by showing fully or partially embedded feldspar laths in the grains of clinopyroxene. They have altered olivine grains in varying proportions. Kumar et al (1992) reported two types of dolerite viz. normal dolerite and granulitic dolerite with porphyritic feldspar from Phenai Mata Igneous Complex adjacent to the percent study area.

Table 17.1 Microprobe analyses of olivine from picrite. C = Core

–	1C	2 C	3 C	4 C	5 C	6 C	7 C
SiO_2	54.15	55.65	40.87	41.03	55.02	40.34	41.20
TiO_2	0.03	0.01	0.00	0.01	0.02	0.00	0.00
Al_2O_3	1.46	1.60	0.02	0.05	1.98	0.03	0.04
Cr_2O_3	0.06	0.05	0.03	0.03	0.05	0.04	0.03
FeO	10.27	10.34	14.52	14.34	10.30	14.50	14.63
MnO	0.04	0.02	0.16	0.17	0.03	0.15	0.17
MgO	21.83	21.98	44.76	44.71	23.62	43.21	42.93
CaO	0.53	0.35	0.39	0.38	0.35	0.41	0.36
Na_2O	0.10	0.11	0.06	0.02	0.07	0.01	0.03
K_2O	0.14	0.17	0.00	0.00	0.18	0.01	0.00
NiO	0.08	0.09	0.33	0.32	0.12	0.30	0.28
Total	88.68	90.36	101.13	101.05	91.73	99.00	99.67
Formula (on the basis of 4 oxygen)							
Si	1.416	1.424	1.012	1.015	1.392	1.020	1.033
Ti	0.001	0.000	0.000	0.000	0.000	0.000	0.000
Al	0.045	0.048	0.001	0.001	0.059	0.001	0.001
Cr	0.001	0.001	0.001	0.001	0.001	0.001	0.001
Fe	0.225	0.221	0.301	0.297	0.218	0.307	0.307
Mn	0.001	0.000	0.003	0.004	0.001	0.003	0.004
Mg	0.851	0.839	1.652	1.649	0.890	1.629	1.604
Ca	0.015	0.009	0.010	0.010	0.009	0.011	0.010
Na	0.005	0.005	0.003	0.001	0.003	0.001	0.001
K	0.005	0.006	0.000	0.000	0.006	0.000	0.000
Ni	0.002	0.002	0.006	0.006	0.002	0.006	0.006
Total	2.565	2.556	2.989	2.984	2.583	2.979	2.967
Fo%	79.11	79.11	84.60	84.74	80.35	84.16	83.94

Table 17.2 Microprobe analyses of pyroxene from picrite. P = Phenocryst; PC = Phenocryst core: PR = Phenocryst rim; GM = Ground mass

–	1 PC	2 PR	3 PC	4 P	5 PC	6 PC	7 PR	8 PC	9 PR
SiO_2	49.65	51.73	52.20	52.35	51.37	51.24	48.84	52.83	50.11
TiO_2	1.77	1.49	1.39	1.47	1.37	1.47	2.34	0.64	1.50
Al_2O_3	3.97	3.00	3.03	3.32	3.06	3.24	4.49	2.79	3.72
Cr_2O_3	0.02	0.01	0.00	0.01	0.01	0.00	0.00	0.50	0.18
FeO	10.74	8.54	7.45	7.68	7.49	9.42	8.01	4.73	8.33
MnO	0.27	0.23	0.09	0.11	0.11	0.27	0.13	0.06	0.12
MgO	10.79	12.56	13.52	13.20	13.56	11.73	12.22	16.55	14.28
CaO	23.47	23.84	23.50	23.59	23.79	22.86	22.61	23.01	21.98
Na_2O	1.07	0.96	0.84	0.85	0.89	1.20	0.99	0.22	0.39
K_2O	0.00	0.00	0.00	0.00	0.00	0.00	0.00	0.00	0.00
NiO	0.00	0.00	0.00	0.00	0.00	0.00	0.00	0.03	0.04
Total	101.76	102.35	102.03	102.57	101.65	101.42	99.61	101.34	100.65
Formula (on the basis of 6 oxygen)									
Si	1.856	1.898	1.906	1.903	1.889	1.902	1.839	1.913	1.861
Ti	0.050	0.041	0.038	0.040	0.038	0.041	0.066	0.018	0.042
Al	0.176	0.130	0.131	0.143	0.133	0.142	0.200	0.119	0.164
Cr	0.000	0.000	0.000	0.000	0.000	0.000	0.000	0.014	0.005
Fe	0.336	0.262	0.228	0.233	0.230	0.292	0.252	0.143	0.259
Mn	0.009	0.007	0.003	0.003	0.004	0.008	0.004	0.002	0.004
Mg	0.601	0.687	0.736	0.715	0.743	0.649	0.686	0.893	0.790
Ca	0.940	0.937	0.919	0.919	0.938	0.909	0.912	0.893	0.875
Na	0.078	0.068	0.060	0.060	0.064	0.087	0.072	0.015	0.028
K	0.000	0.000	0.000	0.000	0.000	0.000	0.000	0.000	0.000
Ni	0.000	0.000	0.000	0.000	0.000	0.000	0.000	0.001	0.001
Total	4.046	4.031	4.021	4.017	4.039	4.030	4.032	4.011	4.028
Ca	50.08	49.69	48.83	49.20	49.05	49.13	49.31	46.28	45.47
Mg	32.02	36.42	39.08	38.30	38.89	35.07	37.06	46.30	41.09
Fe	17.89	13.89	12.09	12.51	12.05	15.80	13.63	7.43	13.45

–	10 GM	11 GM	12 PC	13 PR	14 PC	15 PR	16 PC	17 PR	18 P
SiO_2	52.30	52.67	51.11	50.86	50.18	50.30	54.03	49.72	51.81
TiO_2	1.39	0.67	0.99	1.57	1.27	1.53	0.43	1.57	1.27
Al_2O_3	1.87	2.95	3.88	3.67	4.50	3.64	1.86	3.83	3.36
Cr_2O_3	0.00	0.86	0.48	0.17	0.42	0.16	0.61	0.21	0.28
FeO	10.35	4.00	5.90	8.42	6.66	8.31	3.96	8.44	7.66
MnO	0.23	0.04	0.10	0.14	0.09	0.12	0.04	0.11	0.14
MgO	14.43	16.77	15.53	14.27	16.33	14.10	17.31	14.24	14.97
CaO	20.86	22.46	22.10	22.04	21.04	21.95	22.85	21.89	21.83
Na_2O	0.36	0.29	0.32	0.33	0.36	0.31	0.26	0.37	0.32
K_2O	0.00	0.00	0.00	0.00	0.00	0.01	0.00	0.01	0.00
NiO	0.01	0.06	0.05	0.05	0.06	0.02	0.06	0.04	0.01
Total	101.79	100.76	100.46	101.52	100.90	100.45	101.40	100.42	101.65
Formula (on the basis of 6 oxygen)									
Si	1.924	1.911	1.878	1.870	1.840	1.869	1.945	1.852	1.891
Ti	0.039	0.018	0.027	0.044	0.035	0.043	0.012	0.044	0.035

Table 17.2 (continued)

–	10 GM	11 GM	12 PC	13 PR	14 PC	15 PR	16 PC	17 PR	18 P
Al	0.081	0.127	0.169	0.160	0.196	0.160	0.079	0.169	0.145
Cr	0.000	0.025	0.014	0.005	0.012	0.005	0.018	0.006	0.008
Fe	0.319	0.121	0.181	0.259	0.204	0.258	0.119	0.263	0.234
Mn	0.007	0.001	0.003	0.004	0.003	0.004	0.001	0.004	0.004
Mg	0.791	0.907	0.850	0.782	0.892	0.781	0.929	0.791	0.815
Ca	0.822	0.873	0.870	0.868	0.827	0.874	0.882	0.874	0.854
Na	0.026	0.020	0.023	0.023	0.025	0.023	0.018	0.027	0.023
K	0.000	0.000	0.000	0.000	0.000	0.000	0.000	0.000	0.000
Ni	0.000	0.002	0.002	0.001	0.002	0.001	0.002	0.001	0.000
TOTAL	4.010	4.006	4.017	4.017	4.036	4.018	4.004	4.031	4.010
Ca	42.56	45.92	45.76	45.47	42.98	45.67	45.69	45.33	44.89
Mg	40.96	47.69	44.72	40.96	46.40	40.83	48.14	41.03	42.82
Fe	16.49	6.39	9.52	13.57	10.63	13.50	6.17	13.64	12.29

–	19 P	20 PC	21 P	22 PR	23 PC	24 GM	25 P	26 P
SiO_2	52.18	52.88	51.10	50.59	52.90	49.79	51.55	51.24
TiO_2	1.30	0.63	1.35	1.53	0.55	1.57	1.30	1.18
Al_2O_3	2.71	2.86	1.86	3.70	2.54	3.94	3.35	2.99
Cr_2O_3	0.13	0.51	0.00	0.18	0.60	0.21	0.22	0.18
FeO	8.43	4.80	10.91	8.45	4.54	8.46	8.24	8.17
MnO	0.15	0.07	0.23	0.14	0.06	0.12	0.12	0.13
MgO	15.15	15.87	13.70	13.64	16.11	14.25	13.90	14.43
CaO	21.52	22.84	20.47	21.98	21.94	21.75	21.86	21.62
Na_2O	0.31	0.25	0.35	0.29	0.26	0.31	0.31	0.26
K_2O	0.01	0.00	0.00	0.00	0.00	0.00	0.00	0.00
NiO	0.02	0.03	0.02	0.05	0.09	0.04	0.05	0.05
Total	101.91	100.74	100.00	100.55	99.58	100.43	100.87	100.24
Formula (on the basis of 6 oxygen)								
Si	1.904	1.925	1.922	1.878	1.941	1.853	1.901	1.902
Ti	0.036	0.017	0.038	0.043	0.015	0.044	0.036	0.033
Al	0.117	0.123	0.083	0.163	0.110	0.174	0.146	0.131
Cr	0.004	0.015	0.000	0.005	0.017	0.006	0.006	0.005
Fe	0.257	0.146	0.343	0.262	0.139	0.263	0.254	0.254
Mn	0.005	0.002	0.007	0.004	0.002	0.004	0.004	0.004
Mg	0.824	0.861	0.768	0.755	0.881	0.791	0.764	0.798
Ca	0.841	0.891	0.825	0.874	0.863	0.867	0.864	0.860
Na	0.022	0.018	0.026	0.021	0.018	0.022	0.022	0.019
K	0.001	0.000	0.000	0.000	0.000	0.000	0.000	0.000
Ni	0.001	0.001	0.000	0.001	0.003	0.001	0.001	0.002
Total	4.011	3.999	4.012	4.007	3.990	4.025	3.999	4.008
Ca	43.77	46.94	42.61	46.23	45.81	45.14	45.91	44.97
Mg	42.86	45.36	39.66	39.90	46.78	41.15	40.59	41.76
Fe	13.38	7.70	17.73	13.87	7.41	13.71	13.50	13.27

Table 17.3 Microprobe analyses of feldspar from picrite

–	1	2	3	4	5
SiO_2	67.56	68.11	61.54	51.01	50.42
TiO_2	0.31	0.20	0.16	0.10	0.07
Al_2O_3	20.11	21.36	23.70	31.47	29.53
Cr_2O_3	0.00	0.00	0.00	0.00	0.00
FeO	0.79	0.65	1.50	1.10	1.95
MnO	0.02	0.00	0.01	0.01	0.02
MgO	0.33	0.23	1.08	0.13	1.05
CaO	1.42	1.55	5.19	14.37	13.46
Na_2O	5.55	5.98	6.91	3.51	3.22
K_2O	7.03	6.26	1.61	0.27	0.21
NiO	0.00	0.00	0.00	0.00	0.00
Total	103.12	104.34	101.70	101.97	99.93
Formula (on the basis of 32 oxygen)					
Si	11.769	11.678	10.856	9.175	9.269
Ti	0.040	0.026	0.022	0.014	0.010
Al	4.150	4.340	4.954	6.706	6.433
Cr	0.000	0.000	0.000	0.000	0.000
Fe	0.115	0.093	0.221	0.165	0.300
Mn	0.003	0.000	0.001	0.001	0.003
Mg	0.085	0.059	0.284	0.035	0.287
Ca	0.265	0.284	0.981	2.768	2.652
Na	1.875	1.988	2.364	1.225	1.148
K	1.563	1.369	0.361	0.062	0.049
Ni	0.001	0.000	0.000	0.000	0.000
Total	19.866	19.837	20.045	20.152	20.151
Or	42.20	37.60	9.75	1.53	1.28
Ab	50.63	54.60	63.79	30.21	29.82
An	7.17	7.80	26.47	68.26	68.90

Geochemistry

Twenty representative samples were analyzed for its major elements by XRF and the trace elements by ICP-MS The analyses were carried out at NGRI, Hyderabad and Wadia Institute of Himalayan Geology, Dehra Dun and are presented in Tables 17.5 and 17.6. Considerable variations are noticeable in MgO, Al_2O_3, Ni and Cr contents of these rocks. The chondrite normalized REE pattern (Fig. 17.3) of both the rock types have gentle slope and share a similar pattern pointing a genetic link between them.

Petrogenesis

Various discrimination diagrams and the presence of quartz in the CIPW norm clearly indicate that all these rocks in the present discussion are "tholeiitic" in nature. Vijaya Kumar et al (2006) pointed out that constant (La/Ce)n ratios but variable (La/Yb)n are characteristic geochemical traits of tholeiites. In the present

Table 17.4 Microprobe analyses of opaques from picrite

–	1	2	3	4	5
SiO_2	0.10	0.96	3.66	0.15	0.12
TiO_2	13.87	14.10	24.37	29.14	25.79
Al_2O_3	0.77	1.01	0.88	1.04	1.05
Cr_2O_3	0.02	0.02	0.02	0.03	0.02
FeO	76.29	75.95	63.77	63.09	67.38
MnO	1.14	1.39	0.80	0.97	1.47
MgO	1.14	1.21	0.28	0.06	0.06
CaO	0.15	0.11	1.77	0.17	0.05
Na_2O	0.01	0.42	0.00	0.01	0.07
K_2O	0.00	0.03	0.00	0.00	0.01
NiO	0.00	0.00	0.04	0.03	0.02
Total	93.47	95.21	95.58	94.68	96.03
Formula (on the basis of 32 oxygen)					
Si	0.035	0.333	1.151	0.049	0.037
Ti	3.747	3.690	5.756	7.020	6.287
Al	0.327	0.415	0.327	0.396	0.401
Cr	0.004	0.007	0.004	0.007	0.005
Fe	22.926	22.097	16.746	16.902	18.265
Mn	0.347	0.409	0.211	0.263	0.403
Mg	0.609	0.629	0.130	0.027	0.028
Ca	0.057	0.042	0.594	0.058	0.017
Na	0.004	0.285	0.000	0.006	0.047
K	0.000	0.014	0.000	0.001	0.005
Ni	0.000	0.000	0.011	0.007	0.005
Total	28.057	27.920	24.930	24.736	25.501

case, constant (La/Ce)n and variable (La/Yb)n (Table 17.6) further confirms the tholeiitic nature of these rocks.

Fractional crystallization and crustal contamination are the two important processes in the modification of basaltic rocks. De Paolo (1981) in his classic work proved that concurrent assimilation and fractional crystallization is also an important process in the petrogenesis of basic rocks. It is a common assumption that magma with high $Mg^{\#}$ value (~0.70), Ni (> 400–500 ppm) and Cr (> 1,000 ppm) are considered to be relatively undifferentiated and are primitive (Wilson, 1989, p 22). One of the basaltic dykes (D-21A, D-21B, D-21C, D-21i) in the present area with $Mg^{\#}$ (0.64–0.66), Ni (~ 400 ppm), Cr (~ 1,000 ppm) and Ni/MgO (32–40) exhibit a primitive magma signature. Primitive basalts have been reported from various parts of Deccan Traps earlier also (Beane and Hooper, 1988; Krishnamurthy and Cox, 1977; Peng and Mahoney, 1995; Krishnamurthy et al., 2000; Greenough et al., 1998). Le Bas (2000) argued that a rock with $\geq$ 12 wt% MgO, < 52 wt% SiO_2 and < 3 wt% total alkali is to be termed as "picrite". On the basis of this terminology, the primitive basalt in the present area is to be named as picrite as it contains 12–13 wt% MgO, 47–48 wt% SiO_2 and 2.06–2.19 wt% Na_2O+K_2O.

Table 17.5 Major element analyses in wt% along with CIPW norms. Basalt →D-1A, D-1B, D-6A, D-6B, D-7, D-7i, D-9A, D-9B, D-9i, D-21A, D-21B, D-21C & D-21i; Dolerite → D-2A, D-2B, D-8A, D-8B, D-8i, D-10, D-10i,

Sample	D 1 A	D 1 B	D 2 A	D 2 B	D 6 A	D 6 B	D 7	D 7 i	D 8 A	D 8 B	D 8 i
SiO_2	48.43	48.49	47.34	47.3	48.81	48.83	46.36	49.22	49.21	48.63	48.64
Al_2O_3	16.84	16.83	15.76	15.77	14.53	14.54	15.16	14.15	14.16	15.86	15.86
Fe_2O_3	14.9	14.89	15.44	15.43	15.93	15.94	13.62	13.57	13.54	12.71	12.6
MnO	0.19	0.18	0.2	0.19	0.27	0.27	0.16	0.16	0.15	0.15	0.15
MgO	3.27	3.24	5.28	5.3	3.64	3.67	8.75	7.77	7.69	7.11	7.14
CaO	10.35	10.32	9.35	9.35	10.11	10.07	9.26	9.24	9.19	9.46	9.25
Na_2O	1.77	1.78	1.82	1.82	1.78	1.78	1.76	1.75	1.72	1.71	1.72
K_2O	0.43	0.43	0.38	0.38	0.23	0.24	0.76	0.75	0.8	0.78	0.8
TiO_2	2.51	2.49	2.94	2.93	2.62	2.61	2.03	2.04	1.76	1.85	1.78
P_2O_5	0.26	0.26	0.31	0.31	0.23	0.23	0.29	0.28	0.29	0.27	0.29
Sum	98.95	98.91	98.82	98.78	98.15	98.18	98.15	98.93	98.51	98.53	98.23
Mg#	0.33	0.32	0.43	0.43	0.34	0.34	0.59	0.56	0.56	0.55	0.56
Q	12.96	13.05	10.57	10.49	14.95	14.93	3.915	8.918	9.075	8.153	8.259
Or	2.541	2.541	2.246	2.246	1.359	1.418	4.491	4.432	4.728	4.61	4.728
Ab	14.98	15.06	15.4	15.4	15.06	15.06	14.89	14.81	14.55	14.47	14.55
An	36.73	36.66	33.7	33.74	30.98	30.98	31.22	28.54	28.55	33.30	33.19
Di	3.827	3.791	0.93	0.905	7.48	7.355	4.967	7.001	7.475	4.683	4.041
Hy	6.371	6.313	12.7	12.78	5.599	5.732	19.49	16.11	15.69	15.54	15.91
Il	0.406	0.385	0.428	0.406	0.578	0.578	0.342	0.342	0.321	0.321	0.321
Hm	14.9	14.89	15.44	15.43	15.93	15.94	13.62	13.57	13.54	12.71	12.6
Tn	5.637	5.615	6.665	6.668	5.686	5.661	4.541	4.566	3.906	4.127	3.955
Ap	0.616	0.616	0.734	0.734	0.545	0.545	0.687	0.663	0.687	0.64	0.687

Sample	D 9 A	D 9 B	D 9 i	D10	D 10 i	D 21 A	D 21 B	D 21 C	D 21 i
SiO_2	49.68	49.65	49.7	49.71	49.7	48.36	47.65	47.94	47.14
Al_2O_3	15.81	15.8	15.81	15.82	15.8	9.45	9.98	10.24	10.44
Fe_2O_3	13.23	13.24	13.21	13.27	13.25	15.21	15.09	14.62	14.57
MnO	0.17	0.16	0.17	0.15	0.15	0.18	0.17	0.18	0.18
MgO	6.01	6	5.97	5.99	5.97	12.03	12.3	12.42	13.01
CaO	9.01	9.01	9	9.58	9.68	11.06	11.00	10.53	10.28
Na_2O	1.7	1.7	1.71	1.62	1.63	1.42	1.42	1.49	1.43
K_2O	0.82	0.82	0.83	0.75	0.73	0.64	0.65	0.70	0.71
TiO_2	2.01	2.03	2	1.77	1.77	2.06	2.06	2.25	2.22
P_2O_5	0.31	0.31	0.31	0.29	0.28	0.22	0.23	0.26	0.26
Sum	98.75	98.72	98.71	98.95	98.96	100.62	100.54	100.62	100.22
Mg#	0.50	0.50	0.50	0.50	0.50	0.64	0.64	0.65	0.66
Q	11.33	11.32	11.34	11.36	11.28	4.414	3.031	2.988	1.75
Or	4.846	4.846	4.905	4.432	4.314	3.762	3.857	4.15	4.163
Ab	14.38	14.39	14.47	13.71	13.79	12.00	11.97	12.638	12.135
An	33.09	33.06	33.01	33.68	33.94	17.54	18.945	19.162	19.957
Di	2.533	2.469	2.579	4.964	5.432	22.893	21.512	18.908	17.372
Hy	13.80	13.8	13.67	12.62	12.35	19.352	20.664	22.17	24.347
Il	0.364	0.342	0.364	0.321	0.321	0.373	0.369	0.382	0.378
Hm	13.23	13.24	13.21	13.27	13.25	15.204	15.089	14.615	14.565
Tn	4.464	4.541	4.44	3.931	3.931	4.567	4.576	5.019	4.961
Ap	0.734	0.734	0.734	0.687	0.663	0.529	0.541	0.606	0.606

Table 17.6 Trace element analyses in ppm. Basalt → D-1A, D-1B, D-6A, D-6B, D-7, D-7i, D-9A, D-9B, D-9i, D-21A, D-21B, D-21C & D-21i; Dolerite → D-2A, D-2B, D-8A, D-8B, D-8i, D-10, D-10i

Sample	D 1 A	D 1 B	D 2 A	D 2 B	D 6 A	D 6 B	D 7	D 7 i	D 8 A	D 8 B	D 8 i
Sc	39.71	43.192	46.931	45.57	27.10	26.695	70.825	50.311	68.252	46.22	47.82
V	435.7	471.68	516.63	491.33	242.8	263.95	628.89	442.36	542.76	385.99	374.47
Cr	181.1	183.09	238.76	236.63	134.6	109.30	808.48	533.81	692.66	497.31	520.33
Co	58.25	62.99	59.089	56.086	55.78	56.032	113.18	81.068	99.635	66.493	70.443
Ni	80.4	89.963	81.772	76.521	45.78	43.475	297.43	232.75	218.58	187.42	162.18
Cu	224.3	245.51	282.90	264.87	226.5	227.27	232.94	168.13	200.17	153.86	150.26
Zn	127.1	137.55	159.74	151.32	106.7	129.66	210.60	146.85	185.29	150.94	130.16
Ga	21.73	23.217	26.218	25.31	17.65	17.585	34.74	25.096	37.273	24.513	26.107
Rb	12.25	11.009	14	13.43	10.78	10.775	45.559	33.849	49.924	33.679	35.24
Sr	304.5	318.46	260.58	244.56	464.3	484.62	681.57	497.97	745.42	494.92	522.75
Y	37.09	38.794	51.547	49.761	35.26	35.222	55.244	41.032	56.037	37.262	39.734
Zr	206.6	222.68	264.54	250.99	174.8	202.58	398.05	294.74	387.71	273.38	270.36
Nb	23.83	24.369	19.248	17.312	7.338	28.235	66.791	53.544	33.672	41.559	36.815
Cs	0.2	0.187	0.28	0.266	0.439	0.438	1.974	1.479	1.824	1.253	1.315
Ba	45.24	43.127	35.124	32.28	49.44	54.242	124.53	89.967	128.96	89.236	91.82
La	22.77	22.711	20.535	19.549	28.49	28.813	59.127	44.51	61.317	40.971	41.693
Ce	47.98	48.781	45.869	44.296	56.37	57.474	116.33	86.488	117.48	78.507	82.435
Pr	5.655	5.746	5.732	5.59	6.35	6.541	12.585	9.362	12.575	8.46	9.01
Nd	29.61	30.401	31.726	30.12	32.29	32.851	60.993	45.251	60.774	40.852	43.528
Sm	6.898	7.228	8.13	7.79	7.226	7.209	12.632	9.348	12.443	8.245	8.761
Eu	2.111	2.248	2.497	2.382	2.135	2.155	3.697	2.756	3.773	2.462	2.615
Gd	6.696	7.268	8.183	7.69	7.018	6.974	12.39	9.136	12.845	8.232	8.982
Tb	0.983	1.029	1.246	1.198	0.931	0.934	1.535	1.159	1.594	1.05	1.126
Dy	6.231	6.582	8.261	8.156	5.873	5.847	9.634	7.008	9.72	6.301	6.785
Ho	1.217	1.262	1.668	1.578	1.132	1.112	1.764	1.325	1.852	1.226	1.283
Er	3.06	3.192	4.182	4.071	2.804	2.801	4.569	3.315	4.553	3.035	3.2
Tm	0.333	0.358	0.464	0.457	0.303	0.305	0.485	0.359	0.497	0.328	0.352
Yb	1.93	2.023	2.632	2.554	1.717	1.761	2.712	1.994	2.81	1.84	1.973
Lu	0.413	0.434	0.583	0.549	0.37	0.372	0.58	0.424	0.592	0.396	0.417
Hf	4.817	5.262	6.146	5.82	4.316	4.844	8.997	6.568	8.667	6.192	6.055
Ta	1.877	0.931	0.171	0.467	0.512	1.748	5.343	3.983	1.079	2.443	1.64
Pb	2.074	2.096	1.892	1.929	1.893	2.01	6.137	3.827	5.28	7.84	4.431
Th	1.392	1.45	0.997	0.922	1.654	1.812	3.955	3.015	4.095	2.843	2.705
U	0.343	0.366	0.252	0.279	0.299	0.32	0.795	0.647	0.898	0.747	0.728
(La/Ce)n	1.23	1.2	1.16	1.14	1.31	1.3	1.31	1.33	1.35	1.35	1.31
(La/Yb)n	8.46	8.05	5.6	5.49	11.9	11.74	15.64	16.01	15.65	15.97	15.16
∑REE	369.84	393.89	458.30	435.75	309.81	369.58	718.53	542.56	646.05	487.65	478.59

Sample	D 9 A	D 9 B	D 9 i	D10	D 10 i	D 21 A	D 21 B	D 21 C	D 21 i
Sc	50.6	62.427	45.553	35.924	35.931	*n.d.*	*n.d.*	*n.d.*	*n.d.*
V	473.19	580.63	436.21	307.49	306.02	286.86	296.00	306.03	308.81
Cr	443.15	527.76	420.96	332.43	343.14	1,154.9	1,149.6	997.03	1,036.4
Co	76.298	93.124	69.205	52.766	52.877	67.97	68.88	64.173	61.86
Ni	139.99	169.94	132.2	103.54	104.20	483.69	486.69	395.58	411.2
Cu	176.99	213.78	180.98	113.60	112.62	*n.d.*	*n.d.*	*n.d.*	*n.d.*
Zn	157.34	199.22	157.19	98.854	93.755	*n.d.*	*n.d.*	*n.d.*	*n.d.*
Ga	28.834	35.49	26.473	19.768	19.957	*n.d.*	*n.d.*	*n.d.*	*n.d.*
Rb	38.085	45.934	34.8	24.612	24.168	15.56	15.18	17.68	17.46
Sr	576.29	692.32	522.25	392.45	402.59	246.37	252.07	242.14	247.3
Y	45.641	55.186	40.667	30.379	30.452	19.93	19.918	20.557	20.86
Zr	322.95	396.99	293.68	208.90	210.84	125.99	126.06	134.12	135.3
Nb	54.513	68.087	54.365	35.094	32.87	24.46	25.465	26.62	26.63

Table 17.6 (continued)

Sample	D 9 A	D 9 B	D 9 i	D10	D 10 i	D 21 A	D 21 B	D 21 C	D 21 i
Cs	0.395	0.39	0.303	0.199	0.195	n. d.	n. d.	n. d.	n. d.
Ba	102.55	127.98	96.396	64.42	62.515	238.92	258.97	256.36	281.66
La	48.727	61.565	46.779	33.045	31.78	19	18.1	18.5	20.05
Ce	94.407	118.80	88.743	62.77	62.914	40	39.54	42	44.1
Pr	10.32	12.658	9.424	6.772	6.82	5.3	5.31	5.53	5.82
Nd	49.603	60.118	44.852	32.729	32.947	21.2	21.07	22	23.4
Sm	10.104	12.313	9.227	6.696	6.831	4.85	4.55	5.2	5.1
Eu	2.991	3.689	2.786	2.048	2.052	1.57	1.49	1.63	1.65
Gd	10.266	12.347	9.119	6.909	6.712	4.86	4.71	5.15	5.1
Tb	1.286	1.574	1.161	0.858	0.876	0.71	0.688	0.75	0.75
Dy	7.66	9.397	7.069	5.219	5.247	3.84	3.65	3.89	4
Ho	1.452	1.789	1.309	0.989	0.983	0.75	0.704	0.756	0.787
Er	3.695	4.507	3.285	2.454	2.544	1.85	1.83	1.89	1.94
Tm	0.393	0.494	0.355	0.264	0.27	0.253	0.246	0.26	0.27
Yb	2.236	2.753	1.998	1.505	1.515	1.58	1.523	1.61	1.68
Lu	0.483	0.604	0.432	0.322	0.334	0.237	0.218	0.24	0.248
Hf	7.241	8.874	6.685	4.794	4.798	n. d.	n. d.	n. d.	n. d.
Ta	3.983	5.176	3.846	2.561	1.92	n. d.	n. d.	n. d.	n. d.
Pb	4.302	6.052	4.266	1.96	1.751	n. d.	n. d.	n. d.	n. d.
Th	3.16	4.163	3.211	2.2	2.078	n. d.	n. d.	n. d.	n. d.
U	0.859	0.804	0.671	0.441	0.426	n. d.	n. d.	n. d.	n. d.
(La/Ce)n	1.33	1.34	1.36	1.36	1.3	1.23	1.18	1.14	1.17
(La/Yb)n	15.63	16.04	16.79	15.75	15.05	8.63	8.53	8.24	8.56
∑REE	586.62	719.63	542.36	383.79	382.27	242.45	244.40	257.88	259.89

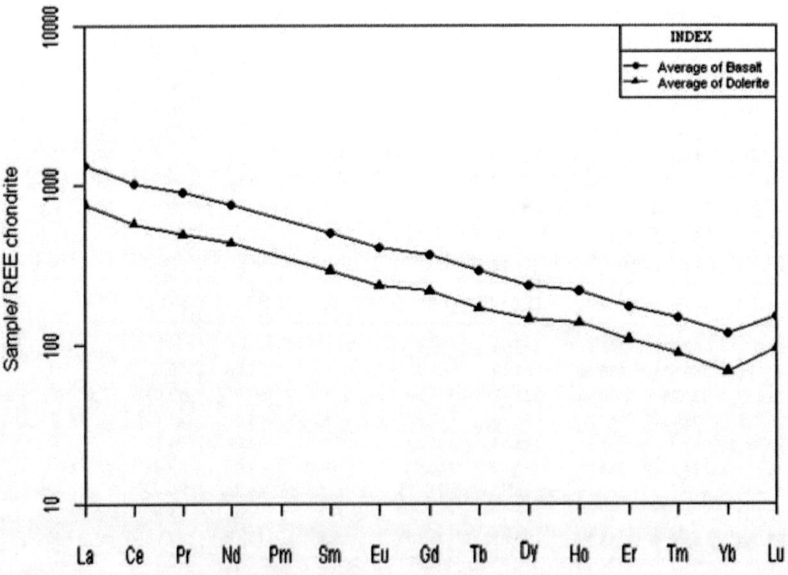

Fig. 17.3 REE chondrite normalized diagram (normalizing values after Nakamura, 1974) of "average basalt" (D-1A, D-1B, D-6A, D-6B, D-7, D-7i, D-9A, D-9B, D-9i, D-21A, D-21B, D-21C & D-21i) and "average dolerite" (D-2A, D-2B, D-8A, D-8B, D-8i, D-10, D-10i)

In some cases, olivine phenocrysts in the primitive rocks can be mantle xenoliths also. However, in the present case, as the CaO content of the olivine grains of picrite are higher (0.35–0.53 wt%) than those of the olivine in mantle peridotite (Fig. 17.4) and slightly lower NiO content than "mantle peridotite" (Fig. 17.5), it is confirmed that these olivine grains are not xenocrysts from the mantle but are derived from a

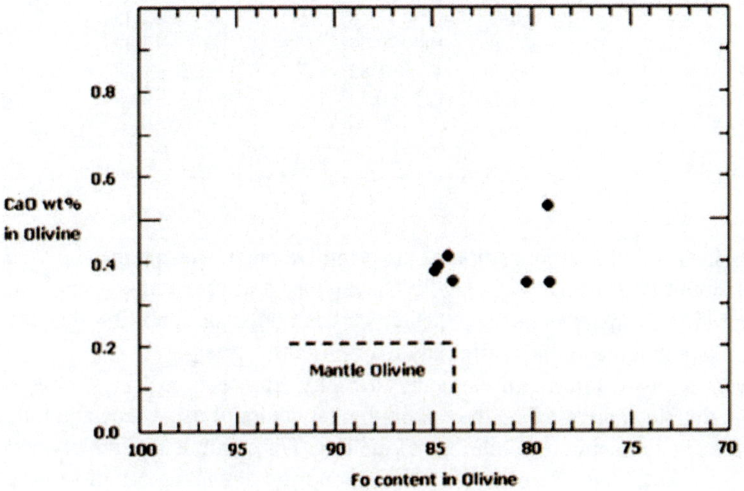

Fig. 17.4 CaO wt% verses mol% Fo in olivine phenocryst from the picrite basalt (field demarcation after Simkin and Smith, 1970)

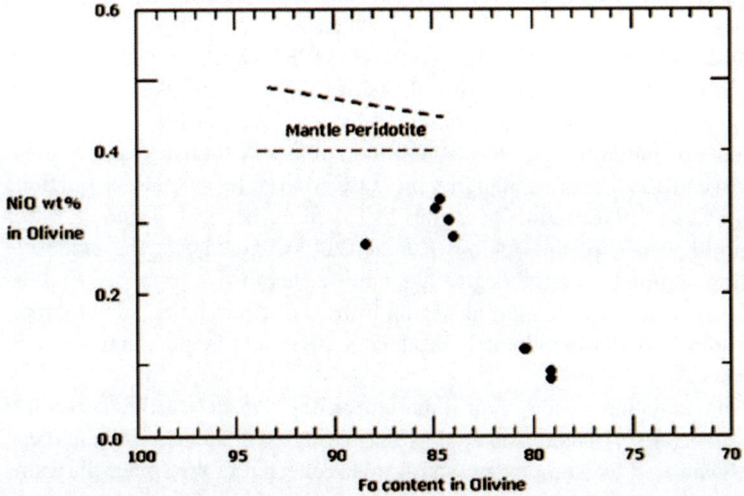

Fig. 17.5 NiO wt% versus mol% Fo in olivine from the picrite basalt. Range of NiO content of olivine in mantle peridotite is after Sato and Tohara (1985)

Table 17.7 Estimated primary magma composition

Oxide	Average of parental magma	Calculated parental melt
SiO_2	45.8	46.24
TiO_2	2.00	2.21
Al_2O_3	11.30	10.29
Fe_2O_3	12.71	15.29
MnO	0.16	0.18
MgO	12.58	12.45
CaO	9.64	10.93
Na_2O	1.82	1.48
K_2O	0.77	0.69
P_2O_5	0.25	0.25

primary magma. The initial composition of the primitive magma of the present area is calculated (Table 17.7) using "Petrolog 2.1.3a version" software and is found that not much variation in major element geochemistry is noticeable pointing that the assumed primitive magma did not suffer any major modification.

The concentration of rare earth elements in rocks representing primary melts often impart valuable constraints on the mineralogical nature of the source and thus on the pressure and temperature conditions of melting. For example a relatively low concentration of HREE (say, two to five times chondritic abundance) indicate the presence of residual garnet at the source, while high HREE concentrations (more than ten times chondrite) rule out the existence of garnet as a residual phase and thereby indicate a lower pressure conditions of melting (Wilson, 1989). Thus the high HREE abundance ($\sim$10 x chondrite) in the primitive rocks of the present area points to a garnet-free source, probably a spinel-lherzolite. Melting experiments on lherzolite done by Jaques and Green (1980) show that tholeiitic magmas can be produced by 20–30% melting at pressures below 15–20 Kb, while at high pressures with the same extents of melting picritic melts are formed; but alkali basalts are generated by smaller percentages of melting. Takahashi and Kushiro (1983) found that the product of initial melting of a spinel-lherzolite was tholeiitic at low pressure ($<$ 15 Kb) and alkali basaltic at higher pressure. Major element based equations developed by Chen (1988) return percentage of melting between 8 and 13% for the most primitive rock of this region. Kumar (2003) opined that the extensive mantle-melting regime by a large degree at shallower depth was probably attained progressively by increasing the melting region from garnet peridotite, spinel-garnet peridotite to spinel peridotite ultimately producing the vast tholeiitic magmatism in Deccan Traps.

Crustal contamination which is an important process in the modification of a magma (De Paolo, 1981; Thompson et al., 1982; Dostal and Dupuy, 1984) and can very well be evaluated by isotopic data, but is also reflected in the spider diagrams with negative Nb and positive La-Th peaks. Due to the lack of isotopic data, we are evaluating the process of crustal contamination based on geochemistry only. Thompson et al (1984) pointed out that fusible crustal rock types are generally much richer in Ba, Rb, Th, K and light REE, but have similar or lower concentrations of Nb, Ta, P, Zr, Hf, Y and middle REE. In the present case, none of the samples exhibit any characteristic enrichment or depletion in the elemental concentrations that signifies crustal contamination in the OIB normalized spider diagram (Fig. 17.6) pointing that this process was not involved in the modification of the magma.

The evolution of modified basalt and dolerite from the primitive picrite rock can be evaluated with the fractional crystallization process. The removal of olivine and clinopyroxene is manifested by this decreasing Ni and Cr with $Mg^{\#}$ value (Fig. 17.7a, b). Various workers have argued that feldspar fractionation is important during the evolution of Deccan tholeiites (e.g. Cox and Hawkesworth, 1985; Cox and Devey, 1987; Lightfoot et al., 1990) even though in these rocks substantial feldspar fractionation does not occur during the earliest stages of differentiation (Sen, 1986). Mass balance calculations based on Stormer and Nicholls (1978) were carried out with "PetroGraph" software (please refer Petrelli et al., 2005 for details). The mineral compositions used for the modeling is given in the Table 17.8. Average of the picrite (D-21A, D-21B, D-21C and D-21i) is taken as the parent and the most evolved rock with the lowest Mg# value (average of D-1A, D-1B, D-6A and D-6B) as the daughter rock. The results (Table 17.9) show a slightly higher

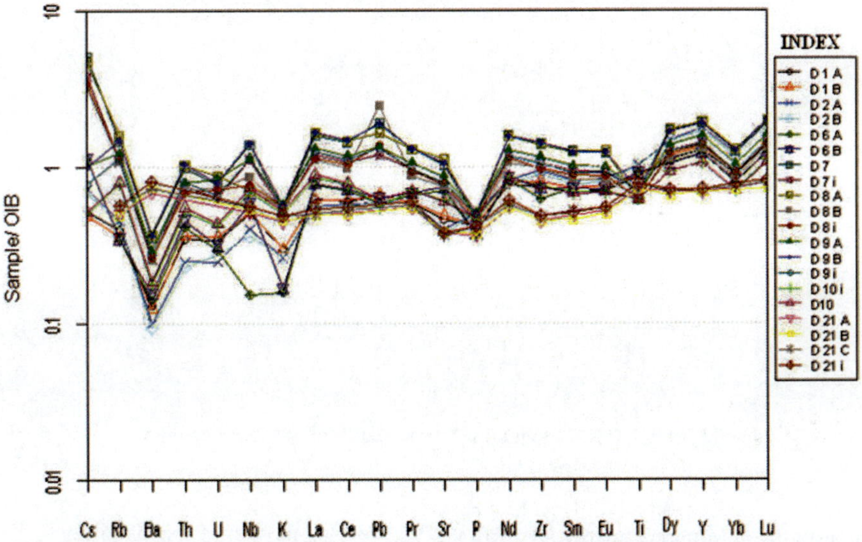

Fig. 17.6 OIB normalized spider diagram basaltic and doleritic rocks. Basalt →D-1A, D-1B, D-6A, D-6B, D-7, D-7i, D-9A, D-9B, D-9i, D-21A, D-21B, D-21C & D-21i; Dolerite → D-2A, D-2B, D-8A, D-8B, D-8i, D-10, D-10i. Normalizing values after Sun and Mc Donough (1989)

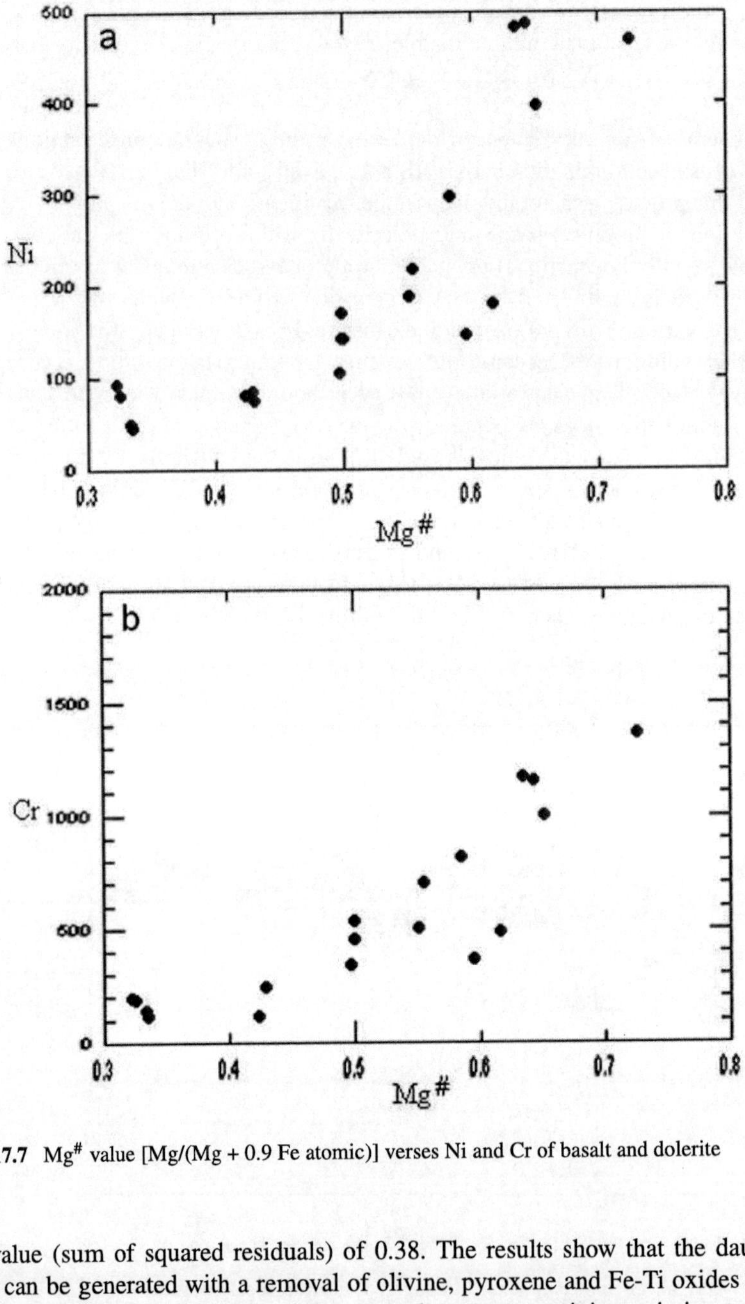

Fig. 17.7 Mg# value [Mg/(Mg + 0.9 Fe atomic)] verses Ni and Cr of basalt and dolerite

R^2 value (sum of squared residuals) of 0.38. The results show that the daughter rock can be generated with a removal of olivine, pyroxene and Fe-Ti oxides in the proportions 44:50:6 with 64% of the original magma remaining pointing towards the importance of olivine and pyroxene fractionation in the petrogenesis of these rocks.

Table 17.8 Mineral compositions used for major element modelling

Oxide	Olivine	Pyroxene	Feldspar	Titano-magnetite
SiO_2	40.74	53.64	50.45	0.12
TiO_2	0	0.43	0.07	26.87
Al_2O_3	0.05	1.84	29.55	1.09
FeOt	14.24	3.93	1.95	70.19
MnO	0.17	0.04	0.02	1.53
MgO	44.40	17.17	1.05	0.06
CaO	0.38	22.68	13.47	0.05
Na_2O	0.02	0.26	3.22	0.08
K_2O	0	0	0.21	0.01

Table 17.9 Major element modeling results. Average of D-21A, D-21B, D-21C and D-21i is taken as the "parent magma" and average of D-1A, D-1B, D-6A and D-6B having lowest $Mg^{\#}$ value is taken as the "evolved magma"

Oxides	Parent magma	Evolved magma
SiO_2	47.65	49.48
TiO_2	2.14	2.6
Al_2O_3	10	15.96
FeO^t	14.83	15.68
MnO	0.17	0.23
MgO	12.41	3.51
CaO	10.69	10.39
Na_2O	1.44	1.81
K_2O	0.67	0.34

RESULTS

Oxides	Bulk composition added or sub	Obs. difference between magmas	Calculated diff. between magmas	Obs. – calc. residuals
SiO_2	44.362	1.831	1.691	0.14
TiO_2	1.932	0.462	0.221	0.241
Al_2O_3	−1.71	5.956	5.835	0.121
FeO^t	13.265	0.851	0.798	0.053
MnO	0.199	0.056	0.011	0.046
MgO	30.703	−8.893	−8.981	0.088
CaO	11.421	−0.301	−0.341	0.04
Na_2O	−0.154	0.372	0.648	−0.276
K_2O	−0.019	0.334	0.118	−0.452

Sum of the squares of the residuals	0.3873
Wt% of substracted phases	
Olivine	43.82
Pyroxene	50.4
Feldspar	0
Fe-Ti oxide	5.78
% crystallized	36.18

Acknowledgements We thank S. F. Sethna and Rajesh Srivastava for their constructive comments for the improvement of the manuscript. We are thankful to N. V. Chalapathi Rao, Banaras Hindu University, India for providing the electron microprobe analyses of minerals. The financial support from Department of Science and Technology, New Delhi in the form of research grant (ESS/16/295/2006) is acknowledged.

References

Beane JE, Hooper PR (1988) A note on the picrite basalts of the Western Ghats, Deccan Traps, India. In: Subbaro KV (ed) Deccan flood basalts. Geol Soc India Memoir 10: 117–133

Chen CH (1988) Estimation of the degree of partial melting by (Na_2O+K_2O) and (Al_2O_3/SiO_2) of basic magmas. Chem Geol 71: 355–364

Courtillot V, Feraud G, Maluski H, Vandamme D, Morean MG, Besse J (1988) Deccan flood basalts and cretaceous/tertiary boundary. Nature 333: 842–846

Cox KG, Devey CW (1987) Fractionation processes in Deccan Traps magmas: Comments on the paper by G. Sen – mineralogy and petrogenesis of the Deccan traps lava flows around Mahabaleshwar, India. J Petrol 28: 235–238

Cox KG, Hawkesworth CJ (1985) Geochemical stratigraphy of the Deccan Traps at Mahabaleshwar, Western Ghats, India. With implications for open system magmatic processes. J Petrol 26: 355–377

De Paolo DJ (1981) Trace element and isotopic effects of combined wallrock assimilation and fractional crystallization. Earth Planet Sci Lett 53: 189–202

Dostal JB, Dupuy C (1984) Geochemistry of the North Mountain Basalts (Nova Scotia, Canada). Chem Geol 45: 245–261

Greenough JD, Hari KR, Chatterjee AC, Santosh M (1998) Mildly alkaline basalts from Pavagadh Hill, India; Deccan flood basalts with an asthenospheric origin. Mineral Petrol 62: 223–245

Gwalani LG, Rock NMS, Chang WJ, Fernandez S, Allegre CJ, Prinzohofer A (1993) Alkaline rocks and carbonatites of Amba Dongar and adjacent areas, Deccan Igneous Province. Gujarat, India: Geology, petrography and petrochemistry. Mineral Petrol 47: 219–253

Hari KR (1988) Mineralogical and Petrological Studies of the Lamprophyres around Chhaktalao area, Madhya Pradesh. J Geol Soc India 51: 28–30

Hofmann C, Feraud G, Courtillot V (2000) $^{40}Ar/^{39}Ar$ dating of mineral separates and whole rocks from the Western Ghats lava pile: Further constraints on duration and age of the Deccan Traps. Earth Planet Sci Lett 180: 13–27

Jaques AL, Green DH (1980) Anhydrous melting of peridotite at 0–15 kb and the genesis of tholeiitic basalts. Contrib Mineral Petrol 73: 287–310

Karkare SG, Srivastava RK (1990) Regional dyke swarms related to the Deccan Trap Alkaline Province, India. In: Parker AJ, Rickwood PC, Tucker DH, (eds) Mafic dykes and emplacement mechanism, A. A. Balkema, Rotterdam: 335–347

Krishnamurthy P, Cox KG (1977) Picrite basalts and related lavas from the Deccan Traps, Rajpipla, India. Contrib Mineral Petrol 62: 53–75

Krishnamurthy P, Gopalan K, Macdougall JD (2000) Olivine compositions in picrite basalts and the Deccan volcanic cycle. J Petrol 41(7): 1057–1069

Kumar S (2003) Variation in the thickness of the lithosphere underneath the Deccan volcanic province, evidence from rare earth elements. Memoir Geol Soc India No. 52: 179–194

Kumar S, Karkare SG, Martin T (1992) Petrology and geochemistry of basalt of Phenai Mata Hill, District of Baroda, Gujarat, Western India. Acta Universitatis Carolinae-Geologica No. 3–4: 313–329

Le Bas MJ (2000) IUGS reclassification of the high-Mg and picritic volcanic rocks. J Petrol 41(10): 1467–1470

Lightfoot PC, Hawkesworth CJ, Devey CW, Rogers NW, Van Calsterem PWC (1990) Source and differentiation of Deccan traps, implications of geochemical and mineral chemical variations. J Petrol 31: 1165–1200

Nakamura N (1974) Determination of REE, Ba, Fe, Mg, Na and K in carbonaceous and ordinary chondrites. Geochim Cosmochim Acta 38: 757–773

Peng ZX, Mahoney JJ (1995) Drillhole lavas from the northwestern Deccan Traps, and the evolution of Reunion hotspot mantle. Earth Planet Sci Lett 134: 169–185

Petrelli M, Poli G, Perugini D, Peccerillo A (2005) Petrograph: A new software to visualize, model and present geochemical data in igneous petrology. Geochem, Geophys, Geosystems 6: Q 07011doi:10.1029/2005GC00932

Sato H, Tohara T (1985) Geochemical characteristics of back-arc basin basalt. In: Nasu N, (ed) Information of active ocean margins, Terra Scientific Publishing Company, Tokyo: 399–410

Sen G (1986) Mineralogy and petrogenesis of the Deccan trap lava flows around Mahabaleshwar, India. J Petrol 27: 627–663

Simkin T, Smith JV (1970) Minor-element distribution in olivine. J Geol 78: 304–325

Srivastava RK (1994) Petrology, petrochemistry and genesis of the alkaline rocks associated with the Ambadungar Carbonatite Complex, Baroda district, Gujarat, India. J Geol Soc India 43: 23–39

Stormer JC, Nicholls J (1978) XLFRAC: A program for the interactive testing of magmatic differentiation models. Comput Geosci 4: 143–159

Sukeshwala RN, Avasia RK (1972) Carbonatite – alkalic complex of Panwad – Kawant, Gujarat and its bearing on the structural characteristics of the area. Bull Volcanol 35: 564–578

Sun SS, McDonough WF (1989) Chemical and isotopic systematic of ocean basalts: Implications for mantle composition and processes. Geol Soc London Spec publ 42: 313–345

Takahashi E, Kushiro I (1983) Melting of dry peridotite at high pressures and basalt magma genesis. Amer Mineral 68: 659–679

Thompson RN, Dickin AP, Gibson IL, Marrison MA (1982) Elemental fingerprints of isotopic contamination of Hebridean Paleocene mantle-derived magmas by Archean sial. Contrib Mineral Petrol 79: 159–168

Thompson RN, Morrison MA, Hendry GL, Parry SJ (1984) An assessment of the relative roles of a crust and mantle in magma genesis: An elemental approach. Philos Trans R Soc Lond A 310: 549–590

Vijaya Kumar K, Narsimha Reddy M, Leelandam C (2006) Dynamic melting of the Precambrian mantle, evidence from rare earth elements of the amphibolites from the Nellore-Khammam Schist Belt, South India. Contrib Mineral Petrol 152(2): 243–256

Viladkar SG, Avasia RK (1992) Pyroxenes from alkaline rocks of the Chhota Udaipur carbonatite-alkaline province, Gujarat, India. J Geol Soc India 39: 313–319

Wilson M (1989) Igneous petrogenesis, Unwin Hyman, London: 465

Chapter 18
Occurrence of Xenoliths in the Lamprophyre and Picrobasalt Dykes of Bakhatgarh – Phulmal Area, Jhabua District, Madhya Pradesh, India

K.R. Randive

Introduction

Xenoliths in lavas and intrusions provide direct evidence on basement composition (e.g. Rudnick, 1992; Rudnick and Fountain, 1995). Dykes carrying xenoliths are rare in the Deccan Traps, but when present they carry xenoliths of crustal rocks such as quartzite, rhyolite, granite, granodiorite, sandstone, calcareous rock, syenite, etc. (Deshmukh and Sehgal, 1988; Karkare and Srivastava, 1990). Nevertheless, there are known occurrences of mantle and crustal xenoliths in the Deccan Traps, namely (1) The granite-quartzite rafts at Mandaleshwar (Duraiswami and Karmalkar, 1996; Subbarao et al., 1999); (2) The granitic and rhyolitic xenoliths near Ahmednagar (Sharma et al., 1999); (3) The mantle derived websterite and pyroxenite, and lower crustal mafic and felsic granulite xenoliths at Murud–Janjira (Dessai et al., 2004); (4) The quartz – alkali feldspar – dolomite – bearing partially associated xenoliths southwest of Mahabaleshwar (Mahoney, 1998); (5) The quartz – alkali feldspar – bearing xenoliths at Susanne (Dessai and Viegas, 1995); and (6) Xenoliths of diverse lithological variety (gneisses, quartzites, granite mylonite, felsic granulite, carbonate rock, tuff) in the Rajmane and Talwade dykes (Ray et al., 2008).

The Chhota Udaipur alkaline – carbonititic Sub-Province (Gwalani et al., 1993), which is a part of Deccan Igneous Province and comprises mantle derived rocks (carbonatites), show paucity of mantle xenoliths. Nevertheless, there are abundant crustal xenoliths in the carbonatite breccias/tuffaceous carbonatites (Gwalani et al., 1993; Srivastava, 1994, 1997). The crustal xenoliths are also abundant in the hybrid "calcareosiliceous" dykes (Gwalani et al., 1993; Hari et al., 1998; Randive, 2005).

K.R. Randive (✉)
Department of Geology, Rashtrasant Tukadoji Maharaj Nagpur University, Nagpur, Maharashtra 440001, India
e-mail: randive101@yahoo.co.in

R.K. Srivastava (ed.), *Dyke Swarms: Keys for Geodynamic Interpretation*,
DOI 10.1007/978-3-642-12496-9_18,

Geological Setup

Bakhatgarh – Phulmal sector (Fig. 18.1) is one of the five sectors of the Chhota Udaipur alkaline-carbonatitic sub-province, identified by Gwalani et al. (1993). The major rock types occurring in this area include Precambrian quartz-mica-schist, Cretaceous Bagh sediments and Upper Cretaceous to Lower Eocene Deccan Trap basaltic lava flows and dykes. The intrusive dykes follow three distinct structural trends viz. EW, NE-SW and ENE-WSW, which are in concurrence with the E-W trend of the Narmada lineament (e.g. Deshmukh and Sehgal, 1988). These dykes may be grouped into four distinct types viz. (1) picrobasalts; (2) lamprophyres;

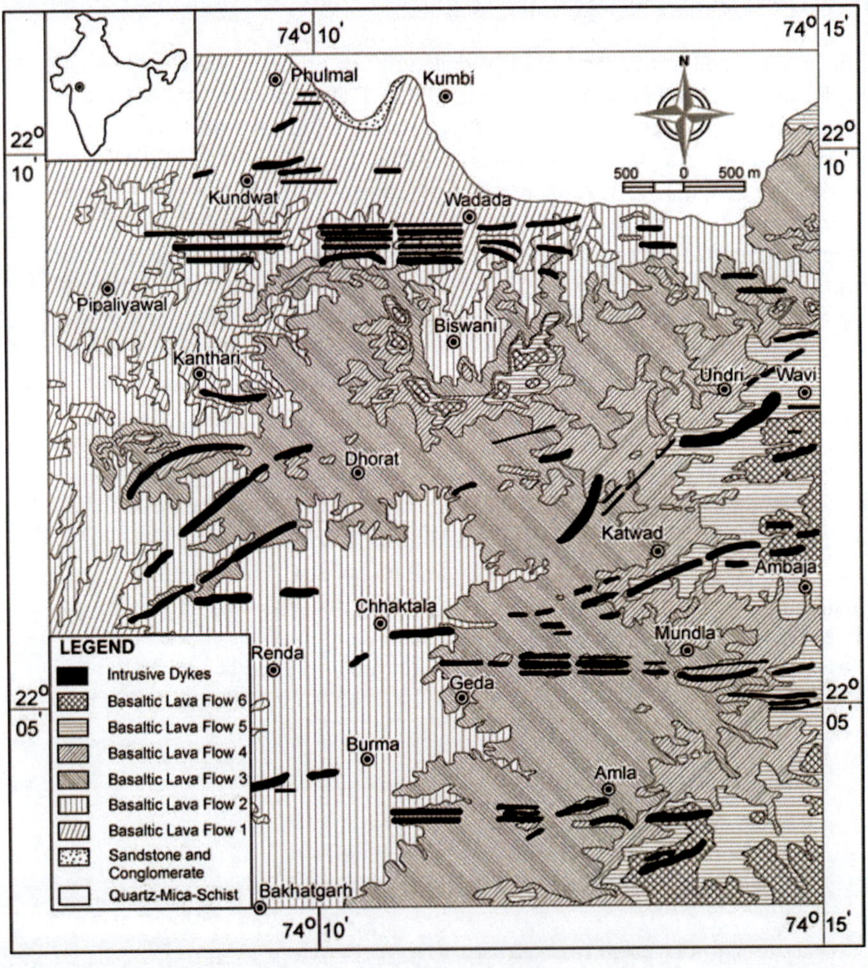

Fig. 18.1 The geological map of the Bakhatgarh – Phulmal area, Jhabua district, Madhya Pradesh, India

(3) tholeiitic comprising of basalt, dolerite, gabbro, porphyritic basalt and giant plagioclase basalt; and (4) Calcareosiliceous dykes represented by late stage carbonate breccias, calcite, and quartz dykes and veins (Randive et al., 2005; Randive, 2008).

Field Occurrence of Xenoliths

During the present study two xenolith bearing dykes (one picrobasalt and lamprophyre each) were noticed towards the north of Kundwat (22°10"N, 74°10"E). The xenoliths are of two types, namely, type-I xenoliths are nodular and rounded, enclosed within the host rock as isolated blobs without showing conspicuous reaction boundaries (Fig. 18.2a). However, surrounding liquid permeate through these in the form of micro veins and along the mineral boundaries. On the contrary the

Fig. 18.2 (**a**) Type-I xenolith in lamprophyre. (**b**) Type-II xenolith in the lamprophyre. (**c**) and (**d**) Type-II xenoliths in the picrobasalt showing distinct reaction boundary with the host rock. (**e**) Photomicrograph of the fine-grained variety showing microveins, individual minerals cannot be identified microscopically. (**f**) Coarse-grained variety of the type-I xenolith showing corrosion and melt formation along the grain boundaries of quartz and feldspar. (**g**) Plagioclase feldspar showing twining and slightly curved twin lamellae due to magmatic stresses, and (**h**) Type-II xenolith showing quartz grains rimmed by anatectic melt films, the melt traversed throughout the rock forming interconnecting network

Table 18.1 Table showing comparison between type-I and type-II xenoliths

Sr. No.	Type-I xenoliths	Type-II xenoliths
1.	Xenoliths vary in size between 3 cm and 5 cm. These are relatively few in number	The xenoliths range in size from 1 cm to more than 10 cm. Frequency of occurrence of xenoliths is higher but within a small area (~50 m^2) of the picrobasalt dyke
2.	Nodular, have acquired appreciable roundedness. Reaction boundary with the host rock is not conspicuous	Irregular in shapes, occasionally rounded, but invariably shows distinct reaction boundary with the host rock
3.	Two types of nodules were distinguished based on their grain size, viz. (i) fine-grained, and (ii) coarse-grained. High temperature varieties of quartz and feldspar, such as quartz, crystoballite and high-albite are present	This type of xenoliths is coarse grained, pinkish coloured; and composed of high temperature minerals such as quartz, tridymite and sanidine
4.	The individual minerals are partially corroded and the interstices are filled with dark coloured magmatic material in coarse grained variety	The fractured quartz grains are rimmed by the thick glassy melt films and contain innumerable fluid inclusions
5.	The xenoliths are impoverished in the incompatible trace elements and the rare earth elements. Shows distinct –ve Th; +ve Sr and Ti; and +ve Eu anomalies. The fine grained xenolith shows strong +ve Nb anomaly	The xenoliths are relatively enriched in the incompatible trace element and rare earth elements. Shows distinct +ve Th; –ve Sr and Ti; and –ve Eu anomalies

type-II xenoliths are irregular in shape and shows distinct reaction boundaries with the host rock (Fig. 18.2b–d). Table 18.1 gives general comparison between type-I and type-II xenoliths. Both types of xenoliths were observed in the lamprophyre dyke, whereas only type-II xenoliths were so far observed in the picrobasalt dyke.

Petrography

Fine as well as coarse grained nodules comprise type-I xenoliths, these are comprised of fine- and coarse-grained varieties. In the fine grained variety mineralogy is not clearly discernible even at higher magnification, although some micro veins traverse throughout the xenoliths. Innumerable fluid inclusions and dusty opaque minerals are concentrated along these veins (Fig. 18.2e). In the coarse grained variety quartz, plagioclase and potassic feldspars are present; quartz is euhedral to sub-hedral, with ubiquitous basal sections. The inter-grain boundaries are almost invariably occupied and often corroded by the liquid media showing earthy-dull-greenish colour, often with innumerable iron-titanium oxides (Fig. 18.2g). Quartz and feldspar both shows magmatic stress twinning, feldspar shows polysynthetic twinning (e.g. Fig. 18.2f). The type-II xenoliths commonly contain anhedral and fractured quartz grains having innumerable fluid inclusions. The inclusions are either grown alongside the fractures, or even disseminated throughout the crystals. It is also observed that the individual grains are rimmed by yellowish-brown isotropic (glassy) material, which at places forms colloform banding (Fig. 18.2h). Presence of normative corundum in xenoliths especially type-II (3.81%, Table 18.3) indicate their peraluminous character.

Geochemistry

The XRD analyses are obtained using XPERT-PRO PANALYTICAL XRD instrument at the Vishweshwaraiah National Institute of Technology, Nagpur (Table 18.2), whereas major, trace, and rare earth elements for host rocks and xenoliths were analysed at National Geophysical Research Institute, Hyderabad using Phillips 2502 X-ray Fluorescence Spectrometer and Perkin Elmer Sciex ELAN II ICP Mass Spectrometer respectively. The results are presented in Tables 18.2, 18.3 and 18.4.

XRD data indicates presence of high temperature minerals such as Tridymite, crystoballite, sanidine, high albite and anorthoclase. The sanidine – high albite series is characteristic of the alkali feldspars of volcanic rocks formed at about 900°C for sanidine and 1,350°C for albite. Monoclinic sanidine and triclinic high-albite form limited solid solution with increasing temperature; the change in symmetry occurs at Ab_{63}, which results into formation of sanidine (Ab_{00}–Ab_{63}), anorthoclase (Ab_{63}–Ab_{90}), and albite (Ab_{90}–Ab_{100}) (Deer et al., 1962; Phillips and Griffen, 1986). From

Table 18.2 X-ray diffraction analysis of the xenoliths

Reference code	Compound name	Chemical formula	Remarks
KR/155A Fine-grained type-I xenolith in lamprophyre			
00-010-0393	Albite, disordered	$Na(Si_3Al)O_8$	Major
01-083-2469	Quartz	SiO_2	Minor
01-082-1233	Cristobalite	SiO_2	Minor
00-010-0353	Sanidine, high	$KAlSi_3O_8$	Minor
KR/155B Coarse-grained type-I xenolith in lamprophyre			
01-078-2315	Quartz	SiO_2	Major
01-071-0261	Tridymite	SiO_2	Minor
01-071-1151	Albite high	$Na(AlSi_3O_8)$	Minor
KR/145A type-II xenolith in picrobasalt			
01-078-2315	Quartz	SiO_2	Major
00-010-0353	Sanidine, high, syn	$KAlSi_3O_8$	Minor
01-083-1609	Albite high	$Na\ (AlSi_3O_8)$	Trace
01-075-1631	Anorthoclase	$(Na_{0.667}K_{0.333})(AlSi_3O_8)$	Trace

Table 18.3 Major oxide concentration of the xenoliths and host rocks

	KR-155	KR-155(A)	KR-155(B)	KR-145	KR-147
		Type-I			
Sample	Lamprophyre	Fine	Coarse	Picrobasalt	Type-II
SiO_2	46.64	69.5	71.32	48.03	71.32
Al_2O_3	11.58	14.85	13.87	12.5	13.55
$Fe_2O_3{}^T$	12.38	0.21	1.8	12.1	2.1
MnO	0.2	0.05	0.09	0.17	0.07
MgO	7.42	2.34	2.47	7.34	2.97
CaO	13.09	0.61	2.84	11	0.21
Na_2O	0.92	6.29	3.02	2.75	1.42
K_2O	2.37	3.6	2.69	1.36	6.69
TiO_2	2.98	1.22	0.42	2.19	0.36
P_2O_5	1.15	0.03	0.09	0.65	0.07
Total	98.73	98.71	98.6	98.09	98.75
Q	2.19	14.87	34.43	0.00	33.20
or	14.30	21.56	16.12	8.26	40.04
ab	7.95	53.93	25.92	23.93	12.17
an	20.90	1.67	13.69	18.25	0.59
C	0.00	0.00	1.06	0.00	3.81
di	30.08	0.00	0.00	26.99	0.00
hy	9.45	5.90	6.24	5.25	7.49
ol	0.00	0.00	0.00	6.00	0.00
mt	6.63	0.00	0.00	5.50	0.00
il	5.78	0.11	0.20	4.28	0.61
hem	0.00	0.21	1.83	0.00	1.88
ti	0.00	0.84	0.00	0.00	0.00
ap	2.72	0.07	0.21	1.55	0.16
%AN	72.45	3.01	34.57	43.27	4.64

the above mineral assemblage, it may be inferred that quartzofeldspathic basement rocks were caught-up by the upwelling magmas at the higher temperatures.

Most of the major elements show a limited compositional range such as SiO_2 (69.5 – 71.32), Al_2O_3 (13.55 – 14.85), total iron (0.21 – 1.22), MgO (2.34 – 2.97), P_2O_5 (0.03 – 0.09), and MnO (0.05 – 0.09). However CaO, Na_2O, and K_2O are comparatively higher in type-I (coarse), type-I (fine), and type-II xenoliths respectively. The reason is not difficult to understand, since their mineralogy shows relative abundance of calcic, sodic, and potassic feldspars. Such influence is also seen on the trace element abundances since Rb, Ba, Th, U concentration is higher in type-II (potash feldspar rich) and Sr is higher in type-I (plagioclase feldspar rich) xenoliths.

The Upper Continental Crust (UCC) normalised trace and rare earth element spidergrams (Figs. 18.3 and 18.4; Taylore and McLennan, 1985) indicate that the fine-grained variety of type-I xenoliths is most impoverished in the incompatible trace elements (ITE), whereas coarse-grained variety is slightly enriched in ITE and type-II xenoliths are even more enriched. Similarly, REE spidergram gives almost

Table 18.4 Trace and rare earth element concentration of the host rocks and xenoliths

	KR-155	KR-155(A)	KR-155(B)	KR-145	KR-147
		Type-I			
Sample	Lamprophyre	Fine	Coarse	Picrobasalt	Type-II
Sc	18.249	1.121	2.17	33.788	9.176
V	355.336	20.405	44.012	332.307	30.566
Cr	179.771	31.615	37.386	119.074	8.677
Co	55.334	14.784	124.325	70.367	154.825
Ni	76.712	14.474	17.143	44.569	4.981
Cu	172.049	24.726	32.668	111.709	26.957
Zn	141.588	124.963	84.035	310.051	260.122
Ga	40.44	48.585	37.023	18.301	19.496
Rb	58.483	60.538	69.245	21.612	201.045
Sr	1007.12	214.805	599.797	808.869	167.835
Y	35.203	0.748	2.959	26.964	38.834
Zr	396.341	12.019	24.006	265.92	93.1
Nb	145.775	349.824	11.145	125.502	37.316
Cs	2.133	0.052	0.169	2.737	0.435
Ba	1,802.401	183.231	574.961	1,300.11	953.068
La	118.934	1.668	12.263	85.997	64.077
Ce	218.157	2.656	19.168	156.744	128.541
Pr	24.64	0.306	1.802	14.056	11.864
Nd	91.244	1.198	6.223	63.561	51.866
Sm	14.883	0.204	0.853	9.531	9.394
Eu	4.053	0.09	0.757	2.956	1.267
Gd	11.301	0.169	0.727	10.744	10.389
Tb	1.488	0.025	0.085	1.147	1.359
Dy	7.806	0.15	0.532	4.851	6.573
Ho	1.258	0.03	0.09	0.891	1.303
Er	3.233	0.075	0.271	2.742	4.154
Tm	0.447	0.013	0.042	0.393	0.685
Yb	2.554	0.078	0.257	2.109	3.708
Lu	0.354	0.009	0.036	0.307	0.539
Hf	8.884	0.24	0.303	5.594	2.571
Ta	8.213	5.813	2.004	16.491	7.159
Pb	29.258	27.976	27.575	16.782	22.566
Th	16.842	1.557	0.811	11.084	28.094
U	3.609	0.48	0.415	1.726	4.686

flat pattern for REEs in xenoliths, whereas slightly LREE enriched – HREE depleted pattern for the host rocks. Other peculiarities are:

(a) Negative Th anomaly for type-I for and positive Th anomaly for and type-II xenoliths,
(b) Positive Sr and Ti anomalies for type-I and negative Sr and Ti anomalies for type-II xenoliths,

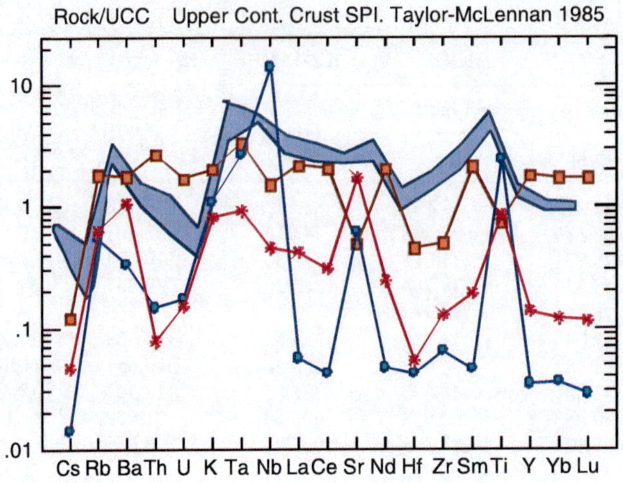

Fig. 18.3 The UCC normalised trace element spidergram. Shaded portion indicates field for host rocks (picrobasalt and lamprophyre) and the curves represent xenoliths. *Circles*: type-I (fine-grained); *stars*: Type-I (coarse-grained); *squares*: type-II xenoliths

(c) Unusually high positive Nb anomaly for the fine-grained variety of the type-I xenolith, and
(d) The positive Eu anomaly for the type-I and negative Eu anomaly for the type-II xenoliths.

All these observed variations are significant and sheds some light on the petrogenesis of these xenoliths.

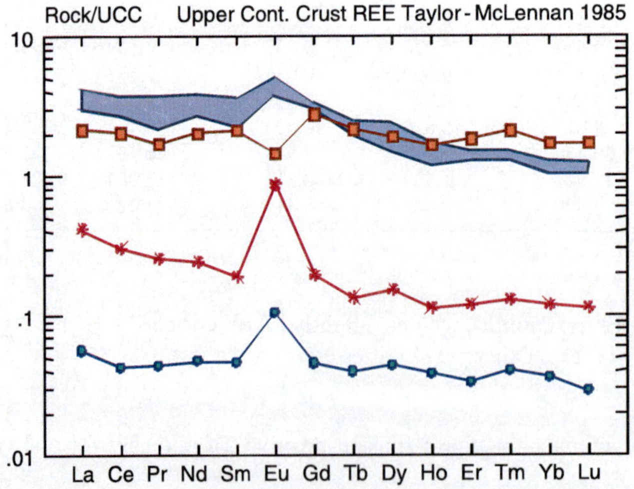

Fig. 18.4 The UCC normalised REE spidergram for the host rocks and xenoliths (symbols as in Fig. 18.3)

Discussion

The mineralogy of all the xenoliths irrespective of their host rock, does not resemble the mantle mineralogy (Olivine ± Orthopyroxene ± Clinopyroxene ± Garnet ± Phlogopite ± Spinels) found in the mantle xenoliths reported worldwide (Nixon, 1987). Therefore, the xenoliths found in the lamprophyre and picrobasalt from Bakhatgarh–Phulmal area are crustal in origin and collected at depth from the rocks underneath.

The positive strontium anomaly in the type-I xenoliths shows feldspar control over their distribution owing to comparable ionic potential of Ca^{+2} and Sr^{+2}. The elevated U and Th concentration in the type–II xenoliths may indicate their abundance in the original protolith in comparison to the upper continental crust. However, the abundance of titanium is unexpected from the type-I xenolith mineralogy, and thought to be due to deposition of solutions brought by the microveins from the host lamprophyres. The Fe-Ti rich solutions were formed due to deuteritic alteration of the ulvospinels from the lamprophyres and distributed to the xenoliths through network of micro-veins and crystallised there forming secondary minerals such as hematite and titanite. CIPW norm (Table 18.3) indicates formation of titanite in fine-grained type-I xenolith. Green and Pearson (1987) calculated high partition coefficient (6.1) of Nb for titanite in the rhyolitic and dacitic magmas; therefore, the high positive Nb anomaly observed in fine grained variety of type-I xenolith is possibly due to accommodation of Nb in the titanite structure.

The Eu anomalies observed in two types of xenoliths can be explained on the basis of the diffusivities of REEs during melting of the crustal rocks. According to Kushiro (1980), the REEs show increasing diffusivity with the increasing degrees of partial melting; however Eu is an exception to this rule, since it shows anomalous and much higher diffusion coefficient than the other REEs (Nakamura and Kushiro, 1998; Mungall, 2002; Frezzotti et al., 2004). Therefore the earlier anatectic melt shows Eu enrichment and hence positive anomaly. However, at the higher degrees of partial melting, melts are much more enriched in ITE due to melting of accessory mineral phases from the original source rocks. These melts migrate freely and also form interconnecting films facilitating melt migration and silica crystallisation such as crystoballite rim over quartz crystals. This silica residue is depleted in REE including Eu, and hence the negative Eu anomaly (Frezzotti et al., 2004). The two types of xenoliths observed in the Bakhatgarh – Phulmal area imply that the xenoliths represents anatectic melts formed due to different degrees of partial melting of the crustal rocks. In the Chhota Udaipur Subprovince, the Precambrian Aravalli rocks (granites, granitic gneisses and other metamorphics), and the Cretaceous Bagh sediments (sandstone and limestone), form the basement rocks. It is likely that the xenoliths of these rocks were collected by the ascending magmas. The carbonatite breccias and calcareosiliceous dykes occurring in the subprovince also contain fragments of the basement rocks mentioned above (e.g. Simonetti et al., 1995, 1998).

The Chhota Udaipur Subprovince records the stupendous magmatism during wanning stages of the Deccan Trap magmatism along the Narmada rift zone.

Unfortunately, no direct geophysical data at Amba Dongar or adjoining areas is available. However, the geophysical studies of the Narmada-Son lineament have indicated its heterogeneous and complicated nature (Kaila, 1988; Reddy; 1999). Dessai et al. (2004) suggested a 3-layered structure of the crust based on P-wave velocities such as: (1) 2–12 km thick layer of granitic gneiss having Vp: 5.7–6.22 km/s (below 0–2 km thick cover of the Deccan Trap Vocanics), (2) 12–25 km thick layer of Felsic granulites having Vp: 5.7–6.22 km/s, and (3) 25–42 km thick layer of mafic granulites having Vp: 6.76–7.26 km/s. Felsic or mafic granulite xenoliths were not so far found in the dykes of Bakhatgarh-Phulmal area, which possibly indicates the removal of granulite layer by thermal and magmatic erosion, as has been proposed at Billimora by Dessai et al. (2004). Present area lies between Mehamadabad – Billimora and Thundara – Sindad DSS Profiles of the Kaila et al. (1981); although no direct geophysical data on the Chhota Udaipur alkaline-carbonatitic sub-province is available, it is possible that the crustal thickness in this part is much lower (<20 km).

Thus, the samples collected by the magma were from the relatively shallow depths, and represent upper part of the continental crust. The above process can be envisaged as follows: initially the hot ascending magma comes in contact with the upper continental crust comprising Aravalli granites, gneisses and other rocks below the cover of Deccan Trap basaltic lava flows. The crust is fragmented and the original mineralogy of the fragmented blocks is subjected to polymorphic transformation and restructuring due to different P-T condition inside the magma. The enclosed crustal fragments also acquire appreciable roundedness (Fig. 18.5a). Subsequently, the nodules having lower residence time in the magma were subjected to lower degrees of partial melting and vice versa, and result into formation of type-I and type-II xenoliths respectively (Fig. 18.5b).

Presence of quartz-rich xenoliths are not uncommon and have been reported worldwide, for e. g. Rocky Mountains, USA. (Doe et al., 1969); Shikoku, Japan (Sato, 1975); Los Encinos Volcanic field, Central Mexico (Luhr et al., 1995);

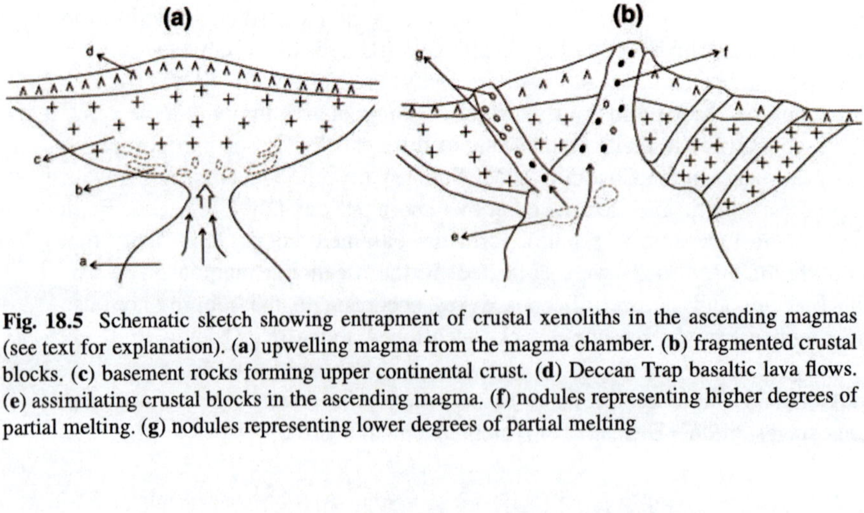

Fig. 18.5 Schematic sketch showing entrapment of crustal xenoliths in the ascending magmas (see text for explanation). (**a**) upwelling magma from the magma chamber. (**b**) fragmented crustal blocks. (**c**) basement rocks forming upper continental crust. (**d**) Deccan Trap basaltic lava flows. (**e**) assimilating crustal blocks in the ascending magma. (**f**) nodules representing higher degrees of partial melting. (**g**) nodules representing lower degrees of partial melting

Nograd – Gomor and other volcanic fields, Hungary (Kovacs et al., 2001); Veliki Majdan, Serbia (2004); Western Yunnan, China (Xu et al., 2007), and others. Their incorporation in the host magmas were interpreted in several ways including following.

(1) Hybridization of magmas such as lamprophyric and granitic, syenitic, trachytic, or dacitic magmas (Nemec, 1988; Branigan, 1989; Venturelli et al., 1991; Sheppard, 1995)
(2) As a product of crustal anatexis resulting into formation of partial melts of felsic composition (Dessai et al., 2004; Dostal et al., 2005; Ray et al., 2008)
(3) Formed by large degrees of batch melting of gneisses and schists, leaving quartz-rich residue (Frezzotti et al., 2004)
(4) In addition to above, an interesting finding of quartz-megacryst in the lamprophyres of Baiya area, Yunnan, China has been argued to represent phenocryst of the orginal magma as against xenocryst collected from the crustal sources. The quartz in the source magma have altered the original basic melts and transformed them into lamprophyric melts (Xu et al., 2007)

Some of the above propositions look reasonable for the xenoliths occurring in the Bakhatgarh – Phulmal area; however, more detailed work will be necessary to comment further on this aspect.

Acknowledgement I acknowledge the Department of Science and Technology, New Delhi for the financial support through the research project ESS/16/301/2006. I am thankful to V. Balaram, M. Satyanarayanan, T. Gnaneshwar Rao, Parijat Roy, Janisar Sheikh and Sachin Pachpor for the help during chemical analyses at N. G. R. I., Hyderabad. The generous help and support provided by A. M. Varade, Manjusha Mahajan, Rajkumar Meshram and Manoj Sahu is also duly acknowledged. Anonymous referee is thanked for the comments that helped for the improvement of the original manuscript. I profusely thank Rajesh Srivastava for his constant encouragement and help; and lastly Chetana, who kept inspiring me all the time.

References

Branigan NP (1989) Hybridization in middle proterozoic high level ring complexes, Aland, SW Finland. Precamb Res 45: 83–96

Deer WA, Howie RA, Zussman J (1962) Rock forming minerals Vol. 4: framework silicates. Wiley, New York

Deshmukh SS, Sehgal MN (1988) Mafic dyke swarm in Deccan volcanic province of Madhya Pradesh and Maharashtra. In: Subbaroa KV (ed) Geol Soc Ind Mem 10: 323–340

Dessai AG, Markwick A, Vaselli O, Downes H (2004) Granulite and pyroxenite xenoliths from the Deccan Trap: Insight into the nature and composition of the lower lithosphere beneath cratonic India. Lithos 78: 263–290

Dessai AG, Viegas AAAA (1995) Multi-generation mafic dyke swarm related to Deccan Magmatism, south of Bombay: Implications on the evolution of the western Indian continental margin. In: Devaraju TC (ed) Dyke swarms of peninsular India. Geol Soc Ind Mem 33: 435–451

Doe BR, Lipman PW, Hedge CE (1969) Primitive and contaminated basalts from the southern rocky mountains, USA. Contrib Mineral Petrol 21: 142–156

Dostal J, Keppie JD, Hamilton MA, Aarab EM, Lefort JP, Murphy JB (2005) Crustal xenoliths in triassic lamprophyre dykes in western Morocco: Tectonic implications for the Rheic Ocean suture. Geol Mag 142: 159–172

Duraiswami RA, Karmalkar NR (1996) Unusual xenolithic dyke at Mandaleshwar and its episodic nature. Gond Geol Mag 11: 1–10

Frezzotti M-L, Peccerillo A, Zanon V, Nikogosian I (2004) Silica-rich melts in quartz xenoliths from Vulcano Island and their bearing on processes of crustal anatexis and crust-magma interaction beneath the Aeolian Arc, Southern Italy. J Petrol 45: 3–26

Green TH, Pearson NJ (1987) An experimental study of Nb and Ta partitioning between Ti-rich minerals and silicate liquids at high pressure and temperature. Geochem Cosmochem Acta 51: 55–62

Gwalani LG, Rock NMS, Chang W-J, Fernandez S, Allegre C-J, Prinzhofer A (1993) Alkaline rocks and carbonatites of Amba Dongar and adjacent areas, Deccan igneous province, Gujarat, India. 1. Geology, petrography and petrochemistry. Mineral Petrol 47: 219–253

Hari KR, Satish Kumar M, Santosh M, Rai Sumit K (1998) Petrology, geochemistry, stable isotope studies of carbonate-rich dyke-like bodies, Chhaktalao area, Madhya Pradesh. J Geol Soc India 52: 591–595

Kaila KL (1988) Mapping the thickness of Deccan Trap flows in India from DSS studies and inferences about a hidden Mesozoic basin in the Narmada – Tapti region. In: Subbarao KV (ed) Deccan Flood Basalts Mem Geol Soc India 20: 91–116

Kaila KL, Krishna VG, Mall DM (1981) Crustal structure along Mehmadabad – Billimora profile in the Cambay basin, India, from deep seismic soundings. Tectonophysics 76: 99–130

Karkare SG, Srivastava RK (1990) Regional dyke swarms related to the Deccan Trap Alkaline Province, India. In: Parker AJ, Rickwood PC, Tucker DH, (eds) Mafic dykes and emplacement mechanism. A.A.Balkema, Rotterdam: 335–347

Kovacs I, Zs N, Kothay K, Bali E, Zajacs Z, Cs S (2001) Quartz and feldspar xenocrysts in mafic lavas from Nograd-Gomar volcanic field, Bakoni-Balaton highland volcanic field and Villany Mountains (Hungary). Mitteilungen der Osterreischischen Mineralogischen Gesellschaft 146: 155–156

Kushiro I (1980) Viscosity, density and structure of melts at high pressure and their petrologic applications. In: Hargraves RB, (ed) Physics of Magmatic Processes. Princeton university Press, New Jersey: 93–120

Luhr JF, Pier JG, Aranda-Gomez JJ, Podosek FA (1995) Crustal contamination in early basin and range hawaiites of the Los Encinos Volcanic field, Central Mexico. Contrib Mineral Petrol 18: 621–642

Mahoney JJ (1988) Deccan Traps. In: Macdougall JD, (ed) Continental flood basalts. Kluwer Academic Publishers, Dordrecht: 151–194

Mungall JE (2002) Emperical models relating viscosity and trace diffusion in magmatic silicate melts. Geochem Cosmochem Acta 66: 125–143

Nakamura E, Kushiro I (1998) Trace element diffusion in jadeite and diopside melts at high pressures and its geochemical implications. Geochem Cosmochem Acta 62: 3153–3160

Nemec D (1988) Origin of syenite porphiries in the central Bohemian Pluton by magma mixing. Ne Jahrb Mineral Abh 159: 59–71

Nixon PH (1987) Mantle xenoliths. Wiley, New York, 844p

Phillips WR, Griffen DT (1986) Optical mineralogy the nonopaque minerals. CBS Publishers and Distributors, Delhi

Prelevic D, Foley SF, Cvetkovi V, Romer RL (2004) Origin of minette by mixing of lamproite and dacite magmas in Veliki Majdan, Serbia. J Petrol 45: 759–792

Randive KR (2005) Intrusive dykes of Chhaktala, Jhabua district, Madhya Pradesh: Field relations, petrography and petrochemical characteristics with reference to their PGE concentration. J Econ Geol Georesource Manag, Hindustan Publishing Corporation of India 2(1-2): 117–137

Randive KR (2008) Compositional variation of micas from the lamprophyre dykes of Bakhatgarh – Phulmal area, Jhabua district, M.P., India. In: Srivastava RK, Sivaji C, Chalapathi Rao NV, (eds)

Indian Dykes: Geochemistry, Geophysics and Geochronology. Narosa publishing House Pvt. Ltd., New Delhi: 133–141

Randive KR, Prasad GJS, Anil Kumar A, Rama Rao R, Bhaskara Rao YJ, Dayal AM (2005) Picrobasalt and lamprophyre dyke swarm of Bakhatgarh-Phulmal area, Chhota Udaipur alkaline complex, Deccan Igneous Province. Indian J Geochem 20: 191–212

Ray R, Shukla AD, Sheth HC, Ray JS, Duraiswami RA, Vanderkluysen L, Routela CS, Mallik J (2008) Highly heterogeneous Precambrian basement under the Central Deccan Traps, India: Direct evidence from xenoliths in dykes. Gond Res 13: 375–385

Reddy PR, Koteshwar Rao P, Kalachand S (1999) Crustal configuration of the Narmada-Son lineament in Central India from deep seismic sounding studies. Mem Geol Soc Ind 43: 353–365

Rudnick RL (1992) Xenoliths – samples of the lower continental crust. In: Fountain DM, Arculus R, Kay RW, (eds) Continental Lower Crust. Developments in Tectonics 23 Elsevier, Amsterdam: 269–316

Rudnick RL, Fountain DM (1995) Nature and composition of the continental crust: A lower crustal perspective. Rev Geophys 33: 267–309

Sato H (1975) Diffusion coronas around quartz xenocrysts in Aadesiter and Basalt from tertiary volcanic region of Northeastern Shikoku, Japan. Contrib Mineral Petrol 50: 49–64

Sharma RK, Pandit MK, Warrier S (1999) Hybrid acid xenoliths in dolerite dykes intruding Deccan Flood Basalts, Pune – Ahmednagar region, Western India. J Geol Soc India 54: 303–308

Sheppard S (1995) Hybridization of shoshonitic lamprophyre and calc-alkaline granite magma in the early Proterozoic Mt. Blundy igneous suite, Northern Territory. Aus J Earth Sci 42: 173–185

Simonetti A, Bell K, Viladkar SG (1995) Isotopic data from the Amba Dongar Carbonatite Complex. West-Central india: Evidence for an enriched mantle source. Chem Geol 122: 185–198

Simonetti A, Goldstein SL, Schmidberger SS, Viladkar SG (1998) Geochemical and Nd, Pb, and Sr isotope data from Deccan Alkaline Complexes – Inferences for mantle sources and plume-lithosphere interaction. J Petrol 39: 1847–1864

Srivastava RK (1994) Petrology, petrochemistry and genesis of the alkaline rocks associated with the Ambadungar Carbonatite Complex, Baroda district, Gujarat, India. J Geol Soc India 43: 23–39

Srivastava RK (1997) Petrology, geochemistry and genesis of rift related carbonatites of Ambadungar, India. Mineral Petrol 61: 47–66

Subbarao KV, Hooper PR, Dayal AM, Walsh JN, Gopalan K (1999) Narmada dykes. In: Subbarao KV (ed) Deccan Volcanic Province, Geol Soc Ind Mem 43(2): 891–902

Taylore SR, McLennan SM (1985) The Continental Crust: its composition and evolution. Blackwell, Oxford

Venturelli G, Toscani L, Salvioli-Mariani E, Capedri S (1991) Mixing between lamproitic and dacitic components in Miocene volcanic rocks of S.E. Spain. Mineral Mag 55: 282–285

Xu X-W, Zhang B-L, Qui K-Z, Mao Q, Cai X-P (2007) Origin of lamprophyres by the mixing of basic and alkaline melts in magma chamber in Baiya area, Western Yunnan, China. Lithos 99: 339–362

Chapter 19
Diorite Vein in Quenched Basalt and Its Implication for the Origin of Late-Granitoid Intrusives in Naga Hills Ophiolite, Northeast India

Naresh C. Ghose and Nilanjan Chatterjee

Introduction

The Upper Cretaceous-Eocene Naga Hills Ophiolite (NHO) occurs as dismembered tectonic slices below accretionary wedge sediments at a convergent boundary where the Indian plate subducted eastward under the Burmese microplate (Curray et al., 1979; Ni et al., 1989). Tectonic slices of such oceanic crustal segments are known to occur all along the eastern fringes of the Indian plate from Myodia in Dibang valley (Ghosh et al., 2007) at the northern end of the Indo-Burman Ranges (IBR), through the Naga-Manipur Hills of India, and extending southward through Chin and Arakan Hills in Myanmar to Andaman–Nicobar islands in Bay of Bengal. A complete set of ophiolitic litho-assemblages as found in many Tethyan ophiolites is exposed at Naga Hills. An important feature at Naga Hills is the occurrence of late-granitoids with base-metal sulphide/oxide mineralisation cross-cutting the youngest basaltic volcanic units of the ophiolite sequence. The late-granitoids are thus the youngest magmatic rocks in the NHO. They occur as minor dykes, apophyses and intrusives within pillow lavas (Singh, 1979), gabbroids and basalts (Agrawal, 1985), and cover sediments of paralic origin (Jopi Formation) overlying the basalts (Venkataramana et al., 1986). Late-granitoids are distinguished from 'oceanic plagiogranite' (Coleman, 1977) associated with cumulate gabbroids of the transitional zone (between the crustal and mantle parts of the ophiolite section) by the presence of two feldspars, viz., plagioclase and potash feldspar, amphibole and base-metal mineralisation.

Analogous to the Newfoundland and Papuan ophiolites (Coleman, 1977), the two distinct felsic rocks in the NHO, e.g., plagiogranite and late-granitoid, belong to separate stratigraphic dispositions and are derived from different melt sources. The internal relationship of gabbro-plagiogranite in the cumulate section

N. Chatterjee (✉)
Department of Earth, Atmospheric and Planetary Sciences, Massachusetts Institute of Technology, Cambridge, MA 02139, USA
e-mail: nchat@mit.edu

N.C. Ghose
G/608, Raheja Residency, Koramangala, Bangalore 560034, India

R.K. Srivastava (ed.), *Dyke Swarms: Keys for Geodynamic Interpretation*,
DOI 10.1007/978-3-642-12496-9_19,

of Troodos ophiolite indicates that "plagiogranites form the roof or screen between the underlying gabbros and the diabase above" (Coleman, 1977). "Plagiogranites are the end product of magmatic differentiation of basalt melt in Troodos and Semail ophiolites" and "the basalt first gave rise to the underlying cumulate section of mafic and ultramafic rocks" (Peterman et al., 1971). Thus development of plagiogranite precedes basaltic volcanism, the latter normally forming a carapace on the ophiolite as the youngest mafic magmatic unit. On the other hand, the late-granitoids postdate basaltic volcanism in the NHO and may have originated through partial melting of (i) ocean floor basalt in the contact aureole of a spreading center (ii) oceanic crust in a subducting slab environment, or (iii) ophiolitic basalt after the emplacement of the ophiolite.

Bulk chemical composition of the NHO basalts indicates their origin in a within-plate or a spreading environment (Ghose et al., 1986; Sengupta et al., 1989; Srikanth et al., 2004). The basalts exhibit greenschist and amphibolite facies metamorphic textures indicating that the felsic veins and dykes perhaps intruded through amphibolite/gabbro partial melting during hydrothermal metamorphism in the contact aureole of the spreading center (cf. Flager and Spray, 1991; Koepke et al., 2004). Another possibility is that the felsic veins formed through slab melting and adakitic magmatism in a subducting plate environment (cf. Drummond and Defant, 1990). In this study, textural criteria are used to demonstrate felsic melt emplacement in microscopic scale in quenched host basalt. Inter-relationship between basalt, plagiogranite and late-granitoid, and source of sulphide mineralsation are evaluated on the basis of petrography, field relationships, and limited geochemical data.

Geology

The Upper Cretaceous-Eocene NHO of Naga-Manipur Hills in the northern part of Indo-Myanmar Ranges is a narrow NNE-SSW trending crescent-shaped belt (Fig. 19.1a), comprised of dismembered and imbricated sheets of mantle-derived serpentinised peridotite tectonite (dunite – harzburgite – lherzolite), mafic-ultramafic cumulates (peridotite – norite – gabbronorite – gabbro – hornblende gabbro – plagiogranite – anorthosite), minor dolerite, volcanics and volcaniclastics dominated by basalt and spilite with intrusive felsic veins, and marine sediments containing radiolaria (Fig. 19.1b) (Agrawal and Kacker, 1980; Chattopadhyay et al., 1983; Agrawal and Ghose, 1986; Venkataramana et al., 1986; Ghose and Agrawal, 1989). The basalts are altered, metamorphosed to greenschist or amphibolite and occasionally deformed. In addition, both the basalts and the ultramafic units are associated with fault-bound tectonic slivers and lenses of Group-C eclogite surrounded by glaucophane schist and greenschist (Fig. 19.2d–f), giving credence of their metamorphism in a subduction zone and subsequent exhumation (Ghose and Singh, 1980, 1981; Ghose et al., 1984, 2010; Vidyadharan et al., 1986) along a clockwise P-T path reaching peak conditions of 1.7–2.0 GPa/ 580–610°C (Chatterjee and Ghose, 2010). However, preservation of glassy and quenched textures with occasional pillow structures indicates the basalts were emplaced under water on the ocean floor (Figs. 19.2a–c and 19.3a–d).

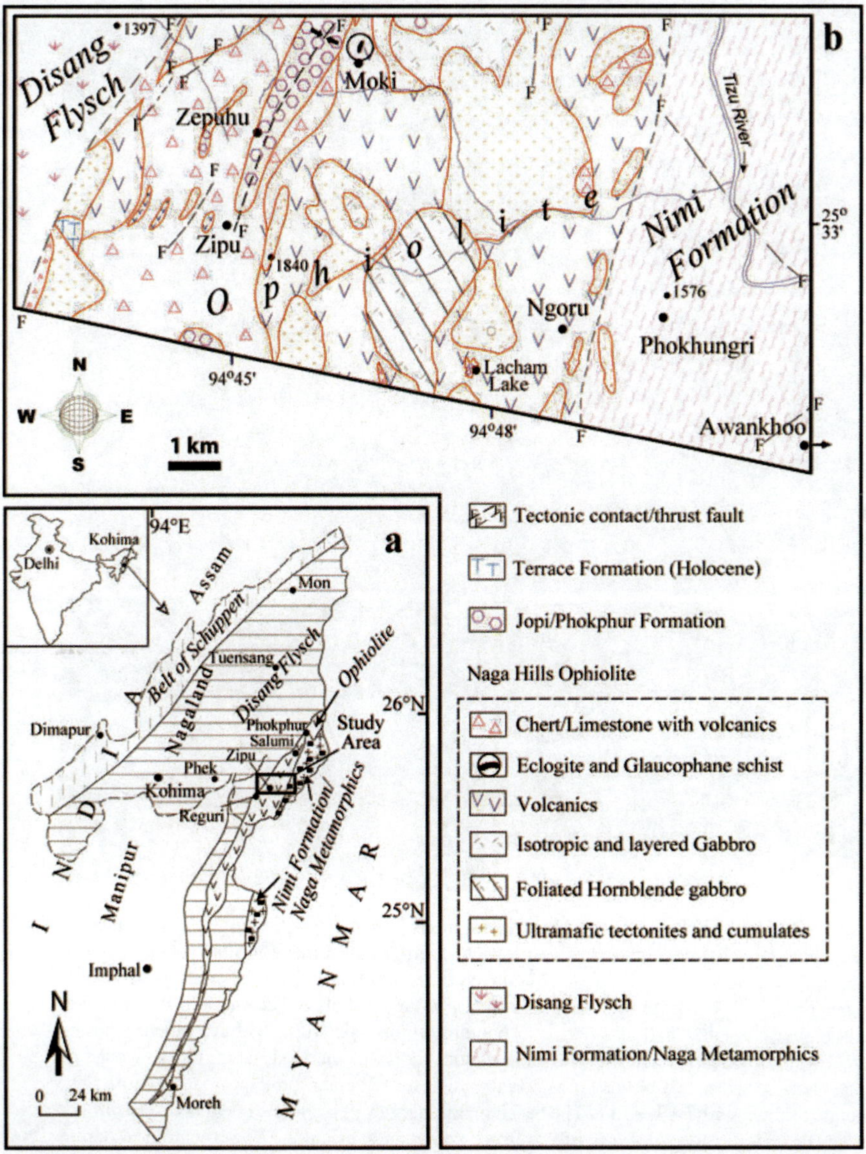

Fig. 19.1 (**a**) Location of the ophiolite belt in the Nagaland-Manipur Hills, Northeast India, showing Naga Hills Ophiolite (NHO) and other major lithological units. The study area is shown by a *bold rectangle*. (**b**) Simplified geological map of the study area in the central part of the ophiolite belt in the Phek district of Nagaland (based on detailed map of Agrawal and Ghose, 1986). Numbers in the map depict elevation (in ft)

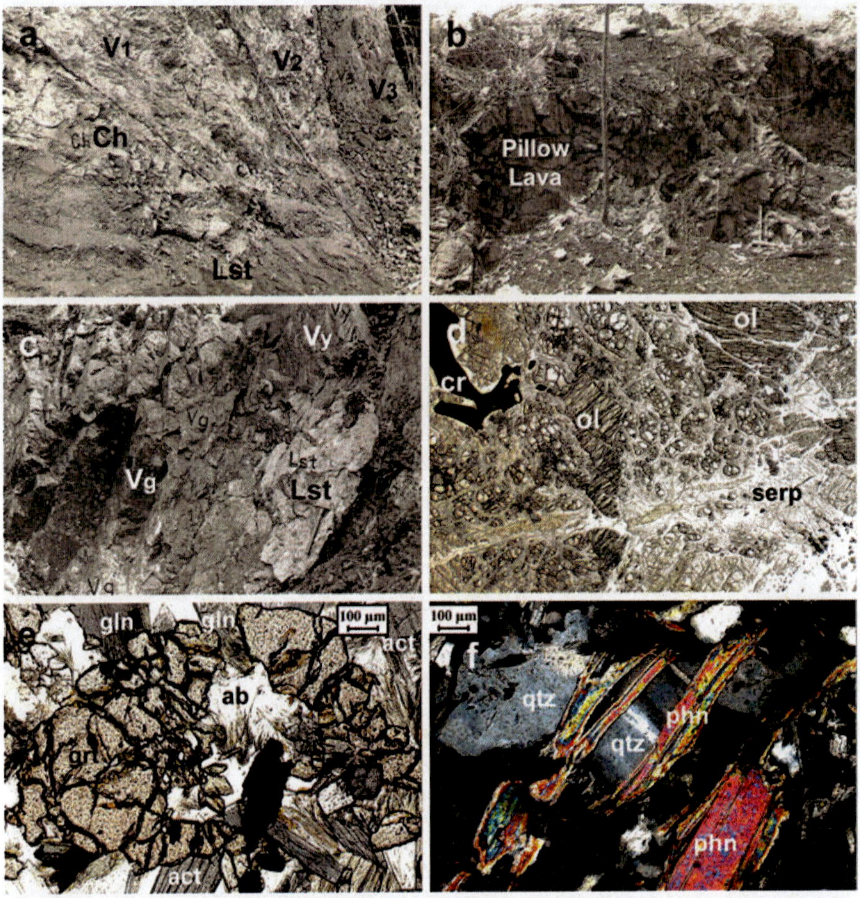

Fig. 19.2 Field photographs (*a–c*) and photomicrographs in plane polarised light (*d*, *e*) and cross polarised light (*f*) of basalt, ultramafics and low- to high-pressure metamorphic rocks in NHO. (**a**) Multiple basalt flows 8 km NNW of Zipu on Waziho-Zipu road. V_1– quenched, glassy basalt overlying a chert bed (Ch) with limestone clast (Lst), V_2-aphyric altered basalt, and V_3- massive, plagioclase phyric basalt. Chert also occurs as boudins within volcanics. (**b**) Lava showing pillow structure 1 km northwest of Salomi, Tuensang district (north of the study area). (**c**) A wedge of limestone (Lst) between two basalt flows 8 km southeast of Waziho on Zipu road section, with sharp contact shown with *broken line*. The basalt on the right (Vy) is glassy, quenched, and *brownish yellow*. The basalt on left (Vg) is fine grained, *green*, and vesicular. (**d**) Serpentinised dunite 108/79 from 1 km west of Zipu, with olivine (ol) porphyroclasts and skeletal chromite (cr). (**e**) Eclogite 139/79 from 3 km ENE of Purr, with garnet (grt), glaucophane (gln), barroisite/actinolite (act) and albite (ab). (**f**) Glaucophane schist ML-225 from 3 km NNW of Chekhung, showing S-C 'mica fish' structure defined by phengitic mica (phn) surrounded by quartz (qtz)

The NHO is tectonically sandwiched between marine sediments in the west (Disang Flysch of Upper Cretaceous–Upper Eocene age) and accretionary prism sediments/crystalline rocks (Nimi Formation/Naga Metamorphics of Mesozoic age) in the east (Fig. 19.1a, b). The low-grade metasediments of Nimi Formation probably representing the accretionary wedge, and its underlying Naga Metamorphics

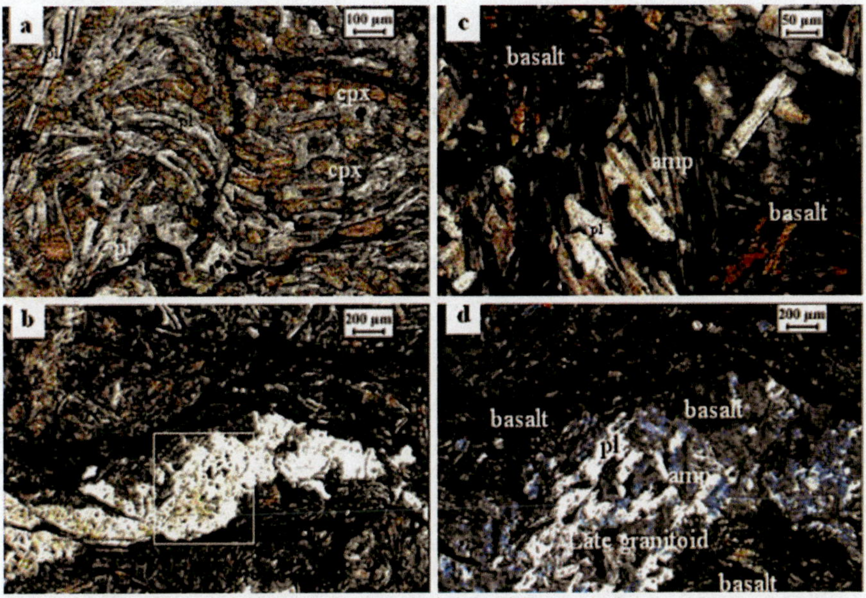

Fig. 19.3 Plane polarised light photomicrographs of quenched basalt 27/79 (olivine tholeiite) from 3 km SSW of Waziho, close to ophiolite-Disang contact. (**a**) Quenched texture consisting of elongated and folded skeletal plagioclase (pl) filled with *brown* clinopyroxene (cpx) and *green* glass. (**b**) Leucocratic diorite vein cutting across basalt. (**c**) Highly magnified view of a part of diorite vein (Fig. 19.3d) showing elongated amphibole (amp) crystals and plagioclase (pl). (**d**) Enlarged view of leucocratic diorite vein (*square area* in Fig. 19.3b) showing panidiomorphic texture with plagioclase (pl) laths and amphibole (amp)

crystalline rocks have thrust over the NHO from the east. The NE to NNE trending major boundary thrust in the east shows moderate to steep southeasterly dips (Fig. 19.1b). The NE-SW contact between the NHO and Disang Flysch in the west is marked by brecciation, silicification, fault gauge, ferrugination and intermixing of litho-units. Thin wedges and slivers of serpentinite, dolerite and crystalline limestone are seen within Disang phyllite and slate. The structural milieu in the western mega-lineament bordering the Disang marine sediments represents a tectonic mélange that shows poly-phase deformation and metamorphism (viz. greenschist, blueschist, and eclogite) and contains S-C mylonite indicating ductile deformation in the shear zone (Fig. 19.2f). The ophiolite is unconformably overlain by ophiolite-derived sediments of paralic origin (Jopi/Phokphur Formation of Late Eocene-Oligocene age) (Agrawal, 1985; Acharyya et al., 1986).

No single litho-unit of the ophiolite suite has long strike length or long tectonic contacts with other members of the ophiolite. Their wedge–shaped linear patterns indicate emplacement along high angle faults. Alternate slivers of serpentinite and volcanics on the Waziho–Zipu track are bounded by NNE trending, 60–80° easterly dipping faults. In general, discordant contacts and absence of large intrusive bodies and thermal aureole suggest emplacement of the ophiolite as tectonic slices in thrust sheets.

Late-Granitoids and Base-Metal Mineralisation

The late-granitoids include quartz porphyry, granite, granodiorite and diorite, associated with Cu-Mo mineralisation (Fig. 19.4a), and occur as minor leucocratic veins, apophyses and dykes of variable width cross-cutting mainly the volcanic (Fig. 19.3b–d), and occasionally the gabbroic and marine volcano-sedimentary units (Agrawal, 1985; Agrawal and Ghose, 1989). The late-granitoids show pronounced

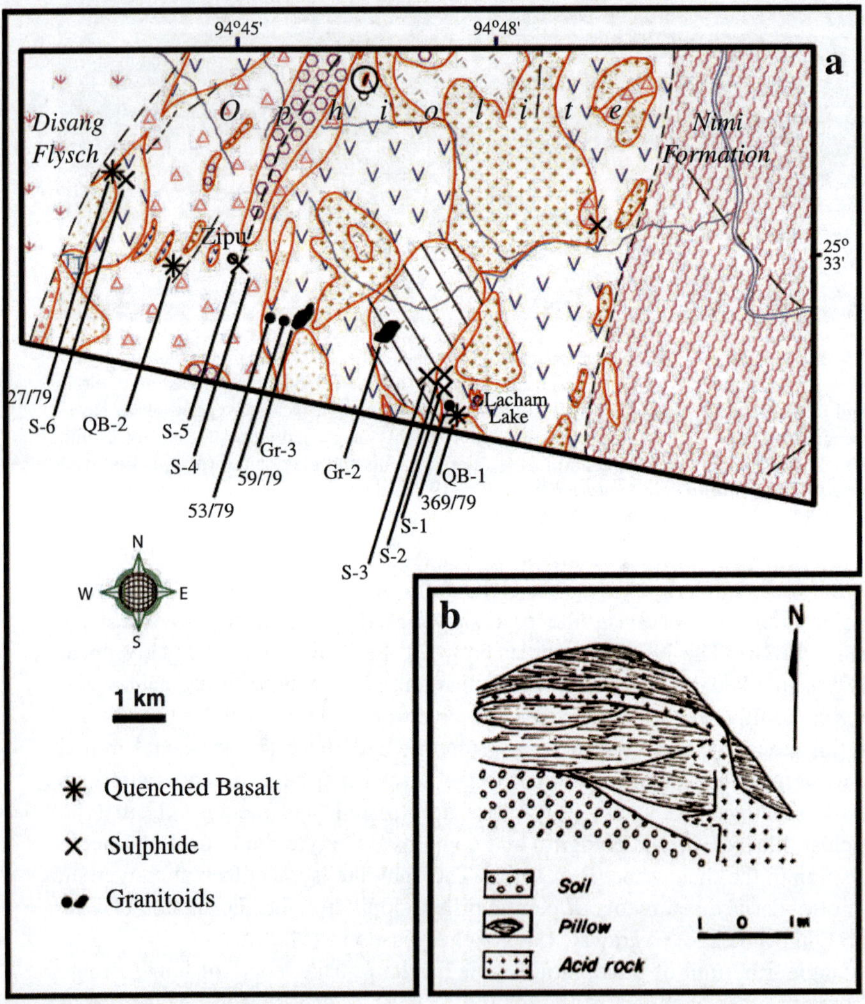

Fig. 19.4 (**a**) Map showing occurrences of intrusive granitoids (*solid dot/oblong*), quenched basalt (*star*) and sulphide mineralisation (*cross*). Lithological symbols are as in Fig. 19.1b. (**b**) Late felsic intrusive cutting across pillow lava (spilite), 1 km north of Salomi, Tuensang district, Nagaland, representing youngest magmatic intrusion

hydrothermal alteration including propylitisation, sericitisation, and brecciation. These intrusives occur along (i) N-S mega-fractures between the ophiolite and Nimi Formation in the east, and (ii) an ESE-WNW trend between Lacham Lake and south of Waziho (WNW of Zipu) in the southern part of the study area (Fig. 19.4a). The latter is marked by occurrences of a string of sulphide mineralisation. These form an array, starting from Lacham Lake and follow westward towards Zipu reaching close to the Disang contact. This zone is also marked with notable occurrences of basalt characterised by quenched texture (samples QB-1 (371/79), QB-2 (37/79), and 27/79; Fig. 19.4a). The field relations of the Lacham Lake-Zipu transect (Agrawal, 1985) are as follows:

(i) In the vicinity of Lacham Lake, volcanics are tectonically juxtaposed with hornblende gabbro showing distinct foliation, 0.5 km west of Lacham (Fig. 19.4a). Both volcanics and gabbros are traversed by dykes and veins comprising quartz diorite, coarse pegmatitic diorite, quartz vein, quartz-feldspar porphyry, and veins of calcite and igneous breccia.
(ii) At Zipu, sulphide mineralisation covering an area of about 0.35 km^2, as specks and disseminations is observed in cherts, volcanics, quartz porphyry and granodiorite. The late-granitoids appear to be a carrier of sulphides in these rocks.

Petrography of Late-Granitoids

The late-granitoid intrusives are medium-to coarse-grained, and primarily composed of variable amounts of quartz, plagioclase and amphibole, and alkali feldspar and minor biotite in some samples (Figs. 19.3 and 19.5). The textures are hypidiomorphic, granular, occasionally mylonitic with bands of recrystallised quartz, and porphyritic (Fig. 19.5a–d). Based on the feldspar varieties and their abundances, the rocks may be classified into quartz porphyry, granite, granodiorite and diorite. Both twinned and untwinned feldspars are present. Graphic intergrowth between sodic plagioclase and potash feldspar is noted. The granulated recrystallised quartzofeldspathic matrix shows elongation along a preferred direction that possibly resulted from ductile deformation (Fig. 19.5c–d).

Plagioclase: Plagioclase is a major component in all felsic varieties except quartz porphyry (Figs. 19.3a–d and 19.5e). Compositionally, it ranges between albite and oligoclase. They occur in two modes, the larger phenocrysts are feebly twinned and marginally altered to sericite, while the smaller crystals are prominently twinned and less altered (Fig. 19.5e).

Alkali feldspar: These are rare in diorite, but significant in granite and granodiorite (Fig. 19.5d, f). They occur mostly as untwinned, subhedral to anhedral grains altering to sericite. Inclusions of spinel and plagioclase in alkali feldspar are noted (Fig. 19.5f).

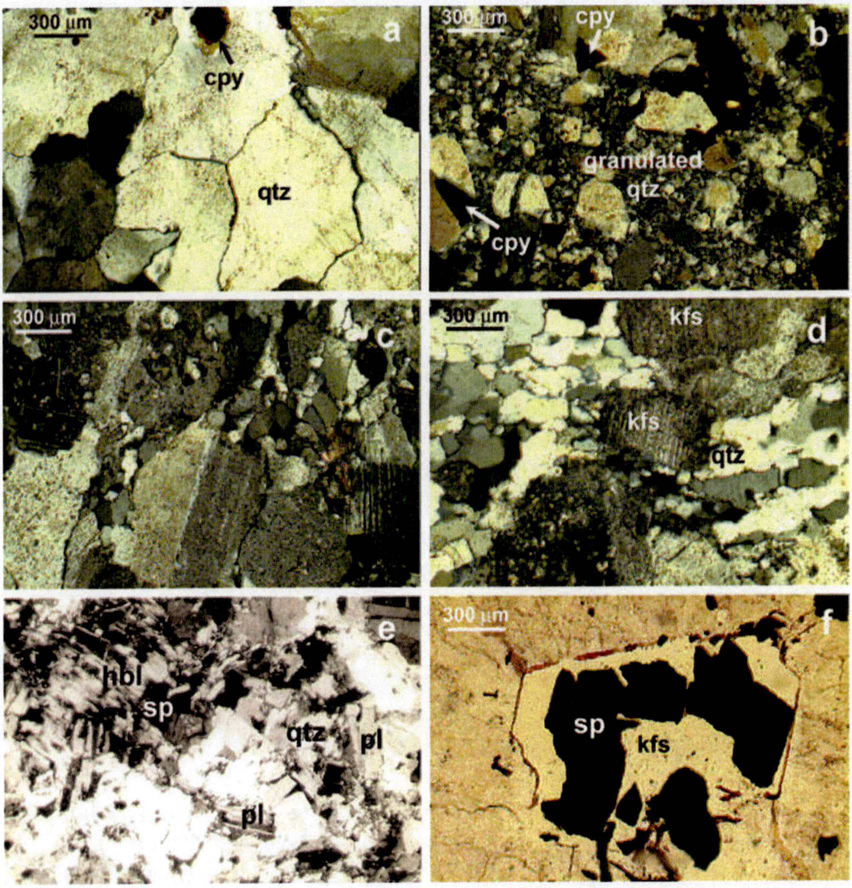

Fig. 19.5 Plane polarised light photomicrographs of granitoids in intrusive veins in NHO basalt. (**a, b**) Porphyritic granite 53/79 from 1.25 km south east of Zipu (Fig. 19.4a), with granulated quartz (qtz) vein showing disseminated chalcopyrite (cpy) at grain boundaries and as inclusions in quartz. (**c, d**) Sheared granite 369/79 from 1 km N60°W of Lacham Lake (Fig. 19.4a), showing preferred orientation of recrystallised quartz (qtz) at K-feldspar (kfs) grain boundaries. (**e**) Medium grained hypidiomorphic diorite Y-38 from 1 km northeast of Yisi containing quartz (qtz), plagioclase (pl), hornblende (hbl) and minor spinel (sp). (**f**) Quartz porphyry 59/79 from 1.5 km southeast of Zipu (Fig. 19.4a), with K-feldspar (kfs)-hosted spinel (sp) inclusion showing atoll structure

Amphibole: The primary amphibole variety is hornblende (Figs. 19.3b–d and 19.5e). It is pleochroic in shades of yellowish green to yellowish brown and contains inclusions of plagioclase. Secondary chlorite and epidote after hornblende and plagioclase are also observed.

Biotite: Green or brown biotites are occasionally present. The former is altered to iron oxide which forms a rim in association with epidote. Pale brown biotite is interleaved with muscovite. Parallel alignment of biotite grains and opaque minerals points to post-crystalline deformation.

Quartz: Quartz grains are of variable sizes. The smaller grains are located at the rims of large plagioclase phenocrysts (Fig. 19.5b–d). Recrystallisation of granulated quartz imparts a vague gneissic fabric. Inclusion of feldspar in larger quartz grain suggests late crystallisation of quartz. Discrete inclusions of opaques in quartz and *vice-versa* suggest simultaneous crystallisation of the two minerals.

Secondary minerals: The granitoids have been overprinted by low temperature hydrothermal alteration giving rise to an assemblage consisting of sericite, calcite, epidote, chlorite, talc and actinolite in decreasing order of abundance.

Base-metal mineralisation: Epigenetic Cu-Mo sulphide/oxide mineralisation associated with late-granitoid intrusives east of Zipu and west of Lacham Lake (Fig. 19.4a) is pyrite, chalcopyrite and magnetite. These occur as disseminated grains or inclusions in quartz and K-feldspar (Fig. 19.5a–b, e–f). Some opaque grains are altered to a reddish brown colour. The Cu and Mo contents of Lacham Lake samples are 0.1–1.3 wt% and 15–150 ppm (Roy et al., 1982), and those of Zipu samples are 0.1–1.8 wt% and 16–25 ppm (Agrawal and Ghose, 1989). The acid magma perhaps acted as a carrier of epigenetic mineralising solutions of hydrothermal origin.

Bulk Chemistry

In order to examine chemical differences between plagiogranite and late-granitoids in NHO, three critical and relatively unaltered samples were selected, one each from quenched basalt (27/79), plagiogranite (16R-80), and late-granitoid (369/79). Samples 27/79 and 369/79 are from an ESE-WNW array that probably constitutes a deep fracture/lineament (Fig. 19.4a). Sample 27/79 is an olivine tholeiite from 3 km SSW of Waziho near the Disang/ophiolite contact. It shows elongated and folded skeletal crystals of plagioclase enclosing clinopyroxene (Fig. 19.3a). Veins of amphibole + plagioclase cut across the sample (Fig. 19.3b–d). Sample 369/79 is a sheared granitoid intrusion from 0.5 km southwest of Lacham Lake, showing preferred orientation of recrystallised quartz at feldspar grain boundaries (Fig. 19.5b–d). The plagiogranite sample 16R-80 is from the ultramafic-mafic cumulate sequence located 1 km northeast of Reguri (Fig. 19.1a), close to the ophiolite-Disang contact, outside and to the southwest of the study area.

Analytical Methods

Chemical analyses of these three samples were carried out at National Geophysical Research Institute (NGRI), Hyderabad, India. Major elements were determined by X-ray fluorescence spectrometry (XRF) on a Philips MAGIX PRO Model 2440 instrument following the pressed pellet method (Roy et al., 2009). Analysis of reference materials JB-2 and JSP-1 yielded results within 3.37% (with a spread of ± 3%) of the certified values (Govindaraju, 1994), and the precision was <3% RSD in majority of the cases (Krishna et al., 2007). Trace elements were analysed by Inductively Coupled Plasma Mass Spectrometry (ICP-MS) on a Perkin Elmer SCIEX ELAN DRC II instrument using the closed acid digestion technique (Roy et al., 2007). JB-2, GSP-1 and BR reference materials were used as external standards, and Rh was used as internal standards to monitor the accuracy and precision of the results. The results for JB-2, GSP-1 and BR were within 1.5, 1.36 and 1.43% respectively (with a spread of ± 3%) of the certified values (Govindaraju, 1994), and the overall precision was <2% RSD for REE and <5% RSD for other trace elements.

Results

The major and trace element compositions of the three samples are given in Tables 19.1 and 19.2. The chondrite-normalised REE and MORB-normalised incompatible element contents are plotted in Fig. 19.6a, b. The patterns for the quenched basalt 27/79 are similar to that of average low-Ti basalt prevalent in the NHO (Sengupta et al., 1989; Srikanth et al., 2004). It shows peaks of Ba, Ta, Pb and Nd, and troughs of Rb, K, Th-U and Zr-Hf (Fig. 19.6b). However, the REE contents of quenched basalt 27/79 are higher than those of the average low-Ti basalt (Fig. 19.6a). Minor peaks of Ta in the basalts may be related to the presence of sulphide, a source for tantalum. The patterns for the plagiogranite 16R-80 and the late intrusive granitoid 369/79 are similar, but their REE patterns cross over (Fig. 19.6a). The late-granitoid contains higher LREE and lower HREE than the plagiogranite, which shows a Eu trough. Both samples show troughs for Ce and Er, and peak for Pb. The plagiogranite shows Cs-Rb enrichment similar to the average low-Ti basalt (Fig. 19.6b). Enrichment of LILE (viz., Cs,Rb,K,Ba) in the plagiogranite and low LREE compared to the late-granitoid suggests that the plagiogranite was probably more contaminated by crust than the late-granitoid. In addition, the late-granitoid contains higher concentrations of base-metals viz. Cu, Zn, Pb and transition metals viz. V, Cr than the plagiogranite 16R-80 (Table 19.1).

The range of Sr/Y ratios of the analysed samples is 1.9–3.2, and the range of La/Yb ratios is 0.8–3.6 (Tables 19.1 and 19.2; Fig. 19.7a, b). These values are much lower than the lowest values for adakites (Sr/Y = 40, La/Yb = 20, Castillo et al., 1999). The Nb/Ta (=3.1–4.5) and Zr/Sm (=5.7–13.3) ratios of the analysed samples are low compared to continental rocks including Archean tonalite-trondjemite-granodiorite (TTG) gneisses, continental crust and continental basalts (approximately 12–20 and 31–45, respectively, Pfander et al., 2007). Compared to the plagiogranite and granitoid of the NHO, Tibetan slab migmatite and Manaslu granite from the collision zone of the Higher Himalayas contain much higher Rb (239–286 ppm), U (1.2–8.0 ppm) and Th (1.6–6.0 ppm), and lower Zr (19–25 ppm), whereas the Sr (76–331 ppm) values overlap (Barbey et al., 1996).

The low Sr/Y and La/Yb values of the samples in this study indicate that they are not adakites produced by slab melting. On the other hand, the Sr/Y and La/Yb ratios of the three samples are similar to the ratios for MORB (Fig. 19.7). Thus all three samples probably formed near an oceanic spreading center, as also concluded

Table 19.1 Bulk compositions of basalt and felsic rocks from Naga Hills Ophiolite

	27/79	369/79	16R-80
	XRF analysis: weight percent		
Major oxides			
SiO_2	47.44[a]	73.68	72.35
TiO_2	1.92	0.17	0.24
Al_2O_3	12.37	13.58	14.55
Fe_2O_3	13.12	2.20	3.04
MnO	0.18	0.02	0.04
MgO	8.52	0.59	1.02
CaO	9.56	1.90	1.26
Na_2O	2.32	5.73	5.86
K_2O	0.20	0.19	0.55
P_2O_5	0.22	0.02	0.05
LOI	4.74	0.89	1.12
Total[b]	99.47	98.78	99.82
Fe_2O_3[b]	1.97	0.33	0.46
FeO[b]	10.03	1.68	2.33
Mg#[b]	60.2	38.5	43.9
	ICP-MS analysis: parts per million (weight)		
Incompatible trace elements			
Cs	0.1	0.1	1.1
Rb	2.4	4.0	15.0
Ba	56.1	96.9	75.9
Th	0.5	2.7	1.1
U	0.1	0.4	0.3
Ta	1.6	0.4	0.7
Nb	6.9	2.0	2.1
Pb	2.1	14.6	7.8
Sr	141.0	160.1	203.0
Zr	39.9	44.0	103.9
Hf	1.1	1.9	5.0
Y	50.5	49.3	103.8
Compatible trace elements			
Sc	47.1	10.0	9.3
V	316.4	35.5	14.0
Cr	70.1	34.3	8.1
Co	51.8	4.7	3.5
Ni	71.3	17.0	19.6
Cu	100.9	37.0	6.3
Zn	89.5	88.6	47.0
Ga	18.6	17.6	14.2

[a] From Ghose et al. (1986).
[b] Calculated based on 15% of total Fe is Fe^{3+}; Mg# = atomic Mg/(Mg+Fe)
Sample locations – 27/79: Quenched basalt 3 km SSW of Waziho (in ophiolite, near Disang contact), 369/79: Late-stage Granitoid 1 km N60°W of Lacham Lake, 16R-80: Plagiogranite 1 km northeast of Reguri (in ophiolite, near Disang contact).

Table 19.2 ICP-MS analysis of rare earth elements in basalt and felsic rocks from NHO

	27/79	369/79	16R-80
	Parts per million (weight)		
Rare earth elements			
La	5.9	17.2	7.7
Ce	17.6	13.2	6.8
Pr	2.5	6.2	3.6
Nd	17.0	28.7	19.3
Sm	5.1	7.7	7.8
Eu	1.8	1.6	1.2
Gd	7.3	4.1	5.6
Tb	1.3	0.9	1.5
Dy	7.4	7.4	15.4
Ho	1.6	1.9	4.0
Er	5.2	4.0	8.4
Tm	0.8	1.0	2.1
Yb	4.7	4.7	9.5
Lu	0.7	0.7	1.3

Sample locations as in Table 19.1

by Sengupta et al. (1989). Furthermore, although limited geochemical data are available, the Sr-Y and La-Yb systematics (Fig. 19.7) are consistent with the origin of the plagiogranite through fractional crystallisation of NHO basalt, whereas the late-granitoid may be a product of basaltic partial melting. The elevated Y and Yb of the plagiogranite may be modelled through 45% low-pressure fractional crystallisation of the basalt, and the slightly higher Sr/Y and La/Yb ratios of the late-granitoid may indicate partial melting of the basalt to form the granitoid (by analogy with the melting curves for primitive mantle) (Fig. 19.7a, b).

Discussion

Greenschist and amphibolite facies assemblages are known to form during hydrothermal metamorphism of ocean-floor basalts in contact aureoles although the igneous textures are not significantly altered (Cann, 1979). The NHO basalts show intergranular (Fig. 19.3a) to intersertal, vitrophyric, variolitic, more commonly porphyritic textures, and occasionally with megacryst of plagioclase (Ghose, 2010) that are typically developed in a mid-ocean ridge lava suite (Smith and Perfit, 2007) and in the Southeast Pacific Rise (Pandey et al., 2009). However, they also show evidence of deformation, cataclysm, and shearing in the form of folded, elongated crystals (Fig. 19.3a). Sericite, calcite, epidote, chlorite, talc and actinolite bearing veins provide evidence of greenschist facies metamorphism (Ghose et al., 2010), whereas presence of euhedral brown amphibole with a core of clinopyroxene in a carbonate clast in quenched basalt (Ghose, 2010) provide evidence of amphibolite facies metamorphism. Such features have previously been observed in ophiolitic

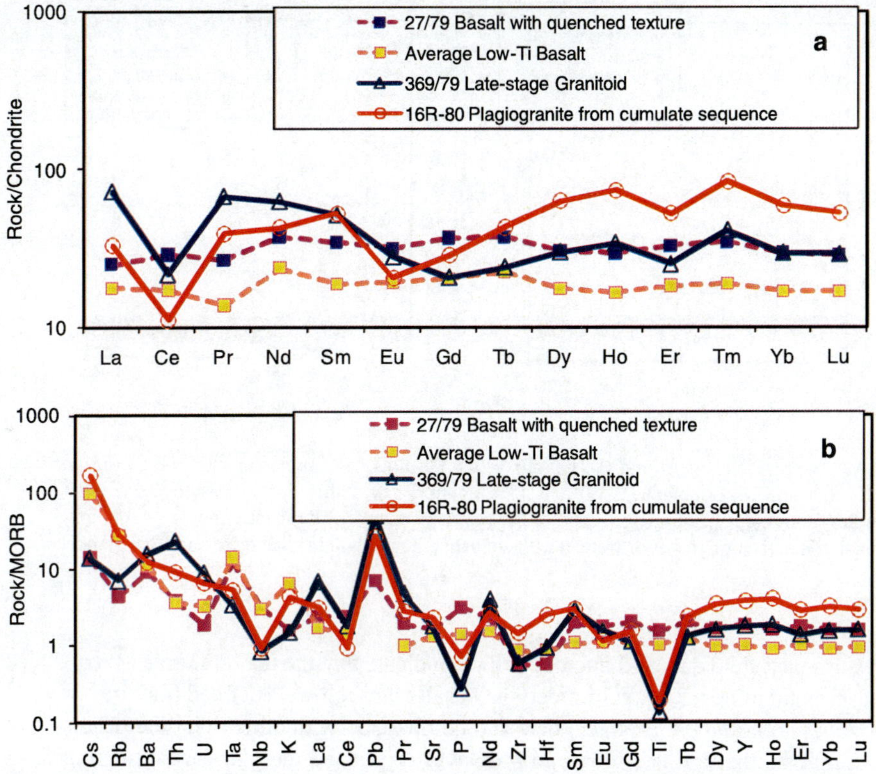

Fig. 19.6 (**a**) Chondrite-normalised (McDonough and Sun, 1995) rare earth element, and (**b**) MORB-normalised (Sun and McDonough, 1989) incompatible element patterns of basalt and granitoids analysed in this study and average low-Ti NHO basalt from Sengupta et al. (1989) and Srikanth et al. (2004)

gabbros of the Canadian Appalachians (Flager and Spray, 1991) and Nidar ophiolite in eastern Ladakh (Rao et al., 2004). We suggest the observed metamorphic textures in NHO basalts (Fig. 19.3b, d) originated through hydrothermal metamorphism in the high thermal gradient of a contact aureole of a magma chamber associated with an oceanic spreading center.

During hydrothermal metamorphism below the surface of oceanic crust, chalcophile elements such as Cu, Zn and Pb are mobilised upward through solution in hot sea-water and deposited on or near the surface. Fractures and shear zones developed parallel to the spreading ridge provide an easy passage for chalcophile element migration. Field data indicates that the intrusive granitoids of the NHO occur along major N–S and possibly ESE-WNW fracture/shear zones (Fig. 19.4a). Disseminated Cu-Mo sulphides and inclusions of spinel with atoll structure in quartz and K-feldspar (Fig. 19.5e, f) in the granitoid intrusives indicate that the granitoids were a major carrier of sulphides and base-metal mineralisation in the NHO.

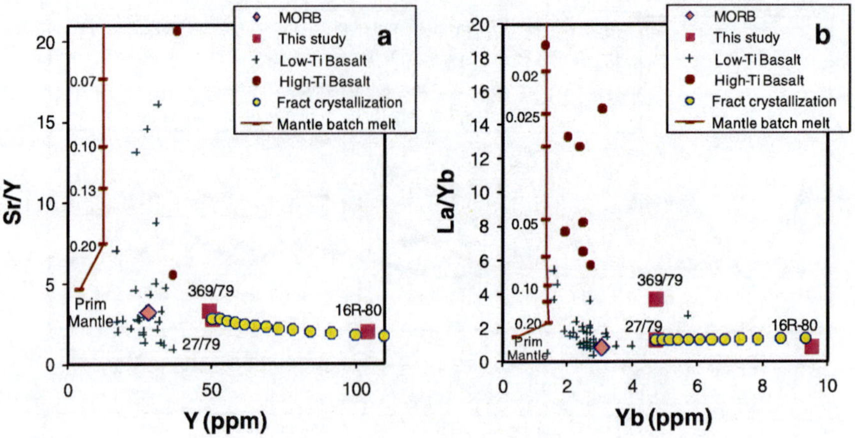

Fig. 19.7 (**a**) Sr/Y versus Y, and (**b**) La/Yb versus Yb plots for samples in this study, average MORB (Sun and McDonough, 1989) and 45% fractional crystallisation of basalt 27/79. Also shown are non-modal batch melting curves for primitive mantle. The mantle melting curve and fractional crystallisation of basalt were calculated using the method in Chatterjee and Bhattacharji (2008)

In summary, field data and microstructures indicate that the late-granitoids carrying base-metal mineralisation intruded the basalts through well-defined fractures or shear zones. Limited geochemical data and comparison with other ophiolite occurrences indicate that the late-granitoid intrusives originated through partial melting of basalt during hydrothermal metamorphism in the contact aureole of an oceanic spreading center.

Acknowledgements The paper emerges out of a collection of samples from the Naga Hills Ophiolite made available to one of the authors (NCG) by O. P. Agrawal, retired Geological Adviser to Govt. of Nagaland. We are thankful to Fareeduddin of Geological Survey of India, Bangalore, D. K. Bhattacharya of Ranchi University for taking photomicrographs, and V. Balaram for kindly undertaking chemical analysis of limited number of samples at the National Geophysical Research Laboratory, Hyderabad.

References

Acharyya SK, Roy DK, Mitra ND (1986) Stratigraphy and palaeontology of the Naga Hills ophiolite belt. Mem Geol Surv India 119: 64–79

Agrawal OP (1985) Geology and geochemistry of the mafic-ultramafic complex of Indo-Burman ranges between Meluri and Awankhoo, Phek district, Nagaland, India. Unpublished Ph D thesis, Patna University, Patna

Agrawal OP, Ghose NC (1986) Geology and stratigraphy of the Naga Hills ophiolite between Meluri and Awankhoo, Phek district, Nagaland, India. In: Ghose NC, Varadarajan S (eds) Ophiolites and Indian Plate Margin. Sumna Publication, Patna: 163–195

Agrawal OP, Ghose NC (1989) Mineral resources in the ophiolite belt of Nagaland, N. E. India. In: Ghose NC (ed) Phanerozoic Ophiolites of India. Sumna Publication, Patna: 245–280

Agrawal OP, Kacker RN (1980) Nagaland ophiolites, India – a subduction zone ophiolite complex in Tethyan orogenic belt. Internatinal Ophiolite Symposium, Cyprus: 454–461

Barbey P, Brouand M, Le Fort P, Pecher A (1996) Granite-migmatite genetic link: The example of the Manaslu granite and Tibetan Slab migmatites in central Nepal. Lithos 38: 63–79

Cann JR (1979) Metamorphism in the ocean crust. In: Talwani M, Harrison CG, Hayes DE (eds) Deep Drilling Results in the Atlantic Ocean: Ocean Crust. Maurice Ewing Series, vol 2. American Geophysical Union, Washington, DC: 230–238

Castillo PR, Janney PE, Solidum RU (1999) Petrology and geochemistry of Camiguin Island, southern Philippines: Insights to the source of adakites and other lavas in a complex arc setting. Contrib Mineral Petrol 134: 33–51

Chatterjee N, Bhattacharji S (2008) Trace element variations in Deccan basalts: Roles of mantle melting, fractional crystallization and crustal assimilation. J Geol Soc India 71: 171–188

Chatterjee N, Ghose NC (2010) Metamorphic evolution of the Naga Hills eclogite and blueschist, Northeast India: Implications for early subduction of the Indian plate under the Burma microplate. J Metamorph Geol 28: 209–225

Chattopadhyay B, Venkataramana P, Roy DK, Bhattacharyya S, Ghosh S (1983) Geology of Naga Hills ophiolites. Geol Surv India Rec 112(2): 59–115

Coleman RG (1977) Ophiolites: Ancient Oceanic Lithosphere. Springer-verlag, Berlin: 229p

Curray JR, Moore DG, Lawyer LA, Emmel FJ, Raitt EW, Henry M, Keikhefer R (1979) Tectonics of the Andaman Sea and Burma. In: Watkins J, Montadert L, Dickinson P (eds) Geological and Geophysical Investigations of Continental Slopes and Rises, vol 29. Memoirs – American Association of Petroleum Geologists, Trondheim : 189–198

Drummond MS, Defant MJ (1990) A model for trondhjemite-tonalite-dacite genesis and crustal growth via slab melting: Archaean to modern comparisons. J Geophys Res 95: 21503–21521

Flager PA, Spray JG (1991) Generation of plagiogranite by amphibolite anatexis in oceanic shear zones. Geology 19: 70–73

Ghose NC, Agrawal OP (1989) Geological framework of the central part of Naga Hills. In: Ghose NC (ed) Phanerozoic Ophiolites of India and Associated Mineral Resources. Sumna Publication, Patna: 165–188

Ghose NC, Agrawal OP, Chatterjee N (2010) A geological and mineralogical study of eclogite and glaucophane schists in the Naga Hills Ophiolite, Northeast India. Island Arc. doi:10.1111/j.1440-1738.2010.00710.x

Ghose NC, Agrawal OP, Singh RN (1986) Geochemistry of the ophiolite belt of Nagaland, N.E. India. In: Ghose NC, Varadarajan S (eds) Ophiolites and Indian Plate Margin. Sumna Publication, Patna: 241–294

Ghose NC, Agrawal OP, Windley BF (1984) Geochemistry of the blueschist-eclogite association in the ophiolite belt of Nagaland, India. Seminar Cenozoic Crustal Evolution on the Indian Plate Margin, Patna University, Abstracts, Patna: 27–30

Ghose NC, Fareeduddin (2010) Textural fingerprints of magmatic, metamorphic and sedimentary rocks associated with the Naga Hills ophiolite, Northeast India. In: Ray J, Sen G, Ghosh B (eds) Topics in Igneous Petrology, DOI10.1007/978-90-481-9600-5_13@Springer Science+ Business Media B.V. 2001, Springer, Berlin, 321–351

Ghose NC, Singh RN (1980) Occurrence of blueschist facies in the ophiolite belt of Naga Hills, east of Kiphire, N.E. India. Geol Rundsch 69: 41–43

Ghose NC, Singh RN (1981) Structure of the Naga Hills ophiolites and associated sedimentary rock in the Tuensang district of Nagaland, N.E. India. Ofioliti 6: 237–254

Ghosh B, Mahoney J, Ray J (2007) Mayodia ophiolites of Arunachal Pradesh, Northestern Himalaya. J Geol Soc India 70: 395–604

Govindaraju K (1994) Compilation of Working Values and sample description for 383 geostandards. Geostan Newsl 18(Special issue): 158p

Koepke J, Feig ST, Snow J, Freise M (2004) Petrogenesis of oceanic plagiogranites by partial melting of gabbros: An experimental study. Contrib Mineral Petrol 146: 414–432

Krishna AK, Murthy NN, Govil PK (2007) Multielement analysis of soils by WD-XRF spectrometry. At spectrosc 28(6): 202–214

McDonough WF, Sun SS (1995) The composition of the Earth. In: McDonough WF, Arndt NT, Shirey S (eds) Chemical Evolution of the Mantle: Chemical Geology, vol 120. Kluwer Academic Publishers, Dodrecht: 223–253

Ni JF, Guzman-Speziale M, Bevis M, Holt WE, Wallace TC, Seager WR (1989) Accretionary tectonics of Burma and the three-dimensional geometry of the Burma subduction zone. Geology 17: 68–71

Pandey SK, Shrivastava JP, Roonwal GS (2009) Petrography and mineral chemistry of neovolcanics occurring between Pacific and Nazca plate boundaries. J Geol Soc India 74: 559–572

Peterman ZE, Coleman RG, Hildreth RA (1971) Sr87/Sr86 in mafic rocks of the Troodos massif, Cyprus. US Geological Survey Professional Papers 750-D: D157–D161

Pfänder JA, Münker C, Stracke A, Mezger K (2007) Nb/Ta and Zr/Hf in ocean island basalts — Implications for crust – mantle differentiation and the fate of Niobium. Earth Planet Sci Lett 254: 158–172

Rao DR, Rai H, Kumar JS (2004) Origin of oceanic plagiogranite in the Nidar ophiolitic sequence of eastern Ladakh, India. Curr Sci 87(7): 999–1005

Roy P, Balaram V, Krishna AK, Singh RS, Chavan CD, Charan SN, Murthy NN (2009) A Simplified and Rapid Method for the Determination of Sulphur in kimberlites and other Geological Samples by Wavelength-Dispersive X-ray Fluorescence Spectrometry. At spectrosc 30(5): 178–183

Roy P, Balaram V, Kumar A, Satyanarayanan M, Gnaneswar Rao T (2007) New REE and Trace Element data on two kimberlitic reference materials by ICP-MS. Geostan Geoanal Res 31: 261–273

Roy RK, Kacker RN, Chattopadhyay B (1982) Geochemical characteristics and tectonic setting of the Naga Hills ophiolite volcanics. Ophioliti 2/3: 479–498

Sengupta S, Acharyya SK, Van der Hul HJ, Chattopadhyay B (1989) Geochemistry of volcanic rocks from Naga hills Ophiolites, North East India and their inferred tectonic setting. J Geol Soc London 146: 491–498

Singh RN (1979) Geochemistry of the ophiolite suite east of Kiphire, Tuensang District, Nagaland. Unpublished Ph.D. Thesis, Patna University, Patna

Smith MC, Perfit MR (2007) Petrography and petrogenesis of a mid-ocean ridge lava suite. Cited 10 Feb 2009. Available from: http://serc.carleton.edu/files/NAGTWorkshops/petrology/teaching_examples/MORB_Petrography_and_Petrology_Key_1.1.pdf

Srikanth B, Subba Rao MV, Rao BV, Nirmal Charan S, Balaram V, Ejung OC (2004) Geochemical signatures in the basaltic rocks of Naga Hills ophiolite belt: Implications for petrogenesis and tectonic environment of emplacement. J Applied Geochem 6: 177–189

Sun SS, McDonough WF (1989) Chemical and isotopic systematics of oceanic basalts: Implications for mantle compostion and processes. In: Saunders AD, Norry DJ (eds) Magmatism in the Ocean Basins, Special Publications, vol 42. Geological Society, London: 313–345

Venkataramana P, Datta AK, Acharyya SK (1986) Petrography and petrochemistry. In: Mitra ND and 12 others (eds) Geology of Nagaland Ophiolite, vol 119. Geological Survey of India, kolkata: 33–63

Vidyadharan KT, Shukla R, Bhattacharyya S (1986) Blueschists, C-type eclogites and garnet-bearing mafic segregations from Naga Hills ophiolite belt. Geol Surv India Rec 114(4): 7–13

Chapter 20
The Parishan Pluton in Qorveh Area, an Example for Magma Mingling Process, Southeastern Sanandaj, Iran

Ashraf Torkian

Introduction

Mafic microgranular enclaves (MMEs) are common in granitoid rocks, and their (MMEs) presence is the most striking feature of granitoid plutons in calc-alkaline suites, originating in subduction zones (Didier, 1973). They have received much more attention by researchers (Barbarin, 1991; Castro et al., 1991; Fershtater and Borodina, 1991; Barbarin and Didier, 1992; Elburg, 1996; Dorais et al., 1990; Dahlquist, 2002, among others). Theories of petrogenesis of enclaves (as summarized) may be divided into four groups: (a) restite; (b) fragments of country rocks picked up during intrusion of the pluton; (c) mafic segregations of early crystallizing phases; and (d) globules of mafic magma that mingles with the host granitic magma (Elburg, 1996). However, mafic microgranular enclaves are generally considered as products of magma mingling (incompletely mixing) between magmas of contrasted compositions, either as cogenetic mafic and felsic magmas or as two coeval magmas (Orsini et al., 1991). In this paper, the term mixing is used to indicate the process of chemical interaction between different magmas to form a completely blended homogenous hybrid. In contrast, mingling is used to indicate the process of physical (mechanical) mixing between different magmas to form a rock product which preserves compositional heterogeneities (Vernon, 1984; Sparks and Marshall, 1986).

The scope this paper is to examine mafic microgranular enclaves as products of magma mingling between the Parishan granodiorite-granite and mafic magma (mafic dykes) in the Qorveh Granitoid Complex (QGC), Iran. Based on petrography, mineral chemistry and whole rock geochemistry of representative samples from MMEs and their host rocks, we discuss magma mingling between the mafic magma and the granodiorite-granite pluton.

A. Torkian (✉)
Department of Geology, Faculty of Sciences, Bu-Ali Sina University, Hamedan, Iran
e-mail: a-torkian@basu.ac.ir

R.K. Srivastava (ed.), *Dyke Swarms: Keys for Geodynamic Interpretation*,
DOI 10.1007/978-3-642-12496-9_20,

Field Relations

Numerous intrusive rocks of various size intrude the Sanandaj-Sirjan Zone, western Iran (e.g. Valizadeh and Cantagrel, 1975; Berberian and King, 1981; Berberian and Berberian, 1981). They show large range of variation in rock types, but are dominated by granodiorite-granite with minor amount of mafic bodies and ideal for studying the evolution of igneous rocks. The Qorveh Granitoid Complex (GGC), as a major plutonic complex in the region, is a typical I-type granitoid. The QGC is an elliptical batholith and is a multi-intrusion structure, containing several rock-types. The Eocene-Oligocene QGC (Bellon and Broud, 1975) is located 70 km SE Sanandaj city in western Iran.

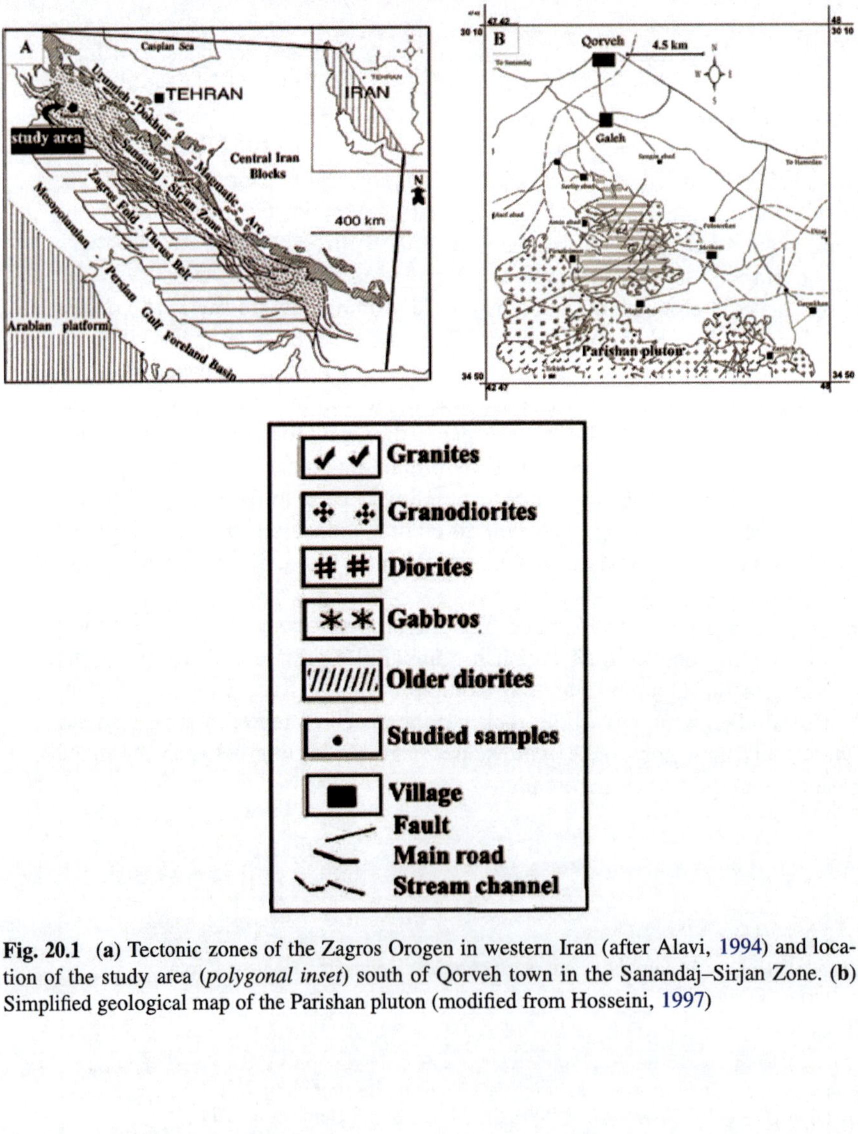

Fig. 20.1 (**a**) Tectonic zones of the Zagros Orogen in western Iran (after Alavi, 1994) and location of the study area (*polygonal inset*) south of Qorveh town in the Sanandaj–Sirjan Zone. (**b**) Simplified geological map of the Parishan pluton (modified from Hosseini, 1997)

The Parishan pluton is apart of the QGC that is composed the granodioritic-granitic and the dioritic units and dykes of dioritic to meladioritic. The dioritic unit occurs as stocks or small masses. These rocks consist mainly of plagioclase (45–55%), hornblende (15–20%), quartz (10–15%), alkali-feldspar (3–5%) and biotite (1–8%), with accessory apatite, zircon and opaque minerals. The grain size is largely invariable, showing textures ranging from fine-grained to porphyritic. Plagioclase occurs as zoned euhedral to subhedral laths that are sometimes altered to sericite, clinozoisite, epidote and calcite. Hornblende occurs as green euhedral to subhedral laths, 0.2–1.5 mm in length. Quartz and alkali-feldspar occur interstitially. For a description of these rocks the reader is referred to Torkian, 2008; Torkian et al., 2008; the present study only deals with the granodiorite-granite unit.

The Parishan granodiorite-granite pluton is massive, and MMEs in it make up no more than 5% of the outcrops, and generally appear as synplutonic dykes. They are typically about 0.4–2.5 m in width. They are subangular to near-oval in shape (Fig. 20.1) and they are fine-grained than the host rock and contain higher proportions of ferromagnesian minerals. They are typically about 10–40 cm in maximum diameter, although some are longer or shorter. Abundance and size of enclaves decrease irregularly with distance from the contact with mafic dykes. Enclaves are pillow-like in shape with planar margins in part. The contacts between large enclaves and the host granodiorite are mainly chilled to fine grained, very less commonly, with lobate crenulations to cuspate margins and convex toward the felsic host. The lobate contact suggests that the enclave and host rock coexisted as contemporaneous magmas and also imply that the mafic magma (synplutonic dykes) from which the MMEs originate was at least partly plastic when it intruded into granitoid (Pitcher, 1991). Centimeter sized clots of mafic minerals (corresponding to amphibole-rich clots of Sial et al., 1998; or aggregates of amphiboles in the sense of Stephens, 2001) are observed in individual MMEs.

Petrography

Granodioritic-Granitic Unit

The Parishan pluton occurs as an elliptical body (Fig. 20.2). By modal mineralogy, the rock varies in composition from granodiorite to granite. In general, these rocks are medium to coarse-grained with granular to porphyritic textures. Mineral composition is normally constituted by plagioclase, K-feldspar (orthoclase and microcline), quartz, hornblende and small amounts of apatite, allanite, sphene, magnetite and zircon. Sometimes K-feldspar occurs as large anhedral patches that engulf quartz, plagioclase and, to a lesser degree, biotite. Plagioclase occurs as zoned euhedral to subhedral tabular laths, up to 2–3 mm in length. Textural relationship indicates that the tabular plagioclase crystals were the first felsic minerals to crystallize. They have zoning and albite twinning.

Many of them have a distinct oscillatory zoned core, which is more An-rich than the rim. Hornblende commonly occurs as subhedral to euhedral grains,

Fig. 20.2 Subangular to near-oval in shaped MMEs is the sharp contacts, near the Majid abad village

2–3 mm in length and it exhibit normal brownish-green pleochroism and fall in the ferro-hornblende to magnesio-hornblende fields with contents of $0.68 > Mg/(Mg + Fe^{+2}) > 0.23$, and $7.5 > TSi > 6.5$ (all of in p.f.u.) (Torkian, 2008). Orthoclase (2–2.5 mm) often has carlsbad twinning and perthitic texture (perthitic exsolution lamellae). Quartz occurs as isolated anhedral grains locally associated with mymekite.

Mafic Dykes and Related MMEs

The mafic dykes are dioritic-meladioritic in composition and mainly composed of plagioclase, hornblende, quartz, K-feldspar and biotite, with accessory apatite, zircon and opaque minerals. The MMEs show different textures including microgranular and microporphyric (Fig. 20.3a). In most mafic microgranular enclaves, plagioclase and, less commonly, green colored euhedral to subhedral hornblende laths are present as large crystals. Plagioclase occurs as zoned euhedral to subhedral laths that are sometimes altered to sericite, clinozoisite, epidote and calcite. Plagioclase crystals reveal discontinuous more sodic rims which are a notable feature. They display normally or complexly zoned structure combination with fine-grained margins, suggesting magmatic origin of enclaves. The compositional range of anorthite contents of plagioclase is higher (An58–73) in comparison to those of the host rocks (Fig. 20.4). Hornblende occurs as green euhedral to subhedral laths, 0.2–1.5 mm in length. Sometimes hornblende, quartz, apatite and Fe-Ti oxides

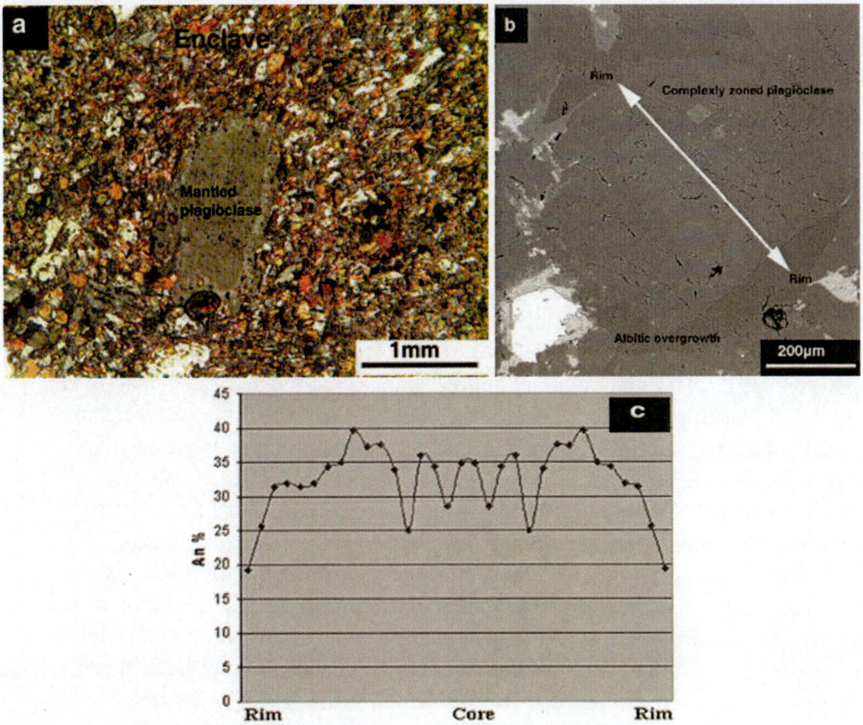

Fig. 20.3 (**a**) Photomicrograph of mantled plagioclase in microporphyric enclaves. (**b**) Back-scattered electron scanning image of plagioclase crystal. The traverse is indicated by the *arrow*. (**c**) Geochemical traverse showing variations in anorthite contents across oscillatory – complexly zoned plagioclase. An contents in rims are very low (albite-rich overgrowth)

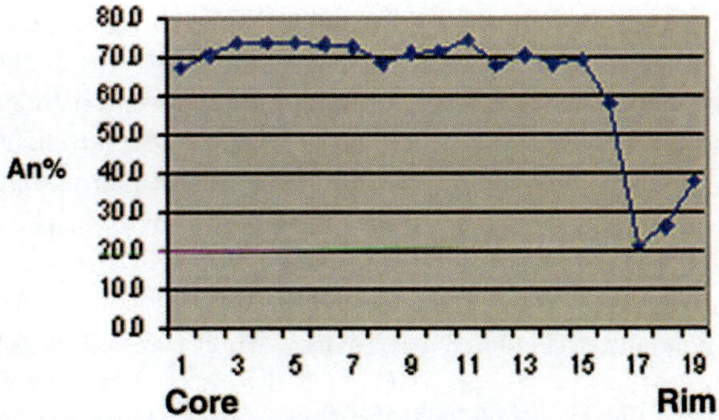

Fig. 20.4 Geochemical traverse showing variations in anorthite contents of core and rim of plagioclase in MMEs. An contents in rim is 20–40

enclose in plagioclases. Quartz is generally anhedral and poikilitic, with all other phases included in it. Apatite appears as long tiny needles scattered throughout the groundmass.

Geochemistry

Sampling and Analytical Methods

Ten fresh samples from the granodioritic-granitic rocks and MMEs were selected for geochemical analyses. The whole-rock major, trace including rare earth element analyses were performed by X-ray fluorescence technique in the Geoanalytical Laboratory of Washington State University, USA. Detection limits ranged from 0.01 to 0.1 wt%, and 0.1 to 10 ppm for major and trace elements, respectively (Table 20.1). Major element oxide abundances were recalculated to 100%, with all the iron expressed as FeO^T. Mineral compositions were determined at the University of Oklahoma, Norman (USA), using a CAMECA SX50 electron microprobe. Chemical microanalysis of minerals was performed by wavelength-dispersive spectrometry using 20 kV acceleration, 20 nA beam current and 2 μm spot size. Matrix corrections employed the PAP algorithm (Pouchou and Pichoir, 1985), with oxygen content calculated by stoichiometry. Standard materials were natural crystalline solids for all elements except Ba, for which NIST glass K309 was used. Data processing was by means of the program Minpet (version 2.02; Richard, 1995).

Whole-Rock Geochemistry

Major element compositions of the granodioritic-granitic unit and the mafic microgranular enclaves are given in Table 20.1. The major elements (TiO_2, MgO, FeO, and Al_2O_3) data for host rocks display a curvilinear trend with increasing SiO_2 content, whereas these variations are little for enclaves and they centered as a cluster (Fig. 20.5). Also, the Alumina Saturation Index (ASI) (molar Al_2O_3/(Na_2O+K_2O+CaO)) varies from 0.9 to 1.18, which indicates that the host rock is I-type in character thus supporting the results of petrographical observations. K_2O is always higher in the granodioritic-granitic rocks than in the enclaves, whereas P_2O_5 seem to be scattered in comparison with MMEs.

Further, MMEs are characterized by low abundances of SiO_2 (49–53 wt%) and high contents of FeO^T (8–∼10 wt%), MgO (4.8–6.2 wt%), CaO (8.5–9.5 wt%), Al_2O_3 (∼16–17 wt%) and TiO_2 (1.1–1.8 wt%). On Hrarker diagrams a compositional gap in SiO_2 contents, in felsic rocks and MMEs could be observe (Fig. 20.5). In enclaves Zr, Ba and Rb increase while Cr, Ni and Sr decrease with increasing SiO_2 (Table 20.1 and Fig. 20.5). The geochemical features combined with the high volume of the granitoid rocks are inconsistent with an origin via differentiation of

Table 20.1 Chemical compositions of rocks from the Parishan Granodiorite Pluton (Granodiorite-granite unit and MMEs), loss of ignition not determined

Rocks	Granodiorite-granite unit						MMEs			
Samples	Pmj-12	Pmj-2	Pmbj-2	Pmi1b-1	Pt-10	Pt-11	Pzre-1	Pzre-2	Pmi-14-3	Pshv-14
SiO_2 (wt%)	67.21	67.99	71.49	63.76	64.27	67.19	49.16	52.04	53.05	52.5
TiO_2	0.37	0.3	0.33	0.61	1.13	0.6	1.85	1.24	1.14	1.52
Al_2O_3	15.17	15.4	12.81	16.36	15.68	15.59	16.13	16.94	15.98	17.14
FeO^T	3.03	2.84	3.28	5.02	4.84	1.92	9.86	7.98	8.13	8.29
MnO	0.06	0.06	0.05	0.11	0.1	0.03	0.18	0.15	0.18	0.16
MgO	0.89	0.31	0.26	0.75	1.45	0.89	6.06	6.16	5.04	4.75
CaO	1.94	1.25	1.24	3.06	3.24	2.39	9.55	8.95	8.55	9.46
Na_2O	4.15	4.49	3.7	4.9	5.05	4.7	3.38	3.8	3.94	3.69
K_2O	4.89	5.28	6.23	3.02	2.94	5.01	1.11	1.2	2.29	1.16
P_2O_5	0.07	0.49	0.06	0.16	0.32	0.1	0.29	0.23	0.16	0.24
Total	97.77	97.95	99.45	97.76	99.02	98.41	97.56	98.69	98.42	98.9
Ba (ppm)	840.6	975.6	786	620.6	690	872.4	193.9	163.5	380.2	194.9
Rb	161.9	166.4	139	89.7	82.8	86.1	41.8	67	133	38.9
Sr	127.1	109.2	81	280	287.8	169.8	306.3	310.1	264.8	339.8
Ga	24.3	27.6	0	29.3	27.6	23.1	24.2	23.3	25.8	22.2
Nb	30.3	43.2	15	39.9	39.8	32	15.4	20.3	18.7	13.4
Hf	8.33	8.13	0	13.1	6.78	–	–	–	4.76	3.33
Zr	432.1	436.7	223	819.9	378.3	1,002.7	203.7	185.1	258.3	162.9
Y	47.2	58.7	25	50.2	53.8	45	41.3	41.7	43.4	35.4
Th	20	11.4	10	15.8	10.7	19.9	1.8	4.1	6	4.7
U	5.5	3	6	2.6	3.7	5.9	2.1	0.7	3	1.5
Cr	30.5	5.1	0	4.4	3.9	13.9	197.5	209.7	193.8	173.3
Ni	15	1.1	8	4	3.1	7.7	73.2	54.7	55.5	20
Sc	13.3	10.3	0	17.9	18.9	8.9	48.6	42.5	46	46.5
V	33	11.2	28	37.7	115.6	60	368.4	272.6	243.2	329.2
Cu	4.6	2.4	1	4.1	8.4	6.3	59.2	27.9	24.7	15.8
Pb	14.3	17.5	18	13.4	6.8	3.9	4	3.9	6.1	3.7
Zn	53.5	58.7	48	90.1	63.9	26.7	113	77.7	123.6	65.6

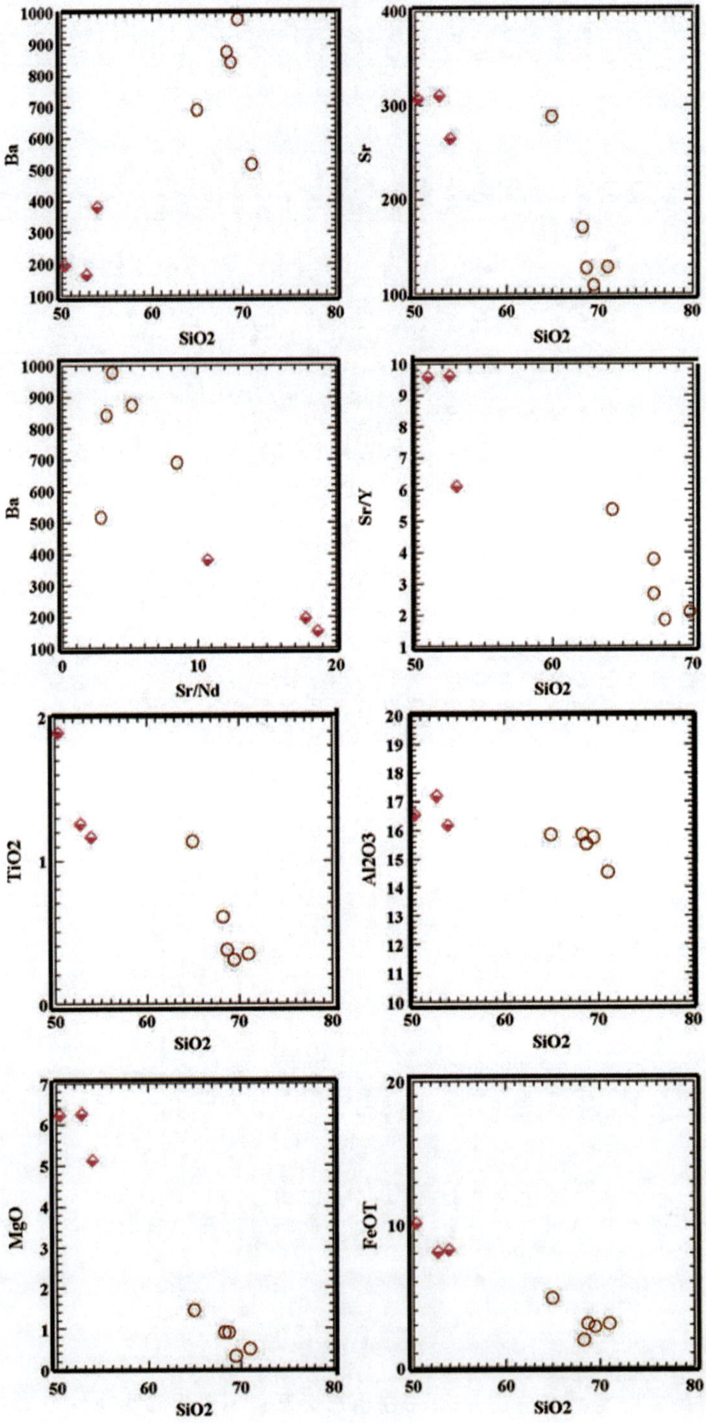

Fig. 20.5 (continued)

mantle-derived basaltic parent magma and assimilation. The granodioritic and the granitic rocks show moderate values of molar $Al_2O_3/(MgO + FeO^T)$ and molar $CaO/(MgO + FeO^T)$ suggesting an origin involving dehydration melting of a metagreywacke source (Torkian et al., 2008)

Discussion

Origin of MMEs

There are two general models that might explain the presence of the mafic microgranular enclaves that are widespread throughout the high levels of granitoid plutons such as the Parishan granodiorite-granite intrusive.

Firstly, the MMEs might represent the remnants of an igneous protolith that was partially melted to yield the host granitic magma (e.g. Chappell and White, 1992). Many of the features of the studied enclaves clearly point to an igneous origin; such as fine-grained margins, microgranular and microporphyritic textures, occasional presence of textural and compositional disequilibrium in plagioclase crystals. The enclaves have higher Ni (55–73 ppm) and Cr (194–210 ppm) concentrations than those of the host granodiorite-granite rock (Ni = 1–15 ppm and Cr = 0–30 ppm). It should be noted that Harkers' pattern can hardly be explained by the restite model proposed for the origin of microgranular enclaves. It is therefore unlikely that the MMEs are restitic in origin.

Secondly, the MMEs and granitoids could have resulted from the mingling of at least two major magma pulses, possibly derived from different sources and/or depths. In such a scenario, the MMEs would represent globules of mafic magma (e.g. Didier and Barbarin, 1991).

Evidence for Magma Mingling in Parishan Pluton

Here the magma mingling process is inferred though field, textural and geochemical evidences.

There is a close spatial and temporal relationship between granitoids and the MMEs as indicated by multi-pulse structures and fine-grained nature of enclaves and less commonly seen crenulated margins. These features are generally indicative of magma mingling (Wiebe, 1991). The sharp contact boundaries (with chilled margins) between the enclaves and the host rock, in combination with a lack of

Fig. 20.5 Harker variation diagrams for major and trace elements in the granodiorite-granite unit and its enclaves. Note that there is a compositional gap between the host rock (granodiorite-granitic unit) and the mafic enclaves implying magma mingling. *Half-full diamond*: enclave; *Circle*: granodiorite-granite compositional or textural zoning parallel to the enclave margins, suggests that the MMEs were not significantly affected by the enclosing host rock; consequently, the strongly chilled margin of mafic enclaves within the granodiorite-granite can be interpreted as a primary criterion for magma mingling (Wiebe, 1991). MMEs formed by crystallization due to the rapid cooling of a mafic magma within a felsic magma and they have little opportunity for chemical interaction between the magmas. As noted by Bea et al. (1977), the preferred interpretation of the origin of MMEs is that they represent fragments of rapidly cooled mafic magma (dykes) that was disrupted and then entrained by the fractionating host magma. Similar process has been invoked for the present study.

Among the microtextural evidence for magma mingling (and mixing), the chemical composition of plagioclase is of particular interest. In the studied MMEs, microprobe analyses of plagioclase crystals within the mafic microgranular enclave reveal that the cores have higher anorthite content than the rims zones (Figs. 20.3 and 20.4). Roberts et al. (2000) suggested that such a trend reflects magma mingling during the evolution of the magmas. The preservation of disequilibrium textures is therefore due to rapid crystallization and can be attributed to magma mingling. It is likely that the relatively small size of the pluton and in the high level intrusion were important factors in terms of its rapid cooling (Elburg, 1996), and it is likely that the residence time of the enclaves (as liquid/melt) in the felsic magma was very short (Tindle, 1991), thereby preventing significant mixing.

Further, major and trace elements trends on Harker variations diagrams do not suggest the mixing of two end members (the mafic enclaves and their felsic host rocks). However, as noted by Clemens (1989) and Wall et al. (1987), however Harker variation diagrams are insufficient tools for the diagnosis of such processes, because linear patterns are equally consistent with magma mixing and closed-system fractional crystallization. As an additional evidence for magma mingling SiO_2 contents is considered here. Considerable difference in the of SiO_2 contents of the granodiriti-granitic host (∼64–67 wt%) and MMEs (∼49–53 wt%) indicates that the felsic and mafic magmas did not evolve to intermediate rocks via a complete mixing process.

In summary, despite the bimodal distribution of plagioclase compositions in the MMEs and the host granodioritic-granitic rock, no convincing evidence exists to suggest that the mafic and felsic magmas mixed homogeneously. The relatively large contrasts in temperature, density, and viscosity between the two magmas would have been a major barrier to complete mixing. Therefore, a plausible model for the generation of MMEs in the granodioritic-granitic rocks of this pluton involves the mingling of the mafic and felsic magmas that played an important role in the occurrence of the enclaves in the host rocks. When two contrasting magmas encounter each other pillows typically develop in a matrix of granodioritic-granitic magma. A similar model has been proposed in many previous studies for comparable rocks in others parts of the world (e.g. Furman and Spera, 1985; Barbarin, 1988; Elburg, 1996; among others).

Conclusions

The Qorveh Granitoid Complex displays field relations and mineralogical and geochemical characteristics typical of calc-alkaline granitoids of the Sanandaj-Sirjan Zone, which is considered to have developed during subduction of the Arabian plate beneath continental crust of the Iranian plate (e.g. Valizadeh and Cantagrel, 1975; Berberian and Berberian, 1981). The MMEs in the felsic host rocks do not show textural or mineralogical evidence of an origin as a refractory residue related to the generation of acidic magma; instead, the coexistence of mafic dykes and felsic magma can be considered as evidence of magma mingling. The granodioritic-granitic magma is possibly derived from an intra-crustal source. The injection and rapid cooling of successive pulses of mafic magma (as dyke) into upward-moving granodioritic-granitic magma resulted in the formation of the observed MMEs. Felsic magma is dominant over mafic magma in the studied area, thereby providing ideal conditions for magma mingling. The mingling involved quenching of the mafic magma in the felsic magma, with consequent "pillowing" of the mafic magma. The compositional and textural disequilibrium shown by large plagioclase crystals is consistent with rapid crystallization resulting from magma mingling rather than magma mixing.

Acknowledgments I express my appreciation to B. C. Prabhakar for constructive comments and helpful suggestions. I also thank G. Morgan, University of Oklahoma in Norman (USA), and to R. Conery, GeoAnalytical Laboratory Washington State University (USA). I express my appreciation to Rajesh K. Srivastava for encouraging me to contribute to the IDC-6 volume.

References

Alavi M (1994) Tectonics of the Zagros Orogenic belt of Iran: New data and interpretations. Tectonophysics 229: 211–238

Barbarin B (1988a) Field evidence for successive mixing and mingling between the Piolard Diorite and the Saint-Julien-la-Vetre Monzogranite (Nord-Forez, Massif Central, France). Can J Earth Sci 25: 49–59

Barbarin B (1991) Enclaves of the Mesozoic calc-alkaline granitoids of the Sierra Nevada Batholiths, California. In: Didier J, Barbarin B (eds) Enclaves and Granite Petrology. Elsevier, Amsterdam: 135–153

Barbarin B, Didier J (1992) Genesis and evolution of mafic microgranular enclaves through various types of interaction between coexisting felsic and mafic magmas. Trans Roy Soc Edinburgh Earth Sci 83: 145–153

Bea F, Fershtater G, Montero P, Smirnov V, Zin'kova E (1997) Generation and evolution of subduction-related batholiths from the central Urals: Constraints on the P-T history of the Uralian orogen. Tectonophysics 276: 103–116

Bellon H, Broud J (1975) Donnes nouvelles sur le domaine metamorphjque du Zagros (zone de Sanandaj-Sirjan) au niveau de Kermanshah-Hamadan (Iran), Nature, age et interpretation des series metamorphiques et des intrusions, evolution structural. Fac Sci Orsay Paris 14

Berberian F, Berberian M (1981) Tectono-plutonic episodes in Iran. Geol Survey Iran 52: 565–592

Berberian M, King GCP (1981) Towards a Paleogeography and Tectonic Evolution of Iran. Can J Earth Sci 18: 210–265
Castro A, Moreno-Ventas I, De La Rosa JD (1991) H-type (hybrid) granitoids: A proposed revision of the granite-type classification and nomenclature. Earth Sci Rev 31: 237–253
Chappell BW, White AJR (1992) I- and S-type granites in the Lachlan Fold Belt. Trans Roy Soc Edinburgh Earth Sci 83: 1–26
Clemens JD (1989) The importance of residual source material (restite) in granite petrogenesis: A comment. J Petrol 30: 1313–1316
Dahlquist JA (2002) Mafic microgranular enclaves: Early segregation from metaluminous magma (Sierra de Chepes), Pampean Ranges, NW Argentina. J South Am Earth Sci 15: 643–655
Didier J (1973) Granites and Their Enclaves. Elsevier, Amsterdam
Didier J, Barbarin B (1991) The different types of enclaves in granites – nomenclature. In: Didier J, Barbarin B (eds) Enclaves and Granite Petrology. Elsevier, Amsterdam: 19–24
Dorais MJ, Whitney JA, Roden MF (1990) The origin of mafic enclaves from the Dinkey Creek Pluton, Central Sierra Nevada Batholith. J Petrol 31: 53–81
Elburg MA (1996) Evidence of isotopic equilibration between microgranitoid enclaves and host granodiorite, Warburton Granodiorite, Lachlan Fold Belt, Australia. Lithos 38: 1–22
Fershtater GB, Borodina NS (1991) Enclaves in the Hercynian granitoids of the Urals Mountains, USSR. In: Didier J, Barbarin B (eds) Enclaves and Granite Petrology. Elsevier, Amsterdam: 83–94
Furman T, Spera FJ (1985) Co-mingling of acid and basic magma with implications for the origin of mafic I-type xenoliths, field and petrochemical relations of an usual dike complex at Eangle Peak Laek, Sequoia National Park, California, USA. J Volcanol Geotherm Res 24: 151–178
Hosseini M (1997) Geological Quadrangle Map of Qorveh, No. 5, Sanandaj (Qorveh), 1:100000. Geol Surv Iran.
Orsini JB, Cocirta C, Zorpi MJ (1991) Genesis of mafic microgranular enclaves through differentiation of basic magmas, mingling and chemical exchanges with their host granitoid magmas. In: Didier J, Barbarin B (eds) Enclaves and Granite Petrology. Elsevier, Amsterdam: 445–464
Pitcher WS (1991) Synplutonic dykes and mafic enclaves. In: Didier J, Barbarin B (eds) Enclaves and Granite Petrology. Elsevier, Amsterdam: 383–392
Pouchou JL, Pichoir F (1985) "PAP" correction procedure for improved quantitative microanalysis. In: Armstrong JT (ed) Microbeam Analysis. San Francisco Press, San Francisco, CA: 104–106
Richard LR (1995) MinPet: Mineralogical and petrological data processing system, version 2.02. MinPet Geological Software, Québec
Roberts MP, Pin C, Clemens JD, Paquette J (2000) Petrogenesis of mafic to felsic plutonic rock associations: The calc-alkaline Querigut Complex, French Pyrenees. J Petrol 41: 809–844
Sial AN, Ferreira VP, Fallick AE (1998) Amphibole-rich clots in calc-alkalic granitoids in the Borborema province, northeastern Brazil. J South Am Earth Sci 11: 457–471
Sparks RSJ, Marshall LA (1986) Thermal and mechanical constraints on mixing between mafic and silisic magmas. J Volcanol Geotherm Res 29: 99–124
Stephens WE (2001) Polycrystalline amphibole aggregates (clots) in granites as potential I-type restite: An ion microprobe study of rare-earth distributions. Aust J Earth Sci 48: 591–601
Tindle AG (1991) Treace elements behavior in microgranular enclaves from granitic rocks. In: Didier J, Barbarin B (eds) Developments in Petrology. Enclaves and Granite Petrology. Elsevier, Amsterdam: 313–332
Torkian A (2008) Magmatism investigation of the South-Qorveh Granodiorite intrusive body (Kurdistan), PhD Thesis, Univ. of Isfahan, Iran
Torkian A, Khalili M, Sepahi AA (2008) Petrology and geochemistry of the I-type calc-alkaline Qorveh Granitoid Complex, Sanandaj-Sirjan Zone, western Iran. N Jb Miner Abh 185/2: 131–142

Valizadeh MV, Cantagrel JM (1975) Premières données radiométriques (K-Ar et Rb-Sr) Sur Les micas du Complexe magmaitque du Mont Alvand, près Hamadan (Iran Occidental), Série D. C.R.A. S, Paris: 1083–1086

Vernon RH (1984) Micro granitoid enclaves in granites globules of hybrid magma quenched in a plutonic environment. Nature 309: 438–439

Wall VJ, Clemens JD, Clarke DB (1987) Models for granitoid evolution and source compositions. J Geol 95: 731–749

Wiebe RA (1991) Commingling of contrasted magmas in the plutonic environment: Examples from the Nain Anorthositic Complex. J Geol 88: 197–209

Chapter 21
Petrology and Geochemistry of Cretaceous Mafic and Silicic Dykes and Spatially Associated Lavas in Central-Eastern Coastal Madagascar

C. Cucciniello, J. Conrad, C. Grifa, L. Melluso, M. Mercurio, V. Morra, R.D. Tucker, and M. Vincent

Introduction and Geological Remarks

Widespread magmatism occurred throughout Madagascar in Late Cretaceous time. It is generally attributed to the arrival of Madagascar over the Marion plume head during separation of Greater India from Africa (Mahoney et al., 1991; Storey et al., 1995, 1997). Evidence of this magmatism is also present in southwestern India, mainly in the form of mafic dyke swarm and silicic lavas (Kumar et al., 2001; Torsvik et al., 2000; Pande et al., 2001; Melluso et al., 2009). Recent age determinations suggest that the main volume of Madagascan igneous rocks was erupted between 92 and 84 Ma (Storey et al., 1995; Torsvik et al., 1998; Melluso et al., 2005; Cucciniello et al., 2010) and that the magmatism possibly migrated from north to south (Storey et al., 1995).

The Madagascan flood basalt province (LIP) consists of lava sequences, swarms of dykes and sills, and intrusive ring complexes. Remnants of the igneous province are well-exposed and preserved in the NW part of the island (Mahajanga and Morondava basins), in the southern Madagascar (e.g. Volcan de l'Androy complex

L. Melluso (✉)
Dipartimento di Scienze della Terra, Università di Napoli Federico II, Napoli, Italy
e-mail: melluso@unina.it

C. Cucciniello and V. Morra
Dipartimento di Scienze della Terra, Università di Napoli Federico II, Napoli, Italy

J. Conrad
USGS, Menlo Park, CA, USA

C. Grifa and M. Mercurio
Dipartimento di Studi Geologici ed Ambientali, Università del Sannio, Benevento, Italy

R.D. Tucker
USGS, Reston, VA, USA

M. Vincent
CNRIT, Antananarivo, Madagascar

R.K. Srivastava (ed.), *Dyke Swarms: Keys for Geodynamic Interpretation*,
DOI 10.1007/978-3-642-12496-9_21, © Springer-Verlag Berlin Heidelberg 2011

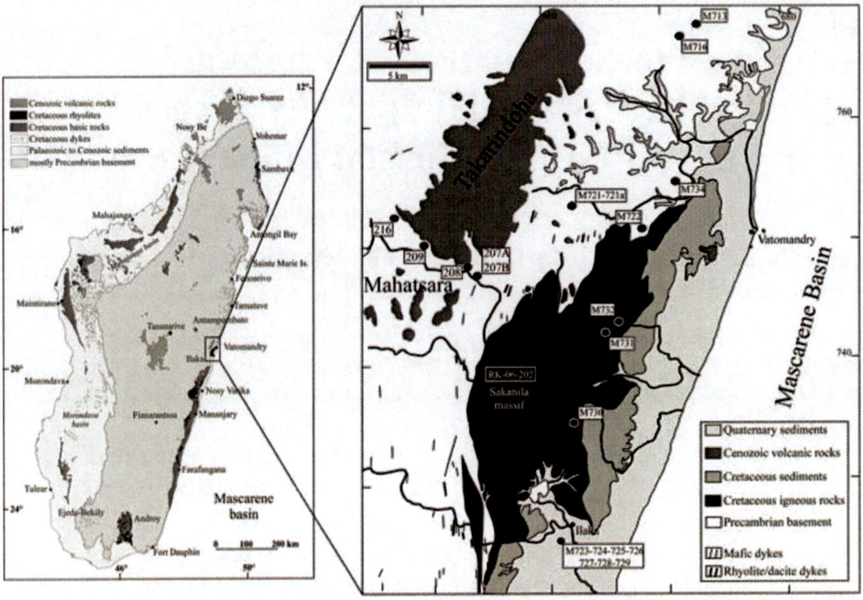

Fig. 21.1 Simplified geological map of central-eastern coast of Madagascar showing the location Cretaceous igneous rocks in this region

and Ejeda-Bekily dyke swarm) and along the eastern coastal area (Sambava, Tamatave and Mananjary districts; Fig. 21.1). The Cretaceous volcanic rocks in the eastern sector overlie Late Archean to Proterozoic basement dominated by granulite and amphibolite facies rocks, migmatites and granites (Windley et al., 1994; de Wit, 2003 and references therein) whereas the lava flows of the western sector cover Paleozoic to Mesozoic sedimentary rocks deposited in the Karoo-age related sedimentary basins of Mahajanga and Morondava (Besairie and Collignon, 1972). The Madagascan LIP is dominated by mafic rocks; silicic rocks represent only a small volume of the province, overall (Melluso et al., 2001, 2005; Mahoney et al., 2008). The mafic rocks are mostly Mg-poor basalts and basaltic andesites with MgO rarely exceeding 8 wt%. The trace element and isotopic composition of mafic products divide them into several groups (Melluso et al., 1997, 2001, 2002, 2003, 2005; Storey et al., 1997; Mahoney et al., 1991, 2008).

The origin of silicic rocks associated with flood basalt lavas (e.g. Paraná-Etendeka, Deccan and Karoo LIPs) is a matter of considerable interest. Two main models have been proposed to explain the generation of silicic in continental setting: (1) prolonged fractional crystallization of a parental basaltic magma (usually with crustal assimilation) (e.g. Sheth and Melluso, 2008; Melluso et al., 2008; Lustrino et al., 2010), and (2) melting of the Precambrian continental crust or its underplated basic igneous rocks (or differentiates) (e.g. Garland et al., 1995; Kirstein et al., 2000).

This paper focuses on a suite of mafic and silicic igneous rocks cropping out in the central-eastern coast (Vatomandry and Mahatsara districts; Fig. 21.1). Here we present new mineral and whole rock chemical data on a dyke swarm and a few spatially associated lavas, in order to assess petrogenetic processes that produced the rhyolitic magmas and investigate the transition from basaltic to silicic magmatism in eastern Madagascar. We also present a ^{40}Ar-^{39}Ar date for rhyolite from the Sakanila massif, southwest of Vatomandry, in order to evaluate the age relationship between basalt and rhyolite in this silicic massif of the east coast.

Sampling and Analytical Techniques

Twenty-two rock samples were analysed in this study (Table 21.1). The mafic dykes are intrusive into the Precambrian crust, and the silicic rocks crop out as minor dykes and extrusive flows erupted on the eroded Precambrian surface. These are overlain by olivine melilititic lava flows of Miocene age (Melluso et al., 2007 and in prep.) and by Quaternary alluvial and beach deposits. Our samples are basaltic dykes and rhyolitic lavas from the Vatomandry district (Fig. 21.1). Stratigraphic relationships between mafic and rhyolitic rocks are unclear, because of heavy vegetation and laterization, but, based on regional considerations, the silicic rocks appear to overlie the mafic dykes. One sample (RK-06-202) was collected from a fresh road cut on Sheet T48 ($X = 65{,}794$, $Y = 73{,}9578$) near the centre of the Sakanila massif (Fig. 21.1). The rock is a feldspar-phyric rhyolite with minor quartz phenocrysts. Banding is defined by cm-thick color bands in the pinkish-white groundmass. The majority of exposed rocks in the Sakanila massif are basalts; the silicic rocks (dacite/rhyolite) tend to weather preferentially and are consequently very poorly exposed. The magnetic imaging suggests that there is interlayering of lavas with different magnetic signatures suggesting that this Cretaceous field comprises interlayered basalt and rhyolite.

Rock chips of the samples, cleaned of visible alteration products, were crushed in low-blank agate mortars. Major and minor oxides and some trace elements (Sc, V, Cr, Ni, Rb, Sr, Y, Zr, Nb and Ba) were analyzed with Axios Panalytical X-ray fluorescence (XRF) instrument at Napoli on pressed powder pellets, following the methods described by Ronga et al. (2010). Relative precision was generally better than 1–2% for the major oxides and better than 5–10% for trace elements. L.O.I. (weight loss on ignition) was measured gravimetrically after igniting the samples at 1,000°C. Other trace elements were analysed by inductively coupled plasma mass spectrometry (ICP-MS) at ACTLABS, Ancaster, Ontario. Mineral compositions were obtained at CISAG (Centro Interdipartimentale Strumentazioni per Analisi Geomineralogiche), University of Napoli utilizing an energy-dispersive microprobe system linked to a JEOL JSM5310 electronic microscope. Detailed information can be found in Melluso et al. (2009, 2010).

Ar-Ar age determination Plagioclase feldspar was prepared as a high-purity (>95%) mineral separate using standard purification techniques such as density separation and magnetic susceptibility. The mineral separate was then placed in an

Table 21.1 Major (wt%) and trace element (ppm) analyses of the igneous rocks of Vatomandry and Mahatsara

Sample	Type	SiO_2	TiO_2	Al_2O_3	Fe_2O_{3t}	MnO	MgO	CaO	Na_2O	K_2O	P_2O_5	LOI	Mg#	Cr	Ni	Sc	V	Rb	Sr	Y	Zr	Nb	Ba
M713	tho bas	49.24	3.75	12.12	17.03	0.24	5.26	9.05	2.41	0.56	0.35	0.36	41	123	54	37	–	8	210	44	235	15	–
M716	tho bas	50.81	3.48	12.82	13.76	0.17	6.48	8.01	2.97	1.10	0.41	2.11	51	43	53	26	–	25	437	29	244	17	88
M721	tho bas	48.55	2.56	14.10	14.83	0.20	6.99	9.90	2.33	0.31	0.23	0.47	51	322	101	41	–	2	207	38	179	10	–
M723	tho bas	48.70	1.73	14.89	12.94	0.19	7.74	11.33	2.15	0.19	0.15	0.35	57	240	82	40	–		187	20	105	5	–
M724	tho bas	49.54	4.14	11.70	17.81	0.23	4.75	8.33	2.30	0.68	0.53	0.99	37	47	46	36	–	14	272	65	404	31	42
M725	tho bas	49.57	4.08	11.69	17.58	0.22	4.65	8.49	2.31	0.91	0.51	0.83	37	39	37	36	–	15	251	58	363	29	66
M729	tho bas	49.14	4.20	11.49	17.97	0.24	4.73	8.68	2.30	0.75	0.51	0.90	37	38	36	34	–	12	245	58	358	27	38
M734	tho bas	49.29	3.39	12.68	16.25	0.24	5.76	9.31	2.25	0.50	0.33	1.00	44	136	54	36	–	4	223	38	227	15	–
RT-06-207B	tho bas	47.77	2.10	14.74	13.88	0.18	7.99	10.76	2.08	0.28	0.22	0.90	56	341	127	37	230	–	196	24	129	6	–
RT-06-208	tho bas	51.64	2.96	13.53	12.97	0.16	5.73	7.92	3.01	1.14	0.94	1.68	50	8	38	20	267	18	687	36	418	21	163
RT-06-216A	tho bas	48.89	2.75	15.16	12.91	0.17	6.37	10.42	2.47	0.50	0.36	0.61	53	276	105	29	198	3	239	33	191	11	30
RT-06-207A	tho bas	48.54	2.32	16.39	11.53	0.16	7.14	10.90	2.37	0.40	0.25	1.20	58	280	128	29	174	1	260	24	156	7	27
RT-06-209	tran bas	46.02	2.41	14.11	16.57	0.23	7.26	9.81	2.62	0.67	0.32	3.51	50	61	57	28	516	7	314	16	123	11	158
M726	dac	68.79	1.35	13.46	5.39	0.15	2.25	1.03	5.34	1.82	0.41	1.64	48	–	–	17	–	32	183	75	515	53	395
M722	dac	66.78	1.10	11.55	7.04	0.12	3.26	3.74	3.46	2.80	0.13	1.33	51	52	20	12	–	77	208	41	321	27	428
M727	rhy	75.56	0.65	12.20	2.23	0.07	0.94	0.30	3.91	4.03	0.12	1.23	49	–	–	8	–	90	123	56	532	52	756
M728	rhy	73.13	0.67	12.34	2.98	0.15	1.10	1.57	4.37	3.55	0.14	1.43	45	–	–	12	–	76	144	55	532	56	615
M730	rhy	74.14	0.72	11.69	2.88	0.14	1.21	1.41	4.21	3.49	0.09	1.53	49	–	–	8	–	76	121	69	623	62	659
M731	rhy	75.25	0.66	11.45	2.70	0.11	0.75	1.19	4.16	3.62	0.10	0.53	38	–	–	7	–	91	116	70	638	64	732
M732	rhy	77.00	0.61	11.07	2.10	0.06	0.17	0.58	3.21	5.12	0.06	1.37	15	–	–	10	–	110	67	64	625	65	684
M721A	rhy	73.02	0.87	12.05	3.74	0.15	1.02	1.28	4.02	3.69	0.16	1.56	38	–	–	8	–	104	346	62	577	56	711

tho bas tholeiitic, *tran bas* transitional basalt, *dac* dacite, *rhy* rhyolite. The major oxide analyses are recalculated to 100 wt% LOI-free; Mg# = molar Mg*100/(Mg+Fe^{2+}).

ultrasonic bath to remove loose grains and fine-grained contaminants (mostly clays), then rinsed in alcohol and deionized water. For final purification plagioclase was hand-picked under a binocular microscope to >99% purity.

Ten (10) mg of sample RK-06-202, together with the TCR-2 sanidine fluence monitor (age is 27.87 Ma), was packaged in Cu foil and sealed in quartz vials and irradiated for 20 h in the U.S. Geological Survey TRIGA reactor in Denver, Colorado. Laboratory standard K-glass and CaF_2 were included in the 20-h irradiation to measure reactor-produced Ca- and K-derived isotopes for samples not shielded with Cd foil (corrections for Cd-shielded samples are already well known by the Menlo Park laboratory).

$^{40}Ar/^{39}Ar$ isotope ratio measurements were conducted at the U.S. Geological Survey, Menlo Park, Geochronology Laboratory using a custom resistance furnace with an online MAP 216 mass spectrometer. Plagioclase was heated by resistance furnace and analyzed by the $^{40}Ar/^{39}Ar$ incremental-heating technique (Table 21.2). The copper foil-wrapped sample was manipulated through a glass sample manifold and dropped into the furnace for analysis using a magnetic slug. In this technique, the isotopic composition of the fractional gas in temperature increments is measured, and the data are presented as either "age spectra" or "isochron" diagrams (Fig. 21.2). Furnace temperatures were monitored by use of an optical fiber thermometer, and about 10–14 temperature increments per sample were used. In the "age spectra" (or plateau) diagram (Fig. 21.2), the apparent age of each temperature increment is plotted as a fraction of the cumulative ^{39}Ar released. A sequence of consecutive, concordant "step ages", comprising more than 50% of the released ^{39}Ar, is called a plateau age. In the "isochron" diagram ($^{40}Ar/^{36}Ar$ vs. $^{39}Ar/^{36}Ar$; Fig. 21.2), the lower intercept is a direct measurement of the initial ratio ($^{40}Ar/^{36}Ar$), and the slope of the regression is proportional to the sample's age. Age spectra were interpreted using commonly accepted criteria for the determination of incremental heating ages, such as that described in McDougall and Harrison (1999). In general, the plateau age is preferred over the isochron age unless (*i*) the age spectra is significantly disturbed (i.e. U-shape or "climbing" spectra), or poorly-defined (MSWD greater than 2.0), or (*ii*) the regression calculation indicates an initial ratio significantly different (at 2-sigma) from the atmospheric $^{40}Ar/^{36}Ar$ value of 295.5. In these cases, the isochron age may be preferred. As a general rule, the cited age is interpreted as the time elapsed since quenching or eruption of the rhyolitic magma.

Eruption Age of the Sakanila Massif

Sample RK-06-202. There is significant disturbance apparent in the $^{40}Ar/^{39}Ar$ release spectrum for this sample (Fig. 21.2). No contiguous portion including at least 50% of the total ^{39}Ar release with concordant ages can be defined (Table 21.2). The shape of the spectrum, however, has a slight U- or "saddle-" shape that is associated with the presence of excess ^{40}Ar (McDougall and Harrison, 1999). Isochron calculations are consistent with this idea, as the 700–1,090°C steps, which comprise

Table 21.2 Results of $^{40}Ar/^{39}Ar$ incremental heating analyses of volcanic rocks from Madagascar

Temp (°C)	Age ± 1σ (Ma)	$\%^{40}Ar^{*}$	K/Ca	K/Cl	moles $^{40}Ar^{*}$	$\sum^{39}Ar$	^{40}Ar	^{39}Ar	^{38}Ar	^{37}Ar	^{36}Ar
RK-06-202					*Plagioclase*			J = 0.0039770 ± 0.0000080			
600	84.234 ± 0.434	65.05	0.35	159	8.648e-14	0.14	5.404695 ± 0.010849	0.292809 ± 0.000595	0.012957 ± 0.000147	0.438840 ± 0.001692	0.006505 ± 0.000032
650	84.228 ± 0.450	81.36	0.68	214	5.802e-14	0.23	2.899038 ± 0.005832	0.196358 ± 0.000399	0.006853 ± 0.000158	0.150819 ± 0.001120	0.001865 ± 0.000033
700	87.526 ± 0.389	79.47	1.48	189	4.823e-14	0.30	2.467020 ± 0.004968	0.156885 ± 0.000319	0.005939 ± 0.000089	0.055521 ± 0.001346	0.001725 ± 0.000018
750	86.793 ± 0.467	78.69	1.68	150	3.182e-14	0.35	1.643728 ± 0.003321	0.104398 ± 0.000212	0.004576 ± 0.000056	0.032584 ± 0.001036	0.001191 ± 0.000018
800	85.920 ± 0.491	76.34	1.57	136	2.799e-14	0.39	1.490695 ± 0.003015	0.092805 ± 0.000189	0.004361 ± 0.000047	0.031083 ± 0.000859	0.001199 ± 0.000017
850	86.305 ± 0.420	82.60	2.01	157	3.349e-14	0.44	1.648013 ± 0.003330	0.110497 ± 0.000225	0.004663 ± 0.000044	0.028864 ± 0.000523	0.000975 ± 0.000016
900	86.907 ± 0.376	85.76	2.45	229	4.397e-14	0.51	2.084292 ± 0.004210	0.144074 ± 0.000293	0.004777 ± 0.000052	0.030797 ± 0.000773	0.001008 ± 0.000017
950	87.415 ± 0.397	80.27	2.14	207	4.794e-14	0.58	2.427572 ± 0.004896	0.156129 ± 0.000317	0.005586 ± 0.000082	0.038251 ± 0.000774	0.001626 ± 0.000019
1,000	88.239 ± 0.621	56.12	1.49	5	3.852e-14	0.64	2.790206 ± 0.005621	0.124262 ± 0.000253	0.115831 ± 0.000301	0.043867 ± 0.000694	0.004152 ± 0.000026
1,050	90.213 ± 0.863	36.83	0.87	91	7.445e-14	0.75	8.217836 ± 0.016476	0.234829 ± 0.000478	0.017503 ± 0.000104	0.142106 ± 0.001120	0.017600 ± 0.000050
1,090	94.019 ± 1.474	23.31	0.65	198	1.282e-13	0.93	10.087835 ± 0.020216	0.174919 ± 0.000357	0.010977 ± 0.000088	0.142033 ± 0.000601	0.026214 ± 0.000063
1,130	103.417 ± 1.980	20.93	0.47	140	3.803e-14	0.98	3.332564 ± 0.006872	0.047058 ± 0.000100	0.003724 ± 0.000036	0.053006 ± 0.000567	0.008930 ± 0.000026
1,160	108.119 ± 3.284	20.11	0.35	111	1.226e-14	0.99	1.118148 ± 0.002444	0.014491 ± 0.000045	0.001313 ± 0.000019	0.021854 ± 0.000268	0.003029 ± 0.000016
1,190	109.209 ± 7.596	19.23	0.22	104	3.472e-15	1.00	0.331118 ± 0.000872	0.004065 ± 0.000029	0.000389 ± 0.000017	0.009622 ± 0.000333	0.000908 ± 0.000011
1,250	119.502 ± 7.068	22.77	0.23	109	2.566e-15	1.00	0.206669 ± 0.000451	0.002737 ± 0.000016	0.000245 ± 0.000013	0.006310 ± 0.000228	0.000542 ± 0.000008
1,400	145.866 ± 19.775	29.03	0.06	29	8.461e-16	1.00	0.053449 ± 0.000149	0.000737 ± 0.000012	0.000140 ± 0.000018	0.006303 ± 0.000250	0.000130 ± 0.000006
Total Gas Age = 88.1 ± 0.2 Ma								Isochron Age = 86.4 ± 0.3 Ma (700–1,090°C)			

Notes: $^{40}Ar^{*}$ is radiogenic argon, isotopes in volts (1.91e-14 mol/V), corrected for blank, background, discrimination, and decay. Plateau ages are inverse variance weighted means; uncertainty in the plateau ages is reported as the standard deviation about the weighted mean.

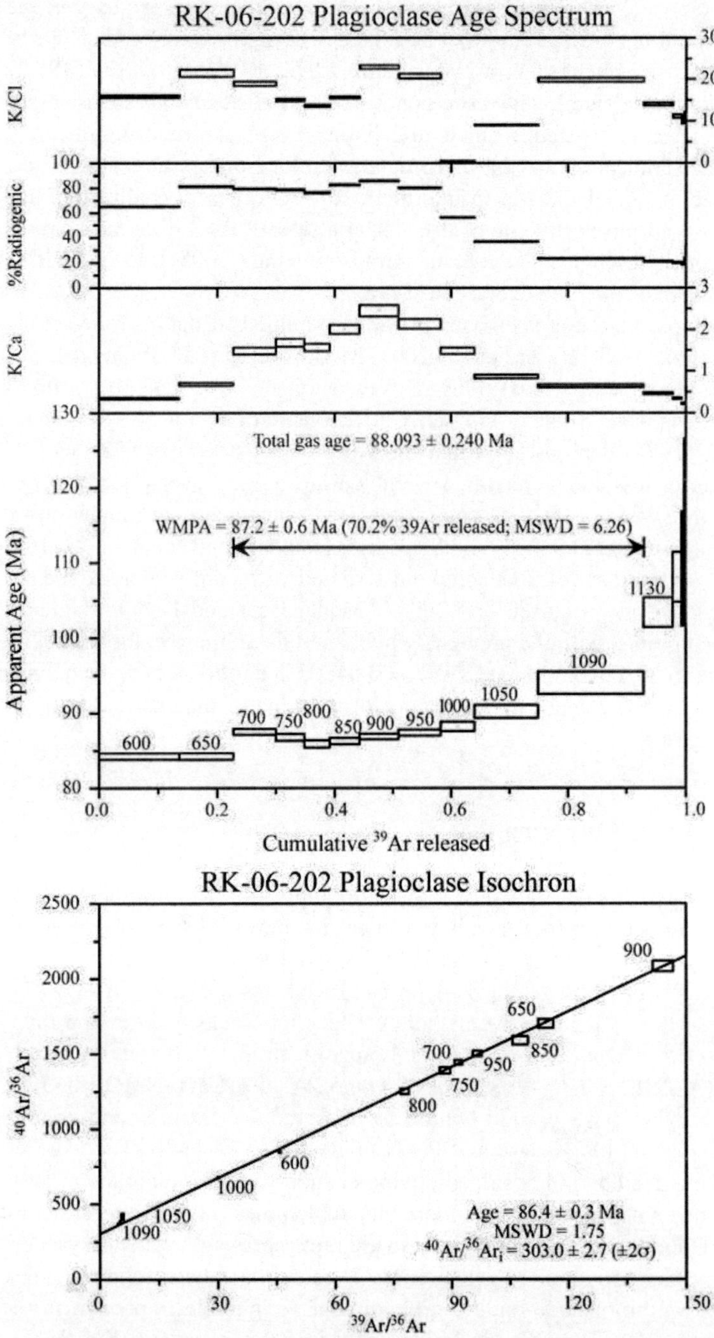

Fig. 21.2 $^{40}Ar/^{39}Ar$ age spectrum and isochron plots for plagioclase from Sakanila rhyolite. Plateau age is weighted mean of concordant, sequential step ages. Isochron age is calculated from the best fit slope of step compositions. All errors are $\pm 1\sigma$

about 70% of the total ^{39}Ar release, give a best fit, least squares regression line that indicates an initial ^{40}Ar/^{36}Ar ratio of 303.0 ± 2.7 (2-sigma error) and an age of 86.4 ± 0.3 Ma. This segment gives a plateau age of 87.2 ± 0.6 Ma but the MSWD value of 6.26 indicates significant discordance. The good fit of the isochron calculation (MSWD = 1.75) and a calculated initial ^{40}Ar/^{36}Ar ratio significantly higher than 295.5 indicates that the discordance results from excess ^{40}Ar. Thus, the isochron age of 86.4 ± 0.3 Ma in interpreted to reflect the crystallization age of the plagioclase porphyry rhyolite of the Sakanila massif. According to the most recent International Geological Time-Scale (Gradstein et al., 2005) this falls within the Coniacian Stage of the Late Cretaceous Period.

Our new isotopic date, together with previously published data (Storey et al., 1995; Torsvik et al., 1998; Kumar et al., 2001; Melluso et al., 2005), suggest that basalt and rhyolite of the east-coast massifs were emplaced over a short period of time within the Coniacian Stage of the Late Cretaceous Period.

Storey et al. (1995) demonstrated that the volcanic rocks of the rifted eastern margin are more-or-less coeval having a weighted mean age (inverse isochron) of 87.6 ± 1.5 Ma (MSWD = 1.09; 95% confidence). Subsequent age determinations in peninsula India (Kumar et al., 2001) and St. Mary's Islands (Pande et al., 2001) lend credence to the notion that volcanic activity associated with continental fragmentation was short-lived and occurred on both the Madagascar and Dharwar shields. Our new date, combined with the previously published data, improve the reliability of the mean age to 87.1 ± 0.3 Ma (MSWD = 0.54; 95% confidence) and establish a precise age of Late Cretaceous (Coniacian) for the time of magmatic activity in India and Madagascar.

Classification and Petrography

The Vatomandry and Mahatsara dykes range in composition from basalt to rhyolite. The distribution in the total alkali vs. silica diagram (T.A.S.; Le Bas et al., 1986; Fig. 21.3) is bimodal. On the basis of their CIPW norms, the basalts are quartz and olivine tholeiites. A single dyke from Mahatsara district (RT-06-209) is transitional basalt (<3% norm. hypersthene). The silicic rocks (dacites and rhyolites) are quartz normative (18.8–32.1%) and range from metaluminous to weakly peraluminous (A/CNK = 0.74–1.08; A/CNK = mol. $Al_2O_3/(CaO+Na_2O+K_2O)$).

The basaltic dykes have pseudo-ophitic to intergranular textures and contain olivine, clinopyroxene, plagioclase and Fe-Ti oxides (Fig. 21.4a–d). The Mg-rich dykes are characterized by iddingsitized olivine phenocrysts and microphenocrysts with rare chrome spinel inclusions. In more altered samples olivine is replaced by clay minerals. Clinopyroxene and plagioclase are rarely present as phenocrysts but represent a large proportion of the groundmass. Pigeonite has been found in a few tholeiitic dykes, within pseudo-ophitic groundmass. Amphibole is present in the transitional basalt RT-06-209 and is interstitial or rim of clinopyroxene (Fig. 21.4c).

The silicic rocks are often altered and sometimes show spherulites. Dacites are weakly to strongly porphyritic, with phenocrysts of plagioclase and alkali

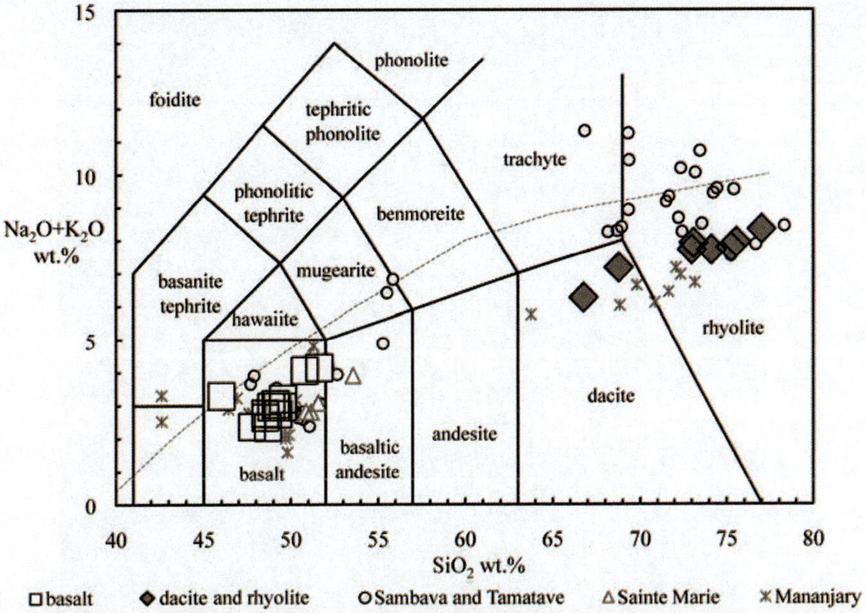

Fig. 21.3 Total alkalis vs. SiO_2 (T.A.S., wt%) classification diagram for the Vatomandry and Mahatsara igneous rocks. The data of other basalts and silicic rocks of eastern Madagascar are from Storey et al. (1997), Melluso et al. (2002, 2003, 2009 and unpublished data). The subalkaline-alkaline line is from Irvine and Baragar (1971)

feldspar set in an altered groundmass of alkali feldspar, quartz and opaque oxides. Phenocrysts of plagioclase and alkali feldspar are generally altered.

Rhyolites are porphyritic and contain phenocrysts and microphenocrysts of plagioclase, clinopyroxene, alkali feldspar, quartz and opaque oxides (Fig. 21.4e–f). The phenocrysts are set in a fine-grained devitrified groundmass composed by quartz, alkali feldspar and opaque oxides. Apatite, titanite and zircon (Fig. 21.4f) are the main accessory phases.

Mineral Chemistry

Representative mineral compositions are given in Tables 21.3, 21.4 and 21.5. *Olivine* phenocrysts from the most Mg-rich dykes show a fairly restricted range of composition (Fo_{63-75}; Fig. 21.5a). The most Fe-rich varieties (Fo_{14-15}; Fig. 21.5a) are present in the groundmass and are characterized by high MnO (0.8–1.1 wt%) content. Calculated distribution coefficients indicate that most olivines are not in equilibrium with host rock (Fe/MgKdOl-liq = 0.30–0.33; Roeder and Emslie 1970; Fig. 21.5b). *Clinopyroxene* in the tholeiitic and silicic rocks is augite-ferroaugite ($Ca_{32}Mg_{47}Fe_{21}$ to $Ca_{42}Mg_{25}Fe_{33}$; Mg# = 75–38, where Mg# = Mg*100/ ($Mg+Fe^{2+}$); Fig. 21.5c) and has low TiO_2 content (<1.9 wt%). Al_2O_3 tend to

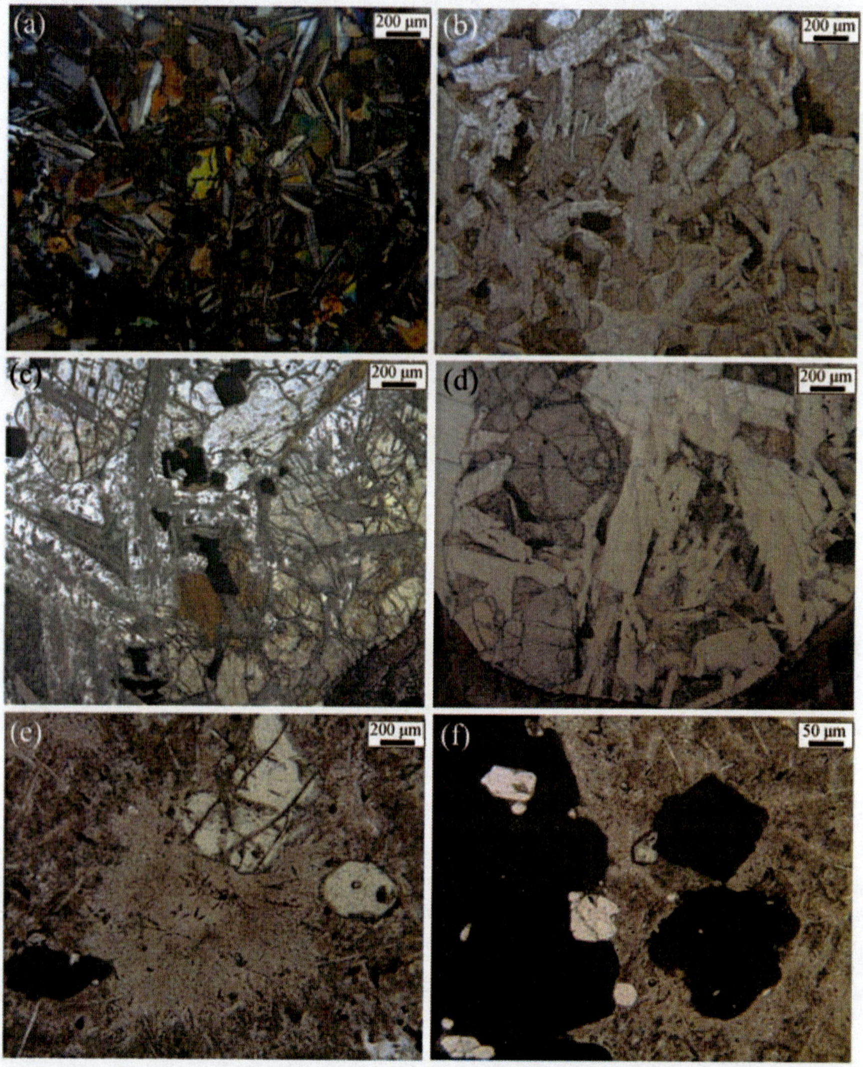

Fig. 21.4 Photomicrographs of Vatomandry and Mahatsara igneous rocks. (**a**) tholeiitic basalt M721, crossed nicols: olivine-phyric dolerite. (**b**) tholeiitic basalt M723, crossed polars; pseudo-ophitic intergrowths of clinopyroxene and plagioclase. (**c**) transitional basalt RT-06-209, crossed polars: subhedral amphibole and clinopyroxene, plagioclase lath. (**d**) Mg-rich tholeiitic basalt RT-06-207A, crossed polars: olivine and plagioclase phenocrysts in a groundmass of plagioclase, clinopyroxene and opaque oxides. (**e**) rhyolite M731, crossed polars: clinopyroxene phenocrysts set in a fine-grained devitrified groundmass composed by quartz, alkali feldspar and opaque oxides. (**f**) rhyolite M731, crossed polars: zircon microphenocryst in a fine-grained devitrified groundmass

Table 21.3 Representative chemical analyses of olivines and oxides

Sample			SiO_2	TiO_2	Al_2O_3	FeO	MnO	MgO	CaO	Cr_2O_3	Sum	Fo	Ulv	Cr#	Ilm
Olivine															
M721	tho bas	gm	36.42	–	–	31.53	0.54	31.43	0.26	–	100.2	64	–	–	–
M721	tho bas	gm	36.69	–	–	31.96	0.7	31.19	0.4	–	100.9	63	–	–	–
M721	tho bas	gm	36.82	–	–	30.3	0.58	31.74	0.37	–	99.8	65	–	–	–
M721	tho bas	gm	37.03	–	–	29.38	0.29	33.04	0.32	–	100.1	66	–	–	–
M721	tho bas	core	38.53	–	–	22.97	0.31	38.44	0.25	–	100.5	75	–	–	–
M721	tho bas	rim	36.97	–	–	30.14	0.42	31.75	0.3	–	99.6	65	–	–	–
M721	tho bas	core	38.01	–	–	23.62	0.13	37.28	0.3	–	99.3	74	–	–	–
M721	tho bas	rim	38.49	–	–	25.18	0.36	37.1	0.38	–	101.5	72	–	–	–
M721	tho bas	rim	37.6	–	–	27.38	0.35	34.64	0.3	–	100.3	69	–	–	–
RT-06-207A	tho bas	core	38.97	–	–	22.84	0.39	36.31	0.50	–	99.0	74	–	–	–
RT-06-207A	tho bas	rim	37.64	–	–	24.10	0.61	36.64	0.27	–	99.3	73	–	–	–
RT-06-207A	tho bas	core	37.33	–	–	24.96	0.47	35.83	0.26	–	98.9	72	–	–	–
RT-06-207A	tho bas	rim	39.31	–	–	23.76	0.33	37.23	0.49	–	101.1	74	–	–	–
RT-06-207A	tho bas	core	37.22	–	–	23.99	0.25	35.85	0.31	–	97.6	73	–	–	–
RT-06-207A	tho bas	rim	38.07	–	–	23.69	0.24	36.75	0.41	–	99.2	73	–	–	–
RT-06-207A	tho bas	interst	30.81	–	–	60.01	0.83	5.53	0.36	–	97.6	14	–	–	–
RT-06-207A	tho bas	interst	30.99	–	–	60.10	1.12	5.97	0.53	–	98.7	15	–	–	-
RT-06-207A	tho bas	interst	29.89	–	–	59.79	0.95	5.64	0.56	–	96.8	14	–	–	–
Cr-spinel															
RT-06-207A	tho bas	incl	–	6.97	14.68	41.61	0.24	7.30	–	25.52	96.3	–	–	54	–
magnetite															
M721	tho bas	gm	–	23.71	1.4	68.51	1.13	0.06	–	0.11	94.9	–	69	–	–
M721	tho bas	gm	–	24.37	1.67	69.07	1.47	0.07	–	0.15	96.8	–	69	–	–
M721	tho bas	gm	–	24.2	1.54	69.11	1.49	0.04	–	–	96.4	–	69	–	–
M721	tho bas	gm	–	31.48	1.47	60.78	1.23	0.34	–	0.19	95.5	–	91	–	–
M721	tho bas	gm	–	23.99	0.76	70.68	1.42	–	–	–	96.9	–	68	–	–
RT-06-207A	tho bas	gm	–	22.39	2.00	69.37	1.46	0.09	–	–	95.3	–	64	–	–

Table 21.3 (continued)

Sample			SiO_2	TiO_2	Al_2O_3	FeO	MnO	MgO	CaO	Cr_2O_3	Sum	Fo	Ulv	Cr#	Ilm
RT-06-207A	tho bas	gm	–	19.22	1.78	67.88	1.20	0.17	–	0.15	90.4	–	58	–	–
RT-06-209	tran bas	gm	–	14.19	0.98	77.34	0.98	0.12	–	–	93.6	–	41	–	–
M722	dac	gm	–	2.70	0.06	88.31	0.97	0.01	–	–	92.1	–	8	–	–
M722	dac	gm	–	5.05	0.34	85.54	1.51	0.15	–	–	92.6	–	15	–	–
M731	rhy	gm	–	9.53	0.88	81.47	3.05	0.08	–	0.16	95.2	–	27	–	–
M731	rhy	gm	–	5.08	1.24	86.88	0.4	–	–	0.09	93.7	–	15	–	–
ilmenite															
RT-06-207A	tho bas	gm		49.19	0.26	43.00	0.32	3.39	–	0.25	96.4	–	–	–	94
RT-06-207A	tho bas	gm		46.96	0.22	47.72	0.38	1.24	–	–	96.5	–	–	–	91
RT-06-207A	tho bas	gm		46.92	0.26	48.49	0.42	0.88	–	0.07	97.0	–	–	–	91
M731	rhy	gm		39.12	0.15	53.17	4.75	0.09	–	0.16	97.4	–	–	–	75
M731	rhy	gm		38.81	0.04	50.94	7.28	0.25	–	0.09	97.4	–	–	–	74
M731	rhy	gm		38.56	0	51.41	7.74	0.11	–	0.18	98.0	–	–	–	74
M731	rhy	gm		37.65	0.13	52.84	6.46	0.05	–	0.07	97.2	–	–	–	72
M722	dac	gm		48.01	0.04	49.43	2.74	0.28	–	–	100.5	–	–	–	90
M722	dac	gm		47.05	0.22	51.05	1.70	0.12	–	–	100.1	–	–	–	89
M722	dac	gm		46.25	0.06	49.29	2.59	0.30	–	–	98.5	–	–	–	88

Fo = atomic 100*Mg/(Mg + Fe + Mn), Cr# = Cr*100/(Cr + Al), Ulv = ulvöspinel (mol%); Ilm = ilmenite (mol%).

Table 21.4 Representative chemical analyses of pyroxenes and amphiboles

Sample			SiO_2	TiO_2	Al_2O_3	FeO	MnO	MgO	CaO	Na_2O	K_2O	Ba	Cr_2O_3	F	Cl	Sum	Ca	Fe*	Mg	Mg#
Pyroxene																				
M725	tho bas		51.86	0.78	1.71	19.67	0.32	13.33	14.80	0.21	–	–	–	–	–	102.7	30.2	31.9	37.9	55
M725	tho bas		50.92	1.39	2.77	12.43	0.35	14.54	19.26	0.38	–	–	–	–	–	102.0	38.9	20.2	40.9	68
M725	tho bas		49.99	0.58	1.04	28.14	0.66	9.87	12.13	0.23	–	–	–	–	–	102.6	25.1	46.5	28.4	38
M725	tho bas		49.32	0.29	0.93	29.79	0.80	9.58	9.75	0.12	–	–	–	–	–	100.6	20.8	50.9	28.4	36
M725	tho bas		50.54	1.09	2.18	15.90	0.36	14.82	14.29	0.30	–	–	–	–	–	99.5	30.0	26.7	43.3	62
M725	tho bas		50.36	0.86	1.99	15.66	0.37	12.64	18.40	0.22	–	–	–	–	–	100.5	37.9	25.8	36.3	59
M734	tho bas		50.33	1.12	2.26	12.30	0.44	14.38	17.99	0.26	–	–	–	–	–	99.1	37.5	20.7	41.7	68
M734	tho bas		51.37	0.42	1.04	22.54	0.37	11.86	14.20	0.23	–	–	–	–	–	102.0	29.2	36.8	34.0	48
M734	tho bas		51.37	0.68	1.23	21.03	0.38	13.52	13.78	0.13	–	–	–	–	–	102.1	27.9	33.9	38.1	53
M734	tho bas		50.22	1.15	2.01	16.21	0.43	15.27	14.31	0.14	–	–	–	–	–	99.7	29.5	26.8	43.8	63
M734	tho bas		50.01	0.75	1.23	20.95	0.49	12.00	14.56	0.26	–	–	–	–	–	100.3	30.3	34.9	34.8	51
M734	tho bas		51.05	0.85	1.55	12.82	0.31	16.37	15.36	0.21	–	–	–	–	–	98.5	31.8	21.2	47.1	69
M734	tho bas		50.38	0.68	1.27	16.03	0.33	13.45	15.61	0.20	–	–	–	–	–	98.0	33.1	27.1	39.7	60
M734	tho bas		47.99	1.26	2.29	9.66	0.22	13.94	17.62	0.23	–	–	–	–	–	93.2	39.4	17.2	43.4	72
M734	tho bas		51.20	1.18	2.06	12.08	0.32	15.42	17.01	0.15	–	–	–	–	–	99.4	35.3	20.1	44.6	69
M721	tho bas	core	49.94	0.68	1.08	21.90	0.39	10.17	16.36	0.37	–	–	0.05	–	–	100.9	34.1	36.3	29.5	45
M721	tho bas	rim	49.04	0.84	1.28	20.47	0.58	11.12	15.62	0.25	–	–	0.10	–	–	99.3	32.9	34.6	32.6	49
M721	tho bas	core	51.78	0.72	2.05	9.01	0.18	16.12	19.25	0.37	–	–	0.55	–	–	100.0	39.4	14.7	45.9	76
M721	tho bas	rim	49.55	1.12	1.78	17.59	0.39	14.67	13.81	0.20	–	–	–	–	–	99.1	28.6	29.1	42.3	60
M721	tho bas	core	49.67	1.91	1.94	18.58	0.41	12.87	15.51	0.25	–	–	0.06	–	–	101.2	32.2	30.7	37.1	55
M721	tho bas	rim	50.97	1.38	2.22	15.49	0.40	13.35	18.34	0.21	–	–	0.00	–	–	102.4	37.2	25.2	37.7	61
M721	tho bas	rim	51.42	1.05	1.70	18.19	0.39	16.50	11.81	0.11	–	–	0.12	–	–	101.3	24.0	29.4	46.6	62
M721	tho bas		50.50	1.32	2.20	13.43	0.37	13.74	19.30	0.23	–	–	–	–	–	101.1	39.2	21.9	38.9	65
M721	tho bas	core	50.91	1.03	3.12	9.43	0.30	15.46	19.61	0.31	–	–	0.65	–	–	100.8	40.3	15.6	44.2	75
M721	tho bas	core	49.92	1.43	2.45	15.11	0.43	13.82	17.83	0.25	–	–	0.02	–	–	101.3	36.2	24.7	39.1	62
RT-06-207A	tho bas	core	49.46	0.85	1.62	17.96	0.53	14.10	13.57	0.36	–	–	0.11	–	–	98.6	28.5	30.3	41.2	58
RT-06-207A	tho bas	rim	49.14	0.79	1.06	21.75	0.54	11.26	14.35	0.31	–	–	0.08	–	–	99.3	30.3	36.7	33.0	48
RT-06-207A	tho bas	core	52.70	1.01	1.20	14.84	0.16	18.50	11.83	0.17	–	–	0.13	–	–	100.5	24.0	23.8	52.2	69

Table 21.4 (continued)

Sample			SiO_2	TiO_2	Al_2O_3	FeO	MnO	MgO	CaO	Na_2O	K_2O	Ba	Cr_2O_3	F	Cl	Sum	Ca	Fe*	Mg	Mg#
RT-06-207A	tho bas	rim	48.43	1.48	2.18	16.54	0.12	13.59	14.60	0.27	–	–	–	–	–	97.2	31.4	28.0	40.7	59
RT-06-207A	tho bas	rim	48.28	0.30	0.80	25.25	0.47	9.37	12.92	0.06	–	–	–	–	–	97.4	28.1	43.6	28.3	40
RT-06-207A	tho bas	interst	50.30	0.50	1.02	19.39	0.45	15.96	9.51	0.13	–	–	0.03	–	–	97.3	20.2	32.8	47.0	59
RT-06-207A	tho bas	core	48.60	1.35	3.37	8.96	0.16	14.28	19.22	0.24	–	–	0.67	–	–	96.8	41.6	15.4	43.0	74
RT-06-207A	tho bas	interst	46.25	0.06	0.86	37.57	0.75	5.48	6.82	0.12	–	–	–	–	–	97.9	15.4	67.4	17.2	21
RT-06-207A	tho bas	interst	46.59	0.40	0.85	37.14	0.94	4.82	6.69	0.03	–	–	–	–	–	97.4	15.5	68.9	15.5	19
RT-06-207A	tho bas	core	48.86	1.53	3.32	8.30	0.09	14.69	18.76	0.18	–	–	0.95	–	–	96.7	41.0	14.3	44.7	76
RT-06-209	tran bas	core	47.40	1.67	7.02	8.68	0.11	12.77	21.74	0.50	–	–	0.03	–	–	99.9	46.9	14.8	38.3	72
RT-06-209	tran bas	rim	50.75	1.16	3.28	8.93	0.14	14.20	21.87	0.38	–	–	0.06	–	–	100.8	44.9	14.5	40.6	74
RT-06-209	tran bas	core	49.94	1.57	4.25	9.06	0.16	13.67	21.61	0.55	–	–	–	–	–	100.8	45.2	15.1	39.8	73
RT-06-209	tran bas	rim	50.60	1.21	4.12	9.40	0.00	14.15	22.14	0.32	–	–	0.09	–	–	102.0	45.0	14.9	40.0	73
RT-06-209	tran bas	rim	49.77	1.21	4.34	9.48	0.20	13.10	21.52	0.57	–	–	–	–	–	100.2	45.5	16.0	38.5	71
RT-06-209	tran bas	rim	48.31	1.40	3.87	8.60	0.13	13.45	21.42	0.44	–	–	0.04	–	–	97.7	45.6	14.5	39.9	74
RT-06-209	tran bas	core	48.08	1.23	4.17	8.24	0.14	12.79	20.85	0.47	–	–	–	–	–	96.0	46.1	14.5	39.4	73
M722	dac		51.26	0.66	1.36	13.05	0.30	13.33	19.54	0.25	–	–	–	–	–	99.8	40.3	21.5	38.2	65
M722	dac		50.47	0.01	0.34	19.22	0.62	8.22	19.76	0.51	–	–	–	–	–	99.1	42.3	33.2	24.5	43
M722	dac		51.00	0.57	1.44	12.70	0.32	13.82	18.72	0.18	–	–	–	–	–	98.8	38.9	21.1	40.0	66
M722	dac		51.36	0.60	1.50	12.99	0.23	13.62	18.93	0.25	–	–	–	–	–	99.5	39.3	21.4	39.3	65
M722	dac		50.81	0.39	0.98	16.77	0.54	10.30	19.49	0.27	–	–	–	–	–	99.6	41.2	28.6	30.3	52
M722	dac		51.25	0.61	1.44	13.87	0.40	13.40	18.18	0.25	–	–	–	–	–	99.4	37.9	23.2	38.9	63
M722	dac		51.38	0.76	1.40	14.13	0.67	12.97	18.50	0.41	–	–	–	–	–	100.2	38.5	24.0	37.5	62
M722	dac		51.29	1.01	1.29	14.85	0.55	12.22	18.81	0.45	–	–	–	–	–	100.5	39.3	25.1	35.5	59
M730	rhy		52.48	0.38	0.92	11.65	1.89	14.06	18.97	0.68	–	–	–	–	–	101.0	38.6	21.6	39.8	68
M730	rhy		52.77	0.48	0.81	11.66	1.97	14.06	19.23	0.61	–	–	–	–	–	101.6	38.9	21.6	39.6	68
M730	rhy		53.03	0.43	0.80	11.16	1.74	14.17	18.97	0.68	–	–	–	–	–	101.0	38.9	20.7	40.4	69
M731	rhy		51.38	0.53	1.05	11.36	1.83	13.97	19.02	0.70	–	–	–	–	–	99.8	39.0	21.1	39.9	69
M731	rhy		51.49	0.59	0.79	11.43	1.62	14.37	19.01	0.55	–	–	–	–	–	99.9	38.6	20.7	40.6	69
M731	rhy		51.83	0.56	0.86	11.64	1.75	13.73	18.77	0.62	–	–	–	–	–	99.8	38.8	21.7	39.5	68
M731	rhy		51.91	0.36	0.81	11.73	1.86	14.02	18.96	0.69	–	–	0.06	–	–	100.4	38.6	21.6	39.7	68

Table 21.4 (continued)

Sample			SiO_2	TiO_2	Al_2O_3	FeO	MnO	MgO	CaO	Na_2O	K_2O	Ba	Cr_2O_3	F	Cl	Sum	Ca	Fe*	Mg	Mg#
M731	rhy		51.80	0.43	0.86	11.71	1.86	13.89	19.23	0.70	–	–	0.13	–	–	100.6	39.1	21.6	39.3	68
M731	rhy		51.84	0.57	1.42	11.85	1.66	13.35	19.62	0.79	–	–	0.07	–	–	101.2	40.2	21.7	38.1	67
M731	rhy		53.04	0.40	0.82	11.67	1.78	13.72	19.44	0.75	–	–	–	–	–	101.6	39.6	21.4	38.9	68
M731	rhy		52.90	0.35	0.71	11.83	2.28	14.29	19.37	0.73	–	–	–	–	–	102.5	38.5	21.9	39.5	68
M731	rhy		51.64	0.46	0.69	11.68	2.01	13.97	18.99	0.59	–	–	–	–	–	100.0	38.7	21.8	39.6	68
amphibole																				
RT-06-209	tran bas	core	43.93	3.04	9.01	14.81	0.70	13.33	10.40	2.92	0.48	–	0.23	–	0.12	99.0	–	–	–	60
RT-06-209	tran bas	rim	47.25	2.76	8.68	14.06	0.55	13.76	10.27	3.19	0.60	0.83	–	0.01	0.07	102.0	–	–	–	63
RT-06-209	tran bas	core	45.14	3.21	9.10	14.89	0.44	12.97	10.61	3.16	0.71	0.34	–	–	0.06	100.6	–	–	–	60

Ca, Fe* (Fe + Mn) and Mg in mol%; Mg# = Mg*100/(Mg + Fe + Mn).

Table 21.5 Representative chemical analyses of feldspars

Sample			SiO_2	Al_2O_3	FeO	CaO	Na_2O	K_2O	BaO	SrO	Sum	An	Ab	Or
M734	tho bas		60.22	23.31	0.93	5.81	7.70	0.76	–	–	98.7	28.1	67.5	4.4
M734	tho bas		55.27	27.17	0.81	10.79	5.40	0.16	–	–	99.6	52.0	47.1	0.9
M734	tho bas		57.13	26.05	0.82	9.16	6.29	0.36	–	–	99.8	43.7	54.3	2.0
M734	tho bas		52.95	28.61	0.98	12.34	4.50	0.16	–	–	99.5	59.7	39.4	0.9
M734	tho bas		53.28	29.55	0.69	12.88	4.53	0.19	–	–	101.1	60.5	38.5	1.1
M734	tho bas		57.17	25.75	0.72	9.25	6.18	0.34	–	–	99.4	44.4	53.6	2.0
M734	tho bas		56.05	26.62	0.57	10.30	5.62	0.28	–	–	99.4	49.5	48.9	1.6
M734	tho bas		55.51	26.74	0.60	10.79	5.54	0.25	–	–	99.4	51.1	47.5	1.4
RT-06-207A	tho bas	core	51.89	28.68	0.62	13.14	3.96	0.06	–	0.23	98.6	64.1	35.6	0.4
RT-06-207A	tho bas	rim	51.13	29.02	0.87	13.66	3.92	0.16	–	0.24	99.0	64.8	34.3	0.9
RT-06-207A	tho bas	core	50.24	29.11	0.60	13.97	3.30	0.21	–	0.40	97.8	68.5	30.3	1.2
RT-06-207A	tho bas	rim	53.06	28.94	0.92	12.96	4.16	0.14	–	0.45	100.6	62.0	37.2	0.8
RT-06-209	tran bas	core	61.83	23.93	0.18	5.64	8.24	0.61	0.15	0.62	101.2	26.0	70.4	3.6
RT-06-209	tran bas	rim	65.65	21.25	0.09	2.75	9.95	0.02	–	0.22	99.9	13.2	86.7	0.1
RT-06-209	tran bas	rim	65.42	19.15	1.19	0.36	10.55	0.50	0.11	0.33	97.6	1.8	95.1	3.1
RT-06-209	tran bas	core	51.60	27.67	0.57	11.68	4.66	0.10	0.23	0.09	96.6	57.3	41.7	1.0
RT-06-209	tran bas	rim	54.64	26.95	0.41	10.20	5.41	0.19	–	0.69	98.5	49.5	49.4	1.1
RT-06-209	tran bas	gm	64.47	17.78	0.73	0.25	0.36	15.84	0.33	0.14	99.9	1.3	3.7	95.1
M722	dac		52.86	28.60	0.54	12.92	4.34	0.13	–	–	99.4	61.7	37.5	0.7
M722	dac		65.10	17.90	0.20	–	0.27	15.89	0.24	–	99.6	0.0	2.5	97.5
M722	dac		59.15	24.34	0.58	7.59	7.15	0.44	0.11	0.05	99.4	36.0	61.4	2.7
M722	dac		64.87	17.80	0.11	–	0.42	16.27	–	0.03	99.5	0.0	3.9	96.1
M722	dac		57.38	25.27	0.76	8.04	6.76	0.50	–	0.60	99.3	37.9	59.2	2.8
M722	dac		61.52	23.77	0.36	6.17	7.95	0.63	–	–	100.4	28.9	67.5	3.5
M722	dac		68.04	19.58	0.02	0.88	10.82	0.05	0.21	0.21	99.8	4.2	95.1	0.7
M722	dac		64.90	18.43	0.06	0.14	0.27	15.56	0.14	–	99.5	0.7	2.6	96.7
M722	dac		68.22	19.04	0.04	0.41	11.28	0.04	–	0.07	99.1	1.9	97.9	0.2
M730	rhy		65.44	23.25	0.31	4.73	8.71	1.06	–	–	103.5	21.7	72.5	5.8
M730	rhy		62.40	22.50	0.37	4.50	8.56	0.98	–	–	99.3	21.3	73.2	5.5

Table 21.5 (continued)

Sample			SiO_2	Al_2O_3	FeO	CaO	Na_2O	K_2O	BaO	SrO	Sum	An	Ab	Or
M730	rhy		62.72	22.46	0.62	4.74	8.54	0.93	–	–	100.0	22.3	72.5	5.2
M731	rhy		62.24	22.87	0.59	4.72	8.46	0.80	–	–	99.7	22.5	73.0	4.5
M731	rhy		67.14	18.86	0.71	1.08	8.28	3.78	–	–	99.9	5.3	72.9	21.9
M731	rhy		63.76	22.22	0.43	3.77	8.72	1.20	–	–	100.1	18.0	75.2	6.8
M721	rhy	core	69.53	19.74	0.02	0.21	11.84	–	–	–	101.3	1.0	99.0	0.0
M721	rhy	rim	51.63	30.05	0.52	14.11	3.78	0.06	–	–	100.2	67.1	32.5	0.3
M721	rhy		53.11	28.36	1.59	12.67	4.40	0.04	–	–	100.2	61.3	38.5	0.2
M721	rhy	gm	55.64	26.97	1.70	10.53	5.58	0.20	–	–	100.6	50.5	48.4	1.1
M725	rhy		60.33	24.96	1.05	7.45	7.61	0.51	–	–	101.9	34.1	63.1	2.8
M725	rhy		53.58	28.41	0.93	12.04	4.91	0.37	–	–	100.2	56.4	41.6	2.1
M725	rhy		55.08	27.30	0.98	10.86	5.64	0.14	–	–	100.0	51.1	48.1	0.8
M725	rhy		54.35	27.97	0.77	11.46	4.94	0.30	–	–	99.8	55.2	43.1	1.7
M725	rhy		57.35	26.25	0.83	9.29	6.20	0.36	–	–	100.3	44.4	53.6	2.0

An, Ab and Or in mol%. *gm* groundmass, *interst* interstitial, *incl* inclusion.

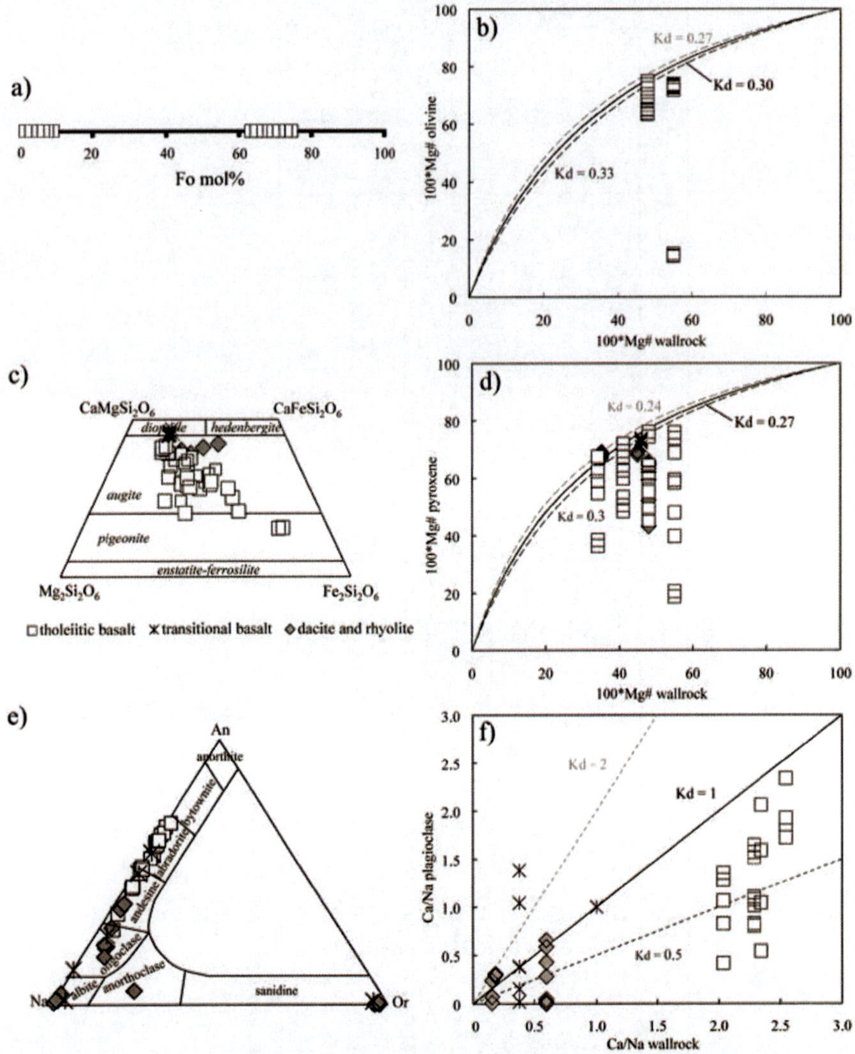

Fig. 21.5 (**a**) olivine compositions. (**b**) Fe-Mg partitioning [$K_d = (Fe/Mg)_{ol}/(Fe/Mg)_{wr}$] between olivine and host rock (Roeder and Emslie, 1970). (**c**) Composition of pyroxene in the rocks of Vatomandry and Mahatsara. (**d**) Fe-Mg partitioning [$K_d = (Fe/Mg)_{px}/(Fe/Mg)_{wr}$] between pyroxene and host rock. (**e**) Composition of plagioclase and alkali feldspar of Vatomandry and Mahatsara rocks. (**f**) Ca/Na (molar) of plagioclase vs. Ca/Na (molar) of the host rock

decrease with decreasing Mg#, whereas MnO increase with decreasing Mg#. Maximum MnO content (1.7–2.3 wt%) has been observed in crystals from two rhyolites (M730 and M731). Similar enrichment in Mn has been found also in Mananjary rhyolites and in the St. Mary's islands dacites, the Indian correlatives of the Madagascan silicic rocks (Melluso et al., 2009). Clinopyroxene in the transitional basalt is diopside ($Ca_{47}Mg_{38}Fe_{15}$ to $Ca_{45}Mg_{40}Fe_{15}$; Mg# = 74–71) and has higher Al_2O_3 (3.3–7.0 wt%) content than augite. Calculated distribution coefficients for the majority of basaltic samples fall outside the range of clinopyroxene-host rock ($^{Fe/Mg}K_d{}^{cpx\text{-}liq} = 0.27 \pm 0.03$; Putirka, 2008). Among rhyolitic rocks, sample M731 contain crystals in chemical equilibrium with the host rock (Fig. 21.5d). The compositional range of *pigeonite* is $Ca_{28}Mg_{28}Fe_{44}$ to $Ca_{15}Mg_{17}Fe_{67}$ (Mg# = 19–40). *Plagioclase* in the tholeiitic dykes is mainly labradorite (An_{67-51}; Fig. 21.5e). It typically shows normal zoning. Only a few compositions are oligoclase (An_{44-28}). Plagioclase phenocrysts in the transitional basalt RT-06-209 show a larger range of compositions (An_{2-57}). Calculated distribution coefficients indicate that most plagioclase phenocrysts are in equilibrium with host rock ($^{Ca/Na}K_d{}^{plag\text{-}melt} = XCa_{plag} * XNa_{liq}/ XNa_{plag} * XCa_{liq} = 0.80$–1.85; Kinzler and Grove, 1992; Fig. 21.5f). Temperatures values calculated using the Putirka (2008) geothermometer range from ~1,231°C for the least evolved basalts to ~1,175°C for the more evolved basalts. Plagioclase of dacites and rhyolites is andesine and oligoclase (Fig. 21.5e). In the dacite M722, some phenocrysts of feldspar are perthitic. Plagioclase-liquid geothermometer (Putirka, 2008) yields values close to 1,000°C, similar to Mananjary rhyolites (Melluso et al., 2009). *Amphibole* in the transitional basalt is a magnesio-hastingsitic hornblende (according to Esawi, 2004) and shows a restricted range in TiO_2 (2.8–3.2 wt%). Amphibole is a relatively rare phase in the Madagascan igneous rocks (Melluso et al., 2005, 2006; Mahoney et al., 2008). *Ti-magnetite* is the dominant spinel of the tholeiitic dykes (rare *Cr-spinels* have been found in the most Fo-rich olivines) and has low Al_2O_3 (<2 wt%) and high ulvöspinel contents (91–58 mol%). *Ilmenite* has low MgO content (0.9–3.4 wt%). Calculated equilibration temperatures and oxygen fugacity (using ILMAT Excel worksheet; Lepage, 2003) plot close to the synthetic quartz-fayalite-magnetite (QFM) buffer, and range from 1,094 to 834°C and from –9.8 to –13.6 log units, respectively. Magnetite in the silicic rocks has low ulvöspinel (27–8 mol%) and Al_2O_3 (<1.5 wt%) contents whereas, ilmenite is low in MgO (<0.3 wt%). The magnetite-ilmenite equilibrium pairs give temperatures and oxygen fugacity between 887 and 716°C, and –11.1 and –14.6 log units, respectively. Equilibration temperature and oxygen fugacity plot above the nickel-nickel oxide (NNO) synthetic buffer.

Geochemistry of the Mafic Dykes

The Vatomandry and Mahatsara basaltic dykes have SiO_2 contents between 46.0 and 51.6 wt% and low compatible trace element concentrations (e.g. Ni = 101–36 ppm). The MgO varies from 4.65 to 7.99 wt% and Mg# (Mg# = Mg*100/(Mg+Fe^{2+})) from 58 to 37, indicating that the samples are evolved. Major and trace element variations using MgO (wt%) as index of differentiation are presented in Fig. 21.6. MgO has negative correlation with Fe_2O_{3t}, TiO_2 and P_2O_5, and positive correlation with Al_2O_3 and CaO. Incompatible trace elements such as Zr, Nb and Rb all show negative correlation with MgO (Fig. 21.6). The chondrite-normalized lanthanide

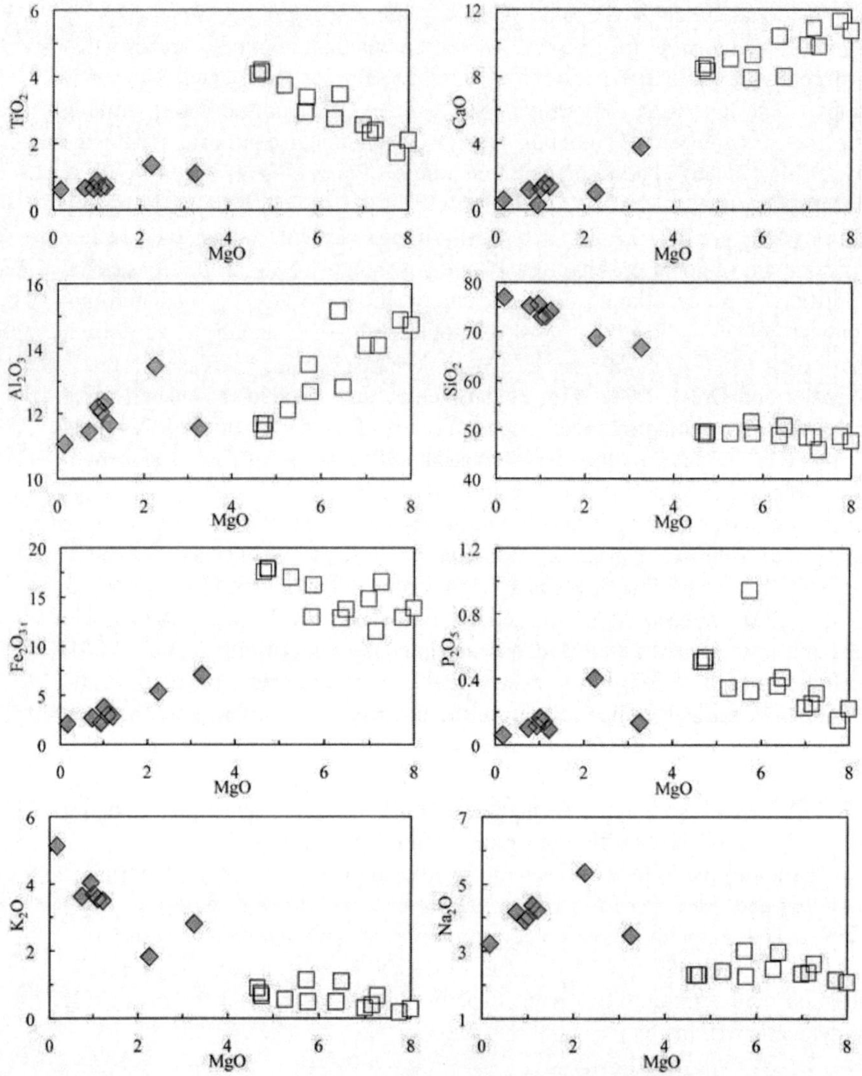

Fig. 21.6 Major (wt%) and trace elements (ppm) of Vatomandry and Mahatsara rocks plotted versus MgO wt% or Zr ppm as differentiation indexes. Fractional crystallization curves for the transition basalt-rhyolite use pure fractional crystallization (*black curve*) and assimilation-fractional crystallization (*grey curve*) models starting from sample M723. Assumed bulk partition coefficients: $D_{Zr} = 0.25$, D_{Nb} and $D_{Rb} = 0$. $r = mass_{assimilated}/mass_{cumulated}$. The crustal contaminant used in the AFC modelling is a felsic granulite sampled in the Vatomandry district. It has a Rb = 152 ppm and a Zr = 635 ppm. Residual liquid fractions (*f*) are indicated on the fractional crystallization curves

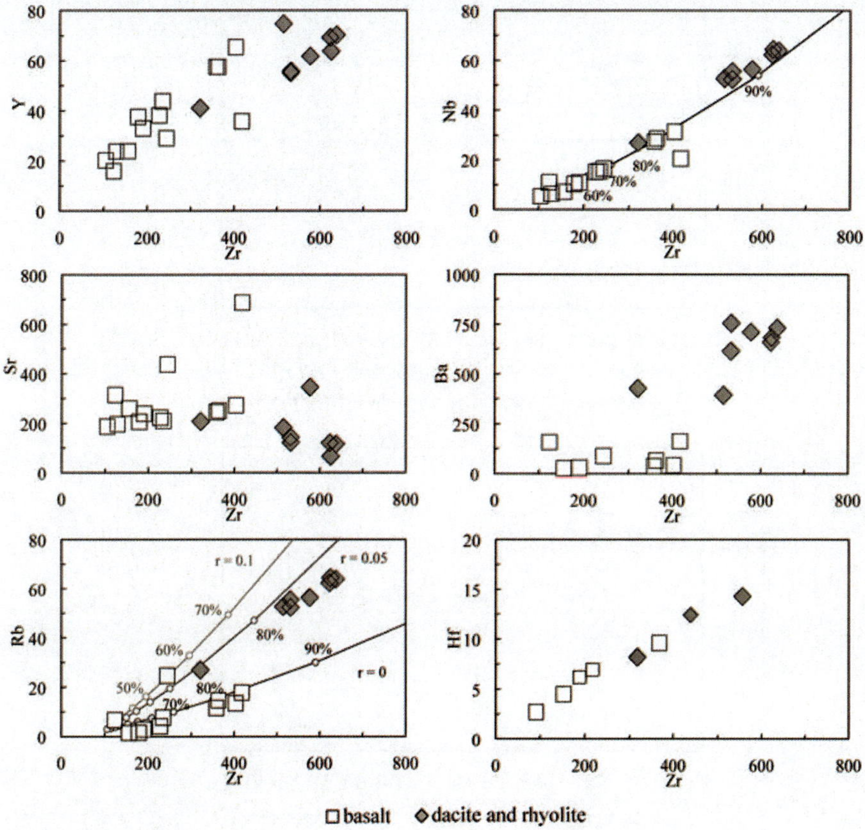

Fig. 21.6 (continued)

patterns of the basaltic dykes (Fig. 21.7) are mildly light lanthanide element (LREE) enriched (La_n = 19 to 96 and La_n/Yb_n from 1.7 to 3.7, the subscript "n" means chondrite normalized; Table 21.6). The most differentiated samples show troughs at Eu (Eu/Eu* = 0.93–0.85). The less evolved basalts have low concentration of incompatible elements (e.g. Nb = 5–9 ppm; Zr = 91–153 ppm) similar to those of Southwest Indian Ridge (SWIR) MORB and show no troughs at Nb and Ti on mantle-normalized incompatible element diagram (Fig. 21.8). The evolved basalts are more enriched in incompatible elements than SWIR MORB and show troughs at Sr and Ti on mantle-normalized incompatible element diagram (Fig. 21.8). In addition, the evolved dykes have an incompatible element pattern very similar to those of Sainte Marie Island and Mg-rich Tamatave dykes (group 1–1a; Melluso et al., 2002; Fig. 21.8). The other Tamatave dykes are characterized by higher incompatible element contents (group 2; Melluso et al., 2002; Fig. 21.8). Melluso et al. (2002) showed that two different basalt groups cannot be produced by closed-system fractional crystallization of plagioclase and pyroxene; rather, they require distinct parental magmas.

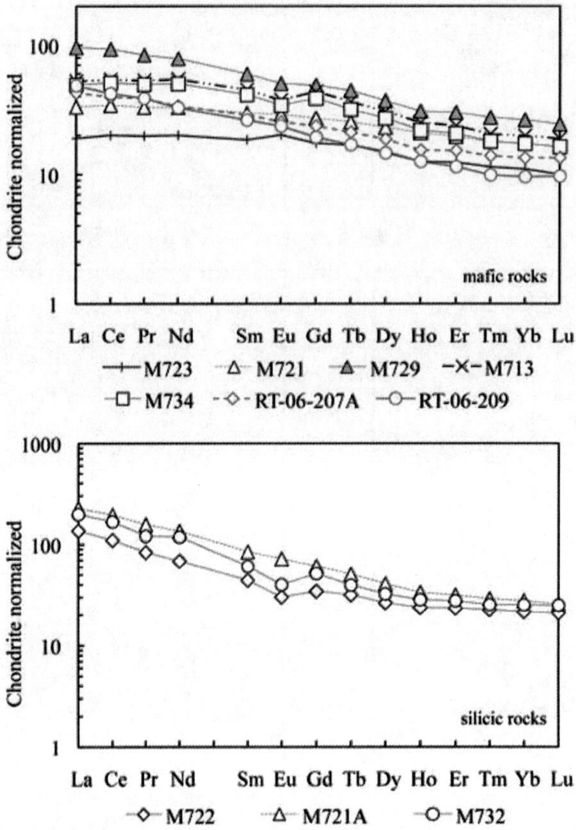

Fig. 21.7 Chondrite-normalized lanthanide element patterns for basalts and silicic rocks of Vatomandry. The chondrite values used for normalization are taken from Boynton (1984)

Geochemistry of the Silicic Rocks

The Vatomandry silicic rocks (SiO_2 = 66.8–77.0 wt%) are characterized by relatively low TiO_2 (<1.3 wt%), P_2O_5 (<0.4 wt%), Th (7–9 ppm) and Nb (27–65 ppm) contents and high Rb/Nb (1.2–2.9) and Ba/Nb (7.5–15.9) ratios (Fig. 21.6; Table 21.1). Dacites and rhyolites show a general decrease in Fe_2O_{3t}, Al_2O_3, CaO, TiO_2 and P_2O_5 with MgO (Fig. 21.6). SiO_2, K_2O and incompatible trace elements as Rb, Ba, Nb and Zr define a nearly linear trend of increasing concentration with decreasing MgO (Fig. 21.6). These trends are generally consistent with fractional crystallization of plagioclase, clinopyroxene, Fe-Ti oxides and apatite. The silica gap between basalts and rhyolites is much less marked when other indicators of differentiation are used (Fig. 21.6). The silicic rocks are moderately LREE-enriched (La_n/Yb_n = 6.3–8.2; Fig. 21.7) and show a progressive increase in all lanthanides with increasing SiO_2. Negative Eu anomaly is noted in the dacites and rhyolite (Eu/Eu* = 0.77–0.72) (Table 21.6). In the primitive mantle-normalized patterns

Table 21.6 ICP-MS trace element contents (ppm) for representative igneous rocks from Vatomandry and Mahatsara districts

Sample Type	M723 tho bas	M721 tho bas	M729 tho bas	M734 tho bas	M713 tho bas	M721A rhy	M732 rhy	M722 dac
V	319	367	454	389	441	25	10	150
Cr	230	260	30	118	90	–	10	50
Co	50	47	44	42	42	2	6	18
Ni	130	120	60	66	63	–	7	50
Cu	140	220	210	–	–	–	–	40
Zn	140	160	220	–	–	140	–	110
Ga	22	24	25	–	–	27	–	21
Rb	3	5	16	10	10	119	110	87
Sr	211	210	285	195	181	383	86	238
Y	25	37	62	37	43	65	52	46
Zr	91	153	368	217	189	556	440	318
Nb	5	9	25	20	19	51	65	24
Cs	0.6	1.3	–	0.22	0.43	1.1	0.59	3.3
Ba	40	63	193	103	95	791	672	473
La	5.9	10.3	29.8	15.2	16.2	70.3	61.3	42.3
Ce	16.2	27.6	75.3	41.1	43.6	157.0	135.4	88.9
Pr	2.4	4.0	10.2	6.0	6.5	19.2	14.8	10.2
Nd	12.0	19.7	47.0	30.0	32.5	81.0	71.1	41.1
Sm	3.6	5.8	11.6	8.0	8.8	16.5	11.8	8.6
Eu	1.5	2.2	3.7	2.5	2.8	5.3	3.0	2.2
Gd	4.5	7.0	12.6	10.0	11.3	15.8	13.3	8.9
Tb	0.8	1.2	2.1	1.5	1.7	2.4	1.9	1.5
Dy	4.8	7.4	11.9	8.7	9.9	13.1	10.4	8.5
Ho	0.9	1.5	2.2	1.6	1.8	2.4	2.0	1.7
Er	2.6	4.1	6.3	4.3	5.0	6.6	5.8	4.9
Tm	0.4	0.6	0.9	0.6	0.7	0.9	0.8	0.7
Yb	2.3	3.6	5.5	3.6	4.1	5.8	5.3	4.5
Lu	0.3	0.5	0.8	0.5	0.6	0.8	0.8	0.7
Hf	2.7	4.5	9.6	6.9	6.2	14.3	12.4	8.2
Ta	0.4	0.7	2.0	3.0	3.1	3.8	8.2	1.9
Pb	–	–	–	1.9	2.0	14.0	15.7	10.0
Th	0.5	0.7	2.1	1.3	1.3	8.8	10.0	6.9
U	0.2	0.2	0.6	–	–	2.5	–	1.6
La_n/Yb_n	2.0	2.1	3.8	3.1	2.9	7.6	6.7	5.3
Eu/Eu*	0.9	1.0	1.0	1.2	1.2	1.0	1.3	1.1

n means chondrite normalized; $Eu/Eu^* = Eu_n/(Gd_n{*}Yb_n)\hat{}0.5$.

(Fig. 21.8) the silicic rocks show weak or no troughs at Nb and marked troughs at Sr and Ti. In addition, the silicic rocks have trace element concentrations very similar to those of the Mananjary rhyolites (Melluso et al., 2009).

Strontium and Neodymium isotope data on silicic rocks were analysed previously by Melluso et al. (2009). The silicic rocks have distinctly more radiogenic Sr ($^{87}Sr/^{86}Sr_i = 0.70551$–$0.70833$) and less radiogenic Nd ($^{143}Nd/^{144}Nd_i = 0.51224$–$0.51240$; $\varepsilon Nd_i = -5.6$ to -2.5) than Mananjary rhyolites (Melluso et al., 2009).

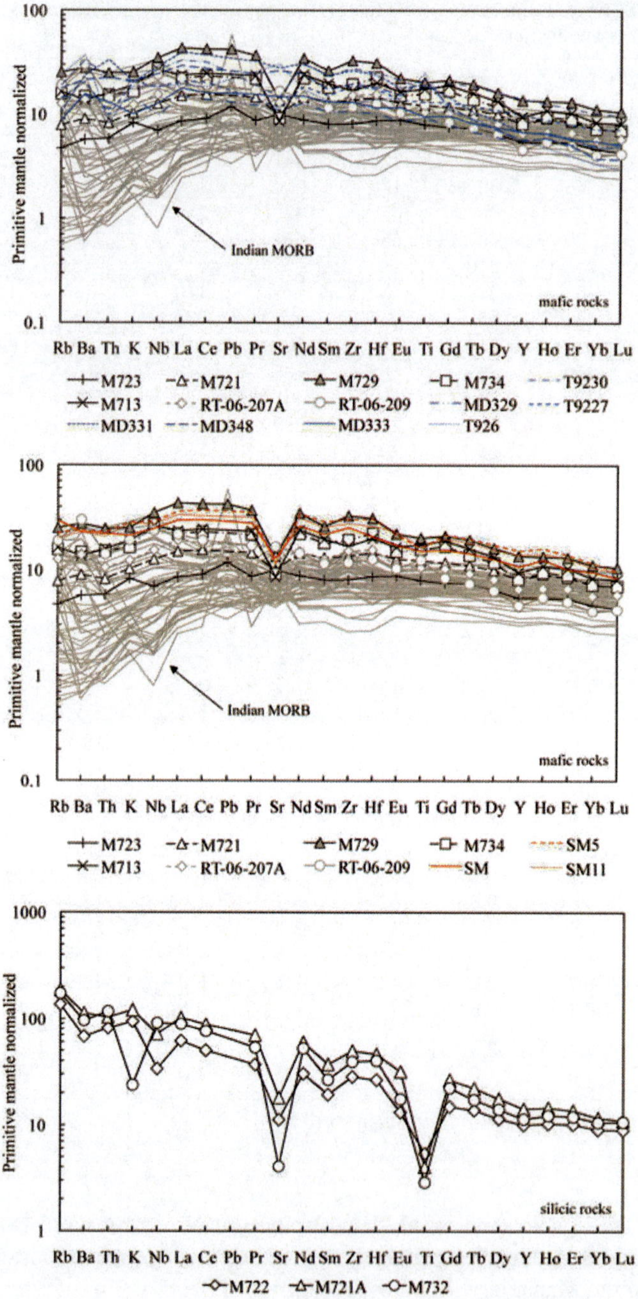

Fig. 21.8 Primitive-mantle normalized incompatible element diagram for basalts and silicic rocks of Vatomandry. The primitive mantle values used for normalization are taken from Sun and McDonough (1989). Indian MORB data are from Rehkämper and Hofmann (1997) and Janney et al. (2005). The data for the Sainte Marie Island and Tamatave dykes are from Melluso et al. (2002, 2003 and unpublished data)

Discussion

Origin and Evolution of the Basaltic Rocks

The Vatomandry basaltic dykes crystallized from evolved magmas as indicated by low MgO (4.7–8.0 wt%), Ni (<105 ppm) and Cr (<350 ppm) concentrations. The mineralogical assemblage and geochemical variations suggest that fractional crystallization of augite, plagioclase, and lesser olivine from more primitive magma compositions was responsible for the evolution of the basaltic rocks. In the pseudoternary diagram Fo-Di-SiO_2 (Grove et al., 1982; Fig. 21.9), the tholeiitic basalts plot near the 1-atm ol+pl+cpx cotectics, suggesting magmatic evolution at shallow depth. The transition from less to more evolved basalts was modelled by 56% fractional crystallization of olivine (17%), augite (29%) and plagioclase (54%; Table 21.7). All tholeiitic dykes have low Ba/Zr (<1) and Th/Ta (<1.4) and high Ce/Pb (20–22) ratios, suggesting minimal crustal contamination. The least evolved basalts are characterized by high Zr/Nb (15–22), Ti/Zr (62–98) and Zr/Hf (31–35) and low Ba/Nb (1.3–5.4) ratios. These values are well within the range of transitional or normal-MORB wordwide (e.g. Sun and McDonough, 1989; le Roex, 1987), and so are the lanthanide chondrite-normalized patterns (Fig. 21.7).

Basaltic rocks with MORB-like composition are well documented in the Madagascan igneous province (Melluso et al., 1997, 2001, 2003, 2005; Storey et al., 1997). The "depleted" geochemical characteristics of the Mailaka and Antanimena basalts are evident from values of high field strength elements (e.g. Zr = 50–230 ppm; Nb = 2–9 ppm) typical of normal-MORB. However, the relatively high

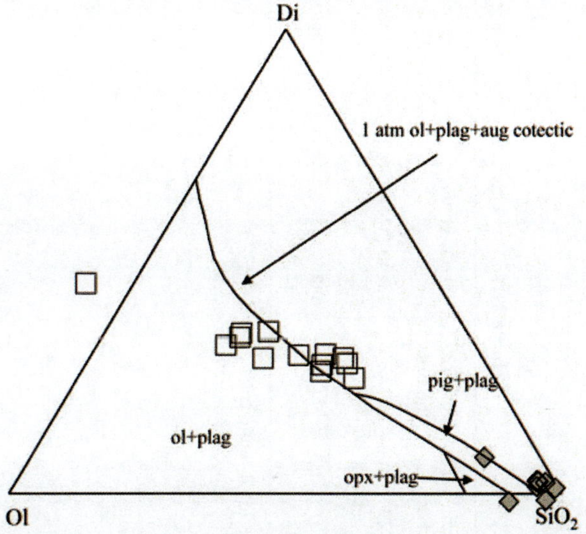

Fig. 21.9 Pseudoternary diagram Fo-Di-Qz (after Grove et al., 1982), with the igneous rocks of Vatomandry and Mahatsara. Symbols as in Figs. 21.3 and 21.5

Table 21.7 Major element mass-balance calculations for the transition from mafic to evolved rock types

From M723	to M729	Olivine (Fo_{69})	augite ($Mg_{44}Ca_{40}Fe_{16}$)	plagioclase (An_{67})		% solid	f (M723)	$\sum R^2$
tho bas	tho bas	–10.1	–15.0	–31.4		–56.5	0.44	0.26
M729	M722	augite ($Mg_{43}Ca_{30}Fe_{27}$)	plagioclase (An_{60})	Ti-magnetite	apatite			
tho bas	dac	–25.2	–22.7	–13.8	–1.1	–62.9	0.16	0.37
M722	M731	augite ($Mg_{43}Ca_{30}Fe_{27}$)	plagioclase (An_{38})	magnetite	ilmenite			
dac	rhy	–15.9	–12.1	–2.0	–0.8	–30.7	0.11	0.47

f, residual liquid fraction from M723.
$\sum R^2$, sum of squares of residuals.

$^{87}Sr/^{86}Sr_i$ ratio and the low Ce/Pb and εNd_i values observed in most Mailaka and Antanimena basalts indicate a crustal input during their genesis.

The tholeiitic lava flows and associated dyke swarm from Sambava, Tamatave and Mananjary transects appear to have been derived by mixing of continental lithospheric mantle and MORB mantle source (Storey et al., 1997). A few basaltic rocks from southeastern coast have Sr-Nd-Pb isotopic compositions resembling those of the present-day Marion hotspot (Storey et al., 1997).

None of the Vatomandry basalts have trace elements concentrations resembling those of present-day Marion hotspot basalts. The observed patterns of the present-day Marion hotspot lavas can be produced by low degrees of partial melting of enriched garnet-bearing source.

Origin and Evolution of the Silicic Rocks

As reported in the Introduction, the origin of silicic rocks in continental flood basalt provinces remains controversial and any model for their genesis must take into account the substantial bimodal composition of the samples and the presence of underlying continental crust. Direct derivation of the Vatomandry dacites and rhyolites by partial melting of the continental crust seems to be precluded by the strong isotopic differences [e.g. the crustal rocks of the Antananarivo Block have lower $^{143}Nd/^{144}Nd$ ratios ($\varepsilon Nd = -18.6$ to -28.6; Kröner et al., 2000) than the Vatomandry silicic rocks ($\varepsilon Nd_i = -5.6$ to -2.5; Melluso et al., 2009)]. Partial melting of a gabbroic (s. l.) crustal source could be a plausible hypothesis for the genesis of the Vatomandry silicic rocks. However, experimental melting of amphibolite assemblage under water-saturated conditions at 5 kbar (Heltz, 1976) produced liquids with higher Al_2O_3 and CaO contents relative to the Vatomandry silicic rocks. In addition, the high K/Rb (302–475) ratio of the silicic rocks rule out the presence of residual amphibole in the source. Silicic magmas derived from partial melting of garnet-granulite or eclogite rocks are characterized by low Y and heavy rare earth element (HREE) contents as consequence of residual garnet (Schnetger, 1994; Wareham et al., 1997). The high Y (>41 ppm) and HREE (e.g. Yb_n >22) contents observed in the silicic rocks rule out the possibility of garnet-granulite (or eclogite) as potential source. A partial melting model assuming a volatile-free source with 40% plagioclase, 40% clinopyroxene, 10% olivine, 10% magnetite (Mahoney et al., 2008) indicate that high degrees of partial melting (>20%) are needed to generate the Vatomandry silicic rocks, therefore a large amount of heat is required. We argue that fractional crystallization model can explain the genesis of Vatomandry silicic rocks and the silica gap between basalts and rhyolites. Mass-balance calculations for major elements using the observed phenocryst compositions indicate that the subtraction of gabbroic or Fe-gabbroic assemblages is required in the transition from basalts to rhyolites (Table 21.7).

Simulation of a fractional crystallization path (Fig. 21.6) starting from tholeiitic basalt (M723) indicates that the high Rb and Ba contents of the Vatomandry rhyolites could not be the result of closed-system fractional crystallization of basaltic melts. In addition, the Sr-Nd isotopic compositions of Vatomandry silicic rocks suggest the involvement of crustal rocks in their genesis. In order to model the trace element variation in the transition from basalt to rhyolite, we used a fractional crystallization + crustal assimilation model (AFC; DePaolo, 1981). AFC processes for silicic suites associated with flood basalts are well documented in other areas showing bimodal magmatism (e.g. Sheth and Melluso, 2008; Lustrino et al., 2010). The results of AFC modelling (Fig. 21.6) indicate that the range of trace element contents requires extensive fractional crystallization ($f = 80$–85%) and low r (mass assimilated/mass cumulated; silicic rocks data are clustered along the model curve at $r = 0.05$); thus the Vatomandry rhyolites can be formed by fractional crystallization starting from basaltic parental melts plus 4% crustal contamination.

Conclusions

The Cretaceous igneous rocks of the central-eastern coastal Madagascar are basaltic dykes and silicic rocks (dacites and rhyolites). An age of 86.4 $\pm$ 0.3 Ma has been determined for one rhyolite from Sakanila massif. The basaltic dykes are chemically evolved and for most major and trace elements show a compositional trend which is consistent with fractional crystallization of plagioclase, clinopyroxene and minor amount of olivine. In addition, the low Ba/Zr and Th/Ta and high Ce/Pb ratios observed in Vatomandry basalts indicate little crustal contamination. The less evolved samples have low incompatible trace element abundances similar to average transitional or normal MORB. The marked geochemical differences between the Vatomandry basalts and the present-day Marion hotspot lavas suggest different mantle sources and depths of melting. The presence of a depleted MORB-like component is well documented throughout the Cretcaeous Madagascan large igneous province.

The Vatomandry silicic rocks have isotopic compositions different from those of crustal basement beneath the Vatomandry district, implying that they did not originate by direct partial melting of old continental crust. In addition, the hypothesis of partial melting of a basic source is ruled out by high degrees of partial melting obtained (>20%). The origin of silicic rocks is more likely to be related to low-pressure fractional crystallization of mafic parental magmas, combined with assimilation of crustal rocks.

Acknowledgements We thank Roberto de' Gennaro, Sergio Bravi, Marianna Marrazzo and Luigi Franciosi for their help with data acquisition and field work. This project has been granted by MURST (COFIN 2008) grants to L. Melluso. The careful review of Hetu Sheth and the editorial comments of Rajesh Srivastava were very useful for the preparation of a revised version.

References

Besairie H, Collignon M (1972) Géologie de Madagascar I – Les terraines sédimentaires. Ann Geol Madagascar 35: 553 pp

Boynton WV (1984) Cosmochemistry of the rare earth elements: Meteorite studies. In: Henderson P (ed), Rare Earth Element Geochemistry. Elsevier, Amsterdam: 63–114

Cucciniello C, Langone A, Melluso L, Morra V, Mahoney JJ, Meisel T, Tiepolo M (2010) U-Pb ages, Pb-Os isotopes ratios, and platinum-group element (PGE) composition of the West-Central Madagascar flood basalt province. J Geol 118: 523–541

DePaolo DJ (1981) Trace element and isotopic effects of combined wallrock assimilation and fractional crystallization. Earth Planet Sci Lett 53: 189–202

de Wit MJ (2003) Madagascar: Heads it's a continent, tails It's an island. Ann Rev Earth Planet Sci 31: 213–248

Esawi EK (2004) AMPH-CLASS: An Excel spreadsheet for the classification and nomenclature of amphiboles based on the 1997 recommendations of the International Mineralogical Association. Comput Geosci 30: 753–760

Garland F, Hawkesworth CJ, Mantovani MSM (1995) Description and petrogenesis of the Paraná rhyolites, southern Brazil. J Petrol 36: 1193–1227

Gradstein FM, Ogg JG, Smith AG (2005) A Geologic Time Scale 2004. Cambridge University Press, UK

Grove TL, Gerlach DC, Sando TW (1982) Origin of calcalkaline series lavas at Medicine Lake Volcano by fractionation, assimilation and mixing. Contrib Mineral Petrol 80: 160–182

Heltz RT (1976) Phase relations of basalts in their melting ranges at $P_{H2O} = 5$ kb. Part II melt compositions. J Petrol 17: 139–193

Irvine TN, Baragar WRA (1971) A guide to the chemical classification of the common rocks. Canadian J Earth Sci 8: 523–548

Janney PE, le Roex AP, Carlson RW (2005) Hafnium isotope and trace element constraints on the nature of mantle heterogeneity beneath the central Southwest Indian Ridge (13°E to 47°E). J Petrol 46: 2427–2464

Kinzler RJ, Grove TL (1992) Primary magmas of mid-ocean ridge basalts 1. Experiments and methods. J Geophys Res 97: 6885–6906

Kirstein LA, Peate DW, Hawkesworth CJ, Turner SP, Harris C, Mantovani MSM (2000) Early Cretaceous basaltic and rhyolitic magmatism in southern Uruguay associated with the opening of the South Atlantic. J Petrol 41: 1413–1438

Kröner A, Hegner E, Collins AS, Windley BF, Brewer TS, Razakamanana T, Pidgeon RT (2000) Age and magmatic history of the Antananarivo Block, Central Madagascar, as derived from zircon geochronology and Nd isotopic systematics. Am J Sci 300: 251–288

Kumar A, Pande K, Venkatesan TR, Bhaskar Rao YJ (2001) The Karnataka Late Cretaceous dikes as products of the Marion hot spot at the Madagascar – India breakup event: Evidence from $^{40}Ar/^{39}Ar$ geochronology and geochemistry. Geophys Res Lett 228: 2715–2718

Le Bas MJ, Le Maitre RW, Streckeisen A, Zanettin B (1986) A chemical classification of volcanic rocks based on the total alkali – silica diagram. J Petrol 27: 745–750

Lepage LD (2003) ILMAT: An Excel worksheet for ilmenite-magnetite geothermometry and geobarometry. Comput Geosci 29: 673–678

le Roex AP (1987) Source regions of mid-ocean ridge basalts: Evidence for enrichment processes. In: Menzies MA, Hawkesworth CJ (eds) Mantle Metasomatism. Academic Press, New York: 389–422

Lustrino M, Marrazzo M, Melluso L, Tassinari CCG, Brotzu P, Gomes CB, Morbidelli L, Ruberti E (2010) Petrogenesis of early cretaceous silicic volcanism in SE Uruguay: The role of mantle and crustal sources. Geochem J 44: 1–22

Mahoney JJ, Nicollet C, Dupuy C (1991) Madagascar basalts: Tracking oceanic and continental sources. Earth Planet Sci Lett 104: 350–363

Mahoney JJ, Saunders AD, Storey M, Randriamanantenasoa A (2008) Geochemistry of the Volcan de l'Androy basalt-rhyolite complex, Madagascar Cretaceous igneous province. J Petrol 49: 1069–1096

McDougall I, Harrison TM (1999) Geochronology and Thermochronology by the $^{40}Ar/^{39}Ar$ Method, 2nd ed. Oxford University Press, New York

Melluso L, Cucciniello C, Petrone CM, Lustrino M, Morra V, Tiepolo M, Vasconcelos L (2008) Petrology of Karoo volcanic rocks in the southern Lebombo monocline, Mozambique. J Afr Earth Sci 52: 139–151

Melluso L, Morra V, Brotzu P, D'Antonio M, Bennio L (2002) Petrogenesis of the late cretaceous tholeiitic magmatism in the passive margins of northeastern Madagascar. In: Menzies MA, Ebinger CJ, Baker J (eds) Volcanic Rifted Margins, Special Paper. Geological Society of America, Boulder, CO vol 362: 83–98

Melluso L, Morra V, Brotzu P, Franciosi L, Grifa C, Lustrino M, Morbidelli P, Riziky H, Vincent M (2007) The Cenozoic alkaline magmatism in central-northern Madagascar: A brief overview. Per Mineral 76: 169–180

Melluso L, Morra V, Brotzu P, Franciosi L, Petteruti Liebercknecht AM, Bennio L (2003) Geochemical provinciality in the Cretaceous magmatism of northern Madagascar, and mantle source implications. J Geol Soc London 160: 477–488

Melluso L, Morra V, Brotzu P, Mahoney JJ (2001) The Cretaceous igneous province of central-western Madagascar: Evidence for heterogeneous mantle sources, crystal fractionation and crustal contamination. J Petrol 42: 1249–1278

Melluso L, Morra V, Brotzu P, Razafiniparany A, Ratrimo V, Razafimahatratra D (1997) Geochemistry and Sr-isotopic composition of the late Cretaceous flood basalt sequence of northern Madagascar: Petrogenetic and geodynamic implications. J Afr Earth Sci 34: 371–390

Melluso L, Morra V, Brotzu P, Tommasini S, Renna MR, Duncan RA, Franciosi L, d'Amelio F (2005) Geochronology and petrogenesis of the Cretaceous Antampombato-Ambatovy complex and associated dyke swarm, Madagascar. J Petrol 46: 1963–1996

Melluso L, Morra V, Fedele L (2006) An overview of phase chemistry and magmatic evolution in the Cretaceous flood basalt province of northern Madagascar. Per Mineral 75: 175–188

Melluso L, Sheth HC, Mahoney JJ, Morra V, Petrone CM, Storey M (2009) Correlations between silicic volcanic rocks of the St Mary's Islands (southwestern India) and eastern Madagascar: Implications for Late Cretaceous India-Madagascar reconstructions. J Geol Soc Lond 166: 1–12

Melluso L, Srivastava RK, Guarino V, Zanetti A, Sinha AK (2010) Mineral compositions and magmatic evolution of the Sung Valley ultramafic-alkaline-carbonatitic complex (NE India). Can Mineral 48: 205–229. doi:10.3749/canmin.48.1.205

Pande K, Sheth HC, Bhutani R (2001) ^{40}Ar-^{39}Ar age of the St. Mary's Islands volcanics, southern India: Record of India-Madagascar break-up on the Indian subcontinent. Earth Planet Sci Lett 193: 39–46

Putirka K (2008) Thermometers and barometers for volcanic systems. In: Putirka K, Tepley F (eds) Minerals, Inclusions and Volcanic Processes. Rev Mineral Geochem, vol 69. American Mineralogical Society, Washington, DC: 61–120

Rehkämper M, Hofmann AW (1997) Recycled ocean crust and sediment in Indian Ocean MORB. Earth Planet Sci Lett 147: 93–106

Roeder PL, Emslie RF (1970) Olivine-liquid equilibrium. Contrib Mineral Petrol 29: 275–289

Ronga F, Lustrino M, Marzoli A, Melluso L (2010) Petrogenesis of a basalt-comendite-pantellerite rock suite: The Boseti volcanic complex (main Ethiopian rift). Mineral Petrol 98: 227–243

Schnetger B (1994) Partial melting during the evolution of the amphibolite- to granulite-facies gneisses of the Ivrea Zone, northern Italy. Chem Geol 113: 71–101

Sheth HC, Melluso L (2008) The Mount Pavagadh volcanic suite, Deccan Traps: Geochemical stratigraphy and magmatic evolution. J Asian Earth Sci 32: 5–21

Storey M, Mahoney JJ, Saunders AD (1997) Cretaceous basalts in Madagascar and the transition between plume and continental lithosphere mantle sources. In: Mahoney JJ, Coffin MF (eds) Large Igneous Provinces: Continental, Oceanic and Planetary Flood Volcanism, Geophysial Monograph, vol 100. Amercian Geophysical Union, Washington, DC: 95–122

Storey M, Mahoney JJ, Saunders AD, Duncan RA, Kelley SP, Coffin MF (1995) Timing of hot spot-related volcanism and the breakup of Madagascar and India. Science 267: 852–855

Sun S-S, McDonough WF (1989) Chemical and isotopic systematics of oceanic basalts: implications for mantle composition and processes. In: Saunders AD, Norry MJ (eds) Magmatism in the Ocean Basins, Special Publication, vol 42. Geological Society of London, London: 313–345

Torsvik TH, Tucker RD, Ashwal LD, Carter LM, Jamtveit B, Vidyadharan KT, Venkataramana P (2000) Late Cretaceous India- Madagascar fit and timing of break-up related magmatism. Terra Nova 12: 220–224

Torsvik TH, Tucker RD, Ashwal LD, Eide EA, Rakotosolofo NA, de Wit MJ (1998) Late Cretaceous magmatism of Madagascar: Paleomagnetic evidence for a stationary hotspot. Earth Planet Sci Lett 164: 221–232

Wareham CD, Millar IL, Vaughan APM (1997) The generation of sodic granite magmas, western Palmer Land, Antarctic Peninsula. Contrib Mineral Petrol 128: 81–96

Windley BF, Razafiniparany A, Razakamanana T, Ackermand D (1994) Tectonic framework of the Precambrian of Madagascar and its Gondwana connections: A review and reappraisal. Geol Rundsch 83: 642–659

Chapter 22
Magnetic Fabric Studies on Mafic Dykes at a Volcanic Rifted Margin in the Henties Bay – Outjo Dyke Swarm, NW Namibia

Miriam Wiegand, Robert B. Trumbull, Reinhard O. Greiling, and Tim Vietor

Introduction

Large-scale mafic dyke swarms are prominent features of the Mesozoic volcanic rifted margins in the South Atlantic and Indian Oceans that relate to Gondwana break up. Examples are (parts of) the Okavango dyke swarm in eastern South Africa which are related to the 180 Ma Karoo large igneous province (LeGall et al., 2002; Watkeys, 2002; Jourdan et al., 2006), and two dyke swarms related to the 130 Ma Paraná-Etendeka large igneous province on the conjugate South Atlantic margins: the Ponta Grossa dyke swarm in Brazil (Raposo & Ernesto, 1995; Renne et al., 1996a, b) and the Henties Bay – Outjo dyke (HOD) swarm in NW Namibia (Trumbull et al., 2004), which is the subject of this paper. Previous AMS work in the Paraná-Etendeka province includes studies from the Ponta Grossa dykes in Brazil by Raposo and Ernesto (1995) and by Glen et al. (1997) who investigated basalt flows and mafic sills from both sides of the South Atlantic.

So far, there has been no comprehensive study on magma flow directions and emplacement mechanisms for the HOD dykes. Our work aims to determine regional patterns of magma flow by AMS studies on dykes located throughout the HOD. The flow patterns will help address questions about the mode of dyke emplacement and magma supply beneath the crust with respect to basement structures and distribution of major intrusive centres. This paper presents the first results and interpretation of magnetic fabric analysis from 165 samples distributed among 16 sites in the HOD.

M. Wiegand (✉)
Institut für Angewandte Geowissenschaften, Karlsruhe Institute of Technology, Karlsruhe, Germany
e-mail: miriam.wiegand@kit.edu

R.B. Trumbull
GFZ German Research Centre for Geosciences, 14473 Potsdam, FR Germany

R.O. Greiling
Institut für Angewandte Geowissenschaften, Karlsruhe Institute of Technology, 76187 Karlsruhe, FR Germany

T. Vietor
NAGRA, CH-5430 Wettingen, Switzerland

R.K. Srivastava (ed.), *Dyke Swarms: Keys for Geodynamic Interpretation*,
DOI 10.1007/978-3-642-12496-9_22, © Springer-Verlag Berlin Heidelberg 2011

Regional Setting

The Henties Bay – Outjo dyke (HOD) swarm was emplaced in the southern Etendeka province of NW Namibia. Associated with the dykes is a series of compositionally diverse intrusive complexes (Damaraland complexes; Trumbull et al., 2000) and erosional remnants of the Etendeka flood basalts (Marsh et al., 2001). The country rocks are tightly folded platformal and geosynclinal metasediments and syn- to post kinematic granites of the Damara Belt, a NE-SW trending Pan-African mobile belt that separates the earlier Proterozoic Angola and Kalahari cratons to the north and south, respectively (e.g. de Wit et al., 2008). During the Early Cretaceous break up of western Gondwana, Damara Belt structures were partly reactivated and the region was the site of extensive bimodal magmatism which makes up the Paraná-Etendeka Large Igneous Province (e.g. Peate, 1997).

Because of its relationship to the submarine Walvis Ridge and Rio Grande Rise hotspot tracks, and supported by geochemical-petrologic evidence, the magmatism is generally attributed to the influence of the Tristan mantle plume (O'Connor and Duncan, 1990; Peate, 1997; Thompson et al., 2001, 2007; Trumbull et al., 2003, 2007). Post-rift denudation and erosion removed on the order of 3–5 km of stratigraphy in the study area (Raab et al., 2005). As a consequence, the Etendeka basalts are preserved only in a few areas (Fig. 22.1b).

The distribution of the HOD dolerite dykes, many of which are geochemically equivalent to the Etendeka basalts (Trumbull et al., 2004), indicates a much larger original extent of the volcanic province.

Mafic Dykes in the HOD Swarm

The distribution of dykes in the HOD swarm was described by Hahne (2004) and Trumbull et al. (2004) based on mapping from high-resolution aeromagnetic data and Landsat ETM+ images. The orientation of mafic dykes within the HOD is generally NE-SW, but coast-parallel and other orientations are also present. The influence of Damara Belt structures on dyke orientation is apparent at a regional scale (see inset to Fig. 22.1a), but locally, dykes often cut across host-rock foliation and lithologic contacts. Hahne (2004) measured 210 dykes from different parts of the HOD and found a range in thickness from about 1.5 to 45 m (average 3 m, 90% of dykes less than 5 m). A few much larger dolerite intrusions, up to 500 m across and 3–4 km in outcrop length, occur in the central part of the HOD (e.g. Okombahe, Schwarze Spitzkoppe). These are compositionally equivalent to the thinner dykes.

Petrographic and geochemical studies of about 100 dykes from the HOD (Trumbull et al., 2004, 2007) found that most are texturally and compositionally simple, fine-to medium grained tholeiitic dolerites. In thin section the dolerites typically display intersertal to subophitic textures with plagioclase and clinopyroxene as the dominant minerals, lesser olivine and minor, fine-grained opaque minerals (Fe-Ti oxides and sulphides, see below) and acicular apatite. Most of the dykes are compositionally similar to the main series of low-Ti lavas in the Etendeka flood

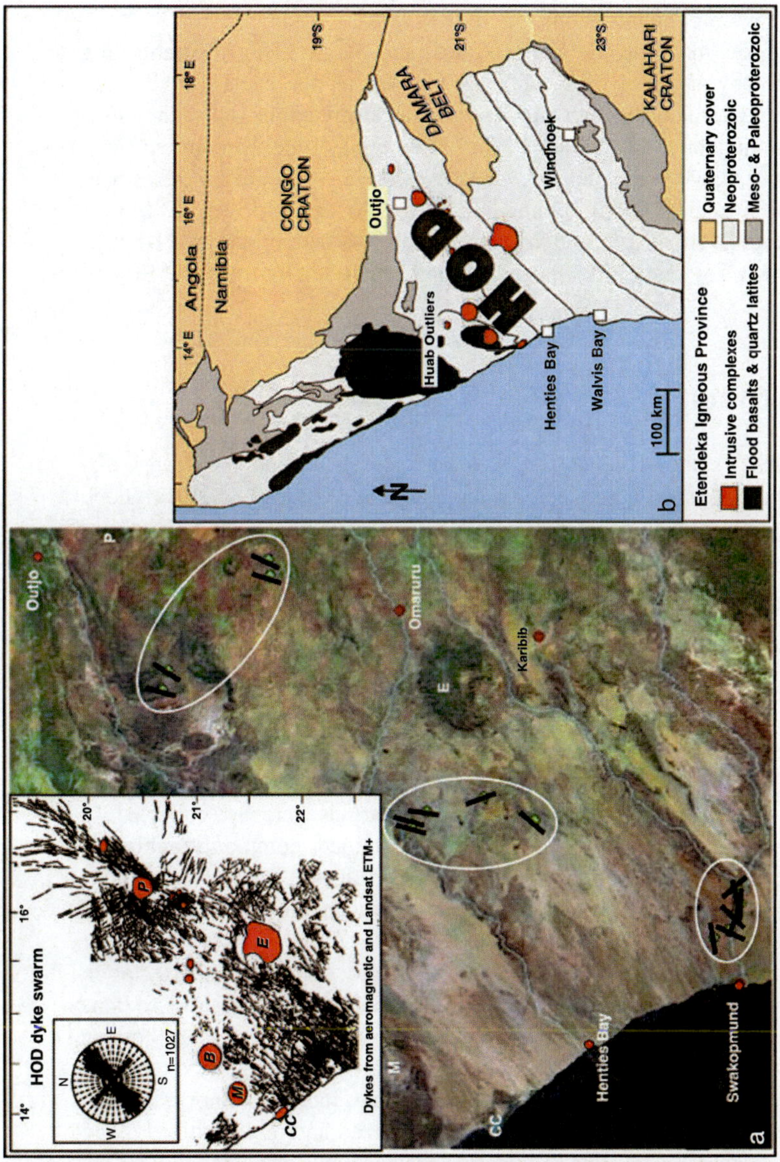

Fig. 22.1 A Satellite image of the study area showing the three areas of the Henties Bay-Outjo dyke (HOD) swarm investigated in this study (NE, central & SW, outlined in *ellipses*). *Green points* mark sample locations. *Black bars* show dyke trends. The *inset* shows distribution of intrusive complexes (*red*) and HOD dykes (*black*) together with a direction rose of dyke trends. B Simplified Map of NW Namibia showing the distribution of Etendeka volcanic and intrusive rocks in the Neoproterozoic Damara Fold Belt (after Trumbull et al., 2007)

basalt sequence but less geochemically evolved. Also present in much lesser amount are alkaline basaltic dykes (trachybasalt, tephrite). Most dykes are unvesiculated, which is consistent with the inferred 3–5 km paleodepth of intrusion (Raab et al., 2005). Chilled margins are typically well-developed and in many cases allow the recognition of dyke-in-dyke intrusions.

Over 6,000 dykes make up the HOD swarm, and the problem of establishing intrusion ages is obvious. However, an Early Cretaceous age can be confidently assigned for the great majority. Existing K-Ar age determinations fall in the range of 116–143 Ma (Siedner and Mitchell, 1976; Erlank et al., 1984; Hunter and Reid, 1987). Erlank et al. (1984) reported $^{39}Ar/^{40}Ar$ ages of 125–130 Ma for dykes near Horingbaai, Renne et al. (1996a) obtained an $^{39}Ar/^{40}Ar$ age of 132 ± 0.7 Ma for gabbroic sills from the Huab area, and Wigand et al. (2004) determined an $^{39}Ar/^{40}Ar$ age of 132 ± 1 Ma for mafic alkaline dykes and small stocks from the Erongo complex.

The Studied Dykes

For the present AMS study of the HOD, dykes were sampled from three clusters distributed along the dyke swarm between Swakopmund and Outjo (Fig. 22.1a). Each sample locality represents a single dyke. One dyke was sampled twice (localities 374 and 375) at a distance of 10 km to test along-strike changes in magnetic fabrics. The orientation of sampled dykes is mostly NE-SW (10 of 16), three strike E–W, two NW–SE and one strikes N–S. Most of the dykes are vertical, three of them are inclined between 49 and 62°. Dyke thickness varies from 1 to 18.3 m (Table 22.1). Although dykes are continuous at a large scale, field observations show that they are segmented. Segments are separated by small "bridges" of country rock.

Dyke orientation was measured by solar compass or by holding the compass well away from the rock surface. Samples were mostly taken from the centre of the dykes. Where possible, samples from near the dyke margins were also collected.

Thin section examination reveals fine to coarse-grained, commonly ophitic to subophitic and intersertal to intergranular textures. Preferred orientation of minerals is generally lacking but for two exceptions: one sample contains bands of plagioclase arranged in a chevron-like pattern together with ophitic to intergranular fabrics and radial arrangements of plagioclase. The plagioclase preferred orientation appears to be an effect of magma flow. In the second sample we observe ribbons of outsize (up to 4 mm) olivine-crystals with sets of parallel microcracks hosting hematite, magnetite, and goethite. Oxy-hornblende partly replacing pyroxene and sericite and/or epidote partly replacing plagioclase may indicate minor alteration in some samples. Some olivine crystals are serpentinized and/or chloritized along fractures, and display dark rims due to very fine replacements of oxide- and/or clay minerals, e.g. smectite.

Whole-rock chemical analyses of the sampled dykes are presented in Table 22.1 and compositions are plotted on the total alkali ($Na_2O + K_2O$) – silica classification diagram in Fig. 22.2, along with published data from the Etendeka low-Ti basalts (Tafelberg type, Marsh et al., 2001). Most dykes classify as basalt on this

Table 22.1 Geochemical analyses of selected dolerite dyke samples from the HOD

Sample	G2	G10	G43	171	372	373	374	375	376	377	415	418	421
SiO_2	49.0	48.5	56.5	50.0	47.6	51.2	49.3	49.2	47.4	47.6	52.5	45.8	44.3
TiO_2	0.94	1.41	1.24	1.00	0.69	1.16	1.07	1.08	0.92	0.68	3.17	1.36	3.27
Al_2O_3	14.5	16.7	15.2	16.0	12.7	15.1	13.6	13.7	12.8	13.0	12.9	13.4	14.2
Fe_2O_3	–	–	–	–	1.57	2.63	2.75	3.04	3.40	2.68	–	2.53	–
FeO	–	–	–	–	7.91	7.18	7.22	7.02	6.72	6.92	–	8.19	–
*FeOtot	9.42	11.4	8.00	9.64	–	–	–	–	–	–	12.6	–	14.7
MnO	0.17	0.18	0.13	0.18	0.16	0.17	0.17	0.17	0.17	0.17	0.18	0.18	0.26
MgO	8.86	5.99	4.42	6.94	14.2	6.86	10.3	10.1	13.0	13.8	3.09	11.4	4.59
CaO	10.7	9.59	7.01	10.1	9.89	9.72	9.87	9.86	9.56	10.6	6.2	12.6	8.3
Na_2O	1.76	2.60	2.72	2.03	1.50	2.48	1.94	1.88	1.62	1.42	3.10	1.83	3.53
K_2O	0.75	1.22	2.07	1.15	0.60	1.05	0.95	1.04	0.74	0.52	2.67	0.54	2.10
P_2O_5	0.10	0.16	0.21	0.12	0.09	0.13	0.14	0.14	0.12	0.09	0.61	0.27	1.48
LOI	2.85	1.00	1.59	1.74	2.30	1.56	1.94	2.20	2.93	1.94	1.44	0.98	1.51
Total	99.0	98.7	99.1	98.9	99.2	99.2	99.3	99.3	99.4	99.3	98.5	99.1	98.2
Cr	431	60	92	29	1,396	118	602	593	1,159	1,301	<10	652	<10
Ni	100	120	48	41	498	58	264	214	391	469	<10	337	34
Rb	17	32	49	30	20	196	27	37	41	19	55	10	38
Sr	185	232	262	188	114	221	162	168	168	116	340	324	680
Ba	201	330	650	253	121	128	248	259	194	96	1,343	228	922
Y	23	31	41	28	22	24	26	25	22	22	42	25	53
Zn	74	91	85	82	62	81	77	84	75	268	121	146	122
Zr	105	150	222	120	73	91	115	117	87	69	259	111	358
V	308	227	201	263	229	317	254	248	237	249	42	301	225

Analyses on whole-rock powders by X-ray fluorescence at GFZ Potsdam, with total iron reported as FeO (*FeOtot), except where noted. Ferrous and ferric iron values were calculated from XRF determination and separate wet-chemical analyses of FeO by titration.

diagram, one is basaltic andesite and two samples plot in the fields of basanite or basaltic trachy-andesite. The comparison of dyke compositions with Etendeka lavas supports the conclusion based on more complete geochemical evidence that the dykes represent feeders to now-eroded basalts. However, the dykes tend to be less geochemically evolved (lower silica) than their erupted derivatives (see also Marsh et al., 2001).

Magnetic Fabric (AMS) Studies

Methods

For the AMS study reported here, fresh samples of well-exposed HOD dykes at 16 localities were cored and oriented in the field using a portable drilling machine with a water-cooled diamond bit. Each locality provided 4–22 standard cylinders, 2.5 cm in diameter and 2.1 cm in height. The anisotropy of magnetic susceptibility of a total of 165 sample cylinders was measured in a low field (300 A/m) with a Kappabridge KLY-2 instrument (Jelinek, 1980; Hrouda, 1994). We also

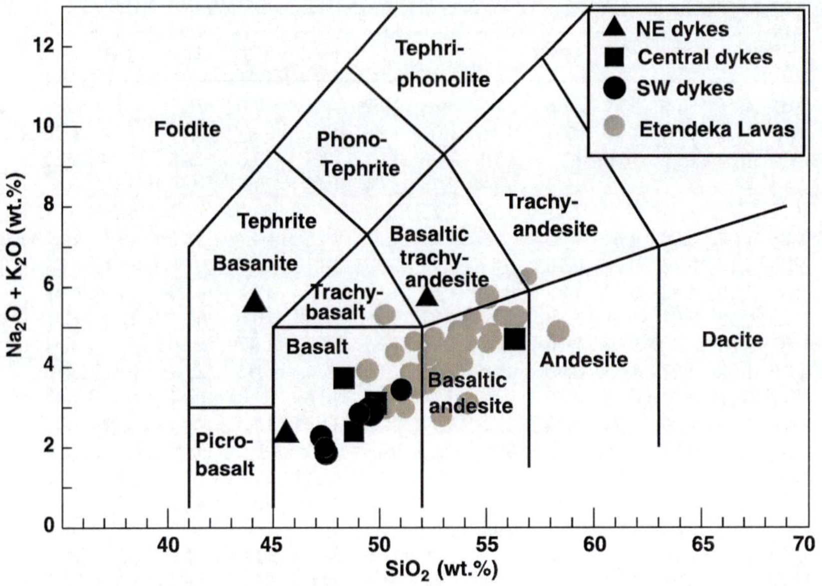

Fig. 22.2 Whole-rock geochemical compositions of the investigated dyke samples plotted on the total alkali – silica (TAS) classification diagram. *Coloured* symbols refer to the three sampled areas displayed in Fig. 22.1

determined the magnetic susceptibility as a function of temperature (kappa-T; Hrouda, 1994) for the temperature range of –192–700°C using a CLS-2 apparatus in combination with the Kappabridge KLY-2. We used liquid nitrogen for cooling, and samples were kept in an argon atmosphere during heating to avoid oxidation.

The principal axes of the AMS ellipsoid (K_{max} or K_1, K_{int} or K_2, K_{min} or K_3) were calculated using the software package ANISOFT (Hrouda et al., 1990). The long axis of the AMS-ellipsoid (K_1) corresponds to the magnetic lineation, and the short axis of the AMS ellipsoid (K_3) represents the pole to the magnetic foliation. The mean susceptibility value (bulk or volume susceptibility) is given by $K_{mean} = (K_1+K_2+K_3)/3$ (e.g. Tarling and Hrouda, 1993, de Wall 2005) in SI units. The intensity of anisotropy or degree of eccentricity of the AMS ellipsoid is expressed by the parameter P_j (Jelinek, 1981), which is defined as: $P_j = \exp\, [2(\ln K_1 - \ln K_{mean})^2 + (\ln K_2 - \ln K_{mean})^2\, (\ln K_3 - \ln K_{mean})^2]^{1/2}$. The shape of the AMS ellipsoid is illustrated by the T parameter, which varies from +1 for a perfectly oblate ellipsoid to –1 for a perfectly prolate ellipsoid (Jelinek, 1981). The shape parameter is defined by: $T = (2\ln K_2 - \ln K_3)/(\ln K_1 - \ln K_3)$. Further details and applications have been discussed by, for example Rochette (1987), Rochette et al. (1992), Borradaille (1988, 1991), Tarling and Hrouda (1993), Siegesmund et al. (1995), Borradaile and Jackson (2004), Martin-Hernandez et al. (2004), de Wall (2005), Dietl (2005) and Greiling et al. (2009).

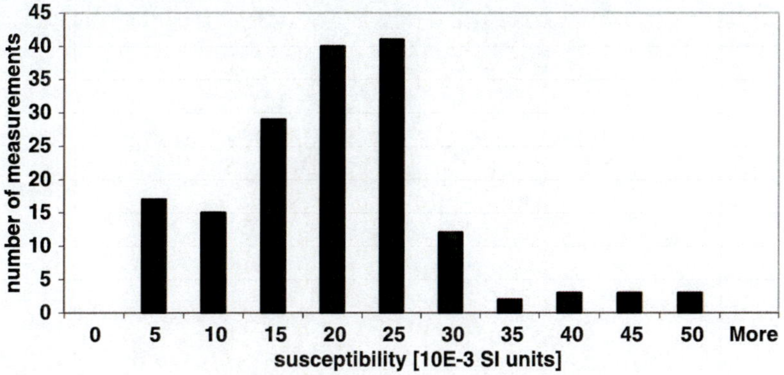

Fig. 22.3 Histogram of magnetic susceptibility distribution in the HOD based on K_{mean} column in Table 22.2

Results

Bulk Susceptibility and Magnetic Mineralogy

The bulk susceptibility values for the samples range from 3 to 46.88 × 10^{-3} SI with most between 10 and 30 × 10^{-3} SI (Table 22.2 and Figs. 22.3 and 22.4). One sample yielded a very low value of 0.7 × 10^{-3} SI. Generally, the susceptibility values of all cylinders from the same sample locality cluster around a common value. The HOD bulk susceptibility values above 3 × 10^{-3} SI units indicate the main contribution to rock magnetism is from ferrimagnetic minerals, as expected for non-metamorphic basaltic rocks.

Reflected-light microscopy revealed a variety of opaque minerals in these samples. Titanomagnetite (tmt), magnetite (mt) and ilmenite (ilm) occur in almost every sample whilst hematite, pyrrhotite, goethite, chromite, covellite, pyrite and chalcopyrite are only present in a few specimens and/or in minor amounts. Because tmt has strong ferrimagnetic properties and ilm is paramagnetic, the bulk susceptibility values identify tmt as the dominant magnetic mineral in our sample set. This is confirmed by measurements of magnetic susceptibility as a function of temperature (kappa-T), which distinguish different ferromagnetic minerals based on their Curie temperatures and other characteristic magnetic transitions. Figure 22.5 illustrates kappa-T curves of three samples, which are representative for low, moderate, and high mean susceptibility values in the sample set.

The kappa-T curve for sample 415–2 (Fig. 22.5a) shows a typical trend indicating tmt and magnetite contributions to the susceptibility. A minor peak at about –160°C is attributed to the Verwey transition in magnetite. With rising temperature the susceptibility increases up to the main peak at about 73°C, the Curie temperature for tmt. According to Lattard et al. (2006) this temperature indicates a tmt solid solution with a composition of 30% magnetite (Fe_3O_4) and 70% ulvöspinel (Fe_2TiO_4) (TM70). A decrease between 530 and 584°C is related to the Curie temperature

Table 22.2 AMS data for rocks from the samples of the mafic dykes

								K_{max}		K_{int}		K_{min}		Dyke			AMS
Site	Sample	K_1	K_2	K_3	K_{mean} [10–3] SI	P'	T	Azim.	Inc.	Azim.	Inc.	Azim.	Inc.	Width (m)	Strike	Dip	Fabric type
NE cluster																	
411	411-1o	0.02161064	0.02146498	0.02118438	21.42	1.02	0.327	106	70	224	10	317	17	?	17	?	I
	411-1U	0.02712473	0.02672896	0.02477631	26.21	1.102	0.676	95	74	221	9	313	13				
	411-2o	0.02612885	0.02596381	0.02506755	25.39	1.095	0.845	84	58	198	14	295	28				
	411-2U	0.02669234	0.02638896	0.02377262	25.93	1.087	0.705	115	60	19	3	287	30				
	411-3o	0.02798784	0.0277763	0.02559586	27.12	1.104	0.83	91	56	204	14	302	30				
	411-3u	0.03324001	0.0327606	0.02861624	31.54	1.18	0.806	107	60	209	7	303	29				
	411-4o	0.03004974	0.02970222	0.02640517	28.72	1.154	0.82	354	63	211	22	115	15				
	411-4U	0.03259311	0.03225691	0.03051363	31.13	1.159	0.843	266	73	21	7	113	15				
415	415-1	0.03993918	0.0392922	0.03911862	39.45	1.022	–0.567	102	25	228	52	358	27	3	15	90	III
	415-2	0.03988885	0.03940707	0.03917408	39.49	1.019	–0.346	103	24	268	65	10	5				
	415-3O	0.04331149	0.04161746	0.04149104	42.14	1.049	–0.857	120	18	218	22	355	61				
	415-3U	0.04845048	0.04621899	0.04597053	46.88	1.06	–0.795	120	17	28	4	284	72				
	415-4O	0.04171559	0.03970263	0.03960178	40.34	1.06	–0.903	99	27	247	59	2	14				
	415-4U	0.04855505	0.04579448	0.04578047	46.71	1.07	–0.989	98	25	297	64	191	8				
418	418-1o	0.0188646	0.01772195	0.01714345	17.91	1.102	–0.307	157	80	301	8	32	6	4	32	90	II
	418-1U	0.00817746	0.00815137	0.00813017	8.153	1.006	–0.106	176	84	294	3	24	5				
	418-2O	0.01917206	0.01818163	0.0173663	18.24	1.104	–0.074	204	70	105	3	14	19				
	418-2U	0.00795774	0.00795218	0.00793708	7.949	1.003	0.512	282	45	134	40	29	17				
	418-3	0.00827888	0.00826069	0.00823094	8.264	1.003	–0.346	239	73	344	5	75	17				
421	421-1O	0.04038266	0.0399577	0.03929622	40.09	1.013	–0.887	290	4	194	57	22	32	4	25	73 SE	III
	421-1M	0.0256437	0.02533103	0.02528273	25.42	1.016	–0.725	286	9	183	54	22	35				
	421-1U	0.02439762	0.02409537	0.02404701	24.18	1.016	–0.716	288	10	182	58	24	30				
	421-2O	0.04576958	0.04548255	0.04542332	45.56	1.008	–0.666	247	20	352	35	133	48				
	421-2U	0.03655751	0.036379	0.03635714	36.43	1.006	–0.78	240	14	356	60	143	26				
	421-3O	0.02566991	0.02550888	0.02550377	25.56	1.007	–0.936	292	3	195	66	23	24				
	421-3U	0.02806345	0.02792365	0.0278929	27.96	1.007	–0.65	116	4	25	16	219	74				

Table 22.2 (continued)

								K_{max}		K_{int}		K_{min}		Dyke			AMS
Site	Sample	K_1	K_2	K_3	K_{mean} [10–3] SI	P'	T	Azim.	Inc.	Azim.	Inc.	Azim.	Inc.	Width (m)	Strike	Dip	Fabric type
Central cluster																	
171	171-1O	0.0161403	0.01604394	0.01599576	16.06	1.009	–0.339	354	10	256	41	95	48	2.4	40	90	I
	171-1m	0.02105622	0.0210184	0.02095537	21.01	1.005	0.272	213	6	309	45	117	44				
	171-1u	0.0212002	0.0211727	0.02110922	21.16	1.004	0.394	222	11	325	51	123	37				
	171-2O	0.01616723	0.01607069	0.01603208	16.09	1.009	–0.433	0	2	268	38	93	52				
	171-2m	0.02094389	0.02090836	0.02084775	20.9	1.005	0.285	212	2	304	40	120	50				
	171-2u	0.02093011	0.02088418	0.01919707	20.88	1.005	0.113	10	12	271	35	116	52				
G02	G2-1	0.00988743	0.00979392	0.00954067	9.74	1.038	0.466	318	33	209	26	90	46	2.1	15	62 E	I
	G2-2	0.01032923	0.01023046	0.00998811	10.18	1.035	0.428	233	4	325	29	135	61				
	G2-3	0.00956428	0.00952002	0.0091706	9.42	1.047	0.779	194	14	288	16	66	69				
	G2-4	0.00981847	0.00969284	0.00948024	9.66	1.036	0.265	203	13	326	67	108	18				
	G2-5	0.01026444	0.01014214	0.00991373	10.11	1.036	0.312	185	4	284	62	93	27				
	G2-6	0.01009769	0.00981416	0.00972623	9.88	1.04	–0.52	359	24	231	54	101	25				
	G2-7	0.00843739	0.00834278	0.00811623	8.30	1.041	0.421	326	73	166	16	74	5				
	G2-8	0.0092739	0.00915345	0.00894813	9.13	1.037	0.266	5	12	226	74	97	10				
	G2-9	0.00786528	0.00779329	0.00756492	7.74	1.042	0.528	328	86	162	4	72	1				
G03	G03-1	0.00065771	0.0006562	0.0006511	0.65	1.011	0.549	176	14	269	15	43	69	1.2	19	49 SE	IV
	G03-2	0.00066508	0.00066396	0.00065945	0.66	1.009	0.595	222	18	317	14	82	66				
	G03-3	0.00067284	0.0006717	0.00066734	0.67	1.009	0.606	207	9	299	13	84	74				
	G03-4	0.00067792	0.00067711	0.00067447	0.68	1.005	0.549	60	3	151	9	310	81				
	G03-5	0.00068059	0.00067984	0.00067564	0.68	1.008	0.685	237	4	145	23	336	66				
	G03-6	0.00067589	0.00067434	0.00067036	0.67	1.008	0.442	229	10	135	15	349	72				
	G03-7	0.00068853	0.0006864	0.0006816	0.69	1.01	0.38	235	5	145	1	42	84				
	G03-8	0.00069022	0.00068891	0.0006843	0.69	1.009	0.567	262	13	171	4	63	76				
G10	G10-1	0.00985296	0.00973173	0.00951156	9.70	1.036	0.294	175	26	36	57	275	19	18.3	8	70 E	I
	G10-2	0.01054049	0.01043943	0.01027481	10.42	1.026	0.246	356	3	97	73	265	17				
	G10-3	0.01201895	0.01198082	0.01174609	11.92	1.025	0.728	159	22	336	68	69	1				
	G10-4	0.01084961	0.0106501	0.01050819	10.67	1.033	–0.161	329	47	100	31	207	26				
	G10-5	0.0103856	0.01012978	0.01005945	10.19	1.034	–0.563	336	38	89	26	203	40				
	G10-6	0.01314633	0.01283625	0.01277811	12.92	1.031	–0.68	312	49	97	36	200	18				
	G10-7	0.01167161	0.01153744	0.01149002	11.57	1.016	–0.469	0	12	166	77	270	3				

Table 22.2 (continued)

Site	Sample	K_1	K_2	K_3	K_{mean} [10–3] SI	P'	T	K_{max} Azim.	K_{max} Inc.	K_{int} Azim.	K_{int} Inc.	K_{min} Azim.	K_{min} Inc.	Dyke Width (m)	Dyke Strike	Dyke Dip	AMS Fabric type
	G10-8	0.01219507	0.01208174	0.01189245	12.06	1.026	0.257	349	1	137	89	260	0				
	G10-9	0.01156065	0.01141402	0.01139111	11.46	1.016	–0.722	3	6	268	41	100	49				
G43	G43-1	0.00383424	0.00380154	0.00380192	3.80	1.01	–1.123	186	26	326	58	87	18	?	160	90	I
	G43-2	0.003587	0.00354187	0.00353228	3.55	1.017	–0.649	188	38	297	22	50	44				
	G43-3	0.00417683	0.00416356	0.00410465	4.15	1.019	0.639	179	42	165	24	54	38				
	G43-4	0.00450774	0.00444972	0.00443142	4.46	1.018	–0.518	186	41	294	19	43	42				
	G43-5	0.00376232	0.00373431	0.00371077	3.74	1.014	–0.086	176	19	275	24	52	58				
	G43-6	0.00428713	0.00425572	0.00418867	4.24	1.024	0.367	197	35	294	9	37	53				
	G43-7	0.00404687	0.00399281	0.00397478	4.00	1.019	–0.491	177	26	290	38	62	40				
	G43-8	0.00449901	0.00443272	0.00441448	4.45	1.02	–0.57	185	32	306	39	70	35				
	G43-9	0.00424437	0.00419776	0.00416333	4.20	1.02	–0.35	240	24	142	17	20	60				
SW cluster																	
G01	G01-1	0.01782792	0.01771849	0.01740782	17.65	1.025	0.484	277	21	55	63	180	17	?	32	80 SE	I
	G01-2	0.01031925	0.0102106	0.00993137	10.15	1.04	0.449	262	71	354	1	85	19				
	G01-3o	0.01173401	0.01163433	0.01140598	11.59	1.029	0.401	197	47	339	36	84	20				
	G01-3	0.01560006	0.01550101	0.01532768	15.48	1.018	0.276	208	60	337	20	75	21				
	G01-4	0.01285451	0.01241385	0.01239502	12.55	1.042	–0.919	280	48	177	12	77	40				
	G01-4u	0.01094197	0.01087511	0.01053327	10.78	1.042	0.677	277	50	177	8	81	38				
	G001-5a	0.01774348	0.01753807	0.01738534	17.56	1.021	–0.138	302	5	35	30	205	60				
	G001-5b	0.01887113	0.01865668	0.01841427	18.65	1.025	0.069	301	12	40	38	196	49				
	G001-5c	0.01267405	0.01251121	0.01239346	12.53	1.023	–0.16	297	41	183	25	71	39				
	G001-5d	0.01254709	0.01240321	0.01226553	12.40	1.023	–0.017	298	40	187	24	75	41				
	G001-6o	0.01576139	0.01557242	0.01552088	15.62	1.016	–0.569	178	55	50	23	309	25				
	G001-6m	0.01638679	0.01615008	0.01610306	16.21	1.019	–0.668	181	27	21	61	275	9				
	G001-6u	0.01599453	0.01577456	0.01570651	15.83	1.019	–0.531	180	26	21	62	274	8				
	G001-7o	0.01619178	0.01603133	0.01604257	16.04	1.011	–1.049	14	84	114	1	204	6				
	G001-7m	0.01606352	0.01587775	0.01570945	15.88	1.023	–0.044	340	83	203	5	113	5				
	G001-7u	0.01604737	0.01583174	0.01568745	15.86	1.023	–0.191	348	84	200	5	110	3				
	G001-8o	0.01100973	0.01087488	0.01074329	10.87	1.025	–0.007	281	66	168	10	74	22				
	G001-8m	0.01236556	0.01222729	0.01211472	12.24	1.021	–0.098	264	66	171	1	81	24				

Table 22.2 (continued)

								K_{max}		K_{int}		K_{min}		Dyke			AMS
Site	Sample	K_1	K_2	K_3	K_{mean} [10–3] SI	P'	T	Azim.	Inc.	Azim.	Inc.	Azim.	Inc.	Width (m)	Strike	Dip	Fabric type
	G001-8u	0.0136809	0.01343411	0.01312294	13.41	1.043	0.127	273	60	82	30	175	5				
G58	G58-1	0.01340796	0.01330445	0.01310272	13.27	1.024	0.331	149	16	314	74	58	4	0.9–1	80	80 NW	IV
	G58-2o	0.01424714	0.01415402	0.01392545	14.11	1.024	0.429	147	8	332	82	237	1				
	G58-2u	0.01496923	0.01488318	0.0146621	14.84	1.022	0.442	144	19	309	71	53	5				
	G58-4	0.01430993	0.01423477	0.01399795	14.18	1.023	0.525	58	79	223	10	314	3				
	G58-5	0.01403817	0.01398528	0.01372782	13.92	1.024	0.659	115	84	317	6	226	2				
	G58-6o	0.01432565	0.01428445	0.01401732	14.21	1.024	0.742	292	4	157	85	22	4				
	G58-6u	0.0137308	0.0136614	0.01343278	13.61	1.023	0.545	183	84	316	4	46	5				
	G58-7o	0.01812279	0.01792069	0.01761128	17.88	1.029	0.214	62	70	305	10	212	18				
	G58-7u	0.01863514	0.01840912	0.0180857	18.38	1.031	0.184	64	72	304	9	212	16				
	G58-8o	0.0182527	0.01805444	0.01776426	18.02	1.028	0.19	46	68	305	4	213	21				
	G58-8u	0.01743315	0.01721134	0.01693796	17.19	1.029	0.107	41	72	303	3	212	18				
	G58-9o	0.01877485	0.01851381	0.01825462	18.51	1.029	0.002	78	74	308	10	215	12				
	G58-9u	0.01801576	0.01776002	0.0175096	17.76	1.029	–0.004	77	76	307	9	215	11				
	G58-10o	0.01811143	0.01788792	0.01764474	17.88	1.026	0.048	85	88	301	2	211	1				
	G58-10u	0.01882774	0.01858796	0.01835189	18.59	1.026	–0.012	328	88	127	2	217	1				
372	372-1O	0.02177161	0.02169132	0.02127034	21.7	1.006	–0.185	301	18	105	71	209	5	11	25	70 SE	III
	372-1U	0.01537532	0.01511405	0.01507911	15.19	1.021	–0.763	313	10	212	47	52	41				
	372-2O	0.025371	0.02482	0.024269	24.82	1.045	0.011	126	10	219	19	9	68				
	372-2M	0.01935021	0.01911146	0.01883833	19.1	1.027	0.069	130	11	224	25	18	63				
	372-2U	0.0182052	0.0180216	0.0177732	18	1.024	0.154	132	13	227	19	10	66				
	372-3O	0.02205301	0.02159487	0.02118212	21.61	1.041	–0.04	288	8	53	76	196	11				
	372-3U	0.01435891	0.01412152	0.01390957	14.13	1.032	–0.051	293	11	93	79	202	4				
	372-4O	0.02021524	0.01984206	0.0194927	19.85	1.037	–0.022	292	4	129	86	22	1				
	372-4U	0.01380944	0.01360136	0.0133892	13.6	1.031	0.02	286	8	116	82	16	1				
373	373-1O	0.00599732	0.00596932	0.00590736	5.958	1.016	0.381	56	18	157	30	300	54	6	44	50 NW	I
	373-1U	0.00730365	0.0072703	0.0071978	7.25	1.019	0.475	52	12	148	29	302	59				
	373-2O	0.00800298	0.00797433	0.00780239	7.96	1.013	0.442	35	14	128	9	250	73				
	373-2U	0.00867919	0.00863696	0.00853785	8.618	1.017	0.397	40	15	133	10	255	72				
374	374-1O	0.02681365	0.02645352	0.02617283	26.48	1.025	–0.113	185	8	278	19	73	69	4	110	90	III
	374-1M	0.02478271	0.02460537	0.02449946	24.63	1.012	–0.252	181	4	273	29	83	61				

Table 22.2 (continued)

Site	Sample	K_1	K_2	K_3	K_{mean} [10–3] SI	P′	T	K_{max} Azim.	K_{max} Inc.	K_{int} Azim.	K_{int} Inc.	K_{min} Azim.	K_{min} Inc.	Dyke Width (m)	Dyke Strike	Dyke Dip	AMS Fabric type
	374-1U	0.02417567	0.02397399	0.02391156	24.01	1.013	–0.366	191	2	282	27	97	62				
	374-2O	0.02380738	0.02344927	0.02342335	23.56	1.018	–0.866	6	11	112	54	269	34				
	374-2U	0.02091124	0.02083417	0.02041757	20.83	1.008	0.082	28	23	134	33	270	47				
	374-3O	0.02460707	0.02434854	0.02421439	24.39	1.016	–0.304	15	11	217	78	106	4				
	374-3M	0.02237359	0.02232007	0.02220634	22.3	1.008	0.366	16	37	201	53	108	3				
	374-3U	0.02234014	0.02228449	0.02215538	22.26	1.009	0.405	17	45	195	45	286	1				
	374-4O	0.02509072	0.02476048	0.02464129	24.83	1.019	–0.462	181	12	296	63	85	24				
	374-4M	0.02194219	0.02183309	0.02168472	21.82	1.012	0.147	180	20	307	59	81	23				
	374-4U	0.02259144	0.02247245	0.02228387	22.45	1.014	0.225	186	26	313	50	81	27				
375	375-1	0.02185277	0.02169466	0.02143257	21.66	1.02	0.25	169	10	74	26	277	62	at least 5.5	92	90	III
	375-2	0.02435305	0.02398673	0.02396263	24.1	1.018	–0.88	9	20	117	41	260	42				
	375-3	0.02159136	0.02136645	0.02130219	21.42	1.014	–0.546	348	2	257	21	84	69				
	375-4	0.02114192	0.02098006	0.02093802	21.02	1.01	–0.587	356	8	265	8	129	79				
	375-5	0.0207249	0.02062587	0.02053923	20.63	1.009	–0.073	343	19	245	22	109	60				
	375-7O	0.01974056	0.01952713	0.01947231	19.58	1.015	–0.591	187	9	85	54	284	34				
	375-7U	0.01942972	0.01928683	0.01921152	19.31	1.012	–0.311	188	7	72	75	280	13				
	375-8O	0.02205455	0.02182284	0.02170261	21.86	1.016	–0.311	0	3	269	25	97	64				
	375-8U	0.02028903	0.02012799	0.0200213	20.13	1.016	–0.006	352	2	261	26	85	64				
	375-9	0.02185821	0.02172565	0.02160614	21.73	1.012	–0.051	164	10	69	24	276	64				
	375-10	0.0211452	0.02104842	0.02062341	21.04	1.011	0.115	17	16	121	39	269	46				
	375-11O	0.02197682	0.02177	0.02156319	21.77	1.019	0.005	1	3	94	49	269	41				
	375-11U	0.02269639	0.02261063	0.02240298	22.57	1.013	0.414	10	14	124	58	272	28				
	375-13o	0.02168628	0.02129948	0.02112638	21.37	1.027	–0.375	182	10	296	66	89	21				
	375-13U	0.02352101	0.02313341	0.02297558	23.21	1.024	–0.41	179	10	292	65	85	23				
	375-14	0.024843	0.02437995	0.02427705	24.5	1.025	–0.635	180	14	281	38	73	49				
	375-15O	0.02503539	0.02458375	0.02442086	24.68	1.026	–0.465	185	4	280	54	93	36				
	375-15U	0.0265488	0.02608	0.0259412	26.19	1.024	–0.537	173	6	273	59	80	30				
	375-16	0.02197338	0.02164183	0.0199234	21.67	1.027	–0.139	111	68	334	17	239	14				
	375-17	0.02163296	0.0214057	0.02128134	21.44	1.017	–0.28	356	4	264	29	93	61				
	375-18O	0.01816776	0.01799297	0.01789927	18.02	1.015	–0.306	345	4	251	50	78	40				
	375-18U	0.01972822	0.01953653	0.01945242	19.56	1.016	–0.229	347	7	250	41	85	48				

Table 22.2 (continued)

Site	Sample	K_1	K_2	K_3	K_{mean} [10–3] SI	P′	T	K_{max} Azim.	K_{max} Inc.	K_{int} Azim.	K_{int} Inc.	K_{min} Azim.	K_{min} Inc.	Dyke Width (m)	Dyke Strike	Dyke Dip	AMS Fabric type
376	376-1O	0.02402754	0.02384374	0.02339737	23.87	1.012	–0.266	161	14	284	66	66	19	2.2	76	80–90 SE	III
	376-1U	0.02364034	0.02344537	0.0233843	23.49	1.011	–0.536	151	13	271	65	56	21				
	376-2O	0.02553989	0.02534193	0.02525818	25.38	1.011	–0.401	160	15	292	69	66	15				
	376-2U	0.02642148	0.02620115	0.02606737	26.23	1.014	–0.235	162	10	281	70	69	17				
	376-4O	0.02087243	0.02072095	0.02065663	20.75	1.011	–0.399	160	2	68	43	253	47				
	376-4M	0.02112345	0.02099739	0.02090915	21.01	1.01	–0.17	344	1	75	50	253	40				
	376-4U	0.02140224	0.02128935	0.02120841	21.3	1.009	–0.168	164	1	73	54	255	36				
377	377-1O	0.0147289	0.0144872	0.01446536	14.56	1.02	–0.833	70	36	216	49	326	17	?	140	85 SW	III
	377-1U	0.01685523	0.01665495	0.01656149	16.69	1.018	–0.358	66	41	230	48	329	8				
	377-2O	0.01655601	0.01627806	0.01618977	16.35	1.022	–0.628	48	6	273	82	138	6				
	377-2M	0.01700485	0.01671959	0.01644776	16.91	1.021	–0.462	46	8	206	81	316	3				
	377-2U	0.01711123	0.01683898	0.01677979	16.78	1.024	–0.646	48	8	189	80	317	6				
	377-3O	0.01624954	0.01581493	0.01569553	15.92	1.037	–0.562	220	5	313	30	121	60				
	377-3M	0.01695051	0.01651366	0.01636417	16.61	1.037	–0.483	220	2	311	24	125	66				
	377-3U	0.01681454	0.01638771	0.01623774	16.48	1.037	–0.477	222	4	314	23	123	66				
	377-4O	0.01779926	0.01729525	0.01722549	17.44	1.036	–0.75	224	5	316	22	121	67				
	377-4M	0.01844169	0.01790199	0.01780633	18.05	1.038	–0.693	224	4	315	12	115	77				
	377-4U	0.01779302	0.01724983	0.01718541	17.41	1.039	–0.785	226	6	318	25	123	64				

K_1 is the long principal axis of the AMS ellipsoid, K_2 and K_3 are the intermediate and the short principal axes, respectively. K_{mean} is the bulk magnetic susceptibility in 10^{-3} SI units, P′ (P_j) is the anisotropy factor. Azimuth (azim.) and inclination (incl.) of the principal axes refer to geographic directions. See text for details.

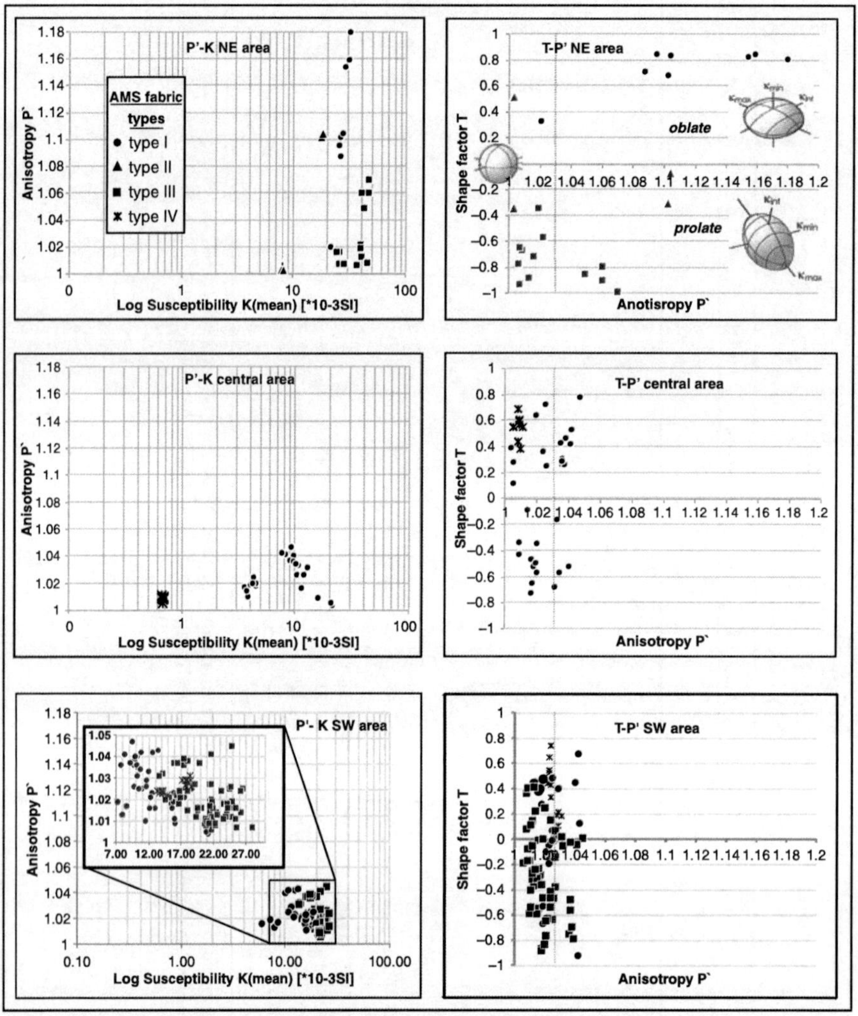

Fig. 22.4 *Left*: Corrected anisotropy factor P′ (P_j) versus bulk susceptibility K (mean) for dyke samples from the HOD. Each symbol represents a specimen. *Right*: Jelinek diagrams, showing the shape factor (T) of the AMS ellipsoids versus the corrected anisotropy P′

of magnetite. In contrast to the heating experiment, a distinct rise in susceptibility during cooling, at ca. 590°C, indicates the formation of magnetite during heating. The reflected light micrograph for this sample (Fig. 22.5a) shows intergrowths of tmt with ilm as well as single grains of both phases. The ratio of tmt to ilm in this thin-section is nearly 1:1. The sample 415–2 contains an unusually high proportion of opaque minerals (10–15 vol%), of which about 1% are sulphides. Sample 411–3u (Fig. 22.5b) shows a Verwey transition of magnetite at about –154°C and a peak at the Curie temperature of tmt (TM70) at about 69°C. At higher temperatures further peaks occur at about 356, 493 and 560°C. The two intermediate Curie points relate

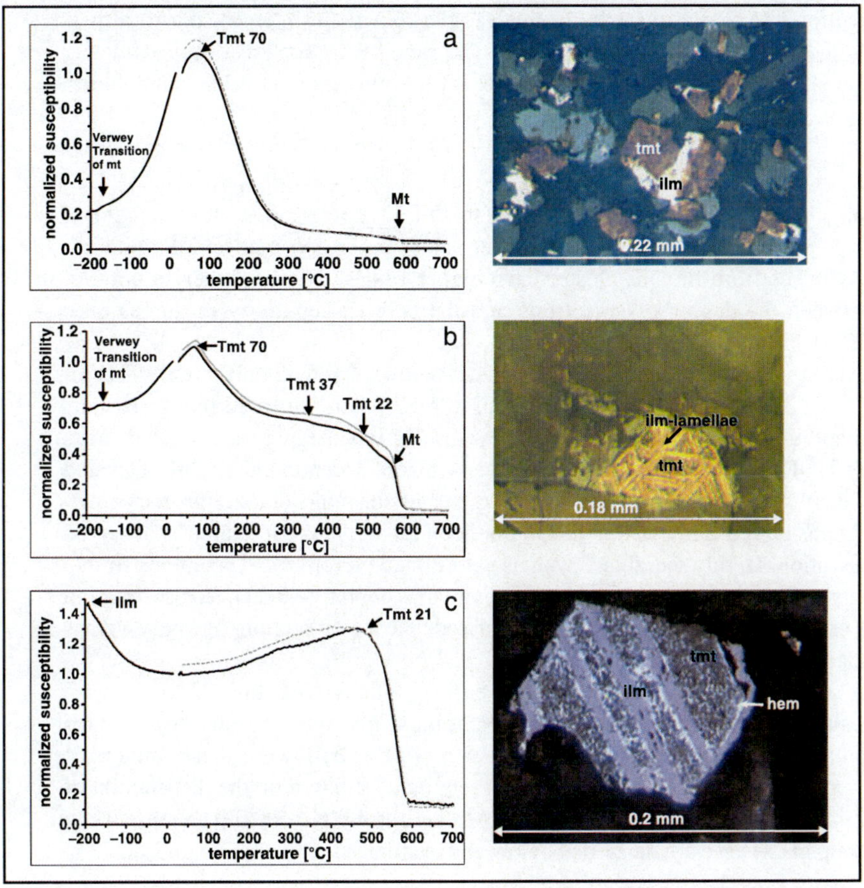

Fig. 22.5 Temperature versus susceptibility (K-T) diagrams. The K-T curve a (sample 415–2) at *left* shows a typical trend indicating titanomagnetite (tmt) and magnetite contributions to the susceptibility. For details see text. *Right*: reflected light micrograph showing intergrowths of tmt with ilmenite (ilm), as well as single grains of both phases. The tmts have been stained red with ferrofluid to distinguish them clearly from the whitish ilm. Sample b (411–3u) K-T curve shows a Verwey transition of magnetite at about –154°C and a peak at the Curie temperature of tmt (TM70) at about 69°C. For details see text. *Right*: reflected light photomicrograph showing reddish tmt containing a set of exsolved whitish ilm lamellae parallel with the grain boundaries. C Left: K-T curve of sample 373–1o with a peak susceptibility at about –192°C representing the Curie temperature of ilm. There is only one other peak at about 470°C, indicating tmt of composition TM21. *Right*: reflected light micrograph showing tmt intergrown with ilm. The *white margins* and the *spots* inside of the magnetite grains are hematite (hem), the oxidation phase of magnetite

to an exsolution of the tmt solid solution, which can also be seen in the micrograph. A reddish tmt grain contains a set of exsolved whitish ilm lamellae parallel with the grain boundaries. But there are also homogeneous tmt and ilm crystals as well as a small amount of sulphides. Sample 373–1o (Fig. 22.5c) shows a peak susceptibility at the temperature of about –192°C representing the Curie temperature of ilmenite. There is only one other peak at about 470°C, which corresponds to tmt of composition TM21. Petrography of this sample shows tmt intergrown with ilm as well as individual grains of both phases. The ratio of tmt to ilm is approximately 3:2. The total opaque fraction in the sample is 3–5 vol%, including 1% of sulphides.

AMS or Magnetic Fabric

The anisotropy (P_j) values vary from 1.009 up to 1.18. There appears to be a regional pattern of anisotropy variations, with lower values in the SW and central domains, whereas dykes from the NE cluster have higher values. There is rarely a correlation between the degree of anisotropy and the bulk susceptibility in the samples (Fig. 22.4).

Most samples from the HOD have P_j values below 1.05, with only a few measurements from 3 different localities above this value. According to Tarling & Hrouda (1993) and de Wall (2005) P_j values below 1.1 for ferrimagnetic-dominated rocks like the HOD dykes may be interpreted as primary. Exceptional, slightly elevated values of P' reflect the presence of aligned clusters of multi-domain primary magnetite grains or of aligned magnetite and/or hematite in olivine fractures as observed in thin section. Highly variable P_j values determined on different specimens from a single dyke locality (e.g. 1.02–1.18) may point to secondary effects. Rock alteration and its effects on the AMS results are discussed below (see section "Primary Versus Secondary Fabrics" in this chapter).

The shapes of AMS ellipsoids displayed in a Jelinek plot (Fig. 22.4) show no dominant fabric type. A wide range of AMS ellipsoid shapes was also reported for the Ponta Gross dykes by Raposo & Ernesto (1995). Following these authors, 4 fabric types can be distinguished in the present data set based on the distribution of major, intermediate and minor axes of the AMS ellipsoid (K_1, K_2 and K_3) relative to the dyke plane. These are described below and examples are illustrated on Fig. 22.6.

Fabric type I or *normal fabric* (Rochette et al., 1991) is defined by the magnetic foliation parallel with the dyke plane and with K_3 nearly perpendicular to it. We allow slight deviations of the K_1-K_2-surface from the dyke plane. K_1 and K_2 in some cylinders are interchanged, consistent with an oblate magnetic fabric. Seven of 16 sample sites represent this type. In four of the localities the axes are clustering well, in three others there is some scatter.

Fabric type II is termed *intermediate fabric* by Rochette et al. (1992, 1999), and characterized by K_1 and K_3 defining a plane parallel with the dyke plane. One site from our set reveals this type of pattern.

Fabric type III: The intermediate and minor axes K_2 and K_3 cluster or form a girdle nearly parallel with the dyke plane, whereas K_1 is perpendicular to it. Compared to type I fabric, K_1 and K_3 are reverse relating to the dyke walls, which Rochette et al. (1999) termed an abnormal fabric. The plane of magnetic foliation and the dyke plane have completely different azimuths, often perpendicular to one another. In several samples, K_2 and K_3 axes are interchanged, which indicates a prolate magnetic texture. Fabric type III is observed at 7 of 17 sites.

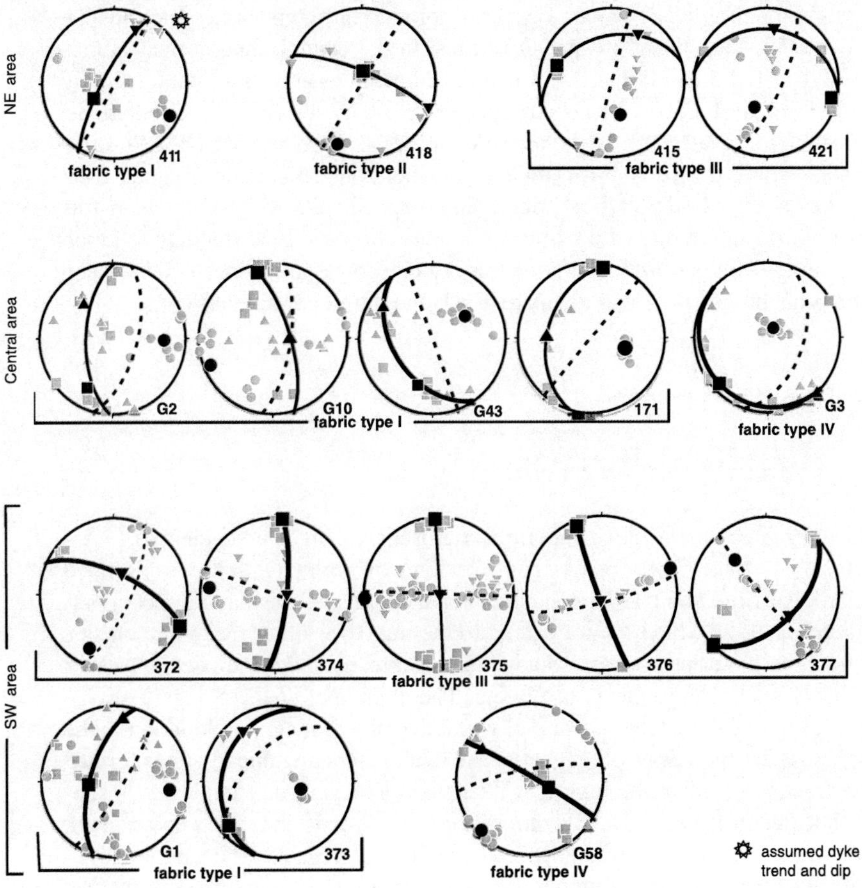

Fig. 22.6 Fabric type distribution in the HOD. Types I to IV according to Raposo and Ernesto (1995), displayed in lower hemisphere of Schmidt net. *Grey squares*, *triangles* and *circles* represent K_1, K_2, K_3 of one specimen, respectively. *Black squares*, *triangles* and *circles* show mean K_1, K_2, K_3 of the sampled locality. *Dashed lines* show dyke planes, *solid lines* show magnetic foliation planes. See text for details

Fabric type IV constitutes a category of samples which do not fit any of the other. AMS axes cluster well but mean AMS axes do not lie within the dyke plane and cannot be related to it, except for steep K_1 axes in one site. Two localities match with this category.

Regional Distribution of Fabric Types

In the SW, fabric type III is the most common. Figure 22.6 shows horizontal magnetic lineations (K_1 axis) and magnetic foliation planes (plane of K_1 and K_2 axes) perpendicular to the dyke planes. Two sites with normal fabrics occur and one with fabric type IV. Only one of the normal fabric type dykes shows a horizontal lineation whereas the foliation plane is flat. The other normal fabric type dykes and the type IV dyke reveal steep lineations and foliations. In the central area, normal fabrics dominate (4 out of 5 localities). Their magnetic lineations are horizontal to slightly inclined while the foliation planes are almost parallel with the dyke orientations. One exception is a site with fabric type IV. In the NE, two sites of NE-SW striking dykes display inverse fabrics, both with a horizontal lineation oriented E-W and a relatively flat N-dipping foliation plane perpendicular to dyke strike. A further locality has an intermediate fabric, a steep lineation and E-W trending foliation plane. Samples from a fourth locality in this region show a steep lineation and steep foliation probably parallel with the dyke trend, and thus a normal fabric.

Discussion

Primary Versus Secondary Fabrics

From petrographic observations under the microscope, there is no evidence of mineral deformation, shearing or other signs of tectonic overprint. One site, as described in section "The Studied Dykes", contains olivine aggregates, displaying sets of parallel microfractures, which do not extend to the matrix or other rock components. These fractures have been mineralized with magnetite, hematite and goethite, which are probably alteration products of olivine. The alignment of the elongate olivine aggregates and their fractures is parallel with the plane of magnetic foliation in the sample, and the platy shapes of magnetite and other magnetic mineral grains formed in these fractures explain the high P' values measured in these samples. It is suggested that fracturing and mineralization occurred prior to magma emplacement, since the fractures do not extend beyond the olivine-aggregates. The parallel alignment of the olivine aggregates within the dyke is interpreted as an effect of magma flow during dyke emplacement. Therefore, the AMS signal is interpreted as a primary flow fabric although in this particular case there is an enhancement due to secondary mineralization.

A possible evidence for secondary alteration affecting the magnetic fabric may be indicated by the variation of anisotropy values within different specimens from localities 411 and 418. Apart from these latter samples, there is no evidence of secondary alteration of the HOD fabrics. Therefore, the fabrics can be interpreted as primary, which originated during magma emplacement and thus represents a flow fabric.

Interpretation of the Fabric Types

The normal type of magnetic fabric (type I in our study) is generally accepted as an indication of magma flow orientation with the longest axis of magnetic susceptibility K_1 parallel with the flow direction (Rochette 1988, 1992, Raposo and Ernesto, 1995). At 5 sample sites showing normal fabric types in this study, the

Fig. 22.7 Interpretation map of dyke fabrics in the HOD investigated in the present study. *White bars* illustrate dyke trends. *Big yellow circles* represent flat-lying magnetic foliations, *yellow bars* steep foliations. The *black dots* indicate plunges of magnetic fabric long axes: shallow at *yellow bars* and steep at *small yellow circles*. *Transparent yellow bars* and *circles* show fabric type II to IV sample sites

inferred magma flow direction is close to horizontal and oriented NE–SW, N–S or NW–SE, parallel with the dyke orientation (Figs. 22.6 and 22.7). The magnetic foliation plane is mostly steep and shows some minor deviation from the orientation of the dyke walls. In two exceptions to this pattern (G1 and 411), the mean K_1 axes plunge more steeply (55–56°) to the west or to the SW.

The development of the inverse magnetic fabric (type III) cannot be explained unequivocally and magma flow directions cannot be deduced directly from the AMS ellipsoid. One of the possible causes for inverse fabric is a higher proportion of single-domain (SD) magnetite whose longest susceptibility axis K_1 (easy axis of spontaneous magnetization; Tarling and Hrouda, 1993) is already saturated and does not respond to the applied field during AMS measurements. The measured susceptibility is thus lowest along K_1 and highest along K_3, a characteristic, which Rochette et al. (1992, 1999) called inverse fabric. However, median destructive field values from alternating field demagnetization of all three fabric types yield values below 10 mT, indicating no major influence of SD grains. Also, preliminary studies of hysteresis did not give indications of SD magnetite. On that account we favour alternative explanations for these inverse fabrics by rotation of elongate magnetic particles during flow with velocity gradients, so that the particle long axes align perpendicular to the flow direction (Dragoni et al. 1997). Ellwood and Ledbetter (1977) showed this effect in sediments and other workers proposed the same for ash flow tuffs (e.g., Hillhouse and Wells, 1991; Lamarche and Froggatt, 1993; Le Pennec et al., 1998). Alternatively, these fabrics may reflect magma cooling progreesively from the wall rock into the dyke, normal to the dyke trend (Khan, 1962).

The intermediate or type II fabric was described by Raposo & Ernesto (1995) in the Ponta Grossa dyke swarm, Brazil. Rochette et al. (1992, 1999) and Ferré (2002) explain the formation of intermediate fabrics theoretically by mixing of normal and inverse magnetic fabrics. Park et al. (1988) described similar fabrics in the Proterozoic Mealy dyke swarm of Labrador and they ascribed this to vertical compaction of a static magma column. The latter explication may be applicable also to the HOD.

Finally, two sites have completely irregular magnetic fabrics (type IV) which do not fit any of the types defined above. We cannot explain the formation of these fabrics at present. Presumably, they represent some mixture of the above fabric types. In one of the sites (Fig. 22.6, G3) none of the AMS ellipsoid axes lies in the dyke plane. However, the shallow plunge and SW-NE orientation of the mean K_1 axis (magnetic lineation) is consistent with other K_1 directions in this central area. The other sample of type IV (G58) shows steep magnetic lineations but none of the other ellipsoid axes is in the dyke plane. The mean K_2 axis is relatively close (40° deviation) to the dyke trace, so this sample may be interpreted as a normal fabric type with some disturbance.

Implications for Magma Flow Directions

The K_1 orientations in normal fabrics are interpreted directly as magma flow orientations (Fig. 22.7). At sites with other fabric types, magma flow is difficult to interpret. These localities are marked with transparent flow symbols.

We recognize relatively consistent flow directions within the central region and varying directions in the NE and SW area. In the central region we find exclusively horizontal magma flow directions, mostly NE–SW oriented. In contrast, in the other areas there is both horizontal and vertical magma flow. Of the horizontal one, E–W directed magma flow occurs in the NE, while in the SW flow orientations trend mostly E–W and NE–SW, depending on dyke orientation.

Flow fabric orientations suggest magma ascent both far inland and in coastal areas. Horizontal flow is more frequent in coastal and central areas. Since dykes are segmented, there is no lateral continuity in dykes. Therefore, a strictly horizontal magma transport is not likely across long distances. The transport has to be more complex, probably including more than one site of magma ascent apart from the perhaps dominant one observed at the inland side of the HOD. This is consistent with the observed steep flow directions in the central and coastal areas.

Conclusions

On the basis of the presented data it is concluded that fabric types similar to those observed in the Ponta Grossa dyke swarm and other major dyke swarms can also be recognized in the HOD.

The present results point to magma ascent both far inland and close to the coast. Subhorizontal magma flow directions can be observed in each area. Since there is no lateral continuity in dyke trends because of segmentation, strictly horizontal transport over long distances is not possible. Consequently, we presume a complex magma flow model, with multiple conduits of magma ascent for the Henties Bay-Outjo dyke swarm.

Acknowledgements This paper is a contribution to the research Priority Programme 1375-SAMPLE. Funding by the German Research Council, DFG (grant GR 682/13-2) is gratefully acknowledged. We thank Oliver Frei (University of Freiberg) for assistance in the field work and the Geological Survey of Namibia for their support. Also we thank Erwin Appel (University of Tübingen) for the use of the Alternating Gradient Magnetometer and Moti Lal Rijal (University of Tübingen) for his engaged assistance. Discussions with Agnes Kontny (KIT) and Jens Grimmer (KIT) were very helpful. Reviews by an anonymous colleague and editor Rajesh Srivastava lead to substantial improvements in the manuscript.

References

Borradaille GJ (1988) Magnetic susceptibility, petrofabrics and strain. Tectonophysics 156: 1–20

Borradaille GJ (1991) Correlation of strain with anisotropy of magnetic susceptibility AMS. Pure Appl Geophys 135: 15–29

Borradaile GJ, Jackson M (2004) Anisotropy of magnetic susceptibility (AMS): magnetic petrofabrics of deformed rocks. In: Martin-Hernandez F, Lüneburg CM, Aubourg C, Jackson M (eds) Magnetic Fabric. Methods and Applications, Special Publications, vol 238. Geological Society, London: 299–360

de Wall H (2005) Die Anisotropie der magnetischen Suszeptibilität - eine Methode zur Gefügeanalyse. Z dt Geol Ges 155/2-4, 287–298

de Wit MJ, de Brito Neves BB, Trouw R, Allard J, Pankhurst RJ (2008) West-Gondwana: Pre-Cenozoic Correlations Across the South Atlantic Region; "the Ties that Bind", Special Publication, vol 294. Geological Society, London: 1–8

Dietl C (2005) Die magnetische Suszeptibilität – eine wertvolle Materialeigenschaft zur Charakterisierung von Granitintrusionen. Z dt Geol Ges 155: 299–309

Dragoni M, Lanza R, Tallarico A (1997) Magnetic anisotropy produced by magma flow: theoretical model and experimental data from Ferran dolerite sill (Antarctica). Geophys J Int 128: 230–240

Ellwood BB, Ledbetter T (1977) Antarctic bottom water fluctuations in the Vema Channel: Effects of velocity changes on particle aligment and size. Earth Planet Sci Lett 35(2): 189–198

Erlank AJ, Marsh JS, Duncan AR, Miller RM, Hawkesworth CJ, Betton PJ, Rex DC (1984) Geochemistry and petrogenesis of the Etendeka volcanic rocks from SWA/ Namibia. In: Erlank AJ (ed) Petrogenesis of Volcanic Rocks of the Karoo Province, Special Publication, vol 13. Geological Society of South Africa, Johannesburg: 195–245

Ferré EC (2002) Theoretical models of intermediate and inverse AMC fabrics. Geophys Res Lett 29, 31-1–31-4. doi: 10.1029/2001GL014367

Glen JMG, Renne PR, Milner SC, Coe RS (1997) Magma flow inferred from anistropy of magnetic susceptibility in the coastal Paraná-Etendeka igneous province: Evidence for rifting before flood volcanism. Geology 25: 1131–1134

Graham JW (1954) Magnetic susceptibility anisptropy, an unexploited petrographic element. Bull Geol Soc America 65: 1257–1258

Greiling RO, Grimmer JC, Kontny A (2009) Interpretation von Strukturen in der Lithosphäre mit Hilfe magnetischer Gefügeuntersuchungen. Zeitschrift für Geol Wissenschaften 37: 361–371

Hahne K (2004) Detektierung eines mesozoischen Gangschwarmes in NW Namibia und Rekonstruktion regionaler Spannungszustände während der Südatlantiköffnung. Ph.D. dissertation, University of Potsdam

Hillhouse JW, Wells RE (1991) Magnetic fabric, flow directions, and source area of the lower Miocene Springs Tuff in Arizona, California and Nevada. J Geophys Res 96: 9947–9954

Hrouda F (1994) A technique for the measurement of thermal changes of magnetic susceptibility of weakly magnetic rocks by the CS-2 apparatus and KLY-2 Kappabridge. Geophys J Int 118: 604–612

Hrouda F, Jelinek V, Hruskova L (1990) A package of programs for statistical evaluation of magnetic data using IBM-PC computers. EOS Trans Am Geophys Union 71: 1289

Hunter DR, Reid DL (1987) Mafic dike swarms in Southern Africa. In: Halls HC, Fahrig WF (eds) Mafic Dike Swarms, Special Paper, vol 34. Geological Association Canada, St. John′s: 445–456

Jelinek V (1980) Kappabridge KLY-2. A precision laboratory bridge for measuring magnetic susceptibility of rocks and its application (including anisotropy). In: Leaflet, Geofyzika, Brno

Jelinek V (1981) Characterization of the magnetic fabrics of rocks. Tectonophysics 79: 63–67

Jourdan FG, Feraud H, Bertrand MK, Watkeys AB, Kampunzu, Le Gall B (2006) Basement control on dike distribution in large igneous provinces: Case study of the Karoo triple junction. Earth Planet Sci Lett 241: 307–322

Khan MA (1962) The anisotropy of magnetic susceptibility of some igneous and metamorphic rocks. J Geophys Res 67: 2873–2885

Lamarche G, Froggatt PC (1993) New eruptive vents for the Whakamaru ignimbrite (Taupo volcanic zone) identified from magnetic fabric study. N Z J Geol Geophys 36: 213–222

Lattard D, Engelmann R, Kontny A, Sauerzapf U (2006) Curie temperatures of synthetic titanomagnetites in the Fe-Ti-O system: Effects of composition, crystal chemistry, and thermomagnetic methods. J Geol Res 111: B12S28

Le Gall B, Tshoso G, Jourdan F, Féraud G, Bertrand H, Tiercelin JJ, Kampunzu AB, Modisi MP, Dyment J, Maia M (2002) $^{40}Ar/^{39}Ar$ geochronology and structural data from the giant Okavango and related mafic dyke swarms, Karoo igneous province, northern Botswana. Earth Planet Sci Lett 202: 595–606

Le Pennec JL, Chen Y, Diot H, Froger JL, Gourgaud A (1998) Interpretation of anisotropy of magnetic susceptibility fabric of ignimbrite in terms of kinematic and sedimentological mechanisms: an Anatolian case study. Earth Planet Sci Lett 157: 105–127

Marsh JS, Ewart A, Milner SC, Duncan AR, Miller R, Mc G (2001) The Etendeka Igneous Province: Magma types and their stratigraphic distribution with implications for the evolution of the Paraná-Etendeka flood basalt province. Bull Volcanol 62: 464–486

Martin-Hernandez F, Lüneburg CM, Aubourg C, Jackson M (2004) Magnetic fabric: methods and applications – an introduction. In: Martin-Hernandez F, Lüneburg CM, Aubourg C, Jackson M (eds), Magnetic Fabric. Methods and Applications, Special Publications, vol 238. Geological Society, London: 1–7

O'Connor JM, Duncan RA (1990) Evolution of the Walvis Ridge-Rio Grande Rise hot spot system; implications for African and South American Plate motions over plumes. J Geophys Res 95: 17475–17502

Park JK, Tanczyk E, Desbarats A (1988) Magnetic fabric and its significance in the 1400 Ma Mealy diabase dikes of Labrador, Canada. J Geophys Res 93: 4301–4319

Peate DW (1997) The Parana-Etendeka Province. In: Mahoney JJ, Coffin ME (Eds.) Large Igneous Provinces; Continental, Oceanic, and Planetary Flood Volcanism. American Geophys Union, Geophys Monographs 100: 217–245

Raab MJ, Brown RW, Gallagher K, Weber K, Gleadow AJW (2005) Denudational and thermal history of the Early Cretaceous Brandberg and Okenyenya igneous complexes on Namibia's Atlantic passive margin. Tectonics 24, TC3006

Raposo MIB, Ernesto M (1995) Anisotropy of magnetic susceptibility in the Ponta Grossa dyke swarm (Brazil) and its relationship with magma flow direction. Phys Earth Planet Inter 87: 183–196

Renne PR, Deckart K, Ernesto M, Féraud G, Picirillo EM (1996a) Age of Ponta Grossa Dike Swarm (Brazil), and implications to Paraná flood volcanism. Earth Planet Sci Lett 144: 199–211

Renne PR, Glen JM, Milner SC, Duncan AR (1996b) Age of Etendeka flood volcanism and associated intrusions in southwestern Africa. Geology 24: 659–662

Rochette P (1987) Magnetic susceptibility of the rock matrix related to magnetic fabrics studies. J struct Geol 9: 1015–1020

Rochette P (1988) Inverse magnetic fabric in carbonate bearing rocks. Earth Planet Sci Lett 90: 229–237

Rochette P, Jenatton L, Dupuy C, Boudier F, Reuber J (1991) Diabase dikes emplacement in the Oman Ophiolite: a magnetic fabric study with reference to geochemistry. In: Peters T (ed) Ophiolite Genesis and Evolution of the Oceanic Lithosphere. Ministry of Petroleum and Minerals, Sultanate of Oman

Rochette P, Jackson M, Aubourg C (1992) Rock magnetism and the interpretation of the anisotropy of magnetic susceptibility. Rev Geophys 30: 209–226

Rochette P, Aubourg C, Perrin M (1999) Is this magnetic fabric normal? A review and case studies in volcanic formations. Tectonophysics 307(1-2): 219–234

Siedner G, Mitchell JG (1976) Episodic Mesozoic volcanism in Namibia and Brazil; K-Ar isochron study bearing on the opening of the South Atlantic. Earth Planet Sci Lett 30: 292–302

Siegesmund S, Ullmeyer K, Dahms M (1995) Control of magnetic rocks fabrics by mica preferred orientations: A quantitative approach. J Struct Geol 17: 1601–1613

Tarling DH, Hrouda F (1993) The Magnetic Anisotropy of Rocks. Chapman and Hall, London

Thompson RN, Gibson SA, Dickin AP, Smith DP (2001) Early Cretaceous basalt and picrite dykes of the southern Etendeka region, NW Namibia: Windows into the role of the Tristan mantle plume in Paraná-Etendeka magmatism. J Petrol 42: 2049–2081

Thompson RN, Riches AJV, Antoshechkina PM, Pearson DG, Nowell GM, Ottley CJ, Dickin AP, Hards VL, Nguno A-K, Niku-Paavola V (2007) Origin of CFB magmatism: Multi-tiered intracrustal picrite-rhyolite magmatic plumbing at Spitzkoppe, western Namibia, during Early Cretaceous Etendeka magmatism. J Petrol 48: 1119–1154

Trumbull RB, Bühn B, Romer RL, Volker F (2003) The petrology of basanite-tephrite intrusions in the Erongo complex and implications for a plume source of Cretaceous alkaline complexes in Namibia. J Petrol 44: 93–112

Trumbull RB, Emmermann R, Bühn B, Gerstenberger H, Mingram B, Schmitt A, Volker F (2000) Insights on the genesis of the Cretaceous Damaraland igneous complexes in Namibia: The Nd- and Sr- isotopic perspective. Commun Geol Surv Namibia 12: 313–324

Trumbull RB, Reid DL, De Beer CH, Romer RL (2007) Magmatism and continental breakup at the west margin of southern Africa: A geochemical comparison of dolerite dikes from NW Namibia and the Western Cape. South Afr J Geol 110: 477–502

Trumbull RB, Vietor T, Hahne K, Wackerle R, Ledru P (2004) Aeromagnetic mapping and reconnaissance geochemistry of the Early Cretaceous Henties Bay-Outjo mafic dike swarm, Etendeka Igneous Province, Namibia. J Afr Earth Sci 40: 17–29

Watkeys MK (2002) Development of the Lebombo rifted volcanic margin of southeast Africa. In: Menzies MA, et al. (eds) Volcanic Rifted Margin. Spec Pap Geol Soc Am 362: 29–48

Wigand M, Schmitt AK, Trumbull RB, Villa IM, Emmermann R (2004) Short-lived magmatic activity in an anorogenic subvolcanic complex: 40Ar/39Ar and ion microprobe U-Pb zircon dating of the Erongo, Damaraland, Namibia. J Volcanol Geothermal Res 130(3-4): 285–305

Chapter 23
Diopsidites and Rodingites: Serpentinisation and Ca-Metasomatism in the Oman Ophiolite Mantle

Marie Python, Masako Yoshikawa, Tomoyuki Shibata, and Shoji Arai

Introduction

The term "rodingite" was originally defined by P. Marshall in reference to grossular-rich dykes cross-cutting the serpentinites of the Dun Mountain in East Nelson, New Zealand (Bell et al., 1911). This term was later extended to describe a certain kind of alteration product of mafic rocks, implying mainly the gain of Ca and the loss of Si. In the resulting lithologies, the primary phases coexist with rodingite typical minerals, i.e. grossular, hydrogarnet, prehnite, wollastonite, zoisite/clinozoisite, epidote, chlorite, diallage (or diopside) and more or less serpentine (Capedri et al., 1978; Kobayashi and Shoji, 1988; Bilgrami and Howie, 1960; Honnorez and Kirst, 1975; Schandl et al., 1989; Dubińska, 1995; Frost et al., 2009). The mineralogical and chemical compositions of rodingites are in general highly variable, and it seems that their global high Ca and low Si contents, and their probable common genesis process are the only parameters in common to all of them (e.g. Kobayashi and Shoji, 1988; Bilgrami and Howie, 1960; Honnorez and Kirst, 1975; Schandl et al., 1989; Dubińska, 1995; Frost et al., 2009; Bach and Klein, 2009). Rodingites occur extremely frequently at the contact between mafic and ultramafic lithologies in various contexts and localities (ophiolites, tectonised serpentinites bodies, present day ocean peridotites, eclogite massifs, etc.; Kobayashi and Shoji, 1988; Mittwede and Schandl, 1992; Dubińska, 1995, 1997; El-Shazly and Al-Belushi, 2004; Li et al., 2007; Frost et al., 2009). Taking account their large variability, a precise definition of

M. Python (✉)
Division of Earth and Environmental Sciences, Graduate School of Natural Science and Technology, Kanazawa University – Kakuma-machi, Kanazawa 920-1192, Japan; Institute for Geothermal Sciences, Graduate School of Science, Kyoto University, Beppu 874-0903, Japan
e-mail: marie@bep.vgs.kyoto-u.ac.jp

M. Yoshikawa and T. Shibata
Graduate School of Science, Institute for Geothermal Sciences, Kyoto University, Beppu 874-0903, Japan

S. Arai
Division of Earth and Environmental Sciences, Graduate School of Natural Science and Technology, Kanazawa University – Kakuma-machi, Kanazawa 920-1192, Japan

R.K. Srivastava (ed.), *Dyke Swarms: Keys for Geodynamic Interpretation*,
DOI 10.1007/978-3-642-12496-9_23,

the term *rodingites*, based on the mineralogy, the texture or the chemistry (like *gabbro*, *dolerite* or *granite*) was not established. Alternatively, this word rather refers to their supposed common genesis process, which, although not totally understood, involves the interaction between a mafic rock and a fluid issued from a serpentinisation reaction. The composition of a fluid reacting with a relatively large amount of ultramafic rock and leading to extensive serpentinisation will be controlled by the serpentinisation reaction (Capedri et al., 1978; Muraoka, 1985; Frost and Beard, 2007; Bach and Klein, 2009). In general, this "serpentinisation-buffered fluid" is under-saturated in Si and rich in Ca coming from the decomposition of the pyroxenes into serpentine (Capedri et al., 1978; Muraoka, 1985; Frost and Beard, 2007; Bach and Klein, 2009). Its interaction with a mafic rock will lead to alteration with Ca enrichment and Si loss, and results in a Si-poor Ca-rich rodingite.

We recently described a new lithology cropping out in the mantle section of the Oman ophiolite. It is present as dykes in the mantle, located a few meters to a few tens of meter below the mantle-crust boundary, and usually shows a diopside monomineralic modal composition, so that we called it *diopsidite* (Python et al., 2007). The composition of the diopsidite is close to the magnesian end-member (Mg# > 0.95) with very low Al_2O_3, TiO_2, Cr_2O_3, and other incompatible element concentrations (Python et al., 2007). In detail, 4 samples contain anorthite as a major phase (up to 55 vol%), which composition is close to the Ca end-member. The textural, mineralogical and chemical characteristics of all these lithologies are in contradiction with a magmatic origin, and we interpreted them as the result of very deep hydrothermal circulation in the oceanic lithosphere (Python et al., 2007). While magmatic dykes are ubiquitous in the mantle of the Oman ophiolite, diopsidites are rather uncommon and represent about 2% of the dykes. The anorthite-bearing "diopsidites" have a gabbroic modal composition, however, in order to make the distinction with the magmatic gabbros and outline their similarities with the diopsidites, we will call them *anorthite-diopsidites* in the rest of this paper.

Rodingites occasionally include high-Mg diopside-rich zones, which composition is close to what we described for the Oman diopsidites (e.g. Dubińska, 1995; 1997; El-Shazly and Al-Belushi, 2004; Python et al., 2007), so that it is temping to consider that diopsidites are a particular kind of rodingites in which only diopside crystallised (Schandl et al., 1989; El-Shazly and Al-Belushi, 2004). The diopsidites would then be the result of the interaction of some of the ubiquitous gabbroic dykes with a serpentinisation issued fluid, and their monomineralic specific modal composition would be the consequence of some particular condition, like the nature of the protolith (ultramafic rather than mafic?) or other chemical parameter (nature of the fluid, f_{O_2}, X_{CO_2}).

In this paper, we present new data on diopsidites and on more classical rodingite found in some particularly altered zone of the Oman ophiolite mantle. We compare their mineralogical and chemical compositions and the state of alteration of their host (more or less altered harzburgites) in order to understand, at least partly, their similarities and differences, their origin and genesis process.

Geological Background and Previous Work

The Oman ophiolite belongs to the ophiolite belt enclosed in the Alpine mountain chain. It is located at the North-eastern extremity of the Arabian Peninsula and represents one of the biggest and best exposed section of oceanic lithosphere in the world. It is not included yet in the Alpine collision belt, so that its internal structures are remarkably well preserved. Although its precise tectonic setting is not clearly understood (mid-ocean ridge, e.g. Coleman, 1981; MacLeod and Yaouancq, 2000; Coogan et al., 2002; or arc-related basin, Pearce et al., 1981; Ishikawa et al., 2002), the mantle structures and the presence of an extensive sheeted dyke complex (Ceuleneer et al., 1988) show that it was a former spreading centre where basaltic MORB-like magmas emplaced on a deep seafloor in an extensional environment.

The Oman ophiolite is globally altered up to a certain point. The upper crust, from the basalts to the upper gabbros, is extensively hydrothermally altered, and the primary lithologies were metamorphosed in the greenschist facies. In the resulting altered rock, an albite + actinolite + chlorite + epidote + quartz ± titanite paragenesis is commonly observed (Nehlig, 1994; Nehlig et al., 1994; Juteau et al., 2000; Kawahata et al., 2001). Recent studies in the deeper gabbroic section also led to the conclusion that interactions at very high temperature between seawater and gabbros may have occurred down to the mantle/crust boundary (Manning et al., 2000; Kawahata et al., 2001; Nicolas et al., 2003; Bosch et al., 2004). On the other hand, the mantle peridotites, even said "fresh" always present at least a low amount of serpentine replacing olivine (Monnier et al., 2006; Nasir et al., 2007), some areas may even be extremely altered with a totally serpentinised, or even lately silicified or carboned, mantle (Stanger, 1985; Combe et al., 2006; Nasir et al., 2007). This serpentinisation, silicification, carbonation events in the mantle section mostly occurred at late stage in the history of the ophiolite (Tertiary, see Stranger, 1985; Nasir et al., 2007), and present days alteration by ground water is still occurring, as shown by the numerous ultra-alkaline sources (Barnes et al., 1978; Neal and Stanger, 1984; Stanger, 1985).

In the mantle section, mafic and ultramafic dykes of magmatic origin are very common (Ceuleneer et al., 1996; Python and Ceuleneer, 2003; Python et al., 2008). They are interpreted as the relics of the magma that circulated in the mantle to reach the surface at the time the ophiolite was a spreading centre. We published systematic studies of large set of these dykes distributed all along the Oman ophiolite (Ceuleneer et al., 1996; Benoit et al., 1996; 1999; Python and Ceuleneer, 2003; Python et al., 2008). The petrographical and chemical (major, trace, isotopes) data showed that most of them belong to two main magma suites, one issued from a MORB-type magma and the second one issued from parent melts richer in silica and ultradepleted in incompatible elements (Benoit et al., 1996; 1999; Python and Ceuleneer, 2003; Python et al., 2008). At the scale of the Oman ophiolite (i.e. about 400×80 km), these two families crop out in different areas which hardly overlap each other. Dykes from both families are frequently altered, leading to two types of secondary lithologies. The first one show a mineralogical association mainly composed of albite, actinolite-tremolite, epidote, serpentine, which is similar to the ones observed in the hydrothermally altered crustal gabbros. The second type of dyke alteration occurs in highly serpentinised areas and led to the formation of rodingites, i.e. low silica highly calcic assemblages dominated by grossular, clinozoisite/zoisite, diopside, chlorite, and prehnite, where the primary texture is generally preserved and the primary lithology is more or less visible according to the degree of alteration. Diopsidite dykes do not have a magmatic origin and crop out only close to the petrological Moho (Python et al., 2007). Their presence or absence is independent of the alteration degree of the peridotite.

Field Occurrences and Petrography

In total, 50 samples, including 7 gabbros, 9 diopsidites, 3 anorthite-diopsidites, 5 rodingites and 26 harzburgites/serpentinites, were analysed. Among the 26 harzburgites/serpentinites, 11 are relatively fresh samples taken away from any magmatic or hydrothermal dykes or serpentinised regions; 10 are the host of diopsidite dykes; and 5 are the host of the rodingites. The mineralogical characteristics of each sample are shown in Table 23.1.

Gabbros

The gabbros chosen for this study are fresh and relatively representative of the mafic magmatic dykes cropping out in the mantle section of the Oman ophiolite. A detailed description of the various facies of gabbros was already given in Python and Ceuleneer (2003) and will not be repeated here. The selected gabbros consist of 4 gabbronorites, 3 olivine gabbros, including 1 amphibole-rich olivine gabbro. The gabbronorites are relatively rich in orthopyroxene (10–20 vol%) and amphibole (up to 10 vol%), they are devoid of accessory minerals except for one sample which contains ilmenite associated with the amphibole. They are coarse-grained to pegmatitic and show a typical cumulate texture. The olivine gabbros are fine grained with adcumulate texture; they often contain chromian-spinel as accessory phase. One sample is rich in late amphibole (≈10 vol%) crystallising at crystal boundaries. Detailed textural, mineralogical and chemical description may be found in Python and Ceuleneer (2003) and Python et al. (2008).

Diopsidites and Anorthite-Diopsidites

Field occurrences of typical diopsidites are shown in Fig. 23.1. They are whitish dykes showing variable properties; the contact with the host may be progressive or sharp; the width of the dyke goes from a few centimetres to more than one metre; a tremolite-rich wall may exist at the dyke rim or not (Fig. 23.1). The cores of the diopsidite dykes themselves however are rather homogeneous, with a monomineralic diopside composition, a coarse grained fibrous to cataclastic texture (Fig. 23.2a and Table 23.1, see Python et al., 2007), and absent accessory phases. Trace amount

Table 23.1 Bulk chemical and modal compositions of the studied 50 samples. The modal composition is given in the abundance order. AP: various accessory phases (see text)

Label	Mineralogy	SiO_2	TiO_2	Al_2O_3	FeO*	MnO	MgO	CaO	Na_2O	K_2O	P_2O_5	Cr (ppm)	LOI	Total
Diopsidites														
07DIOP16	100% diopside	55.30	0.01	0.35	1.07	0.02	19.37	23.40	0.00	0.00	0.00	2,066	0.71	99.41
07DIOP20	100% diopside	55.46	0.00	0.55	1.20	0.02	20.51	22.19	0.00	0.00	0.00	2,112	1.18	99.83
05SDM9a	100% diopside	55.87	0.01	0.24	0.81	0.02	19.85	23.15	0.00	0.00	0.00	277	0.73	99.84
07DIOP19	100% diopside	55.92	0.01	0.18	0.91	0.02	19.56	23.41	0.00	0.00	0.00	132	0.77	99.89
07DIOP21	100% diopside	55.66	0.01	0.23	0.91	0.02	20.38	22.84	0.00	0.00	0.00	948	1.13	99.92
07DIOP3a	100% diopside	55.72	0.00	0.35	1.43	0.04	18.97	23.64	0.00	0.00	0.00	176	0.68	100.05
07DIOP8	100% diopside	55.96	0.01	0.15	0.93	0.02	19.72	23.42	0.00	0.00	0.00	42	0.63	100.11
07DIOP3b	100% diopside	55.22	0.01	0.51	1.84	0.05	20.41	22.41	0.00	0.00	0.00	2,261	1.38	100.36
Anorthite-diopsidites														
07HLHL5a	Diopside + Anorthite ± AP	50.02	0.08	15.24	0.94	0.01	10.84	22.43	0.48	0.38	0.01	8	2.39	100.43
07HLHL5f1	Diopside + Anorthite ± AP	49.15	0.10	18.17	1.87	0.03	10.78	19.95	0.59	0.06	0.00	117	1.41	100.68
07HLHL5b	Diopside + Anorthite ± AP	46.60	0.09	21.59	2.16	0.02	11.46	16.98	1.65	0.25	0.01	15	4.43	100.81
Gabbros														
07HLHL5f2	Plg + Opx + Cpx +Amph	45.15	0.02	31.42	0.28	0.00	2.77	21.31	0.31	0.06	0.00	2	0.88	101.31
07HLHL5g	Plg + Opx + Cpx +Amph ± Ilm	46.39	0.26	24.80	2.06	0.04	6.34	20.03	1.12	0.05	0.01	62	2.70	101.09
07HSIN5g	Plg + Olv + Cpx ± Amph	46.58	0.31	23.03	3.04	0.05	8.36	16.54	0.88	0.05	0.00	310	2.05	98.82
07HLHL5h	Plg + Opx + Cpx ± Amph	48.65	0.04	20.64	4.12	0.08	12.45	14.24	1.00	0.05	0.00	564	2.42	101.24

Table 23.1 (continued)

Label	Mineralogy	SiO_2	TiO_2	Al_2O_3	FeO*	MnO	MgO	CaO	Na_2O	K_2O	P_2O_5	Cr (ppm)	LOI	Total
07HLHL5i	Plg + Opx + Amph ± Cpx	48.30	0.05	22.02	3.35	0.07	12.17	14.40	1.06	0.06	0.00	306	2.07	101.46
07HSIN5h	Plg + Cpx + Olv + Amph ± Spl	47.81	0.24	17.15	7.23	0.12	12.38	13.92	1.60	0.00	0.00	515	0.44	100.41
07HSIN5c	Plg + Olv + Cpx ± Spl	46.52	0.20	16.97	8.18	0.13	13.44	12.55	1.52	0.00	0.00	402	0.62	99.44
Harzburgites														
07DIOP24	Olv + Opx+Ser ± Cpx ± Spl	44.49	0.00	0.66	9.06	0.13	43.52	1.87	0.00	0.00	0.00	3,149	5.58	99.59
07DIOP25	Olv + Opx+Ser ± Cpx ± Spl	43.41	0.02	0.61	9.40	0.14	44.77	1.80	0.00	0.00	0.00	3,274	7.67	99.98
07DIOP15	Olv + Opx+Ser ± Cpx ± Spl	43.45	0.01	0.62	9.19	0.14	44.33	2.00	0.00	0.00	0.00	2,921	4.82	99.61
07DIOP14	Olv + Opx+Ser ± Cpx ± Spl	39.42	0.01	1.19	10.68	0.14	46.05	1.86	0.00	0.00	0.00	3,130	9.11	99.21
07DIOP6	Olv + Opx+Ser ± Cpx ± Spl	40.09	0.01	2.63	10.43	0.15	43.39	2.95	0.00	0.00	0.00	1,824	8.39	99.49
07HLHL5c	Olv + Opx+Ser ± Cpx ± Spl	44.66	0.01	0.53	8.96	0.13	44.66	0.91	0.05	0.00	0.00	3,281	5.66	99.88
07HLHL5d	Olv + Opx+Ser ± Cpx ± Spl	45.96	0.01	0.65	8.96	0.13	43.40	1.20	0.01	0.00	0.00	2,925	3.44	100.29
07HLHL5e	Olv + Opx+Ser ± Cpx ± Spl	44.64	0.01	0.49	9.02	0.13	45.54	0.89	0.04	0.00	0.00	2,674	4.90	100.74
07DIOP18	Olv + Opx+Ser ± Cpx ± Spl	44.66	0.00	0.66	9.01	0.13	43.69	1.89	0.00	0.00	0.00	2,559	9.99	99.88

Table 23.1 (continued)

Label	Mineralogy	SiO_2	TiO_2	Al_2O_3	FeO*	MnO	MgO	CaO	Na_2O	K_2O	P_2O_5	Cr (ppm)	LOI	Total
Nephrites (diopsidites host)														
05SDM9e	Olv + Ser + Trem ± Spl	47.47	0.01	0.73	5.11	0.07	41.67	4.52	0.04	0.00	0.00	2,369	8.26	99.53
07DIOP17	Olv + Trem + Ser ± Spl	49.71	0.01	0.50	5.60	0.07	34.61	10.29	0.00	0.00	0.00	1,973	8.24	100.69
07DIOP12	Olv + Ser + Opx ± Trem ± Spl	46.27	0.01	0.79	6.08	0.08	42.72	3.74	0.08	0.00	0.00	2,090	8.71	99.69
07DIOP13	Olv + Ser + Opx ± Trem ± Spl	44.68	0.01	0.64	6.87	0.09	43.58	3.53	0.04	0.00	0.00	2,213	8.78	99.36
07DIOP23	Ser + Olv ± Opx ± Trem ± Spl	42.21	0.00	0.58	7.84	0.08	43.95	4.48	0.00	0.00	0.00	3,199	9.10	99.00
07DIOP5	Olv + Ser + Opx ± Trem ± Spl	43.14	0.04	1.62	9.66	0.13	40.98	3.95	0.00	0.00	0.00	2,740	8.87	99.39
05SDM2a	Trem + Ser ± Olv ± Spl	59.02	0.01	0.72	1.07	0.02	25.96	13.63	0.12	0.00	0.00	1,748	2.51	100.49
07DIOP9	Trem + Ser ± Olv ± Spl	58.46	0.01	0.88	0.94	0.02	25.69	14.10	0.21	0.00	0.00	233	2.55	100.26
07DIOP11	Trem + Ser ± Olv ± Spl	54.25	0.01	1.24	1.51	0.03	28.05	15.22	0.07	0.00	0.00	1,758	4.33	100.33
Serpentinites (rodingites host)														
07ROD2	Ser + Mgn ± Olv ± Spl	46.65	0.01	0.92	7.72	0.09	43.89	0.49	0.00	0.00	0.00	4,155	12.23	99.66
07HLHL7d	Ser + Mgn ± Olv ± Spl	45.67	0.01	0.58	9.41	0.12	43.87	0.45	0.00	0.00	0.00	3,369	11.51	99.95
07HLHL7c	Ser + Mgn ± Olv ± Spl	46.71	0.00	0.41	9.13	0.13	43.25	1.03	0.00	0.00	0.00	2,828	9.41	100.52

Table 23.1 (continued)

Label	Mineralogy	SiO_2	TiO_2	Al_2O_3	FeO*	MnO	MgO	CaO	Na_2O	K_2O	P_2O_5	Cr (ppm)	LOI	Total
07ROD3	Ser + Mgn ± Olv ± Spl	47.86	0.00	0.87	7.74	0.07	44.07	0.16	0.00	0.00	0.00	3,511	12.02	100.65
07HLHL8a2	Ser ± Olv ± Trem ± Spl	43.55	0.04	3.33	9.50	0.12	40.15	3.01	0.00	0.00	0.00	3,712	11.79	99.54
Rodingites (type-I)														
07ROD1a	Czo + Chl + Grt ± Dio ± Trem	38.91	0.22	23.33	4.53	0.12	16.17	16.91	0.00	0.00	0.00	151	6.28	100.04
07ROD1b	Grt + Chl ± Czo ± Dio ± Trem	37.10	0.30	22.87	5.86	0.53	11.34	22.13	0.00	0.00	0.00	120	4.97	99.95
Rodingites (rodingitised diopsidites, type-II)														
07HLHL8a1	Dio + Grt + Chl ± Cal ± Tit ± Pre ± Trem ± Zr	49.39	0.08	11.41	0.82	0.02	12.81	24.86	0.00	0.00	0.00	7	2.82	99.32
07HLHL7a	Grt + Dio + Chl + Trem ± Cal ± Pre ± Tit ± Zr	46.59	0.26	12.48	0.48	0.01	11.65	27.99	0.00	0.00	0.00	36	4.17	99.39
07HLHL7b	Dio +Grt + Hgr + Cal ± Chl ± Tit	40.15	0.12	17.26	0.62	0.02	9.09	32.30	0.00	0.00	0.01	70	7.12	99.48

Plg: plagioclase; *Opx*: orthopyroxene; *Cpx*: clinopyroxene; *Amph*: amphibole; *Ilm*: ilmenite; *Olv*: olivine; *Spl*: spinel; *Ser*: serpentine; *Trem*: tremolite; *Mgn*: magnetite; *Czo*: clinozoisite; *Chl*: chlorite; *Grt*: garnet; *Hgr*: hydrogarnet; *Pre*: prehnite; *Dio*: diopside; *Cal*: calcite; *Tit*: titanite; *Zr*: zircon

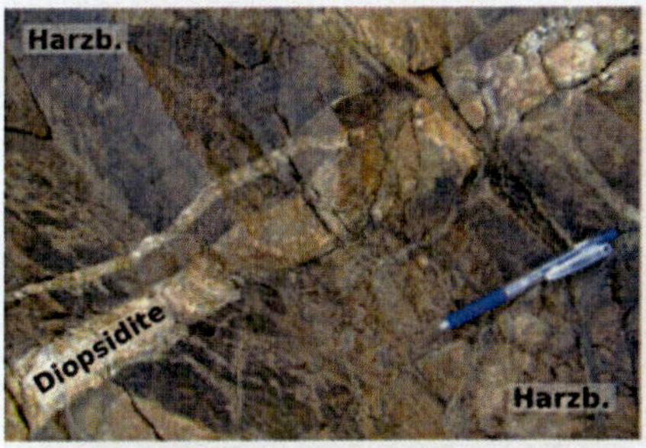

(a) Diopsidite dyke in fresh peridotite

(b) Two types of contact with the host

(c) Dyke with tremolite-rich rim

(d) Transformed harzburgite

(e) Anorthite-diopsidite cross-cut by a gabbronorite dyke

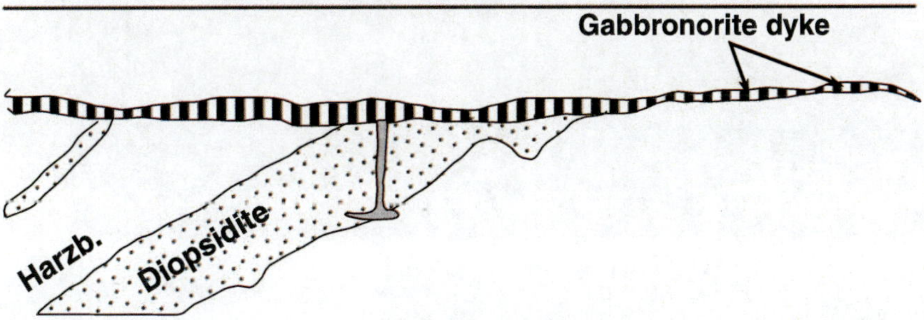

(f) Sketch of *(a)*

Fig. 23.1 Field views of various diopsidite dykes and their host. (**a**) General view of a diopsidite dyke showing clear contact with its host. The harzburgite is fresh up to the contact with the dyke. (**b**) Detail of a dyke with two types of contacts with its host. One is progressive and the other sharp with fibrous antigorite outlining it. (**c**) Diopsidite dyke with a highly tremolitised host. (**d**) Relics of partially altered harzburgite in a matrix of tremolite at the contact with a diopsidite dyke (**e**) and (**f**) Anorthite-bearing diopsidite cross-cut by a gabbronorite dyke

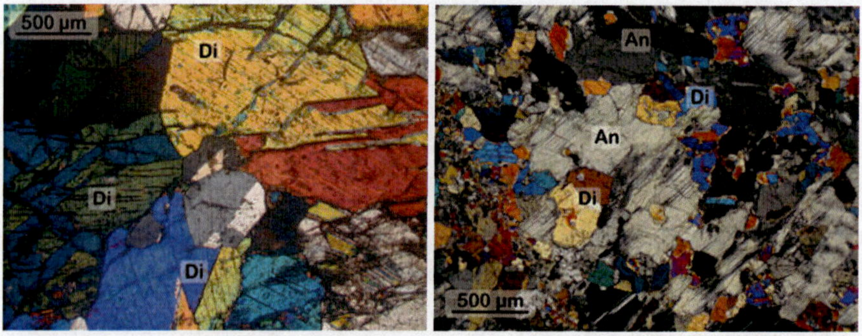

(a) Diopsidite (b) Anorthite-diopsidite

Fig. 23.2 Thin section (crossed Nichols) of the two types of diopsidities. (**a**) Diopsidite with a monomineralic diopside composition; (**b**) anorthite-diopsidite. *Di:* Diopside; *An:* Anorthite

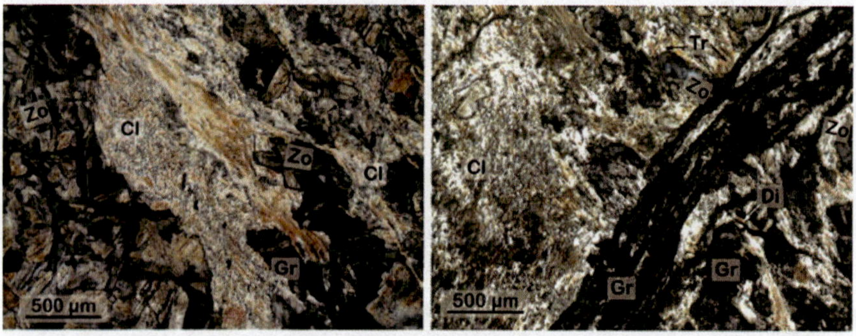

(a) Rodingites

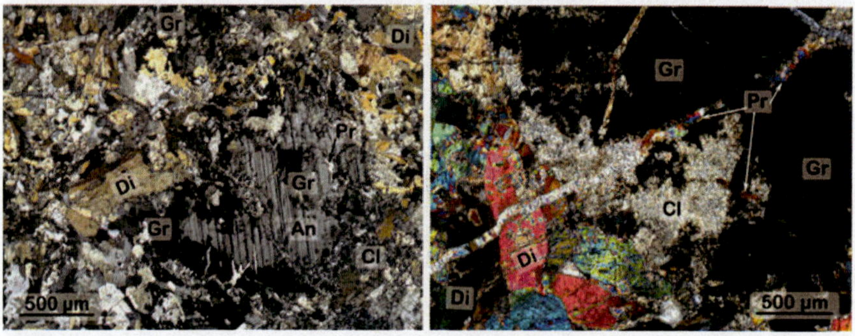

(b) Rodingitised An-diopsidite

Fig. 23.3 Photomicrographs (crossed Nichols) of Oman rodingites. In the totally rodingitised rock, the original texture is not visible. (**a**) rodingitised gabbros: Chlorite (*Cl*) and zoisite/clinozoisite (*Zo*) are the main constituents, grossular (*Gr*) is late and crystallise in vein cross-cutting the others minerals, low amount of diopside (*Di*) and tremolite (*Tr*) may appear in the space left between the other minerals. (**b**) rodingitised diopsidites: Anorthite (*An*) is progressively transformed into grossular; diopside is stable in the rodingitised product but may react with grossular to form chlorite, late prehnite (*Pr*) crystallise in veins of olivine and chromian-spinel may be included at the rim of some diopsidite dyke, close to the host.

The anorthite-diopsidites always crop out in a very fresh peridotite and may be cross-cut by magmatic dykes (gabbronorite, see Fig. 23.1e, f). Their field occurrences are more systematic than the diopsidites but their microscopic characteristics show a stronger variability. The main constitutive phases are diopside and anorthite at rather equivalent proportions (Fig. 23.2b and Table 23.1). A very low abundance of a great diversity of small accessory phases is observed, with the sporadic presence of titanite, rutile, zircon, corundum, phlogopite, etc. The textures are fine grained and more similar to that of the olivine gabbro adcumulates (Python and Ceuleneer, 2003) than to that of the diopsidites. The presence of subhedral diopside in inclusion within large anorthite suggests a rather contemporaneous crystallisation with a slightly early diopside compared to the anorthite (Fig. 23.3b).

Rodingites

Rodingites in Oman appear as very altered whitish dykes in totally serpentinised harzburgite. In the partially rodingitised dyke, a part of the primary textures and mineralogies are preserved. In Oman, we could distinguish two types of rodingite according to their protolith:

1. Type-I rodingites derive from the alteration of gabbroic dykes in the mantle and are mainly composed of chlorite and zoisite with minor grossular (Fig. 23.3a);
2. Type-II rodingites stem from the transformation of anorthite-diopsidite into a rodingite-like rock. Garnet and chlorite are their main constituents (Fig. 23.3b).

These two types of rodingites show different mineralogical composition and textures (Fig. 23.3). The studied five samples include 2 type-I and 3 type-II rodingites.

In type-I rodingites, zoisite and clinozoisite are subhedral and may form a mosaic cut by veins bearing an assemblage of late chlorite and grossular (Fig. 23.3a). Low amount of small diopsides and tremolites is observed at the grain boundaries between zoisite, grossular and chlorite, and traces of serpentine appear when approaching the contact with the host. In the 2 samples studied here, the primary adcumulate texture is still observable from place to place, but no relics of the primary minerals could be analysed. The original texture however shows that the protolith was a magmatic dyke.

Type-II rodingites contain abundant coarse grained euhedral diopsides and grossular (Fig. 23.3b). In one dyke of anorthite-diopsidite cropping out in a particularly serpentinised host, the progressive transformation of the primary lithology into a rodingite-like assemblage is clearly observable (Fig. 23.3b). Grossular first appears as coronæ around anorthite, which it tends to totally replace as the rodingitisation progress (Fig. 23.3b). Diopside may be partially transformed into tremolite and chlorite but seems to be generally stable in rodingite. It breaks into smaller grains, and remains relatively abundant in the final product. The degree of rodingitisation decreases from the rim of the dyke to its core, at the contact with the serpentinised harzburgite, no anorthite is left and the large grains of grossular and diopside are partly decomposed to form chlorite and late veins of prehnite. The core of the dykes may be hardly affected by rodingitisation, in that case, a fine grained assemblage of garnet and chlorite appears sporadically at the grain boundary between anorthite and diopside.

Harzburgites and Serpentinites

The ultramafic rocks used for this study are of 3 types according to their degree of alteration and their position and relationship with diopsidite or rodingite dykes.

1. Fresh harzburgites were sampled far (more than 2 m) from any magmatic, hydrothermal dyke or serpentinised area;
2. Serpentinites were taken close to or at the contact with rodingite dykes, and include (for one sample) the chloritic blackwall frequently observed between rodingites and their host (Frost, 1975; Dubińska, 1997; Frost et al., 2009);
3. Nephrites resulting from a particular kind of alteration of harzburgite, which is not serpentinisation, at various degrees were sampled close to or at the contact with diopsidite dykes.

Fresh Harzburgites

The Oman harzburgites were described in several studies (e.g. Monnier et al., 2006). The harzburgites studied here show the same characteristics as the average Oman ones. They mostly show a porphyroclastic texture, contain 10–25 vol% of orthopyroxene, up to 3 vol% of chromian-spinel, less than 5 vol% of clinopyroxene and are serpentinised to a certain degree. They were taken away from the large altered zones associated to the rodingites and their amount of serpentine does not exceed 20 vol%. Orthopyroxenes and chromian-spinels are generally fresh, clinopyroxenes are often absent or present at very low abundance as exsolution within orthopyroxenes (Fig. 23.4a). Amphibole is very rare in harzburgite but may appear in a few samples as an alteration phase of the pyroxenes.

Serpentinites (Host of the Rodingites)

Rodingites always occur in highly serpentinised harzburgite (LOI over 10 wt% in most of the cases, see Table 23.1). The Fig. 23.4b shows the typical rodingite host, i.e. a serpentinite containing less than 5 vol% of relics of fresh olivine, abundant mesh-textured serpentine and bastite. The original porphyroclastic texture is preserved and totally serpentinised orthopyroxene (bastite) is easily recognizable in thin section (Fig. 23.4b). Magnetite is abundant at the external rim of the mesh or as inclusion in bastite. Chromian-spinels show preserved cores and magnetite alteration coronæ. One sample contains the chloritic blackwall frequently observed at rodingite dyke rims. At the thin section scale the contact between the

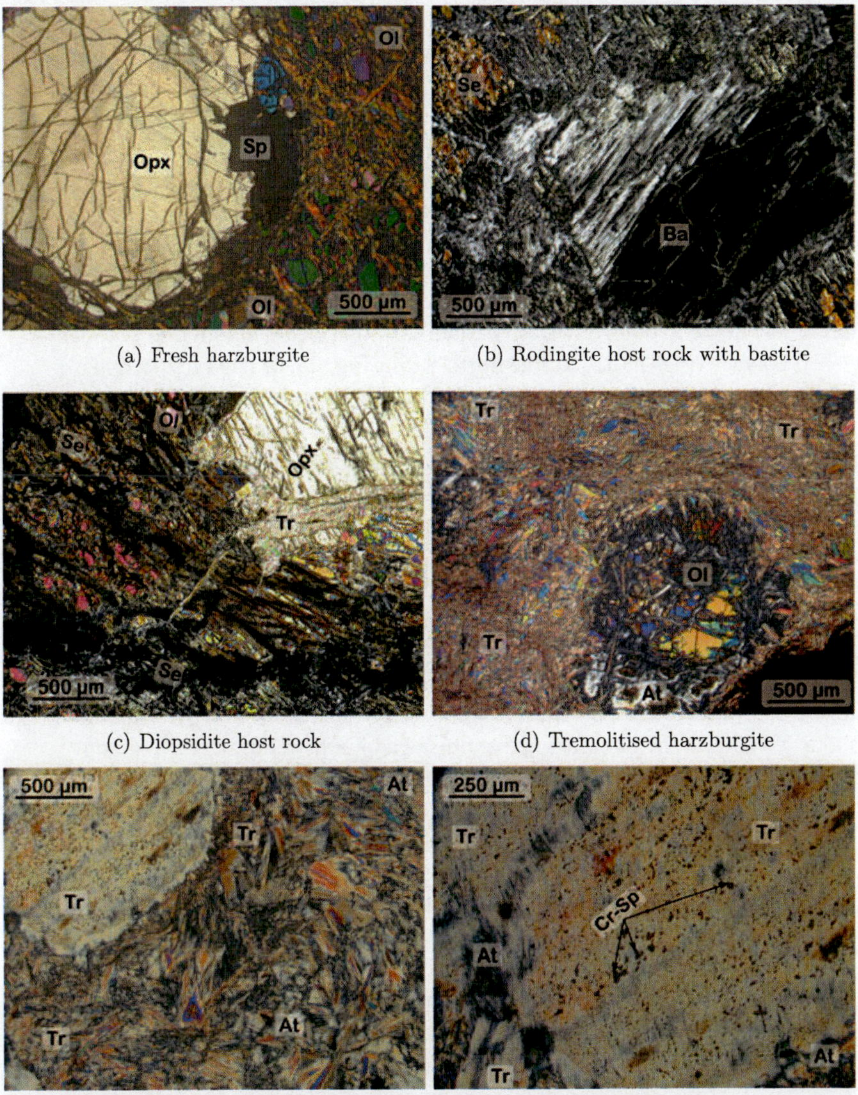

(a) Fresh harzburgite

(b) Rodingite host rock with bastite

(c) Diopsidite host rock

(d) Tremolitised harzburgite

(e) Tremolitised harzburgite with Opx pseudomorph and inclusions of Cr-rich spinel

Fig. 23.4 Photomicrographs (crossed Nichols) of rodingites and diopsidites host rocks. (**a**) Fresh harzburgite with a coarse grained porphyroclastic texture. (**b**) Totally serpentinised rodingite host with orthopyroxene pseudomorphs (Bastite: *Ba*). (**c**) Partially altered diopsidite host with orthopyroxene (*Opx*) partly transformed into tremolite (*Tr*) and relatively serpentinised olivine (*Ol*; serpentine: *Se*). (**d**) Host of diopsidite, largely transformed into tremolite, with relics of olivine. Olivine is altered into antigorite (*At*) and tremolite needles replace the antigorite. (**e**) Nephrite close to the contact with a diopsidite dyke with preserved coarse-grained porphyroclastic texture. Pseudomorphs of orthopyroxene, totally transformed into tremolite, include numerous small grains of Cr-rich spinel (*Cr-Sp*). Needles of tremolite crystallise in the serpentine mesh formed after olivine serpentinised host and the chloritic zone is sharp and outlined by a progressive increase of the grain size from the external to the internal rim of the dyke.

In Oman, rodingites occur in large scale serpentinisation zones. The host at the contact with the rodingite is almost totally altered, and the serpentinisation degree seems to be independent of the distance to the dyke. The serpentinised area may abruptly end at a few centimetres to a few metres from the rodingite, and be in direct contact (generally through a fault) with less altered harzburgite.

Nephrites (Host of the Diopsidites)

Diopsidites may occur in relatively fresh peridotite with a clear, although sometimes progressive, contact (Fig. 23.1a, b); the host rock in Fig. 23.1a is particularly hard and fresh and does not show higher serpentinisation degree than the average harzburgite of Oman taken far from any dyke (Fig. 23.4a). The host of the diopsidites presents various degree of alteration depending on the location and the proximity to the dyke (LOI varying from 2.5 to 9 wt% depending on the proportion of serpentine, see Table 23.1). In the example in Fig. 23.1a, the host is fresh up to the contact with the dyke; in (Fig. 23.1b), there is a few centimetres wide altered zone at one of the rim; in Fig. 23.1c, this altered zone reaches a few tens of centimetre and the Fig. 23.1d shows a detailed view of an about 80 cm large altered zone, mainly composed of whitish minerals including a few relics of serpentinised olivine.

The alteration of the host, i.e. the amount of serpentine and whitish minerals, tends to increase with the diopsidite proximity (Fig. 23.1b, c). At low alteration level, about 40–60% of the olivine is serpentinised and large relics of fresh orthopyroxene are surrounded by coronæ of tremolite (Fig. 23.4c). The modal proportion of tremolite increase with the alteration degree and needles of this mineral progressively replace the serpentine that formed from olivine (Figs. 23.1c and 23.4d, e). Relics of more or less serpentinised olivine appear from place to place (Figs. 23.1d and 23.4c). When totally altered, the primary lithology is completely transformed into tremolite. The original porphyroclastic texture is preserved with large orthopyroxene pseudomorphs surrounded by a matrix of tremolite (≈90 vol %) and serpentine (≈10 vol %) showing a mesh texture inherited from the serpentinised original rock (Fig. 23.4e). The tremolitised orthopyroxene pseudomorphs exhibit numerous of small Cr-rich spinel inclusions (Fig. 23.4e), the surrounding mesh-textured matrix does not contain any opaque or oxide minerals of any sort, except for rare relics of the original chromian-spinels, which is partially decomposed into an assemblage of ferritchromite and chlorite. Magnetite is in particular completely absent of the diopsidite host.

Chemistry

Analytical Methods

The bulk composition of the 50 samples was measured using the X-ray fluorescence analyser (Rigaku ASC48) at the Institute for Geothermal Sciences in Beppu. Each sample was crushed into a fine power, and then heated at 900°C during more than 24 h. The weight before and after heating was measured and gives the LOI. The heated sample powder was then mixed in a proportion of 1 g of sample for 4 g of Li-borate and melted into a glass bead, following the method described by Sugimoto et al. (2007). The glass bead was irradiated by X-ray for analysis.

All the samples were analysed for single mineral chemistry using the electron-microprobe of the Kanazawa University (JEOL 8800). A standard program with an acceleration voltage of 20 KV, a probe current of 20 nA and a probe diameter of 3 μm was used. Counting times varied from 10 to 50 s on the peak, and from 5 to 25 s on the background, the internal program of the probe was used for ZAF corrections. The chemical characteristics of the main mineral phases are presented in the Tables 23.2, 23.3, 23.4, 23.5, and 23.6.

Table 23.2 Selected compositions of clinopyroxenes in the various lithologies related to the diopsidites

Label	SiO_2	TiO_2	Al_2O_3	Cr_2O_3	FeO	MgO	CaO	Na_2O	Total	Mg#
Diopsidites										
05SDM08a	55.31	0.01	0.16	0.00	0.52	18.18	26.09	0.00	100.28	98.42
05SDM08a	55.40	0.01	0.14	0.02	0.43	18.31	25.90	0.00	100.22	98.69
05SDM08a	55.16	0.00	0.16	0.03	0.34	18.55	25.68	0.01	99.92	98.98
05SDM08a	55.49	0.00	0.17	0.00	0.34	18.59	25.74	0.01	100.34	98.99
05SDM08b	55.58	0.02	0.14	0.00	0.25	18.62	25.61	0.02	100.25	99.25
05SDM08b	55.42	0.00	0.05	0.02	0.10	18.68	25.87	0.00	100.17	99.69
05SDM09a	55.18	0.00	0.03	0.03	0.52	18.09	25.84	0.01	99.70	98.43
05SDM09a	55.67	0.00	0.09	0.00	0.44	18.48	25.70	0.05	100.43	98.69
06HLHL02c	55.47	0.01	0.32	0.02	0.27	19.25	24.60	0.03	99.98	99.23
06HLHL02c	55.59	0.03	0.34	0.14	0.13	18.68	25.79	0.02	100.74	99.62
Anorthite-diopsidites										
07HLHL05a	54.87	0.14	1.03	0.00	0.68	17.63	26.08	0.04	100.56	97.88
07HLHL05a	54.52	0.14	1.07	0.00	0.27	17.76	25.90	0.04	99.83	99.17
07HLHL05f	54.77	0.00	1.61	0.00	0.55	17.81	25.90	0.02	100.72	98.30
07HLHL05f	54.70	0.00	1.41	0.01	1.06	17.61	25.77	0.03	100.67	96.75
07HLHL05j	54.90	0.04	1.05	0.00	0.62	17.87	25.75	0.04	100.36	98.08
07HLHL06b	55.65	0.00	0.13	0.06	0.84	18.05	25.98	0.02	100.82	97.45
07HLHL06b	53.83	0.19	2.02	0.00	0.49	17.14	25.99	0.06	99.72	98.43
07HLHL06b	54.71	0.02	1.94	0.00	0.30	17.63	26.05	0.06	100.84	99.06
07HLHL06b	54.79	0.10	1.33	0.00	0.20	17.86	26.21	0.04	100.59	99.36
Gabbros										
05SDM09h	52.77	0.13	2.94	1.21	2.92	17.37	22.25	0.21	99.90	91.38
2005SDM8-1	53.41	0.15	2.91	0.90	4.72	22.67	14.59	0.16	99.69	89.54
2005SDM8-1	53.31	0.10	3.03	0.82	4.27	21.60	16.34	0.12	99.78	90.02
2005SDM8-2	52.45	0.08	3.38	1.29	2.52	16.54	23.73	0.01	100.12	92.13
07HLHL05f	53.07	0.17	2.14	0.09	5.50	18.72	19.37	0.09	99.36	85.84
07HLHL05f	53.41	0.18	1.83	0.03	4.03	16.55	23.64	0.09	100.00	87.97
07HLHL05g	52.52	0.29	2.29	0.00	5.84	15.94	22.50	0.09	99.70	82.96
07HLHL05h	52.94	0.13	2.30	0.04	4.41	16.53	22.70	0.12	99.39	86.99
07HLHL05i	54.13	0.13	2.02	0.07	4.95	19.54	19.45	0.15	100.62	87.55
07HLHL05i	53.82	0.13	1.33	0.07	3.85	16.83	24.05	0.06	100.37	88.62

Table 23.2 (continued)

Label	SiO_2	TiO_2	Al_2O_3	Cr_2O_3	FeO	MgO	CaO	Na_2O	Total	Mg#
Nephrites (diopsidites host)										
05SDM09b	55.75	0.02	0.21	0.00	0.52	18.85	25.38	0.02	100.80	98.48
05SDM09b	55.17	0.00	0.02	0.00	0.31	18.75	25.52	0.00	99.84	99.07
05SDM08c	55.50	0.03	0.24	0.25	0.35	18.78	25.60	0.02	100.76	98.98
05SDM08c	55.75	0.00	0.10	0.04	0.26	18.74	25.86	0.02	100.78	99.21
Harzburgites										
06HLHL04d	54.15	0.06	1.70	0.72	1.95	17.69	24.14	0.11	100.64	94.17
07HLHL05d	54.41	0.00	1.58	0.83	2.09	17.80	24.27	0.14	101.26	93.83
07HLHL05d	54.80	0.03	1.56	0.80	2.12	17.93	24.23	0.16	101.73	93.78
07HLHL05d	55.25	0.05	1.51	0.53	2.08	17.98	24.50	0.12	102.16	93.90
07HLHL05d	52.28	0.01	1.61	0.73	2.88	19.89	20.96	0.09	98.60	92.49
07HLHL05d	54.44	0.04	1.59	0.61	2.24	17.87	23.97	0.15	100.99	93.44
07HLHL05d	54.34	0.02	1.91	0.85	2.33	17.77	23.84	0.15	101.33	93.16
06HLHL02b	53.91	0.06	2.44	1.03	1.94	17.43	24.09	0.16	101.21	94.12
06HLHL04d	54.15	0.06	1.70	0.72	1.95	17.69	24.14	0.11	100.64	94.17
06HLHL04d	55.05	0.03	1.56	0.61	3.51	25.11	14.96	0.10	101.07	92.74
Rodingites (type-I)										
06HLHL01a	54.52	0.01	0.32	0.05	1.25	18.37	24.86	0.03	99.44	96.32
06HLHL01a	55.53	0.02	0.10	0.03	0.93	18.65	25.25	0.00	100.56	97.27
07HLHL07e	54.93	0.03	0.90	0.00	2.16	16.78	25.79	0.07	100.81	93.26
07HLHL07e	55.12	0.00	0.33	0.00	2.12	17.01	25.93	0.01	100.66	93.47
07HLHL07e	53.24	0.11	2.99	0.00	1.83	17.71	24.04	0.05	100.07	94.52
07HLHL07e	54.98	0.01	0.32	0.00	1.69	17.37	25.95	0.05	100.45	94.83
07ROD1a	54.68	0.06	0.29	0.00	2.80	16.22	26.13	0.06	100.35	91.16
07ROD1a	53.61	0.08	1.72	0.07	2.42	16.40	25.65	0.06	100.15	92.36
07ROD1a	54.61	0.09	0.91	0.00	2.36	16.97	25.56	0.08	100.66	92.75
Rodingites (rodingitised diopsidites, type-II)										
07HLHL07a	55.05	0.11	0.86	0.00	0.72	17.88	26.08	0.03	100.84	97.78
07HLHL07a	55.05	0.02	0.54	0.00	0.30	18.11	26.41	0.03	100.51	99.09
07HLHL07a	53.79	0.33	2.25	0.00	0.57	17.59	25.70	0.05	100.31	98.22
07HLHL07a	52.99	0.47	2.45	0.00	0.54	17.21	25.84	0.03	99.54	98.28
07HLHL07b	55.04	0.15	0.55	0.00	0.73	17.72	26.15	0.03	100.45	97.75
07HLHL07b	55.41	0.00	0.51	0.00	0.14	18.18	26.12	0.03	100.44	99.56

Table 23.3 Selected compositions of olivine in the various lithologies related to the diopsidites

Label	SiO_2	FeO	MnO	MgO	CaO	NiO	Total	Fo%
Diopsidites								
05SDM08c	41.68	3.63	0.05	55.50	0.04	0.45	101.36	96.46
05SDM08c	41.55	2.86	0.03	56.45	0.03	0.51	101.43	97.24
05SDM08c	41.64	3.49	0.07	55.85	0.01	0.46	101.54	96.61
05SDM08c	41.70	3.01	0.00	55.79	0.01	0.47	100.98	97.06
05SDM08c	41.17	3.32	0.03	54.67	0.01	0.47	99.68	96.71

Table 23.3 (continued)

Label	SiO_2	FeO	MnO	MgO	CaO	NiO	Total	Fo%
Gabbros								
05SDM09h	40.14	10.49	0.15	49.59	0.02	0.33	100.77	89.40
2005SDM8-1	40.18	10.14	0.21	49.09	0.03	0.44	100.11	89.61
2005SDM8-2	40.97	9.07	0.11	50.34	0.02	0.36	100.87	90.82
2005SDM8-2	40.96	8.94	0.11	49.80	0.04	0.40	100.25	90.85
2005SDM8-2	40.85	8.87	0.17	50.01	0.02	0.40	100.33	90.95
07HLHL05f	40.60	9.99	0.13	49.43	0.02	0.34	100.51	89.82
Nephrites (diopsidites host)								
05SDM09c1	41.43	4.71	0.06	54.30	0.01	0.40	100.93	95.36
05SDM09e	40.97	8.44	0.14	50.72	0.02	0.41	100.69	91.46
05SDM09e	40.82	7.31	0.11	50.96	0.02	0.34	99.57	92.55
05SDM09e	41.59	5.74	0.10	52.60	0.01	0.39	100.47	94.23
05SDM09e	41.82	4.89	0.11	53.41	0.00	0.40	100.66	95.12
05SDM09e	41.35	4.45	0.08	53.13	0.01	0.39	99.44	95.51
05SDM09e	41.96	3.28	0.04	54.25	0.02	0.44	99.99	96.72
05SDM09c1	40.96	4.50	0.07	53.66	0.00	0.44	99.64	95.50
05SDM09c1	41.37	5.08	0.06	53.90	0.03	0.42	100.89	94.98
Serpentinites (rodingites host)								
07ROD2	41.26	8.83	0.13	50.17	0.02	0.31	100.72	91.01
07ROD2	41.02	8.50	0.14	49.90	0.02	0.31	99.91	91.28
07HLHL07d	41.81	8.77	0.14	50.27	0.02	0.39	101.41	91.08
07HLHL07c	41.28	8.40	0.12	50.90	0.02	0.35	101.08	91.53
Harzburgites (gabbros host)								
07HLHL05g	40.28	9.00	0.10	49.77	0.01	0.38	99.57	90.79
07HLHL05h	40.28	9.37	0.18	49.42	0.01	0.33	99.58	90.39
07HLHL05h	40.67	9.11	0.16	49.90	0.02	0.33	100.20	90.71
07HLHL05h	40.68	9.04	0.11	49.99	0.02	0.36	100.19	90.79
07HLHL05i	40.80	9.35	0.13	50.13	0.02	0.38	100.85	90.53
07HLHL05i	40.56	8.93	0.16	50.46	0.02	0.36	100.50	90.97
Harzburgites								
05SDM08d	40.97	8.45	0.14	50.49	0.00	0.42	100.50	91.42
05SDM09f	40.49	8.70	0.09	50.53	0.04	0.39	100.35	91.19
05SDM09f	40.61	8.57	0.12	50.60	0.02	0.37	100.33	91.32
06HLHL02b	41.02	8.78	0.10	50.33	0.00	0.39	100.64	91.08
06HLHL04d	41.00	8.92	0.12	50.33	0.03	0.35	100.75	90.95
07HLHL05c	40.74	8.51	0.11	50.47	0.03	0.39	100.25	91.36
06HLHL04c	40.80	9.64	0.19	49.69	0.01	0.46	100.80	90.18
06HLHL04c	40.98	9.35	0.19	50.02	0.06	0.28	100.87	90.51

Mineral Chemistry

The chemical composition of clinopyroxenes, olivines, plagioclases, amphiboles, serpentines and chlorites in the various studied lithologies are summarised in Tables 23.2, 23.3, 23.4, 23.5, 23.6 and Figs. 23.5, 23.6, 23.7 and 23.8. In the

Table 23.4 Selected compositions of plagioclase in diopsidites, gabbros, and gabbros which reacted with a diopsidite

Label	SiO_2	TiO_2	Al_2O_3	CaO	Na_2O	K_2O	Total	An%
Anorthite-diopsidites								
07HLHL05a	43.21	35.77	0.03	20.09	0.13	0.01	99.25	98.75
07HLHL05a	43.15	35.98	0.03	20.24	0.02	0.01	99.46	99.71
07HLHL05b	43.07	36.44	0.04	19.87	0.08	0.06	99.56	98.93
07HLHL05b	42.73	36.32	0.05	20.18	0.03	0.00	99.34	99.77
07HLHL05f	43.23	36.31	0.08	20.21	0.13	0.00	100.00	98.83
07HLHL05f	43.27	35.61	0.00	20.38	0.02	0.00	99.31	99.79
07HLHL05j	43.64	36.42	0.02	20.26	0.07	0.00	100.43	99.42
07HLHL05j	43.33	36.34	0.03	20.37	0.02	0.00	100.09	99.80
07HLHL06a	43.75	36.40	0.03	20.27	0.12	0.01	100.58	98.87
07HLHL06b	43.72	36.03	0.02	20.05	0.19	0.00	100.02	98.33
07HLHL06b	43.42	35.85	0.02	20.10	0.12	0.00	99.51	98.94
Gabbros								
07HLHL05f	44.37	35.37	0.17	19.39	0.63	0.00	99.96	94.49
07HLHL05f	44.26	35.89	0.08	19.62	0.51	0.00	100.39	95.48
07HLHL05g	44.31	35.17	0.15	19.26	0.63	0.01	99.56	94.34
07HLHL05h	43.86	35.56	0.14	19.08	0.54	0.01	99.27	95.06
07HLHL05h	43.89	35.60	0.12	19.29	0.54	0.00	99.46	95.14
07HLHL05i	44.57	34.69	0.17	19.28	0.70	0.02	99.44	93.75
07HLHL05i	44.83	34.93	0.11	19.18	0.67	0.01	99.76	94.04

chemical diagrams (Fig. 23.5), the two trends corresponding to the two magmatic dyke families are clearly visible. As for their petrographic characteristics, the gabbros show chemical compositions similar to the other gabbros (Ceuleneer et al., 1996; Python and Ceuleneer, 2003) and will not be detailed here.

Diopsidites and Anorthite-Diopsidites

The chemical composition of the various minerals in diopsidites was precisely described in (Python et al., 2007). Diopside show a very high Mg# (**=Mg/(Mg+Fe)**) and low concentrations in all incompatible elements (Table 23.2 and Fig. 23.5a). Its composition is clearly out of the magmatic domain given by the gabbros (Fig. 23.5a).

Olivines are quite infrequent in diopsidites and only a few grains could be analysed (Table 23.3 and Fig. 23.7). Their composition is close to the forsterite end-member (96.5 < Fo% < 98), with relatively high NiO (0.4–0.5 vol%) and low MnO (<0.07) contents (Table 23.3 and Fig. 23.7). Plagioclase is present only in the anorthite-diopsidite, its composition is close to the anorthite end-member and presents a strong contrast with the anorthitic plagioclase present in the gabbros (Table 23.4).

Amphiboles are scarce in the core of diopsidite dykes, when present, they are always of the tremolite type. They tend to be more abundant, in equilibrium with diopsides, at the rim of the dyke with a large "tremolitised" zone in the host (like

Table 23.5 Selected compositions of amphiboles in the various lithologies related to the diopsidites

Label	Type	SiO_2	TiO_2	Al_2O_3	Cr_2O_3	Fe_2O_3	FeO	MnO	MgO	CaO	Na_2O	K_2O	NiO	$H_2O_{(c)}$	Total
Diopsidites															
05SDM02b	Cannilloite	53.72	0.00	0.45	0.01	0.00	0.95	0.04	20.00	22.32	0.07	0.00	0.05	2.18	99.79
05SDM02b	Cannilloite	54.53	0.00	0.19	0.05	0.00	0.81	0.05	16.92	25.05	0.04	0.01	0.02	2.15	99.82
05SDM08a	Cannilloite	54.00	0.01	0.10	0.08	0.00	0.76	0.03	16.98	25.63	0.02	0.00	0.00	2.15	99.75
05SDM08a	Cannilloite	54.58	0.02	0.15	0.04	0.00	0.67	0.00	18.23	24.05	0.02	0.00	0.00	2.17	99.92
05SDM08b	Cannilloite	50.57	0.08	0.23	0.21	0.00	4.02	0.01	16.27	26.28	0.02	0.00	0.01	2.14	99.84
05SDM09a	Cannilloite	54.23	0.01	0.03	0.06	0.00	0.67	0.01	18.70	23.66	0.00	0.00	0.00	2.16	99.51
05SDM09a	Cannilloite	54.71	0.01	0.15	0.07	0.00	0.45	0.02	17.48	25.35	0.03	0.01	0.00	2.17	100.43
Gabbros															
06HLHL04e	Magnesiohornblende	53.93	0.59	4.71	0.65	0.00	3.56	0.06	21.38	12.39	0.64	0.22	0.10	2.18	100.41
07HLHL05c	Tremolite	54.50	0.03	3.11	1.00	1.36	0.87	0.04	22.13	13.43	0.54	0.03	0.07	2.17	99.27
07HLHL05f	Anthophyllite / Cummingtonite	58.60	0.00	0.17	0.00	0.59	7.51	0.27	29.35	0.65	0.04	0.00	0.07	2.21	99.46
07HLHL05f2	Tremolite	56.55	0.14	1.65	0.03	0.72	3.23	0.14	22.56	12.59	0.36	0.02	0.04	2.19	100.22
07HLHL05g	Magnesiohornblende	53.39	0.51	4.44	0.00	0.62	5.08	0.11	19.73	12.55	0.36	0.11	0.13	2.14	99.17
07HLHL05h	Magnesiohornblende	49.10	0.70	8.83	1.88	0.00	2.58	0.04	19.70	12.85	1.15	0.12	0.11	2.15	99.20
07HLHL05i	Magnesiohornblende	51.70	0.13	6.79	0.04	1.33	2.50	0.08	20.70	12.58	0.85	0.29	0.09	2.15	99.24
07HLHL05i	Magnesiohornblende	50.01	0.10	8.83	0.04	1.91	2.73	0.05	19.93	12.24	1.13	0.28	0.26	2.15	99.65
Nephrites (diopsidites host)															
05SDM09c	Tremolite	58.31	0.00	0.46	0.02	0.31	0.53	0.02	24.39	13.38	0.16	0.00	0.11	2.21	99.90
05SDM09c	Tremolite	57.82	0.00	0.42	0.03	0.00	0.64	0.00	24.29	14.32	0.21	0.00	0.00	2.21	99.93
05SDM09c	Magnesiohornblende	53.04	0.09	4.43	0.73	1.63	0.43	0.01	22.65	12.89	1.09	0.02	0.10	2.16	99.26
05SDM09c	Tremolite	59.11	0.02	0.32	0.00	0.66	0.16	0.02	24.52	13.74	0.13	0.00	0.00	2.23	100.92

Table 23.5 (continued)

Label	Type	SiO_2	TiO_2	Al_2O_3	Cr_2O_3	Fe_2O_3	FeO	MnO	MgO	CaO	Na_2O	K_2O	NiO	$H_2O_{(c)}$	Total
Harzburgites															
05SDM02e	Tremolite	58.04	0.00	0.60	0.09	0.55	0.26	0.03	24.28	13.61	0.20	0.02	0.08	2.21	99.97
05SDM08d	Cannilloite	56.90	0.02	1.95	0.48	0.00	3.27	0.11	25.98	8.28	0.68	0.00	0.07	2.20	99.95
05SDM08d	Tremolite	55.41	0.01	2.09	1.09	1.89	0.24	0.04	22.68	13.12	0.71	0.01	0.06	2.18	99.53
05SDM09d	Magnesiohornblende	53.19	0.06	4.00	0.01	2.59	0.15	0.01	22.70	13.08	1.06	0.01	0.13	2.16	99.14
05SDM09d	Magnesiohornblende	54.22	0.12	3.90	0.00	1.88	0.32	0.02	23.10	13.21	0.99	0.01	0.09	2.19	100.01
05SDM09d	Tremolite	55.50	0.02	2.04	0.26	1.58	0.21	0.05	23.75	12.88	0.63	0.00	0.07	2.18	99.17
06HLHL04d	Tremolite	55.43	0.09	3.48	0.54	0.00	2.10	0.07	22.44	13.11	0.59	0.01	0.05	2.20	100.11
06HLHL04d	Tremolite	59.03	0.01	0.71	0.34	0.00	5.00	0.21	25.78	7.29	0.25	0.01	0.08	2.22	100.92
Rodingites (type-I)															
07ROD1a	Cannilloite	48.73	0.14	7.46	0.06	0.00	3.50	0.10	19.68	17.95	0.10	0.00	0.04	2.08	99.83
06HLHL01a	Tremolite	57.22	0.04	0.73	0.00	0.48	1.67	0.06	23.22	13.33	0.21	0.01	0.00	2.18	99.15
06HLHL01a	Tremolite	58.60	0.00	0.30	0.00	0.08	1.71	0.02	23.83	13.51	0.14	0.00	0.00	2.21	100.39
07HLHL07e	Cannilloite	50.81	0.01	7.62	0.00	0.00	1.56	0.00	16.55	22.04	0.03	0.00	0.08	2.19	100.90
Rodingites (rodingitised diopsidites, type-II)															
07HLHL07a	Cannilloite	49.67	0.21	7.44	0.00	0.00	0.64	0.00	13.49	26.90	0.02	0.00	0.04	2.16	100.57

Table 23.6 Selected compositions of serpentines and chlorites in diopsidites, rodingites, and their related hosts

Label	Variety	SiO_2	Al_2O_3	Cr_2O_3	FeO*	MnO	MgO	CaO	NiO	Total
Diopsidites										
05SDM02b1	Clinochlore	31.42	13.33	4.47	3.29	0.03	31.95	0.11	0.48	85.08
05SDM02b1	Serpentine	43.02	0.17	0.00	1.76	0.03	39.87	0.31	0.00	85.17
05SDM08a	Serpentine	44.42	0.90	0.06	2.22	0.03	37.31	0.08	0.04	85.11
05SDM08b	Serpentine	43.45	0.93	0.63	2.92	0.01	37.75	0.03	0.05	85.78
05SDM09a	Talc-chlorite	41.18	5.25	0.56	1.15	0.00	36.51	0.19	0.12	85.03
Rodingites (type-I)										
07ROD1a	Sheridanite	28.74	22.43	0.00	6.94	0.05	27.06	0.01	0.09	85.34
07ROD1a	Clinochlore	28.88	21.35	0.04	6.97	0.09	27.24	0.02	0.07	84.67
06HLHL01b	Sheridanite	28.73	22.38	0.00	6.51	0.09	28.21	0.01	0.09	86.02
06HLHL01b	Clinochlore	28.58	21.31	0.08	8.29	0.14	27.06	0.03	0.15	85.66
Rodingites (rodingitised diopsidites, type-II)										
07HLHL06b	Sheridanite	27.34	17.40	5.83	6.58	0.09	27.43	0.03	0.26	84.96
07HLHL06b	Clinochlore	28.67	20.76	0.91	5.52	0.01	28.85	0.01	0.25	85.04
07HLHL07a	Clinochlore	29.00	21.33	0.00	8.59	0.10	27.18	0.03	0.15	86.43
07HLHL07b	Talc-chlorite	39.31	1.80	0.00	4.74	0.07	38.56	0.04	0.20	84.73
Serpentinites (rodingites host)										
06HLHL01c	Serpentine	43.48	0.56	0.00	2.27	0.03	38.68	0.07	0.19	85.28
06HLHL01c	Serpentine	43.68	0.52	0.00	2.16	0.01	38.82	0.03	0.22	85.44
07HLHL08b	Penninite	34.80	12.10	0.00	7.60	0.09	30.66	0.17	0.14	85.60
07HLHL07c	Talc-chlorite	41.12	1.02	0.45	4.61	0.10	37.06	0.06	0.12	84.59
Nephrites (diopsidites host)										
05SDM09e	Serpentine	41.89	0.40	0.02	3.46	0.05	38.83	0.08	0.32	85.06
05SDM09c2	Serpentine	43.43	0.25	0.01	2.29	0.03	39.01	0.04	0.00	85.08
05SDM03	Serpentine	44.02	0.10	0.03	6.38	0.08	36.05	0.05	0.13	86.83

The variety column gives the type of mineral: *clnchl*: clinochlore; *talc-chl*: talc-chlorite; *serp*: serpentine; *sherid*: sheridanite; *pen*: penninite

in Fig. 23.1c, d). Their chemical compositions are rich in Mg and poor in Fe and incompatible elements like Na and K (Table 23.5 and Fig. 23.6a). With decreasing Si content, Mg# and Na+K in A site do not change and stay quite high for the first and close to 0 for the later (Fig. 23.6a). The Fe_{total} in the amphibole is on average lower than 1 wt%, and ferric iron represent only a low proportion or may even be absent (Table 23.5). Amphiboles in diopsidites and gabbros show different compositions and evolutions.

Rodingites

The mineral assemblage of the rodingite shows a great variability. It notably includes numerous Al and Ca-rich phases like zoisite, garnet, and prehnite. However, in order to focus on the comparison between diopsidite and rodingite, only the chemical composition of the mineral present in both lithologies, i.e. diopside and amphibole, will be detailed here.

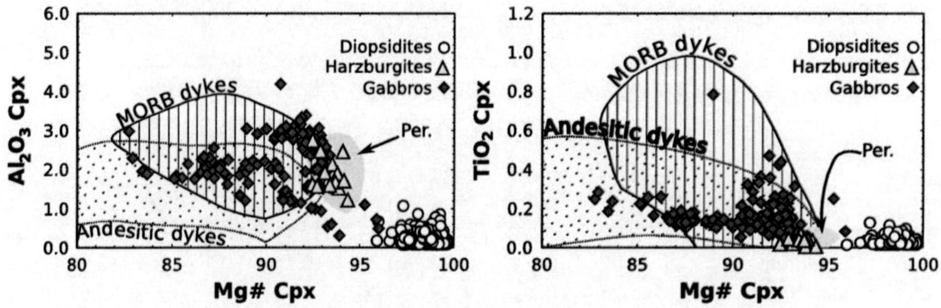

(a) Clinopyroxene in diopsidite, peridotites & gabbros

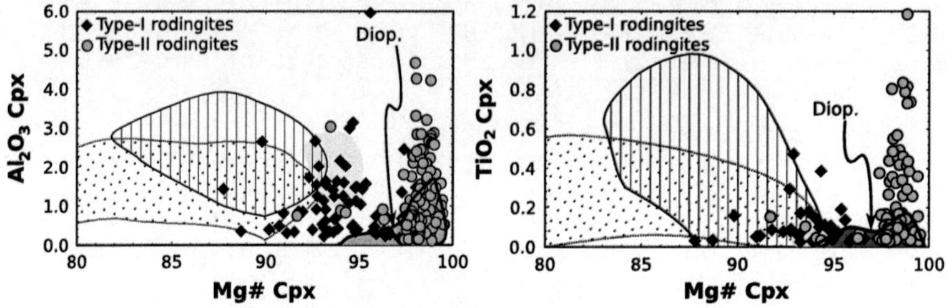

(b) Clinopyroxene in rodingite

Fig. 23.5 Al_2O_3 and TiO_2 vs Mg# for clinopyroxene (see Table 23.2). The *grey*, *hatched* and *dotted* areas represent the chemical domains for, respectively, the peridotites (*Per.*, Monnier et al., 2006) and the two series of magmatic dykes (Python and Ceuleneer, 2003) in Oman. (**a**) Clinopyroxenes in the diopsidites (*white circles*); their host (*grey triangles*), and gabbroic dykes (*grey diamonds*). (**b**) Clinopyroxenes in rodingites with gabbroic (*black diamonds*) and diopsiditic (*grey circles*) protoliths. The chemical domain for diopsidites (*Diop.*, Python et al., 2007) is given in *dark grey*

The compositions of these two minerals show a large scatter (Mg# from 87 to 99%, TiO_2 between 0 and 1 wt% and Al_2O_3 up to 5 wt% for the diopside; Mg# 0.9 to 1, with great variation of the Si and Fe contents for the amphibole, see Tables 23.2 and 23.5, Figs. 23.5b and 23.6b). Type-I and Type-II rodingites occupy two different chemical fields that hardly overlap (Figs. 23.5b and 23.6b). Type-I rodingites show lower Mg# and higher Fe_{total} content, with a greater proportion as Fe_2O_3, than type-II rodingites for a similar chemical range of Al_2O_3 and TiO_2 in diopside. Amphibole in these two types of rodingite plot in very different zone, and the Si content in type-II rodingite is sensibly lower than in type-I (7.6-8 atom per formula unit for type-I and 6.6-7.1 for type-II rodingite).

Fresh and Altered Harzburgites

Olivine is the only phase common to every kind of harzburgites, it is dominant in fresh harzburgites and frequently appears as relics in the altered ones. In fresh harzburgites and serpentinites (i.e. in the samples taken far from any dykes and the

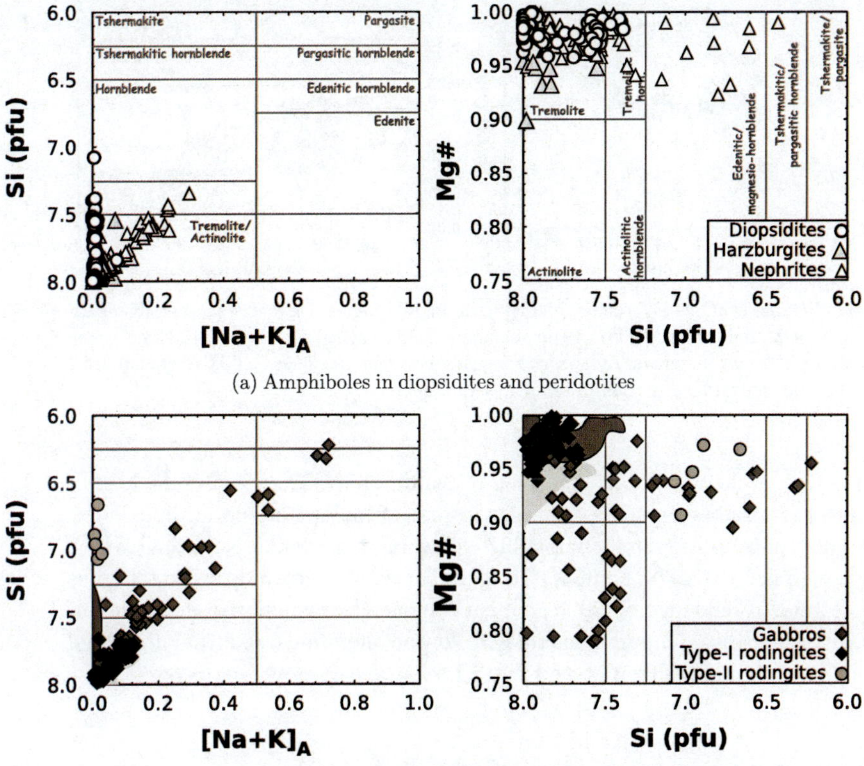

(a) Amphiboles in diopsidites and peridotites

(b) Amphiboles in Rodingites and gabbros

Fig. 23.6 Chemical composition of the amphiboles in the various studied lithologies. *Left*: Si, in atom per unit formula, vs Na+K in the A site. *Right*: Mg# (**Mg/(Mg+Fe)**) vs Si, in atom per unit formula. *Symbols* are the same as in Fig. 23.5: chemical domains for the harzburgites and the diopsidites are given for comparison

host of rodingite), its chemical composition is similar to that of the other Oman peridotites (Fig. 23.7), with a forsterite content ranging from 90 to 92%, a NiO content between 0.32 and 0.55 wt%, and a MnO content between 0.1 and 0.2 wt% (Table 23.3 and Fig. 23.7). The olivine in the nephrites (i.e. the host of diopsidites) show intermediate compositions between the peridotite chemical field (Monnier et al., 2006) and the diopsidite one (Python et al., 2007): $90<Fo<96.5$; $0.3<NiO<0.5$; $MnO<0.2$ (Table 23.3 and Fig. 23.7).

Clinopyroxene is present only in the fresh harzburgites, its chemical composition show no dissimilarity with the average peridotite of Oman (Monnier et al., 2006; Table 23.2 and Fig. 23.5a), with a Mg# ranging from 92 to 95%, an Al_2O_3 content comprised between 1.5 and 3 wt%, and a TiO_2 concentration lower than 0.1 wt%.

Amphiboles are abundant only in the nephrites host of diopsidite and sporadically appear as clinopyroxene alteration product in fresh harzburgites. They have a tremolite to tremolitic-hornblende composition (Table 23.5 and Fig. 23.6a) and

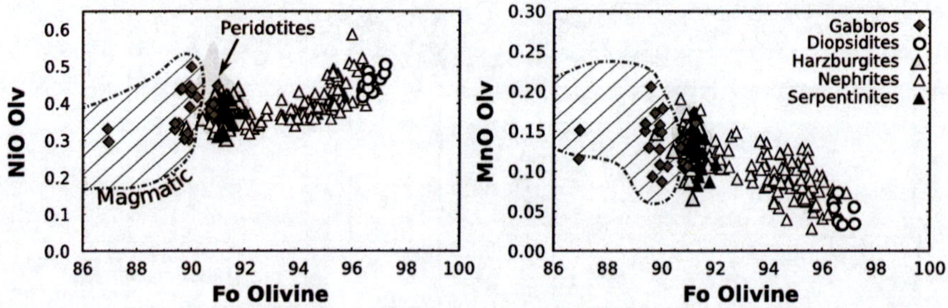

Fig. 23.7 NiO and MnO vs Fo content of olivine in the lithologies where it is present. The peridotites (*in grey*, Monnier et al., 2006) and the magmatic field (hatched surface enclosed by a *dotted line*) given by the two magmatic dyke series (see Python and Ceuleneer, 2003 for details) are compared to the diopsidites and rodingites related lithologies

occupy different chemical field according to the lithology. Tremolite in the host of diopsidite has a rather high Mg#, similar to that of the amphibole in diopsidites, but a higher concentration in incompatible elements (in particular Na and K) which increase with decreasing Si content. On the other hand, tremolites in harzburgites show a lower Mg# and incompatible element content. The evolution of these parameters with decreasing Si suggests that they follow an alteration trend from high Mg# tremolite to actinolite (Table 23.5 and Fig. 23.6a).

Whole Rock Chemistry

The whole rock compositions are given in Table 23.1 and represented in Figs. 23.8 and 23.9. The ACF and SCF diagrams (Fig. 23.8a) show the inter evolution of Al_2O_3, CaO and F (=MgO+FeO*, with FeO*=FeO_{total}) for the ACF diagram; and of SiO_2, CaO and F for the SCF one. Figure 23.8b shows the variation of SiO_2 with CaO and F. The various lithologies occupy different chemical domains. The harzburgites, the fresh ones as well as the serpentinised host of rodingites, are purely ferro-magnesian, their SiO_2 content only slightly vary according to the hydration level, they contain virtually no CaO, Al_2O_3 or other elements (see Table 23.1). The nephrites present various concentrations of SiO_2, CaO, FeO and MgO depending on their modal amount of tremolite. Their CaO and SiO_2 contents increase with decreasing MgO+FeO* (Fig. 23.8). The gabbros are rich in plagioclase, and then in CaO and Al_2O_3. The scatter of the data from gabbros, mostly along a line in the ACF and SCF diagrams, reflects the variations in the modal composition of the rock: Al_2O_3 and CaO decrease together with increasing F, i.e. with increasing modal proportion of ferro-magnesian minerals in the rock (olivine and pyroxenes). Type-I rodingites (formed after gabbros) show a composition similar to that of the gabbros except for a significantly lower content in Si (Fig. 23.8). On the other hand, type-II rodingite (from anorthite-diopsidite) have a very calcic chemical composition, close to that of the most calcic rodingite described in the literature (Coleman, 1977)

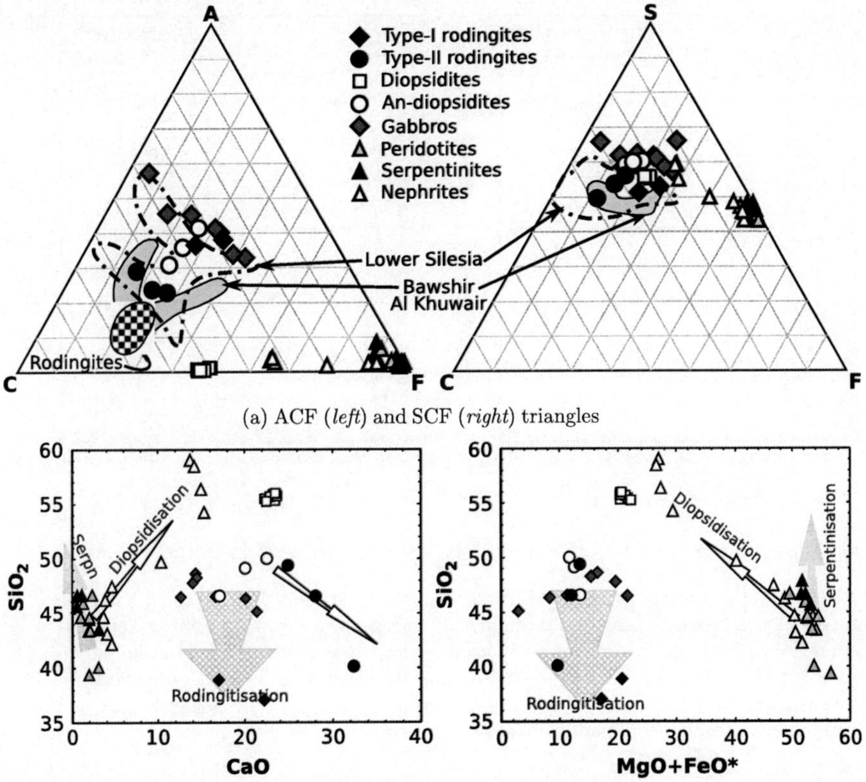

(a) ACF (*left*) and SCF (*right*) triangles

(b) Evolution of SiO_2 with CaO and MgO+FeO_{total}

Fig. 23.8 Bulk composition of the various studied facies. **ACF**: Al_2O_3–CaO–MgO+FeO* diagram; *SCF*: SiO_2–CaO–MgO+FeO* diagram, with FeO*=FeO_{total}. (**a**) The chemical domains for rodingites in ophiolites as defined by Coleman (1977), Lower Silesia rodingites (Dubińska, 1995), Oman metasomatised mafic and ultramafic rocks (El-Shazly and Al-Belushi, 2004) are given respectively by the *squared, white* and *grey surfaces*. (**b**) Symbols are the same as in a. The *grey arrow* shows the chemical evolution during serpentinisation of harzburgites. The *squared arrow* shows the evolution during the rodingitisation of gabbros. The *white arrow* shows the chemical evolution of harzburgites during "diopsiditisation" process, i.e. the progressive transformation of the primary lithology into tremolite ± antigorite. At close contact with the diopsidite, the host is almost totally composed of tremolite and show extremely high Si content

and their composition noticeably differ from that of their protolith or the type-I rodingites. Anorthite-diopsidites have a rather gabbroic composition and diopsidites exhibit a pure diopside composition in agreement with their monomineralic nature (Fig. 23.8).

Figure 23.9 show two profiles across two diopsidite dykes and their host (two cases in two different wadis in the northern part of the Oman ophiolite), one with a large zone of nephrite between the diopsidite and the host (wadi Sudum), and one with rather sharp contacts with the host (wadi Hilti). The CaO and SiO_2 contents

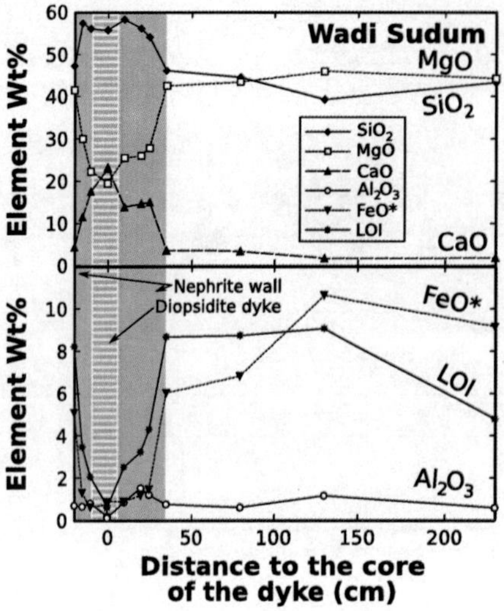

(a) Diopsidite dyke in the wadi Sudum

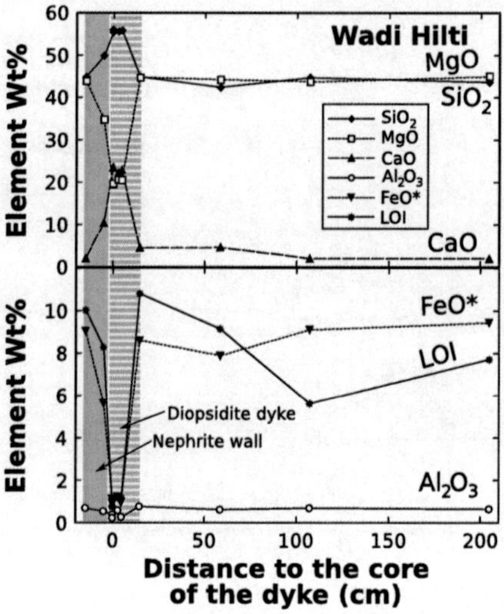

(b) Diopsidite dyke in the wadi Hilti

Fig. 23.9 Profiles across two diopsidite dykes in the Hilti block (northern part of the Oman ophiolite). The centre of each dyke was chosen as origin and samples taken at various distance from this point. The region in *grey* represents the zone affected by the "diopsiditisation", i.e. the diopsidite itself and its nephrite wall (area containing whitish minerals mixed with the primary lithology in Fig. 23.1c). The area hatched in *white* represents the diopsidite dyke itself. The other part (in *white background*) show the evolution of peridotite far from the dyke

in the harzburgite starts to progressively increase at about one metre from the dyke itself and becomes very high in the nephrite (whitish "tremolitised" zone, Fig. 23.1c). The bulk FeO concentration shows an inverse progression from the same distance. The LOI first increases from the "fresh" harzburgite value to "serpentinised" value when approaching diopsidite dyke, and then abruptly drops in the nephrite. Within the nephrite and the diopsidite dyke itself, the other elements show a progression mostly related to the evolution of the mineralogy, i.e. the general transformation of the harzburgite into a nephrite, and the abrupt change into pure diopsidite within the dyke (Fig. 23.9).

Discussion

Alteration Processes in Harzburgite

The observation of the fresh and altered harzburgites suggests that two types of alteration occurred in the mantle of the Oman ophiolite. The first type of alteration concerns the serpentinites hosting the rodingites. It results in serpentinisation in a rather closed system. Olivines and pyroxenes are decomposed into serpentine and magnetite (Fig. 23.4b) and the bulk composition of the resulting rock is globally the same as that of the primary rock, except for the addition of water and Si, and removal of Ca (see Table 23.3, Fig. 23.10a). This kind of serpentinisation, and its association with rodingite formation, was largely described in various contexts (Muraoka, 1985; Frost and Beard, 2007; Pomonis et al., 2008; Frost et al., 2009; Bach and Klein, 2009) and probably result of the interaction of harzburgite with seawater or groundwater (Barnes et al., 1978; Neal and Stanger, 1984; Stanger, 1985; Nasir et al., 2007; Pomonis et al., 2008).

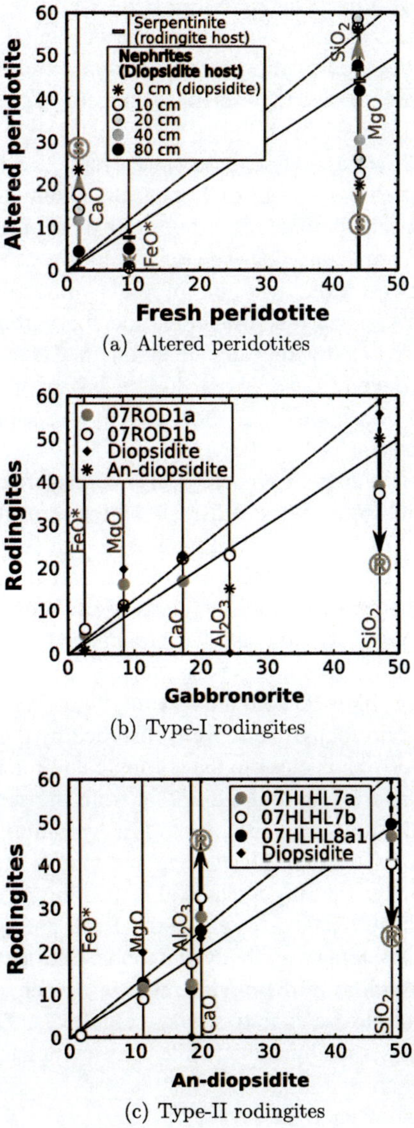

(a) Altered peridotites

(b) Type-I rodingites

(c) Type-II rodingites

Fig. 23.10 Major-element isocon diagrams for the rodingites (Type-I and Type-II) and the altered peridotites (serpentinites and nephrites) plotted against the composition of their respective probable protoliths. (**a**) The *arrow* marked with the *M* shows the compositional evolution of the peridotites by Ca-metasomatism close to a diopsidite dyke. The distance to the diopsidite dyke (from 0 cm in the core of the dyke, to 80 cm from the dyke boundary) is given in the legend. (**b**) and (**c**) The *arrow* marked by *R* shows the chemical evolution of an original rock (gabbronorite for type-I rodingites and anorthite-diopsidite for type-II rodingites) by rodingitisation. All values are given in wt%, the *solid line* represents constant mass and the *dashed line* constant volume metasomatism

The second type of alteration concerns the harzburgites close to the diopsidite dykes. It results in the transformation of the primary lithology into a nephrite. At low alteration stage, relatively far from the diopsidite (Fig. 23.4c), only orthopyroxene is partly altered into tremolite and olivine is serpentinised to a higher degree than observed in the average Oman peridotite. At higher alteration stage, tremolite entirely replaces orthopyroxene and also crystallise in equilibrium with minor antigorite in place of olivine (Fig. 23.4d, e). The alteration process is not homogeneous and "pockets" of serpentine containing relics of fresh olivine are observed (Fig. 23.4d). These olivines are richer in Mg and Ni, and poorer in Mn than the average Oman harzburgite (Table 23.3, Fig. 23.7). The chemical transformation becomes stronger with the degree of alteration, i.e. the forsterite content increases with the modal proportion of tremolite and the reduction of the olivine relics size. The progressive bulk chemical transformations observed during "diopsiditisation" are illustrated in the Figs. 23.8 and 23.9, the SiO_2 and CaO contents increase as the "F" (=MgO+FeO*) decreases, reflecting the expanding tremolite crystallisation. The profiles in Fig. 23.9 show the evolution of the composition of the harzburgite when approaching a diopsidite dyke. The diopsidite-related alteration affects a larger area that what we could think by simply looking to the outcrops. At about 80 cm from the dyke, the progressive transformation of olivine into forsterite, of orthopyroxene into tremolite, and the partial and increasing serpentinisation are pointed out by the augmentation in CaO, LOI, and the decreasing FeO* content. Beyond these about 80 cm, harzburgites are not affected any more by the "diopsiditisation" and show exactly the same composition as the average Oman peridotite.

This alteration process is also characterised by the absence of magnetite or any other Fe sink co-crystallising with the serpentine and the tremolite. Isocon diagram in Fig. 23.10a shows that progressive but substantial addition of CaO and SiO_2, and loss of FeO* and MgO occurred during the transformation, suggesting that the fluid responsible of alteration process itself was rich in the two first elements and undersaturated in the two last. The pronounced enrichment in MgO of olivine relics in the nephrites and diopsidites and the absence of Fe sink in the final product especially point to high Fe solubility in the fluid. The few Wt% of Cr_2O_3 contained in the primary orthopyroxene were absorbed by the crystallisation of Cr-rich spinel with composition is close to the chromite end-member (Fig. 23.4e). This second type of alteration is then quite different from the serpentinisation observed in association with the rodingites, it leads to less hydration, to a strong general enrichment in Ca and Si, and to the loss of Mg and almost the whole Fe. By contrast, Ca tends to be removed during a classical serpentinisation process (Muraoka, 1985; Frost and Beard, 2007; Frost et al., 2009; Bach and Klein, 2009) as iron and other major elements remain globally immobile (see the horizontal bars in Fig. 23.10a). The compositions of the olivine relics in the serpentinites host of rodingite and in fresh harzburgite are exactly similar (Fig. 23.7). The olivine chemistry does not change during serpentinisation while it is completely transformed by diopsidite-related alteration.

Formation of Rodingites in Oman

Whatever their protolith (gabbro or anorthite-diopsidite), the Oman rodingites petrographical and chemical characteristics favours the hypothesis made by the previous workers for rodingites genesis in general (Coleman, 1977; Capedri et al., 1978; Kobayashi and Shoji, 1988; Mittwede and Schandl, 1992; Dubińska, 1997; Bach and Klein, 2009). They appear only in a highly serpentinised host, they are rich to extremely rich in Ca and relatively poor in Si. The serpentinisation degree of the surrounding harzburgite seems to be independent of the presence or the absence of dyke and of the distance to the rodingite (when they are present). The same gabbroic dyke may be strongly rodingitised when its host is serpentinised and fresh out the serpentinisation zone. The isocon diagrams in Fig. 23.10b and 10c show that loss of Si and gain of Ca are the two main consequences of the rodingitisation.

Consequently, the Oman rodingites do not differ from those that were described in other place and contexts. They form as a consequence of the extensive serpentinisation of the mantle, by reaction of a gabbroic protolith with a fluid which composition is controlled by the serpentinisation reaction in the host (Coleman, 1977; Capedri et al., 1978; Kobayashi and Shoji, 1988; Mittwede and Schandl, 1992; Dubińska, 1997; Bach and Klein, 2009). The chemical and textural scatter observed in the Oman rodingite is related to some variety in the protolith, when derived from gabbro, they inherited the gabbroic texture and chemical composition; with a diopsiditic protolith, the absence, or exceptional abundance, of some elements will prevent the crystallisation of some rodingite characteristic phases.

About the Diopsidite Genesis

In our previous study (Python et al., 2007), we suggested that seawater penetrated down to the mantle, after crossing the whole oceanic crust. Reactions with the crustal lithologies may cause the dissolution of large amount of Ca in the fluid, and if such a fluid reach the mantle, chemical reactions leading to Ca-metasomatism of the mantle lithologies are most likely. Large deposits of chlorite rock with a preserved magmatic texture resulting from the alteration of gabbroic lithologies by hydrothermal fluid were discovered in the crustal section of the Oman ophiolite (Yoshitake et al., 2009). The fluid issued from this reaction should be extremely rich in Ca, Si, poor in Fe and Mg. This fluid is a good candidate for mantle metasomatism and diopsidite formation. The crystallisation of large amount of tremolite during peridotite alteration, the fibrous texture of the diopsidites and the nephrites, the variation of the alteration level with the distance to the dyke, all these data are compatible with our hypothesis. The field relationships between the diopsidites and the magmatic gabbros, the second cross-cutting the first (Fig. 23.1e, f), show in addition that diopsidites are not related to late alteration of the ophiolite but formed during the magmatic activity time, i.e. at the time the Oman ophiolite was a spreading centre.

Diopsidite genesis is related to alteration of harzburgite in a Ca-rich environment. The isocon diagrams (Fig. 23.10) show the chemical evolutions during the transformation of peridotite into nephrite and the rodingitisation of gabbros and anorthite-diopsidites. The partially preserved porphyroclastic texture, the transformation of the primary olivine into an assemblage composed of forsterite, antigorite and tremolite, the cracking of olivine relics observed in the nephrite suggest that the alteration occurred with volume augmentation at rather constant mass. Thus, nephrites and diopsidites likely result from the transformation of harzburgite by addition of Ca, Si and removal of Fe, Mg at constant mass (Fig. 23.10a). The drop in density and the volumetric expansion resulting from this transformation was probably adjusted by general cracking leading to the genesis dykes. Diopsidites may have precipitated directly from the fluid, or have replaced the tremolitised harzburgite in a kind of final alteration stage. Diopsidites and anorthite-diopsidites cannot be the result of the rodingitisation of gabbroic dykes. Even if we assume that they resulted from the transformation of some gabbroic protolith (Fig. 23.10b), then the "diopsiditisation" would imply gain of or approximately immobile Si depending on the conditions (constant mass or volume, see Fig. 23.10). Whatever the conditions of a hypothetical transformation, CaO and MgO increase and Al_2O_3 and FeO* decrease from the gabbro to the diopsidite (Fig. 23.10b). Among all these chemical changes, only the gain of Ca is in accordance with a rodingitisation process. Al is usually considered as immobile during alteration, which seems confirmed by the relatively similar Al contents of the rodingites and their protoliths (see Figs. 23.8a and 23.10). Considering the chemistries of the diopsidites and gabbros, it is very unlikely that the firsts derives from the alteration of the seconds. If it was however the case, the process leading to the formation of the diopsidites would have nothing in common with rodingitisation, except for slight enrichment in Ca.

The "tremolitisation" of the harzburgite seems to be strictly related with the formation of diopsidite dykes, as shown by the variation in the degree of alteration in the host with the distance to the diopsidite. The presence of the antigorite ± tremolite assemblage in the nephrite between diopsidite and its host shows that diopsidites with a nephrite wall form out of the forsterite + diopside stability field. Diopside and tremolite coexist in a large domain of temperature, but the equilibrium forsterite + antigorite + tremolite in the nephrite suggest a temperature of formation around 550–600°C at relatively high X_{CO_2} (Trommsdorff and Evans, 1972; Trommsdorff and Connolly, 1990). Conversely, the existence of diopsidite dykes at direct contact with harzburgite without any tremolitised border shows that the genesis of the diopsidites is independent of the nephrite formation. The assemblage forsterite + diopside is stable at various temperature but the absence of serpentine and the presence of traces of tremolite suggest, in the absence of nephrite, a temperature over 650°C with a relatively high X_{CO_2} (Trommsdorff and Evans, 1972; Trommsdorff and Connolly, 1990). In the case of anorthite-diopsidite, the equilibrium diopside-anorthite suggests a *magmatic* temperature, i.e. over 900°C. This high-T assumption is also supported by the absence of high grade contact metamorphism neither in the host nor in the anorthite-diopsidite at the contact with cross-cutting magmatic dykes (Fig. 23.1f). The conditions of diopsidite genesis are then quite variable. They probably first depend on the local temperature of the mantle when the metasomatic event happened.

Are the Diopsidites a Kind of Rodingite?

A clear mineralogical or chemical definition of the term *rodingite* does not exist. Rodingites were first defined as garnet-rich rock from the Roding River in New Zealand (Bell et al., 1911), but as new rodingites were discovered all around the world, the initial meaning of the word changed. Nowadays, the most commonly accepted signification defines rodingite as a mafic rock (gabbro, diabase, basalt, etc.) transformed by Ca metasomatism in a serpentinite environment. The invoked source of Ca is usually the breakdown of pyroxenes into serpentine within the host, which release this element in the fluid during alteration. The resulting Ca-rich fluid may then react with a gabbro or any mafic lithology and lead to its Ca enrichment and the rodingite formation (Muraoka, 1985; Kobayashi and Shoji, 1988; El-Shazly and Al-Belushi, 2004; Bach and Klein, 2009). The Ca-rich composition of the rodingite is probably its most characteristic property. However, Bach and Klein (2009) and Frost et al. (2009) argued that during rodingitisation a high Ca activity in the fluid is not necessary and the Si activity gradient between the serpentinising peridotite and the mafic lithology is the key factor for rodingite genesis.

The mineralogical, textural and chemical characteristics of the diopsidites are in contradiction with a magmatic origin (see Figs. 23.5, 23.6 and 23.7; Python et al., 2007). Diopsidites are, like rodingites, characterised by a Ca-rich chemical composition. Moreover, in some heterogeneous rodingite bodies, zones entirely made of purely magnesian diopsides may be observed (Dubińska, 1995; Dubińska, 1997; El-Shazly and Al-Belushi, 2004), it is then tempting to consider that the Oman diopsidites are actually a kind of rodingite. Considering that diopsidites probably result from genuine Ca-metasomatism (as shown by the invasive crystallisation of tremolite during the alteration of their host), we could say that the diopsidite is a kind of rodingite, which differ from more "classical" rodingite (as described in other places) by some chemical characteristics probably related to a different, non-gabbroic, protolith (the harzburgite itself or some other ultramafic rocks present in the mantle?).

However, diopsidites show significant differences with rodingites. The alteration of their host harzburgite results in a lithology richer in Ca than the primary one, showing that the fluid responsible of the alteration itself was probably rich in Ca. In consequence, the source of Ca for the diopsidite genesis cannot be the breakdown of pyroxenes into serpentinite. In addition, the high bulk SiO_2 content of the resulting nephrites and diopsidites seems to be incompatible with a very low Si activity in the fluid. Although rodingitisation is closely related to serpentinisation, without which it does not happen, serpentinisation responsible of the formation of rodingite does not depend on rodingitisation and rodingite themselves. In absence of any mafic lithology, no rodingite will form but the serpentinisation may takes place. In the diopsidite case, alteration and diopsidite formation are intimately linked and the one will not occur without the other.

In the rodingite case, we are in presence of simple serpentinisation with migration of the Ca from the altered harzburgite to the rodingite, and the Si from the rodingite to the fluid. The diopsidite case implies a real metasomatism, with invasion of the mantle by a Ca-rich fluid of external origin. Pure serpentinisation does not even occur during diopsidite genesis. Serpentinites are observed only in association with rodingitised diopsidites and are related to the rodingitisation and not the diopsidite genesis. We can then say that diopsidite is not some kind of rodingite but a specific lithology with a specific genesis process.

Conclusion

There were two types of alteration in the Oman mantle, related to the genesis of rodingites for the one and to the formation of diopsidites for the other:

1. A classical serpentinisation process led to slight increase in SiO_2 and decrease in CaO, by transformation of orthopyroxenes into serpentine, in the resulting lithology. The rodingites form in association with this serpentinisation, probably by reaction of the serpentinisation-resulting fluid with a mafic dyke in the mantle.
2. Diopsidite-related alteration led to strong increases in SiO_2 and CaO, and loss of Fe and Mg, resulting from the transformation of orthopyroxene and olivine into tremolite. The remaining olivine lost its Fe and became forsteritic. Diopsidites form in close association with this kind of alteration.

Two types of rodingites formed from two types of protoliths (gabbros and anorthite-diopsidites). Most of their chemical variations may be explained by differences in the primary mineralogy and chemistry of their respective protoliths.

Taking account the chemical variations between diopsidites and gabbros, it is very unlikely that the diopsidites stem from some transformation of gabbro. If it is anyway the case, the phenomenon responsible of the alteration was not a rodingitisation process. The diopsidites most probably formed by genuine Ca metasomatism of the mantle. They share some petrological, textural and chemical characteristics with rodingites. However, the source of Ca and the genesis process are rather different for these two lithologies.

Acknowledgements We are grateful to the staff of the Ministry of Commerce and Industry in Oman for help at any aspect of the field work; to Satoko Ishimaru and Akihiro Tamura for assistance and advice at the microprobe of Kanazawa University. We thank Andrew McCraig whose comments greatly contributed to improve this manuscript. This work was partially supported by Grant-in-Aid for Creative Scientific Research (19GS0211).

References

Bach W, Klein F (2009) The petrology of seafloor rodingites: Insights from geochemical reaction path modelling. Lithos 112(1–2): 103–117. doi: 10.1016/j.lithos.2008.10.022

Barnes I, O'Neil JR, Trescases JJ (1978) Present day serpentinization in New Caledonia, Oman and Yugoslavia. Geochim Cosmochim Acta 42(1): 144–145. doi: 10.1016/0016–7037(78)90225–9

Bell JM, Clarke EdC, Marshall P (1911) The geology of the Dunn mountain subdivision, Nelson: NZ Geol. Surv Bull 12: 78

Benoit M, Ceuleneer G, Polvé M (1999) The remelting of hydrothermally altered peridotite at mid-ocean ridges by intruding mantle diapirs. Nature 402(6761): 514–518. doi: 10.1038/990073

Benoit M, Polvé M, Ceuleneer G (1996) Trace element and isotopic characterization of mafic cumulates in a fossil mantle diapir (Oman ophiolite). Chem Geol 134(1–3): 199–214. doi: 10.1016/S0009-2541(96)00087-3

Bilgrami SA, Howie RA (1960) Mineralogy and petrology of a rodingite dike, Hindubagh, Pakistan. Am Mineral 45(7–8): 791–801

Bosch D, Jamais M, Boudier F, Nicolas A, Dautria JM, Agrinier P (2004) Deep and high-temperature hydrothermal circulation in the Oman ophiolite – Petrological and isotopic evidence. J Petrol 45(6): 1181–1208. doi: 10.1093/petrology/egh010

Capedri S, Garuti G, Rossi A (1978) Rodingites from Pindos. Constraints on the "rodingite" problem. Neues Jb Miner Abh 132(3): 242–263

Ceuleneer G, Monnereau M, Amri I (1996) Thermal structure of a fossil mantle diapir inferred from the distribution of mafic cumulates. Nature 379(6561): 149–153. doi: 10.1038/379149a0

Ceuleneer G, Nicolas A, Boudier F (1988) Mantle flow patterns at an oceanic spreading centre: The Oman peridotites record. Tectonophysics 151(1–4): 1–26. doi: 10.1016/0040-1951(88)90238-7

Coleman RG (1977) Ophiolites: Ancient Oceanic Lithosphere? No. 12 in Minerals and rocks. Springer, Berlin. doi: 10.1016/0037-0738(79)90041-1

Coleman RG (1981) Tectonic setting for Ophiolite Obduction in Oman. J Geophys Res 86(B4): 2497–2508. doi: 10.1029/JB086iB04p02497

Combe JP, Launeau P, Pinet P, Despan D, Harris E, Ceuleneer G, Sotin C (2006) Mapping of an ophiolite complex by high-resolution visible-infrared spectrometry. Geochem Geophys Geosyst. doi: 10.1029/2005GC001214

Coogan LA, Thompson GM, MacLeod CJ (2002) A textural and geochemical investigation of high level gabbros from the Oman ophiolite: Implications for the role of the axial magma chamber at fast-spreading ridges. Lithos 63(1–2): 67–82. doi: 10.1016/S0024-4937(02)00114-7

Dubińska E (1995) Rodingites of the eastern part of the Jordanow-Gogolow serpentine massif, Lower Silesia, Poland. Can Mineral 33(3): 585–608

Dubińska E (1997) Rodingites and amphibolites from the serpentinites surrounding Sowie Góry block (Lower Silesia, Poland): Record of supra-subduction zone magmatism and serpentinization. Neues Jahrb Mineral Abh 171(3): 239–279

El-Shazly AEDK, Al-Belushi M (2004) Petrology and Chemistry of Metasomatic Blocks from Bawshir, Northeastern Oman. In Ernst WG (ed) Serpentine and Serpentinites: Mineralogy, Petrology, Geochemistry, Ecology, Geophysics, and Tectonics, vol 8. Bellwether Publcation, Geological Soceity of America, Columbia, MD: 388–422. doi: 10.2747/0020-6814.46.10.904 *International Book Series*

Frost R (1975) Contact metamorphism of serpentinite, chloritic blackwall and rodingite at Paddy-Go-Easy Pass, central Cascades, Washington. J Petrol 16(2): 272–313. doi: 10.1093/petrology/16.2.237

Frost BR, Beard JS (2007) On silica activity and serpentinization. J Petrol 48(7): 1351–1368. doi: 10.1093/petrology/egm021

Frost BR, Beard JS, McCaig A, Condliffe E (2009) The Formation of Micro-Rodingites from IODP Hole U1309D: Key to understanding the process of Serpentinization. J Petrol 49(9): 1579–1588. doi: 10.1093/petrology/egn038

Honnorez J, Kirst P (1975) Petrology of rodingites from the equatorial Mid-Atlantic fracture zones and their geotectonic significance. Contrib Mineral Petrol 49(3): 233–257. doi: 10.1007/BF00376590

Ishikawa T, Nagaishi K, Umino S (2002) Boninitic volcanism in the Oman ophiolite: Implications for thermal condition during transition from spreading ridge to arc. Geology 30(10): 899–902. doi: 10.1130/0091-7613(2002)03

Juteau T, Manac'h G, Moreau O, Lécuyer C, Ramboz C (2000) The high temperature reaction zone of the Oman ophiolite: New field data, microthermometry of fluid inclusions,

PIXE analyses and oxygen isotopic ratios. Mar Geophys Res 21(3–4): 351–385. doi: 10.1023/A:1026798811446

Kawahata H, Nohara M, Ishizuka H, Hasebe S, Chiba H (2001) Sr isotope geochemistry and hydrothermal alteration of the Oman ophiolite. J Geophys Res 106(B6): 11083–11099. doi: 10.1029/2000JB900456

Kobayashi S, Shoji T (1988) Metasomatic process in the formation of rodingite in Boso Peninsula, Chiba, Japan. J Mineral Petrol Econ Geol 83(12): 514–526

Li XP, Zhang LF, Wang ZL (2007) Petrology of rodingite derived from eclogite in western Tianshan, China. J Metamorph Geol 25(3): 363–382. doi: 10.1111/j.1525–1314.2007.00700.x

MacLeod CJ, Yaouancq G (2000) A fossil melt lens in the Oman ophiolite: Implications for magma chamber processes at fast spreading ridges. Earth Planet Sci Lett 176(3–4): 357–373. doi: 10.1016/S0012-821X(00)00020–0

Manning CE, MacLeod CJ, Weston PE (2000) Lower-crustal cracking front at fast-spreading ridges: Evidence from the East Pacific Rise and the Oman Ophiolite. In Dilek YR, Moores EM, Elthon D, Nicolas A (eds) Ophiolites and oceanic crust: new insights from field studies and the Ocean Drilling Program, Special Publication, vol 349 Geological Society of America, Boulder, CO: 261–272, doi: 10.1130/0–8137–2349–3.261

Mittwede SK, Schandl ES (1992) Rodingites from the southern Appalachian Piedmont, South Carolina, USA. Euro J Mineral 4(1): 7–16

Monnier C, Girardeau J, Le Mée L, Polvé M (2006) Along-ridge petrological segmentation of the mantle in the Oman ophiolite. Geochem Geophys Geosyst. doi: 10.1029/2006GC001320

Muraoka H (1985) Serpentinization reaction responsible for rodingite formation of the Ashidachi ultramafic complex, Southwest Japan. J Jpn Assoc Mineral Petrol Econ Geol 80(10): 413–428

Nasir S, Al Sayigh AR, Al Harthy A, Al Khirbash S, Al Jaaidi O, Musllam A, Al Mishwat A, Al Bu'saidi S (2007) Mineralogical and geochemical characterization of listwaenite from the Semail Ophiolite, Oman. Chem Erde-Geochem 25(3): 213–228. doi: 10.1016/j.chemer.2005.01.003

Neal CR, Stanger G (1984) Calcium and magnesium hydroxide precipitation from alkaline groundwaters in Oman, and their significance to the process of serpentinization. Mineral Mag 48(347): 237–241

Nehlig P (1994) Fracture and permeability analysis in magma-hydrothermal transition zones in the Samail Ophiolite (Oman). J Geophys Res 99(B1): 589–601. doi: 10.1029/93JB02569

Nehlig P, Juteau T, Bendel V, Cotten J (1994) The root zones of oceanic hydrothermal systems: Constraints from the Samail Ophiolite (Oman). J Geophys Res 99(B3): 4703–4713. doi: 10.1029/93JB02663

Nicolas A, Mainprice D, Boudier F (2003) High-temperature seawater circulation throughout crust of oceanic ridges: A model derived from the Oman ophiolites. J Geophys Res 108(B8): 2371. doi: 10.1029/2002JB002094

Pearce JA, Alabaster T, Shelton AW, Searle MP (1981) The Oman ophiolite as a cretaceous arc-basin complex: evidence and implications. Philos Trans R Soc A 300 (1454): Extensional Tectonics Associated with Convergent Plate Boundaries

Pomonis P, Tsikouras B, Karipi S, Hatzipanagiotou K (2008) Rodingite formation in ultramafic rocks from the Koziakas ophiolite, western Thessaly, Greece: Conditions of metasomatic alteration, geochemical exchanges and T-X(CO) evolutionary path. Can Mineral 46(3): 569–581. doi: 10.3749/canmin.46.3.569

Python M, Ceuleneer G (2003) Nature and distribution of dykes and related melt migration structures in the mantle section of the Oman ophiolite. Geochem Geophys Geosyst. doi: 10.1029/2002GC000354

Python M, Ceuleneer G, Arai S (2008) Chromian spinels in mafic – ultramafic mantle dykes: Evidence for a two-stage melt production during the evolution of the Oman ophiolite. Lithos 106(1–2): 137–154. doi: 10.1016/j.lithos.2008.07.001

Python M, Ceuleneer G, Ishida Y, Barrat JA, Arai S (2007) Oman diopsidites: A new lithology diagnostic of very high temperature hydrothermal circulation in mantle peridotite below oceanic spreading centres. Earth Planet Sci Lett 255(3–4): 289–305. doi: 10.1016/j.epsl.2006.12.030

Schandl ES, O'Hanley DS, Wicks FJ (1989) Rodingites in serpentinized ultramafic rocks of the Abitibi greenstone belt, Ontario. Can Mineral 27(4): 579–591

Stanger G (1985) Silicified serpentinite in the Semail nappe of Oman. Lithos 18: 13–22. doi: 10.1016/0024-4937(85)90003-9

Sugimoto T, Shibata T, Yoshikawa M (2007) Procedure of making a fused glass bead for whole rock major elements analyses by X-ray fluorescence spectrometer RIGAKU SYSTEM3270. Annual report of Institute for Geothermal Sciences, Graduate School of Science, Kyoto University FY2006: 44–47

Trommsdorff V, Connolly JAD (1990) Constraints on phase diagram topology for the system CaO-MgO-SiO-CO-HO. Contrib Mineral Petrol 104(1): 1–7. doi: 10.1007/BF00310641

Trommsdorff V, Evans BW (1972) Progressive metamorphism of antigorite schist in the Bergell tonalite aureole (Italy). Am J Sci 272: 423–437

Yoshitake N, Arai S, Ishida Y, Tamura A (2009) Geochemical characteristics of chloritization of mafic crust from the northern Oman ophiolite: Implications for estimating the chemical budget of hydrothermal alteration of the oceanic lithosphere. J Mineral Petrol Sci 104: 156–163. doi: 10.2465/jmps.081022b

Chapter 24
Petrogenetic Comparison of the Mafic Dykes in the Kohistan Paleo-Island Arc-Back-Arc System, Himalayas of North Pakistan

Tahseenullah Khan, Mamuro Murata, Mohammad Zafar, and Hafiz Ur Rehman

Introduction

Mafic dykes are common expression of crustal extension and represent a major avenue by which basaltic magma is transferred from the mantle to the upper crust and represent short-lived magmatic events that carry important temporal and chemical constraints for the evolution of the lithospheric mantle (Maurice et al., 2009). Mafic dykes occur in the Kohistan paleo-island arc, are generally, less abundant as compared to the back-arc, where, these dykes occur as swarm (Khan et al., 1992, 2007). Previously, the dykes of the area were discussed in terms of their origin and the timing of their intrusions (Petterson and Windley, 1992; Sawada et al., 1993; Khan et al., 1992, 2007). We reproduce our previous individual data with reference to recent research activities carried out on the Kohistan island arc (e.g. Jagoutz et al., 2009; Khan et al., 2009) and compare the mafic dykes of the area spatially, temporally and petrogenetically.

Tectonic Setting of the Area

The Kohistan paleo-island arc and back-arc terrane covers about 40,000 km^2 area and at present squeezed between the collided Indian and Karakoram-Asian plates

T. Khan (✉)
Department of Earth and Environmental Sciences, Bahria University, Islamabad, Pakistan
e-mail: bangash1444@hotmail.com

M. Murata
Department of Geosciences, Faculty of Science, Naruto University of Education, National University Corporation Naruto, Tokushima 772-8502, Japan

M. Zafar
Department of Earth and Environmental Sciences, Bahria University, Islamabad, Pakistan

H.U. Rehman
Department of Earth and Environmental Science, Faculty of Science, Kagoshima University, Kagoshima 890-0065, Japan

R.K. Srivastava (ed.), *Dyke Swarms: Keys for Geodynamic Interpretation*,
DOI 10.1007/978-3-642-12496-9_24, © Springer-Verlag Berlin Heidelberg 2011

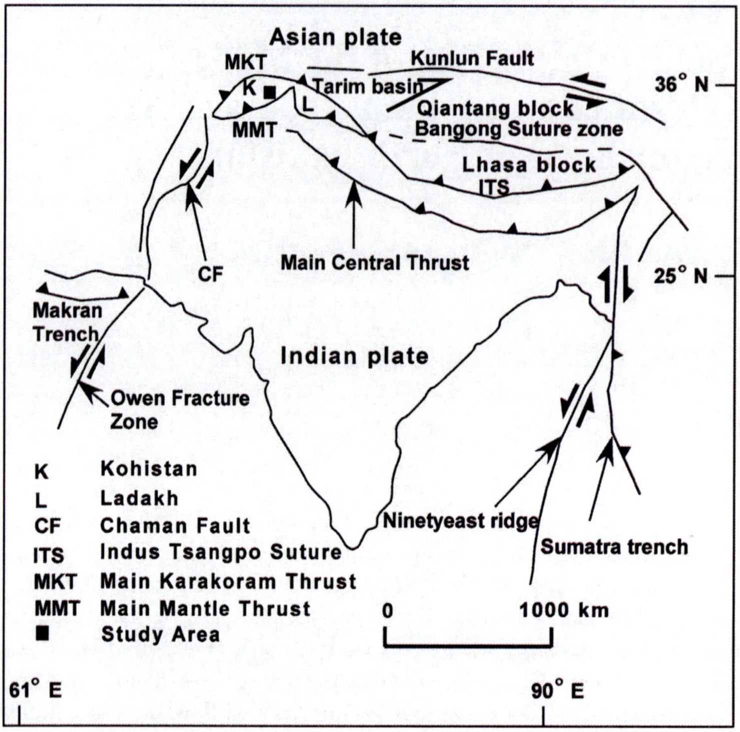

Fig. 24.1 Sketch map of the Indian plate and its margins, showing the tectonic position of the Kohistan paleo-island arc-back-arc terrane

(Fig. 24.1). The paleomagnetic and isotopic age reveal the paleoposition of Kohistan island arc to the south of the equator in Cretaceous time (Yoshida et al., 1996; Khan et al., 1997b; 2009). There is constraint on the timing of collision of the island arc with the Indian continental plate and the ages assigned for collision and complete suturing range between ~65 Ma and ~34 Ma (Patriat and Achache, 1984; Dewey et al., 1989; Klootwijk et al., 1991; Le Pichon et al., 1992; Yoshida et al., 1996; Rowley, 1996; Yin and Harrison, 2000; Najman et al., 2001, 2003; Zhu et al., 2005; Ding et al., 2005; Aichison et al., 2007). Bard (1983) considered closure of the Kohistan-Karakoram Suture at 60–50 Ma, which is supported by paleomagnetic data (Khan et al., 2009). However, some other workers assigned Late Cretaceous; 100–75 Ma for the Northern Suture on the basis of isotopic and paleomagnetic data (Petterson and Windley, 1985; Yoshida et al., 1996). The paleomagnetic direction of Kohistan paleo-island arc rocks is similar to that of Albian-Aptian (?) components maintained by the rocks in the immediate northern blocks belonging to the Eurasian continent (Klootwijk et al., 1994; Zaman and Torii, 1999), which indicates the paleoposition of the Kohistan island arc very close to the northern margin of the Eurasian continent in Cretaceous (Yoshida et al., 1996). The present position of the arc is due to the continued northward movement of the Indian plate since Jurassic after it separated from Gondwanaland.

Outline of Geology

The Kohistan paleo-island arc comprises from south to north the Jijal complex (Jan and Howie, 1981; Jan and Windley, 1990; Burg et al., 1998) showing Sm-Nd modal age of 104 Ma (Coward et al., 1986), Sm-Nd isochron cooling age of 91 $\pm$ 3 Ma (Yamamoto and Nakamura, 1996) and 96 $\pm$ 3 Ma U-Pb ages (Anczkiewicz and Vance, 1997). The Jijal complex consists of ultramafic rocks overlain by granulite facies gabbros.

The Kamila amphibolites, which tectonically overlay the Jijal complex, are composed mainly of undeformed amphibolitised gabbronorites to the north and variably deformed and sheared amphibolites, hornblendites, tonalites, pillow basalts, gabbros, diorites and granites to the south (Jan 1979; Treloar et al., 1990). The rocks were considered to be arc- related by Jan (1977, 1990) and Schaltegger et al. (2002).

The Chilas complex comprises layered ultramafic, mafic rocks and quartz diorites (Jan et al., 1984; Khan et al., 1989, 1994; Treloar et al., 1996; Fig. 24.2). Quartz diorite of the Chilas complex intrudes the paragneiss of the Gilgit formation of the Jaglot group whereas the lower contact of the complex is marked with the Kamila amphibolites and seems to be intrusive. Zeitler et al. (1980) reported 84 Ma (U-Pb) crystallisation age for gabbronorite, diorite and gabbroic dyke of the Chilas complex, which is consistent with ^{40}Ar– ^{39}Ar ages of 82 $\pm$ 5 Ma and 83 $\pm$ 3 Ma from hornblende of the Chilas complex (Treloar et al., 1989). Jagoutz et al. (2004) assigned 102 $\pm$ 5 Ma crystallisation age of the Chilas complex using Sm-Nd whole-rock clinopyroxene-plagioclase isochron.

The Kohistan batholith, which constitutes a 300 km long and 60 km broad belt, intrudes the Jaglot group and the Chalt volcanic group due to the continued subduction of the oceanic crust beneath the Andean type margin (Treloar et al., 1996). The batholith consists of numerous large to small plutons, sills, dykes and sheets emplaced over a time span ranging between 102 and 30 Ma (Petterson and Windley, 1991). A wide range of rocks including gabbro, hornblendite, diorite, quartz diorite, monzogranite, granodiorite, granite, tonalite, trondhjemite, aplite and pegmatite constitute the batholith (Petterson and Windley, 1991; Treloar et al., 1996). The Matum Das trondhjemite in the arc gives $\sim$154 Ma age (Schaltegger et al., 2004 cited in Jagoutz et al., 2009).

The Jaglot group occurs as septum within the Kohistan batholith and has been considered to represent rock assemblages of back-arc basin origin (Fig. 24.3) (Khan et al., 1994, 1997a; Treloar et al., 1996). The group includes Thelichi formation, Gashu-confluence volcanics and the Gilgit formation (Khan et al., 1994). The exact age of the Jaglot group is not known, however, mid Jurassic to Cretaceous is suggested on the basis of Rhabdophyllian fauna found in limestone from Kalam area (Bender and Raza, 1995).

The volcanic rocks run from Hunza in the east to Chitral in the west along northern margin of the arc (Tahirkheli, 1982; Coward et al., 1986; Pudsey, 1986). Treloar et al. (1996) divided the northern Kohistan paleo- arc volcanic rocks into Cretaceous Chalt volcanics including the Yasin volcanics and Eocene Shamran volcanics. These volcanic rocks range from basalt to rhyolite in composition with varying grade of

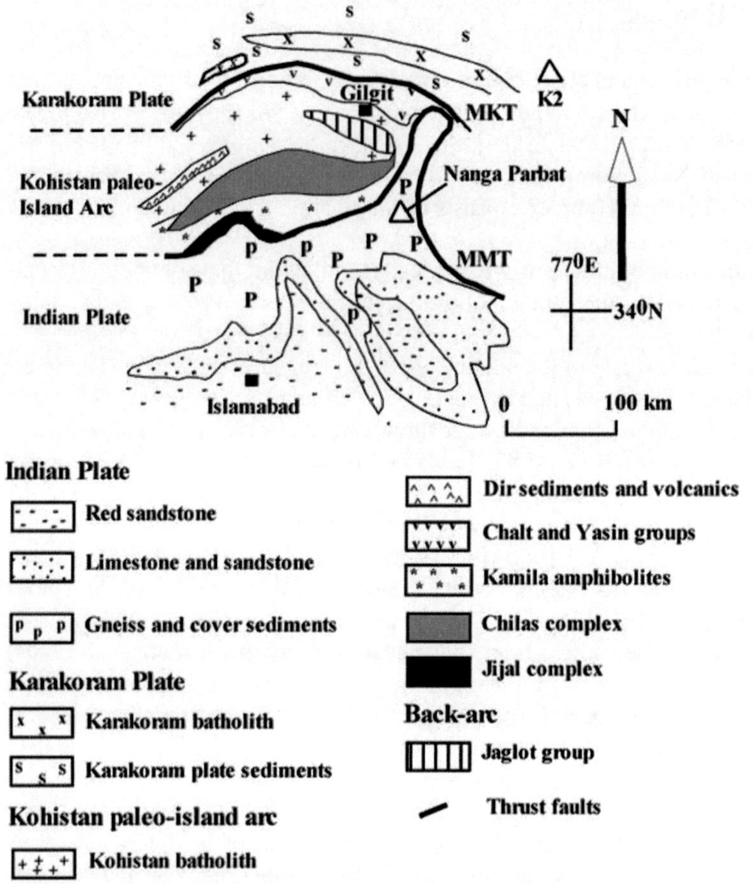

Fig. 24.2 A simplified geological map of North Pakistan showing the position of Kohistan paleo-island arc back-arc terrane

metamorphism. The Chalt and the Yasin groups are considered to be Albian-Aptian in age (Tahirkheli, 1982; Pudsey et al., 1985; Petterson et al., 1990; Khan et al., 1996; Treloar et al., 1996). On the other hand, the Shamran and Teru volcanics yield Ar-Ar hornblende age of 58 ± 1 Ma and 61 Ma U-Pb age respectively (Treloar et al., 1989; Khan et al., 2009).

Mafic Dykes

Mafic dykes of the Kohistan paleo-island arc between Gilgit and Chilas belong to the Chilas complex and Kohistan batholith (Petterson and Windley, 1992) whereas the mafic dykes of the Jaglot group occur as separate entity in Majne and Chileli Gah areas of Kar Gah Valley in Gilgit (Fig. 24.3).

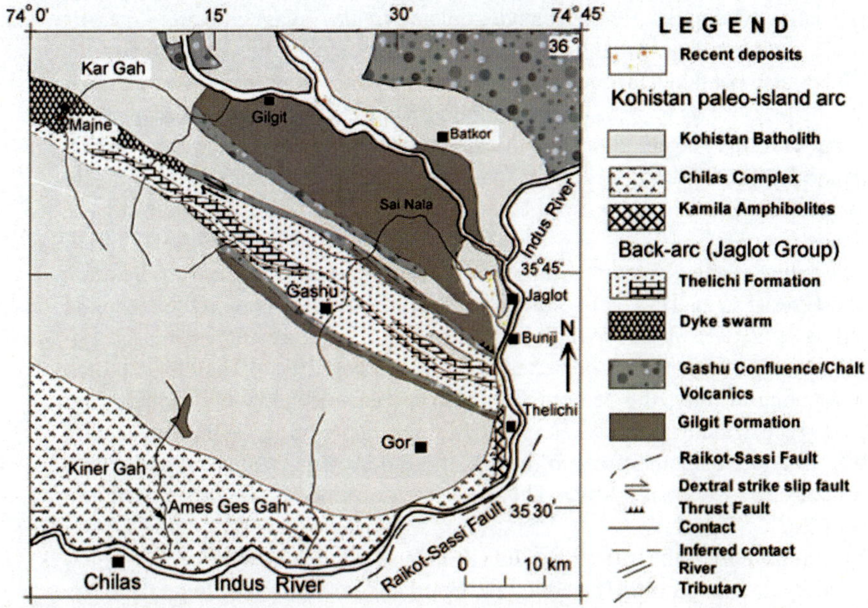

Fig. 24.3 Geological map of a part of the Kohistan paleo-island arc-back-arc terrane, exposed in Gilgit and Chilas areas of Gilgit-Baltistan

Mafic dykes occur sparsely in the Chilas complex. These are more common in the ultramafic – mafic association of the complex at Thak-Indus confluence and in Thurli Gah, and are mostly oriented NS but locally make networks. The dykes are greenish black, fresh looking, fine-grained porphyritic to aphyric and some are foliated. Chilling effects have been noticed in a few dykes intruding the ultramafics and gabbros in Thurly Gah. Many of the dykes are pinched but some are truncated. Jointing is a common feature. The mafic dykes are a few cm to 3 m thick and most are less than 10 m long, but some extend for several 10 to a few 100 m. For example, 1 m thick porphyritic dyke, trending N10°W, extends for 200 m along the contact of dunite and neighbouring rocks along KKH near Thurli Gah. The margins of this dyke are aphyric but the core is porphyritic, possibly due to flowage differentiation. Dyke swarms also intrude the orthopyroxene quartz diorite in the southernmost part of Darel Gah near Shatial where the dykes are recrystallised and the host rock is converted to amphibolite. This part of the Chilas complex represents a ductile shear zone. In Kiner Gah and Ames Ges Gah, the mafic dykes intrude gabbronorite and pyroxene diorite and cut the NW foliation of the host rocks. These range from 10 cm to 3 m in thickness, but their length is difficult to ascertain due to glaciofluvial cover on the rocks.

In thin sections, dykes are fine-grained, with or without igneous texture, i.e. phenocrysts, ophitic, poikilitic or intergranular mosaic. Many of the rocks are equi- to subequigranular and display granoblastic texture but some are foliated. The dykes consist mainly of plagioclase, hornblende, opaque oxides (magnetite and ilmenite) ± pyrite and biotite. Epidote and apatite are common accessories, but some of the mafic dykes also contain quartz. A few contain relics of orthopyroxene and clinopyroxene. The plagioclase is fresh, displays albite twinning and ranges in compositions from An_{40} to An_{60}. The amphibole is bluish green and at places replaced by biotite. Zircon and apatite occur typically in the form of tiny inclusions in plagioclase and hornblende. Quartz occurs as independent grains but some tiny inclusions in hornblende.

Mafic dykes of the Kohistan batholith are exposed in near Chalt, Kar Gah, Sai Nala and Huddar Gah. The total volume of the dykes, as elsewhere in the batholith, is much less (Petterson and Windley, 1992). The dykes are intrusive in multi-episodic plutons of the batholith (Petterson and Windley, 1992). The oldest pluton of calc-alkaline affinity, the Matum Das pluton, is ca 154 Ma old (Schaltegger et al., 2004) in which mafic dykes of 75 Ma age are intruded. These dykes are generally thin (<2 m) and strike northwest. In Huddar Gah, within the tonalite, a mafic dyke shows isoclinal folding. The dykes are deformed and recrystallised into amphibolites.

In the middle part of the batholith at Jut (Kar Gah), a 2 m thick dyke cuts the foliation of its host quartz diorite. The contact is sharp and no visible chilling effects are observed. The dyke is fine grained and weakly porphyritic containing hornblende and plagioclase as recognizable minerals. The mafic dykes in the Kohistan batholith near the junction of Shinghai Gah with Kar Gah and at Hurkus, Sai Nala intrude the diorites, which are metamorphosed into amphibolites. They are foliated, fine grained and non porphyritic. The apparent thickness of these dykes may reach up to 1 m. On the basis of petrography, the mafic dykes are grouped into two types. The first one is the recrystallised (amphibolite) with fine-grained and foliated textures. Plagioclase (An_{45}) is the common mineral. Biotite replaces hornblende. Epidote symplectites are also developed. Quartz, zircon, apatite and opaque oxides are present in traces. The dykes of second type contain plagioclase and hornblende phenocrysts in the trachytic and intersertal groundmass. Plagioclase is albite twinned and its anorthite content is also up to An_{45}. Hornblende is brown and twinned, marginally altered to chlorite. Biotite is developed at the expense of hornblende. Anhedral quartz grains are present but in traces and may be secondary.

Mafic dykes of the Jaglot group occur as dyke swarm in an approximately 100–4,000 m wide and 15 km long belt (Fig. 24.3). The dykes are basaltic and doleritic and strike northwest-southeast with high angle dip. Their orientation in general is north-south. Within the dykes, the individual dykes range in thickness between 10 cm to 5 m. The country rocks for the dykes are not clearly exposed in the area. In the northeastern part of the area, the dykes have intrusive contact with gabbros whereas in the southwestern part, the contact with the Thelichi formation is marked with a 2 m wide breccia zone. Individual dykes are also present within the Thelichi formation in Chileli Gah area.

In thin sections, the dykes exhibit ophitic to sub-ophitic textures. Plagioclase and ± diopsidic-augite (relics) are the primary mineral phases whereas titanomagnetite,

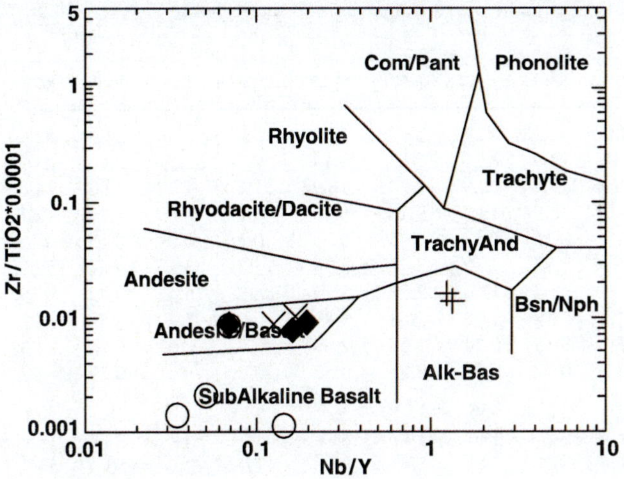

Fig. 24.4 Rock classification diagram showing mafic dykes of the Kohistan paleo-island arc-back-arc after Winchester and Floyd (1977). Symbols: *open circle*, mafic dykes of the Chilas complex; *cross* and *plus*, mafic dykes of the Kohistan batholith; *filled diamond*, mafic dykes of the Jaglot group

magnetite, ilmenite, sphene, apatite ± rutile occur as minor and accessory minerals. Actinolite-tremolite, chlorite, epidote, calcite ± quartz and ± biotite developed as secondary minerals. Hornblende seems to develop from diopsidic-augite? Titanomagnetite, magnetite, ilmenite and ± rutile are disseminated and occasionally form pseudomorphs after amphiboles and/or pyroxenes. Sphene is present as disseminated subhedral grains, veinlets and pseudomorphs after amphibole and/or pyroxene and composite grains with ilmenite.

Geochemistry

Major and trace elements of the mafic dykes were obtained in the Geoscience Advance Research Laboratories, Geological Survey of Pakistan, Islamabad, and the Department of Geosciences, Naruto University of Education, Tokushima, Japan

For major elements analysis using XRF, fused disks were prepared in a 1:10 rock-powder: lithium tetraborate ratio. Thirty-two USGS and GSJ standard reference samples were used for the calibration of the instrument. The analyses for the major elements were obtained using the Fundamental Parameter method in which the results were normalized on a volatile free basis with the total iron expressed as Fe_2O_3 (Murata, 1993). For trace elements analysis viz., Ba, Rb, Sr, Nb, Zr, Y, Cr, Ni, Co and V, powder samples were pressed. The same standard samples, which were used for major elements, used also for the empirical method with matrix and overlapping corrections.

Table 24.1 Comparison of mafic dykes composition of the Kohistan paleo-island arc back arc terrane

Sample	KT33	KT26	KT35	AGT6	KGT1	KGT59	ST45	ST30	M1	M2	M3
SiO_2	47.64	44.66	42.18	42.11	53.25	57.7	53.55	57.47	51.95	52.04	49.71
TiO_2	1.16	1.31	1.09	1.36	0.87	1.14	0.79	1.21	2.81	2.05	3.25
Al_2O_3	15.16	20.46	19.54	20.97	18.31	16.94	19.06	17.39	13.7	14.86	12.88
Fe_2O_3	11.34	14.06	12.98	13.75	9.24	6.59	9.92	6.59	12.92	12.13	15.5
MnO	0.18	0.17	0.15	0.13	0.16	0.09	0.18	0.09	0.15	0.15	0.26
MgO	8.14	5.65	8.24	7.57	3.74	3.34	3.31	3.25	4.81	5.39	3.88
CaO	13.79	10.99	12.45	12.66	7.86	5.75	8.83	5.45	9.42	9.35	7.85
Na_2O	2.14	2.15	2.12	1.15	1.58	3.89	2.75	3.7	3.49	3.22	4.49
K_2O	0.41	0.34	0.15	0.24	1.58	2.59	0.72	2.61	0.14	0.23	0.39
P_2O_5	0.12	0.15	0.02	0.02	0.21	0.52	0.26	0.61	0.29	0.27	0.43
V	241	310	346	436	165	131	145	139	362	304	422
Cr	343	20	24	19	19	81	52	94	97	148	40
Co	46	53	57	59	58	36	12	20	53	51	97
Ni	147	19	59	31	12	47	21	36	24	40	23
Zn	86	95	85	88	95	78	77	104	0	0	0
Y	29	21	20	29	24	16	24	15	37	31	88
Zr	99	15	23	19	88	184	69	167	215	186	287
Nb	2	3	1	1	4	20	3	20	6	6	6
Rb	12	7	12	5	42	65	17	91	3	7	4
Sr	190	497	346	355	433	802	471	1,024	246	328	129
Ba	130	18	0	63	364	588	450	731	43	19	49
Zr/Nb	49.50	5.00	23.00	19.00	22.00	9.20	23.00	8.35	35.83	31.00	47.83
Zr/Y	3.41	0.71	1.15	0.66	3.67	11.50	2.88	11.13	5.81	6.00	3.26
Nb/Y	0.07	0.14	0.05	0.03	0.17	1.25	0.13	1.33	0.16	0.19	0.07

Key: KT, AGT, KGT, ST and M rock samples are from Kiner Gah, Ames Ges Gah, Kar Gah Sai Nala and Majne area. Fe_2O_3 measured as total iron.

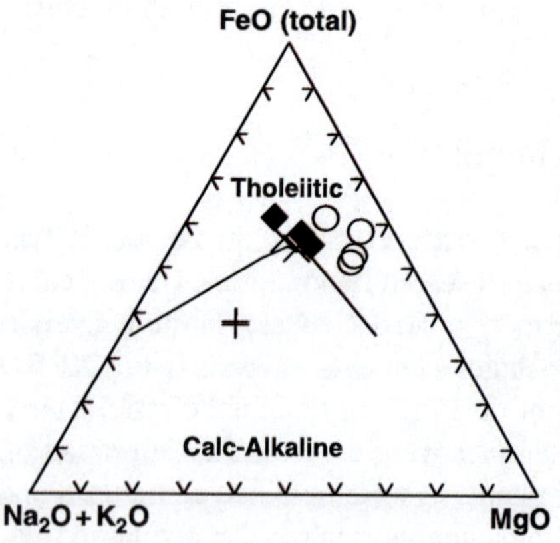

Fig. 24.5 AMF diagram showing classification of the mafic dykes after Irvine and Baragar (1971). *Symbols* as in Fig. 24.4

Geochemical data on the mafic dykes are presented in Table 24.1. Mafic dykes of the Chilas complex are low in SiO_2 (42.11–47.64 wt%), Na_2O, K_2O, P_2O_5, high

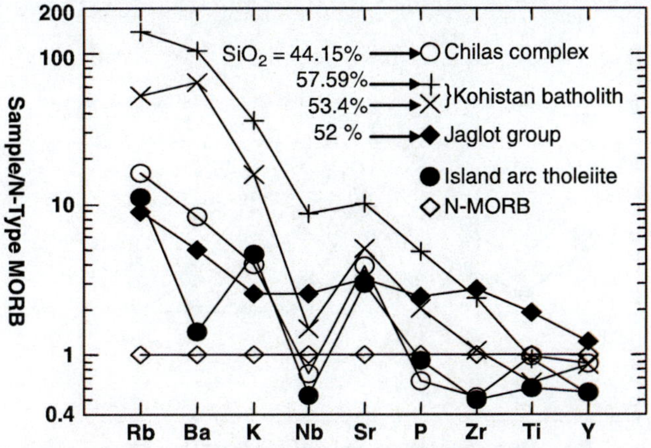

Fig. 24.6 Average data of mafic dykes of the Kohistan paleo-island arc-back-arc (N-MORB normalized diagram after Sun and McDonough, 1989). For comparison, island arc tholeiite (*filled circle*) and N-MORB (*open diamond*) are used (data source Sun, 1980)

in TiO_2, Al_2O_3 (up to 20.97 wt%), Fe_2O_3, MgO and CaO and resemble compositionally the high alumina basalts because of their high Al_2O_3 content except sample KT33 with 15.16% Al_2O_3. The analyses classify as subalkaline basalts on Nb/Y versus Zr/TiO_2 diagram of Winchester and Floyd (1977) and tholeiites on AMF diagram of Irvine and Baragar (1971) (Figs. 24.4 and 24.5). Trace elements geochemistry shows variations in all the mafic dykes. Sample KT33 contains comparatively high contents of Cr (343 ppm) and Ni (147 ppm; Table 24.1). The rest of the trace and minor elements including K, Ti and P are shown in N-MORB and primitive mantle-normalized diagrams (Figs. 24.6 and 24.7). The average composition when plotted shows depletion in HSF elements compared to the LIL elements with strong negative and positive spikes at Nb and Sr respectively.

Mafic dykes of Kohistan batholith are high in SiO_2 (53.25–57.7 wt%), Na_2O, K_2O and P_2O_5, and are low in Al_2O_3 (16.94–19.06 wt%), TiO_2, Fe_2O_3, MgO and CaO. Dykes KGT1A and ST45 have similar chemical composition and KGT59 and ST30 have close chemical similarities, thus making two distinct groups. These dykes plot as andesite/basalt and alkali basalt (Fig. 24.4). The AFM diagram distinguishes the dykes as tholeiitic and calc-alkaline.

Mafic dykes of the Kohistan batholith depict high values of Rb (17–91 ppm), Sr (433–1,024 ppm) and Ba (364–731 ppm). Samples KGT59 and ST30 contain high Nb (20 ppm), Zr (167–184 ppm), Sr (802–1,024 ppm) and Ba (588–731 ppm) as compared to other mafic dykes of the area. The average composition of calc-alkaline mafic dykes on spider diagrams show relatively more enrichment in both the LIL and HFS elements with negative spike at Nb and positive spike at Sr. The LIL/HFS elements ratio is relatively high in these dykes as compared to the tholeiitic dykes of the batholith which show almost similar geochemical patterns (Figs. 24.6 and 24.7).

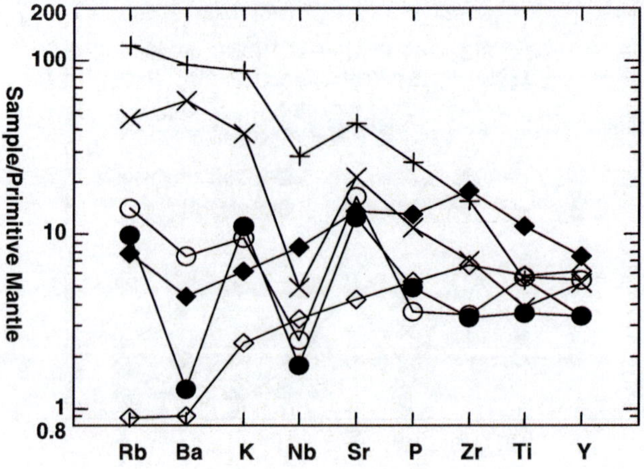

Fig. 24.7 Average data of the mafic dykes of the Kohistan paleo-island arc-back-arc (Primitive-mantle-normalized diagram after Sun and McDonough, 1989). *Symbols* as in Fig. 24.6

Mafic dykes of the Jaglot group contain SiO_2 (49.71–52.04 wt%), TiO_2 (2.05–3.2 wt%), Fe_2O_3 (12.13–15.5 wt%), Al_2O_3 (12.88–14.86 wt%) and Na_2O (3.22–4.49 wt%) and low MgO contents (3.88–5.39 wt%). The K_2O contents (0.07–0.19 wt%) are higher than the mafic dykes of Chilas complex and lower than the mafic dykes of Kohistan batholith.

These dykes contain high values of V (304–422 ppm), Y (31–88 ppm), Zr (187–287 ppm) and Nb (6 ppm). As shown in the N-MORB- normalized diagram, the average composition of the dykes show nearly flat pattern with slightly high HFS/LIL elements ratio whereas on primitive mantle normalized diagram, these are slightly enriched in Rb and Ba as compared to HFS elements (Figs. 24.6 and 24.7). In these dykes, negative and positive spikes at Nb and Sr are absent.

Petrogenesis

It is generally understood that the partial melting of mantle lherzolite and peridotite results in the formation of basaltic magma which is injected through tensional fissures into narrow zones as pillow basalts on the surface, and dykes and layered intrusions as greater depths (Wilson, 1989). The partial melts from the deformed peridotites may be evacuated through dykes at maximum depth of 50 km (Nicholas, 1986). Partial melting of a metasomatised mantle source has been suggested to have produced the mafic dykes of the Chilas complex and Kohistan batholith (Petterson and Windley, 1992).

Mafic dykes of the Chilas complex are characterised mainly by low SiO_2 and high Al_2O_3, CaO and MgO. The low silica contents may be attributed to (i) primitive melt composition to intrude the cumulate phase of the ultramafic-mafic association and the main gabbronorites/gabbros and pyroxene diorites of the Chilas complex and/or recrystallisation of these rocks in amphibolite grade? There are also many possibilities to define low absolute values of Y, Zr, P and Nb in the dykes than N-MORB, which may possibly be due to (i) crystallisation of mafic phases (Pearce and Norry, 1979); (ii) retention of these elements at source (e.g. rutile, zircon and sphene); (iii) contamination by fluids enriched in LIL elements; (iv) high degree of partial melting of the mantle source; (v) stability of the minor residual phases, and (vi) the re-melting of the already depleted mantle source (Pearce, 1982). High values of Ti may possibly due to presence of ilmenite in these dykes. The relatively higher values of MgO (8.24 wt%), Ni (147 ppm) and Cr (343 ppm) in sample KT33 may indicate low degree of partial melting or less fractional crystallisation. Crystal-liquid distribution coefficient data indicate that Ni partitions mainly into olivine during partial melting and fractional crystallisation process, whilst Cr enters chrome spinel and clinopyroxene (Wilson, 1989). These incompatible elements depleted dykes have major element characteristics between alkali olivine basalts and picrites (Khan et al., 1993), which are consistent with melting of hornblende-bearing sub-arc mantle source (Khan et al., 1989) before the closure of the Northern suture. The geochemical patterns of the mafic dykes of the Chilas complex are more or less horizontal approaching those found in the low-K island arc tholeiites. These dykes having tholeiitic affinities intrude at places the calc-alkaline hosts indicating reversal in magmatism. The basaltic liquid thus produced, migrated from the source, and solidified as mafic dykes during the subduction related processes within the root zone of magma chamber of the Chilas complex or magma generated by diapirism in early stages of intra-arc rifting during formation of back-arc basin (Khan et al., 1989; Schaltegger et al., 2002; Burg et al., 2006; Jagoutz et al., 2006, 2007).

A general lack of igneous texture, the relict nature of pyroxene, equi- to subequigranular granoblastic mosaic, and local foliation suggest metamorphism in amphibolite facies for the mafic dykes of the Chilas complex. In some dykes (e.g. KT26), the orthopyroxene has sharp boundaries against hornblende, possibly suggesting recrystallisation under hornblende granulite facies conditions (Jan et al., 1984).

Dykes in the Kohistan batholith are generally thin (<2 m), tholeiitic to calc-alkaline nature and may have been derived from the partial melting of the heterogeneous mantle source. Their production may be at shallow depths within the arc crust. The dykes intrude gabbroic-diorite. These are amphibolitised basaltic andesite and show similar geochemical patterns but grade towards calc-alkaline. High contents of biotite and amphibole and low concentration of the opaque oxides signify them to be derived under high volatile pressure. The dykes have coherent geochemical patterns with the host rocks, which are enriched in the incompatible elements and might be the result of the partial melting of the more metasomatised mantle source at shallow depth and/or formed due to amphibole fractionation from parent magma from which the mafic rocks of the Chilas complex were formed (Jagoutz et al., 2009). The calc-alkaline/alkaline basaltic dykes of the Kohistan batholith show trachytic-type texture and contain primary brown hornblende. The dykes of the Kohistan batholith are produced after the closure of the Northern suture (Petterson and Windley, 1985; Treloar et al., 1989). The tholeiitic dykes of the batholith are the components of one igneous event dated at 75 Ma (Treloar et al., 1989).

Mafic dykes of the Jaglot group show petrological and geochemical characteristics opposite to a typical island arc, as evidenced by its mode of occurrence, mineral assemblage and trace elements data. The dykes show basaltic and doleritic textures and contain sphene, ilmenite, titanomagnetite, magnetite, zircon and apatite as accessory mineral phases. Such minerals phases are not found commonly in island arc rocks except magnetite as the minor accessory mineral phase (Wilson, 1989).

High TiO_2 content of the dykes may suggest a composition transitional between tholeiitic and alkaline basaltic types. The increasing of TiO_2 and Fe_2O_3 and decreasing of Al_2O_3 with decreasing MgO suggest early onset of plagioclase fractionation (e.g. Harper, 2003). Their high contents of Ti, V, Y, Zr and Nb suggest their origin from enriched mantle.

Discussion

Mafic dykes of the Kohistan paleo-island arc-back arc system in Chilas and Gilgit areas show distinct petrological and geochemical variations. Jagoutz et al. (2009) described the origin of the Kohistan batholith as the derivative of mantle derived melts that evolved through amphibole dominated fractionation and intra-crustal assimilation. Mafic dykes, which belong to Chilas complex and Kohistan batholith differ petrologically and geochemically. Mafic dykes of the Chilas complex intrude cumulates of ultramafic-mafic association, gabbronorites and pyroxene diorites in N–S and NW–SE directions. These are mainly amphibolitised but fresh dykes apparently retain flowage differentiation features. These dykes contain hornblende, plagioclase ± orthopyroxene, epidote ± zircon, apatite, magnetite and ilmenite. The low concentrations of both the LIL and HFS elements show that the mafic dykes of the Chilas complex are the products of mantle source by high degree of partial melting and/or retention of these elements in rutile, zircon and sphene at source and/or the re-melting of the already depleted mantle source. The sharp negative spike at Nb and positive spike at Sr reveal island-arc type signatures for these dykes, which correspond to low-K island arc tholeiites commonly found in island arcs (Figs. 24.6, 24.7, 24.8 and 24.9).

Mafic dykes of Kohistan batholith intrude gabbroic-diorites, quartz diorites, tonalites and granodiorites in NW-SE directions. These dykes contain hornblende, plagioclase, biotite, apatite, magnetite, ilmenite and pyrite. Hornblende is olive green to green and metamorphic, but in one dyke it is brown and primary. The tectono-magmatic diagrams reveal these dykes to represent island arc, ocean floor and continental margin types tectonic configurations (Figs. 24.8 and 24.9).

The NW–SE strike of the mafic dykes of Jaglot group is analogous to the strike of other rock units of the area, running almost parallel to the E-W trend of the Main Karakoram Thrust (MKT). These dykes which show enriched MORB-type

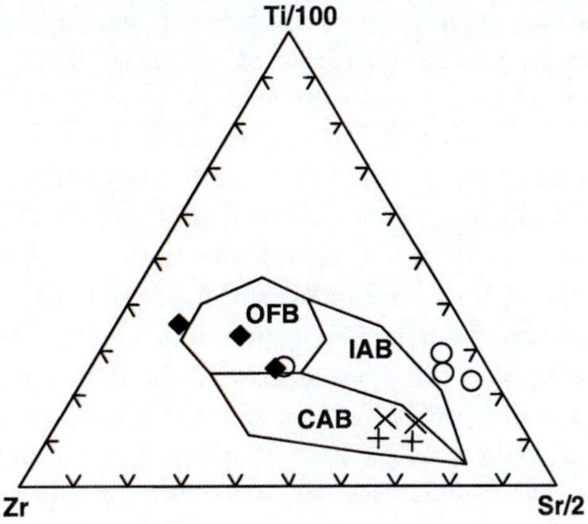

Fig. 24.8 Ti/100-Zr-Sr/2 tectono-magmatic discrimination diagram classifying mafic dykes of the Kohistan paleo-island arc-back arc (after Pearce and Cann, 1973). Symbols as in Fig. 24.6. Key: OFB, Ocean floor basalts; IAB, Island arc basalt; CAB, Calc-alkali basalt

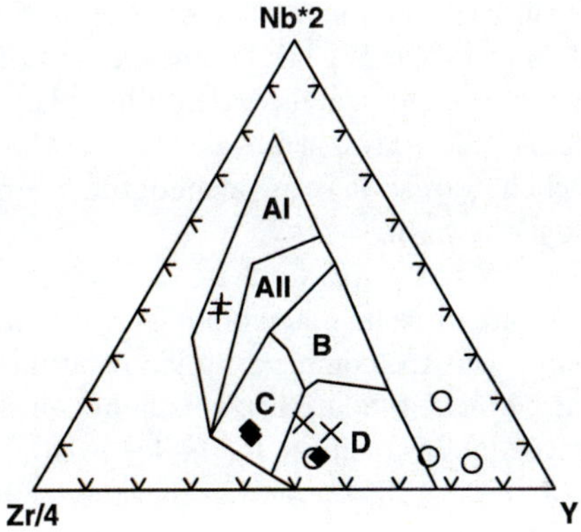

Fig. 24.9 2Nb-Zr/4-Y tectono-magmatic discrimination diagram classifying mafic dykes of the Kohistan paleo-island arc-back arc (after Meschede, 1986). *Symbols* as in Fig. 24.6. Key: AI, Within plate alkali; AII, Within plate tholeiite; B, P-MORB; C, Within plate tholeiite and Volcanic arc basalt; D, N-MORB and Volcanic arc basalt

signatures are generated in back-arc basin set up (Khan et al., 2007). The high concentration of LIL and HFS elements in these dykes may be due to heterogeneity of the mantle.

The tectonic scenario of the investigated area indicates a 120 Ma age for the separation and northward movement of Indian plate from Gondwana continent (Powell, 1979), and 150–130 Ma age for drifting of the Karakoram plate ahead of the Indian plate. The Karakoram plate was separated by the remnant paleo- or meso-Tethys from the Eurasian plate and by neo-Tethys from the Indian plate. During this period an intra-oceanic Kohistan-Ladakh island arc was formed due to the northward subduction of the neo-Tethyan oceanic lithosphere (Honegger et al., 1982). This arc was subsequently accreted to the Karakoram plate at the site of the Shyok (MKT) or Northern suture.

On the basis of the present research, a modified tectonic model is proposed for the Kohistan paleo-island arc terrane. This model covers a span more than 90 Ma from Early Cretaceous to Eocene. According to this model, the Kamila amphibolite belt is considered to represent an arc magmatism related to the northward subduction of the neo-Tethyan lithospheric plate. In the process a marginal or back-arc basin developed between the Kamila amphibolites and the Karakoram plate. In this basin, two stages of turbiditic deposition took place, together with volcanics derived from the subduction-modified mantle wedge (Chalt volcanics) followed by MORB type volcanics and dyke swarm (Jaglot group). The Chilas complex formed from a magma that was probably derived from a metasomatised mantle diapir, and intruded the back-arc basin assemblages of the Jaglot group. All these lithologies are folded and imbricated due to the closure of the back-arc basin and, subsequently, collision of the Kohistan paleo-island arc with the Indian plate.

It is proposed that the 36,000 km^2 Kohistan paleo-island arc can be divided into three tectonic zones: (i) the southern Kamila amphibolites which correspond to Kohistan paleo-island underplated by basal cumulates of the Sapat Gali and Jijal complexes. These complexes, comprising ultramafic rocks at the base and gabbroic on top, occupy the hanging wall of the Indus suture (MMT); (ii) the central part comprises gabbronorites, pyroxene diorites and ultramafic rocks, forming the ~8,000 km^2 coherent Chilas complex; and (iii) the northern Jaglot group to comprise Gilgit formation at the base, followed by the Chalt volcanic group (commonly containing pillow lavas) equivalent of the Gashu-confluence volcanics, and finally, the Thelichi formation (equivalent of the Yasin group). This group also includes the mafic dykes swarm.

The jaglot group was welded to the Karakoram plate ~100–75 Ma ago along the Shyok suture or Main Karakoram Thrust (Petterson and Windley, 1985). This and the central part are complexly folded and imbricated, with a persistent southward vergence. Much of the Kohistan batholith, principally formed in Paleocene-Eocene, cuts across the structures in the Jaglot group.

Khan et al. (1989) proposed that the Chilas complex was probably emplaced during an episode of back-arc rifting. The present research has resulted in the recognition of ocean crust assemblage with two stage turbidite deposition intervened by basaltic volcanisms and dykes swarm (Jaglot group) an assemblage formed very likely in a back-arc basin (cf. Kohn et al., 1993). This back-arc basin, initially a narrow zone of marginal basin, formed within the neo-Tethyan lithospheric plate between the Kohistan island arc and the Karakoram continental plate. The possibility that this back-arc basin formed within the neo-Tethyan lithosphere, is supported by the absence of remnant arc and the alkali basalts of continental affiliation in the area. The presence of dykes swarm in the Jaglot group clearly indicates extensional origin, typically of enriched MORB composition.

The 134 ± 3 Ma K-Ar date of the mafic dyke of the Jaglot group (Khan et al., 2007) reveals that this back-arc basin opened in early Cretaceous. The N–S orientation of the dykes may suggest E–W spreading axis of the back-arc. After the closure of the back arc, and due to continued subduction of the neo-Tethyan oceanic plate, the Kohistan paleo-island arc changed to continental margin type island arc until 55–40 Ma, which is the closure and formation date of subduction and the MMT, respectively (Fig. 24.10).

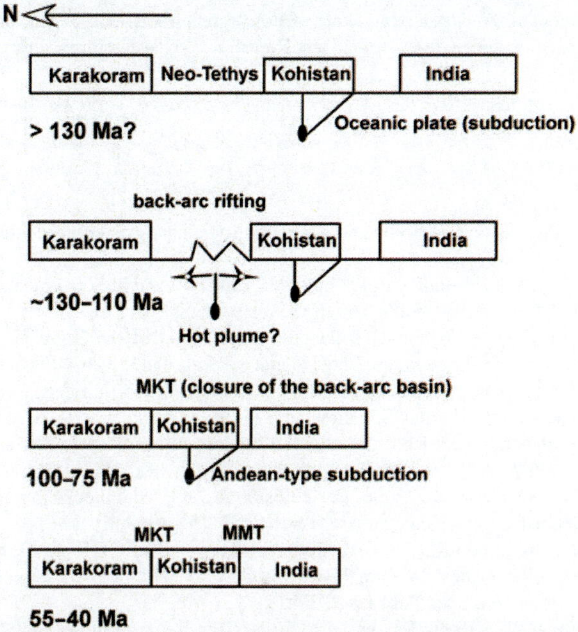

Fig. 24.10 A simplified model illustrating the tectonic evolution of the Kohistan paleo-island arc-back arc over a span of >130 to 40 Ma (modified after Khan et al., 2007)

Conclusions

On the basis of field, petrological and geochemical data, we conclude that:

1. Mafic dykes of the Chilas complex and the Kohistan batholith show island arc and continental margin signatures. Mafic dykes of the Chilas complex are LIL and HFS elements depleted whereas mafic dykes of the Kohistan batholith are LIL elements enriched. Both these dykes show similar geochemical patterns marked with negative and positive anomalies at Nb and Sr respectively. Such geochemical patterns are formed in the collisional tectonics both in island arc and continental margin-type settings.
2. Mafic dykes of the Jaglot group show enriched MORB-type affinity of back-arc basin origin. Their mode of occurrence, mineralogy and geochemistry reveal that these dykes originated due to extensional tectonics.
3. Mafic dykes of the Chilas complex and Kohistan batholith are derived by the partial melting of depleted and heterogeneous mantle sources and the Jaglot group enriched mantle sources during island arc, continental margin and back-arc tectonic settings.

Acknowledgements We are thankful to Imran Khan, Director General, Geological Survey of Pakistan and Tahir Karim, Project Director of the Geoscience Advance Research Laboratories,

Islamabad and the President of the Naruto University of Education, Tokushima, Japan for facilitating this study. We are also indebted to Talat Ahmad and Rajesh K. Srivastava for the critical review of the manuscript.

References

Aitchison JC, Ali JR, Davis AM (2007) When and where did India and Asia collide? J Geoph Res 112: B0542

Anczkiewicz R, Vance D (1997) Chronology of subduction, collision and regional metamorphism in Kohistan, vol 331, NW Himalaya, Pakistan. Terra Abstracts

Bard JP (1983) Metamorphism of an obducted island arc: Example of the Kohistan sequence (Pakistan) in the Himalayan collided range. Earth Planet Sci Lett 65: 133–144

Bender FK, Raza HA (1995) Geology of Pakistan. Gebrunder Borntragaer. Stuttgart, Berlin

Burg JP, Bodinier J, Chaudhry MA, Hussain SS, Dawood H (1998) Intra-arc mantle-crust transition and intra-arc mantle diapers in the Kohistan complex (Pakistani Himalaya): Petro-structural evidence. Terra Nova 10: 74–80

Burg JP, Jagoutz O, Hamid D, Hussain S (2006) Pre-collision tilt of crustal blocks in rifted island arcs: Structural evidence from the Kohistan Arc. Tectonics 25(5): 1–13

Coward MP, Windley BF, Broughton RD, Luff IW, Petterson MG, Pudsey CJ, Rex DC, Khan MA (1986) Collision Tectonics in the NW Himalayas. In: Coward MP, Ries AC (eds) Collision Tectonics. J. Geol Soc London, Spec Publ 19: 203–219

Dewey JF, Cande S, Pitman WC (1989) Tectonic evolution of the India – Eurasia collision zone. Eclogae Geologicae Helvetiae 82: 717–734

Ding L, Kapp P, Wan XQ (2005) Paleocene – Eocene record of ophiolite obduction and initial India – Asia collision. Tectonics, south central Tibet: 24

Harper GD (2003) Fe-Ti basalts and propagating rift tectonics in the Josephine ophiolite. Geol Soc Am Bull 115: 771–787

Honegger K, Dietrich V, Frank W, Gansser A, Thoni M, Trommsdorff V (1982) Magmatism and metamorphism in the Ladkh Himalayas (the Indus-Tsangpo suture zone). Earth Planet Sci Lett 60: 253–292

Irvine TN, Baragar WRA (1971) A guide to the classification of the common volcanic rocks. Canad J Earth Sci 8: 523–549

Jagoutz OE, Burg J-P, Hussain S, Dawood H, Pettke T, Iizuka T, Maruyama S (2009) Construction of the granitoid crust of an island arc part I: Geochronological and geochemical constraints from the plutonic Kohistan (NW Pakistan). Contrib Mineral Petrol 158: 739–755. doi: 10.1007/s00410-009-0408-3

Jagoutz O, Burg JP, Jagoutz E, Muntener O, Petteke T, Ulmer P (2004) Zoned ultramafic intrusions of the Chilas complex in Kohistan (NE Pakistan): Mantle diapers and km-scale melt conduits in extending island arcs. Himalayan J Sci 2(4) Special Issue Extended Abstracts: 166

Jagoutz O, Müntener O, Burg JP, Ulmer P, Jagoutz E (2006) Lower continental crust formation through focused flow in km-scale melt conduits: The zoned ultramafic bodies of the Chilas Complex in the Kohistan Island arc (NW Pakistan). Earth Planet Sci Lett 242(3–4): 320–342

Jagoutz O, Müntener O, Ulmer P, Burg J-P, Pettke T (2007) Petrology and mineral chemistry of lower crustal intrusions: The Chilas complex, Kohistan (NW Pakistan). J Petrol 48(10): 1895–1953

Jan MQ (1977) Petrography of the amphibolites of Swat and Kohistan. Geol Bull Univ Peshawar 11: 51–64

Jan MQ (1990) Petrology and geochemistry of the southern amphibolites of the Kohistan arc, N. Pakistan. In: Sharma KK (ed) Geology and geodynamic evolution of the Himalayan collision zone. Phy Chem Earth 17: 71–92

Jan MQ, Howie RA (1981) The mineralogy and geochemistry of the metamorphosed basic and ultrabasic rocks of the Jijal Complex, Kohistan, N.W. Pakistan. J Petrol 22: 85–126

Jan MQ, Pervez MK, Khattak MUK, Windley BF (1984) The Chilas stratiform complex: Field and mineralogical aspects. Geol Bull Univ Peshawar 17: 153–169

Jan MQ, Windley BF (1990) Chromium Spinel-silicate chemistry in ultramafic rocks of the Jijal complex, Northeast Pakistan. J Petrol 31: 667–715

Khan T, Jan MQ, Khan MA, Kausar AB (1997a) High-grade metasedimentary rocks (Gilgit Formation) in the vicinity of Gilgit, Kohistan, northern Pakistan. J Mineral Petrol Econ Geol 92(11): 465–479

Khan MA, Jan MQ, Weaver BL (1993) Evolution of the lower arc crust in Kohistan, N. Pakistan; temporal arc magmatism through early, mature and intra-arc rift stages. In: Treloar PJ, Searle MP (eds), Himalayan tectonics. Geological Society Special Publications. Geol Soc London, London: 123–138

Khan MA, Jan MQ, Windley BF, Tarney J, Thirlwall MF (1989) The Chilas mafic-ultramafic igneous complex of the Kohistan island arc in the Himalaya of north Pakistan. Geol Soc Am Spec Paper 232: 75–94

Khan T, Khan MA, Aziz A (1992) Petrogenetic comparison between the mafic dykes of Chilas complex and Kohistan batholith, Northern Pakistan. Acta Mineralogica Pakistanica 6: 118–125

Khan T, Khan MA, Jan MQ (1994) Geology of part of the Kohistan terrane between Gilgit and Chilas, Northern areas of Pakistan. Geol Bull Univ Peshawar 27: 99–112

Khan T, Khan MA, Jan MQ, Naseem M (1996) Back-arc basin assemblages in Kohistan, Northern Pakistan. Geodinamica Acta 9(no 1): 30–40

Khan T, Murata M, Karim T, Zafar M, Ozawa H, Rehman H (2007) A Cretaceous dike swarm provides evidence of a spreading axis in the back-arc basin of the Kohistan paleo-island arc, northwestern Himalaya, Pakistan. J Asian Earth Sci 29: 350–360

Khan MA, Stern RJ, Gribble RF, Windley BF (1997b) Geochemical and isotopic constraints on subduction polarity, magma source, and paleogeography of the Kohistan intra-oceanic arc, northern Pakistan Himalaya. J Geol Soc London 154: 935–946

Khan SD, Walker DJ, Hall SA, Burke KC, Shah MT, Stockli L (2009) Did the Kohistan-Ladakh island arc collide first with India? Geol Soc Am Bull 121: 366–384

Klootwijk CT, Conaghan PJ, Nazirullah R, De Jong KA (1994) Further palaeomagnetic data from Chitral (eastern Hindukush); evidence for an early India-Asia contact. Tectonophysics 237 (1–2): 1–25

Klootwijk CT, Gee JS, Peirce JW, Smith GM (1991) Constraints on the India-Asia convergence: Paleomagnetic results from Ninetyeast Ridge. In: Weissel J, Peirce J, Taylor E, Alt J., et al., Proceedings. ODP, Sci. Results 121: College Station, TX (Ocean Drilling Program): 777–882

Kohn MJ, Speer FS, Dalziel IWD (1993) Metamorphic P-T-paths from Cordillera Darwin, a Core Complex in Teirra del Fuego, Chile. J Petrol 34: 519–542

Le Pichon X, Fournier M, Jolivet L (1992) Kinematics, topography, shortening, and extrusion in the India – Eurasia collision. Tectonics 11: 1085–1098

Maurice C, David J, O'Neil J, Francis D (2009) Age and tectonic implications of Paleoproterozoic mafic dyke swarms for the origin of 2.2 Ga enriched lithosphere beneath the Ungava Peninsula, Canada. doi:10.1016/j.precamres.2009.07.007

Meschede M (1986) A method of discriminating between different types of mid ocean ridge basalts and continental tholeiites with the Nb-Zr-Y diagram. Chem Geol 56: 207–218

Murata M (1993) Major and trace component analysis of Korean Institute of Energy and Resources igneous rock reference samples using X-ray fluorescence spectrometer. Res Bull Nat Sci Naruto Univ of Education 8: 37–49

Najman Y, Garzanti E, Pringle M, Bickle M, Stix J, Khan I (2003) Early – middle Miocene paleodrainage and tectonics in the Pakistan Himalaya. Geol Soc Am Bull 115: 1265–1277

Najman Y, Pringle M, Godin L, Oliver G (2001) Dating of the oldest continental sediments from the Himalayan foreland basin. Nature 410: 194–197

Nicholas A (1986) A melt extraction model based on structural studies in mantle peridotites. J Petrol 27: 999–1022

Patriat P, Achache J (1984) India – Eurasia collision chronology has implications for crustal shortening and driving mechanism of plates. Nature 311: 615– 621

Pearce JA (1982) Trace element characteristics of lavas of destructive plate boundaries. In: Thorpe SS (ed), Andesites: orogenic andesites and related rocks. Wiley Press, New York: 525–548

Pearce JA, Cann JR (1973) Tectonic setting of basic volcanic rocks determined using trace element analyses. Earth Planet Sci Lett 19: 290–300

Pearce JA, Norry MJ (1979) Petrogenetic implications of Ti, Zr, Y and Nb variations I volcanic interpretation. Contr Mineral Petrol 69: 217–235

Petterson MG, Windley BF (1985) Rb-Sr dating of the Kohistan arc batholith in the Transhimalayan of the North Pakistan and tectonic implications. Earth Planet Sci Lett 74: 45–57

Petterson MG, Windley BF (1991) Changing source region of magma and crustal growth in the Trans-Himalayas: Evidence from Chalt volcanics and Kohistan batholith, Kohistan, northern Pakistan. Earth Planet Sci Lett 102: 326–341

Petterson MG, Windley BF (1992) The field relations, geochemistry and petrogenesis of the Cretaceous basaltic Jutal dyke-suite, Kohistan, Northern Pakistan. J Geol Soc London 149: 107–114

Petterson MG, Windley BF, Luff IW (1990) The Chalt volcanics, Kohistan, N. Pakistan: High-Mg tholeiites and low Mg calc-alkaline volcanism in a Cretaceous island arc. In: Sharma KK (ed) Geology and geodynamic evolution of the Himalayan collision zone. Phys Chem Earth 17: 19–30

Powell CMcA (1979) A speculative tectonic history of Pakistan and surroundings: Some constraints from Indian ocean. In: Farah A, Dejong K (eds), Geodynamics of Pakistan. Geological Survey Pakistan, Pakistan: 5–24

Pudsey CJ (1986) The Northern Suture, Pakistan: Margin of a Cretaceous island arc. Geol Magazine 123: 405–423

Pudsey CJ, Coward MP, Luff IW, Shackleton RM, Windley BF, Jan MQ (1985) The collision zone between the Kohistan arc and the Asian plate in NW Pakistan. Trans Royal Soc Edinburg 76: 464–479

Rowley DB (1996) Age of collision between India and Asia: A review of the stratigraphic data. Earth Planet Sci Lett 145: 1–13

Sawada Y, Kubo K, Kausar AB, Takashi Y (1993) Primary magma of the Chilas complex in the Kohistan arc, Northern Pakistan. Proc Geosci Colloq, Geoscience Laboratory, Geol Surv Pakistan 7: 97–119

Schaltegger U et al (2004) Crust-mantle interaction during Karakoram- Kohistan accretion (NW Pakistan). Goldschmidt 2004. Geochim Cosmochim Acta 68(11) Copenhagen

Schaltegger U, Zeilinger G, Frank M, Burg JP (2002) Multiple mantle sources during island arc magmatism: U-Pb and Hf isotopic evidence from the Kohistan arc complex, Pakistan. Terra Nova 14: 461–468

Sun SS (1980) Lead isotope study of young volcanic rocks from mid ocean ridges and island arcs. Phil Trans R Soc London, A 297: 409–445

Sun SS, McDonough WF (1989) Chemical and isotopic systematic of ocean basalts: Implication for mantle composition and processes. J Geol Soc London Spec Publ 42: 313–345

Tahirkheli RAK (1982) Geology of the Himalaya, Karakorum and Hindukush. Geol Bull Univ Peshawar 13: 1–151

Treloar PJ, Broadie KH, Coward MP, Jan MQ, Knipe RJ, Rex DC, Williams MP (1990) The evolution of the Kamila shear zone, Kohistan, Pakistan. In: Salibury MH, Fountain DM (eds), Exposed cross sections of the continental crust. Kluwer Academy Press, Amsterdam: 175–214

Treloar PJ, Petterson MG, Jan MQ, Sullivan MA (1996) A re-evaluation of the stratigraphy and evolution of the Kohistan arc sequence, Pakistan Himalaya: Implications for magmatic and tectonic arc-building processes. J Geol Soc London 153: 681–693

Treloar PJ, Rex DC, Guise PG, Coward MP, Searle MP, Windley BF, Petterson MG, Jan MQ, Luff IW (1989) K/Ar and Ar/Ar geochronology of the Himalayan collision in NW Pakistan:

Constraints on the timing of suturing, deformation, metamorphism and uplift. Tectonics 4: 881–909
Wilson M (1989) Igneous petrogenesis, a global tectonic approach. Unwin Hyman Press, London: 1–466
Winchester JA, Floyd PA (1977) Geochemical discrimination of different magma series and their differentiation products using immobile elements. Chem Geol 20: 325–343
Yamamoto H, Nakamura E (1996) Sm-Nd dating of garnet granulites from the Kohistan complex, northern Pakistan. J Geol Soc London 153: 965–969
Yin A, Harrison TM (2000) Geologic evolution of the Himalayan-Tibetan orogen. Annual Review of Earth Planet Sci Lett 28: 211–280
Yoshida M, Zaman H, Ahmad MN (1996) Paleoposition of Kohistan arc and surrounding terranes since Cretaceous time: The paleomagnetic contraints. In: Kausar AB, Yajima J (eds) Geology, Geochemistry, Economic Geology and Rock Magnetism of the Kohistan arc. Proce Geosci Colloq, Geoscience Laboratory, Geol Surv Pakistan: 83–101
Zaman H, Torii M (1999) Palaeomagnetic study of Cretaceous red beds from the eastern Hindukush ranges, northern Pakistan; paleoarc construction of the Kohistan-Karakoram composite unit before the India-Asia collision. Geophys J Int 136(3): 719–738
Zeitler PK, Tahirkheli RAK, Naesser C, Johnson N, Lyon J (1980) Preliminary fission-track ages from Swat valley, northern Pakistan. Geol Bull Univ Peshawar 13: 23–29
Zhu B, Kidd WSF, Rowley DB, Currie BS, Shafique N (2005) Age of initiation of the India – Asia collision in the east central Himalaya. J Geol 113: 265–285

Chapter 25
A Comparison of Chronometers Applied to Monastery Kimberlite and the Feasibility of U-Pb Ilmenite Geochronology

A.K. Noyes, L.M. Heaman, and R.A. Creaser

Introduction

The discovery of economic diamond-bearing rocks of deep-seated origin in North America (e.g. the Lac de Gras kimberlite field in northern Canada) and the recognition of many new kimberlite fields of uncertain economic significance have substantially increased the number of known kimberlite occurrences in North America to more than 1,000. In addition to their potential economic importance, the timing and composition of kimberlites can reveal information regarding the origin and emplacement mechanisms of these deep-seated mantle magmas. Various tectonic scenarios have been proposed for the origin of kimberlite magmatism (see summary by Mitchell, 1986) including a link to subduction (e.g., Sharp, 1974; McCandless, 1999), rifting (e.g., Taylor, 1984; Phipps, 1988) and mantle plumes (e.g., Crough, 1981; Sleep, 1990; Heaman and Kjarsgaard, 2000). However, a single mechanism cannot explain the origin of all kimberlites (Helmstaedt and Gurney, 1984; Mitchell, 1986). Based on the precise and accurate age determinations for more than 100 kimberlites in North America (Heaman et al., 2003, 2004), there are a number of patterns of kimberlite emplacement that require quite distinct tectonic scenarios for their generation. For example, the narrow NW–SE trending corridor of Triassic to Jurassic kimberlite magmatism in eastern North America display a progressive SE younging that is best interpreted as forming along the continental extension of a known mantle plume hotspot track associated with the opening of the North Atlantic Ocean (Heaman and Kjarsgaard, 2000). In contrast, the belt-like

L.M. Heaman (✉)
Department of Earth and Atmospheric Sciences, University of Alberta, Edmonton, AB, Canada T6G 2E3
e-mail: larry.heaman@ualberta.ca

R.A. Creaser and A.K. Noyes
Department of Earth and Atmospheric Sciences, University of Alberta, Edmonton, AB, Canada T6G 2E3

R.K. Srivastava (ed.), *Dyke Swarms: Keys for Geodynamic Interpretation*,
DOI 10.1007/978-3-642-12496-9_25,

corridors of kimberlite magmatism that parallel and young towards continental margins (e.g., the 95–105 Ma Cretaceous corridor of kimberlite magmatism that extends from Somerset Island to the mid-continental U.S.A.) is most likely linked to extensional tectonics accompanying changes in Farallon plate geometry (Heaman et al., 2003, 2004). In order to differentiate between these different tectonic environments for kimberlite genesis, it is imperative to obtain precise and accurate age constraints for the kimberlite emplacement history both within individual fields and on a more regional scale.

Many North American kimberlites and other kimberlites worldwide do not contain primary minerals such as phlogopite, perovskite, and mantle zircon; minerals commonly used to determine kimberlite emplacement ages. For example, in a recent study of 125 kimberlite samples from 38 separate kimberlites on the Diavik property in northern Canada, mantle zircon and kimberlitic perovskite were not identified in any samples and unaltered phlogopite megacrysts were identified in only two samples (Zurevinski, 2009). Therefore, only a small percentage of known kimberlites have accurate emplacement ages (~20% in North America) so there is a need to discover additional approaches to address the geochronology of kimberlites and related rocks. Ilmenite is widely used as an exploration tool for kimberlites because of its diagnostic Mg-rich character, physical resilience and presence in most kimberlites worldwide. Extensive research on ilmenite mineral chemistry and relationship to kimberlite petrology indicates that ilmenite is genetically linked to direct crystallization in kimberlitic or proto-kimberlitic magmas (Mitchell, 1973; Schulze et al., 1995) so obtaining a reliable radiometric date for ilmenite formation could reflect the approximate crystallization age of a kimberlitic magma. Kimberlitic ilmenite is known to contain small amounts of uranium (0.1–0.2 ppm; Kresten, 1974), indicating that U-Pb geochronology might be feasible, however there is currently no information on the U-Pb systematics of kimberlitic ilmenite. Lanzirotti and Hanson (1995) report the U-Pb systematics for two ilmenite fractions from the Straits Schist, Connecticut and these analyses indicate an order of magnitude higher uranium contents (3.7–4.5 ppm) for these relatively low-MgO (~0.05 wt%) metamorphic ilmenites compared to the concentrations Kresten (1974) reported for kimberlitic ilmenite.

The objectives of this study are twofold. The first is to attempt a comprehensive study that applies many of the currently used radiometric dating techniques on a single well-studied kimberlite intrusion. The Monastery kimberlite was considered ideal for this purpose because previous attempts to establish the emplacement age using K-Ar, Rb-Sr, and U-Pb methods all yield similar results. Furthermore, this kimberlite contains all the minerals currently used to determine kimberlite emplacement ages (e.g., phlogopite, mantle zircon and perovskite) so it should be possible to provide a robust chronology of kimberlite emplacement at Monastery, which is critical when assessing the veracity of a new dating tool. The second objective is to investigate the U-Pb systematics of ilmenite and evaluate the feasibility of using ilmenite as a new chronometer for determining the emplacement age of kimberlites and kimberlite indicator minerals from exploration samples. Of particular importance to this study is the fact that the chemistry and petrogenesis of Monastery ilmenite is well established and has been the subject of numerous studies (e.g., Boyd and Nixon, 1975; Mitchell, 1977, 1986; Haggerty et al., 1979; Gurney et al., 1979, 1998; Moore et al., 1992; Griffin et al., 1997; Wyatt et al., 2004). We provide detailed imaging of the selected ilmenite megacrysts to evaluate whether any uranium or thorium might reside in microscopic mineral inclusions and attempt to relate the ilmenite mineral chemistry to the U-Pb systematics.

Geological Setting of the Monastery Kimberlite

Regional Setting

The Monastery kimberlite is located in the Eastern Free State of South Africa, close to the border of Lesotho (Fig. 25.1a). Geologically, the Monastery kimberlite is situated within the Kaapvaal craton along with numerous other diamond-bearing kimberlite fields (e.g., Kimberley field; Fig. 25.1a). The Kaapvaal craton extends in an east-west direction from the Lebombo Mountains to the Orange River at Prieska, and in a north-south direction from the Limpopo Belt to Lesotho (Cahen et al., 1984). The Kaapvaal craton is composed primarily of Archean granite-greenstone belts, Archean basement gneisses and to a lesser extent sedimentary and volcanic rocks of Proterozoic age. It is bounded on the south by the ∼1.4 Ga Namaqua-Natal Mobile Belt and the Archean Limpopo Mobile Belt to the north (Fig. 25.1a). Clifford (1970) noted that diamond-bearing kimberlites are typically located in cratonic areas in Africa that have not been affected tectonically since about 1,850 Ma (e.g., the Eburnian and Huabian orogenies). Except for the northern region, most of the African continent has been stable since the Damaran-Katangan Orogeny at ∼550 Ma (Truswell, 1977).

The majority of South Africa is covered by the Karoo supracrustal sequence that ranges in age from 300 to 160 Ma (Truswell, 1977). The Karoo sedimentary units have been divided into three groups; the Ecca, Beaufort and the Stormberg Groups. The Monastery kimberlite intrudes the Red Bed or Elliot Formation of the Stormberg Group (Whitelock, 1973). The end of the Karoo sedimentation period (190–160 Ma) was punctuated by a spectacular volcanic event in South Africa producing extensive lava flows and dolerite dikes and sills of the Drakensberg large igneous province. The Drakensberg volcanism spanned a period ranging from 184 to 179 Ma with the majority of volcanic activity (lava flows, dolerite dikes and sills) occurring at 183±1 Ma (Duncan et al., 1997). The majority of kimberlites in South Africa were emplaced post-Karoo sedimentation and volcanism as they intrude these units and many kimberlites contain xenoliths of basement gneiss, sandstone, siltstone, dolerite, gabbro and Drakensberg volcanics (Allsopp and Barrett, 1975; Truswell, 1977; Allsopp and Roddick, 1984).

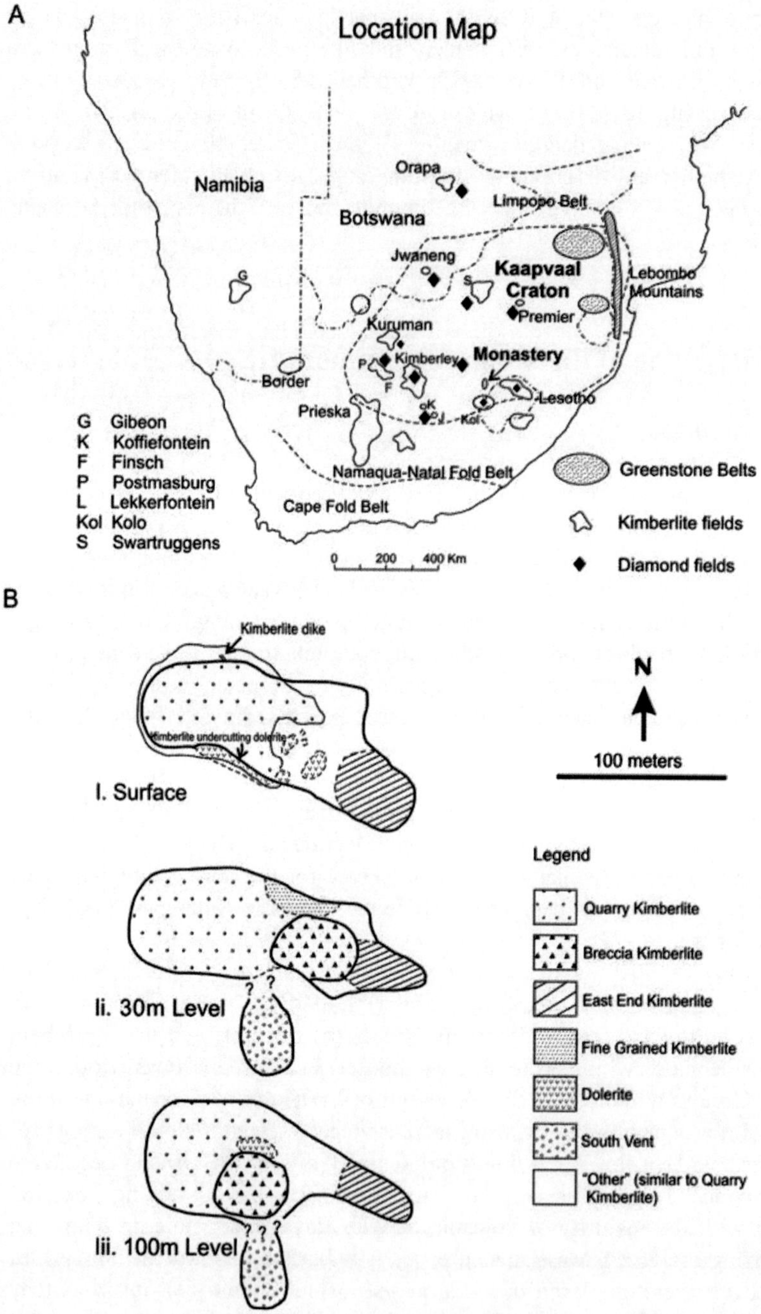

Fig. 25.1 (**a**) Location map of South Africa outlining various kimberlite and diamond fields (after Nixon, 1987). (**b**) Schematic diagram of the Monastery kimberlite pipe at various mine levels (after Whitelock, 1973)

Local Setting

Figure 25.1b is a geological map of the Monastery kimberlite pipe with schematic representation of different kimberlite types and intrusion shape at various levels (after Whitelock, 1973). The Monastery intrusion is irregular in shape with dimensions approximating 180 m in length by 70 m in width (Whitelock, 1973). Whitelock (1973) identified four kimberlite types at the Monastery mine based on colour and mineralogy: the East End Type, Quarry Type, Breccia Type and Fine-Grained Type (see petrographic description below). The East End Type kimberlite is located at the eastern-most tip of the intrusion and was termed the "eastern blow" by Whitelock (1973). This kimberlite type narrows to a dike-like structure, inclining at shallow depths. The Quarry Type kimberlite is the most abundant and most diamondiferous rock type at the Monastery mine (Whitelock, 1973). Dolerite occurs in surface exposures along the southern rim of the Quarry Type kimberlite and also occurs as xenoliths scattered throughout the intrusion (Fig. 25.1b). The Breccia Type kimberlite forms a column in the central region of the intrusion and has a westerly dip of 70° (Whitelock, 1973). A subsurface "South Vent" (Fig. 25.1b Ii&Iii) was discovered through a series of bore-holes and Whitelock (1973) proposed that it may be connected to the main body of the kimberlite.

Previous Geochronology

South African kimberlites have been subdivided into two groups based on their emplacement ages, isotopic compositions and mineralogical assemblages (Smith, 1983). The Group 1 kimberlites contain little to no late crystallizing groundmass mica and typically have emplacement ages ranging from 80 to 114 Ma. Group 2 kimberlites or orangeites are mica-rich and have emplacement ages ranging between 114 and 150 Ma.

A summary of radiometric ages obtained previously for the Monastery kimberlite is shown in Fig. 25.2 and range between 83.0±3.0 and 90.0±4.0 Ma. The U-Pb perovskite date of 83.0 Ma determined by Kramers and Smith (1983) represents the youngest age determination for the Monastery kimberlite. The range of ages obtained for the Monastery kimberlite is slightly smaller (88.1–90.4 Ma) if this perovskite date is omitted. Zartman et al. (1998) obtained a very precise weighted average $^{206}Pb/^{238}U$ date of 90.1±0.5 Ma from ten crystal fragments analysed from a single mantle zircon megacryst, consistent with the ages obtained by other techniques. Allsopp and Barrett (1975) obtained a Rb-Sr isochron date of 90.0±4.0 Ma derived from nine out of ten analyses of fresh, unaltered phlogopite isolated from five separate kimberlite samples. Smith and Barton (1996) obtained a very precise Rb-Sr date of 88.1±0.6 Ma from a single phlogopite megacryst. In the latter study these authors analysed a number of phlogopite pieces from a single 4 cm diameter megacryst and selected portions of the megacryst to specifically identify isotopic and age variations from core to rim; none were found.

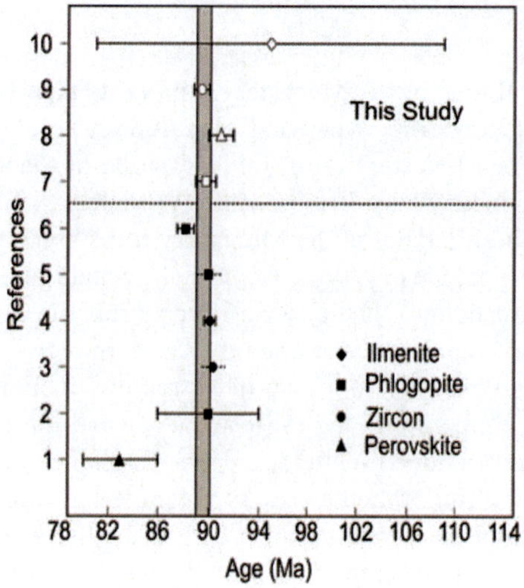

Fig. 25.2 Summary of Monastery kimberlite ages from previous studies and this study. References are as follows: 1. Kramers and Smith (1983) U-Pb on perovskite (82.5±2.5 Ma); 2. Allsopp and Barrett (1975) Rb-Sr on phlogopite (90.0±4.0 Ma); 3. Davis et al. (1976) U-Pb mantle zircon (90.4 Ma); 4. Zartman et al. (1998) U-Pb mantle zircon (90.1±0.5 Ma); 5. McIntyre and Dawson (1976) K-Ar phlogopite (90 Ma); 6. Smith and Barton (1996) Rb-Sr phlogopite (88.1±0.6 Ma); 7. This study, Rb-Sr phlogopite (89.4±0.8 Ma); 8. This study, U-Pb perovskite (91.0±1.0 Ma); 9. This study, U-Pb zircon (88.8±0.5 Ma); 10. This study, U-Pb ilmenite (95.0±14.0 Ma). Samples with dashed error bars are given an estimated minimum of ±1 Ma error

The best estimate for the emplacement age of the Monastery kimberlite prior to this study is approximately 90±2 Ma based on a compilation of K-Ar/Rb-Sr phlogopite and U-Pb mantle zircon age determinations (Allsopp and Barrett, 1975; McIntrye and Dawson, 1976; Davis et al., 1976; Davis, 1977; Smith and Barton, 1996; Zartman et al., 1998) and is slightly older but within error of the proposed Group 1 kimberlite age range for the Kimberley and Jagersfontein areas (86–87 Ma; Smith and Barton, 1996).

Material Analysed in This Study

Petrography

Samples of three main Monastery kimberlite types, Quarry Type, East End Type and Breccia Type, were selected for this study and are denoted M1, M2, and M3, respectively. A fourth kimberlite type (Fine-Grained Type) recognized by Whitelock (1973) was not investigated. These three main kimberlite types possess minor differences in colour, degree of alteration, mineralogy and modal mineral content.

The Quarry Type kimberlite (M1) is dark-blue in colour and contains visible ilmenite megacrysts and macrocrysts. Rounded, olivine macrocrysts range in size from 0.5 cm to 1 mm and are noticeably altered to serpentine. The East End Type kimberlite (M2) is characterized by a pale bluish-grey colour and by an abundance of carbonate in the groundmass. Altered reddish-brown olivine grains range in size from 0.4 cm to 2 cm in diameter and often possess kelyphite rims. Ilmenite megacrysts and macrocrysts are also abundant (Plate 25.1a). The Breccia Type kimberlite (M3) is typically weathered to a brown colour and contains a higher abundance of megacrysts. Garnet and phlogopite (Plate 25.1e) megacrysts can be as large as 12 cm in diameter, often with kelyphite rims.

Apart from the differences noted above, all three samples investigated in this study are petrographically quite similar. The olivine macrocrysts (0.5–12 mm; modal abundance 35–40%) are completely altered to serpentine. Similarly, some phlogopite macrocrysts (~30%) are extensively altered to serpentine and calcite (M1 and M2 only) and others have chloritized rims. The rims of some phlogopite macrocrysts can contain ilmenite (Plate 25.1f) and perovskite. Ilmenite macrocrysts (<1 mm; ~20–25%) are typically subhedral to anhedral in shape. A variety of textures are preserved in the ilmenite macrocrysts, including polycrystalline grains, grains with Mg-rich reaction rims, pitted grains and single homogeneous grains. The association of perovskite grains along with Fe-rich ilmenite around the rims of ilmenite macrocrysts (Plate 25.1b) is common in the Monastery kimberlite. The groundmass is comprised of subhedral olivine grains that are typically <0.5 mm in diameter, rounded phlogopite grains (<1 mm), up to 0.2 mm perovskite crystals with a cubic habit (Plate 25.1g) and anhedral ilmenite fragments (0.1 to 0.3 mm). Perovskite occurs as tan and typically zoned crystals. A nice example of intergrowth between a smaller ilmenite crystal with a larger zoned perovskite grain is shown in Plate 25.1h. The alternating light and dark colour bands in the perovskite crystal correspond to differing chemical and isotopic compositions, which will be discussed later.

Mineral Chemistry

All mineral chemical analyses were obtained using the JEOL 8900 electron microprobe at the University of Alberta. Operating conditions were 15.0 kV accelerating voltage, 1–2 μm beam diameter, 15 nA probe current, 20 s count time on peaks and 10 s count times on background positions, per element. Raw data were corrected using the ZAF correction scheme.

Ilmenite – Discrete Nodules and Megacrysts

This study includes ilmenite megacrysts (>1 cm), macrocrysts (<1 cm but >2 mm) and groundmass ilmenite (<0.1 mm) from the Monastery kimberlite. All ilmenite textures described below are representative of ilmenite typically found within the

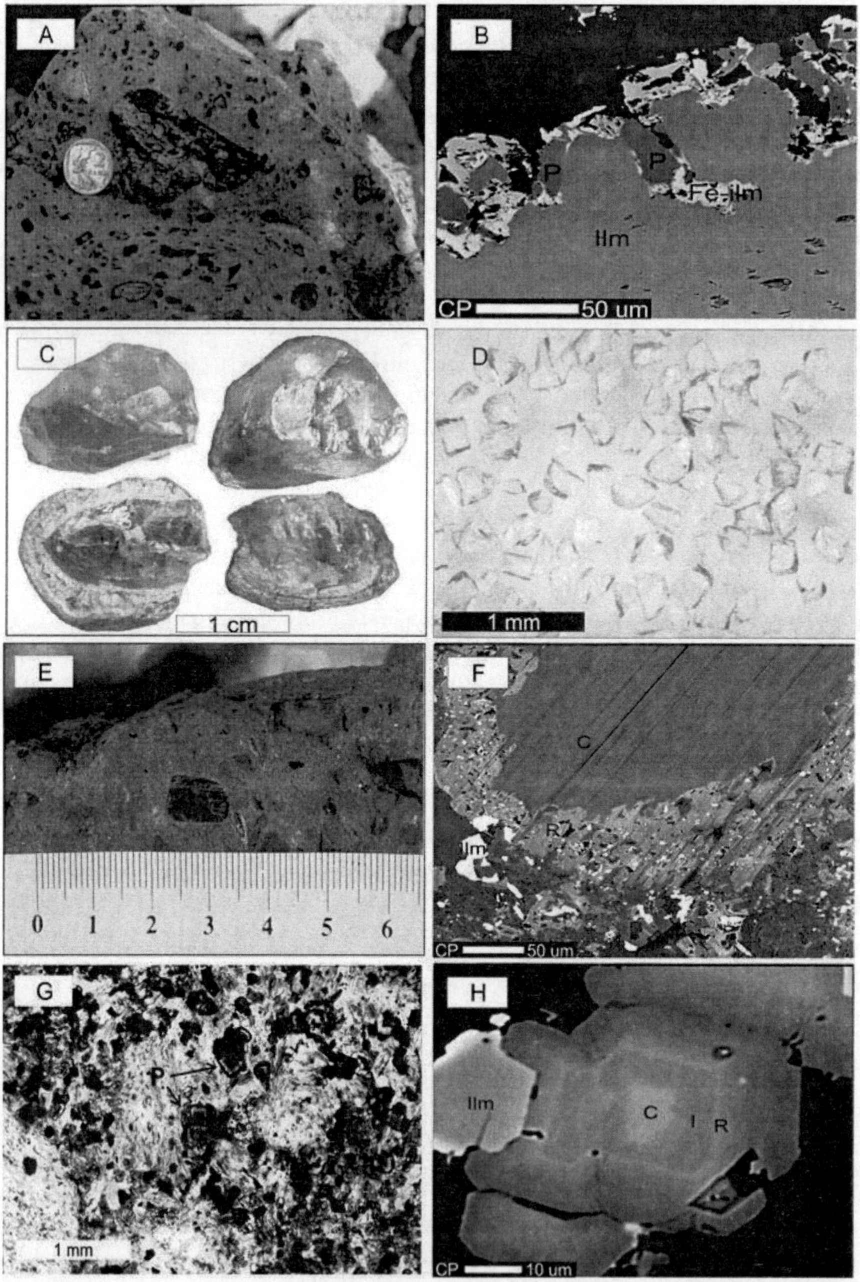

Plate 25.1 (**a**) Ilmenite megacryst within East End Type kimberlite matrix. (**b**) Back-scatter electron image of the rim of an ilmenite macrocryst from the Breccia Type kimberlite. Fe-ilmenite and perovskite (P) association commonly found along rims of ilmenite macrocrysts. (**c**) Photograph of a variety of mantle zircon megacrysts from the Monastery kimberlite. The bottom two crystals contain baddeleyite rims. (**d**) Photograph of a selected fraction of mantle zircon fragments from a

three Monastery kimberlite samples selected for this study and are also representative of the ilmenite selected for U-Pb isotopic analyses below.

Representative chemical analyses of Monastery ilmenite studied here including megacrysts and macrocrysts are presented in Table 25.1. Ilmenite megacryst and macrocryst core analyses (including both discrete nodules and polygranular megacrysts) exhibit a limited range in MgO (7.1–12.5 wt%), TiO_2 (45.7–53.2 wt%), Mn (0.2–0.3 wt%), and Al_2O_3 (0.1–0.9 wt%). The majority of the ilmenite cores have low Cr_2O_3 contents (<0.9 wt%) and plot within the "Kimberlite Reaction Trend" of Haggerty et al. (1979) with some slight overlap with the "Magmatic Trend" (Fig. 25.3). The ilmenite compositions determined in this study agree with results obtained for Monastery ilmenite previously reported by Mitchell (1977). Surface pitting is a distinctive and quite common feature of ilmenite megacrysts in kimberlite and may be the result of alteration (Pasteris et al., 1979). Ilmenite megacrysts from the East End Type kimberlite sample M2 are typically pitted with compositions similar to the majority of ilmenite core analyses in this study. Some Monastery ilmenite megacryst rims are enriched in Mn (10–13 wt%).

Groundmass Ilmenite

Ilmenite grains <100 μm in diameter, categorized as groundmass ilmenite, were analysed from the Quarry Type kimberlite (M1) and the Breccia Type kimberlite (M3). Groundmass ilmenite is generally homogeneous and rarely displays core/rim chemical zoning (Mitchell, 1986). In previous studies, groundmass ilmenite from the Monastery kimberlite has similar compositions to the ilmenite macrocryst rims (Mitchell, 1977; Pasteris, 1980; Boctor and Boyd, 1980; and Agee et al., 1982). A few zoned groundmass ilmenite crystals were identified in this study and they are depleted in MgO and correspondingly enriched in FeO along the rims. Groundmass ilmenite from sample M3 (Table 25.1) has compositions very similar to the megacryst suite.

Phlogopite

Many phlogopite megacrysts consist of homogeneous cores with alteration rims typically containing ilmenite inclusions (Plate 25.1f). The homogeneous phlogopite cores have 0.1–4.0 wt% TiO_2, 10.8–14.4 wt% Al_2O_3, 2.4–7.2 wt% FeO, 0.1–0.6 wt% BaO, and <0.8 wt% Cr_2O_3 (Table 25.1). Examples of zoned phlogopite megacrysts are reported in Table 25.1 (columns 12 and 13). The phlogopite rims

Plate 25.1 (continued) similar crystal to that shown in Plate *(c)*. (**e**) Photograph of a phlogopite megacryst in Breccia Type kimberlite matrix similar to the phlogopite selected for Rb-Sr analyses. (**f**) Back-scatter electron image of a phlogopite from the Breccia Type kimberlite with ilmenite (Ilm) association on rim (R). (**g**) Photomicrograph of zoned perovskite (P) in the groundmass of the East End Type kimberlite. (**h**) Back-scatter electron image of a zoned perovskite crystal from the Quarry Type kimberlite with ilmenite (Ilm) association (C: core; I: intermediate; R: rim)

Table 25.1 Representative ilmenite, phlogopite and perovskite analyses from the Monastery kimberlite, South Africa

	1 M-1-1 (Core)	2 M-1-2 (Core)	3 M-1-3 (Core)	4 M-2-1 (Core)	5 M-2-2 (Core)	6 M-3a-1 (Core)	7 M-3a-2 (Core)	8 M-3b-1 (Core)	9 M-3b-2 (Core)	10 M-3a-2 (GM)	11 M-3a-3 (GM)	12 M-3c-2 (Core)	13 M-3c-2 (Rim)	14 M-2-2
F	–	–	–	–	–	–	–	–	–	–	–	0.49	0.34	–
Na_2O	–	–	–	–	–	–	–	–	–	–	–	0.18	0.28	–
Cl	–	–	–	–	–	–	–	–	–	–	–	0.05	0.03	–
BaO	–	–	–	–	–	–	–	–	–	–	–	0.04	0.12	–
K_2O	–	–	–	–	–	–	–	–	–	–	–	10.23	10.01	–
MgO	12.47	12.39	7.10	8.52	10.20	8.83	10.81	7.63	8.42	9.94	12.81	24.50	21.52	0.05
V_2O_3	0.13	0.19	0.11	0.12	0.18	0.03	0.12	0.13	0.14	0.20	0.09	–	–	0.00
CaO	0.06	0.00	0.02	0.00	0.02	0.01	0.02	0.00	0.00	0.05	0.10	0.22	0.01	39.77
MnO	0.30	0.29	0.21	0.20	0.30	0.19	0.22	0.29	0.32	0.26	0.49	0.01	0.01	0.01
Al_2O_3	0.36	0.29	0.29	0.48	0.15	0.63	0.86	0.11	0.14	0.71	0.61	11.46	14.35	0.18
ZnO	0.00	0.00	0.06	0.16	0.00	0.05	0.00	0.16	0.04	0.00	0.00	–	–	0.18
TiO_2	52.27	53.24	45.73	48.23	47.74	48.55	49.59	47.73	48.84	48.24	52.50	0.97	3.99	56.33
FeO_T	31.07	31.93	44.01	40.56	39.06	4.050	35.94	41.27	39.31	38.45	31.81	6.50	6.30	1.05
SiO_2	0.00	0.01	0.02	0.03	0.02	0.02	0.01	0.00	0.01	0.02	0.00	42.34	40.04	0.00
Cr_2O_3	0.89	0.93	0.72	0.00	0.63	0.00	0.19	0.66	0.84	0.00	0.04	0.10	0.27	0.04
Total	97.53	99.28	98.27	98.30	98.29	98.82	97.76	97.97	98.06	97.87	98.44	96.86	97.14	97.61

1–3 Core analyses of a macrocryst from Quarry Type kimberlite
4–5 Core analyses of a macrocryst from East End Type kimberlite
6–9 Core analyses of a macrocryst from Breccia Type kimberlite
10–11 Groundmass ilmenite from Breccia Type kimberlite
12–13 Core/rim analyses of phlogopite from Breccia Type kimberlite
14 Perovskite analysis from East End Type kimberlite

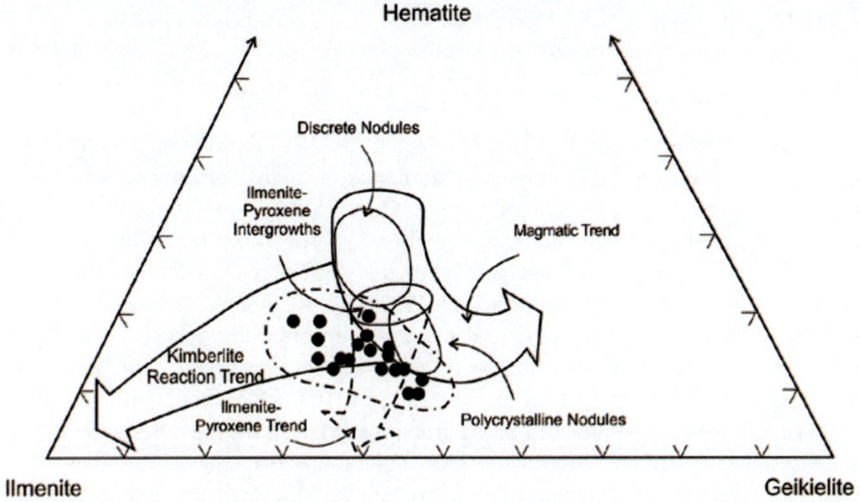

Fig. 25.3 Ternary plot displaying the various ilmenite compositional trends outlined by Haggerty et al. (1979). Envelop of data points represent ilmenite megacryst and macrocryst core compositions from this study

show an increase in BaO, TiO_2, Al_2O_3 and Cr_2O_3 with a very slight decrease in MgO and K_2O. These phlogopite compositions are very similar to those reported from the Fayette County kimberlite, Pennsylvania (Hunter et al., 1984), Dutoitspan kimberlite (Jones and Smith, 1984) and the DeBeers kimberlite (Farmer and Boettcher, 1981). The phlogopite compositions reported here fall into the Cr-poor field outlined by Carswell (1975) for phlogopite megacrysts and are comparable with other phlogopite analyses from South African kimberlites reported by Dawson and Smith (1975).

Perovskite

A representative analysis of Monastery perovskite is presented in Table 25.1. Based on eight analyses, groundmass perovskite from Monastery kimberlite have relatively homogeneous compositions: 37.2–39.8 wt% CaO, 0.05–0.08 wt% MgO, 55.2–56.3 wt% TiO_2, and 0.9–1.4 wt% FeO. Mitchell (1972) noted that there is little chemical variation in perovskite within a kimberlite pipe or between perovskite populations from a variety of kimberlites. This conclusion is valid for the Monastery groundmass perovskite despite the fact that some crystals are spectacularly zoned (Plate 25.1h). In addition to the isolated groundmass crystals, Monastery perovskite commonly occurs rimming ilmenite macrocrysts and these also do not show any significant major or minor element chemical differences from groundmass perovskite (Mitchell, 1972; Agee et al., 1982). Perovskite was not identified in backscatter imaging as inclusions in any ilmenite megacrysts or macrocrysts.

Rb-Sr and U-Pb Analytical Techniques

Rb-Sr Phlogopite

The Breccia Type kimberlite sample M3 was selected for a Rb-Sr phlogopite study because it contains large fresh phlogopite books resulting in ease of mineral separation and minimizing the potential of accidental incorporation of xenocrystic material. Phlogopite megacrysts chosen for analysis in this study range in size from 1.5 to 2.5 cm in length and are similar to the one shown in Plate 25.1e. The margins of this megacryst are deformed and a thin light coloured reaction rim is present. Phlogopite flakes were separated from two different megacrysts (AN1 and AN3) directly from the kimberlite matrix using a pair of tweezers. The first phlogopite cleavage layer from each 'book' was cleaved off and discarded to reduce the amount of surface-altered material in the fractions. Megacryst was chosen for non-leached/leached paired analyses and the phlogopite flakes from megacryst AN1 were leached. A total of seven fractions of phlogopite flakes were analysed and some (e.g. AN3L#2 and AN3L#3) represent duplicate analyses of the same megacryst (see Table 25.2). Although most of the phlogopite analysed here represent pure phlogopite aliquots, chlorite alteration was noted at the margins of some flakes and a small amount of host kimberlite was present in the fraction isolated from megacryst AN1.

Table 25.2 Rb-Sr results for phlogopite fractions from the Monastery Breccia Type kimberlite, South Africa

Sample	Sample weight (mg)	Spike weight (mg)	Rb (ppm)	Sr (ppm)	$^{87}Rb/^{86}Sr$	$^{87}Sr/^{86}Sr$ ± 2 sigma	Age (Ma)[a]
AN1L#3	9.95	5.00	630.6	3.04	647.5	1.52161 ± 0.00006	89.9
AN1L#4	23.48	12.00	616.1	3.27	584.9	1.43924 ± 0.00010	89.3
AN3L#2	26.24	25.00	740.2	21.92	98.88	0.83028 ± 0.00002	91.0
AN3NL#2	25.03	35.00	746.9	37.22	58.48	0.77921 ± 0.00002	92.3
AN3L#3	39.51	35.00	740.8	14.10	154.9	0.90047 ± 0.00002	90.0
AN3NL#3	23.33	20.00	759.1	50.99	43.29	0.75845 ± 0.00002	91.0
AN3L#4(Ac)	14.99	15.00	694.2	26.41	76.77	0.80257 ± 0.00002	91.8

Notes: Concentration data for both non-leached (NL) and leached (L) samples are based on sample weights prior to leaching.

[a]Model ages calculated assuming an initial $^{87}Sr/^{86}Sr$ ratio of 0.7025.

AC acetic acid leach.

Uncertainty in the $^{87}Rb/^{86}Sr$ ratio is 1%.

The seven phlogopite fractions were weighed into acid-washed 15 ml PFA Teflon vessels. The phlogopite leaching method of Brown et al. (1989) was adopted to dissolve carbonate material along fractures and cleavage planes of the phlogopite flakes for the five fractions in Table 25.2. The phlogopite fractions were leached with 3 ml of 0.75 N HCl and left in an ultrasonic bath for 30 min. One fraction (AN3L#4) was given and acetic acid leach. The leachate was discarded and the phlogopite flakes were rinsed twice with Millipore water to ensure complete removal of leachate. All phlogopite fractions were then completely spiked with a measured amount of mixed ^{84}Sr-^{87}Rb tracer solution. The spiked leached and non-leached fractions were dissolved in 3 ml of 24 N HF and 1 ml of 16 N HNO_3 on a hot plate at 150°C for ~12 h. The fractions were then evaporated to dryness. The fluoride precipitate was then converted to a chloride form using 3 ml of 6 N HCl at 80°C for 12 h. In preparation for column chemistry, the samples were evaporated to dryness then re-dissolved in 3 ml of mixed oxalic and hydrochloric acid. Rb and Sr were separated using conventional cation exchange chromatography (Creaser et al., 2004).

Purified Rb and Sr aliquots were re-dissolved in Millipore water and loaded onto single rhenium filaments and coated with a mixture of H_3PO_4 and tantalum gel. The Rb aliquots were analysed on a VG Micromass® 30 thermal ionization mass spectrometer in single collector mode and the Sr isotopic compositions were determined on the VG 354 thermal ionization mass spectrometer in multi-collector, dynamic mode. ^{86}Sr/^{88}Sr ratios were normalized to 0.1194 and replicate analyses of the NBS 987 standard during this study yielded an average ^{87}Sr/^{86}Sr ratio of 0.710271±15 ($n = 20$). Rubidium and strontium blanks are typically low (<100 pg Rb; <400 pg Sr), therefore, only a slight blank correction was required for extremely low-Sr phlogopite analyses. The Rb-Sr age calculation is based on a two-error linear regression treatment (York, 1969) using the Isoplot software of Ludwig (1998). The ^{87}Rb decay constant used was 1.42×10^{-11} yr^{-1} (Davis et al., 1977), and age uncertainties are reported at 2 sigma. A blanket error of 1%, derived by replicate analyses of standards, was applied as the uncertainty for ^{87}Rb/^{86}Sr ratios.

U-Pb Perovskite, Zircon and Ilmenite

Perovskite was isolated from the East End Type kimberlite (M-2-2) by standard crushing and mineral separation techniques (Heaman, 1989). The crushed kimberlite powder was separated into 'heavy' and 'light' mineral fractions using a Wilfley Table. The heavy concentrate was passed through a 70 mesh nylon sieve and the < 70 mesh fraction was passed through the Frantz Isodynamic magnetic separator for an initial separation up to a current of 0.6 amperes. The 0.6 A non-magnetic material from the initial Frantz stage was further separated using the heavy liquid methylene iodide and the "heavy" (>3.2 g/cc^3) fraction was further separated using the Frantz magnetic separator at higher currents. Perovskite is usually magnetic between 0.6

and 1.4 A. Each perovskite grain was hand-picked under a binocular microscope to ensure 100% purity of fractions. There is some variation in crystal colour, degree of rounding, and proportion of alteration component in this perovskite population although there is very little range in chemical composition. To test for possible age variations, four different perovskite fractions were selected that encompassed the range in colour and included one fraction with partial leucoxene coatings (fraction D in Table 25.3).

Mantle zircon grains or fragments were not identified in mineral separates from the East End Type kimberlite, therefore, a mantle zircon xenocryst (~1 cm in diameter; similar to mantle zircons shown in Plate 25.1c) collected from the final concentrates at the Monastery mine, was selected for analysis. The selected crystal did not contain a baddeleyite-rich reaction rim often present on mantle zircons (Heaman and LeCheminant, 1992). This mantle zircon crystal was crushed with an agate mortar and pestle and sieved through a 70 mesh nylon sieve. The <70 mesh fragments were then hand-picked under a binocular microscope selecting only clear, inclusion-free pieces (Plate 25.1d). A total of eight, multi-fragment, mantle zircon fractions were analysed in this study.

Ilmenite selected for this U-Pb study included macrocrysts from crushed East End Type kimberlite (Plate 25.1a) and megacrysts that were collected at the Monastery mine dumps. For identification, the ilmenite fractions from the East End Type kimberlite are labeled as M-2-2 (#2,4,5) and the three megacrysts are labeled as Mon-1, 2, and 3. The ilmenite macrocrysts were hand-picked under a binocular microscope excluding fragments with leucoxene coatings or kimberlite matrix rinds. The ilmenite megacrysts were crushed with an agate mortar and pestle and were sieved with a 70 mesh nylon sieve. These <70 mesh fragments were also hand-picked under a binocular microscope for the same reasons given above for zircon. A total of nine ilmenite fractions were selected for this feasibility study; three ilmenite fractions selected from crushed East End Type kimberlite macrocrysts and six fractions selected from three different ilmenite megacrysts.

Prior to weighing, the perovskite and ilmenite fractions were washed in a solution of 2N HNO_3 and the zircon grains were washed in a solution of 4N HNO_3. All samples were left on a hot plate for 1.5 h at 80°C to dissolve any impurities or difficult to detect adhering material, such as apatite. The grains were then thoroughly washed several times with Millipore water and distilled acetone. The grains were weighed using a UTM2 microbalance located in an HEPA-filtered cleanhood. The perovskite and ilmenite fractions were then transferred to acid-washed 15 ml Savillex digestion vessels and mantle zircon fractions were transferred to TFE Teflon dissolution capsules. The mineral fractions were subsequently spiked with a mixed ^{205}Pb-^{235}U tracer solution and dissolved in 20 drops (one drop corresponds to approximately 0.15 ml) of 48% HF and two drops of 7 N HNO_3 on a hot plate at 80°C (perovskite dissolution was achieved in 50:50 mixture of 48% HF and 7 N HNO_3). Perovskite required 5 days and ilmenite required 10 days at this temperature to dissolve completely. Mantle zircon fractions were sealed in Teflon sleeves, encased in Monel jackets and assembled in a metal carousel that houses 8 samples. They were left in

Table 25.3 U-Pb results for perovskite fractions from the Monastery kimberlite, South Africa

Sample	Description	Sample[a] weight	U (ppm)	Th (ppm)	Pb (ppm)	Th/U	Common Pb (pg)	$^{206}Pb/^{204}Pb$ ± abs	$^{238}U/^{204}Pb$ ± abs	$^{206}Pb/^{238}U$ ± abs	$^{207}Pb/^{204}Pb$ ± abs	$^{206}Pb/^{238}U$ Age (Ma)
M-2-2-A	270 mixed grains	87	119.8	321.3	12.3	2.681	634	34.13 ± 0.223	1040.3 ± 14.5	0.01438 ± 0.00038	16.383 ± 0.058	92.1 ± 2.4
M-2-2-B	140 light grains	43	76.9	94.12	9.5	1.224	308	29.10 ± 0.30	696.0 ± 19.1	0.01424 ± 0.00056	16.146 ± 0.058	91.2 ± 3.6
M-2-2-C	170 dark grains	50	114.7	348.9	10.9	3.042	299	36.59 ± 0.53	1223.4 ± 34.8	0.01422 ± 0.00032	16.501 ± 0.078	91.0 ± 2.0
M-2-2-D	205 leucoxene coating	89	112.6	471.3	10.6	4.187	496	37.40 ± 0.34	1273.5 ± 21.8	0.01430 ± 0.00032	16.521 ± 0.062	91.5 ± 2.0

Notes: Atomic ratios corrected for fractionation, blank (8 pg Pb; 2 pg U), spike and common Pb (Stacey and Kramers, 1975). Uncertainties in atomic ratios are quoted at 2 sigma.
Th concentration determined from amount of ^{208}Pb and $^{206}Pb/^{238}U$ age.
[a]Sample Weight in micrograms; uncertainty is ±2 micrograms.

an oven at 215°C for 4–5 days. After this dissolution period, the samples were evaporated to dryness, and converted to chloride form by adding 15 drops (five drops for the zircon samples) of 3.1 N HCl and returning the digestion vessels to the hot plate (or in the oven in the case of mantle zircon) for 24 h. During this study it was noticed that large ilmenite fractions are difficult to completely dissolve on the hotplate so fractions larger than 1 mg are now dissolved in TFE Teflon digestion vessels in an oven at 215°C for 5 days.

Uranium and lead were separated from the perovskite and ilmenite fractions using a larger column HBr anion exchange chromatography procedure similar that that outlined for titanite and apatite by Heaman and Machado (1992). Minor changes to the procedure were necessary to enhance the removal of Fe^{2+}, Mg, and Ti^{4+} from the ilmenite samples. An extra column volume (~5 ml) of 7 N HNO_3 was added to ensure complete extraction of Fe^{2+} and Ti^{4+} and during the final purification of U^{4+}, an extra column volume of H_2O was added to ensure maximum recovery of uranium. For ilmenite fractions weighing > 5 mg, the entire chromatography procedure was performed twice as the fraction weights are larger than normal and one pass through chromatography was insufficient for purifying the uranium in particular. Uranium and lead from the zircon samples were separated using the HCl procedure outlined by Krogh (1973).

The purified uranium and lead were loaded onto outgassed Re filaments with a mixture of silica gel and phosphoric acid (Cameron et al., 1969) and were analysed on a VG 354 Mass Spectrometer in single Daly photomultiplier detector mode. All isotopic data were corrected for Daly bias by a factor of 0.13%/amu for lead and 0.15%/amu for uranium and were also corrected for mass discrimination (0.09%/amu for lead and 0.16%/amu for uranium). For both perovskite and ilmenite analyses, total procedural blanks for lead and uranium are 8 pg (±50%) and 2 pg (±50%), respectively (blank corrections for ilmenite samples represented by "a", are 3 pg (± 50%) and 1 pg (± 50%), respectively). The total Pb and U procedural blanks for the zircon chromatography were generally in the 2–6 pg (±50%) and 1–5 pg (±20%), respectively. The ilmenite corrected atomic ratios in Table 25.5 have larger than normal analytical uncertainties, which can essentially be attributed to the magnitude of the blank corrections for these analyses. Therefore, it was necessary to reduce the overall procedure blank during this study and continuously monitor the blank concentrations during these experiments as Monastery ilmenite and mantle zircon contain ultra-low concentrations of radiogenic lead (typically 10–200 ppb).

The ^{238}U and ^{235}U decay constants used for age calculations are $1.55125 \times 10^{-10} y^{-1}$ and $9.8485 \times 10^{-10} y^{-1}$, respectively (Jaffey et al., 1971; Steiger and Jäger, 1977). Initial lead isotopic compositions were estimated using the Stacey and Kramers (1975) two-stage growth model. Corrections to the raw isotopic data and numerical propagation of all known sources of uncertainty were calculated using in-house software. All errors are reported at the 95% confidence level. Weighted mean and isochron age calculations were determined using the program Isoplot (Ludwig, 1998).

Geochronology Results

Rb-Sr Phlogopite

The Rb-Sr results for two phlogopite megacrysts, AN1 (solid squares) and AN3 (solid circles) from the Monastery kimberlite are summarized in Table 25.2 and an isochron plot displaying the data is shown in Fig. 25.4a. The Rb and Sr concentrations for phlogopite range from 616 to 759 ppm and 3.0 to 51.0 ppm, respectively. The Monastery phlogopite Sr concentrations obtained in this study are generally lower than those reported by Allsopp and Barrett (1975; 36–403 ppm), but close to those (12.2–29.9 ppm) reported by Smith and Barton (1996). This likely reflects the removal of high Sr minerals, such as carbonate and/or apatite, during the HCl leaching step (Brown et al., 1989). The calculated isochron age (Fig. 25.4a) using all seven phlogopite analyses, two from AN1 and five from AN3, is 89.4±0.8 Ma with an initial $^{87}Sr/^{86}Sr$ value of 0.704±0.001 (MSWD=3.0), consistent with a Group I kimberlite designation.

U-Pb Perovskite

The U-Pb perovskite data from the East End Type Monastery kimberlite (M-2-2) are summarized in Table 25.3 and a compilation of weighted mean $^{206}Pb/^{238}U$ dates with corresponding 2σ uncertainties is shown in Fig. 25.4b. Four different multi-grain (between 140 and 270 grains) fractions were analysed: a mixed fraction comprising both light and dark perovskite grains, a light coloured fraction, a dark coloured fraction and a fraction of perovskite grains possessing leucoxene ($CaFeTiO_5$) coatings. The ranges for $^{206}Pb/^{204}Pb$ and $^{238}U/^{204}Pb$ ratios (29.102–37.400 and 696.0 –1,273.5, respectively) are all very high and typical for groundmass perovskite in kimberlite (Heaman, 1989; Heaman et al., 2003).

The Monastery perovskite fractions analysed in this study have variable U and Th concentrations (77–120 and 94–471 ppm, respectively) and a smaller range in Pb concentrations (9.5–12.3 ppm). Common Pb contents are typically high, ranging from 299 to 634 pg which represents 35–43% of the total Pb present within the perovskite grains. These high common Pb contents require a significant correction in the age calculations (largest impact on the $^{207}Pb/^{235}U$ date) and the uncertainty in this correction is propagated in the error calculation. The fraction of light coloured perovskite grains (M-2-2-B) contains the lowest U, Th and Pb contents and has a distinctly lower $^{238}U/^{204}Pb$ ratio (696) compared to the other fractions suggesting that there is compositional variation at the trace element level, however, this seems to have no affect on the $^{206}Pb/^{238}U$ apparent dates as they are identical (91.0–91.5 Ma) within error (Table 25.3). One perovskite fraction was selected specifically to contain some leucoxene coating and although it has a slightly higher Th concentration (471 ppm), $^{206}Pb/^{204}Pb$, and $^{238}U/^{204}Pb$ isotopic ratios, the $^{206}Pb/^{238}U$ apparent age is identical to the other fractions (Table 25.3). The leucoxene coating can contribute to the observed compositional variation however this does not appear to affect the

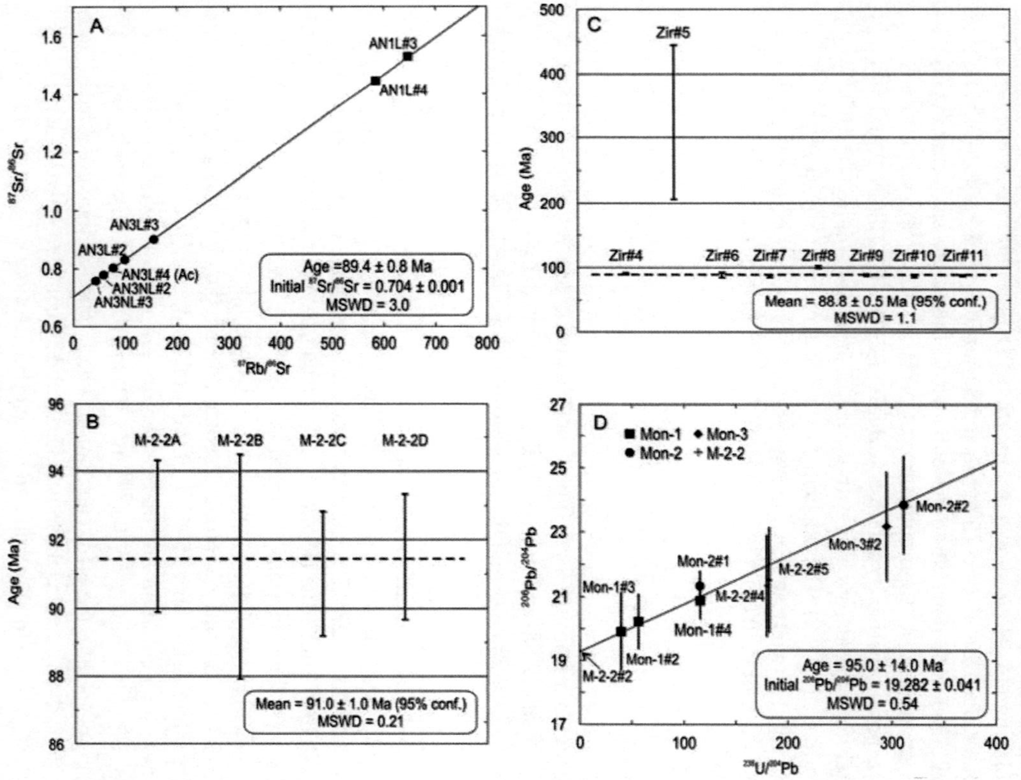

Fig. 25.4 (**a**) Isochron plot showing the distribution of phlogopite megacryst samples analysed from the Monastery kimberlite. (**b**) Weighted mean $^{206}Pb/^{238}U$ age plot of perovskite fractions using the initial $^{206}Pb/^{204}Pb$ ratio of 19.19 for the common Pb correction. (**c**) Weighted mean $^{206}Pb/^{238}U$ diagram showing the distributions of zircon ages about the calculated mean $^{206}Pb/^{238}U$ age line (using Stacey and Kramers (1975) common Pb correction). (**d**) Isochron plot of ilmenite populations from the Monastery kimberlite

apparent age calculation. This could be explained if alteration to form leucoxene occurred slightly later (<1 m.y.) than but close in time to perovskite crystallization.

These four groundmass perovskite fractions yielded an isochron date of 91.9±6.2 Ma and an initial $^{206}Pb/^{204}Pb$ ratio of 19.19 ± 0.97 (MSWD = 0.19). This initial $^{206}Pb/^{204}Pb$ ratio of 19.19 overlaps within analytical uncertainty Stacey and Kramers (1975) average crustal model Pb at 90 Ma ($^{206}Pb/^{204}Pb = 18.57$) but is distinctly more radiogenic than any estimate for depleted mantle Pb at 90 Ma ($^{206}Pb/^{204}Pb = 17.62$). It is also similar to the average $^{206}Pb/^{204}Pb$ of ~19 obtained for Group 1 kimberlites (Smith, 1983). Therefore, the initial $^{206}Pb/^{204}Pb$ ratio of 19.19 from the regression calculation was considered a more accurate estimate and used to recalculate the perovskite $^{206}Pb/^{238}U$ apparent ages (Table 25.3) yielding a weighted average $^{206}Pb/^{238}U$ age of 91.0±1.0 Ma with an MSWD of 0.21 (Fig. 25.4b). This date is more precise and is interpreted to be the best estimate of the crystallization age for the Monastery groundmass perovskite in this sample.

U-Pb Mantle Zircon

The U-Pb results for eight fractions of zircon fragments selected from the same mantle zircon crystal are presented in Table 25.4 and a weighted average $^{206}Pb/^{238}U$ age calculation for the fractions is shown in Fig. 25.4c. Selected grains were devoid of visible inclusions and coatings (Plate 25.1d). The total common Pb contents are typically quite low (2–20 pg) with one exception; Zir#6 contains 141 pg. The common Pb content represents 10–48% of the total lead present in these analyses.

The U (0.93–4.05 ppm), Pb (0.03–0.47 ppm) and Th (0.50–1.12 ppm) concentrations obtained in this study are some of the lowest reported for mantle zircons (e.g., Davis et al., 1976; Zartman et al., 1998). The Th/U ratios obtained for this mantle zircon are < 1 and five analyses indicate a narrow range (0.195–0.209) which is typical for kimberlitic zircons (Ahrens et al., 1967). Such ratios are at the low Th/U end member for primary zircon that crystallized from magmas (Heaman et al., 1990).

The $^{206}Pb/^{238}U$ dates obtained for several fractions of fragments from this mantle zircon megacryst vary between 86.5 and 325.4 Ma, a range that is quite large compared to other mantle zircon studies (e.g., Zartman et al., 1998). Excluding the two extreme outliers (#5 and #8) the weighted mean $^{206}Pb/^{238}U$ date is 89.0 ± 1.7 Ma (MSWD = 10.5). Four fractions (#6,7,9,11) yield identical results with a weighted mean $^{206}Pb/^{238}U$ date is 88.8 ± 0.5 Ma (MSWD = 1.1), which we interpret as the best estimate for the age of closure to Pb diffusion in this crystal and approximate age for kimberlite emplacement. This date is slightly younger than previous U-Pb studies on Monastery zircon megacrysts of 90.4 and 90.1±0.5 Ma reported by Davis et al. (1976) and Zartman et al. (1998), respectively.

U-Pb Ilmenite

This is the first comprehensive U-Pb study on kimberlitic ilmenite extending the early geochemical study of Kresten (1974) who analysed uranium concentrations in

Table 25.4 U-Pb results for mantle zircon fractions from the Monastery kimberlite, South Africa

		Concentrations					Atomic Ratios				Apparent Age	
Sample	Sample weight (μg)	U (ppm)	Th (ppm)	Pb (ppm)	Th/U	Common Pb (pg)	$^{206}Pb/^{204}Pb$	$^{206}Pb/^{238}U$ ± abs	$^{207}Pb/^{235}U$ ± abs	$^{207}Pb/^{206}Pb$ ± abs	$^{206}Pb/^{238}U$ (Ma)	2σ
Zir#4	383	3.77	0.79	0.05	0.209	2	548	0.01420 ± 0.00006	0.0951 ± 0.0034	0.0486 ± 0.0016	90.9	± 0.4
Zir#5	692	0.93	0.70	0.05	0.732	7	1531	0.05178 ± 0.01950	0.3426 ± 0.1304	0.0480 ± 0.0026	325.4 ± 118.4	
Zir#6	327	4.05	< 0.10	0.47	0.008	141	27	0.01378 ± 0.00066	0.0693 ± 0.0462	0.0364 ± 0.0239	88.2 ± 4.2	
Zir#7	357	3.39	0.67	0.04	0.198	3	329	0.01367 ± 0.00022	0.0852 ± 0.0058	0.0452 ± 0.0030	87.5 ± 1.4	
Zir#8	597	3.87	1.12	0.06	0.289	8	1159	0.01574 ± 0.00010	0.0967 ± 0.0052	0.0445 ± 0.0023	100.7 ± 0.6	
Zir#9	772	2.53	0.50	0.03	0.199	6	22832	0.01394 ± 0.00006	0.0852 ± 0.0061	0.0443 ± 0.0031	89.2 ± 0.4	
Zir#10	753	3.50	0.68	0.06	0.195	20	181	0.01351 ± 0.00008	0.0844 ± 0.0050	0.0453 ± 0.0026	86.5 ± 0.5	
Zir#11	719	3.85	0.75	0.06	0.196	10	595	0.03181 ± 0.00006	0.0864 ± 0.0043	0.0454 ± 0.0022	88.4 ± 0.4	

Notes: Atomic ratios corrected for fractionation, blank (2–6 pg Pb; 1–3 pg U), spike and initial common Pb (Stacey and Kramers, 1975).
Uncertainties in atomic ratios are quoted at 2 sigma.
Th concentration determined from amount of ^{208}Pb and $^{206}Pb/^{238}U$ age.

Table 25.5 U-Pb results for ilmenite fractions from the Monastery kimberlite, South Africa

Sample	Description	Sample weight (μg)	U (ppm)	Th (ppm)	Pb (ppm)	Th/U	Common Pb (pg)	$^{206}Pb/^{204}Pb$ ± abs	$^{238}U/^{204}Pb$ ± abs	$^{206}Pb/^{238}U$ ± abs	$^{207}Pb/^{204}Pb$ ± abs	$^{206}Pb/^{238}U$ Age (Ma)
Mon-1#2	50 grains	1150	0.17	0.03	0.20	0.169	230	20.220 ± 0.120	56.4 ± 2.4	0.029 ± 0.007	15.753 ± 0.080	186.1 ± 42.1
Mon-1#3	50 grains	1072	0.09	0.14	0.15	1.485	165	19.901 ± 0.144	39.8 ± 2.4	0.034 ± 0.010	15.663 ± 0.090	212.8 ± 60.4
*Mon-1#4	302 grains	7128	0.03	0.03	0.02	0.808	137	20.869 ± 0.092	115.5 ± 3.2	0.020 ± 0.003	15.659 ± 0.058	127.3 ± 20.4
*Mon-2#1	297 grains	5959	0.05	0.07	0.03	1.252	172	21.334 ± 0.090	115.2 ± 2.5	0.024 ± 0.003	15.717 ± 0.058	153.0 ± 20.4
*Mon-2#2	284 grains	5582	0.05	0.02	0.01	0.317	61	23.845 ± 0.390	311.4 ± 20.0	0.017 ± 0.001	15.803 ± 0.118	108.4 ± 8.0
*Mon-3#2	299 grains	5662	0.04	0.00	0.01	0.027	53	23.174 ± 0.370	294.9 ± 21.8	0.016 ± 0.001	15.725 ± 0.080	99.9 ± 8.2
M-2-2#2	50 grains	732	0.02	0.07	1.32	3.510	943	19.296 ± 0.044	1.0 ± 0.1	0.701 ± 0.072	15.722 ± 0.048	–
*M-2-2#4	339 grains	7413	0.02	–	0.01	–	53	21.330 ± 0.240	179.9 ± 13.4	0.015 ± 0.002	15.490 ± 0.078	98.3 ± 13.4
*M-2-2#5	383 grains	7010	0.02	–	0.01	–	52	21.517 ± 0.290	182.1 ± 13.9	0.016 ± 0.002	15.500 ± 0.122	103.6 ± 13.8

Notes: Atomic ratios corrected for fractionation, blank (8 pg Pb; 2 pg U) or (3 pg Pb, 1 pg U)[a], spike and initial common Pb (Stacey and Kramers, 1975).
Th concentration determined from amount of ^{208}Pb and $^{206}Pb/^{238}U$ age.
Uncertainties in atomic ratios are quoted at 2 sigma.

ilmenite nodules from a few South African kimberlites. In this study, both ilmenite megacrysts and macrocryst/groundmass (East End Type kimberlite) were analysed to test specifically for trace element and isotopic variations within and between megacrysts. Polygranular and deformed ilmenite megacryst types, such as those shown in Figs. 25.4f and 25.5a, c from Noyes (2000), were not analysed.

The U-Pb results for nine Monastery ilmenite fractions are presented in Table 25.5 and on an isochron diagram in Fig. 25.4d. Eight fractions from a total of three ilmenite megacrysts (similar to the one shown in Plate 25.1a) and five macrocryst/groundmass fractions were isolated from a crushed sample of East End Type kimberlite. The U, Th and Pb concentrations (ppm) in kimberlitic ilmenite are very low and generally less than 0.2 ppm: 0.02–0.17, 0.02–0.14 and 0.01–0.20, respectively. These concentrations overlap but tend to be lower than those reported by Kresten (1974) for kimberlitic ilmenite. Common Pb contents range from 52 to 230 pg for fractions weighing between 0.7 and 7.4 mg; representing 48–53% of the total lead present. This proportion of common Pb overlaps with but is at the high end

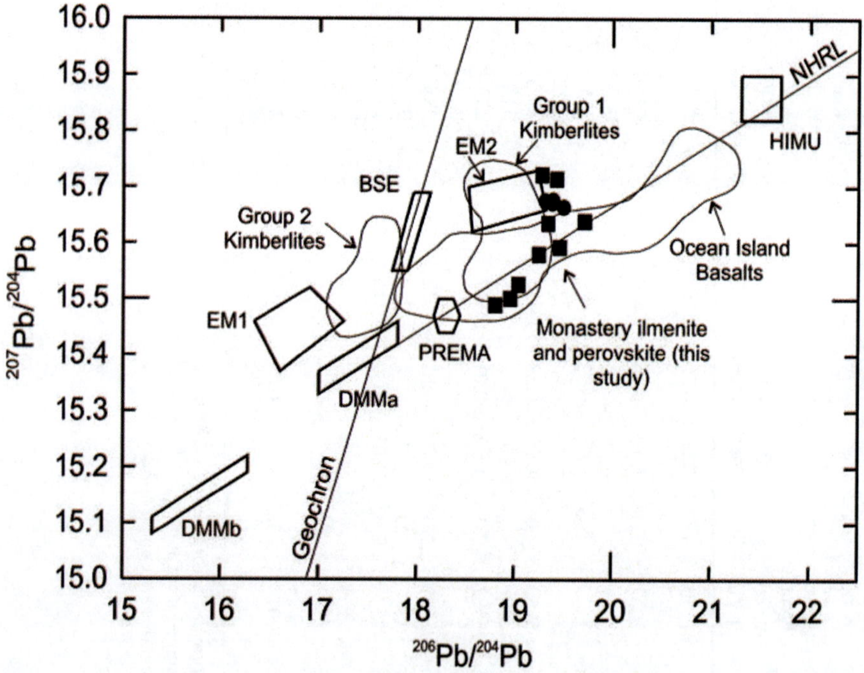

Fig. 25.5 $^{207}Pb/^{204}Pb$ vs $^{206}Pb/^{204}Pb$ diagram displaying various mantle components. The Pb isotopic composition for Monastery ilmenite (*squares*) and perovskite (*circles*) (initial Pb compositions calculated at 90 Ma) are plotted for comparison with Group 1 and Group 2 kimberlites (Smith, 1983), and ocean island basalts (Hart, 1984). *DMM* depleted MORB mantle, *PREMA* prevalent mantle, *EM1* enriched mantle type 1, *EM2* enriched mantle type 2, *BSE* bulk silicate earth, *HIMU* high MU. The Northern Hemisphere Reference Line (NHRL) and the Geochron at 4.55 Ga are shown for reference. After Zindler and Hart (1986)

of what is typically measured for groundmass perovskite (Heaman et al., 2003). The $^{238}U/^{204}Pb$ ratios are variable (40–311) but considerably lower than most perovskite (usually >1,000). The $^{206}Pb/^{204}Pb$ ratios are consistently low but somewhat radiogenic (19.9–23.8). One ilmenite fraction in Table 25.5 (M-2-2#2) was not included in this summary because it contains significantly more common Pb (943 pg), which results in an artificially low $^{238}U/^{204}Pb$ ratio of 1 and a quite high $^{206}Pb/^{238}U$ ratio of 0.7 corresponding to a model age of >3.0 Ga.

Perhaps the most important aspect of the U-Pb data obtained for these ilmenite fractions (except for M-2-2#2) is the range in $^{238}U/^{204}Pb$ ratios. This ratio is sufficiently high enough that it may be feasible to consider U-Pb ilmenite geochronology as and additional tool for determining the emplacement ages of kimberlites. If all nine ilmenite fractions are used in a regression calculation then a date of 95.0±14.0 Ma with an initial $^{206}Pb/^{204}Pb$ initial ratio of 19.282±0.041 (MSWD = 0.51) is obtained. This age is in excellent agreement with the ages determined by several different techniques applied to the Monastery kimberlite (composite age of 89.7±0.3 Ma is considered the current best estimate for the emplacement of the Monastery kimberlite) indicating the geologically meaningful U-Pb ilmenite ages may be achievable. Within the quoted uncertainties, there is no difference between the formation age of ilmenite megacrysts and macrocryst/groundmass ilmenite. It is also significant that the $^{206}Pb/^{204}Pb$ initial ratio of 19.282 ± 0.041 based on the ilmenite regression treatment (Fig. 25.4d) is identical to the $^{206}Pb/^{204}Pb$ initial ratio of 19.19±0.97 obtained for groundmass perovskite, possibly hinting that both minerals formed from a similar Pb reservoir. Monastery ilmenite has significantly lower U and Th contents compared to mantle zircon, suggesting that the partition coefficient (K_D) for U and Th are much lower for mantle ilmenite compared to mantle zircon.

Discussion

Previously, establishing the time of kimberlite emplacement involved the use of one or more of the following chronometers: Rb-Sr phlogopite, $^{40}Ar/^{39}Ar$ phlogopite, U-Pb perovskite, and U-Pb mantle zircon. These minerals are present in a number of kimberlites however many kimberlites worldwide are devoid of these minerals so alternative isotopic dating techniques are required to fully characterize the emplacement patterns of kimberlite magmatism. A recent study by Blackburn et al. (2008) explored the possibility of (U-Th)/He dating of minerals such as apatite, titanite and zircon in kimberlites and showed this technique holds much promise for obtaining accurate age results that are in agreement with more conventional Rb-Sr phlogopite dates. In this study we have continued the search for new chronometers that can constrain the emplacement age of kimberlites by investigated the feasibility of U-Pb ilmenite geochronology because ilmenite is abundant in most kimberlites worldwide and contains trace amounts of uranium, albeit at low concentrations. The discussion initially focuses on the results of more traditional geochronometers

(Rb-Sr phlogopite, U-Pb perovskite, U-Pb mantle zircon) to establish a temporal framework with which to compare the new U-Pb ilmenite results. A total of 10 individual radiometric age dates have been determined for the Monastery kimberlite (Fig. 25.2) and includes 4 techniques: K-Ar phlogopite, Rb-Sr phlogopite, U-Pb mantle zircon and U-Pb perovskite. An obvious outlier is the U-Pb perovskite date of 83.0±3.0 Ma reported by Kramers and Smith (1983). Excluding the U-Pb ilmenite date obtained in this study and the U-Pb perovskite date reported by Kramers and Smith (1983), a weighted mean date for eight independent age determinations is 89.4±0.3 Ma (MSWD = 6.7). Significantly less scatter is achieved if Rb-Sr phlogopite date 88.1±0.6 Ma is also omitted (Smith and Barton, 1996) yielding a weighted mean date of 89.7±0.3 Ma (MSWD = 4.1; $n = 7$); this is our preferred date for the emplacement age of the Monastery kimberlite (corresponds to the shaded band in Fig. 25.2).

Kimberlite Geochronology – Traditional Techniques

Rb-Sr Phlogopite

Rb-Sr phlogopite geochronology has been one of the primary age dating techniques for decades and has been shown to yield accurate and precise emplacement ages for kimberlites (e.g., Allsopp and Barrett, 1975; Smith et al., 1985; Allsopp and Roddick, 1984; Smith et al., 1985; Allsopp et al., 1986; Brown et al., 1989). When present, phlogopite is typically abundant in kimberlites, can occur as megacrysts and these can be easily extracted from the kimberlite matrix. Alteration of phlogopite to chlorite, occurrence of calcite/apatite inclusions or the presence of xenocrystic mica could result in excessive scatter on an isochron diagram potentially producing kimberlite emplacement ages that are too old (Allsopp et al., 1986). Some of these problems can be minimized by selectively hand-picking the phlogopite grains under a binocular microscope and by using an acid leaching technique (Brown et al., 1989) to remove calcite/apatite along cleavage planes, fractures, and grain boundaries.

The trend of higher $^{87}Sr/^{86}Sr$ and $^{87}Rb/^{86}Sr$ ratios in phlogopite for leached fractions compared to non-leached fractions was interpreted by Brown et al. (1989) to be a result of removal of unradiogenic strontium hosted in carbonate. A similar pattern exists for Monastery phlogopite analysed in this study as can be seen by comparing the non-leached/leached pairs for megacryst AN3 seen in Fig. 25.4a. Phlogopite megacryst AN3 contains numerous apatite inclusions that could be the primary source of unradiogenic Sr. The removal of apatite and/or carbonate during leaching would explain the significant reduction of total Sr in the leached fractions (Table 25.2) with Rb concentrations remaining relatively constant.

The Rb-Sr phlogopite date of 89.4 ± 0.8 Ma obtained here is in excellent agreement with the Rb-Sr phlogopite dates of 90.0 ± 4.0 Ma and 88.1 ± 0.6 Ma reported by Allsopp and Barrett (1975) and Smith and Barton (1996), respectively. The latter date is based on the analysis of six leached fractions from a single phlogopite megacryst from the Monastery kimberlite (Smith and Barton, 1996). The Rb-Sr phlogopite age determined in this study is also in good agreement with the ages obtained by other radiometric techniques (Fig. 25.2), such as U-Pb mantle zircon (Davis et al., 1976; Zartman et al., 1998) and U-Pb perovskite (Kramers and Smith, 1983).

U-Pb Perovskite

U-Pb perovskite geochronology was first attempted Kramers and Smith (1983) on concentrates and is advantageous because it is a primary groundmass phase crystallizing directly from the kimberlitic magma, contains high concentrations of uranium (11–348 ppm; Heaman, 1989; Heaman et al., 2003), and is abundant within many kimberlites. U-Pb perovskite dating by thermal ionization mass spectrometry has been shown to yield precise (±2–4 Ma) kimberlite emplacement ages (e.g., Heaman, 1989; Heaman and Kjarsgaard, 2000; Heaman et al., 2003). The small grain size of perovskite (<50 μm) can be a disadvantage for conventional isotope dilution thermal ionization dating as isolating minuscule crystals is challenging and selection of individual grains is time consuming. Perovskite also can contain quite high contents of initial common lead (up to 40% of the total lead) so model ages are sometimes quite sensitive to the choice of the initial common lead isotopic composition.

The similar $^{206}Pb/^{238}U$ ages for the different perovskite fractions determined here (Table 25.3) indicates that there is no obvious age difference between the varieties of perovskite fractions selected. The fraction of darker perovskite grains has higher concentrations of U, Th and Pb than the lighter grains, consistent with the results from previous U-Pb studies of zoned perovskite grains (e.g., Heaman, 1989). The colour zoning observed in most Monastery perovskite (Plates 1G and 1H) with dark cores and light rims likely correspond to a decrease in U and Th content of the magma during perovskite crystallization.

The precise weighted average $^{206}Pb/^{238}U$ date of 91.0 ± 1.0 Ma obtained using all four perovskite fractions (Fig. 25.4b) is interpreted as the best estimate for the age of groundmass perovskite crystallization in the East End Type kimberlite. This U-Pb perovskite age is slightly older than the 82.5±2.5 Ma age obtained on a leach-residue study of perovskite-enriched concentrate from a sample of slightly serpentinized, macroporphyritic kimberlite (Kramers and Smith, 1983). It is unclear why there is a discrepancy in these perovskite age determinations. It is possible that the leach component in the Kramers and Smith (1983) study contains a relatively radiogenic Pb isotopic composition from an unknown older mineral component in this concentrate or the dissolved carbonate component may have a more radiogenic Pb isotopic composition than the primary kimberlite magma and this may contribute to the slightly higher initial $^{206}Pb/^{204}Pb$ isotopic ratio (19.27±0.04) obtained from the leach-residue reference line compared to 19.19 obtained from pure perovskite fractions in this study. Another possibility is the presence of multiple kimberlite intrusions at Monastery with a narrow range (~8 m.y.) of emplacement ages. Multiple kimberlite intrusion events over a narrow time interval of <10 m.y. has been detected in a few kimberlite bodies such as the Buffonta kimberlite in the Kirkland Lake kimberlite field in eastern Canada (Heaman and Kjarsgaard, 2000).

U-Pb Mantle Zircon

The U-Pb mantle zircon dating technique has been widely used to determine kimberlite emplacement ages (e.g., Davis et al., 1976; LeCheminant et al., 1998; Davis, 1977, 1978). Mantle zircon is a rare accessory mineral in kimberlites, but is commonly recovered in heavy mineral concentrates along with diamonds. Mantle zircon can occur as isolated megacrysts (Davis et al., 1976), as inclusions in diamond (Kinny and Meyer, 1994) or as an accessory mineral in mantle material entrained in kimberlite, such as eclogite (Chen et al., 1994; Heaman et al., 2002), metasomatized harzburgite (Kinny and Dawson, 1992) or MARID (Konzett et al., 1998; Hamilton et al., 1998) xenoliths. It is more common to find mantle zircon megacrysts when bulk sampling of kimberlite has taken place. These large (up to several cm) mantle zircon crystals are interpreted to coexist with diamond at great mantle depths and are entrained during kimberlite ascent. They commonly contain a diagnostic baddeleyite-dominated reaction rim that forms from desilicification once entrained in kimberlite magma (Heaman and LeCheminant, 1992) and the presence of these baddeleyite rims, large crystal size (mm to cm) and corresponding chemical composition (i.e. very low U and Th concentrations) help to distinguish mantle from crustal zircon xenocrysts in kimberlite.

There are two main limitations of U-Pb mantle zircon geochronology. The first is that the concentrations of trace elements, in particular uranium and radiogenic lead (often <10 ppm: Heaman et al., 1990), are quite low. As a consequence, this relatively pure natural form of zircon can be difficult to completely dissolve in inorganic acids. For many "young" mantle zircons, there is negligible radiogenic ^{207}Pb produced so many researchers report only ^{206}Pb/^{238}U dates (e.g., Davis et al., 1976; Davis, 1977, 1978). The second limitation relates to the fact that these crystals have resided at high temperatures possibly for many millions of years and the U-Pb system may be disturbed by diffusion of daughter elements out of the crystal. In addition, great care must be taken while selecting zircon grains for U-Pb geochronology; for example, selecting crystals devoid of inclusions, fractures and/or turbidity as these features can either contribute unwanted common Pb (i.e. mineral/fluid inclusions) or can be associated with Pb-loss linked to fluid interaction and alteration.

The uranium content of eight fractions of fragments from the Monastery mantle zircon megacryst analysed in this study (0.9–4.1 ppm) represents some of the lowest concentrations ever reported. In two previous studies of mantle zircon chemistry (Ahrens et al., 1967; Kresten and Berggren, 1975), a much larger range in uranium (7–66 ppm) content was reported. The majority of these mantle zircons typically contained < 30 ppm uranium (Kresten and Berggren, 1975) and these low values were considered to represent primary uranium contents within mantle zircon. Davis et al. (1976) also reported a low U concentration of 6.1 ppm for a zircon crystal isolated from a peridotite nodule from the Monastery kimberlite. The previous lowest published uranium concentration in a mantle zircon is 3 ppm (Zartman et al., 1998) for a zircon crystal from the Monastery kimberlite. In summary, the Monastery mantle zircons have some of the lowest uranium contents known for any zircon, combined with its overall low trace element content (Heaman et al., 1990), possibly represents the most chemically pure natural terrestrial zircon known. Konzett et al. (1998) summarize the U and Th concentration in mantle zircon and clearly show that mantle zircon megacrysts have the lowest uranium contents (up to 80 ppm), MARID zircons are much higher (20–400 ppm) and the zircon inclusion in diamond the highest (360–660 ppm; Kinny and Meyer, 1994).

For the most part, mantle zircon is considered to reside in the lithospheric mantle at temperatures that exceed its closure temperature for Pb diffusion (>900°C; Lee et al., 1997). If true, then these mantle xenocrysts would begin to accumulate radiogenic lead once they are transported by kimberlite magma to lower temperature regions, recording an age that should be close to the time of kimberlite emplacement. It is likely that this assumption is often valid and has guided the application of U-Pb mantle zircon geochronology to obtain ages of kimberlite emplacement. However, the closure temperature to Pb diffusion in zircon is not well established, although it is generally considered to be high (>900°C). All occurrences of mantle zircon indicate that some pre-kimberlite age information can be preserved in the U-Pb system. Kinny et al. (1989) documented two age populations of mantle zircon xenocrysts in the Permian Jwaneng DK2 kimberlite; one similar to the emplacement age of the kimberlite and one that showed slightly disturbed U-Pb systematics of primarily Archean mantle zircon. The Hf isotopic composition of these two age populations confirms the presence of two distinct generations of mantle zircon. Kinny and Meyer (1994) report a concordant 628 Ma U-Pb SHRIMP age for a zircon inclusion in diamond from the Cretaceous Mbuji Mayi kimberlite, Zaire; indicating that mantle zircon can preserve its original formation age, at least when encapsulated in diamond. Zircon in mantle xenoliths can also record pre-kimberlite emplacement ages and may document the time of metasomatism (Konzett et al., 1998; Hamilton et al., 1998) or eclogite metamorphism (Heaman et al., 2002). In all examples, there is now convincing evidence that certain mantle zircons can preserve pre-kimberlite age information and is not always completely reset under high pressure and temperature conditions of the subcontinental mantle.

There are two possibilities to explain the 'old' ages occasionally reported for mantle zircon (Kinny and Meyer, 1994; Schärer et al., 1997; Konzett et al., 1998; Klötzli, 1999). The first is the preservation of older zircon age domains with only partial resetting of the U-Pb system (Kinny and Meyer, 1994; Schärer et al., 1997). The second is a complex growth history for a mantle zircon crystal where new zircon growth occurs around an older core; such growth structure, including oscillatory zoning interpreted a primary magmatic feature, is often revealed by cathodoluminescence imaging (Konzett et al., 1998; Klötzli, 1999). In both cases, the implication is that mantle zircons could have extremely high closure temperatures for Pb diffusion (>900°C).

These examples illustrate some of the possible explanations that could account for the older Monastery U-Pb mantle zircon ages obtained in this study, such as the 325.4 and 100.7 Ma ages obtained for fractions Zir#5 and Zir#8, respectively. Interestingly, the oldest zircon fraction Zir#5 also has the lowest uranium content (0.93 ppm) of the eight analyses. One possible interpretation is that the 325.4 Ma age represents a minimum estimate for the time of original xenocryst formation. In support of this interpretation, Griffin et al. (2000) have reported $^{176}Hf/^{177}Hf$ model ages for some Monastery mantle zircon xenocrysts. The emplacement age obtained for the Monastery kimberlite is consistent at 89.7 Ma whereas, the Hf model ages range from 684 to 789 Ma (Griffin et al., 2000).

A total of three U-Pb dates have been determined for Monastery mantle zircons (Fig. 25.2), 90.4 Ma (Davis et al., 1976), 90.1±0.5 Ma (Zartman et al., 1998) and 88.8±0.5 Ma (this study). These ages are in general agreement and overlap with the 89.7±0.3 Ma preferred age of Monastery kimberlite emplacement.

U-Pb Ilmenite Geochronology – a New Chronometer?

Geochemistry

The major- and trace element geochemistry of ilmenite isolated from Monastery kimberlite has been studied since the early 1900s. Monastery is well known for its ilmenite-pyroxene intergrowths and it is this aspect that has generated much interest, starting with the study of Williams (1932). It was recognized early that the chemistry of ilmenite nodules differs from ilmenite within ilmenite-pyroxene intergrowths even though it was interpreted that they originated from the same source in the upper mantle (Dawson and Reid, 1970).

In an attempt to understand the paragenesis of ilmenite and define the processes responsible for the formation of megacrysts, numerous authors have grouped ilmenite based on mineral associations and chemical compositions (Haggerty et al., 1979; Moore et al., 1992; and Gurney et al., 1998). Haggerty et al. (1979) recognized two major ilmenite suites at Monastery: a discrete nodule suite and an ilmenite-pyroxene intergrowth (symplectite) suite. On an ilmenite-geikielite-hematite ternary plot (Fig. 25.3), Haggerty et al. (1979) demonstrated that there is a chemical difference between the homogeneous and polycrystalline megacrysts of the discrete nodule suite. In addition, Haggerty et al. (1979) identified four distinct compositional trends recorded in individual Monastery ilmenite megacrysts (Fig. 25.3): (1) "Magmatic Reaction Trend" (rim compositions trend towards higher MgO contents); (2) "Ilmenite-Pyroxene Trend" (trend towards lower Fe_2O_3 compositions); (3) "Kimberlite Reaction Trend" (rim compositions trend towards higher FeO contents) and (4) a trend resulting from exsolution and subsolidus reduction (not shown). The mineral chemistry for ilmenite examined in this study demonstrate that the ilmenite megacrysts and macrocrysts selected for U-Pb geochronology have compositions similar to those previously reported for Monastery "Magmatic" and "Kimberlite Reaction" trends shown on Fig. 25.3 and as proposed by Haggerty et al. (1979). These Monastery ilmenites plot within the field for On-craton Group I kimberlites summarized by Wyatt et al. (2004).

Compared to the whole-rock uranium concentration of 4.1 ppm for the Monastery kimberlite (Kresten, 1974), the uranium contents of ilmenite from the Kao, Lemphane and Sekameng/Kolo kimberlites, South Africa, also determined by

Kresten (1974), are an order of magnitude lower (0.1–0.2 ppm). The range in uranium concentrations for Monastery ilmenite obtained here overlaps (0.02–0.17 ppm; Table 25.5) the data range of Kresten (1974) but the Monastery ilmenite generally has even lower uranium contents (typically in the 20–50 ppb range). The Th/U ratios obtained for Monastery ilmenite are quite variable (0.03–1.48) but some fractions (Mon-1#3, Mon-2#1, M-2-2#2) have Th/U greater than unity, indicating that Th may be preferentially incorporated into ilmenite relative to uranium or that some invisible perovskite or other high Th/U material is present. The ilmenite analyses with high Th/U do not have correspondingly high $^{238}U/^{204}Pb$ ratios, which would be expected if some perovskite is present so the variable Th/U ratios is considered a primary chemical feature of ilmenite.

An important factor to address before interpreting the U-Pb results (next section) is the large variation in $^{238}U/^{204}Pb$ ratios of ilmenite within individual megacrysts. Kresten (1974) has suggested that these variations could be the result of either the presence of perovskite or other high U/Pb mineral inclusions. Kresten (1974) noted that primary ilmenite nodules contain approximately 0.1 ppm uranium (slightly higher than most ilmenite analyses reported in this study, Table 25.5), however, he also reported that calcite filled fractures contained 0.4 ppm of uranium (Kresten, 1974). An important finding of Kresten (1974) is that fission track patterns indicate that the majority of uranium is not located within the ilmenite or calcite grains but rather is concentrated at the contact between calcite/ilmenite and perovskite/ilmenite grains. However, the ilmenite megacrysts used in this study did not contain visible reaction rims (such as the one shown in Plate 25.1b), the ilmenite fractions were hand-picked at high magnification in an effort to avoid visible mineral inclusions, and the grains were washed in HNO_3 to remove such contaminants as carbonate along fractures or grain exteriors, prior to dissolution. These precautions were employed to eliminate potential inclusions or contaminants in the ilmenite fractions. As mentioned above, there is no correlation between Th/U and $^{238}U/^{204}Pb$ in the ilmenite analyses, which would be expected if a high Th/U mineral like perovskite (Th/U typically >10) were present. We conclude that it is unlikely that inclusions of a uranium-bearing mineral are present in the Monastery ilmenite megacrysts selected for U-Pb geochronology. However, there is some major element chemical variation within individual megacryst (e.g. M1-2 and M3-2 have higher-MgO and $-Al_2O_3$ rims; Table 25.1) so we cannot rule out the possibility that some of the selected ilmenite fractions may reflect this primary chemical heterogeneity and may be responsible for the variation in $^{238}U/^{204}Pb$ within an individual megacryst.

U-Pb Ilmenite Geochronology

Despite this extensive research on ilmenite mineral chemistry, there is still considerable debate regarding the origin of ilmenite megacrysts in kimberlites. Mitchell (1973, 1977 and 1986) and Pasteris (1980) advocate that ilmenite is a phenocryst phase in kimberlites, whereas others (Gurney et al., 1979; Moore et al., 1992; Griffin et al., 1997) support the hypothesis that ilmenite is a xenocryst phase within kimberlites. U-Pb ilmenite geochronology could potentially resolve this debate depending on the closure temperature of uranium and lead diffusion in ilmenite. If the U-Pb ilmenite dates were older than the kimberlite emplacement age, then this would signify that ilmenite is a mantle xenocryst. In contrast, if the ilmenite dates were the same as the kimberlite emplacement age then this could indicate that ilmenite is a primary phenocryst, crystallizing directly from the kimberlite or proto-kimberlite magma. One caveat however is a similarity in ages doesn't preclude the possibility that ilmenite megacrysts could still be mantle xenocrysts. For this to be the case, ilmenite xenocrysts must be susceptible to complete radiogenic Pb diffusion out of the crystal until entrained in Monastery kimberlite; the same argument used for interpreting mantle zircon megacryst ages. The closure temperature to Pb diffusion in ilmenite is unknown so it is difficult to assess this condition. Considering the large size of the ilmenite megacrysts (up to several centimeters), it seems unlikely that complete loss of radiogenic Pb by volume diffusion alone would occur so the similarity between U-Pb ilmenite and matrix perovskite or phlogopite ages more likely reflects a cognate origin.

One of the most encouraging results in this study is the potential for obtaining geologically meaningful U-Pb ilmenite ages. The ilmenite megacryst fractions shown on the isochron diagram in Fig. 25.4b, show significant variation in $^{238}U/^{204}Pb$ and $^{206}Pb/^{204}Pb$ ratios as well as significant variation within a single megacryst. This is highly promising as such isotopic variations are essential for obtaining precise isochron ages. For example, megacrysts Mon-1 and Mon-2 (square and circle symbols, respectively, Fig. 25.4d) exhibit a significant variation in the $^{238}U/^{204}Pb$ ratios (39.8–115.5 and 115.2–311.4, respectively). Together with the major element core-to-rim chemical zoning, this is interpreted as evidence for a primary internal geochemical variation within ilmenite megacrysts and indicates the great potential to obtain U-Pb ages from a single ilmenite megacryst.

Most important is the fact that the U-Pb composite ilmenite age (95.0±14.0 Ma) obtained here is within error of previously published ages for Monastery kimberlite (Fig. 25.2) and the composite age of 89.7±0.3 Ma, interpreted here as the best estimate for the timing of Monastery kimberlite emplacement, and is in remarkable agreement with Rb-Sr phlogopite, U-Pb perovskite and U-Pb mantle zircon ages obtained for the same samples in this study. If one considers the ilmenite megacrysts on their own (Mon-1, 2, 3) then an age of 87.0±32.0 Ma (MSWD = 0.51) is obtained, which is similar to the composite age although having a larger uncertainty.

It is the relatively poor precision of U-Pb ilmenite age determinations that requires refinement. It is apparent from the ages obtained for the Monastery kimberlite in this study that the errors associated with the $^{206}Pb/^{204}Pb$ and $^{238}U/^{204}Pb$ ratios for ilmenite need to be reduced. The largest contribution to these errors is the uncertainty in the Pb blank correction. An attempt to reduce these errors by carefully evaluating and reducing the Pb blanks in this study resulted in some improvement, however the errors are still significantly large enough to yield relatively imprecise ages. The uranium and lead concentrations are quite low as well (20–50 ppb range) which means that ion beam intensities for ^{206}Pb and ^{238}U during mass spectrometer analysis can be quite low (<10 mV using a Daly photomultiplier detector). However, selecting larger fractions can compensate this for. The requirement to analyse large (~1–5 mg) ilmenite fractions does not pose a serious problem, as ilmenite is quite abundant within most kimberlites. In the case where kimberlites do not possess abundant phlogopite, mantle zircon or perovskite, geologically meaningful emplacement ages with slightly larger errors can be established using U-Pb ilmenite geochronology.

Nature of the Monastery Source Region

Lead Isotopes

The initial $^{207}Pb/^{204}Pb$ and $^{206}Pb/^{204}Pb$ ratios for individual ilmenite and perovskite analyses calculated at 90 Ma are plotted in Fig. 25.5 along with other mantle reservoirs and fields for South African kimberlites. The ilmenite and perovskite analyses plot adjacent to the enriched Type 2 (EM2) mantle component, which also overlaps the isotopic signatures for ocean island basalts (Hart, 1984) and Group 1 South African kimberlites (Smith, 1983). In detail, the initial $^{206}Pb/^{204}Pb$ ratios for perovskite and ilmenite obtained from regression line calculations in this study (19.19 ± 0.97 and 19.28 ± 0.04, respectively) are quite radiogenic and most resemble the more radiogenic ocean island basalt ratios (17.6–21.1) summarized by Hart (1984). These initial $^{206}Pb/^{204}Pb$ ratios are identical to the Monastery whole rock kimberlite result obtained for sample RJ53/54 K of 19.09 (Smith, 1983) and fall within the range reported for Group 1 kimberlites (18.5–20.1). The Monastery Pb isotope data are clearly outside the range reported for Group 2 kimberlites (17.2–17.7; Smith, 1983). The mantle source for Group 1 kimberlites in Pb isotope space is slightly enriched and is characterized by high $^{238}U/^{204}Pb$ ratios (high mu = HIMU), whereas, the mantle source for Group 2 kimberlites is slightly depleted relative to bulk earth and is characterized by lower $^{238}U/^{204}Pb$ ratios (Smith, 1983). The initial Pb isotopic data reported here and the whole rock kimberlite data of Smith (1983) provide compelling evidence for the involvement of a HIMU mantle source region in the origin of the Monastery kimberlite (also proposed by Hatton, 1998). The similarity between the ilmenite and perovskite initial Pb isotopic data is consistent with derivation of both Main Silicate Suite megacrysts of Hatton (1998) and groundmass perovskite from the same kimberlite magma.

Strontium Isotopes

The Rb-Sr phlogopite age calculated for the Monastery kimberlite in this study (89.4 ± 0.8 Ma; Fig. 25.4a) and the initial $^{87}Sr/^{86}Sr$ ratio (0.7044 ± 0.0013) is in agreement within error of results reported in previous studies. Smith (1983) obtained a similar initial $^{87}Sr/^{86}Sr$ ratio of 0.70330 ± 0.00011 for Monastery kimberlite whole rock. The initial strontium isotopic composition reported for a single phlogopite megacryst (Smith and Barton, 1996) including the kimberlite whole rock analysis of Smith (1983) yields the preferred initial $^{87}Sr/^{86}Sr$ ratio of 0.7034 ± 0.0002.

The Monastery kimberlite is considered to be a Group 1 kimberlite based on mineralogical and emplacement age criteria (Smith, 1983). Group 1 kimberlites contain little to no groundmass mica and have emplacement ages ranging from 80 to 114 Ma, whereas, Group 2 kimberlites are micaceous and have emplacement ages ranging from 114 to 150 Ma (Smith, 1983). Smith (1983) has determined initial $^{87}Sr/^{86}Sr$ ratios for fresh Jurassic to Cretaceous Group 1 kimberlites in South Africa to be typically 0.7033–0.7049 and for Group 2 kimberlites to be >0.7074. Both the Sr and Pb initial isotopic composition for Monastery whole rock, megacrysts and groundmass minerals are consistent with this Group 1 kimberlite classification.

Conclusions

It has been demonstrated that multiple age dating techniques can yield consistent and reliable kimberlite emplacement ages. Rb-Sr phlogopite megacrysts (89.4±0.8 Ma), U-Pb mantle zircon megacryst (88.8±0.5 Ma) and U-Pb groundmass perovskite (91.0±1.0 Ma) all from the Monastery kimberlite yield ages that are in excellent agreement both in this study and with age determinations from previous studies. This is the first study to attempt all three radiometric techniques on the same kimberlite host and provides confidence that geologically meaningful and identical ages can be obtained by all techniques.

The ilmenite mineral chemistry determined in this study is similar to results previously reported for Monastery "Magmatic" and "Kimberlite Reaction" Trends shown on Fig. 25.2 and as proposed by Haggerty et al. (1979). They are considered to have formed by primary crystallization in a kimberlitic magma and some ilmenite megacrysts have been influenced by late-stage interaction with a carbonatitic fluid. A preliminary study of U-Pb ilmenite geochronology has yielded very promising results and demonstrates for the first time that ilmenite megacrysts can yield kimberlite emplacement ages. The composite U-Pb ilmenite age of 95.0±14.0 Ma is in agreement with the ages obtained by three different radiometric techniques in this study and the preferred composite emplacement age of 89.7±0.3 Ma based on seven independent age dates determined for the Monastery kimberlite. The U-Pb date obtained here for Monastery ilmenite is the first attempt at determining whether the U-Pb system can be applied to ilmenite megacrysts and whether the age results could provide some constraint for the time of kimberlite emplacement. The general agreement with other more traditional chronometers (Rb-Sr phlogopite, U-Pb perovskite etc.) demonstrates the feasibility of U-Pb ilmenite geochronology applied to kimberlites. This study has also demonstrated that the large variations in ilmenite $^{238}U/^{204}Pb$ ratios are primary and indicates the possibility of dating a single ilmenite megacryst. This could have importance for evaluating the provenance of ilmenite from indicator mineral suites during diamond exploration.

Acknowledgments This research formed the basis of an MSc thesis by Andrea Noyes. It was funded in part by a University of Alberta graduate student assistantship to A. Noyes and by NSERC Discovery grants awarded to L. Heaman and R. Creaser. We would like to thank the technical staff at UofA; M. Labbe and D. Resultay (thin section preparation), B. Böhm (rock crushing), L. Shi (microprobe facilities), O. Levner and L. Raynor (isotope geochemistry) and S. Hagen (mass spectrometer). The authors greatly appreciate the comments and suggestions by journal reviewer N. V. Chalapathi Rao and Editor Rajesh Srivastava.

References

Agee JJ, Garrison JR, Taylor LA (1982) Petrogenesis of oxide minerals in kimberlite, Elliott County, Kentucky. Am Mineral 67: 28–42

Ahrens LH, Cherry RD, Erlank AJ (1967) Observations on the Th-U relationship in zircons from granitic rocks and from kimberlites. Geochim Cosmochim Acta 31: 2379–2387

Allsopp HL, Barrett DR (1975) Rb-Sr age determinations on South African kimberlite pipes. Phys Chem Earth 9: 605–617

Allsopp HL, Bristow JW, Smith CB, Brown R, Gleadow AJW, Kramers JD, Garvie OG (1986) A summary of radiometric dating methods applicable to kimberlites and related rocks. Proceedings of 4th International Kimberlite Conference, Perth, Australia. Cambridge, Victoria Australia 1: 343–357.

Allsopp HL, Roddick JC (1984) Rb-Sr and ^{40}Ar-^{39}Ar age determinations on phlogopite micas from the Pre-Lebombo Group Dokolwayo kimberlite pipe. Geol Soc S Africa Spec Publ 13: 267–271

Blackburn TJ, Stockli DF, Carlson RW, Berendsen P (2008) (U-Th)/He dating of kimberlites – A case study from north-eastern Kansas. Earth Planet Sci Lett 275: 111–120

Boctor NZ, Boyd FR (1980) Oxide minerals in the Liqhobong kimberlite, Lesotho. Am Mineral 65: 631–638

Boyd FR, Nixon PH (1975) Origins of the ultramafic nodules from some kimberlites of northern Lesotho and the Monastery Mine, South Africa. Phys Chem Earth 9: 431–454

Brown RW, Allsopp HL, Bristow JW, Smith C (1989) Improved precision of Rb-Sr dating of kimberlitic micas: An assessment of a leaching technique. Chem Geol 79: 125–136

Cahen L, Snelling NJ, Delhal J, Vail JR (1984) The Geochronology and evolution of Africa. Clarendon Press, Oxford: 512 pp

Cameron AE, Smith DH, Walker RL (1969) Mass spectrometry of nanogram-size samples of lead. Anal Chem 41: 525–526

Carswell DA (1975) Primary and secondary phlogopites and clinopyroxenes in garnet lherzolite xenoliths. Phys Chem Earth 9: 417–430

Chen YD, O'Reilly SY, Kinny PD, Griffin WL (1994) Dating lower crust and upper mantle events: an ion microprobe study of xenoliths from kimberlitic pipes, South Australia. Lithos 32: 77–94

Clifford TN (1970) The structural framework of Africa. In: Clifford TN, Gass IG (eds) African Magmatism and Tectonics. Hafner Publishing Company, Darien, CT: 1–26

Creaser RA, Grutter H, Carlson J, Crawford B (2004) Macrocrystal phlogopite Rb-Sr dates for the Ekati property kimberlites, Slave Province, Canada: Evidence for multiple intrusive episodes in the Paleocene and Eocene. Lithos 76: 399-414

Crough WT (1981) Mesozoic hotspot epeirogeny in Eastern North America. Geology 9: 2–6

Davis GL (1977) The ages and uranium contents of zircons from kimberlites and associated rocks. Carnegie Institute of Washington Yearbook 76: 631–635.

Davis GL (1978) Zircons from the mantle. Carnegie Institution of Washington Yearbook 78: 86–88

Davis DW, Gray J, Cumming GL, Baadsgaard H (1977) Determination of the ^{87}Rb decay constant. Geochim Cosmochim Acta 41: 1745–1749

Davis GL, Krogh TE, Erlank AJ (1976) The ages of zircons from kimberlites from South Africa. Carnegie Institution of Washington Yearbook 75: 821–824.

Dawson JB, Reid AM (1970) A pyroxene-ilmenite intergrowth from the Monastery Mine, South Africa. Contrib Mineral Petrol 26: 296–301

Dawson JB, Smith JV (1975) Chemistry and origin of phlogopite megacrysts in kimberlite. Nature 253: 336–338

Duncan RA, Hooper PR, Rehacek J, Marsh JS, Duncan AR (1997) The timing and duration of the Karoo igneous event, southern Gondwana. J Geophys Res 102: 18127–18138

Farmer GL, Boettcher AL (1981) Petrologic and crystal-chemical significance of some deep-seated phlogopites. Am Mineral 66: 1154–1163

Griffin WL, Moore RO, Ryan CG, Gurney JJ, Win TT (1997) Geochemistry of magnesian ilmenite megacrysts from southern African kimberlites. Russian Geology and Geophysics 38: 421–443. In: NV Sobolev and RH Mitchell (eds) Proceed of 6th International Kimberlite Conference. Allerton Press Inc., New York.

Griffin WL, Pearson NJ, Belousova E, Jackson SE, van Achterbergh E, O'Reilly SY, Shee SR (2000) The Hf isotope composition of cratonic mantle: LAM-MC-ICPMS analysis of zircon megacrysts in kimberlites. Geochim Cosmochim Acta 64: 133–147

Gurney JJ, Jakob WRO, Dawson JB (1979) Megacrysts from the Monastery kimberlite pipe, South Africa. Proceedings 2nd International Kimberlite Conference, Santa Fe. Am Geophys Union, Washington 2: 265–278.

Gurney JJ, Moore RO, Bell DR (1998) Mineral associations and compositional evolution of Monastery kimberlite megacrysts. Proceedings of 7th International Kimberlite Conference, Cape Town. Extended Abstracts: 290–292.

Haggerty SE, Hardie IIIRB, McMahon BM (1979) The mineral chemistry of ilmenite nodule associations from the Monastery diatreme. Proceeding of 2nd International Kimberlite Conference, Santa Fe. Am Geophys Union, Washington 2: 249–256.

Hamilton MA, Pearson DG, Stern RA, Boyd FR (1998) Constraints on MARID petrogenesis: SHRIMP II U-Pb zircon evidence for pre-eruption metasomatism at Kampfersdam. Proceedings of 7th International Kimberlite Conference, Cape Town. Extended Abstract: 296–298.

Hart SR (1984) A large-scale isotope anomaly in the Southern Hemisphere mantle. Nature 309: 753–757

Hatton CJ (1998) The kimberlite-megacryst link at Monastery Mine. Proceedings 7th International Kimberlite Conference, Cape Town. Extended Abstract: 314–316.

Heaman LM (1989) The nature of the subcontinental mantle from Sr-Nd-Pb isotopic studies on kimberlitic perovskite. Earth Planet Sci Lett 92: 323–334

Heaman LM, Bowins R, Crocket J (1990) The chemical composition of igneous zircon suites: Implications for geochemical tracer studies. Geochim Cosmochim Acta 54: 1597–1607

Heaman LM, Creaser RA, Cookenboo HO (2002) Extreme enrichment of high field strength elements in Jericho eclogite xenoliths: A cryptic record of Paleoproterozoic subduction, partial melting, and metasomatism beneath the Slave craton, Canada. Geology 30: 507–510

Heaman LM, Kjarsgaard BA (2000) Timing of eastern North American kimberlite magmatism: Continental extension of the Great Meteor hotspot track? Earth Planet Sci Lett 178: 253–268

Heaman LM, Kjarsgaard BA, Creaser RA (2003) The timing of kimberlite magmatism in North America: Implications for global kimberlite genesis and diamond exploration. Lithos 71: 153–184

Heaman LM, Kjarsgaard BA, Creaser RA (2004) The temporal evolution of North American kimberlites. Lithos 76: 377–397

Heaman LM, LeCheminant AN (1992) Paragenesis and U-Pb systematics of baddeleyite (ZrO_2). Chem Geol 110: 95–126

Heaman LM, Machado N (1992) Timing and origin of midcontinent rift alkaline magmatism, North America: Evidence from the Coldwell Complex. Contrib Mineral Petrol 110: 289–303

Helmstaedt H, Gurney JJ (1984) Kimberlites of southern Africa – are they related to subduction processes? Proceedings of 3rd International Kimberlite Conference, France. Elsevier, Amsterdam 1: 425–434.

Hunter RH, Kissling RD, Taylor LA (1984) Mid- to late-stage kimberlitic melt evolution: Phlogopites and oxides from the Fayette County kimberlite, Pennsylvania. Am Mineral 69: 30–40

Jaffey AH, Flynn KF, Glendenin LE, Bentley WC, Essling AM (1971) Precision measurement of half-lives and specific activities of ^{235}U and ^{238}U. Physical Review C 4: 1889–1906

Jones AP, Smith JV (1984) Ion probe analysis of H, Li, B, F and Ba in micas, with additional data for metamorphic amphibole, scapolite and pyroxene. Neues Jahrbuch fuer Mineralogie Monatshefte 5: 228–240

Kinny PD, Compston W, Bristow JW, Williams IS (1989) Archean mantle xenocrysts in a Permian kimberlite: Two generations of kimberlitic zircon in Jwaneng DK2, southern Botswana. Geol Soc Austr Spec Publ 14: 833–842

Kinny PD, Dawson JB (1992) A mantle metasomatic injection event linked to late Cretaceous kimberlite magmatism. Nature 360: 726–728

Kinny PD, Meyer HOA (1994) Zircon from the mantle: A new way to date old diamonds. J Geol 102: 475–481

Klötzli US (1999) Th/U zonation in zircon derived from evaporation analysis: A model and its implications. Chem Geol 158: 325–333

Konzett J, Armstrong RA, Sweeney RJ, Compston W (1998) The timing of MARID metasomatism in the Kaapvaal mantle: An ion probe study of zircons from MARID xenoliths. Earth Planet Sci Lett 160: 133–145

Kramers JD, Smith CB (1983) A feasibility study of U-Pb and Pb-Pb dating of kimberlites using groundmass mineral fractions and whole rock samples. Isotope Geosci 1: 23–38

Kresten P (1974) Uranium in kimberlites and associated rocks, with special reference to Lesotho occurrences. Lithos 7: 171–180

Kresten P, Berggren G (1975) Kimberlitic zircons – a possible aid in prospecting for kimberlites. Mineral Dep 10: 47–56

Krogh TE (1973) A low-contamination method for hydrothermal decomposition of zircon and extraction of U and Pb for isotopic age determinations. Geochim Cosmochim Acta 37: 485–494

Lanzirotti A, Hanson GN (1995) U-Pb dating of major and accessory minerals formed during metamorphism and deformation of metapelites. Geochim Cosmochim Acta 59: 2513–2526

LeCheminant AN, Heaman LM, Kretschmar U, LeCouteur PC (1998) Complex origins and multiple ages of mantle zircon megacrysts from Canadian and South African kimberlites. Proceedings of 7th International Kimberlite Conference, Cape Town. Extended Abstracts: 486–488.

Lee JKW, Williams IS, Ellis DJ (1997) Pb, U and Th diffusion in natural zircon. Nature 390: 159–161

Ludwig KR (1998) On the treatment of concordant uranium-lead ages. Geochim Cosmochim Acta 62: 665–676

McCandless TE (1999) Kimberlites: Mantle expressions of deep-seated subduction. In: JL Gurney, MD Pascoe, SH Richardson (eds) Proceedings of 7th International Kimberlite Conference 2: 545–549.

McIntyre RM, Dawson JB (1976) Age and significance of some South African kimberlites. 4th European Colloquium on Geochronology, Cosmochronology and Isotope Geology, Abstracts. Elsevier, Amsterdam: 66

Mitchell RH (1972) Composition of perovskite in kimberlite. Am Mineral 57: 1748–1753

Mitchell RH (1973) Magnesian ilmenite and its role in kimberlite petrogenesis. J Geol 81: 301–311

Mitchell RH (1977) Geochemistry of magnesian ilmenites from kimberlites in South Africa and Lesotho. Lithos 10: 29–37

Mitchell RH (1986) *Kimberlites: Mineralogy, Geochemistry, and Petrology*. Plenum Press, New York: 442 pp

Moore RO, Griffin WL, Gurney JJ, Ryan CG, Cousens DR, Sie SH, Suter GF (1992) Trace element geochemistry of ilmenite megacrysts from the Monastery kimberlite, South Africa. Lithos 29: 1–18

Nixon PH (1987) Kimberlitic xenoliths and their cratonic setting. In: Nixon PH (ed) Mantle Xenoliths. Wiley, Chichester

Noyes AK (2000) A feasibility study of U-Pb ilmenite geochronology, Monastery Kimberlite, South Africa. M.Sc. dissertation, University of Alberta, Edmonton, Alberta, 88 pp.

Pasteris JD (1980) The significance of groundmass ilmenite and megacryst ilmenite in kimberlites. Contrib Mineral Petrol 75: 315–325

Pasteris JD, Boyd FR, Nixon PH (1979) The ilmenite association at the Frank Smith mine, R.S.A. Proceedings 2nd International Kimberlite Conference, Santa Fe. Am Geophys Union Washington 2: 265–278.

Phipps SP (1988) Deep rifts as sources for alkaline intraplate magmatism in eastern North America. Nature 334: 27–31

Schulze DJ, Anderson PFN, Hearn BC Jr, Hetman CM (1995) Origin and significance of ilmenite megacrysts and macrocrysts from kimberlite. Internat Geol Rev 37: 780–812

Schärer U, Corfu F, Demaiffe D (1997) U-Pb and Lu-Hf isotopes in baddeleyite and zircon megacrysts from the Mbuji-Mayi kimberlite: Constraints on the subcontinental mantle. Chem Geol 143: 1–16

Sharp WE (1974) A plate tectonic origin for diamond-bearing kimberlite. Earth Planet Sci Lett 21: 351–354

Sleep NH (1990) Monteregian hotspot track: A long-lived mantle plume. J Geophys Res 95: 21983–21990

Smith CB (1983) Pb, Sr and Nd isotopic evidence for sources of southern African Cretaceous kimberlites. Nature 304: 51–54

Smith CB, Allsopp HL, Kramers JD, Hutchinson G, Roddick JC (1985) Emplacement ages of Jurassic-Cretaceous South African kimberlites by the Rb-Sr method on phlogopite and whole-rock samples. Trans Geol Soc S Afr 88: 249–266

Smith CB, Barton ES (1996) Rb-Sr mica age of the Monastery kimberlite phlogopite megacryst sample ROM98. AARL Internal Report, DeBeers.

Stacey JS, Kramers JD (1975) Approximation of terrestrial lead isotope evolution by a two-stage model. Earth Planet Sci Lett 26: 207–221

Steiger RH, Jäger E (1977) Subcommission on geochronology; convention on the use of decay constants in geo- and cosmochronology. Earth Planet Sci Lett 36: 359–362

Taylor LA (1984) Kimberlitic magmatism in the Eastern United States: Relationship to mid-Atlantic tectonism. Proceedings of 3rd International Kimberlite Conference 1: 417–424.

Truswell JF (1977) The Geological Evolution of South Africa. Purnell and Sons, Cape Town: 218 pp

Whitelock TK (1973) The Monastery mine kimberlite pipe. In: Nixon PH (ed) Lesotho Kimberlites. Lesotho National Development Corporation, Maseru: 214–219

Williams AF (1932) The Genesis of the Diamond, vol 1. Ernest Benn Ltd., London: 352 pp

Wyatt BA, Baumgartner M, Anckar E, Grütter H (2004) Compositional classification of "kimberlitic" and "non-kimberlitic" ilmenite. Lithos 77: 819–840

York D (1969) Least squares fitting of a straight line with correlated errors. Earth Planet Sci Lett 5: 320–324

Zartman RE, Richardson SH, Gurney JJ, Moore RO (1998) U-Th-Pb ages of megacrystic zircon from the Monastery kimberlite, Free State, South Africa. Proceedings of 7th International Kimberlite Conference, Cape Town. Extended Abstracts: 989–991.

Zindler A, Hart S (1986) Chemical Geodynamics. Ann Rev Earth Planet Sci 14: 493–571

Zurevinski S (2009) The origin of North American kimberlites. Ph.D. Dissertation, University of Alberta. 180 pp.

Part II
Dykes in Laurasia

Chapter 26
Precise U-Pb Age for the Great Whin Dolerite Complex, N.E. England and Its Significance

M. A. Hamilton and D.G. Pearson

Introduction

The Whin Sill is probably the largest hypabyssal igneous intrusion in Britain and represents an important time marker in the regional stratigraphy of NE England. As noted by Francis (1982) and Johnson and Dunham (2001), the Whin Sill, intrusive into its host Carboniferous sedimentary succession, became known as the first archetype example of a "sill", even though its largest representative exposures define a largely discordant intrusive sheet. The intrusion is more correctly referred to as the Whin Sill complex, comprising a series of sills of varying thickness together with a suite of dykes (Dunham and Strasser-King, 1982). The main sill intrusion is known as the Great Whin Dolerite Complex (Johnson and Dunham, 2001), referred to hereafter as the GWDC. Individual sill-like phases of the intrusion reach 90 m in thickness (Johnson, 1995). Its widespread and transgressive nature, combined with separation into numerous (up to 5) discrete intrusive pulses make it a valuable regional time marker for a variety of geological and tectonic events. For instance, accurate knowledge of the age(s) of this body provides important constraints on the timing of regional tectonic movements, together with the evolution of the regional stress field (Johnson, 1995). Furthermore, the intrusion of the GWDC provides an important datum in the history of the North Pennine orefield where it post-dates regional subsidence but pre-dates the mineralisation (Dunham, 1990). The growing importance of the sill complex in terms of regional European studies of Permo-Carboniferous magmatism and tectonics (e.g., Wilson et al., 2004) highlights the need for a refined, accurate and precise age for the intrusion.

The GWDC has been a reference point for many pioneering geochronological studies, including one of the first attempts, by Arthur Holmes, to date an intrusion

M.A. Hamilton (✉)
Jack Satterly Geochronology Laboratory, Department of Geology, University of Toronto, Toronto, ON, Canada M5S 3B1
e-mail: mahamilton@geology.utoronto.ca

D.G. Pearson
Department of Earth Sciences, Durham University, Durham DH1 3LE, UK

R.K. Srivastava (ed.), *Dyke Swarms: Keys for Geodynamic Interpretation*,
DOI 10.1007/978-3-642-12496-9_26,

using the U-He method (Dubey and Holmes, 1929). It also featured in the earliest assessment of the K-Ar decay scheme to determine the age of basic rocks (Miller and Mussett, 1963). Here we use U-Pb geochronology on the mineral baddeleyite (ZrO_2), separated from a pegmatite phase within the main sill, to provide the best available age on this intrusion and the most precise age for any magmatic body within the European Permo-Carboniferous magmatic event.

Geological Setting and Petrology

The GWDC of northern England is part of a widespread and voluminous magmatic suite located in the foreland of the Variscan Orogen (Wilson et al., 2004). The intrusive complex crops out in a semi-continuous and semi-circular arcuate pattern in the area between Holy Island in the north, Hartlepool in the south and Melmerby – upper Teesdale in the south-west, and underlies an area of approximately 5,000 km^2, with a volume estimated at >200 km^3 (Thorpe and Macdonald, 1985) although this does not take into account its probable extension beneath the North Sea, towards which it dips and thickens (Fig. 26.1). Sheets of Great Whin dolerite are found restricted to Carboniferous sedimentary rocks, from stratigraphic levels at least as old as Asbian, and likely even Holkerian in Upper Teesdale (Johnson and Dunham, 2001), up through the Westphalian Coal measures. Details of the geology and emplacement of the complex have been described by Francis (1982), and more recently by Johnson and Dunham (2001). Liss et al. (2004) define three distinct sill units within the GWDC, the Holy Island Sill, the Alnwick Sill and the Hadrian's Wall-Pennines Sill. Goulty (2005) suggests that intrusion of the sills forming the GWDC was probably accompanied by extrusion of flood basalts. Differences in paleomagnetic directions between different sill phases, combined with variations in magma flow indicators suggest that the Holy Island Sill and Hadrian's Wall-Pennines Sill may represent separate intrusion events (Liss et al., 2004), opening up the possibility that the Whin "event" may have a range of ages. In this study, we have focused only on the Hadrian's Wall-Pennines Sill which has been the target of previous geochronology studies.

The predominant lithology within the sill complex is quartz dolerite with a grain size of 0.5–1 mm. Plagioclase of broadly labradoritic composition together with augite (locally subcalcic augite or pigeonite) form a sub-ophitic texture. Other common minerals include orthopyroxene, titano-magnetite, alkali feldspar and quartz (locally micropegmatite; Johnson and Dunham, 2001; Thorpe and Macdonald, 1985; Dunham et al., 1972). Apatite is an abundant accessory mineral. This study is the first to record the additional presence of baddeleyite in the gabbroic pegmatites of the Whin dolerite. At the High Force quarry [NY881283] in Teesdale the GWDC is over 70 m thick, intruding into Brigantian Tyne Bottom Limestone. Here, well developed coarse-grained pegmatite bands varying from 1 cm to 2 m thick occur in the upper part of the main sill, occurring as flat lenticular masses parallel to the upper contact. Pegmatitic gabbro lenses, sampled at this locality (Fig. 26.1) comprise blades of pyroxene up to 5 cm in length along with plagioclase and minor but abundant intergrowths of interstitial baddeleyite, apatite and oxides.

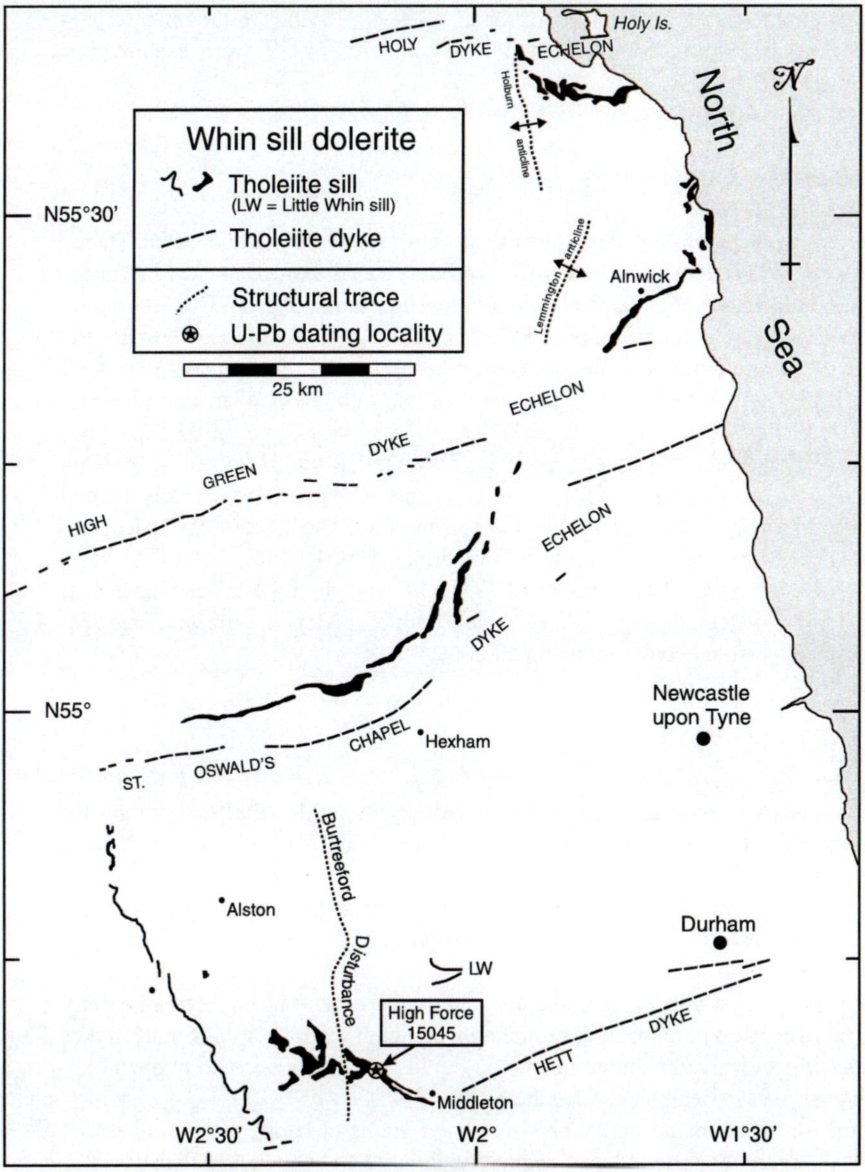

Fig. 26.1 Sketch map showing the outcrop pattern of the Whin Dolerite Complex in northeast England. Principal associated dykes are also highlighted. Modified after Francis (1982) and Thorpe and Macdonald (1985)

Similar coarse-grained to pegmatitic dolerites have been reported from the Barrasford area, where the dolerite reaches a maximum thickness of approximately 40 m (Randall, 1989), intruding Brigantian age rocks. In this region, normal dolerite, dominated by subophitic augite with plagioclase and minor pigeonite, contains thin (1–20 cm) concordant pegmatitic lenses with similar modal mineralogy, that also occur parallel to the upper contacts of the sill. The similarities to the dolerite pegmatites described from the Teesdale region by Tomkeieff (1929) were also noted by Randall (1989).

Previous Age Constraints

Pioneering work by Holmes and co-workers (Holmes and Harwood, 1928; Dubey and Holmes, 1929) using the U-He method produced a Jurassic age (192 Ma) for the GWDC. On the basis of geological field relationships alone, the GWDC demonstrably intrudes Upper Carboniferous rocks but not those of Permian age. Furthermore, eroded clasts of Whin dolerite have been recognized in Lower Permian-aged Upper Brockram continental breccias interbedded with the aeolian Penrith sandstone (Dunham, 1932). Wholerock K-Ar determinations by Fitch and Miller (1967) produced an age reported as 295 $\pm$ 6 Ma, a result that yields an age of 302 Ma when re-calculated using modern decay constants. Despite being widely quoted (e.g., Timmerman, 2004) we are unable to reproduce the uncertainty calculated on this age. More recently, however, Timmerman (2004) using ^{40}Ar–^{39}Ar step-heating methods gave a best-estimate of 294 $\pm$ 2 Ma (2-sigma weighted mean age, not including systematic errors) but no well defined plateaux were obtained (M. Timmerman, personal communication 2003).

Analytical Methods

There has been a growing utilization of accessory minerals other than zircon for U-Pb dating especially in magmatic rocks where zircon is absent, or is expected to be complicated by inheritance. Primary baddeleyite is one of the more useful minerals in the context of dating basic and often zircon-free igneous rocks due to its comparable levels of U, typically low abundance of common Pb, and because it does not appear to suffer from inheritance (Heaman and LeCheminant, 1993). This mineral occurs in a variety of mafic and ultramafic rocks but is most commonly found in mafic pegmatoid patches or intercumulus melt regions of thick mafic dykes and sills where evolved residual magmatic liquids may have become trapped (e.g. Krogh et al., 1987; Heaman and LeCheminant, 1993).

In this study, abundant honey-brown striated grains of baddeleyite were recovered from approximately 12 kg of pegmatitic dolerite from the High Force Quarry following standard crushing, grinding, heavy liquid and magnetic separation procedures. The largest euhedral blade-like grains had length:breadth dimensions on the order of 150:30 microns, reaching a maximum thickness of probably <15 microns. Isotope dilution thermal ionization mass spectrometry (ID-TIMS) U-Pb analysis was carried out at the Geological Survey of Canada (Ottawa), and followed the methods described by Hamilton et al. (1998). All uncertainties are quoted at the 95% confidence level. Further procedural details are also provided in Table 26.1.

Table 26.1 U-Pb isotopic data for baddeleyite from the Great Whin Dolerite Complex

Fraction	Weight (μg)	U p.p.m.	Pb* p.p.m.	$^{206}Pb/^{204}Pb^{a}$	Pb^{b} (pg)	$^{208}Pb/^{206}Pb^{c}$	$^{206}Pb/^{238}U^{c}$	$^{207}Pb/^{235}U^{c}$	$^{207}Pb/^{206}Pb^{c}$	Apparent Age (Myr) $^{206}Pb/^{238}U$
15045 Pegmatitic dolerite, High Force Quarry, Teesdale										
A	20	315	14	1,177	16	0.031	0.047231 ± 0.09	0.341068 ± 0.18	0.05237 ± 0.13	297.5 ± 0.5
B	18	331	17	1,407	15	0.034	0.047286 ± 0.29	0.344532 ± 1.42	0.05287 ± 1.33	297.8 ± 1.7
C	11	627	28	1,034	20	0.033	0.047157 ± 0.16	0.341422 ± 0.27	0.05251 ± 0.19	297.0 ± 0.9
D	9	436	21	688	20	0.031	0.047221 ± 0.14	0.340370 ± 0.30	0.05228 ± 0.24	297.4 ± 0.8

Ratio errors are 1 standard error of the mean, in %. $^{206}Pb/^{238}U$ age errors are 2 standard errors, in Myr. Pb*, radiogenic Pb.

[a]Measured ratio, corrected for mass fractionation and spike.

[b]Total common Pb in analysis (in picograms), corrected for fractionation and spike.

[c]Corrected for spike, blank and common Pb.

Analytical techniques: Mineral separation protocols generally followed the procedures described in detail in Hamilton et al. (1998). Samples were processed using conventional crushing, grinding, Wilfley table and heavy liquid concentration methods. Frantz isodynamic magnetic separation techniques were used for isolation of the least magnetic grains; from these, the best quality, clear, translucent pale to dark honey-brown striated blades and blade fragments were hand-picked under alcohol using a binocular microscope and selected for analysis. Samples were dissolved in concentrated HF in Teflon bombs, together with a mixed $^{205}Pb/^{233-235}U$ spike. Isotopic analysis generally followed the procedures outlined in Hamilton et al. (1998). U and Pb isotopic ratios were measured statically on a Finnigan MAT 261 mass spectrometer (GSC, Ottawa).

Age calculations were carried out using the decay constants 1.55125×10^{-10} year^{-1} (^{238}U) and 9.8485×10^{-10} year^{-1} (^{235}U) after Jaffey et al. (1971).

Results

U-Pb analytical results for four fractions of baddeleyite from the GWDC are provided in Table 26.1. Least paramagnetic baddeleyites (A-D) have U concentrations between 315 and 627 ppm, within the range typically observed for baddeleyite (Heaman and LeCheminant, 1993). Most of the fractions analysed fall on or near concordia (Fig. 26.2), with little sign of discordance. Although fraction B has larger analytical errors than the others, all fractions have overlapping Pb/U and Pb/Pb ratios, and a weighted mean $^{206}Pb/^{238}U$ age of all four fractions yields a precise age of 297.4 ± 0.4 Ma (2σ), with a low MSWD (0.30) and a high probability of fit (p = 0.83). This result is equivalent to a Concordia Age calculation (Ludwig, 1998, 2003) excluding decay-constant errors. If the uncertainties for the U decay-constants are taken into account, a Concordia Age of 297.5 ± 0.5 Ma (2σ) is obtained. Although the difference in minimal, in order to facilitate comparison with other regional U-Pb datasets (below), we prefer to quote 297.4 ± 0.4 Ma as representing the best estimate of the age of emplacement and crystallization of the Great Whin Dolerite at this locality.

A weighted mean of the 13 samples used by Fitch and Miller (1967) to produce their age (recalculated using the decay-constant given by Renne et al., 1998)

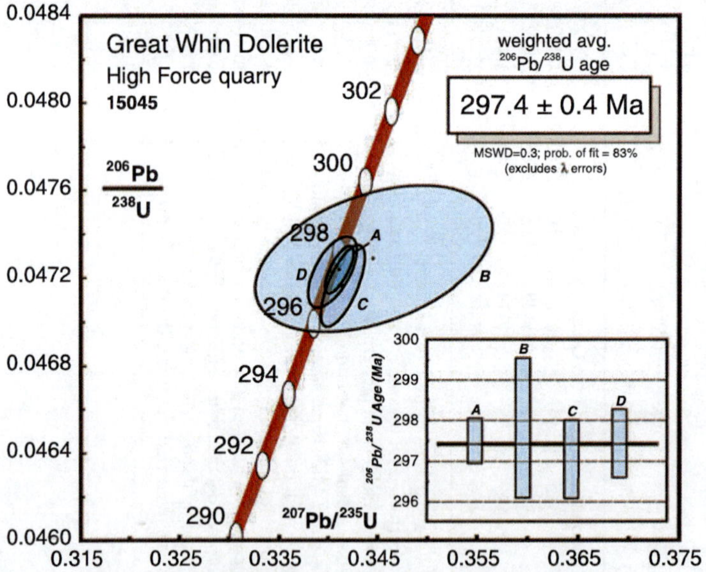

Fig. 26.2 Concordia diagram showing U-Pb isotopic results for baddeleyite analyzed from Great Whin Dolerite Complex, sampled at High Force Quarry. Concordia "band" reflects the uncertainties in the U decay-constants. *Inset* shows the $^{206}Pb/^{238}U$ ages for each fraction. The weighted average $^{206}Pb/^{238}U$ age calculation ignores decay constant errors (but see text). Concordia ellipses and weighted mean error bars (*inset*) are all shown at the 2σ (95% confidence) level of uncertainty

yields an age of 297.6 ± 8.4 Ma, remarkably close to our U-Pb age. However, we note that this mean comprises individual age determinations that span a range from 275 to 323 Ma, with uncertainties as large as ±18 Ma (1 sigma errors). Hence, the coincidence of these ages by different methods is probably somewhat serendipitous. The new age presented by us is an order of magnitude more precise than previous estimates and can be used with confidence for correlation purposes.

Implications

Whin Sill Magmatic Event

Although the previous K-Ar and $^{40}Ar/^{39}Ar$ determinations for the GWDC are within error of the new baddeleyite U-Pb age, the latter is considerably more precise and avoids the additional uncertainties currently recognised as hampering age determinations based on $^{40}Ar/^{39}Ar$ systematics (e.g. ^{40}K decay constant uncertainties, accuracy of ages of neutron flux monitors, etc.; Renne et al., 1998). The $^{40}Ar/^{39}Ar$ age of 294 ± 2 Ma (2σ) presented by Timmerman (2004) was determined on plagioclases from the Holy Island dyke. Although this age falls just outside of error (at 95% confidence) of our U-Pb age (Fig. 26.3), we note that Johnson and Dunham (2001) suggest that there were likely two separate periods of Whin dolerite emplacement and, citing Smythe (1930), observe that there is little supportive evidence that the dykes within the complex actually fed sill-like bodies. Smythe (1930) believed that the dyke phase at Holy Island could be slightly later than the intrusive sheets. Consequently, as Timmerman (2004) acknowledges, the $^{40}Ar/^{39}Ar$ age for the Holy Island dyke likely provides a minimum age for the broader sill complex. Further work remains to done to define any resolvable age differences between the different intrusive units of the Whin Sill complex.

Carboniferous Time Scale and the Age of the Permo-Carboniferous Boundary

The Carboniferous time scale and the age of the Permo-Carboniferous boundary of central and western Europe are subjects of considerable controversy that are poorly constrained from existing geochronology, with a paucity of high-precision TIMS U-Pb ages, particularly in the upper Carboniferous (Menning et al., 2000). Recent advances in high precision U-Pb dating of zircon from Permo-Carboniferous volcanic ashes from marine sections in the Urals has led to the establishment of a Carboniferous-Permian Global Stratotype (GSSP) date of 299 ± 1 Ma (Davydov et al., 2004). This age has since been refined to 298.9 + 0.31/–0.15 Ma (Ramezani et al., 2007). The precise age presented here for the Great Whin Sill complex is consistent with stratigraphic observations of intrusion into Wesphalian B–C units, determined to be < ~309 Ma according to a variety of low-precision age constraints (Menning et al., 2000). The observation of clasts of Whin dolerite within the basal

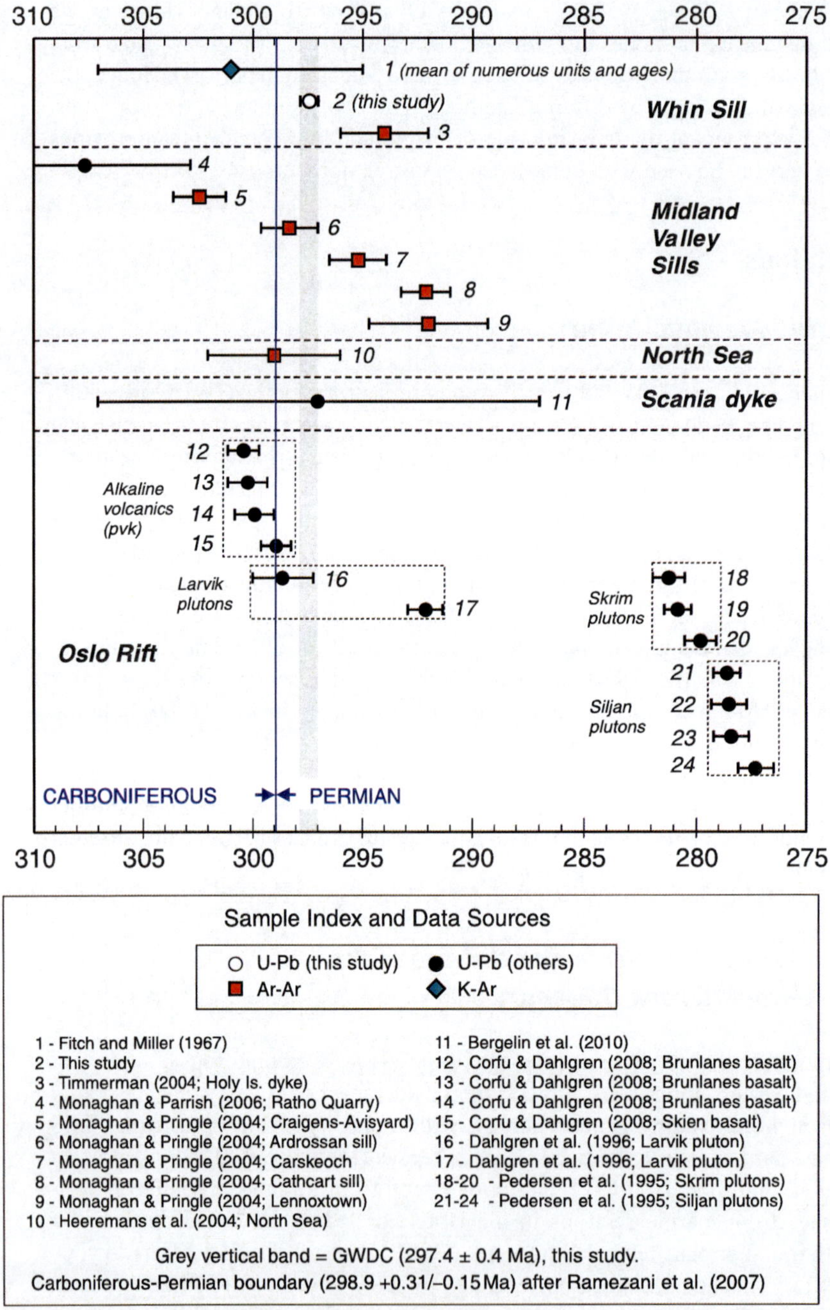

Fig. 26.3 Comparison of published U-Pb, Ar-Ar and K-Ar ages for volcanics, sills and dykes in NW Europe during the late Carboniferous – early Permian. *Grey vertical band* represents age and errors of GWDC sill dated in this study. Carboniferous-Permian boundary after Ramezani et al. (2007)

Permian Brokram formation (Holmes and Harwood, 1928; Dunham, 1932) provides a maximum age for this stratum and hence a minimum estimate for the age of Permo-Carboniferous boundary within the UK. The new datum we present for the Whin Sill (297.4 ± 0.4 Ma) agrees with the less precise age estimates for the Permo-Carboniferous boundary in Europe between approximately 295 and 305 Ma (Fig. 26.3; see also summary by Ramezani et al., 2007). The reversed magnetic polarity of the GWDC (Storetvedt and Gidskehaug, 1969; Liss et al., 2004) is in accord with intrusion during the Kiaman reversed superchron and provides firm geochronological support for establishment of this long-lived reversed interval before 297.4 Ma (*c.f.* Opdyke et al., 2000).

Permo-Carboniferous Rifting and Magmatism in Europe

The new 297.4 ± 0.4 Ma age for the Whin sheet provides a key piece of evidence in unravelling the history of Permo-Carboniferous rifting and magmatism that was widespread across Europe (Wilson et al., 2004 and references therein) in the context of possible plume (Ernst and Buchan, 1997; Torsvik et al., 2008) or extensional models (e.g. Neumann et al., 2004). In Fig. 26.3 we compare the GWDC age with precise U-Pb ages from the Oslo rift area (Pedersen et al., 1995; Dahlgren et al., 1996; Corfu and Dahlgren, 2008) as well as ^{40}Ar–^{39}Ar and U-Pb ages for the Midland Valley sills (MVS) of Scotland (Monaghan and Pringle, 2004; Monaghan and Parrish, 2006), and recent U-Pb baddeleyite results for a Scania dyke in southern Sweden (Bergelin et al., 2010). The numerous earlier Rb-Sr, K-Ar and ^{40}Ar–^{39}Ar ages for this region, summarised by Timmerman (2004) are essentially too imprecise and unreliable to provide rigorous time constraints on these units, both in terms of absolute ages and for understanding the true duration of magmatism in any given igneous province.

As Fig. 26.3 illustrates, the age for GWDC falls approximately in the middle of a 10 m.y. span of ^{40}Ar–^{39}Ar ages determined for a suite of Midland Valley Sills (Monaghan and Pringle, 2004), which range from ~302–292 Ma. Some central North Sea basalt flows also appear to be of similar age (c. 299 Ma, Ar-Ar; Heeremans et al., 2004). A single U-Pb zircon age for a tholeiitic Midland Valley Sill (Ratho Quarry) – at 307.6 ± 4.8 Ma – is distinctly older than the Ar-Ar ages measured on alkaline members of the MVS (Monaghan and Parrish, 2006), and these authors ascribed this to a time gap coupled to a change in tectonic regime between the late Carboniferous and the early Permian in the Midland Valley region. The tholeiitic members of the MVS have long been thought to be contemporaneous with the Whin Sill magmas. However, the two tholeiitic units that are currently dated accurately by U-Pb methods – the Ratho sill (307.6 Ma) and the Whin dolerite (297.4 Ma), demonstrate that the timing of emplacement of similar magma compositions between the two regions differed, possibly by as much as 10 m.y. Unless the Ar-Ar ages for the MVS alkaline magmatism are grossly inaccurate, tectonomagmatic models relating the two areas require an explanation for the contemporaneous nature of alkalic MVS and tholeiitic Whin magmatism over a relatively small area in northern Britain, as alluded to by Monaghan and Parrish (2006).

Results from previous geochronological studies of the WNW- to NW-trending Scania mafic dykes of southern Sweden, which converge to a point near the Skagerrak graben south of the Oslo Rift, have suggested ages near 294 Ma (K–Ar; Klingspor, 1976) and ca. 285 Ma (Timmerman et al., 2009). These have broadly supported the notion that Scania dyke magmatism was essentially coeval with other Permo-Carboniferous igneous activity in northern Europe, though the issue of whether it was precisely coeval with the earliest Oslo Rift magmatism and the Whin (and MVS) sill event has remained unclear. Recently, Bergelin et al. (2010) have identified an older pulse of Scania dykes with Ar-Ar ages clustered near 305–298 Ma (a second pulse occurs near 290 Ma). Moreover, these authors have carried out a parallel U-Pb study of some of the same dykes; preliminary baddeleyite dating of one of these dykes has yielded an age of 297 ± 10 Ma (Bergelin et al., 2010), identical (though with large uncertainty) to the GWDC result presented here (Fig. 26.3). A second Scania dyke has given a preliminary U-Pb age of ca 303 Ma and both dates agree well with the timing of initial magmatism in the Oslo Rift (see below). It seems plausible, therefore, that some of these sill and dyke complexes could have been emplaced during a relatively brief pulse of mafic magmatic activity, over a timespan of only a few million years.

Following this regional perspective, the U-Pb age for the GWDC reported here is most similar to a series of reliable U-Pb ages determined on perovskite (and titanite) from earliest alkalic basalts formed during initial extensional magmatism within the Oslo Rift (Fig. 26.3). Corfu and Dahlgren (2008) established precise ages for four early volcanic members of "stage 2" basalts in the Brunlanes and Skien series, bracketing magmatism narrowly between 300.4 ± 0.7 and 298.9 ± 0.7 Ma. The GWDC age, at 297.4 ± 0.4 Ma, overlaps more directly with ages defined by nested intrusive members of the Larvik plutonic complex (and much of the likely coeval, voluminous rhomb porphyry lavas) that followed these early Oslo Rift basalt sequences, at 298.6 ± 1.4–292.1 ± 0.8 Ma (Dahgren et al., 1996). In contrast, the GWDC is distinctly older than syenite intrusions associated with the Skrim and Siljan plutonic complexes of the southern Oslo Rift, which define a separate, younger episode of rift-related magmatic activity between 281.2 ± 0.6 Ma to 279.8 ± 0.7 Ma (3 ages), and 278.6 ± 0.8 to 277.3 ± 0.8 Ma (4 ages; Pedersen et al., 1995) (Fig. 26.3).

For the Oslo Rift, tectonic models have been advanced in support of limited lithospheric extension as a consequence of (post-Variscan) orogenic collapse following Pangean collisional assembly. Geochemical arguments (e.g. Kirstein et al., 2006) have also been forwarded to support this hypothesis over a model involving a thermally anomalous mantle (i.e. plume) as a mechanism for driving extension and aborted rifting. However, the potentially short period of emplacement, and large volumes of magma associated with the main phase of the Oslo Rift, Midland Valley of Scotland and Whin Sill, together with the Scania dykes of southern Sweden, and the converging triple-junction (rift-rift-rift) geometry of all of these at 300 Ma offer compelling reasons why this magmatism may have a mantle plume origin (Ernst and Buchan, 1997; Neumann et al., 2004; Torsvik et al., 2008). The new U-Pb GWDC age puts this intrusion firmly within this brief pulse of volcanism and suggests that most or all of the magma associated with this event formed in a period of 10 m.y. or less. It is notable that most of the less precise Rb-Sr determinations of dykes and lavas associated with the plutonic rocks yield ages within this time period but with large uncertainties. The widespread nature and timing of the regional magmatism described here could be consistent with the Ernst and Buchan (1997) model of formation in a giant radiating dyke swarm focussed on a plume head north of Denmark, referred to as the "Jutland event" by these authors, or as the Skagerrak-Centered Large Igneous Province of Torsvik et al. (2008). In this scenario, the GWDC would represent a sill fed distally from a central plume source some 500 km to the east. If such a model is applicable, then it becomes inappropriate to use the magma composition of the GWDC to infer lithospheric extension directly beneath northern England.

Implications Regarding Local Basin Evolution and Post-Whin Mineralization

In terms of the local evolution of the Northumbrian sedimentary basin, Johnson and Dunham (2001) concluded from structural observations that the timing of GWDC emplacement was bracketed narrowly between a period of WSW-ENE directed folding responsible for the "Burtreeford Disturbance" and Holburn – Lemmington anticlines (Fig. 26.1), and a later, subsequent phase of tilting linked to uplifts of the Teesdale and Cheviot domes. Hence, the intrusion provides an important time marker for the regional deformation events involved in the evolution of this sedimentary basin, e.g., Corfield et al. (1996).

Finally, the cross-cutting, post-Whin nature of veins associated with the North Pennine Orefield (Johnson and Dunham, 2001) implies that 297.4 ± 0.4 Ma is an effective maximum for the age of this widespread mineralization, an important event which remains otherwise poorly constrained geochronologically.

Acknowledgements We thank G.A.L. Johnson and K. McCaffrey for comments on earlier versions of the manuscript. We also thank A. Monaghan and M. Timmerman for early discussions regarding correlations with Midland Valley sills and other Permo-Carboniferous rocks in NW Europe. MAH gratefully acknowledges numerous enlightening discussions with K. Buchan, and with I. Bergelin for providing details of Scania dykes. We are indebted to Henry C. Halls for an insightful review, and to R. Srivastava for constructive editorial handling of this manuscript.

References

Bergelin I, Obst K, Hamilton M, Söderlund U (2010) $^{40}Ar/^{39}Ar$ whole rock and U-Pb baddeleyite dating of Permian and Carboniferous dolerite dykes in Scania (southern Sweden), and their relation to Pangea post-collisional processes. 6th Int Dyke Conf, Varanasi, India, Program with Abstracts:13

Corfield SM, Gawthorpe RL, Gage M, Fraser AJ, Besly BM (1996) Inversion tectonics of the Variscan foreland of the British Isles. J Geol Soc London 153: 17–32

Corfu F, Dahlgren S (2008) Perovskite U-Pb ages and the Pb isotopic composition of alkaline volcanism initiating the Permo-Carboniferous Oslo Rift. Earth Planet Sci Lett 265: 256–269

Dahlgren S, Corfu F, Heaman LM (1996) U-Pb isotopic time constraints and Hf and Pb isotope source characteristics of the Larvik Plutonic Complex, Oslo paleorift. Geodynamic and geochemical implications for the rift evolution. V. M. Goldschmidt Conf, Jour Conf Abstracts: 120, Cambridge Publ

Davydov VI, Wardlaw BR, Gradstein FM (2004) The Carboniferous period. In: Gradstein FM, Ogg JG, Smith AG, (eds) A Geologic Time Scale 2004. Cambridge University Press, Cambridge (UK): 222–248

Dubey VS, Holmes A (1929) Estimates of the ages of the Whin sill and Cleveland Dyke by the helium method. Nature 124: 447

Dunham KC (1932) Quartz-dolerite pebbles (Whin Sill type) in the Upper Brockram. Geol Mag 69: 425–427

Dunham KC (1990) Geology of the northern Pennine orefield (1st Edition). Vol. 1, Tyne & Stainmore (2nd Edition). Mem Geol Surv Great Britain

Dunham AC, Copley PA, Strasser-King VEH (1972) Submicroscopic exsolution lamellae in pyroxenes of the Whin Sill, northern England. Contrib Mineral Petrol 37: 211–220

Dunham AC, Strasser-King VEH (1982) Later carboniferous intrusions of northern Britain. In: Sutherland DS, (ed) Igneous Rocks of the British Isles. Wiley, New York: 277–283

Ernst RE, Buchan KL (1997) Giant radiating dyke swarms: Their use in identifying pre-Mesozoic large igneous provinces and mantle plumes. In: Mahoney JJ, Coffin MF (eds) Large Igneous Provinces, Geophysical Monograph 100: 297–334

Fitch FJ, Miller J (1967) The age of the Whin Sill. Geol J 5: 233–250

Francis EH (1982) Magma and sediment I. Emplacement mechanism of late Carboniferous tholeiite sills in northern Britain. J Geol Soc London 139: 1–20

Goulty NR (2005) Emplacement mechanism of the Great Whin and Midland Valley dolerite sills. J Geol Soc London 162: 1047–1056

Hamilton MA, Pearson DG, Thompson RN, Kelley SP, Emeleus CH (1998) Rapid eruption of the Skye lavas inferred from precise U-Pb and Ar-Ar dating of the Rum and Cuillin plutonic complexes. Nature 394: 260–263

Heaman LM, LeCheminant AN (1993) Paragenesis and U-Pb systematics of baddeleyite (ZrO_2). Chem Geol 110: 95–126

Heeremans M, Timmerman MJ, Kirstein LA, Faleidel JI (2004) New constraints on the timing of late Carboniferous-early Permian volcanism in the central North Sea. In: Wilson M, Neumann ER, Davies GR, Timmerman M, Heeremans M, Larsen B (eds) Permo-Carboniferous Rifting and Magmatism in Europe, Geol Soc Spec Publ 223: 177–193

Holmes A, Harwood HF (1928) The age and composition of the Whin Sill and related dykes of the north of England. Mineral Mag 21: 493–552

Jaffey AH, Flynn KF, Glendenin LE, Bentley WC, Essling AM (1971) Precision measurement of half-lives and specific activities of ^{235}U and ^{238}U. Phys Rev 4: 1889–1906

Johnson GAL (1995) Robson's geology of northeast England. Trans Nat Hist Soc Northumbria 56: 226–391

Johnson GAL, Dunham KC (2001) Emplacement of the great Whin Dolerite Complex and the Little Whin Sill in relation to the structure of northern England. Proc Yorkshire Geol Soc 53: 177–186

Kirstein L, Davies GR, Heeremans M (2006) The petrogenesis of Carboniferous-Permian dyke and sill intrusions across northern Europe. Contrib Mineral Petrol 152: 721–742

Klingspor I (1976) Radiometric age determination of basalts, dolerites and related syenite in Skåne, southern Sweden. Geol Föreningens Stockholm Förhandlingar 98: 195–216

Krogh TE, Corfu F, Davis DW, Dunning GR, Heaman LM, Kamo SL, Machado N, Greenough JD, Nakamura E (1987) Precise U-Pb isotopic ages of diabase dykes and mafic to ultramafic rocks using trace amounts of baddeleyite and zircon. In: Halls HC, Fahrig WF (eds) Mafic Dyke Swarms, Geol Assoc Canada Spec Pap 34: 147–152

Liss D, Owens WH, Hutton DHW (2004) New palaeomagnetic results from the Whin Sill complex: Evidence for a multiple intrusion event and revised virtual geomagnetic poles for the late Carboniferous for the British Isles. J Geol Soc London 161: 927–938

Ludwig KR (1998) On the treatment of concordant uranium-lead ages. Geochim Cosmochim Acta 62: 665–676

Ludwig KR (2003) Isoplot 3.00: A Geochronological Toolkit for Microsoft Excel. Berkeley Geochronology Center, Special Publication No. 4 (rev. 3.00)

Menning M, Weyer D, Drozdzewski G, van Ameron HWJ, Wendt I (2000) A Carboniferous Time Scale 2000: Discussion and use of geological parameters as time indicators from central and western Europe. Geol Jahrb A 156: 3–44

Miller JA, Mussett AE (1963) Dating basic rocks by the potassium-argon method: The Whin Sill. Geophy J 7: 547–550

Monaghan AA, Parrish RR (2006) Geochronology of Carboniferous-Permian magmatism in the Midland Valley of Scotland: Implications for regional tectonomagmatic evolution and the numerical timescale. J Geol Soc London 163: 15–28

Monaghan AA, Pringle MS (2004) $^{40}Ar/^{39}Ar$ geochronology of Carboniferous-Permian volcanism in the Midland Valley, Scotland. In: Wilson M, Neumann ER, Davies GR, Timmerman M, Heeremans M, Larsen B (eds) Permo-Carboniferous Rifting and Magmatism in Europe, Geol Soc Spec Publ 223: 219–242

Neumann E-J, Wilson M, Heeremans M, Spencer EA, Obst K, Timmerman MJ, Kirstein L (2004) Carboniferous-Permian rifting and magmatism in southern Scandinavia, the North Sea and northern Germany: A review. In: Wilson M, Neumann ER, Davies GR, Timmerman M, Heeremans M, Larsen B (eds) Permo-Carboniferous Rifting and Magmatism in Europe, Geol Soc Spec Publ 223: 11–40

Opdyke ND, Roberts J, Claoue-Long J, Irving E, Jones P (2000) Base of the Kiaman: Its definition and global stratigraphic significance. GSA Bull 112: 1315–1341

Pedersen LE, Heaman LM, Holm PM (1995) Further constraints on the temporal evolution of the Olso Rift from precise U-Pb zircon dating in the Siljan-Skrim area. Lithos 34: 301–315

Ramezani J, Schmitz MD, Davydov VI, Bowring SA, Snyder WS, Northrup CJ (2007) High-precision U–Pb zircon age constraints on the Carboniferous–Permian boundary in the southern Urals stratotype. Earth Planet Sci Lett 256: 244–257

Randall BAO (1989) Doleriite-pegmatites from the Whin Sill near Barrasford, Northumberland. Proc Yorkshire Geol Soc 47: 249–265

Renne PR, Swisher CC, Deino AL, Karner DB, Owens TL, DePaolo DJ (1998) Intercalibration of standards, absolute ages and uncertainties in $^{40}Ar/^{39}Ar$ dating. Chem Geol 145: 117–152

Smythe JA (1930) A chemical study of the Whin Sill. Trans Nat Hist Soc Northumberland, Durham and Newcastle upon Tyne 7: 16–150

Storetvedt KM, Gidskehaug A (1969) The magnetisation of the Great Whin Sill, northern England. Phys Earth Planet Interiors 55: 93–105

Thorpe RS, Macdonald R (1985) Geochemical evidence for the emplacement of the Whin Sill complex of northern England. Geol Mag 2: 105–114

Timmerman MJ (2004) Timing, geodynamic setting and characteristics of Permo-Carboniferous magmatism in northern Europe: A review. In: Wilson M, Neumann ER, Davies GR, Timmerman M, Heeremans M, Larsen B (eds) Permo-Carboniferous Rifting and Magmatism in Europe, Geol Soc Spec Publ 223: 41–74

Timmerman MJ, Heeremans M, Kirstein L, Larsen BT, Spencer-Dunworth E-A, Sundvoll B (2009) Linking changes in tectonic style with magmatism in northern Europe during the late Carboniferous to latest Permian. Tectonophysics 473: 375–390

Tomkieff SI (1929) A contribution to the petrology of the Whin Sill. Mineral Mag 22: 100–120

Torsvik TH, Smethurst MA, Burke K, Steinberger B (2008) Long term stability in deep mantle structure: Evidence from the ~300 Ma Skagerrak-centered large igneous province (the SCLIP). Earth Planet Sci Lett 267: 444–452

Wilson M, Neumann ER, Davies GR, Timmerman M, Heeremans M, Larsen B, (eds) (2004) Permo-Carboniferous Rifting and Magmatism in Europe. Geol Soc Spec Publ, London (UK): 223

Chapter 27
The Melville Bugt Dyke Swarm of Greenland: A Connection to the 1.5-1.6 Ga Fennoscandian Rapakivi Granite Province?

H.C. Halls, M.A. Hamilton, and S.W. Denyszyn

Introduction

The continental fit between Laurentia and Fennoscandia during the Precambrian has been hampered by the lack of definitive geological correlations and by paleomagnetic data that are either imprecise or missing at critical times in the geological record. A major time period from about 1.65 to 1.45 Ga is well represented both paleomagnetically and geochronologically in Fennoscandia and particularly by the rapakivi granite province (Rämö and Haapala, 1995). In Laurentia this time period is weakly represented although the 1.59 Ga Western Channel Diabase in NW Canada is an exception (Hamilton and Buchan, 2010; Irving et al., 1972).

Current fits for Laurentia – Fennoscandia employing both geological correlations during the time period both before and after that under consideration suggest that there has been a long-lived connection between the two continental entities between about 1.8 and 1.3 Ga. To date most fits on paleomagnetic grounds are ones in which Laurentia sits in northern latitudes and is rotated clockwise with respect to its present position, so that northern Greenland becomes juxtaposed with northern Norway (e.g. Buchan et al., 2000; Pesonen et al., 2003).

In this paper we present paleomagnetic, U-Pb age and geochemical data from the Melville Bugt (Bay) dyke swarm that trends NNW along the western coast of Greenland (Nielsen, 1990) for about 1,000 km (Fig. 27.1). Prior to our study, only a single paleomagnetic site had been collected from the Melville Bugt dykes (Hamilton et al., 2004) but the paleomagnetic results could not be readily reconciled

H.C. Halls (✉)
Department of Geology, University of Toronto, Toronto, ON, Canada M5S 3B1;
Department of Chemical and Physical Sciences, University of Toronto Mississauga, Mississauga, ON, Canada L5L 1C6
e-mail: henry.halls@utoronto.ca

M.A. Hamilton
Jack Satterly Geochronology Laboratory, Department of Geology, Earth Sciences Centre, Toronto, ON, Canada M5S 3B1

S.W. Denyszyn
Berkeley Geochronology Center, Berkeley, CA 94709, USA

R.K. Srivastava (ed.), *Dyke Swarms: Keys for Geodynamic Interpretation*,
DOI 10.1007/978-3-642-12496-9_27, © Springer-Verlag Berlin Heidelberg 2011

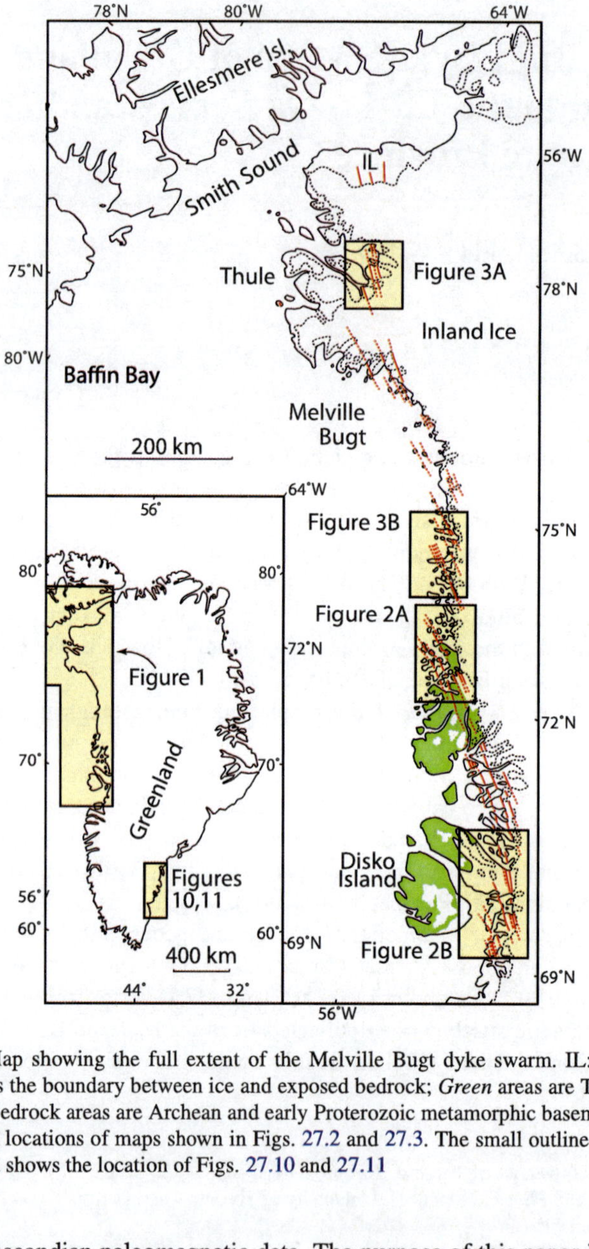

Fig. 27.1 Map showing the full extent of the Melville Bugt dyke swarm. IL: Inglefield Land. Dotted line is the boundary between ice and exposed bedrock; *Green* areas are Tertiary volcanics, uncoloured bedrock areas are Archean and early Proterozoic metamorphic basement rocks. *Boxed areas* are the locations of maps shown in Figs. 27.2 and 27.3. The small outlined area in the map of Greenland shows the location of Figs. 27.10 and 27.11

with Fennoscandian paleomagnetic data. The purpose of this paper is to throw further light upon Fennoscandia-Greenland connections during a critical time period for which paleomagnetic data, particularly in Laurentia, are sparse.

Towards the north the Melville Bugt swarm appears to die out; it is present in Inglefield Land (IL in Fig. 27.1) where dykes are relatively narrow (generally less than 15 m wide, Dawes, 2004) compared to dykes averaging about 100 m across at the eastern end of Inglefield Bredning (fjord) (northernmost inset map in Fig. 27.1). The swarm cannot be traced into Canada as it passes under Paleozoic strata on Inglefield Land (Dawes, 2004). However if Ellesmere Island has been sinistrally offset during Mesozoic time about 200 km from Greenland along the hypothesized Nares Strait fault (e.g. Denyszyn et al., 2009), then the swarm might be expected to cross Ellesmere Island, but no dykes of the expected trend and width are encountered (Frisch, 1984a, b; Denyszyn, 2008 and helicopter reconnaissance by HCH and SD in 2005 and 2006). To the south the swarm passes under the Greenland ice cap (Fig. 27.1). The western margin of the swarm is exposed but the eastern is not so that neither the true width nor length of the swarm are known. Individual dykes can be up to 100–200 m wide and one extends for at least 400 km and shows a remarkably constant geochemistry over this entire distance (Kalsbeek and Taylor, 1986). The dykes are described by Nielsen (1990) as alkaline trachybasalts.

Field Surveys

In the Summer of 2008 fieldwork by two of us (HCH and MAH) resulted in the collection of 11 paleomagnetic sites from the Melville Bugt swarm (Figs. 27.2 and 27.3), using a portable field drill. A few hand samples were also collected. The collection strategy was to collect most samples from both chilled margins if available because margins, if fresh, tend to preserve primary remanences more so than interiors due to the smaller grain size of remanence carriers (in this case magnetite as deduced from thermomagnetic data by Denyszyn, 2008). Samples were collected from each dyke interior for paleomagnetic comparison but also for U-Pb geochronology and petrography. It has been shown that where the host rock has remained un-melted, chilled margins of a diabase dyke carry a primary geochemical signature which is diagnostic of that dyke and different from other members of the swarm (Halls, 1986; Halls et al., 2008). Therefore major, minor and trace element analyses have been obtained on all chilled margin samples, so that any dykes that have been sampled by more than one paleomagnetic site might be recognized and paleomagnetic statistics adjusted to allow for this duplication.

This collection augments a previous one by HCH and SD in the Thule area of Greenland (Fig. 27.3) where three sites were collected in 2004 and 2005, and a single site collected in the Disko Island area ("Torsukattak") in the 400 km long dyke at the southern end of the swarm for which a U-Pb age also exists (Fig. 27.2, Hamilton et al., 2004).

Most paleomagnetic sites were collected using both a magnetic and sun compass. The only sites lacking sun orientations, due to cloudy weather, were sites MB6-9, MB 11, 12 and the one north of Disko Island. The presence of sun measurements is extremely important at latitudes of 73–77°N because the steep inclination of the earth's field necessitates that the rocks be sufficiently non-magnetic that they do not perturb the relatively weak horizontal field that determines the magnetic compass reading.

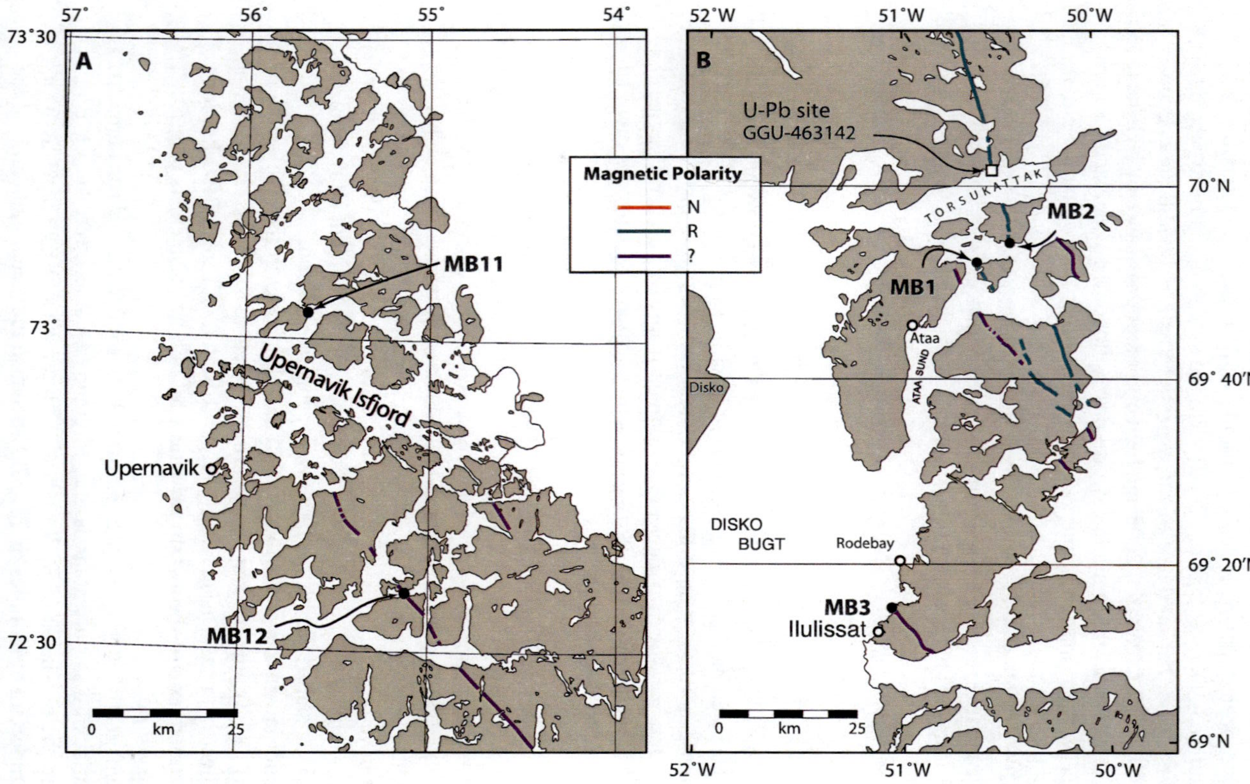

Fig. 27.2 Maps of the southern part of the sampling region showing location of paleomagnetic sites in Melville Bugt dykes: (**a**) Sites MB11 and 12 (**b**) Sites MB1 to 3. *Brown areas* represent-ice-free land; *blue areas* are ice covered

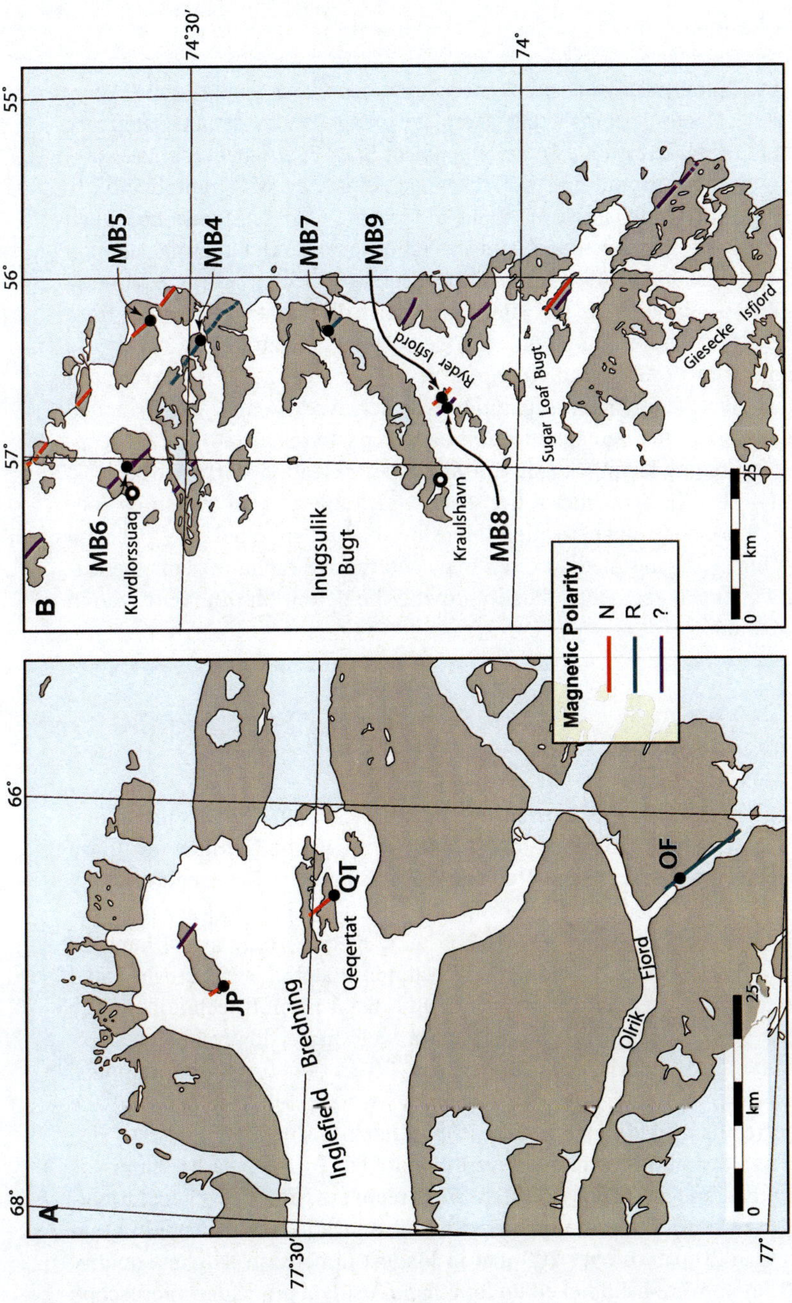

Fig. 27.3 Maps from the northern half of the study area, showing the location of paleomagnetic sites in Melville Bugt dykes. **(a)** Sites OF, QT and JP, **(b)** Sites MB4 to 9. Colours as in Fig. 27.2

Laboratory Methods

Paleomagnetism

Cores or blocks ("samples") were prepared as cylinders, 2.45 × 2.45 cm ("specimens") and at least one specimen per sample was subject to detailed stepwise alternating field demagnetization using a Schonstedt SD-1 instrument and the magnetization measured on a modified DIGICO magnetometer with reproducibility down to magnetization intensities of about 10^{-3} Am^{-1}. At least one specimen per site was thermally demagnetized using a Schonstedt TSD-1 thermal demagnetizer and if the demagnetization proved more successful than AF at separating components, further samples in the site were thermally treated. Samples from one site (MB8) were subject to liquid nitrogen demagnetization (Ozima et al., 1964; Borradaile et al., 2004) in an attempt to remove a strong PEF-like overprint prior to AF treatment. Component separation was achieved by applying Principal Component Analysis to the demagnetization data (Kirschvink, 1980), with acceptance of linear segments having maximum angles of deviation <10° and at least four data points. The AF demagnetization data were subject to an averaging routine (see Halls,1986), once magnetization intensities had fallen to below about 10% of the NRM. This averaging method often extended the AF range of remagnetization circles and/or linear segments, thus improving the definition of magnetization component directions.

U-Pb Geochronology

Most U-Pb analytical work was carried out in the Jack Satterly Geochronology Laboratory (JSGL) at the University of Toronto, with methods following those described in detail by Hamilton and Buchan (2010). Samples for geochronology described here were collected from the freshest, coarsest-grained portions of dykes that ranged in thickness between 100 and 200 m. Generally, only a few hundred grams total of either fresh drill core or hand sample material were crushed and then finely ground using standard jaw crusher and disk/ring mill techniques. The resulting coarse powder was then spoon-fed in a wet slurry slowly over a shaking (Wilfley) table (see Söderlund and Johansson, 2002), left for several minutes, and then a thin band of heavy mineral concentrates were pipetted directly off the table and inspected using a binocular microscope. Using this method, no heavy liquids or magnetic separation techniques were employed. In most cases, hundreds of good quality baddeleyite (ZrO_2) crystals were recovered for U-Pb geochronology from the small sample volumes processed, and showed a serial range in grain size from less than 20 μm to over 200 μm in longest dimension. Highest quality grains of baddeleyite were handpicked under ethanol using a binocular microscope and selected for isotopic analysis. Prior to dissolution, weights of selected fractions were estimated by use of a scaled digital photographic measurement of the length and breadth of each component grain, and an estimate of the maximum thickness, together with the known density.

Baddeleyite grains were prepared for isotope dilution – thermal ionization mass spectrometry (ID-TIMS) analyses at the JSGL by first washing the fractions in 7 N HNO_3, Milli-Q water and acetone on Parafilm, then loading into a Teflon bomb in a drop of HNO_3. The samples were dissolved in concentrated HF mixed with $^{205}Pb/^{235}U$ spike, and, following conversion to chloride, the samples were loaded, without column chemistry, directly onto Re filaments with Si-gel and phosphoric acid. Isotope ratios were measured on a VG354 mass spectrometer using a Daly detector equipped with digital ion counting. System dead-time corrections during this period were 21 ns for Pb and U. Detector characteristics were monitored during the analytical period using the SRM-982 Pb and CBNM 072/6 U standards. Corrections were 0.05%/amu for Daly mass discrimination and 0.10%/amu for thermal mass discrimination. JSGL procedural blanks are normally at the 0.5 pg and 0.1 pg level for Pb and U, respectively. The measured total common Pb in the samples was generally low and was assigned the common Pb composition of the lab blank (see footnote to Table 27.2).

Sample GGU-463142 was processed and analyzed at the Geological Survey of Canada in Ottawa in 2002. Methods for heavy mineral separation at that time involved crushing and grinding much larger samples (approximately 10 kg), followed by conventional Wilfley table and heavy liquid techniques, and isolation of least paramagnetic heavy minerals using a Frantz isodynamic magnetic separator. Details of this, and the GSC isotope dilution – thermal ionization mass spectrometry (ID-TIMS) methods for U-Pb geochronology, are provided in Hamilton et al. (2001, 2002). Pb and U from the larger fractions prepared for sample GGU-463142 were purified using ion exchange chromatography. Isotopic ratios were measured on a Finnigan-MAT 261 mass spectrometer, using Faraday detectors and a secondary electron multiplier. Lead fractionation was approximately 0.09%/amu. All common Pb was assumed to have the composition of the laboratory blank (see footnote to Table 27.2).

Comprehensive error estimation for each sample in this study was made by propagating all known sources of analytical error, and uncertainties in the composition of the laboratory blanks. Decay constants used are those of Jaffey et al. (1971). Age regressions and errors were calculated, and U-Pb Concordia diagrams were generated using the Microsoft Excel Add-in IsoPlot/Ex v.3.0 (Ludwig, 2003). Concordia ellipse errors and calculated ages are shown at the 2σ (95% confidence) level of uncertainty.

Results

Dyke Petrography

Petrographically all Melville Bugt dykes are virtually identical, being extremely fresh olivine basalt with an equigranular to sub-ophitic texture. They contain about

50% plagioclase with up to about 10% saussuritisation, 25% of a pink clinopyroxene, 10–15% fresh olivine with magnetite-rich cracks, 5% magnetite often associated with a deep brown biotite, and accessory minerals, developed especially where saussuritization is greatest: apatite, quartz, chlorite, epidote. Hornblende as an alteration product of clinopyroxene is largely absent and primary grains are rare. Only samples at site MB1 lack olivine.

Paleomagnetism

The full paleomagnetic results are presented in Table 27.1 and specific aspects of specimen behavior under demagnetization are illustrated in Fig. 27.4. More or less antipodal directions are present (Fig. 27.4a, b, f): the swarm can be divided into four dykes that have a down to the SW characteristic direction and five dykes that have intermediate negative inclinations towards the NNE. Those dykes having the steepest negative inclinations (see Fig. 27.4a, b) generally have the least overprinting by present Earth's field (PEF)-like components. Occasionally a specimen with a strong PEF overprint will attain a stable end-point which is shallower (Fig. 27.4c). Those samples that have a progressively larger PEF component fail to achieve end-points but move towards the NE assuming shallow inclinations of ±5° at the limit of AF demagnetization (e.g. Fig. 27.4d). Thermal demagnetization appears incapable of improving component separation (Fig. 27.4e). One site (MB7) attains stable end points at a shallow inclination for both AF and thermal treatment (Fig. 27.4g, h). Another site (MB8) despite having several beautifully exposed narrow fine-grained satellite dykes has an extremely intense NRM ($\sim10^2$ Am^{-1}) overprint that was almost vertical and suspected to be a PEF component. Samples from this site were cooled in field-free space down to liquid nitrogen temperatures (120°K) in an effort to remove the PEF component prior to AF treatment (see Borradaile et al., 2004 for an example of applying this method to diabase). The results (Fig. 27.4i, j) show treatments for two specimens from the same sample. In both cases the low temperature treatment removes a significant fraction (>60%) of the vertical overprint, but the remanence remaining has widely different directions. After step-wise AF demagnetization this remanence gradually moves back to the PEF one. The method did not result in the recovery of a coherent component underlying the PEF and therefore was discontinued.

A summary of sample paleomagnetic directions at each site is given in Fig. 27.5. Note that despite most sites moving from a steep down component, only site MB2 is streaked in such a way to suggest the incomplete removal of the steep component and hence an underestimate of the negative inclination. Two further sites (MB1 and 7) also showed an unexplained streaking in the direction of declination, but one that would not affect the inclination.

In terms of paleomagnetic components, the two antipodal ones (A and B) are considered to be primary because each characterises different dykes that are geographically intermixed (Fig. 27.3b). In addition the chilled margins show the least contamination with the PEF and other secondary components. The host rocks,

Table 27.1 Paleomagnetic results

Site	LAT (°N)	LON (°W)	Width (M)	Strike (°)	DIP	NT	Primary component N	DEC (°)	INC (°)	α_{95} (°)	k	PLAT (°N,–°S)	PLON (°E)	dp (°)	dm (°)	λ (°)	DEC′	INC′	SITE	Secondary component N	DEC (°)	INC (°)	α_{95} (°)	k
MB1	69.8669	50.6201	7.5	345	90	8	8	39.3	–44.5	7	64	10.1	274.1	5.5	8.8	–26	33.1	–40.1	MB1					
MB2	69.8984	50.4479	110	NW	>80°SW	6	4	26.9	–14.9	11.7	62	–10.4	282.5	6.2	12.0	–8	22.2	–7	MB2					
MB3	69.2668	51.0375	50	330	90	7													MB3	6	33.7	47.5	13.5	26
MB4	74.4829	56.3348	60	NW	90	10	5	30.3	–29.3	9.1	72	2.2	274.6	5.5	10.0	–16	31.6	–28.3	MB4					
MB5	74.5579	56.2202	50	308	80°SW	7	6	209.1	46.6	4.9	186	14.2	277.5	4.1	6.3	28	210.3	46	MB5					
MB6	74.5834	57.0260	55	340	90	8				Erratic									MB6	3	204.7	–70.7	15.3	66
MB7	74.2886	56.2729	70	NW	90	9	5	51.9	–8.5	9.6	64	–5.4	251.7	4.9	9.7	–4	53.1	–7.2	MB7					
MB8	74.1068	56.6731	45	315	90	17													MB8	6	154.9	89.2	5.2	166
MB9	73.9129	55.9418	135	320	90	10	6	201.2	50.1	5.6	142	15.7	285.2	5.0	7.5	31	202.1	48.8	MB9					
MB11	73.0373	55.6619	45	325	80°SW	12													MB11	6	226.2	–41.4	17.5	16
MB12	72.5981	55.1217	210	NW	~90	8				Erratic									MB12					
QT	77.4781	66.4256	100	307	80°NE	9	8	195.9	28.8	8.8	41	3.3	278.2	5.3	9.7	15	208.9	30.9	QT					
JP	77.5933	66.8817	70	325	NO CM	9	7	209.1	44.2	3	40	15	266.2	2.4	3.8	26	222.5	44.8	JP					
OF	77.1058	66.2670	200	330	85°SW	9	7	13.1	–6.7	10.5	25	–9.2	280.5	5.3	10.5	–3			OF					
OF	Margin					9	5	19.8	–23.2	7.1	118	–0.2	274.4	4.0	7.6	–12	31.5	–24.2	OF					
Mean pole position							9			8.7		5.0°N	273.8°E				33.1	–31.2	Mean					

Notes: North poles are given for SW-directed remanences but South poles for NE-directed ones. DEC′, INC′: Declination, Inclination for common site locality 75°N, 55°W. NT/N are number of samples collected per site/providing data. For the pole position, N is the number of sites and the confidence limit is A_{95}.

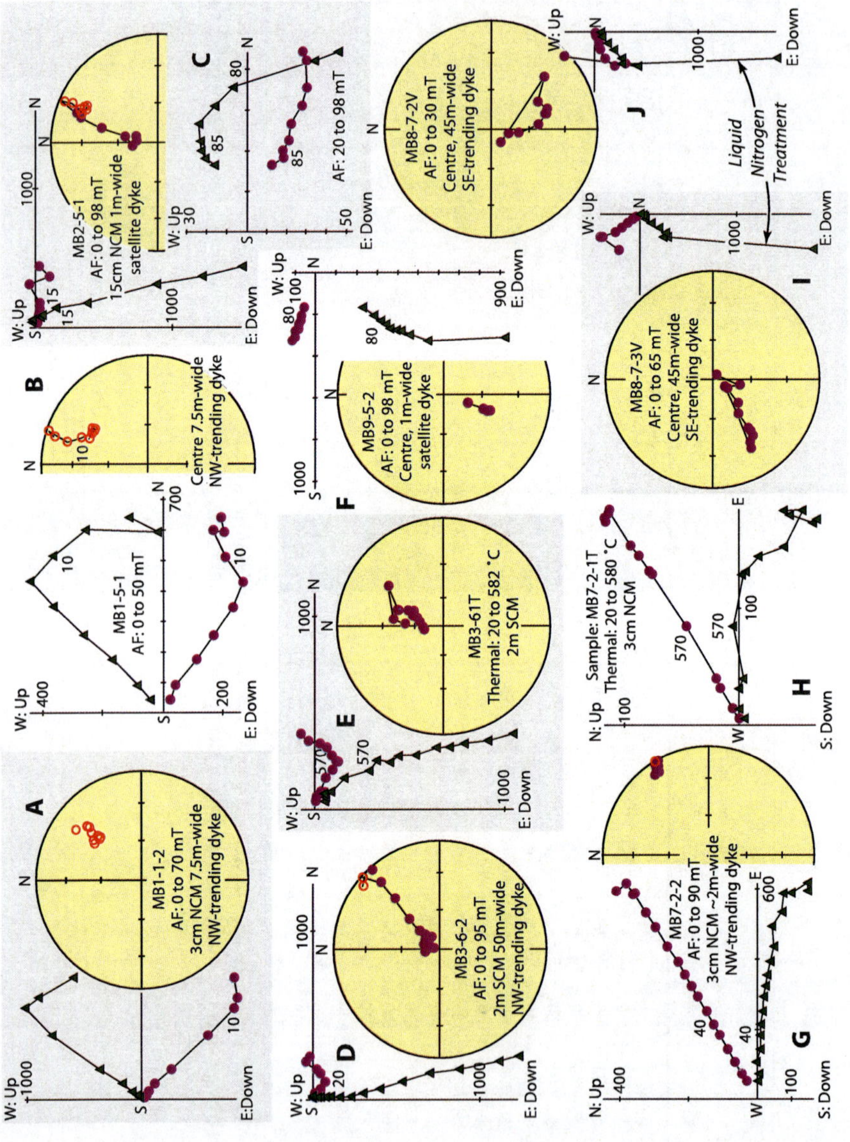

Fig. 27.4 (continued)

coarse-grained granitic gneisses in amphibolite facies, were not sampled (previous experience with such rocks leads to an expectation of weak and unstable remanences) so a baked contact test, needed to conclusively prove a primary origin for the remanence, was not attempted. However, the ca. 720 Ma Franklin dyke swarm is in close proximity to the Melville Bugt swarm in the Inglefield Bredning area and its paleomagnetic direction, shown to be primary by a positive baked contact test (Denyszyn et al., 2009) is distinct from that of the Melville Bugt dykes. This observation rules out any regional overprinting of Melville Bugt dykes after 720 Ma. Apart from the PEF-like remanence, defined at site MB8, the only other components found in Melville Bugt dykes were one of unknown origin (34°/48°, N = 6, α_{95} = 14°) found at site MB3, and a steep negative one to the SW (205°/–71°, N = 3, α_{95} = 15°) from site MB6 which may represent an overprint from the nearby Tertiary volcanic province (see Fig. 27.1 for the distribution of flood basalts and picrites) which gives a mean paleomagnetic direction of 163°/–70°, N = 44, α_{95} = 4° with some sites having a SW declination (Riisager et al., 2003). Therefore despite its proximity and the thickness of the volcanic sequence (5–6 km) the Tertiary volcanic province has had a minimal effect upon the magnetism of the Melville Bugt dykes.

Equal area stereonets of the paleomagnetic data for each Melville Bugt site are given in Fig. 27.5, and in Fig. 27.6 as a stereonet summary of the final Melville Bugt data, using site directions with one polarity reversed (Fig. 27.6a), with polarities separated (Fig. 27.6b), and with directions expressed as virtual geomagnetic poles (Fig. 27.6c).

Fig. 27.4 (continued) Equal area and vector diagrams showing the response of selected Melville Bugt specimens to AF and thermal demagnetization. On stereoplots *solid/open symbols* are downward/upward-pointing magnetizations. On vector diagrams *circles* represent the projection of the tip of the remanence vector, after each AF or thermal demagnetization step, onto the *horizontal plane* and *triangles* represent the projection onto a vertical W-E plane, except for diagrams g and h where the projection plane is N-S. SCM and NCM are, respectively, the southwestern and northeastern dyke margins (e.g. "3 cm NCM..." implies the sample was collected 3 cm inside the northeastern chilled margin). Diagrams (**a**) and (**b**): two samples from the same site that yield a well-defined stable end-point corresponding to an interpreted primary component. (**c**), (**d**) and (**e**) are examples of the response under AF and thermal demagnetization of samples that are strongly overprinted by a PEF-like component but which move to shallow inclinations with NE declination without attaining a stable end-point. (**f**) is an example of a specimen that attains a stable end point corresponding to a suspected primary component that is antipodal to that in (**a**) and (**b**). (**g**) and (**h**) are examples of AF and thermal demagnetization on the same sample which shows that in some sites like MB7 the shallow NE component appears to be real. (**i**) and (**j**) represent an attempt, using liquid nitrogen cooling in field-free space, to eliminate a strong vertical component at site MB8, in an attempt to recover an underlying older component. Note that while the specimens were strongly demagnetized by the technique, the residual remanence direction was quite different, and moreover that subsequent AF demagnetization merely returned the specimens to their initial NRM direction

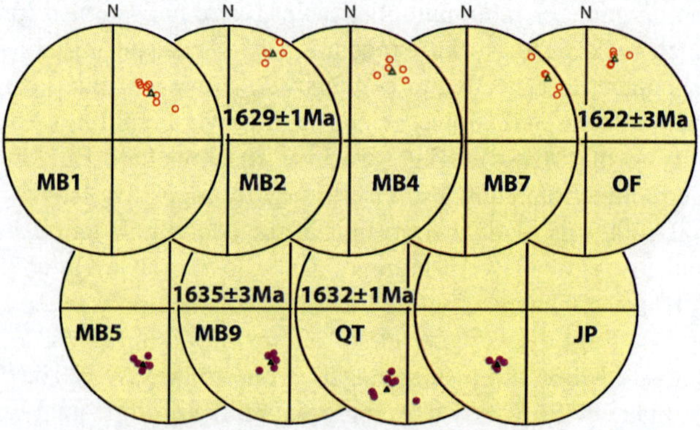

Fig. 27.5 Summary equal area stereonets showing sample directions for each of the Melville Bugt sites that are interpreted to be carrying a primary remanence. Note that only MB2 shows streaking in the expected direction between the primary and the PEF-like component. Sites MB1 and MB7 show streaking but not in a way that could account for error in the inclination of the primary direction due to unresolved components

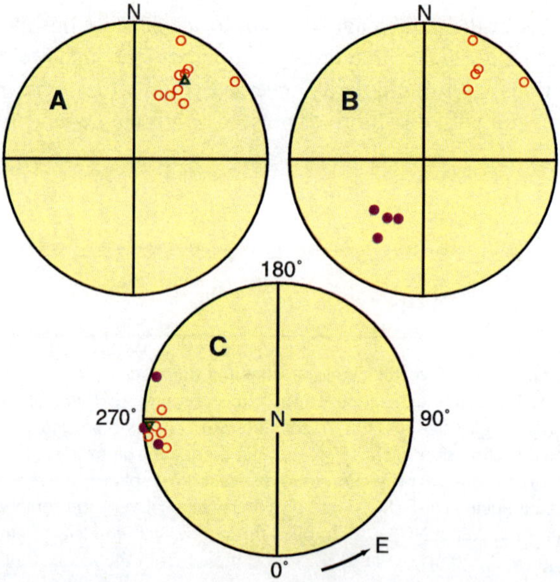

Fig. 27.6 Summaries of site mean data for Melville Bugt dykes: (**a**) Site directions with antipodal directions reversed; (**b**) Site directions without reversing data. *Solid/open* symbols are downward/upward pointing magnetizations; directions corrected to a common site at 77°N, 55°E, the *triangle* in A is the mean direction. (**c**) Virtual geomagnetic poles for each site, showing them to be clustered about the equator. View of earth is looking down from the N pole, with longitudes marked in degrees east. *Solid/open points* are poles in the southern/northern hemisphere. *Triangle* is the mean pole position

U-Pb Geochronology

We present the results of U-Pb isotopic dating of four Melville Bugt diabase dykes in Table 27.2, with graphic representations of the data in Concordia diagrams in Fig. 27.7(a–d). These four individual dykes span almost the entire known width of the swarm as recognized in west and northwest Greenland, as well as the north-south distribution from Inglefield Bredning to Ilulissat (a distance of roughly 1,000 km). Two of these are from dykes that yielded the down-to-the-SW remanence (MB9, QT), and two are from dykes that gave the up-to-the-NNE remanence (MB2, OF), both of which are interpreted to be primary.

Baddeleyite was successfully recovered from all four dykes. Invariably, these consisted of striated, pale to medium (and rarely dark) brown euhedral blades and blade fragments ranging up to ~150 μm in the longest dimension (but typically <60 μm). Fractions chosen for analysis normally comprised between 2 and 20 grains, but single grain fractions were also analyzed (Table 27.2). Uranium concentrations in the baddeleyite crystals are relatively high, mostly between 400 and 1,000 ppm, but reach up to ~2,200 ppm U. Estimated Th/U ratios for all fractions are low (<0.06) and typical of baddeleyite crystallizing in mafic or gabbroic magmas (Heaman and LeCheminant, 1993). Data from most of the samples plot close to the Concordia curve (all <2.5% discordant; Table 27.2, Fig. 27.7).

Site MB9 (sample HMB08-013): This NW-trending (320°) dyke reached up to 135 m in thickness where it was sampled on a small island (Paornarqutût) inside Ryder Isfjord, east of Kraulshavn – in the approximate geographic centre of the known Melville Bugt swarm (Figs. 27.1 and 27.3b). The middle of the dyke at this site was characterized by roughly 50 cm-diameter ovoidal patches of coarse to pegmatitic, locally ophitic gabbro rich in ilmenite and apatite. Baddeleyite was abundant in these segregations, which are interpreted to represent pockets of evolved, trapped melt. Zircon was not observed.

Data for four fractions, comprising between 4 and 6 grains of baddeleyite each, are variably discordant and regress to yield an upper intercept age of 1,635.0 ± 2.7 Ma (2σ, 76% probability of fit; Fig. 27.7a). This result is interpreted to represent the age of emplacement and crystallization of the Melville Bugt dyke at this site, and constrains the remanent magnetization direction here – SW declination and down ("normal" polarity) – to have been characteristic at this time. Although poorly constrained, the lower intercept of the U-Pb array, at ~215 Ma, may reflect disturbance during Mesozoic opening of the Labrador Sea.

Site QT (sample QT-1): The central portion of Qeqertat Island, at the head of Inglefield Bredning in northwestern Greenland, is cut by a conspicuous NW-trending (307°) dyke with a thickness reaching approximately 100 m (Fig. 27.3a). This region encompasses the most northern and eastern limits of demonstrable Melville Bugt dykes. The Qeqertat dyke may continue along strike as the same dyke sampled at site JP as both also exhibit normal magnetic polarity. However, their paleomagnetic directions are distinct at the 95% confidence level and the geochemistry of their chilled margins is dissimilar, suggesting two separate intrusions.

Table 27.2 U-Pb analytical results for baddeleyite from Melville Bugt diabase dykes, west Greenland

Sample fraction	Analysis no.	Description	Weight (µg)	U (ppm)	Th/U	Pb* (pg)	Pb_C (pg)	$^{206}Pb/^{204}Pb$	$^{206}Pb/^{238}U$	±2σ	$^{207}Pb/^{235}U$	±2σ	$^{207}Pb/^{206}Pb$	±2σ	$^{206}Pb/^{238}U$ Age (Ma)	±2σ	$^{207}Pb/^{235}U$ Age (Ma)	±2σ	$^{207}Pb/^{206}Pb$ Age (Ma)	±2σ	Disc. (%)	Corr. Coeff.
HMB08-013 (site MB9) – Ryder Isfjord; 320° strike																						
Bd-1	MAH8163	4 mbr long prismatic	0.2	428	0.009	23	0.5	3,189	0.28533	0.00063	3.9557	0.0128	0.10055	0.00019	1,618.1	3.1	1,625.2	2.6	1,634.3	3.6	1.1	0.8173
Bd-2	MAH8164	5 mbr short blades & frags	0.2	495	0.015	27	0.4	4,530	0.28593	0.00061	3.9613	0.0115	0.10048	0.00015	1,621.2	3.1	1,626.3	2.3	1,633.0	2.8	0.8	0.8617
Bd-3	MAH8165	6 mbr blades & frags	0.2	1,268	0.010	69	0.4	12,194	0.28611	0.00053	3.9661	0.0093	0.10054	0.00009	1,622.1	2.7	1,627.3	1.9	1,634.0	1.7	0.8	0.9273
Bd-4	MAH9012	6 mbr blades & frags	0.2	1,039	0.007	55	0.9	4,205	0.28092	0.00056	3.8888	0.0110	0.10040	0.00016	1,596.0	2.8	1,611.4	2.3	1,631.5	2.9	2.5	0.8443
QT-1 (site QT) - Qeqertat Island, inner Inglefield Fjord; 307° strike																						
Bd-1	MAH0100na	2 mbr long prismatic	0.5	537	0.040	54	0.2	5,332	0.28150	0.00068	3.8994	0.0118	0.10047	0.00014	1,598.9	3.4	1,613.6	2.5	1,632.7	2.5	2.3	0.9011
Bd-2	MAH0101	2 pbr thin blades	0.5	383	0.021	52	0.3	11,943	0.28324	0.00055	3.9220	0.0094	0.10043	0.00009	1,607.6	2.8	1,618.2	1.9	1,632.0	1.8	1.7	0.9270
Bd-3	MAH0102na	3 mbr blades & frags	0.5	601	0.047	82	0.3	15,743	0.28413	0.00056	3.9334	0.0092	0.10040	0.00010	1,612.1	2.8	1,620.6	1.9	1,631.6	1.8	1.3	0.9105
GGU-463142[a] (same dyke as site MB2) – Torsukattak; 356° strike																						
Bd-A	Z7428-A	14 m-dbr, large blades	4.2	1,049	0.037	1200	10.0	7808	0.28524	0.00056	3.9431	0.0097	0.10026	0.00010	1,617.7	2.8	1,622.6	2.0	1,628.9	1.8	0.8	0.9286
Bd-B	Z7428-B	5 m-dbr, blades and frags	4.2	970	0.028	1110	4.6	15,850	0.28608	0.00056	3.9561	0.0097	0.10030	0.00009	1,621.9	2.8	1,625.3	2.0	1,629.6	1.7	0.5	0.9404
Bd-C	Z7428-C	13 mbr, large blades	8.8	607	0.036	1460	10.5	9,028	0.28560	0.00058	3.9522	0.0100	0.10036	0.00009	1,619.5	2.9	1,624.4	2.0	1,630.8	1.7	0.8	0.9420

Table 27.2 (continued)

Sample fraction	Analysis no.	Description	Weight (μg)	U (ppm)	Th/U	Pb* (pg)	Pb_C (pg)	$^{206}Pb/^{204}Pb$	$^{206}Pb/^{238}U$	±2σ	$^{207}Pb/^{235}U$	±2σ	$^{207}Pb/^{206}Pb$	±2σ	$^{206}Pb/^{238}U$ Age (Ma)	±2σ	$^{207}Pb/^{235}U$ Age (Ma)	±2σ	$^{207}Pb/^{206}Pb$ Age (Ma)	±2σ	Disc. (%)	Corr. Coeff.
Bd-D	Z7428-D	20 mbr, large blades	5.6	678	0.037	1040	5.0	13,650	0.28580	0.00059	3.9495	0.0100	0.10023	0.00009	1,620.5	3.0	1,623.9	2.1	1,628.3	1.7	0.5	0.9400
OF-2 (site OF) - Olrik Fjord; 330° strike																						
Bd-1	DWD4671	1 mbr short prismatic	0.6	975	0.060	156	4.8	2,247	0.28765	0.00065	3.9625	0.0158	0.09991	0.00028	1,629.8	3.3	1,626.5	3.2	1,622.4	5.2	–0.5	0.7356
Bd-2	DWD4672	1 mbr short prismatic	0.9	406	0.039	98	2.0	3,453	0.28683	0.00081	3.9493	0.0156	0.09986	0.00024	1,625.7	4.1	1,623.8	3.2	1,621.5	4.4	–0.3	0.8036
Bd-3	DWD4673	1 mbr long prismatic blade	0.6	2,173	0.011	351	1.9	12,609	0.28519	0.00050	3.9379	0.0361	0.10014	0.00089	1,617.5	2.5	1,621.5	7.4	1,626.8	16.5	0.6	0.2393

Notes: All analyzed fractions represent hand-picked, highest quality baddeleyite crystals free of inclusions or adhered/coating phases. Abbreviations: pbr – pale brown; mbr – medium brown; dbr – dark brown; fl – flat; frag - fragment. Pb* is total amount (in picograms) of radiogenic Pb. Pb_C is total measured common Pb (in picograms) assuming the isotopic composition of laboratory blank: ROM [GSC]- 206/204 – 18.221 [18.2]; 207/204 – 15.612 [15.37]; 208/204 – 39.360 [37.4] (errors of 2%). Pb/U atomic ratios are corrected for spike, fractionation, blank, and, where necessary, initial common Pb; 206Pb/204Pb is corrected for spike and fractionation. Th/U is model value calculated from radiogenic 208Pb/206Pb ratio and 207Pb/206Pb age assuming concordance. Disc. (%) – per cent discordance for the given 207Pb/206Pb age. [a]Analyzed at the GSC Geochronology Lab, Ottawa, Canada.

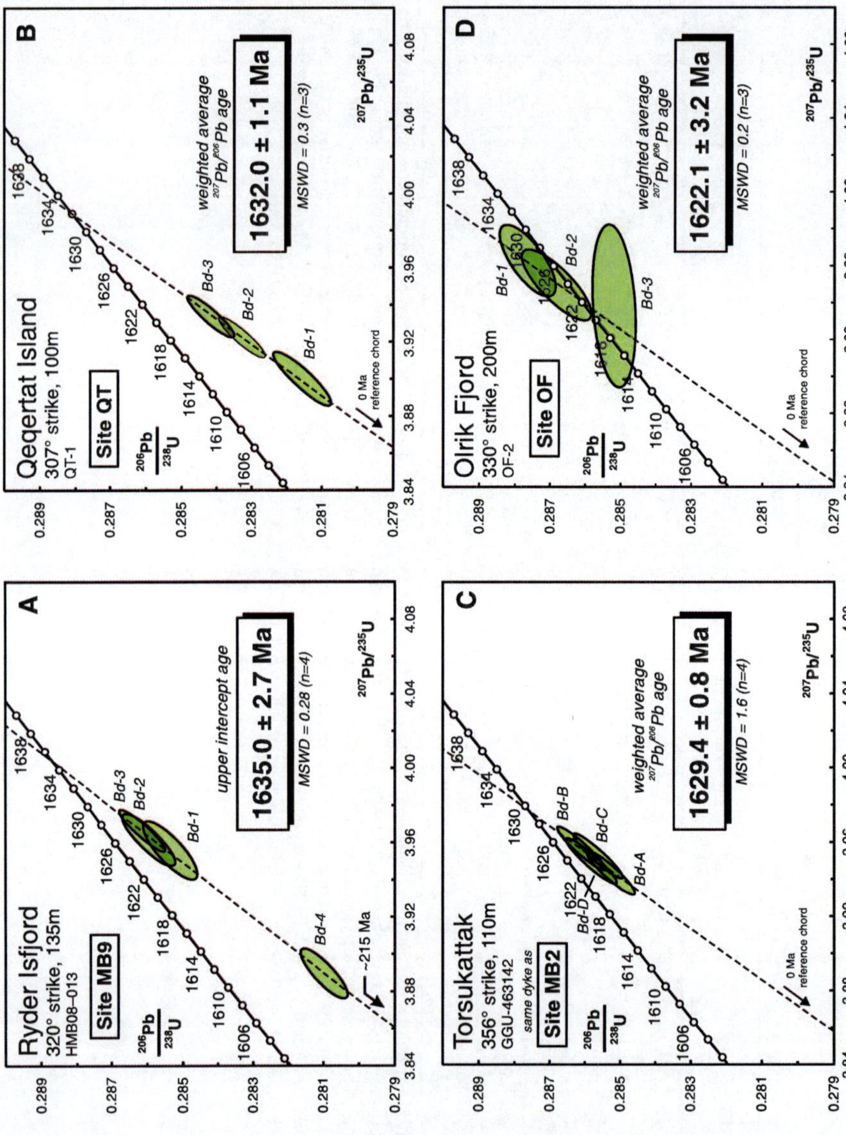

Fig. 27.7 Concordia diagrams showing U-Pb isotopic data for baddeleyite from four widely-distributed Melville Bugt dykes

Three fractions of baddeleyite from sample QT-1 are 1.3–2.3% discordant but are collinear and yield uniform $^{207}Pb/^{206}Pb$ ages – 1,631.6–1,632.7 Ma (Table 27.1; Fig. 27.7b). A weighted average $^{207}Pb/^{206}Pb$ age for all three fractions yields 1,632.0 ± 1.1 Ma (2σ, 74% probability of fit), which we interpret to represent the intrusion and crystallization age of the Melville Bugt diabase dyke at this locality. Like the 1,635 Ma MB9 dyke, the QT dyke yields a magnetic remanence direction, interpreted to be primary, that is oriented to the SW and down (Fig. 27.5), suggesting that this polarity persisted from at least 1,635–1,632 Ma.

Site MB2 (sample GGU-463142): Due east of the large island of Disko, and approximately 80 km north of Ilulissat, the southern end of the large, semi-continuous 400 km-long dyke is exposed on the shores of the fjord Torsukattak (Fig. 27.2b). Further south, its likely extension is covered by the Inland Ice. Sample GGU-463142 was collected from this dyke on the north shore of Torsukattak in 2001 by GEUS (Denmark and Greenland Geological Survey) geologists. Here, the dyke is approximately 70 m thick and strikes almost N–S (354°). Initial dating and paleomagnetic results from this sampling were presented by Hamilton et al. (2004). This dyke undoubtedly continues south along strike across the fjord where it was sampled for paleomagnetic study by us (HCH and MAH) in 2008 (Fig. 27.2b). At site MB2, the dyke also strikes NNW and widens to roughly 100 m in thickness.

U-Pb data for four multigrain fractions of baddeleyite (5–20 grains each) from this sample are tightly clustered and only 0.5–0.8% discordant (Fig. 27.7c). $^{207}Pb/^{206}Pb$ ages are very uniform, ranging narrowly between 1,628.3 and 1,630.8 Ma (Table 27.2). A simple weighted average $^{207}Pb/^{206}Pb$ age for all four fractions is 1,629.4 ± 0.8 Ma (2σ). We interpret this result to represent a rigorous estimate of the age of emplacement of the large Melville Bugt dyke at this locality. Because the MB2 site paleomagnetic data records a remanence magnetization with a NE (and up) declination, by extension we also interpret this age to constrain the timing of a magnetic reversal between that recorded in the QT dyke (1,632 ± 1 Ma) and the MB2 dyke (1,629 ± 1 Ma).

Site OF (sample OF-2): At the head of the long E–W Olrik Fjord in the Thule-Inglefield Bredning region, a wide Melville Bugt dyke (~200 m thick) parallels the eastern wall of a NW-trending glacier-fed valley (Figs. 27.1 and 27.3a). Here the dyke strikes 330°, and a representative coarse-grained piece from paleomagnetic site OF was collected for geochronology (Denyszyn, 2008). Abundant baddeleyite was recovered from sample OF-2.

Three single grains of fresh, tan-coloured, striated and prismatic baddeleyite from this sample were analyzed individually (Table 27.2). Although the results generally straddle concordia (0.6 to –0.5% discordant; Fig. 27.7d), the sample:blank ratios for these analyses are slightly higher than in the other samples and thus individual analytical uncertainties are also greater. $^{207}Pb/^{206}Pb$ ages for these fractions range from 1,621.5 to 1,626.5 Ma. Nonetheless, a relatively precise weighted average $^{207}Pb/^{206}Pb$ age, controlled by Bd-1 and Bd-2, yields a date for the timing of emplacement and crystallization of this large dyke at 1,622.1 ± 3.2 Ma (2σ, 82% probability of fit). Melville Bugt diabase from site OF also carries the upward,

NE-directed remanence, consistent with the result from the slightly older dyke at MB2 (1,629 ± 1 Ma).

In summary, the precise U-Pb geochronological results presented here demonstrate that the timing of emplacement of the widespread Melville Bugt dyke swarm lasted from at least 1,635 to 1,622 Ma, spanning some 13 Myr. Detailed results from within the geographic distribution of the swarm (as currently recognized) reveals no obvious pattern of intrusion. Dykes appear to be intermingled in terms of age and magnetic polarity, though the trends are all subparallel, and we have seen no evidence of cross-cutting relationships within the swarm. The geochronology results (Table 27.2 and Fig. 27.7) show conclusively that the remanent magnetization with SW declination, as exemplified by the dated sites MB9 and QT, is older than the NE-directed remanence found at dated sites MB2 and OF. Because of the small temporal uncertainties on the nearest dykes (age-wise) recording a reversal in remanence direction (i.e. QT & MB2), this polarity switch must have happened within a narrow window between 1 and 5 Myr, centred on ca. 1,630 Ma.

Interpretation

An alternative model to the conventional paleomagnetic one of placing Laurentia in the northern hemisphere for the time period 1,450–1,600 Ma (Buchan et al., 2000, Pesonen et al., 2003, Lubnina et al., 2010) is to reverse the poles and place it in the southern one. The effect of this is to return Laurentia to a position not unlike that existing today with the tip of Greenland pointing southwards, rather than northwards in present day co-ordinates. In Figs. 27.8 and 27.9 configurations for Laurentia versus Fennoscandia are shown, one at ~1.5 Ga and one at ~1.6 Ga respectively. What becomes apparent is that the Melville Bugt dyke swarm with some adjustments of

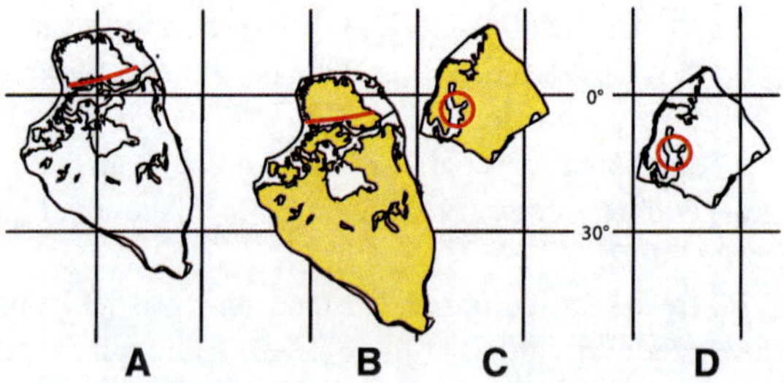

Fig. 27.8 Positions of Laurentia and Fennoscandia for the time ~1.5 Ga based on the average mean poles given by Salminen and Pesonen (2007), together with perturbations in the position of the mean poles allowed by their uncertainty that permit a direct linkage of the Melville Bugt swarm (*red line*) with the Fennoscandian rapakivi province (*red circle*). (**a**) Mean pole for Laurentia (Salminen and Pesonen 2007); (**b**) Mean pole for Laurentia with a displacement allowed by the A95 value (13°); (**c**) mean pole for Fennoscandia with displacement allowed by A95 value (11°), (**d**) mean pole for Fennoscandia (Salminen and Pesonen, 2007)

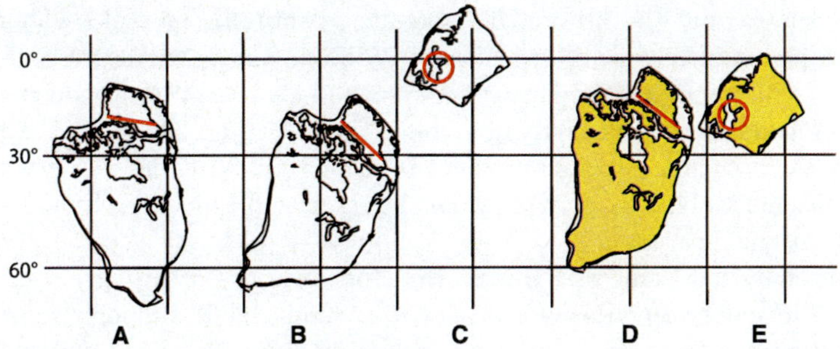

Fig. 27.9 Similar to Fig. 27.8, except that the time period is for ~1.6 Ga. (**a**) position of Laurentia using paleomagnetic data from the 1.59 Ga Western Channel Diabase (Irving et al., 1972; Hamilton and Buchan, 2010); (**b**) the position of the Melville Bugt dykes based upon the pole data in Table 27.1 (**c**) the position of Fennoscandia based on the pole for the Sipoo dykes (Mertanen and Pesonen,1995) and shows that correlation of the ~1630 Ma Melville Bugt dykes (*red line*) with the Fennoscandian rapakivi province (*red circle*) is poor. (**d**) location of Laurentia based on an adjustment of the Melville Bugt pole position that lies within its 95% confidence circle. (**e**) an adjusted position of Fennoscandia based on the premise that the magnetization of the 1,633 Ma Sipoo dykes has an inclination of –30° (see text). A comparison of figures (**d**) and (**e**) shows that the correlation between Melville Bugt dykes and Fennoscandian rapakivi province is now improved but not as close if only the data from the Western Channel Diabase are used to define the position of Laurentia

the paleomagnetic data (see below) approximately points towards the Scandinavian rapakivi granite province, and furthermore that the configuration of Greenland and Fennoscandia may have remained the same throughout the lifetime of rapakivi magmatism. The Melville Bugt dyke swarm has been an enigma because it occurs in isolation with no obvious connection to other coeval magmatic provinces nor to plume centres, a common feature of other major dyke swarms.

If the paleomagnetic configurations shown in Figs. 27.8 and 27.9 are allowable, it means that the source for the Melville Bugt swarm may be represented by the Fennoscandian 1.5–1.6 Ga rapakivi granite province. In this province diabase dykes are associated with rapakivi plutons and generally trend away from them. Cross-cutting relations suggest an order of emplacement, from older to younger, of diabase dykes, rapakivi granite and finally quartz porphyry dykes. It is generally accepted that the rapakivi plutons represent melting of the lower crust, perhaps by a local mantle-derived basaltic melt (e.g. Rämö and Haapala, 1995). The availability of basaltic magma associated with these intrusions is exemplified by the Breven-Hällefors dyke swarm in Sweden. This swarm is composed of over 850 dykes and is at least 150 km wide (Fig. 27.1 of Risku-Norja, 1992). Within the swarm are at least four major dykes several 100 m thick, one of which is about 1 km wide and on the basis of aeromagnetic data, at least 100 km long. The western extent of the swarm is obscured beneath Paleozoic strata and ultimately by the Protogine zone. However the enormous size of these master dykes indicates the capability of rapakivi – associated magmatism to rapidly produce copious amounts of basaltic magma as dykes. In many cases magma-mixing observations suggest the basaltic magma to have been coeval with felsic magma. The Breven-Hällefors dyke swarm lies on strike with a major rapakivi pluton in Estonia. Other plutons (Wiborg, Åland) are also associated with dykes of more modest proportions. A system of diabase dykes (Sipoo and Häme swarms) lie adjacent to the Wiborg granite. The total age range from U-Pb dating for the Fennoscandian rapakivi province is about 150 Myr (from ~1,650 to 1,500 Ma) so that the Melville Bugt dykes would be associated with the early stages of the rapakivi magmatic event.

Paleomagnetically the Häme and Breven-Hällefors swarms are strongly overprinted by a PEF component (Neuvonen, 1967; Bylund, 1985) although some diabase dykes from the latter study and quartz-porphyry dykes associated with the Wiborg granite give a streaked distribution from a PEF direction to a shallow up, N to NE component (Neuvonen, 1986; Mertanen and Pesonen, 1995). One important observation is that for the Sipoo swarm, the diabase dykes record a remanence which is both older and of opposite polarity to the NE component recorded by the quartz-porphyry dykes. The evidence is that the freshest dyke, site SB, which is farthest from the Wiborg granite, records the SW component, whereas in another more altered dyke, closer to the granite, both the SW *and* NE components are present, suggesting that the alteration was produced during the intrusion of the younger Wiborg granite and associated quartz porphyry dykes during a time when the field pointed towards the NE (Mertanen and Pesonen, 1995).

In our study of the Melville Bugt dykes, almost exactly the same observations hold: the dykes were intruded on either side of a reversal and the precise U-Pb data show convincingly that the NE-directed component is the younger, same as for the Sipoo dykes. Although the age data for the Sipoo dykes are less precise, they allow the same reversal to be recorded in the two studies. If so, then we have persuasive evidence that both Fennoscandia and Laurentia remained at 1.63 Ga in orientations not unlike their present ones and therefore that reconstructions that involve placing northern Norway and the Kola Peninsula adjacent to northern Greenland are incorrect. This interpretation can be tested by more precise age data and by more detailed paleomagnetic work on the Häme and Sipoo dykes.

The Melville Bugt and Sipoo swarms both have the same problem, namely that the NE-directed component may have a steeper negative inclination because of the difficulty in removing the strong PEF-like component. However the Melville Bugt dyke population shows several dykes (MB1, MB9, QT and JP) with steeper inclinations that are virtually devoid of PEF overprinting, suggesting that an underestimate of the inclination is minimal. However PEF contamination is more severe in the Fennoscandian dykes so that the shallow NE directions found in the Häme, Sipoo and Breven-Hällefors swarms, are more likely to be underestimating the inclination through failure to achieve a true end point. For the Sipoo dykes a further complication is the presence of a high coercivity remanence with a direction similar to that found as a secondary component at site MB3 (Table 27.1), so that in terms of coercivity the NE component is sandwiched between two remanences which have steep to moderate down inclinations with NE declinations.

The Fennoscandian paleomagnetic data closest in age to the Melville Bugt dykes are those from the 1,633 Ma Sipoo quartz porphyry and diabase dykes (Mertanen and Pesonen, 1995). Specifically, we compare here the Sipoo pole position (26.4°N, 180.4°E) with that for Laurentia from the Melville Bugt dykes (pole 5°N, 273°E, Table 27.1) and also with that from the slightly younger 1,592 Ma Western Channel diabase (pole 6.8° N, 245°E, Irving et al., 1972; Hamilton and Buchan, 2010). The craton positions are shown in Fig. 27.9a–c where Fennoscandia, based on the Sipoo dykes data, occupies more equatorial positions than the two positions of Laurentia and therefore offers little support for a connection between the Melville Bugt dykes and the rapakivi province in Fennoscandia.

However in Fig. 27.9d and e we show a comparison that is closer. The position of Laurentia is adjusted using a pole (2°S, 267°E) that lies within the A_{95} of the Melville Bugt pole position in Table 27.1 (Fig. 27.9d). The streaking of site directions for Sipoo quartz porphyry dykes has a declination range of about 10° and an inclination range of ~30° from about +16 to –16° (Mertanen and Pesonen, 1995), inviting speculation that the true NE component is steeper. In fact one specimen upon thermal demagnetization (Fig. 27.5a of Mertanen and Pesonen, 1995) reaches an inclination of about –30°. If we assume a direction of 22°/–30° as closer to the true component direction, then a new pole position (11.5°N, 183.5°E) places Fennoscandia as shown in Fig. 27.9e. A further 10° increase in inclination to –40° (a few specimens attain these values, Satu Mertanen, personal communication 2010) would place the rapakivi province more or less in line with the Melville Bugt dykes of Fig. 27.9d. Although the correlation is far from perfect, a supportive observation is that the Western Channel Diabase data place Laurentia in a position such that the dykes trend more directly towards the Fennoscandian rapakivi province (c.f. Fig. 27.9a, e). The key element in our reconstruction is the assumption that the NE younger component in the Sipoo dykes has a steeper inclination. However if the rapakivi province extends beyond Fennoscandia with a southerly extension into Amazonia as shown in Fig. 27.1 of Johansson (2009) and the Melville Bugt dykes are fed from there, the positions of Laurentia and Fennoscandia as shown in Fig. 27.9b, c might require little adjustment.

Although the southern end of the Melville Bugt swarm disappears under ice, in SE Greenland high resolution satellite maps, courtesy Google Earth© (Figs. 27.10 and 27.11), image a NNW-trending series of dykes of unknown age, with widths up to 50 m, cutting basement gneisses, and approximately on strike with the Melville Bugt swarm. The dykes have a slightly reddish tinge with respect to the basement, with dyke-normal jointing suggesting that the dykes, like those of the Melville Bugt swarm, have not been metamorphosed. An investigation is needed to verify this correlation, but if true, it suggests that the Melville Bugt swarm could be continuous across southern Greenland and therefore allows it to be fed from the south.

That dykes can propagate laterally for over 2,000 km has been recognized by Ernst and Baragar (1992) from magnetic fabric studies on the giant 1.27 Ga Mackenzie swarm in Canada. Geochemical studies of Mesozoic Farrar diabase sills also suggest a lateral flow of about 2,500 km across Antarctica (Leat et al. 2006), so that a comparable distance of propagation that we advocate for the Melville Bugt dykes is not unreasonable.

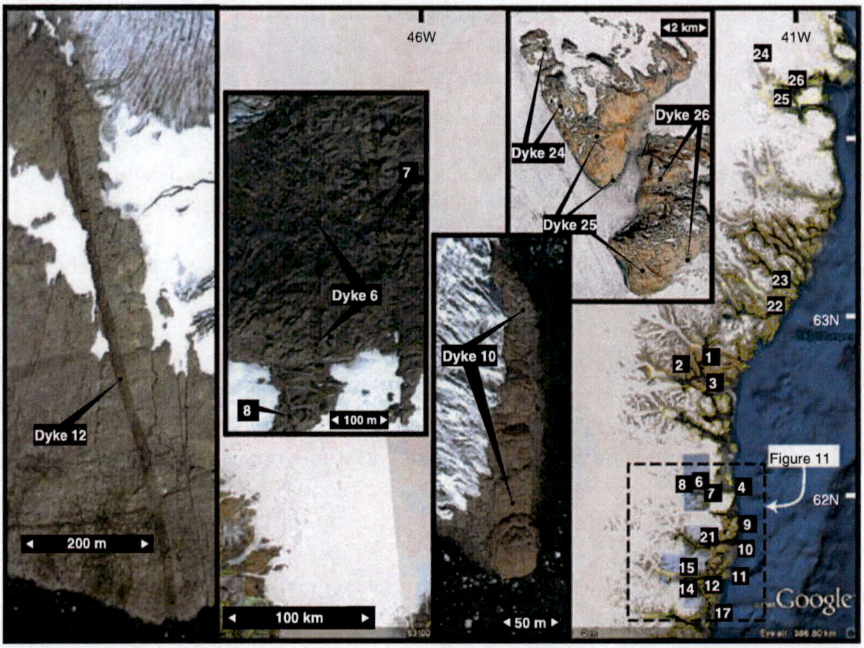

Fig. 27.10 Google Earth© images of SE Greenland showing evidence of NNW-trending dykes that may represent the southerly continuation of the Melville Bugt swarm. Most data come from relatively small survey areas of higher resolution (shown in *blue*). Overall coverage has relatively low resolution but in a few cases (sites 24 and 25) NW-trending dykes are visible (see farthest *inset* picture on the *right*). Lineaments with similar orientations occur at sites 22, 23 and 1–3, but their interpretation as dykes is less secure. However if the dykes exist a dyke swarm trending N to NW is more than 200 km wide, comparable to the minimum estimate for the Melville Bugt swarm

The geochemical study of the Melville Bugt master dyke (Kalsbeek and Taylor, 1986) shows that composition remains constant for 100's of km along strike and therefore that it may be similar to mafic dykes of the Fennoscandian rapakivi province, if we are correct in surmising that the two are genetically linked. We rely for geochemical estimates across the swarm on the work by Nielsen (1990), Denyszyn (2008), and on our unpublished major and trace element data from chilled margins of 10 Melville Bugt dykes. Melville Bugt dykes have P_2O_5 and TiO_2 compositions that average 0.50 ± 0.1 and 2.13 ± 0.2 wt% respectively (H.C. Halls, unpublished data from sites 1–9 and OF). Zr/Y values average 4.8 ± 0.5. One unusual aspect is that Ba concentrations average 870 ± 130 ppm. In Fennoscandia comparatively little work has been done on rapakivi-related diabase. For two Häme dykes P_2O_5 and TiO_2 average 0.50 and 2.3 wt% respectively and Zr/Y averages 7.9; Ba contents average about 500 (Rämö, 1990). However for the nearby (~100 km away) Suomenniemi swarm, also proximal to the Wiborg granite, analyses on five diabase dykes give the following: $P_2O_5 = 0.56$, $TiO_2 = 2.6$, Zr/Y = 4.0 and Ba averages

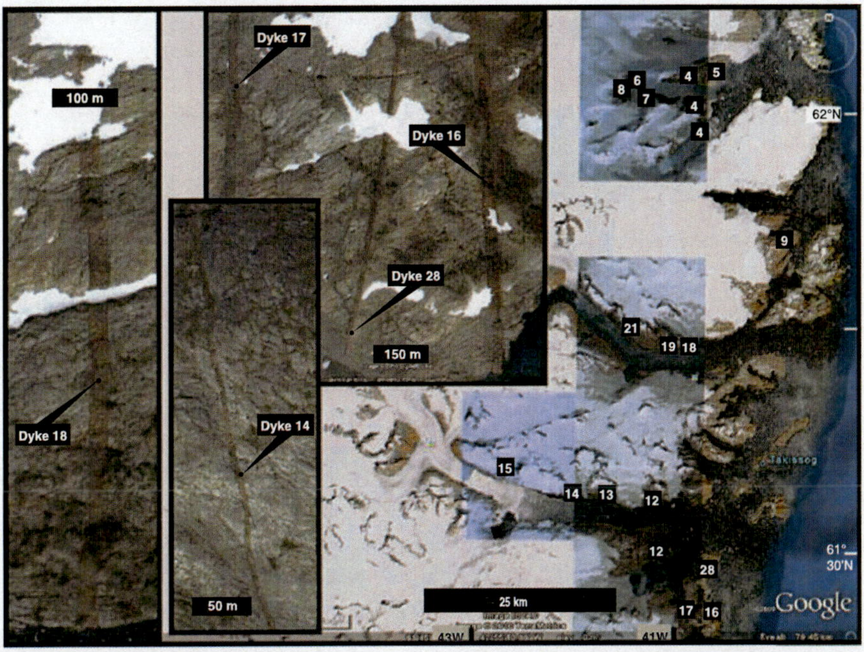

Fig. 27.11 Google Earth© images focusing more on the southern part of the dyke distribution shown in Fig. 27.10 Images are taken only from the higher resolution survey areas shown in *blue*

583 ppm with a maximum value of 809 ppm (Rämö, 1990). For the four major dykes of the Breven-Hällefors swarm, geochemical data reported by Risku-Norja (1992) give the following averages: $P_2O_5 = 0.36$, $TiO_2 = 2.1$ and $Zr/Y = 3.55$, with the Ba content remaining unmeasured. Therefore geochemically the dykes are surprisingly similar, being basalts that are all relatively rich in Ti and P and with Zr/Y ratios that mostly lie between 4 ± 1. However comparisons are flawed owing to the lack of correspondence between analysed and unmeasured elements, to inadequate sampling, and to differences in sampling philosophy (chilled margins in our study versus dyke interiors in the that by Risku-Norja).

The proposed reconstruction does not violate previous trans-Atlantic correlations of basement rocks older than 1.6 Ga (e.g. Zhao et al., 2004). In their reconstruction of the supercontinent Columbia Zhao et al. (their Fig. 17) show collisional orogens at 2.1–1.8 Ga and at 1.8–1.3 Ga which they correlate between Fennoscandia and Laurentia; both of these remain easily correlated in the configuration shown in Fig. 27.9. The configuration shown in Fig. 27.9 merely serves to reconfigure the Archean-Proterozoic margins and the 1.7 Ga Gothian margin with older rocks. In fact our reconstruction also resembles that proposed in Fig. 2 of Johansson (2009) for the time period 0.9 Ga, whereby the Gothian rocks exposed in Laurentia and Fennoscandia represent the *opposing* sides of the orogen. One puzzle is that during the lifetime of rapakivi magmatism (about 150 Myr), subduction-related accretionary magmatism was also taking place along the west coast of Norway representing the later stages of the Gothian orogen (Åhäll et al., 2000) and signifying an ocean farther to the west. If, to the north, Greenland was separated from Fennoscandia by this ocean then a physical connection of the Melville Bugt swarm with the rapakivi igneous province would be invalidated. At the present time evidence for such a northward extension is lacking, thus allowing for dyke injection into Greenland from Fennoscandia.

Conclusions

Paleomagnetic and U-Pb geochronology demonstrate the sense of a magnetic polarity reversal found in the Melville Bugt dyke swarm and precisely date magmatic activity to span a time period of at least 13 million years as determined by maximum and minimum dates of 1,622 ± 3 and 1,635 ± 3 Ma. The paleomagnetic data strongly resemble those reported for the Sipoo dykes in Finland with an estimated U-Pb age of about 1,633 Ma (Mertanen and Pesonen, 1995), without any need for significant rotation of Laurentia; a NNE-N directed component with negative inclination and an older antipodal one are present in both data sets. If the same reversal is being recorded, it implies that Fennoscandia and Laurentia can be juxtaposed such that the southern end of the Melville Bugt dyke swarm can be projected to the proximity of the Fennoscandian rapakivi province. Paleomagnetic data can only be made to support this idea if magnetic overprinting of the Sipoo dykes has led to an underestimation of the negative inclination of the NE directed component. The correlation is best when Laurentia is positioned according to the 1,592 Ma Western Channel Diabase data rather than by those of the Melville Bugt dykes themselves. Diabase dykes associated with the rapakivi plutons have similar geochemical compositions to the Melville Bugt dykes. The enormous size of some of these dykes in Sweden indicates the capability of rapakivi magmatism to produce prodigious quantities of mafic magma and therefore our speculation that the entire Melville Bugt swarm is fed from a basaltic magma source below the continental crust that has ultimately produced the rapakivi granites, may be a reality, especially since the dykes may continue across Greenland. If so then the distance over which magma has been laterally injected, of the order of 2,000 km, offers support to other workers who also have proposed similar distances of basaltic magma transport. Finally our correlation of the Melville Bugt dykes with the Fennoscandian rapakivi granite province represents the first hypothesized Precambrian piercing point between Laurentia and Fennoscandia which is related to a specific magmatic event. The hypothesis is testable by further mapping in southeast Greenland and by more precise U-Pb dating, paleomagnetism and geochemistry of the Fennoscandian and Amazonian rapakivi units.

Acknowledgments For outstanding and friendly logistical support in Ilulissat and Upernavik, we thank Flemming Bisgaard, Flemming Lassen and Ole Sørensen and the crew of M/V Duda Lasøe. Field work by HCH and SWD in the Inglefield Bredning area was supported by the 2001 Nares Strait GeoCruise and by the Canadian Polar Continental Shelf Project. Subsequent field work in the Melville Bugt region was supported by NSERC Discovery grants to HCH and MAH. We are grateful for discussions with Ken Buchan and Tom Frisch (Geological Survey of Canada), Troels

Nielsen (GEUS, Copenhagen), Tapani Rämö (Univ. of Helsinki), and Satu Mertanen (GSF, Espoo, Finland). MAH thankfully acknowledges Jim Connelly (Geol. Museum, Copenhagen), Adam Garde and the 2001 GEUS field party for sample collection at Torsukattak. Continental reconstructions shown in Figs. 27.8 and 27.9 were prepared using GMAP, courtesy of Trond Torsvik (NGU, Trondheim). HCH thanks Alan Lovette for making many of the measurements at the UTM Paleomagnetism Laboratory, and Alison Dias (UTM Computing Services) for expert digital graphics support. MAH and SWD gratefully acknowledge the assistance of colleagues and staff of the Jack Satterly Geochronology Laboratory.

References

Åhäll K-I, Connelly JN, Brewer TS (2000) Episodic rapakivi magmatism due to distal orogenesis? Correlation of 1.69-1.50 Ga orogenic and inboard "anorogenic" events in the Baltic Shield. Geology 28: 823–826

Borradaile GJ, Lucas K, Middleton RS (2004) Low-temperature demagnetization isolates stable magnetic vector components in magnetite-bearing diabase. Geophys J Intern 157:526–536

Buchan KL, Mertanen S, Park RG, Pesonen LJ, Elming S-A, Abrahamsen N, Bylund G (2000) Comparing the drift of Laurentia and Baltica in the Proterozoic: the importance of key paleomagnetic poles. Tectonophysics 319:167–198

Bylund G (1985) Paleomagnetism of Middle Proterozoic basic intrusive in central Sweden and the Fennoscandian apparent polar wander path. Precamb Res 28: 283–310

Dawes P (2004) Explanatory notes, Geological map of Greenland, 1:500 000, Humboldt Gletcher; sheet 6, Geological Survey of Denmark and Greenland Map Series 1, 48p

Denyszyn SW (2008) Paleomagnetism, geochemistry and U-Pb geochronology of Proterozoic mafic intrusions in the high Arctic: relevance to the Nares Strait Problem. Ph.D Dissertation, 197 p, University of Toronto, Canada

Denyszyn SW, Halls HC, Davis DW, Evans DAD (2009) Paleomagnetism and U-Pb geochronology of Franklin dykes in High Arctic Canada and Greenland: a revised age and paleomagnetic pole constraining block rotations in the Nares Strait region. Can J Earth Sci 46: 689–705

Ernst RE, Baragar WRA (1992) Evidence from magnetic fabric for the flow pattern of magma in the Mackenzie giant radiating dyke swarm. Nature 356: 511–513

Frisch T (1984a) Geology, Prince of Wales Mountains, District of Franklin, Northwest Territories. Geol Surv Canada Map 1572A, scale 1:250 000

Frisch T (1984b) Geology, Makinson Inlet, District of Franklin, Northwest Territories. Geol Surv Canada Map 1573A, Scale 1:250 000

Halls HC (1986) Paleomagnetism, structure and longitudinal correlation of middle Precambrian dykes from northwestern Ontario and Minnesota. Can J Earth Sci 23:142–157

Halls HC, Davis DW, Stott GM, Ernst RE, Hamilton MA (2008) The Paleoproterozoic Marathon large igneous province: New evidence for a 2.1 Ga long-lived mantle plume event along the southern margin of the North American Superior province. Precamb Res 162: 327–353

Hamilton MA, Buchan KL (2010) U-Pb geochronology of the western Channel Diabase, northwestern Laurentia: Implications for a large 1.59 Ga magmatic province, Laurentia's APWP and paleocontinental reconstructions of Laurentia, Baltica and Gawler craton of southern Australia. Precamb Res 183: 464–473

Hamilton MA, Buchan KL, Garde AA, Connelly JN (2004) U-Pb age and preliminary paleomagnetism of a Melville Bugt diabase dyke, west Greenland, and implications for mid-Proterozoic Laurentia-Baltica reconstructions. AGU-CGU Joint Assembly, 17-21 May, Montreal, Canada. Eos Trans. AGU, 85(17), Jt. Assem. Suppl., Abstract GP31B–03 (2004)

Hamilton MA, Davis DW, Buchan KL, Halls HC (2002) Precise U-Pb dating of reversely magnetized Marathon diabase dykes and implications for emplacement of giant dyke swarms along the southern margin of the Superior Province, Ontario. Geol Surv Canada Current Res 2002-F6, 8p

Hamilton MA, Goutier J, Matthews W (2001) U-Pb baddeleyite age for the Proterozoic Lac Esprit dyke swarm, James Bay region, Quebec. Radiogenic Age and Isotopic Studies: Report 14. Geol Surv Canada Current Res 2001-F5, 6p

Heaman LM, LeCheminant AN (1993) Paragenesis and U-Pb systematics of baddeleyite (ZrO2). Chem Geol 110: 95–126

Irving E, Donaldson JA, Park JK (1972) Paleomagnetism of the Western Channel diabase and associated rocks, Northwest Territories. Can J Earth Sci 9:960–971

Jaffey AH, Flynn KF, Glendenin LE, Bentley WC, Essling AM (1971) Precision measurement of half-lives and specific activities of 235U and 238U. Phys Rev 4: 1889–1906

Johansson A (2009) Baltica, Amazonia and the SAMBA connection: 1000 million years of neighbourhood during the Proterozoic? Precamb Res 175:221–234

Kalsbeek F, Taylor PN (1986) Chemical and isotopic homogeneity of a 400 km long basic dyke in central West Greenland. Contrib Mineral Petrol 93: 439–448

Kirschvink JL (1980) The least squares line and plane and the analysis of paleomagnetic data. Geophys J R Astron Soc 62: 699–718

Leat PT, Luttinen AV, Storey BC, Millar IL (2006) Sills of the Theron Mountains: evidence for long-distance transport of mafic magmas during Gondwana break-up. In: Hanski E, Mertanen S, Ramo T, Vuollo J (eds) Dyke swarms- time markers of crustal evolution, Taylor and Francis Pub: 183–199

Lubnina NV, Mertanen S, Soderlund U, Bogdanova S, Vasilieva TI, Frank-Kamenetsky D (2010) A new key pole for the east European craton at 1452 Ma: Paleomagnetic and geochronological constraints from mafic rocks in the Lake Ladoga region (Russian Karelia). Precamb Res. doi:10.1016/j.precamres.2010.02.014

Ludwig KR (2003) Isoplot 3.00: A Geochronological Toolkit for Microsoft Excel. Berkeley Geochronology Center, Special Publication No. 4 (rev. 3.00)

Mertanen S, Pesonen LJ (1995) Paleomagnetic and rock magnetic investigations of the Sipoo Subjotnian quartz porphyry and diabase dykes, southern Fennoscandia. Physics Earth Planet Interiors 88:145–175

Neuvonen KJ (1967) Paleomagnetism of the dike systems in Finland III. Remanent magnetization of diabase dikes in Häme, Finland. Comptes Rendus de la Societé géologique de Finlande 39: 87–94

Neuvonen KJ (1986) On the direction of remanent magnetization of the quartz porphyry dikes in SE Finland. Bull Geol Soc Finland 58 (I): 195–201

Nielsen TFD (1990) Melville Bugt dyke swarm: a major 1645 Ma alkaline magmatic event in west Greenland. In: Parker AJ, Rickwood PC, Tucker DH (eds) Mafic dykes and emplacement mechanisms, A.A. Balkema Publ, Rotterdam 497–505

Ozima M, Ozima M, Akimoto S (1964) Low temperature characteristics of remanent magnetization of magnetite self-reversal and recovery phenomena of remanent magnetization. J Geomag Geoelectr 16: 165–177

Pesonen LJ, Elming S-A, Mertanen S, Pisarevsky S, D'Agrella-Filho MS, Meert JG, Schmidt PW, Abrahamsen N, Bylund G (2003) Palaeomagnetic configuration of continents during the Proterozoic. Tectonphysics 375: 289–324

Rämö OT (1990) Diabase dyke swarms and silicic magmatism – evidence from the Proterozoic of Finland. In: Parker AJ, Rickwood PC, Tucker DH (eds) Mafic dykes and emplacement mechanisms, A.A. Balkema Publ, Rotterdam 185–199.

Rämö OT, Haapala I (1995) One hundred years of rapakivi granite. Mineral Petrol 52:129–185

Riisager J, Riisager P, Pedersen AK (2003) Paleomagnetism of large igneous provinces: case-study from West Greenland, North Atlantic igneous province. Earth Planet Sci Lett 214:409–425

Risku-Norja H (1992) Geochemistry of the dolerite dykes of Södermanland, east-central Sweden. Geol Föreningens i Stockholm Förhandlingar 114: 67–91

Salminen J, Pesonen LJ (2007) Paleomagnetic and rock magnetic study of the Mesoproterozoic sill, Valaam island, Russian Karelia. Precamb Res 159: 212–230
Söderlund U, Johansson L (2002) A simple way to extract baddeleyite (ZrO2); Geochem. Geophys. Geosyst, 3 (2), 10.1029/2001GC000212
Zhao G, Sun M., Wilde SA, Li S(2004) A Paleo-mesoproterozoic supercontinent: assembly, growth and breakup. Earth Sci Rev 67: 91–123

Chapter 28
Mafic Melt Emplacement During the Shock Deformation in the Subvolcanic Environment: Example from Tastau Volcanoplutonic Ring Complex, Eastern Kazakhstan

Ksenia Dokukina and Alexander Konilov

Introduction

The Tastau volcanoplutonic complex of the Char transpressional belt, Eastern Kazakhstan, contains the mafic microintrusions (main in size about 10–30 cm) in the form of strings of globules and other bodies that resemble stretched and boudinaged dykes. This paper describes these structures as result of primary magma emplacement rather than postconsolidation stretching. We believe that the formation of these mafic microintrusions was controlled by compressive shear deformations. Progressive stretching at a high-speed deformation during decompression of metasedimentary rock was accompanied by intrusion of mafic melt. Magma fragmentation due to the significant decrease in matrix viscosity took place during high-velocity shearing and cataclastic flow. We propose the synkinematic model of the mafic melt fragmentation and mingling between magma and deforming rock on the hypabyssal level of the crust.

Geological Background

The Char shear belt is the axial zone of the Hercynian Zaisan orogen (Fig. 28.1) that formed during the Late Carboniferous – Permian collision between the Kazakhstan and Siberian continents (Belyaev, 1985; Buslov et al., 2004) and represents sinistral displacement system of the northwest trend parallel to the main structure of the Zaisan orogen. The present-day structural framework of the Zaisan orogen indicates that it belongs to a single Paleozoic accretion-collision complex which is characterized by a collage of different tectonic blocks various in composition and age.

K. Dokukina (✉)
Geological Institute of Russian Academy Science, Moscow 119017, Russia
e-mail: ksdokukina@gmail.com

A. Konilov
Institute of Experimental Mineralogy of the Russian Academy Science, Institutskaya ulitsa, Moscow 142432, Russia

R.K. Srivastava (ed.), *Dyke Swarms: Keys for Geodynamic Interpretation*,
DOI 10.1007/978-3-642-12496-9_28, © Springer-Verlag Berlin Heidelberg 2011

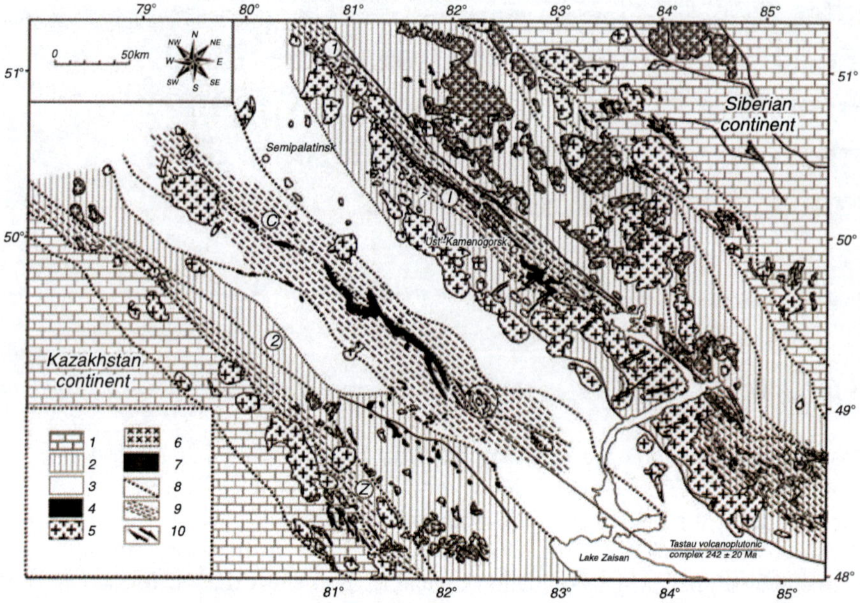

Fig. 28.1 Structural sketch of Zaisan orogen, Eastern Kazakhstan (modified from Vladimirov et al., 2008). 1 – Caledonian complexes of Siberian and Kazakhstan continents; 2 – Hercynian complexes of continental margins; 3 – Char paleooceanic terrane; 4 – Hercynian basite magmatic complexes (Early Permian – Early Jurassic); 5 – Hercynian granitoid magmatic complexes (Early Permian – Early Jurassic); 6 – Devonian – Early Carboniferous batholites; 7 – Char ophiolite suture, 8 – regional faults; 9 – main shear zones (Capital letters in circles: C – Char, I – Irtysh, Z – Zharma); 10 – kinematic of shear deformation. *Number in circles*: 1 – Kalba-Narym accretion prism, 2 – Zharma-Saur island arc

Geological and structural evidences indicate that the Char belt is a good example of a transpressive belt. Thrusting and sinistral transpression along the Char belt were contemporaneous with southeast directed shortening in the Altai orogen during Late Carboniferous-Permian (Ermolov et al., 1981, 1983; Dobretsov et al., 1979; Berzin et al., 1994; Berzin, Kungurtsev, 1996; Buslov et al., 2003, 2004). The Char belt is spatially associated with the Char ophiolite suture and includes three main tectonic units that differ in structure, age, and geodynamic environment (Polyanskii et al., 1979; Dobretsov et al., 1979; Buslov et al., 2003, Vladimirov et al., 2003, Shcherba et al., 1998).

(1) Subduction mélange in the southeastern part of the zone. It is made up of blocks of high-grade metamorphics (garnet amphibolites, eclogites, glaucophane and garnet-glaucophane blue schist), which were brought to the surface in the Late Ordovician-Early Silurian (Dobretsov et al., 1979, Buslov et al., 2001). The subduction-related rocks are Cambrian-Early Ordovician (Dobretsov and Ponomareva, 1969).

(2) Ophiolitic mélange imbricated with blocks of mantle and oceanic crust of different sizes, is composed of massive harzburgite, lherzolite, dunite and pyroxenite, gabbro, amphibolite, and amphibole-plagioclase schist.
(3) Polymictic mélange consisting of the fragments of first (subduction-related) and second (ophiolitic) types of mélange. It marks numerous young fault bands that bound the Char ophiolite suture. Along the northeastern side of the Char shear band the mélange is imbricated with Early Carboniferous turbidite and olistostrome, Late Devonian-Early Carboniferous island arc rocks, and Devonian-Early Carboniferous limestone that fills the Kalba-Narym accretionary prism. Imbricates along the southwestern limits of the zone include fragments of an Early Carboniferous accretionary wedge of the Zharma-Saur island arc. Serpentinitic mélange in the interior Char zone is most often oriented northwestward following zonal strike, as well as interthrust of various compositions.

The Char ophiolitic belt occurs among Devonian-Early Carboniferous volcanoclastics, which are likely fragments of accretionary prisms, separated from Kazakhstan and Siberia during formation of the Char shear band. The mélange units are sealed by Late Carboniferous volcanoclastic wedge and dykes. Younger Permian and Early Triassic faults occur locally.

The Char shear belt (Fig. 28.1) consists of voluminous Permian-Triassic volcanic and plutonic rocks emplaced during regional extension affecting the Zaisan orogen. Coeval mafic-felsic magmatism in the Zaisan orogen mostly consists of anatectic felsic magmas (e.g. Ermolov et al., 1983; Titov et al., 2001) with associated mantle-derived magmas of calc-alkaline affinity.

Tastau Volcanoplutonic Complex

The Tastau volcanoplutonic ring complex (Fig. 28.2) is the largest multiphase intrusive massif 242 ± 20 Ma (U-Pb zircon TIMS, Dokukina et al., 2010) in the Zaisan Magmatic Province. The Tastau volcanoplutonic complex intrudes the Early Carboniferous metasedimentary rock of the Char transpressional zone. Host rocks are low-grade polymictic siltstone and sandstone of greywacke composition (deposit of the Serpukhovian stage of the Early Carboniferous). They were metamorphosed to greenschist (mineral associations with biotite, chlorite, actinolite) and up to hornblende hornfels facies during intrusion of the Tastau complex (Ermolov et al., 1983). Metagreywacke is characterized by variations in chemical compositions (SiO_2 55.1–64.5, TiO_2 0.6–1.0, Al_2O_3 14.0–18.5, ΣFe_2O_3 5.6–9.6, MnO 0.1–0.2, MgO 1.9–4.8, CaO 1.5–7.8, Na_2O 1.8–4.5, K_2O 1.2–3.2 wt%). Metasedimentary rocks are deformed into large linear folds (Fig. 28.2). The metasandstone predominates over the metasiltstone. The rock has a subvertical dip and a northwestern strike.

Close to the central granitoid stock the metasedimentary rocks are intensively brecciated. Breccias are composed of angular ungraded siltstone fragments in a cataclastic sandstone matrix (Fig. 28.3a–e). The fragments also display chaotic

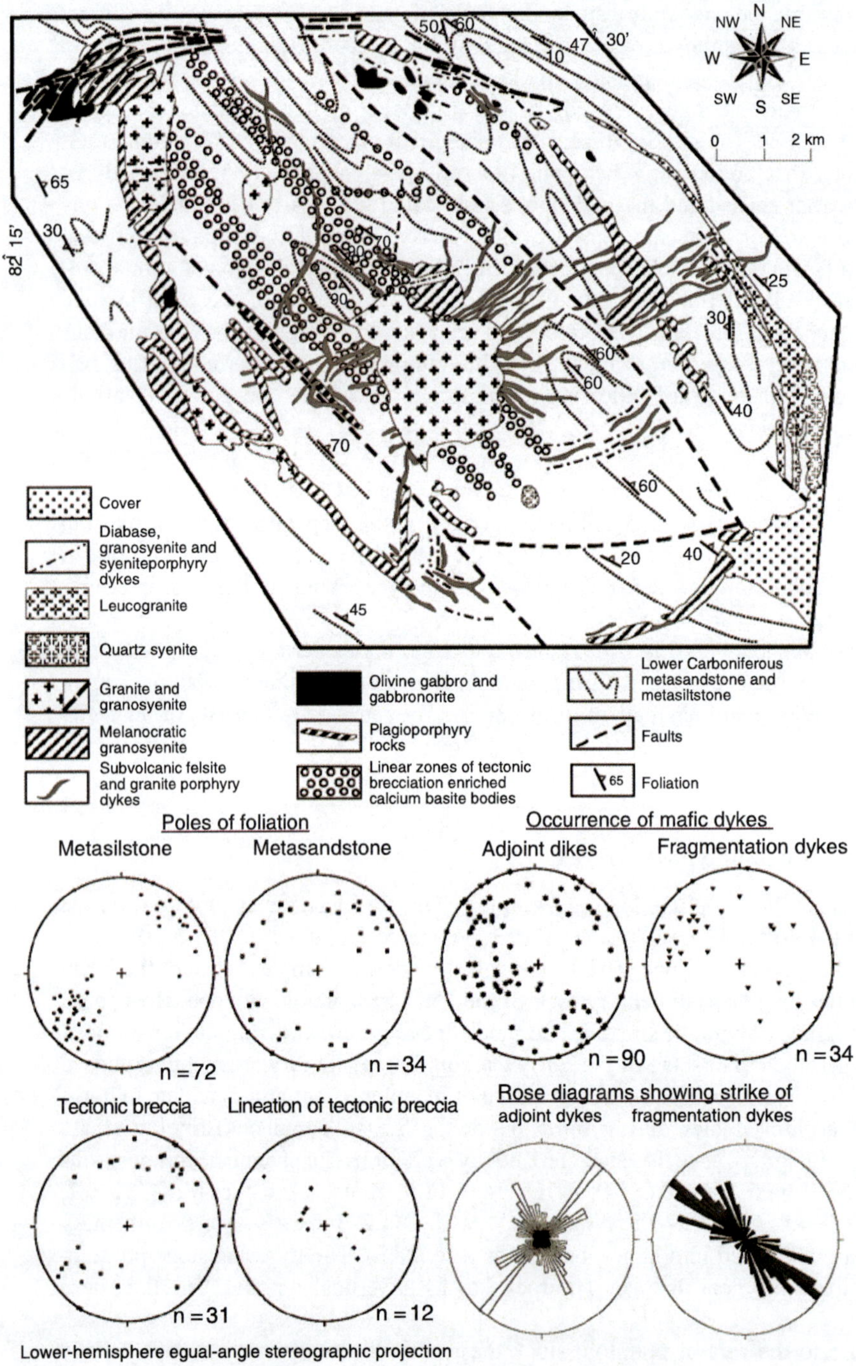

Fig. 28.2 Geological sketch map of the Tastau volcanoplutonic complex (after Ermolov et al., 1983; Dokukina, Vladimirov, 2008)

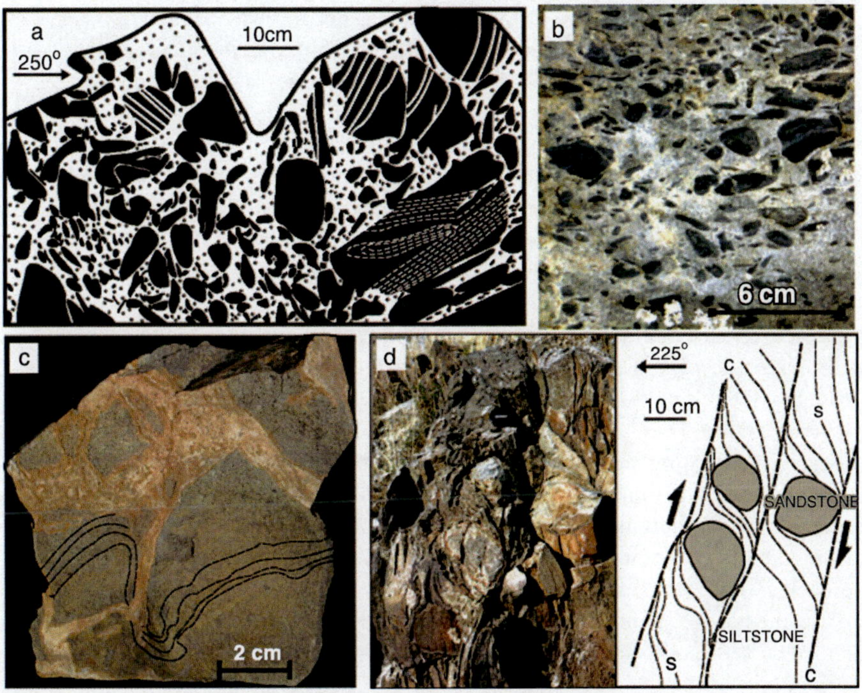

Fig. 28.3 (**a–b**) Sketch and outcrop photo of tectonic breccia and cataclasite of metasandstone and metasiltstone. (**c**) Tectonic crushed sandstone with sign of primary sediment bedding. Note absence of rotation of separate fragments. (**d**) Photographs and interpretation of deformation structures in metasedimentary rocks

orientation and traces of abrasion (Fig. 28.3a). Compressive shear deformation in metasedimentary rocks are fixed on C/C' and C/S tectonites (Fig. 28.3d) and sheath folds.

The volcanoplutonic complex has an ellipse shape (13 × 18 km). Arc-shaped belts of composite dykes intrude the metasedimentary rocks. A wide variety of magmatic rock is represented within the Tastau volcanoplutonic complex: leucogranite, granite, granosyenite, gabbronorite and gabbro-diorite and calcium basite. The granitoid stock and swarm of radial felsic dykes occupy center of the volcanoplutonic complex. Numerous small various intrusive calcium basite bodies of different morphology are restricted to the linear zones of the tectonic brecciation of the metasedimentary rock (Fig. 28.2). The calcium basite dykes gave birth to other bodies of globular and irregular morphologies. The calcium basite will be described in detail in a next part of the article.

Gabbronorite and gabbro-diorite occur as intrusive bodies in the northeastern and northwestern parts of the Tastau complex.

The bulk of the complex is made up of granosyenite and granite rock. Hypabyssal melano- and leucocratic granosyenite, granite and leucogranite are holocrystalline rocks, sometimes quenched in endocontact zones. Melanocratic granosyenite compose central part of the stock and outer intrusive ring. They are not deformed and cut felsic dykes and linear zone, which is rich in calcium basite bodies (Fig. 28.2). The transition between each granitoid phases is gradual and without sharp contacts. Metasedimentary xenoliths (ranging in size from 1 to 20 cm) are abundant within the Tastau complex. They mostly consist of metasedimentary enclaves and foliated gneissic fragments. They are randomly distributed throughout the intrusion. Another important feature of the Tastau volcanoplutonic complex is the extensive occurrence of mafic magmatic enclaves (MME) (Fig. 28.4a). MMEs range in size from tens of centimeters to several meters in some places. They are always darker than the host, with a fine-grained matrix and a porphyritic appearance. The shape, chemical composition, mineralogy and texture of MME undoubtedly support a magmatic origin as a result of repeated interactions between acid and basic magmas (Dokukina et al., 2010). MME mostly have scalloped contacts with the host (Fig. 28.4b). They are sometimes elongated parallel to the flow direction of the granite, suggesting plastic deformation in a partially liquid state (Vernon et al., 1988). MME are sometimes partially or completely disaggregated to form of schlieren structures and heterogeneous hybrid rocks. The composition of MME varies from gabbro, granodiorite to monzodiorite and monzogranite. This suggests that the enclave represents a gabbroic fragment that was incorporated and partly reacted with the host felsic magma.

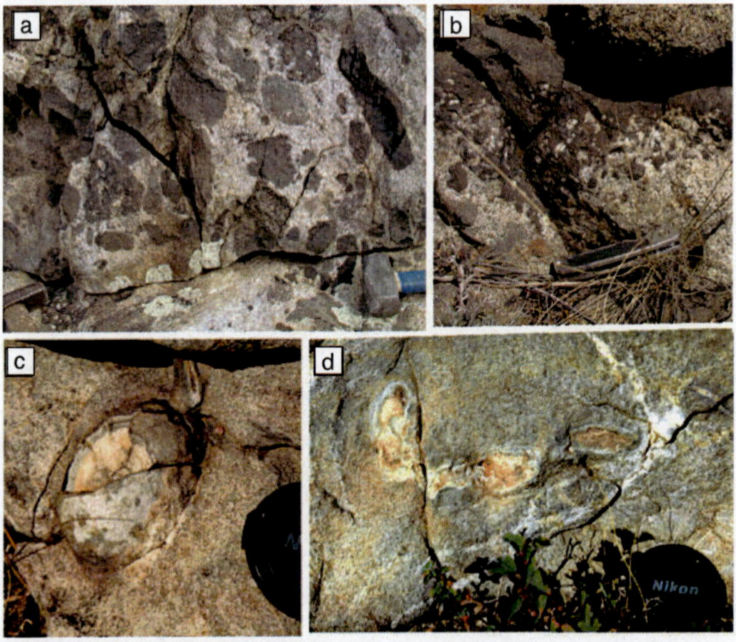

Fig. 28.4 Field photographs showing (**a**) small mafic magmatic enclaves (MME) in the host leucocratic granite. (**b**) Scalloped contact between MME and the host leucocratic granite. (**c**) Calcium basite nodule within a diorite. (**d**) Calcium basite nodules in granite elongated along the flow direction of the granite magma

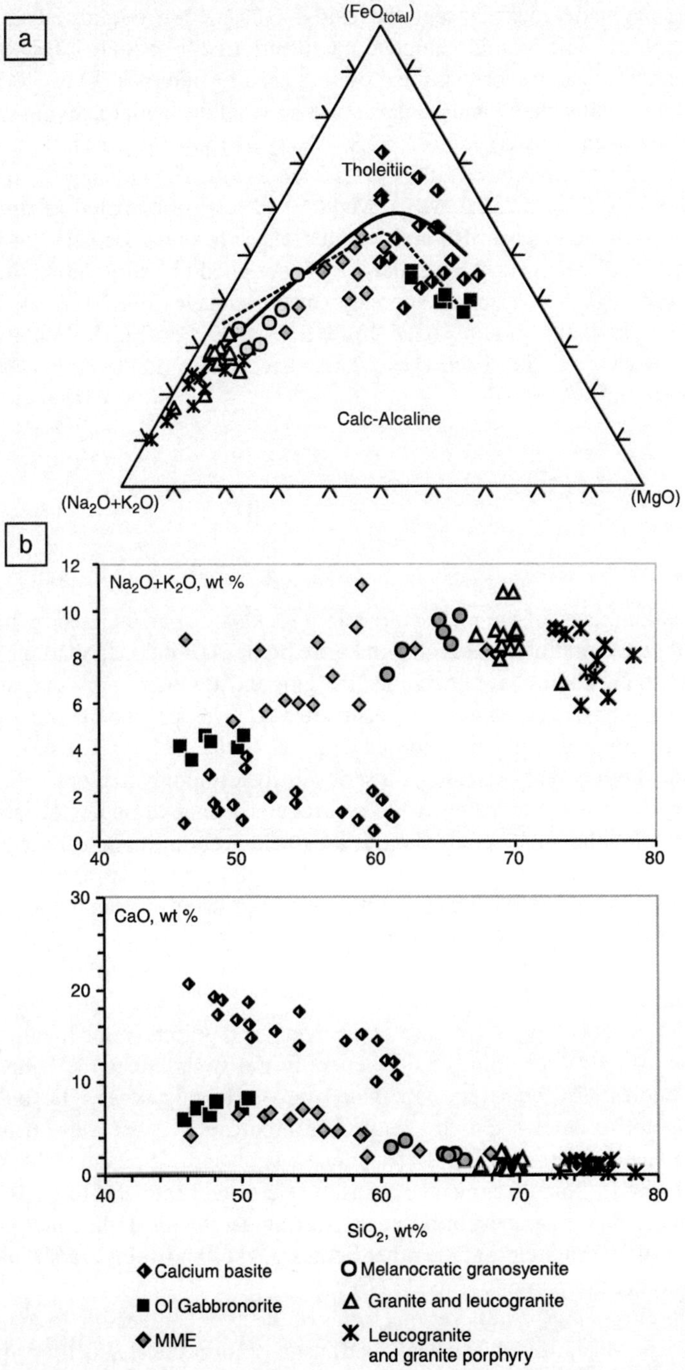

Fig. 28.5 (**a**) AFM diagram for magmatic rock of the Tastau volcanoplutonic complex; (**b**) variational diagrams (Na_2O+K_2O) *vs* SiO_2 and CaO *vs* SiO_2 for magmatic rock of the Tastau volcanoplutonic complex

The Tastau magmatic rocks except the calcium basite are coeval calc-alkaline series with wide variations in chemical compositions (46 < SiO_2 < 78 wt%) (Fig. 28.5). Nonhybrid gabbronorite is the most magnesium-rich rock (#Mg = 0.50–0.55). Synplutonic mafic dikes and enclaves are subalkaline gabbro, monzonite, syenite and quartz syenite (SiO_2 = 46.2–62.8, Al_2O_3 = 15.8–19.6, TiO_2 = 0.75–2.22, FeO_{tot} = 4.7–11.5, MgO = 1.8–5.4, CaO = 2–7 wt%) with high content of alkalies (Na_2O+K_2O = 5.2–9.3 wt%). Mafic enclaves are depleted relatively the gabbro in Sr, Ca, and rich in Rb and K. The felsic rocks are depleted in Ba, Sr, and rich in U. All rocks have a similar LREE-enriched chondrite-normalized REE patterns and negative europium anomaly (mafic enclaves $(Ce/Yb)_N$ = 4.99–5.54, Eu/Eu* = 0.68–0.78; granite $(Ce/Yb)_N$ = 5.48–8.59, Eu/Eu* = 0.24–0.64; felsic veins $(Ce/Yb)_N$ = 2.50, Eu/Eu* = 0.006), except the nonhybrid gabbro ($(Ce/Yb)_N$ = 5.34, Eu/Eu* = 1.02).

Tastau Calcium Basite

Morphology

Tectonic brecciation zones of the metasedimentary rocks are characterized by high (up to 30–40%) concentration of the calcium basite bodies (Dokukina, Vladimirov, 2005, 2008). All mafic bodies have got chilled margins and are located inside central part of the volcanoplutonic complex. The basite are also involved into the mingling process coupled with granitoids and diorites (Fig. 28.4c, d). The typical shape of mafic intrusions in horizontal exposure is a series of dykes, elongate globules, and pods varying in size from 2 to 70 cm wide or irregularly shaped bodies of larger size sometimes. Globular and irregular bodies are exclusively in the linear zone of tectonic brecciation of metagreywacke host rock (Fig. 28.2). In outcrop, the mafic bodies thus often look like beads that connected by a bead string. Tastau mafic bead strings can be tens of meters long and more.

Isolated mafic bodies can be subdivided into several classes, according to their shape and contacts with host rocks.

Dykes and Veins. This type is represented as dykes and adjoint veins mainly of 2–10 cm (up to 70 cm) thick (Fig. 28.6). Generally the veins are found coupled with all other bodies. The veins are subdivided into following types: (1) folded veins, including those deformed into sheath folds indicating shear deformation (Fig. 28.7a) (Passchier, Trouw, 1996); (2) fragmented veins indicating ductile deformation (Fig. 28.7b–d). Some fragmented veins are deformed into sheath folds, at the same time individual fragments of basite are surrounded by the chilled margins; (3) adjoint veins often cross-cutting the other bodies. Dykes and veins are a source of the bodies of other morphology (Fig. 28.6b).

Irregular type (Figs. 28.7, 28.8). Various forms of this type (amoeboid, irregular, interconnected, drop-like, fluidal, and scalloped) usually form the chains following directions of veins and dykes, which were sources of matter for these fragments (Figs. 28.6b, 28.7b). Irregular bodies have scalloped, fluidal, fine-fiber, flame-like

Fig. 28.6 Field photographs showing dyke (**a**) and adjoint veins (**b**). Note adjoint veins are source of globular bodies

and diffusive boundaries (Fig. 28.7b–d). In area where the host rock was the most disintegrated the mafic melt impregnate and assimilate the metasediment (Fig. 28.8).

Globular type (Fig. 28.9). On the weathered surface rounded egg-like forms represent this type (Fig. 28.8a). Sections of pillow unveiling their inner structure demonstrate pronounced exocontact and endocontact zones (Fig. 28.9b, c). Spheroid fragments usually are scattered within the outcrops and form the globules, elongate globules, "hollow" (Fig. 28.9d) and bag-like bodies (Fig. 28.9e). Metasedimentary fragments sometimes penetrated into the basite bodies and chilled margins were formed around them (Fig. 28.9d, e).

All described types of mafic bodies can be found in one outcrop where their combinations look fantastic scenes. Within the outcrop it is possible to see dykes and veins, which take a stand of brittle fractures. These dykes and veins simultaneously feel ductile deformation and folding and are a source of the globular and irregular bodies (Figs. 28.6–28.9).

Comparative Structural Analysis of Mafic Veins and Tectonites

Schistosity of metasiltstone generally is characterized by northwest strike with southwest and northeast subvertical dips (Fig. 28.2). Tectonic breccias and cataclasites have a northwest-southeast strike with southwest and northeast dips.

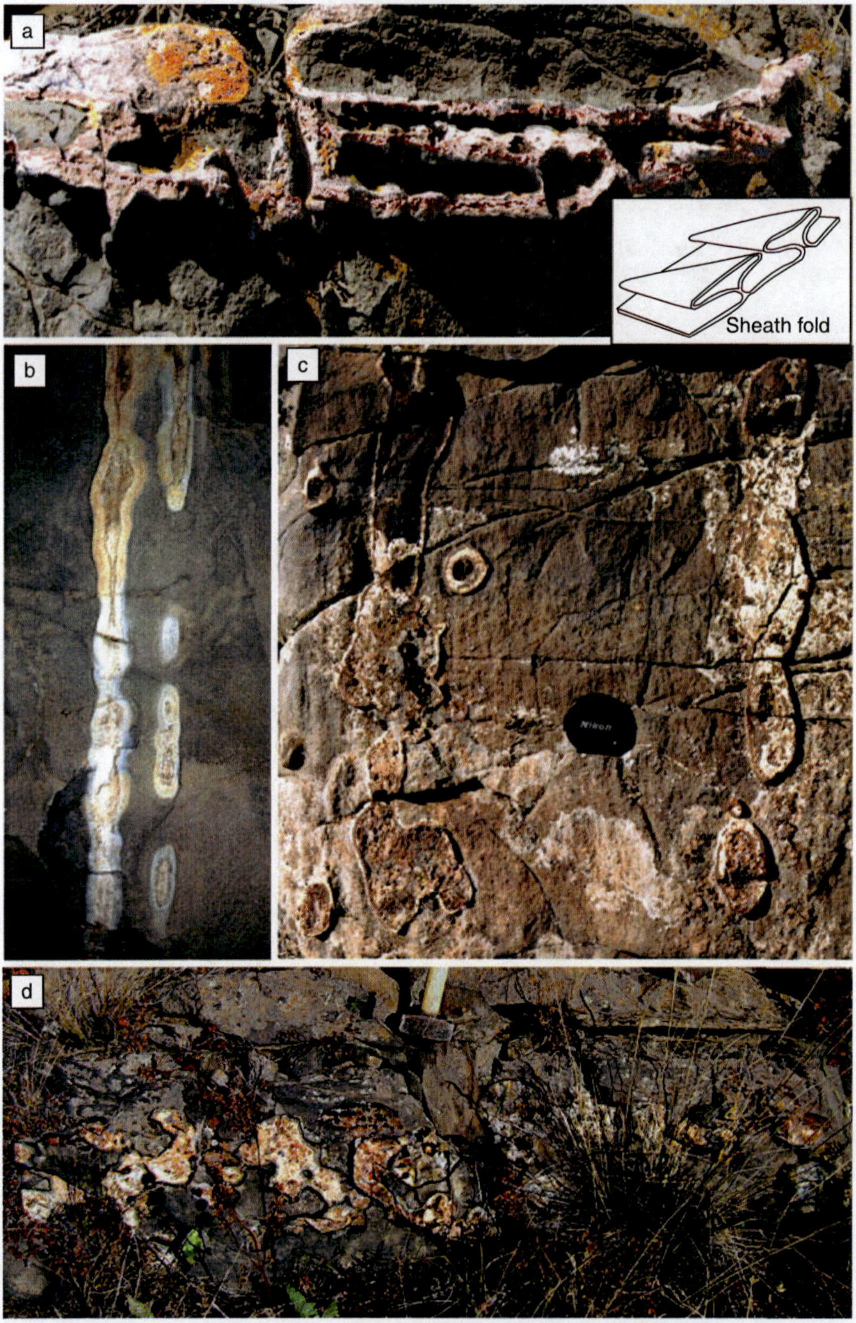

Fig. 28.7 Field photographs showing irregular mafic bodies. (**a**) Folded vein. (**b**) Fragmented veins "bead string". (**c**) Combinations of irregular and steroidal mafic bodies. (**d**) Echelon irregular bodies

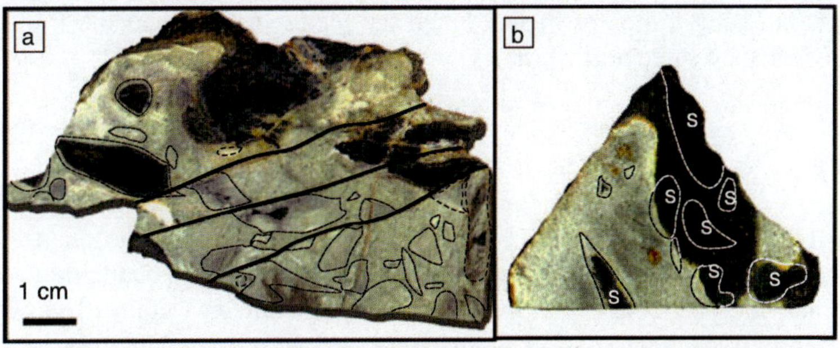

Fig. 28.8 Photographs of sample Bc721-6. (**a–b**) Capturing and assimilation of brecciated metasedimentary rock by the calcium basite melt. In example (**a**) the cleavage structure inherits by disposition of the assimilated fragments of siltstone

Fig. 28.9 Field photographs showing globular mafic bodies. (**a**) Mafic globules in outcrop. (**b**) Large mafic globule. (**c**) Zonal composition of mafic bodies: 1 – central part, 2 – chilled margin, 3 – hornfels zone, 4 – low-grade sediment. Enlarged fragment from (b). (**d**) "Hollow" globular mafic bodies. Note the chilled margin forms from the outside of body and inside around metasiltstone xenolith. Here the internal chilled margin did not prove a metasomatic changing in contrast to the external margin. (**e**) Bag-like mafic body

Stretching lineation of tectonites generally trends northwest with southwest and northeast steep dips.

Fragmented basite veins have two trends of stretch. First is well-marked northwest strike. Second strike is less-marked north-northeast. Bedding of fragmented basite veins swarm is consistent with strike of the linear zone of tectonic breccias. Undeformed basite veins have different trends (Fig. 28.2). However most of undeformed veins have northeast strike with gentle and steep dips.

Thus the dominating trends of metasedimentary rock tectonites and the fragmented basite veins are same northwest.

Mineralogy

The calcium basite is a massive green-gray, dark-gray, black rock. The quartz-clinopyroxene–plagioclase basite is the prevailing rock type, while quartz-clinopyroxene-hornblende-plagioclase varieties are rare. These are fine-grained rocks composed essentially of plagioclase, clinopyroxene, sometimes amphibole (green hornblende), quartz, and subordinate apatite, allanite, ilmenite, titanite, pyrite and zircon. Typical secondary minerals are chlorite, actinolite and epidote. Plagioclase, clinopyroxene and quartz occur as resorbed fission-fragment phenocrysts and in the fine-grained matrix. Phenocryst and groundmass minerals have similar composition. Relatively large grains of quartz are often corroded and granular, that supposes their xenogenic nature. Pnenocrysts and their fragments make up 15–40% of the rock.

Plagioclase

Representative compositions of plagioclase are shown in Tables 28.1–28.3. Plagioclase exhibits bytownite-anorthite ($An_{78–98}$) composition within the calcium basite, and an anorthite ($An_{90–96}$) prevails. Two broad types of plagioclase have been recognized: (1) Small tabular or fission-fragment and resorbed phenocrysts or glomerocrysts in the fine-grained matrix. (2) Tabular crystallites in a groundmass structure coupled with quartz and clinopyroxene (Fig. 28.10a, b). In the matrix, plagioclase is the most abundant mineral with tabular habit and the same calcium composition.

The plagioclase in narrow transition band on contact between basite and host rock tends change it compositions with decreasing of calcium content ($An_{40–75}$) (Fig. 28.10c) that apparently it is connected to a diffusion exchange between sedimentary and mafic magmatic components. An example such interaction is in Fig. 28.10d: a bytownite core (An_{89}) gradually changes to andesine rim ($An_{40–50}$).

Clinopyroxene

Representative clinopyroxene compositions are shown in Tables 28.1–28.3. Salitic clinopyroxene occurs in the matrix along with plagioclase and quartz, and also forms fine-grained aggregates (Fig. 28.10b). Rare resorbed relics of primary zoned tabular clinopyroxene are in basite (Fig. 28.11a). A core of clinopyroxene has higher magnesium content than its rim at equal calcium content. Probably at brusque decompression, which accompanied of basite melt intrusion, the crystallized phenocrysts of clinopyroxene and plagioclase collapsed with new middle composition clinopyroxene forming (Fig. 28.11b). In chilled margins of basite bodies the clinopyroxene forms fine-grain aggregate at the border with coeval magmatic rock (Fig. 28.12a); and fine-grain aggregate with titanite at the border with metasedimentary rock (Fig. 28.11b, 28.12b).

Table 28.1 Representative electron microprobe analyses of minerals from sample kt225/2 in chilled margin of calcium basite globule with primary clinopyroxene and secondary clinoamphibole[a]

	Cpx12	Cam15	Pl 14	Tnt13
Wt% oxides				
SiO_2	51.71	51.85	46.16	30.68
TiO_2	0	0.36	0.00	40.06
Al_2O_3	0.21	4.46	34.20	1.45
Cr_2O_3	0	0	0.04	0
FeO*	15.79	16.54	0.05	0.32
MnO	1.79	0.89	0	0.21
MgO	7.80	12.09	0	0
CaO	23.71	12.27	17.70	28.56
Na_2O	0.05	0.14	1.65	0.08
K_2O	0.02	0.12	0	0
Total	101.08	98.72	99.80	101.36
Cations				
Si	1.994	7.543	2.129	0.988
Al IV	0.006	0.457	1.859	0
Al VI	0.003	0.308	0	0.055
Ti	0	0.039	0	0.970
Cr	0	0	0.001	0
Fe^{3+}	0	0	0	0
Fe^{2+}	0.509	2.012	0.002	0.009
Mn	0.058	0.110	0	0.006
Mg	0.448	2.622	0	0
Ca	0.980	1.913	0.875	0.985
Na	0.004	0.039	0.148	0.005
K	0.001	0.022	0	0
Total	4.004	15.066	5.014	3.017
O	6	23	8	5
X′	0.532	0.434	0.856	–

Notes: * – all iron as FeO. X' = Fe/(Fe+Mg) for clinopyroxene and amphibole. X' = Ca/(Ca+Na+K) for plagioclases.

[a]Microprobe analyses of coexisting minerals (Tables 28.1, 28.2 and 28.3) were performed on polished thin sections following petrographic examination. CamScan X-ray microprobe equipped with a Link EDS analytical system (Institute of Experimental Mineralogy. Russian Academy of Sciences) was used for the analyses. The operating conditions were: 20 kV accelerating voltage. 1.2 mA beam current, 70 s counting time, and a beam 1–5 μm in diameter or, sometimes, analysis in 100–150 μm raster mode, when it was necessary to determine the composition of cryptocrystalline mineral intergrowth.

Amphibole

Amphibole is formed along the edge of calcium basite, occasionally in matrix coupled with clinopyroxene. Near the contact, the mineral size diminishes; amphibole appears instead of clinopyroxene (Fig, 28.12c, Table 28.1) and as split crystallites

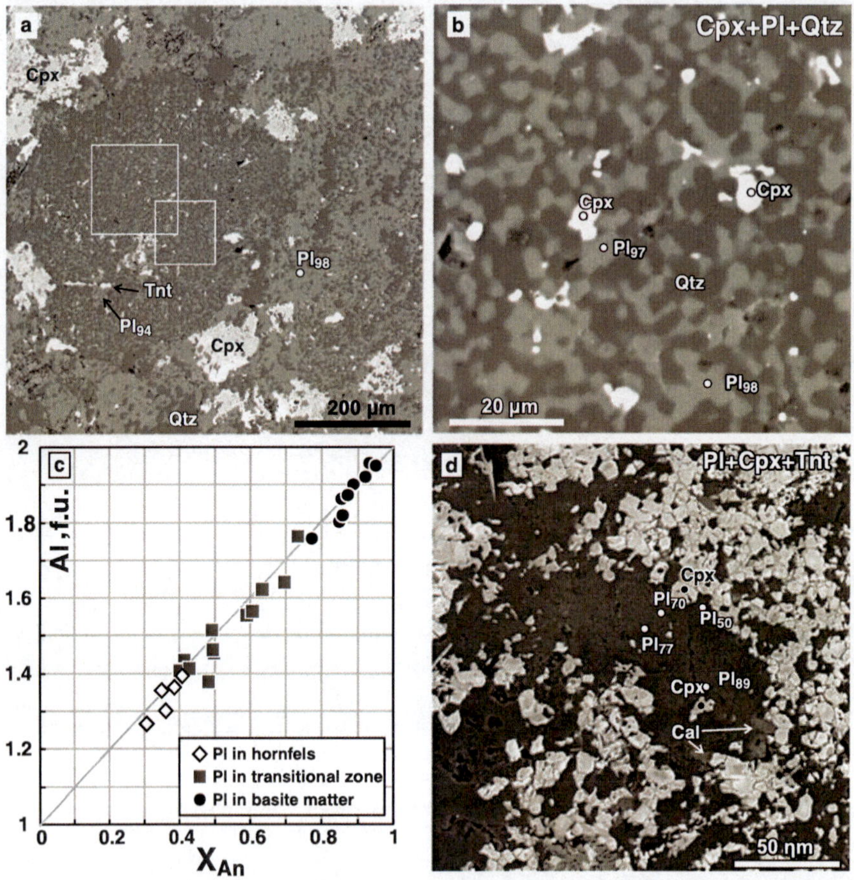

Fig. 28.10 (**a–b**) BSE images of calcium basite with clinopyroxene-plagioclase-quartz assemblage, sample kt140. (b) Enlarged fragment from (a). Note kotectic intergrowths of salite, anorthite and quartz. (**c**) Plagioclase composition at the border between the calcium basite and metasedimentary rock. (**d**) Zoned plagioclase at the border between the calcium basite and metasedimentary rock with clinopyroxene and titanite, sample kt88-3. There and everywhere mineral abbreviations after Kretz (1983), letters for clinopyroxene are iron numbers (100*Fe/(Fe+Mg)), for plagioclase – mole percent of anorthite

upon cryptocrystalline or vitreous basite (Fig. 28.12d). Amphibole morphology indicates the rapid growth of minerals at supercooling.

Xenogenic amphibole inclusions are in basite, which are the most crust contaminated. Example such amphibole (Fig. 28.12e) has varying composition and appreciably differ from all other calcium basite amphiboles by higher aluminum and titanium content (Fig. 28.13a). Around of the amphibole the clinopyroxene prograde formed (Fig. 28.12e) probably at chilling of a captured grain. Amphibole contains numerous inclusions of barium-potassium feldspar (Fig. 28.12e, increased fragment in inset) with different content of barium (Fig. 28.13b).

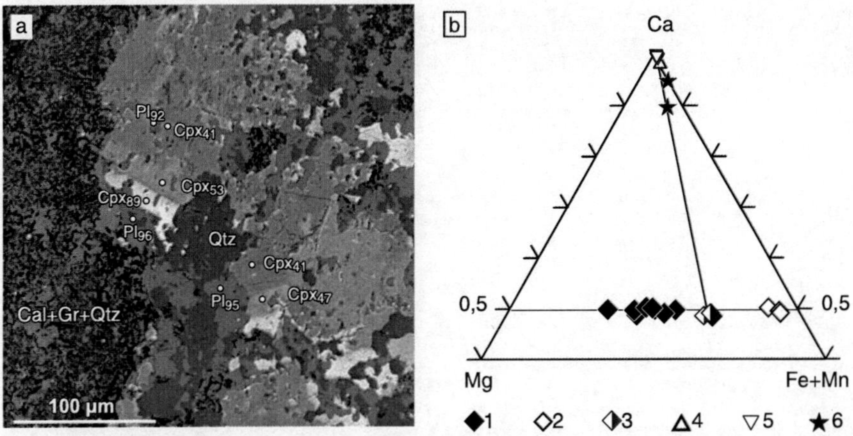

Fig. 28.11 (**a**) BSE image, sample kt140. Primary salitic clinopyroxene phenocryst probably formed in deeper magmatic camera and altered at decompression and emplacement (see text). Note symplectite-like intergrowths of calcite, graphite and quartz on the left. (**b**) Ternary Mg-Ca-(Fe+Mn) diagram showing chemical compositions of rock and minerals from sample kt140: 1 – cores of clinopyroxene phenocrysts, 2 – rims of clinopyroxene phenocrysts, 3 – clinopyroxene from fine-grain matrix, 4 – plagioclase, 5 – plagioclase from Cpx-Pl-Qtz kotectic intergrowths, 6 – Re-integrated bulk composition of fine-grain matrix showed on Fig. 28.10a, b

Calcite and Graphite

Calcite and graphite occur as well as individual grains both symplectite-like intergrowths of calcite, graphite and quartz. Individual grains of calcite formed as primary crystals in equilibrium with Cpx-Pl-Qtz assemblage or as secondary minerals after basite matter. Calcite-graphite-quartz intergrowths formed in interstitial (Fig. 28.11a) and have a wide enough distribution.

Accessory Minerals

Ilmenite, titanite, pyrite, apatite, epidote, clinozoisite, allanite (Ce-La mineralization) and zircon are routine accessory minerals. Ilmenite, titanite, pyrite are the most prevalent ore mineral. Apatite is fluoric (F 2.6 wt%; $X_F = 0.7$) and REE-bearing titanite has been found as a single grain with changing from core to rims of rare-earth element oxides content from 4.94 to 0.2 wt% (La_2O_3 0.13–1.00, Ce_2O_3 0.2–1.84, Nd_2O_3 0.21–1.65 wt%).

Exotic minerals – chromium spinel ($Y_{Cr} = 0.89 \div 0.76$, $Y_{Al} = 0.09 \div 0.24$, $X_{Mg} = 0$), pyrope garnet (0.55 pyrope, 0.17 almandine, and 0.28 grossular moles) and moissanite were found in the crushed rock heavy fraction. These minerals can interpreted as mantle xenocrysts and are evidence of a deep origin of the calcium melt.

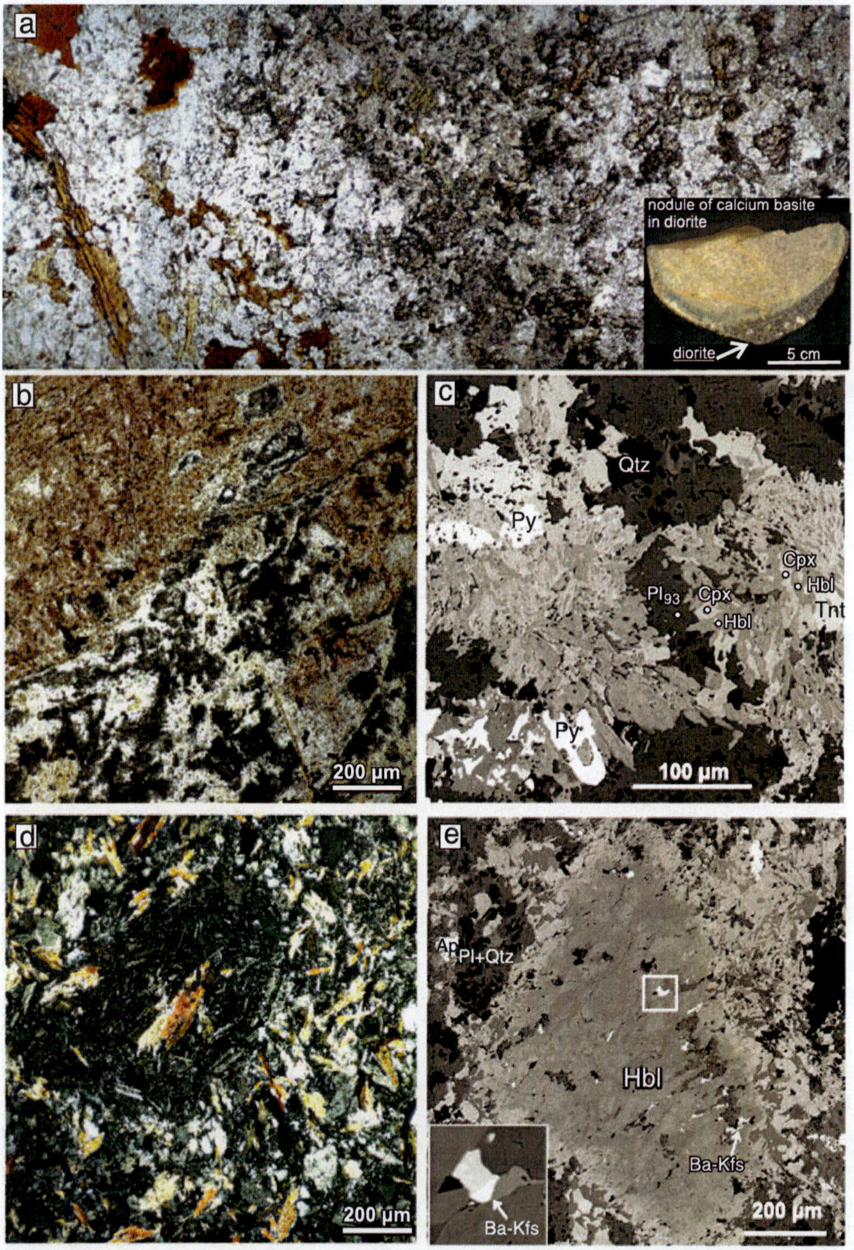

Fig. 28.12 (**a**) Photomicrograph of thin section of the calcium basite nodule in diorites, sample kt122-4, plane polarized light. Photo of this nodule is in an *inset*. Note decreasing of a clinopyroxene grain size at the border between the calcium basite and diorite. (**b**) Photomicrograph of thin section, sample kt88-3, plane polarized light. Flame-like contact of basite matter and hornfels of metasiltstone. Mineral assemblage is fine grain intergrowths of salite and titanite, intermediate plagioclase, rare rhombic amphibole and quartz (see Fig. 28.10d). Hornfels compose by

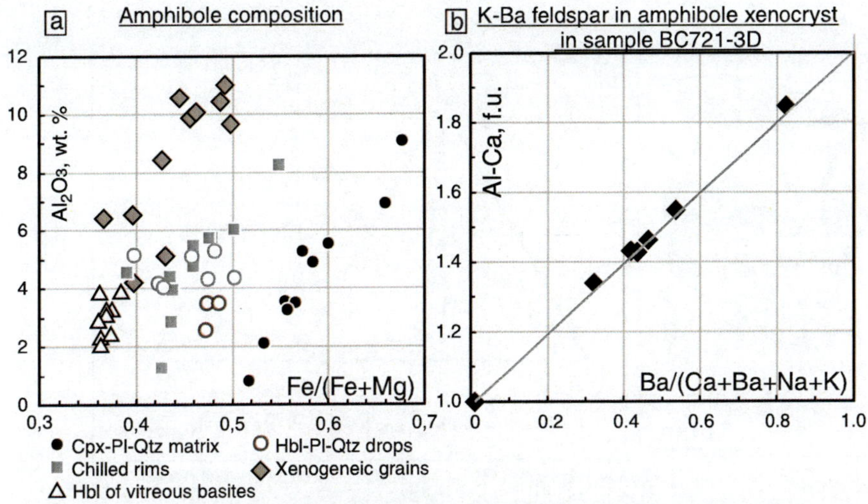

Fig. 28.13 (**a**) Composition diagram Fe/(Fe+Mg) *vs* Al_2O_3 for amphiboles of calcium basite; (**b**) composition diagram Ba/(Ca+Ba+Na+K) *vs* Al-Ca for hyalophane inclusions in xenogenic hornblende

Liquation Structures

Rounded black nodules of quartz-amphibole-plagioclase compositions sometimes are in the calcium basites (Fig. 28.14, Table 28.2). These drops consist of rhombic prismatic amphibole crystals in fine-grain quartz-plagioclase matrix with numerous accessory pyrites (Fig. 28.14c, d). A margin between clinopyroxene-plagioclase and amphibole-plagioclase parts of basite is sharp without a contact changes. Plagioclase composition in nodules is identical with plagioclase in the clinopyroxene-plagioclase part. In the Fig. 28.14 Area 1 corresponds to silicate-oxide host of silicate-sulfide nodule (area 2). Area 1 represents the typical mineralogical assemblage and textures of silicate-oxide globules. Note in the Area 2 the abundance of pyrite (white). The elements distribution maps (Fig. 28.15) illustrate the increase of iron and sulfur in nodule. Ternary MFA and MFC diagrams display the phase and bulk chemistry relationships of silicate-oxide assemblage in area 1 and silicate-sulfide assemblage in Area 2 (Fig. 28.14b). We think these silicate-sulfide nodules can be liquation structures inside calcium basite bodies.

Fig. 28.12 (continued) fine-grain aggregate of biotite, potassium feldspar, intermediate plagioclase and quartz. (**c**) BSE image of neogenic amphibole after clinopyroxene in endocontact of calcium basite globule from metasiltstone, sample kt225-2. (**d**) Split amphibole crystallites upon cryptocrystalline or vitreous basite body, sample kt139, transmitted cross-polarized light. (**e**) BSE image of xenogenic hornblende with hyalophane inclusions with different celsian content (enlarged fragment in *inset*), sample Bc719-3D. Note prograde growth of salite clinopyroxene around amphibole inclusion

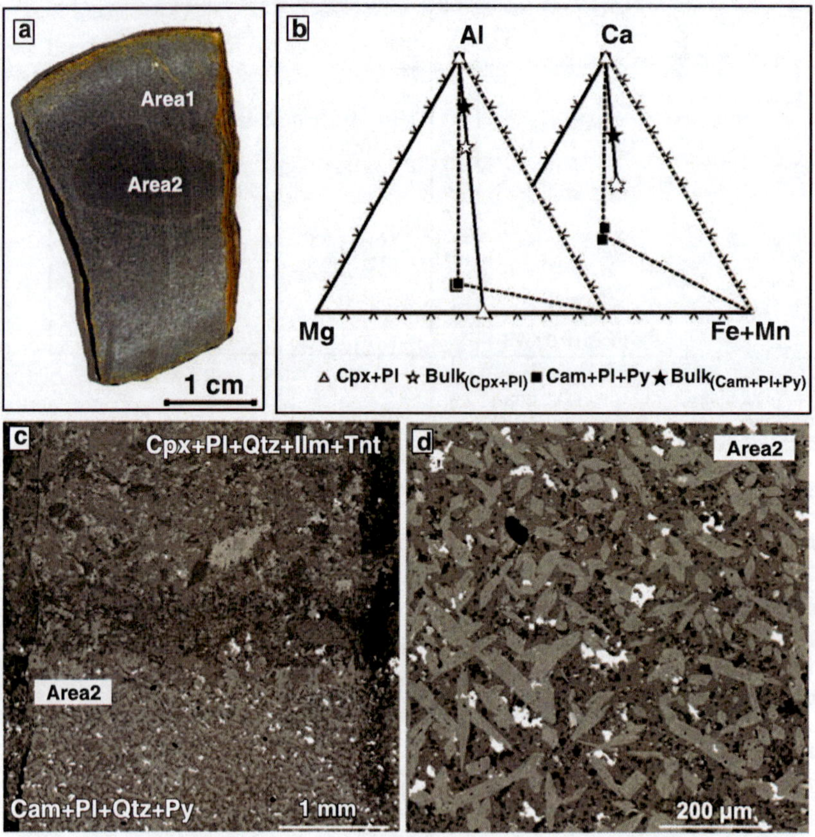

Fig. 28.14 (**a**) Liquation structure (dark amphibole-plagioclase-quartz-pyrite drop) inside of a clinopyroxene-plagioclase-quartz basite globule, sample Bc719-1; (**b**) Ternary MFA and MFC diagrams display the phase and bulk chemistry relationships of a silicate-oxide assemblage in area 1 (*white star and triangles*) and a silicate-sulfide assemblage in area 2 (*black star and boxes*). (**c**) BSE image of margin between the Hbl-Pl-Qtz-Py drop and Cpx-Pl-Qtz matrix. Area 1 corresponds to the silicate-oxide host of the silicate-sulfide nodule (area 2). Area 1 represents the typical mineralogical assemblage and textures of the silicate-oxide globules. Note in the area 2 the abundance of pyrite (*white*). (**d**) Enlarged BSE image of the silicate-sulfide nodule (area 2)

Intergranular Glass

Sometimes a magmatic glass is in interstices of basite (Fig. 28.16, Table 28.3). The glass composition is $SiO_2 = 38.8–43.4$, $Al_2O_3 = 30.5–35.9$, $TiO_2 = 0.00–0.25$, $FeO^{tot} = 0.03–1.37$, $MnO = 0.0–0.1$, $MgO = 0.10–0.42$, $Na_2O = 0–0.24$, $K_2O = 0.23–1.02$, $CaO = 0.47–1.59$ wt% at low cation totals (81–84.5 wt%). We do not know which elements determined so high losses, but suppose it can be carbon. In basite often graphite is found along with calcite and quartz (Fig. 28.11a). We think

Table 28.2 Representative electron microprobe analyses of minerals from sample Bc719-1 of calcium basite with liquation drop

	Cpx 3	Cpx18	Pl 2	Ilm 1	Cam26	Cam25	Pl 28	Scp27	Tnt29
Type	–	Host	–	–	Core	Rim	–	Incl	–
Location	Area 1				Area 2				
Wt.% oxides									
SiO_2	50.17	50.48	44.11	0.06	49.11	48.90	46.84	46.00	29.61
TiO_2	0.05	0.10	0.04	53.35	0.09	0.24	0.05	0.25	36.81
Al_2O_3	0.05	0.40	34.00	0.05	4.02	4.17	33.64	27.58	1.70
Cr_2O_3	0	0.11	0	0.02	0.19	0.11	0	0.01	0
FeO*	15.83	13.15	0.18	35.74	15.56	16.94	0.19	1.32	0.56
MnO	1.58	1.59	0	11.64	1.09	1.20	0	0.11	0.28
MgO	6.86	9.34	0	0	12.01	10.57	0	0.62	0.07
CaO	22.74	22.34	18.53	0.43	11.71	11.75	17.58	14.83	27.83
Na_2O	0.12	0.11	0.93	0	0.26	0.42	1.70	2.44	0.04
K_2O	0	0.03	0	0	0.23	0.16	0	0.09	0
Total	97.40	97.65	97.79	101.29	94.27	94.46	100.00	93.25	96.90
Cations									
Si	2.010	1.989	2.084	0.001	7.507	7.513	2.155	7.031	0.999
Al IV	0	0.011	1.894	0.001	0.493	0.487	1.824	4.969	0
Al VI	0.002	0.008	0	0	0.231	0.268	0	0	0.068
Ti	0.002	0.003	0.001	0.997	0.010	0.028	0.002	0.029	0.934
Cr	0	0.003	0	0	0.023	0.013	0	0.001	0
Fe‴	0	0	0	0.001	0	0	0	0	0
Fe″	0.530	0.433	0.007	0.742	1.989	2.177	0.007	0.169	0.016
Mn	0.054	0.053	0	0.245	0.141	0.156	0	0.014	0.008
Mg	0.410	0.549	0	0	2.737	2.421	0	0.141	0.004
Ca	0.976	0.943	0.938	0.011	1.918	1.934	0.866	2.429	1.006
Na	0.009	0.008	0.085	0	0.077	0.125	0.151	0.723	0.003
K	0	0.002	0	0	0.045	0.031	0	0.018	0
Total	3.992	4.002	5.010	2.000	15.170	15.153	5.007	15.524	3.035
O	6	6	8	3	23	23	8	24,698	5
X′	0.564	0.441	0.917	–	0.421	0.473	0.851	0.656	–

Notes: * – all iron as FeO; X' = Fe/(Fe+Mg) for clinopyroxene and amphibole, X' = Ca/(Ca+Na+K) for plagioclase, X' = AnEq (Al-3)/3 for scapolite. Ilmenite formula was calculated on 2 cation total and 3 oxygen atom basis. Scapolites formula was calculated on the basis of Si+Al = 12.

the basic melt was concentrated of CO_2, which is evidence of a basite melt waterless. Therefore the amphibole growth in margins of basite bodies in metasedimentary rock probably is coherent with a dry basite melt hydration by a water fluid from metasedimentary rock.

PT Estimation

Precise *PT*-parameters of crystallization the basite to determine it was not possible. Rough estimate of crystallization condition of clinopyroxene-amphibole-plagioclase-quartz equilibrium assemblage from the chilled margins of mafic

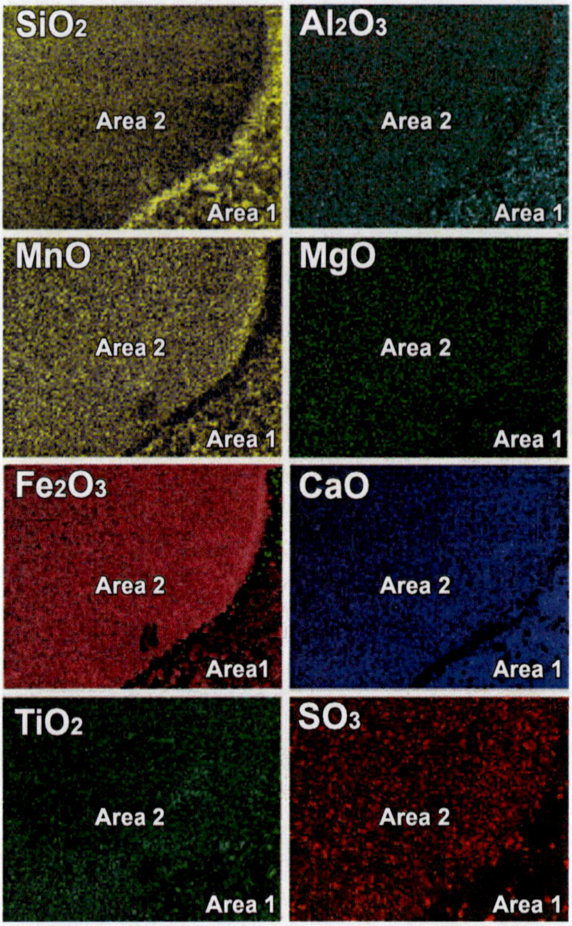

Fig. 28.15 The elements distribution maps illustrate the increase of iron and sulfur in the silicate-sulfide nodule, sample Bc719-1. The bulk composition and element distribution maps in thin section was determined by X-ray fluorescence analysis with an Eagle III-Probe micro-XRF elemental analyzer at the Kurnakov Institute of General and Inorganic Chemistry of the Russian Academy of Sciences, Moscow

bodies corresponds to low pressure about 1 kb and 600–700°C by using Hbl-Pl-Qtz geothermometer (Blundy, Holland, 1990) by hornfels and CaTs-An-Qtz geobarometer (McCarthy, Patiño Douce, 1998) by basites.

Geochemistry

The whole-rock composition of the calcium basite is characterized by wide variations of all major elements (SiO_2 = 46.2–61.2, Al_2O_3 = 12.6–17.7, TiO_2 =

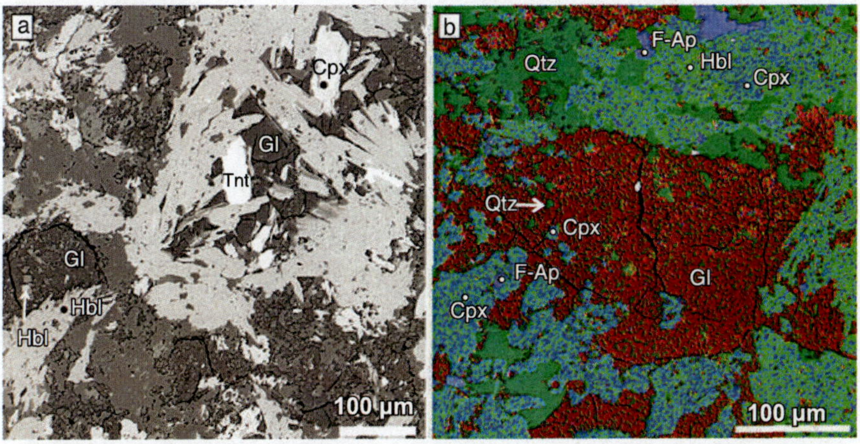

Fig. 28.16 (**a**) BSE image showing the nature of the devitrified intergranular glass with desiccation cracks. Gl – glass, sample kt139. (**b**) False-colored BSE image of the intergranular glass illustrates distribution of elements: Al (*red*), Si (*green*) and Ca (*blue*), sample kt139

0.55–0.85, FeO_{tot} = 3.77–6.87, MnO = 0.35–0.68, MgO = 2.0–5.64, LOI = 0.59–11.5 wt%), low alkali content ($Na_2O + K_2O$ = 0.78–2.9 wt%) and high content of CaO (10.8–20.7 wt%), Table 28.4. Calcium content has inverse relationship with silica content but for all that a dependence of these oxides with rock iron content. The silica content in some analyses can be top-heavy than in primary melt since a xenogenic quartz content is very high. Chemical and mineral composition of the basite from metasedimentary rock is the same as the basite from granite and diorite (Table 28.4). Calcium basite is depleted relatively the nonhybrid gabbronorite and MME in LILE and enriched in HFSE (Fig. 28.17). The chondrite-normalized REE pattern of calcium basite is characterized by low enrichment in LREE ($(La/Lu)_N$ = 2–12, ΣREE 75–185 ppm) and a feebly marked europium anomaly (Eu/Eu* = 0.84–1.06). Thus chemical composition of calcium basite fundamentally differs from compositions everything else subalkaline Tastau magmatic rock (Figs. 28.5, 28.17).

Discussion

Origin of Calcium Basites

Tastau calcium basite bodies have untypical composition for basalt magma derivations. High calcium content controlled by main rock-forming minerals: salite, anorthite and calcite. Such magmatic extreme calcium mineral paragenesis (salite+anorthite) meets in a cumulate of mafic layered intrusions, but are a rarity in subvolcanic condition and are abundant in metasomatic rock, for example in a calcic skarn. However in Tastau volcanoplutonic complex a carbonate

Table 28.3 Representative electron microprobe analyses of minerals and melt from sample kt139 of calcium basite

	Glass 1	Glass 4	Qtz 3	Cpx 2	Hbl 5	Hbl 6	Pl 30
Location	–	–	–	Core	In glass	In matrix	Core
Wt.% oxides							
SiO_2	45.66	43.31	98.60	51.18	51.07	52.28	45.85
TiO_2	0.05	0.22	0.14	0	0.01	0.17	0.07
Al_2O_3	37.07	35.89	0.05	1.20	5.13	4.38	34.46
Cr_2O_3	0.18	0.27	0.07	0.02	0.03	0.07	0
FeO*	0.24	0	0.06	13.87	13.98	12.83	0
MnO	0	0	0	1.43	0.92	0.70	0.13
MgO	0.24	0.28	0	9.08	13.03	13.80	0
CaO	0.84	0.67	0	23.17	12.58	12.59	18.28
Na_2O	0.11	0.02	0	0.05	0.37	0.24	1.22
K_2O	0.19	0.32	0	0	0.19	0.06	0
Total	84.58	80.98	98.92	100	97.32	97.12	100
Cations							
Si			0.998	1.972	7.476	7.596	2.113
Al IV			0.001	0.028	0.524	0.404	1.871
Al VI			0	0.026	0.362	0.346	0
Ti			0.001	0	0.001	0.019	0.003
Cr			0.001	0.001	0.003	0.008	0
Fe^{3+}			0	0	0	0	0
Fe^{2+}			0.001	0.447	1.711	1.559	0
Mn			0	0.047	0.114	0.086	0.005
Mg			0	0.521	2.843	2.989	0
Ca			0	0.957	1.973	1.960	0.902
Na			0	0.004	0.105	0.068	0.109
K			0	0	0.036	0.011	0
Total			1.001	4.003	15.148	15.046	5.003
O			2	6	23	23	8
X′			–	0.462	0.376	0.343	0.892

Notes: * – all iron as FeO. X' = Fe/(Fe+Mg) for clinopyroxene and amphiboles, X' = Ca/(Ca+Na+K) for plagioclases. Compositions of Ca-clinoamphibole included in the melt and matrix grain one are represented. Analytical data of quartz is given for comparison of oxides total in the melt and of anhydrous mineral.

rock (limestone, marl or other) misses, i.e. there is not a sedimentary component for metasomatic interaction with any magmatic intrusion. Also metasomatic columns and typical calcic skarn mineral (so as grossular, wollastonite and other) are not discovered in the calcium basite bodies. Sharp intrusive contact between calcium basite dykes and host metasedimentary rock, regular mineral composition and other features described above suggest the certain magmatic origin of basites. Other open question is how and at what conditions was generated such unusual mafic melt.

We can offer at least two possible scripts of an origin of the calcium mafic melt.

First preferred scenario is carbonatite-like melt forming. Calcium carbonate melt could interact and assimilate a tholeitiic basalt melt in an intermediate magmatic chamber at ascent from a mantle to upper level of crust. The facts answering to

Table 28.4 Whole-rock oxide (wt.%) and trace element (ppm) data for the calcium basites of the Tastau volcanoplutonic complex

Sample	kt 122-2	kt 122/5	kt 122/6	Bc 721/15	Bc 721/15-2	kt 108	kt 131/2	kt 134	kt 134/1	kt 139	kt 140
Host rock	Diorite	Diorite	Diorite	Ms	Ms	Ms	Ms	Ms	Ms	Ms	Ms
wt%											
SiO_2	54.2	57.5	58.7	48.6	50.4	48	60.4	61	46.2	61.2	59.8
Al_2O_3	14.8	15.9	14.9	16.2	17.7	14.2	15.1	15.5	13	12.6	15.4
TiO_2	0.72	0.69	0.66	0.83	0.88	0.72	0.69	0.7	0.55	0.63	0.71
Fe_2O_3t	5.44	4.87	5.22	7.25	666	6.8	5.13	4.85	4.19	6.28	4.94
MnO	0.28	0.33	0.43	0.35	0.57	0.37	0.3	0.31	0.59	0.34	0.43
MgO	2.26	2.33	2.27	3.75	2.42	3.31	2.23	2.24	2.17	5.64	2
CaO	17.60	14.4	15.2	18.9	18.7	19.3	12.3	12.1	20.7	10.8	14.4
Na_2O	0.94	0.75	0.45	1.12	0.83	2.73	1.42	0.77	0.53	0.75	0.36
K_2O	0.71	0.52	0.44	0.18	0.11	0.17	0.32	0.32	0.25	0.29	0.12
P_2O_5	0.22	0.2	0.21	0.38	0.23	0.26	0.2	0.23	0.19	0.23	0.23
LOI	3.41	2.06	1.53	2.34	0.59	3.87	1.73	1.91	11.5	1.11	0.97
Total	101	99.5	100	100	100	99.8	99.8	100	99.9	100	99.3
ppm											
V	100	108	90	155	153	125	123	119	97.8	99	112
Cr	110	92.5	81	99.6	95.8	82	109	117	90.3	107	114
Ba	200	416	90.5	112	101	394	96.7	350	181	159	23.2
Co	18.3	15.7	11	23	26.6	18.1	21.6	20.6	11.7	13.7	14.9
Ni	26.6	31.1	26.9	36.5	37	32.6	38.8	42.6	27.1	41.8	42.8
Rb	18.7	10.6	6.49	0.36	0.28	0.95	5.83	5.76	3.17	4.71	1.16
Sr	513	593	505	603	682	545	488	478	426	346	351
Y	20.7	23.3	18.7	19.1	25.6	15	25.4	22.5	19.4	21	23.8
La	20.9	23.1	18	17.3	38.3	13.9	21.7	19.6	19.3	19.8	24.8
Ce	42.1	46.3	36	36	75.9	29.6	44.1	41.4	39.2	39.9	47.2
Pr	4.99	5.52	4.37	4.58	9.02	3.66	5.12	5.08	4.65	4.98	5.37
Nd	19.7	21	16.5	18.9	36.1	14.7	20.3	20.4	17.9	20.3	21.9
Sm	4.09	4.24	3.43	3.98	6.89	3.01	4.16	4.3	3.49	4.25	4.3
Eu	1.26	1.19	0.97	1.24	1.5	1.09	1.47	1.4	1.23	1.29	1.47
Gd	4.02	4.41	3.47	4.18	6.04	2.97	4.55	4.15	3.62	4.52	4.39
Tb	0.57	0.7	0.55	0.58	0.84	0.49	0.7	0.68	0.56	0.63	0.65
Dy	3.33	3.81	3.02	3.29	4.55	2.62	3.99	3.59	3.07	3.19	3.72
Ho	0.68	0.75	0.6	0.67	0.84	0.53	0.8	0.7	0.7	0.68	0.73
Er	1.97	2.38	1.72	1.85	2.37	1.42	2.39	2.07	1.92	1.91	2.31
Tm	0.28	0.34	0.27	0.29	0.36	0.24	0.36	0.31	0.28	0.28	0.3
Yb	1.85	2.23	1.84	1.78	2.17	1.51	2.41	2.02	1.83	1.91	2.11
Lu	0.30	0.36	0.28	0.29	0.33	0.25	0.36	0.32	0.29	0.29	0.33
Zr	134	136	98.9	160	150	118	135	126	112	111	115
Hf	–	3.71	2.7	3.98	3.66	3.09	3.58	3.17	3.1	3	3.05
Ta	0.45	0.43	0.34	0.38	0.38	0.3	0.49	0.39	0.51	0.36	0.4
Nb	9.76	7.17	4.98	6.68	7.66	5.41	7.74	7.26	6.27	7.23	8.59
Th	4.33	3.77	2.81	2.75	2.99	2.2	3.89	3.38	3.14	2.97	3.3
U	1.53	1.35	0.96	0.79	0.92	0.78	1.26	1.22	1.06	0.98	1.14

Ms – metasediments, Fe_2O_3t: total Fe as Fe_2O_3. The chemical composition was determined according to standard procedures by the XRF and atomic absorption spectroscopy as well as by the ICP-MS method at the Karpinskii All-Russia Institute of Geological Research (CII VSEGEI, St. Petersburg).

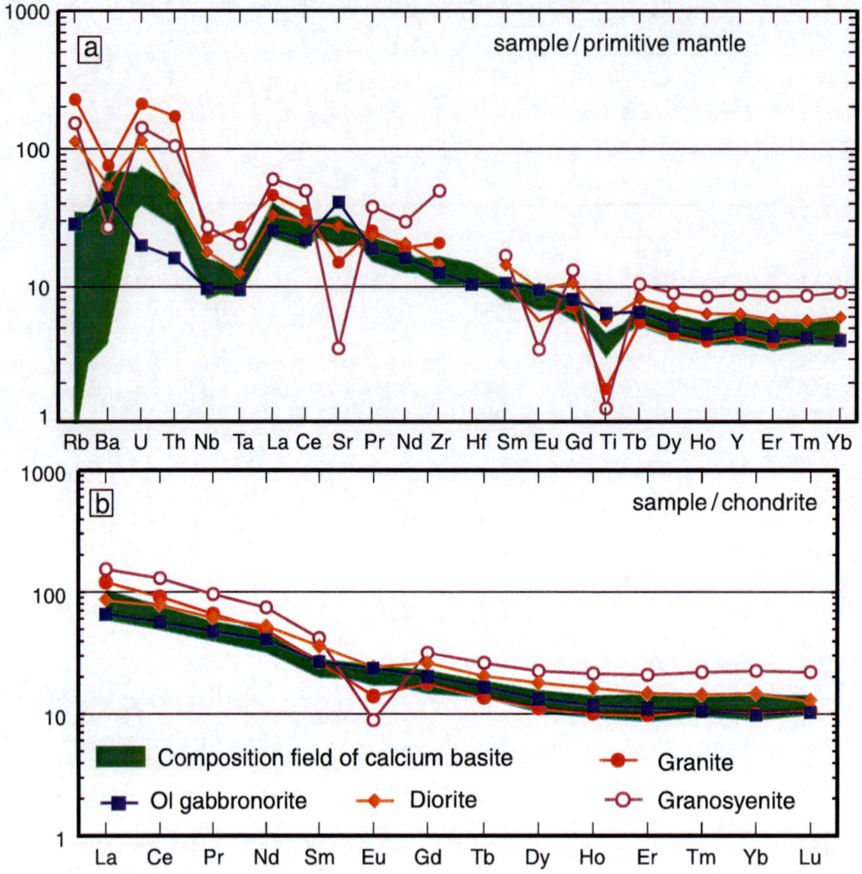

Fig. 28.17 (**a**) Primitive mantle-normalized trace element spider diagrams for the calcium basite. Primitive mantle normalization value is from Hofmann (1988). (**b**) Chondrite-normalized REE patterns for the calcium basite. Chondrite value is from Sun and McDonough (1989). For comparison some compositions of the Tastau coeval magmatic rock are given

this scenario are (1) co-existence the calcium basite bodies with coeval intrusions of gabbro and granite, and participation of the calcium basite in magma mingling structures; (2) inclusions of chromite, pyrope garnet and moissanite as mantle xenocrysts. However the low contents of alkalis in calcium basite forces to concern carefully to this scenario.

Second scenario is saturation of tholeitiic magma by crust carbonate rock (for example calcic marble) at lower level crust. It is possible to present a hypothetical situation when the intermediate magmatic chamber is formed as variant in a stratum of calcic marble. The basalt melt became saturated with calcium and the calcium mafic melt was formed.

Calcium Basite Emplacement

The melt intrusions under hypabyssal conditions generally appear as large plutons or dykes with parallel sides. Magmatic dykes form by the intrusion of magma into a planar fracture and subsequent freezing of the magma inside the open fracture (Spence, Turcotte, 1985; Rubin, 1993; Petford et al., 2000; Bons et al., 2001). After solidification, the dykes normally have a different rheology from their host rocks. Tectonic extension of dykes that are more competent than their host rocks can lead to boudinage with the formation of a series of blocks or lenses (Cloos, 1947; Ramberg, 1955; Ramsay, 1967; Hanmer, 1986; Goscombe, Passchier, 2003). However the basite bodies of the Tastau volcanoplutonic complexes are not boudins because each body has the undeformed chilled margin, i.e. they are microintrusions. Undeformed chilled margins indicate that the melt was fragmented prior to consolidation. Globular and ductile deformed irregular bodies with scalloped and flame-like contacts are typical of an interaction of low-viscosity matters (Furman, Spera, 1985; Cook, 1988, Lowell, Young, 1999 etc.).

There is an empirically determined sequence of mafic bodies emplacement (Fig. 28.18a): (1) dykes, veins; (2) ductile deformed, fragmented veins and globular and irregular bodies; (3) adjoint veins that often cut all mafic bodies of other forms. Basite bodies occur in linear zones of tectonic milling of the folded metasediment, which are boundaries between less deformed rigid blocks. Tectonic zones of the brecciation keep evidences of sedimentary layers boudinage and brittle-ductile deformations of metasedimentary rock (Fig. 28.18b). The fragmentation of the basite veins, orientation axis of sheath folds, lineation of a stretching at a boudinage of sedimentary layers are oriented in the northwest direction reflecting tectonic movements in regional transpressive conditions within the Char transpressive shear zone during the Late Hercynian time (Vladimirov et al., 2003).

Small size of basite fragments (on average 5–15 cm) points to an intensity and short-time fragmentation (Fig. 28.18c) of the melt because the small hot melt portions quick-freeze in the relatively cold hypabyssal environment. Fluorine and carbon in the basite should lower viscosity and solidus temperature (Persikov et al., 1990) of melt and provide high reactivity and high mobility of the calcium basite magma. The melt fragmentation occurred in the moment of a host rock milling and decreasing of metasedimentary matrix viscosity (Fig. 28.18d). Only in this case mechanical instability could develop and as result of surface tension between two viscous mediums the single melt portions could obtain round morphology (Snyder et al., 1997). Thus an interaction between mafic melt and cataclastic flow matrix can be defined the hydrodynamic mingling which occurs exclusively between two immiscible liquids. The result of a mingling process will be a dispersion of domains of one liquid in the other liquid (Zimanowski, Büttner, 2002). The size distribution of the dispersed liquid reflects the physical parameters of both liquids and the mingling energy, i.e. the shear rate or differential flow speed. Under otherwise constant physical parameters the mean size of the dispersed liquid domains will decrease with increasing differential flow speed (Fernandez, Gasquet, 1994; Zimanowski, Büttner, 2002). The formation of mafic microintrusions is complicated

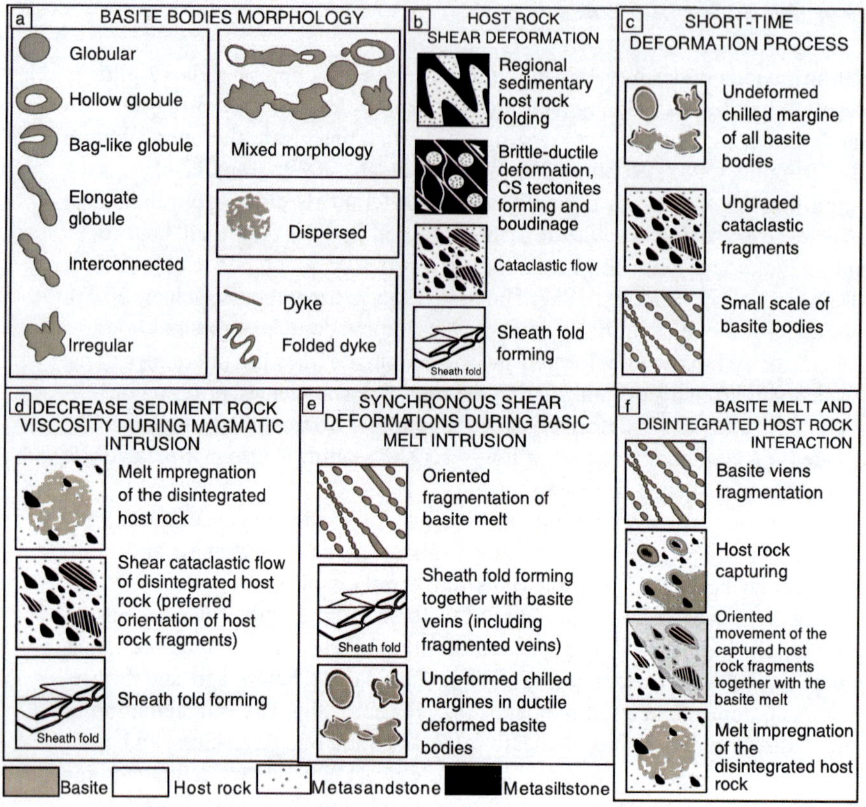

Fig. 28.18 Summary of gross characteristics of interaction between the calcium basite microintrusions and metasedimentary country rock. (**a–f**) are explained in text

by the extreme temperature contrast between the magmatic melt and the hypabyssal metasedimentary rock. As the superheat of basaltic magmatic melts (i.e. the temperature above the solidus) normally does not exceed 150 K (Zimanowski et al., 1998), the time window for hydrodynamic mingling processes is strongly limited by the cooling of the magmatic melt. Once the magma approaches its glass transition temperature due to heat loss, hydrodynamic mingling will be hindered by rising viscosity and finally be stopped when the magma reaches solid-state behaviour. In the case of magmatic melt dispersed in disintegrated sediment this cooling is a critical feature of the system because of its limited heat content. Consequently hydrodynamic mingling of mafic magma and cataclasite must be very rapid to form mixtures with domain sizes about 10 cm in diameter (Zimanowski, Büttner, 2002).

Kinematic indicators (asymmetric folds and boudinage structures) in metasedimentary rock characterize the brittle-ductile shear deformation (Passchier, Trouw, 1996), which changed into cataclastic flow in fault zones with sheath folds forming and milling. Angular and ungraded metasiltstone fragments in cataclasite also point to the short-time period of deformation of the metasedimentary rock. Intersection of basite bodies of different interconnected morphological types is possible only during the short-term change of rheological conditions of metasedimentary rock deformation during the melt intrusion. Thus all evidences indicate the melt emplacement; the melt fragmentation and the cataclastic flow were practically simultaneously and (Fig. 28.18e), in terms of deformation rate, were comparable to crustal earthquakes.

It is possible to reconstruct a relative temporal sequence of rheological conditions of deformation of the rock described by corresponding kinematic indicators. The early brittle-ductile deformation gave way to the ductile deformation with cataclastic flow and, finally, by brittle deformation (Fig. 28.19). Decreasing of the metasedimentary matrix viscosity occurred within very short-time period. On the one hand this short-time intensive tectonic event provoked melt fragmentation, and,

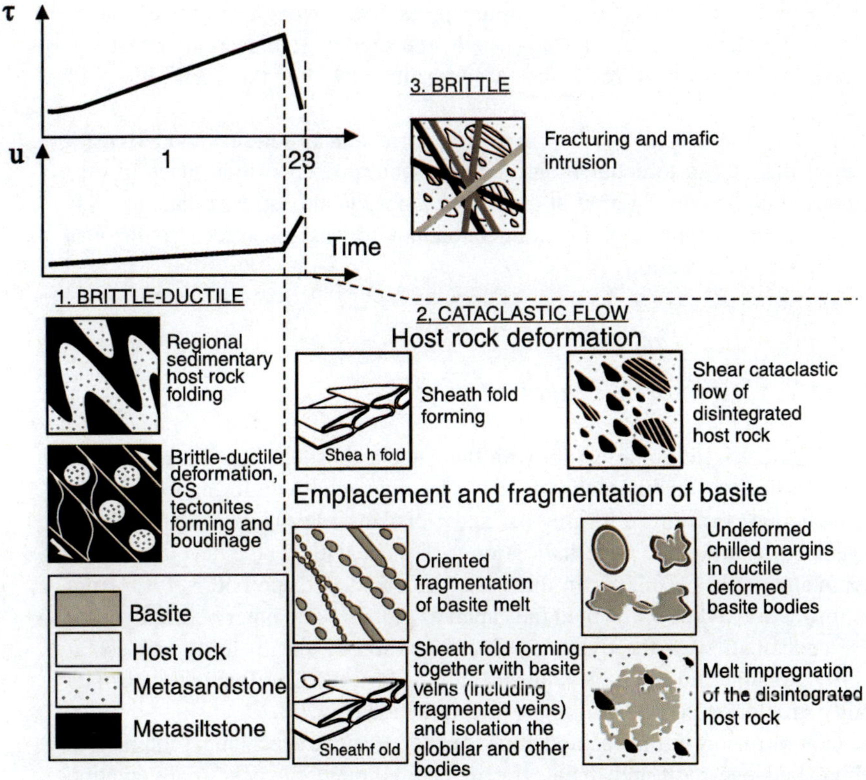

Fig. 28.19 Hypothetical stress-time and displacement-time relationships (from Sibson, 1986) at emplacement of calcium basite microintrusions. τ – stress; u – velocity of deformation 1 – steady increasing of stress with insignificant increasing of velocity of brittle-ductile deformation; 2 – stress drop with the cataclastic flow; 3 – finish of a deformation history with brittle fracturing

on the other hand, provided preservation of already consolidated melt as globules and other plastic bodies with undeformed chilled margins. Next passive intrusion of Tastau magmatic complex was in extension behaviour.

Mafic melt were emplaced syntectonically along dip strike-slip shear zones in Tastau area and were transported from their site generation through the crust to their site of emplacement. Alternating dilation and compression within the anastomosing shear zone system drove migration and ascent of magma. Magma entered extensional jogs within shear zones from a magma chamber and was expelled to other (extensional) parts shear zones during subsequent compression, upward movement being aided by buoyant nature of the magma.

In the Tastau area the melt fragmentation structures and tectonites are evidence of dramatic changes in the tectonic regime from compression to extension and again to compression. Two main factors determined such inversion: (1) regional tectonics and (2) magmatic fluid pressure, which could operate simultaneously.

The model assumes a relative long preservation of regional compressive conditions. Impenetrable areas impeded the magmatic melt motion and produced a fluid overpressure. At excess of limiting pressure a deviator stress relaxation and a hydrofracturing were realized which appeared in the tectonic crushing. The disintegration of host rock was accompanied by a mafic melt filling of fractures.

Softening of the crushed matrix, a loss of mechanical solidity and viscosity decrease of matrix (up to acquisition of quasi-liquid property) took place in ruptured zones with the participation of melt fluid. Consequently stress relaxation was realized with cataclastic flow and melt injection in northwest linear local fault zones in the compression conditions.

Model of the Mafic Melt Fragmentation

Disintegrated matrix in combination with the cataclastic shear flow set conditions for a melt dissipation mainly along the linear tectonic zone with formation of various basite bodies: from dyke to irregular and globular bodies (Fig. 28.20). Subject to the vein's orientation the melt dissipation took place with (1) the development of pseudoboudinage (Fig. 28.20a). In this case, the rounded shape bodies form from an intruding mafic dyke, which then falls apart as globules and other bodies by local collapse and inflation of the dyke. (2) The formation of sheath folds with isolation of the globular and irregular bodies (Fig. 28.20b). (3) Development of frontal instability structures according to shear-flow direction (Fig. 28.20c).

The melt intrusion in the sheared zones led to an active mechanical interaction (mingling) between melt and disintegrated metasedimentary rock: fragmentation of basite veins, capturing of host rock fragments, oriented movement of the captured host rock fragments together with mafic melt, and a melt impregnation of disintegrated host rock (Fig. 28.18f).

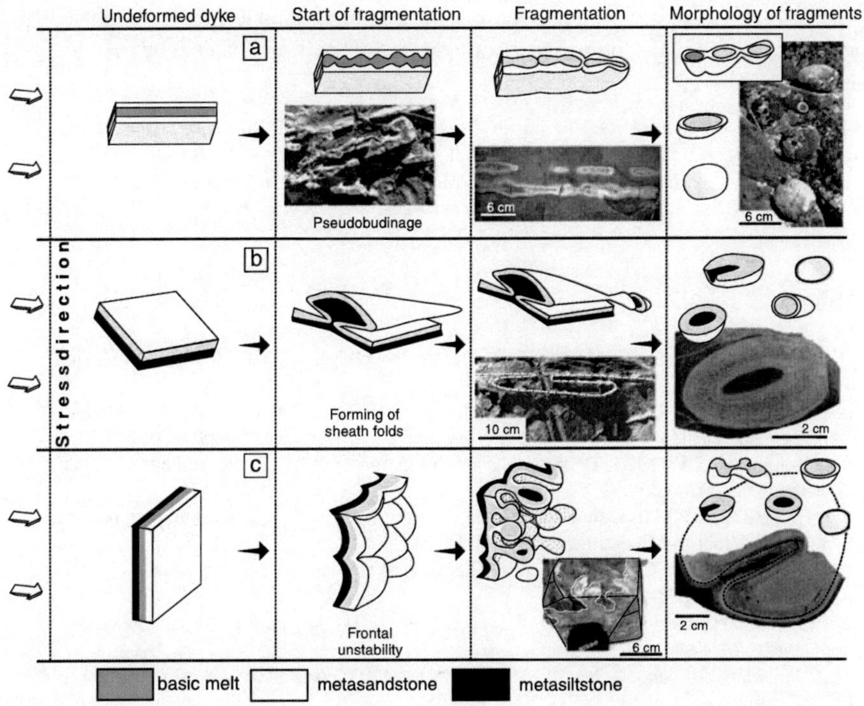

Fig. 28.20 Model of fragmentation of the basite melt in the Tastau volcanoplutonic complex. (**a-c**) are explained in text

Conclusions

Strings of mafic globules and irregular bodies that are found in the Tastau volcanoplutonic complex at first seemed to have formed by boudinage of basite sheets. They are, however, original emplacement structures that experienced little or no stretching. Thus, the major form and orientation of fragmented mafic bodies was established during the intrusive stage. The structures arose from redistribution of solidifying magma along intruded sheets, causing local expansion and collapse that was accommodated by cataclastic flow of the metasedimentary host rock. *In statu* disintegration the host rock were instable and interacted with low-viscosity mafic melt as with immiscible liquid with surface tension between them. Mechanical interaction between melt and disintegrated metasedimentary rock included fragmentation of basite veins, capturing of a host rock fragments in melt, oriented movement of the captured host rock fragments together with mafic melt, melt impregnation of disintegrated host rock. Hydrodynamic mingling processes were strongly limited by the cooling of the mafic melt. The mafic melt intrusion and the cataclastic flow of the disintegrated host rock can comparable in time with upper crustal super earthquake. Correct interpretation of boudin-like structures of melt fragmentation at hypabyssal depth is of major importance for tectonic analysis and paleogeodynamic reconstruction.

Acknowledgements We thank V. G. Vladimirov, M. V. Mints, and O. M. Rosen for critical comments and helpful suggestion. We also thank official reviewer Boris Belyatsky for his valuable suggestions for improving the earlier version of the MS. This research was financially supported by grant from the FADS.

References

Belyaev SYu (1985) Tectonics of the Char Zone, Eastern Kazakhstan UIGGM, Novosibirsk (in Russian)

Berzin NA, Coleman RG, Dobretsov NL, Zonenshain LP, Xuchang X, Chang EZ (1994) Geodynamic map of Paleo Asian Ocean: Eastern segment. Russian Geol Geophys 35: 5–23

Berzin NA, Kungurtsev LV (1996) Geodynamic interpretation of Altai-Sayan geological complexes. Russian Geol Geophys 37: 56–73

Blundy JD, Holland TJB (1990) Calcic amphibole equilibria and a new amphibole-plagioclase geothermometer. Contrib Mineral Petrol 104: 208–224

Bons PD, Dougherty-Page J, Elburg MA (2001) Stepwise accumulation and ascent of magmas. J Metamorph Geol 19: 627–633

Bons PD, Druguet E, Hamann I, Carreras J, Passchier CW (2004) Apparent boudinage in dykes. J Struct Geol 26: 625–636

Buslov MM, Fujiwara Y, Iwata K, Semakov NN (2004) Late Paleozoic-Early Mesozoic geodynamics of Central Asia. Gondwana Res 7/3: 791–808

Buslov MM, Saphonova IYu, Watanabe T, Obut OT, Fujiwara Y, Iwata K, Semakov NN, Sugai Y, Smirnova LV, Kazansky AYu, Itaya T (2001) Evolution of the Paleo-Asian Ocean (Altai-Sayan Region, Central Asia) and collision of possible Gondwana-derived terranes with the southern marginal part of the Siberian continent. Geosci J 5: 203–224

Buslov MM, Watanabe T, Smirnova LV, Fujiwara Y, Iwata K, de Grave I, Semakov NN, Travin AV, Kirinova AP, Kokh DA (2003) Role of strike-slip faults in Paleozoic-Early Mesozoic tectonics and geodynamics of Altai-Sayan and East Kazakhstan folded zone. Russian Geol Geophys 44: 49–75

Cloos E (1947) Boudinage. Transactions of the American Geophysical Union 28: 626–632

Cook NDJ (1988) Diorites and associated rocks in the Anglem complex at the neck, northeastern Stewart Island, New Zeeland: An example of magma mingling. Lithos 21: 247–262

Dobretsov NL, Ermolov PV, Khomyakov VD (1979) Ophiolites and basement composition in the axial part of the Zaisan geosyncline. In: Kuznetsov YuA, (ed) Mafic and ultramafic complexes of Siberia. Nauka, Novosibirsk: 196–219 (in Russian)

Dobretsov NL, Ponomareva LG (1969) Basement composition and age of the Zaysan geosynclines: New data. Russian Geol Geophys 3: 121–125

Dokukina KA, Konilov AN, Kaulina TV, Vladimirov VG (2010) Interaction between mafic and felsic magmas in subvolcanic environment (Tastau igneous complex, eastern Kazakhstan). Russian Geol Geophys 51: 625–643

Dokukina KA, Vladimirov VG (2005) Tectonic fragmentation of basaltic melt. Doklady Earth Sciences 400/5: 660–665

Dokukina KA, Vladimirov VG (2008) Tectonic control of mafic intrusions in the Tastau volcanoplutonic complex (eastern Kazakhstan). Russian Geol Geophys 49: 582–593

Ermolov PV, Polyanskiy NV, Dobretsov NL, Klyopinam LN (1981) Ophiolite of char zone. In: Abdulin AA, Patalakha EI, (eds) Ophiolites. Nauka KazSSR, Alma-Ata: 103–178 (in Russian)

Ermolov PV, Vladimirov AG, Izokh AE, Polyanskiy NV, Kuzebnyi VS, Revyakin PS, Borcov VD (1983) Orogenic magmatism of ophiolite belts Nauka, Novosibirsk (in Russian)

Fernandez AN, Gasquet DR (1994) Relative rheological evolution of chemically contrasted coeval magmas: Example of the Tichka plutonic complex (Morocco). Contrib Mineral Petrol 116: 316–326
Furman T, Spera FJ (1985) Co-mingling of acid and basic magma with the implications for the origin of mafic I-type xenoliths: Field and petrochemical relations of an unusual dike complex at Eagle Lake, Sequoia National Park, California, USA. J Volcanol Geotherm Res 24: 151–178
Goscombe B, Passchier CW (2003) Asymmetric boudins as shear sense indicators — an assessment from field data. J Struct Geol 25: 575–589
Hanmer S (1986) Asymmetrical pull-aparts and foliation fish as kinematic indicators. J Struct Geol 8: 111–122
Hofmann AW (1988) Chemical differentiation of the Earth: The relationship between mantle continental crust and oceanic crust. Earth Planet Sci Lett 90: 297–314
Kretz R (1983) Symbols of rock-forming minerals. Amer Mineral 68: 277–279
Lowell GR, Young GJ (1999) Interaction between coeval mafic and felsic melts in the St. Francois Terrane of Missouri, USA. Precamb Res 95: 69–88
McCarthy TC, Patiño Douce AE (1998) Empirical calibration of the silica–Ca-tschermak's–anorthite (SCAn) geobarometer. J Metamorph Geol 16: 675–686
Passchier CW, Trouw RAJ (1996) Microtectonics. Springer, Berlin Heidelberg
Persikov ES, Zharikov VA, Bukhtiyarov PG, Pol'skoy SF (1990) The effect of volatiles on the properties of magmatic melts. Eur J Mineral 2: 621–642
Petford N, Cruden AR, McCaffrey KJW, Vigneresse JL (2000) Granite formation, transport and emplacement in the Earth's crust. Nature 408: 669–673
Polyanskii NV, Dobretsov NL, Ermolov PV, Kuzebnyi VS (1979) The structure and history of the Char ophiolite belt. Russian Geol Geophys 16: 52–62
Ramberg H (1955) Natural and experimental boudinage and pinch-andswell structures. J Geol 63: 512–526
Ramsay JG (1967) Folding and Fracturing of Rocks. McGraw Hill, New York
Rubin AM (1993) Dikes vs. diapirs in viscoelastic rock. Earth Planet Sci Lett 117: 653–670
Shcherba GN, Bespaev KA, D'yachkov BA, Mysnik AM, Gangenkov GD, Sapargaliev EM (1998) Great Altai (Geology and Metallogeny) Almaty, Gylym (in Russian)
Sibson RH (1986) Earthquakes and rock deformation in crustal fault zones. Ann Rev Earth Planet Sci 14: 149–175
Snyder D, Ch C, Tait S, Wiebe RA (1997) Magma mingling in dikes and sills. J Geol 105: 75–86
Spence DA, Turcotte DL (1985) Magma-driven propagation of cracks. J Geophys Res 90: 575–580
Sun SS, McDonough WF (1989) Chemical and isotopic systematics of oceanic basalts: implications for mantle composition and processes. In: Saunders AD, Norry MJ (eds) Magmatism in the Ocean Basins, Geol Soc Lond Spec Publ 42, 313–345
Titov AV, Khromykh SV, Vladimirov AG, Pospelova N (2001) Melt inclusions in garnet and quartz from dacite porphyries of the Aktobe structure, Kazakhstan: Estimation of generation conditions and composition of primary melts. Doklady Earth Sciences 377/2: 229–233
Vernon RH, Etheridge MA, Wall VJ (1988) Shape and microstructure of microgranitoid enclaves: indicators of magma mingling and flow. Lithos 22: 1–11
Vladimirov AG, Kruk NN, Khromykh SV, Polyansky OP, Chervov VV, Vladimirov VG, Travin AV, Babin GA, Kuibida ML, Khomyakov VD (2008) Permian magmatism and lithospheric deformation in the Altai caused by crustal and mantle thermal processes. Russian Geol Geophys 49: 468–479
Vladimirov AG, Kruk NN, Rudnev SN, Khromykh SV (2003) Geodynamic and granitoid magmatism of collision orogens. Russian Geol Geophys 44: 1275–1292
Zimanowski B, Büttner R (2002) Dynamic mingling of magma and liquefied sediments. J Volcanol Geotherm Res 114: 37–44
Zimanowski B, Büttner R, Lorenz V (1998) Premixing of magma and water in MFCI experiments. Bull Volcanol 58: 491–495

Chapter 29
Small Dacite Dyke, Southern Urals, Russia: Rapidly Quenched Liquid or Fine-Grained Cumulate?

S.Yu. Chistyakova and R.M. Latypov

Introduction

We have recently documented a well-developed internal chemical zonation in small, 5–20 cm thick, mafic dykes from the Åland-Åboland dyke swarm, Finland (Chistyakova and Latypov, 2009a, b; 2010). The feature is puzzling since the dykes appear to be macro- and microscopically quite homogeneous so that one would not expect them to exhibit any systematic chemical zonation. Information about chemical zonation in such cm-scale mafic dykes is very scarce in the petrological literature (Kretz et al., 1985; Brouxel, 1991; Hoek, 1995). Although it is clear that several different mechanisms were involved in the origin of the chemical zonation (e.g. Chistyakova and Latypov, 2010), many aspects of this phenomenon remain unknown. In particular, we have little information as to whether the development of such chemical zonation in small mafic dykes is dominated by rapid quenching of liquids or by *in situ* growth cumulate processes. Similarly, we are unaware of how common chemical zonation is in small mafic dykes from other regions of the world, and whether such zonation is characteristic of only mafic dykes or may also occur in small dykes of other compositions (e.g. granitic, alkaline). The study of internal compositional variations in small mafic dykes (<0.5 m) from different geological provinces is therefore required to characterize general features of their chemical zonation and to improve our understanding the physico-chemical processes operating during emplacement of mafic magma bodies. To start filling this gap, we have undertaken a detailed study of chemical zonation in a small dacite dyke from Southern Urals, Russia. This paper presents results of this study that provides several important constraints on the origin of chemical zonation in small dykes.

S.Yu. Chistyakova (✉)
Department of Geosciences, University of Oulu, FIN-90014 Oulu, Finland
e-mail: sofya.chistyakova@oulu.fi

R.M. Latypov
Department of Geosciences, University of Oulu, FIN-90014 Oulu, Finland

R.K. Srivastava (ed.), *Dyke Swarms: Keys for Geodynamic Interpretation*,
DOI 10.1007/978-3-642-12496-9_29, © Springer-Verlag Berlin Heidelberg 2011

Samples and Methods

We have studied whole-rock major and trace elements in two cross-sections through the 16 cm thick dacite dyke from the Southern Urals, Russia. The most representative of them is presented here. For this study a specific sampling technique was adopted. We found that the internal compositional structure of dacite dyke can be best revealed by sampling on a 1–2 cm-scale in the laboratory using a saw machine. The major and trace elements were determined by X-ray fluorescence (XRF) using a Siemens SRS-303AS spectrometer with Rh tube at the University of Oulu, Finland. Working curves were determined by analysis of the geochemical rock standards NIM-N, NIM-P, NIM-D, BE-N, BIR-1, W-2, MRG-1, data for each having been compiled by Govindaraju (1994). Analyses of standards suggest that relative accuracy for all major elements is within 5% and for most trace elements analyses it is within 15%. Detection limits are 61.3 ppm (F), 3.6 ppm (Sc), 2.6 ppm (Cr), 7.1 ppm (V), 3.7 ppm (Ni), 0.8 ppm (Cu), 1.5 ppm (Zn), 1.6 ppm (Rb), 1.1 ppm (Sr), 1.5 ppm (Zr), 9.8 ppm (Ba). Table 29.1 presents whole rock major and trace elements as well as CIPW normative compositions for one section across the dacite dyke.

Geological Background

The study area is located within the Tagil-Magnitogorsk zone of the Southern Urals (Fig. 29.1). This zone is mostly composed of sub-volcanic and volcanic basaltic to acidic rocks of the Baimak-Buribai formation of Devonian age. In the study area (Fig. 29.2) the Baimak-Buribai sequence is represented from the base upwards by basaltic gialoclastite, basaltic to andesitic pillow lavas, sandstone and acidic volcanites. The basaltic lavas show individual pillows that range from 0.6 to 1.7 m in diameter and are commonly aphyric to finely porphyritic in texture. Chemically they are classified as boninitic rocks (SiO_2, 53–65 wt%; MgO, 6–12 wt%) (Spadea et al., 1998). At this locality the pillow lavas are cut by numerous dyke swarms. The individual dykes have chilled margins and most typically vary in thickness from 0.2 to 0.5 m. They trend between east-north-east and south-east, and north-north-west to north-north-east, with no apparent dependence of their composition on the trend directions. The dykes commonly show medium-grained dolerite centres and aphyric or spherulitic margins. The dyke textures are generally porphyritic with typically fine-grained groundmasses and, often, amygdaloidal texture. When present the amygdales compose up to 40% of the rocks. From the base of the stratigraphy upwards the dykes tend to gradually pass into sill-like bodies that in turn pass into acidic lavas. On this basis the dykes are commonly considered as feeder channels for overlying acidic volcanic lavas. More detailed descriptions of the geology, petrology and petrography of the rocks in the study area can be found in Spadea et al. (1998) and Seravkin et al. (2003).

Table 29.1 Major, trace and CIPW normative compositions for rocks from dacite dyke

Sample		Ur-12	11/1-1	11/1-2	11/1-3	11/2-1	11/2-2	11/2-3	11/3-1	11/3-2	11/3-3	11/4-1	11/4-2	11/5-1	11/5-2	11/5-3	Ur-12
Rock		Pillow lava	Dyke	Dyke	Dyke	Dyke	Dyke	Dyke	Dyke	Dyke	Dyke	Dyke	Dyke	Dyke	Dyke	Dyke	Pillow lava
Distance (cm)		0	2	3.3	4.6	5.9	6.3	7.6	8.9	10.3	11.6	12.9	14.2	15.5	16.8	18	20
SiO_2	wt%	69.20	67.39	65.46	65.49	64.16	64.64	64.02	62.93	63.80	62.47	63.81	64.48	65.47	65.61	66.08	69.20
TiO_2	wt%	0.31	0.59	0.59	0.60	0.59	0.57	0.58	0.60	0.59	0.60	0.57	0.59	0.61	0.59	0.58	0.31
Al_2O_3	wt%	10.23	12.81	14.14	14.38	14.65	14.22	14.36	15.02	14.48	15.09	14.27	14.60	14.50	14.08	13.60	10.23
FeO (total)	wt%	4.14	4.16	5.71	6.39	7.16	7.10	7.09	7.57	7.31	7.68	7.19	7.05	6.39	5.73	4.16	4.14
MnO	wt%	0.10	0.09	0.09	0.09	0.10	0.11	0.11	0.11	0.12	0.11	0.11	0.10	0.09	0.09	0.08	0.10
MgO	wt%	7.06	4.47	4.58	4.65	5.30	5.14	5.10	5.50	5.29	5.68	5.24	4.97	4.39	4.47	4.40	7.06
CaO	wt%	3.65	2.39	2.83	2.79	2.77	2.86	2.94	2.61	2.93	2.71	2.98	2.75	2.70	2.69	2.18	3.65
Na_2O	wt%	1.53	3.52	2.53	2.25	2.02	2.01	2.12	2.10	2.13	2.16	2.04	2.15	2.48	2.65	3.54	1.53
K_2O	wt%	2.16	2.15	1.98	1.98	2.09	2.06	2.04	2.23	2.02	2.20	2.18	2.22	1.97	1.93	2.60	2.16
P_2O_5	wt%	0.07	0.08	0.08	0.08	0.08	0.08	0.08	0.08	0.08	0.08	0.08	0.08	0.08	0.08	0.08	0.07
LOI	wt%	2.61	1.95	na	2.41	na	na	2.77	na	2.78	na	2.89	na	2.35	na	2.10	2.61
Total	wt%	101.06	99.60	97.99	101.11	98.92	98.79	101.21	98.75	101.52	98.78	101.36	98.99	101.03	97.92	99.40	101.06
Sc	ppm	24	33	33	32	32	32	31	33	31	35	31	35	34	32	32	24
V	ppm	105	160	267	316	348	331	336	361	337	362	337	334	320	270	176	105
Cr	ppm	557	30	24	21	17	26	24	21	27	24	16	22	33	15	24	557
F	ppm	200	300	300	100	200	200	300	300	200	300	100	200	100	100	400	200
Ni	ppm	113	23	18	12	13	13	9	10	11	11	10	15	15	19	23	113
Cu	ppm	59	153	134	139	138	131	133	142	135	139	136	137	135	128	122	59
Zn	ppm	33	63	115	132	149	130	119	127	113	129	132	139	120	111	68	33
Rb	ppm	32	24	30	33	34	34	34	35	33	34	35	35	33	30	26	32
Sr	ppm	240	362	279	259	252	244	255	257	259	265	255	255	261	273	366	240
Zr	ppm	68	70	62	62	60	59	61	62	62	62	61	61	62	62	68	68
Ba	ppm	968	1353	1026	680	565	533	533	567	542	587	619	690	678	975	1530	968

Table 29.1 (continued)

Sample		Ur-12	11/1-1	11/1-2	11/1-3	11/2-1	11/2-2	11/2-3	11/3-1	11/3-2	11/3-3	11/4-1	11/4-2	11/5-1	11/5-2	11/5-3	Ur-12
Rock		Pillow lava	Dyke	Dyke	Dyke	Dyke	Dyke	Dyke	Dyke	Dyke	Dyke	Dyke	Dyke	Dyke	Dyke	Dyke	Pillow lava
Distance (cm)		0	2	3.3	4.6	5.9	6.3	7.6	8.9	10.3	11.6	12.9	14.2	15.5	16.8	18	20
Qtz	wt%	33.23	25.81	26.89	27.55	25.27	26.37	25.23	22.56	24.45	21.37	24.66	25.13	26.85	26.99	22.77	33.23
Als	wt%	0.00	0.90	4.81	6.07	7.00	6.10	5.85	7.65	6.08	7.36	5.55	6.39	5.93	4.90	2.00	0.00
Or	wt%	13.20	13.33	12.28	12.21	12.89	12.71	12.64	13.79	12.48	13.61	13.50	13.68	12.16	11.98	16.17	13.20
Pl+Or	wt%	41.76	56.45	48.93	45.99	44.51	44.74	46.18	45.39	45.98	46.25	46.53	46.31	47.51	49.00	58.53	41.76
Opx	wt%	22.55	16.83	19.36	20.39	23.22	22.79	22.74	24.40	23.49	25.02	23.26	22.17	19.71	19.12	16.70	22.55
Mg number	at.%	75.29	65.71	58.86	56.50	56.90	56.37	56.21	56.46	56.37	56.90	56.53	55.73	55.08	58.19	65.35	75.29
An(norm)	mol.%	51.66	26.40	37.31	39.74	42.17	43.10	42.50	39.73	42.30	39.99	43.79	40.47	36.64	35.03	24.47	51.66

Distance is shown from north-western contact. Mg number = 100*Mg/(Mg+Fetotal), An(norm) = 100*An/(An+Ab), Qtz – quartz, Als – Al_2SiO_5, Or – orthoclase, Pl – plagioclase, Opx – orthopyroxene, LOI – loss on ignition, na – not analysed.

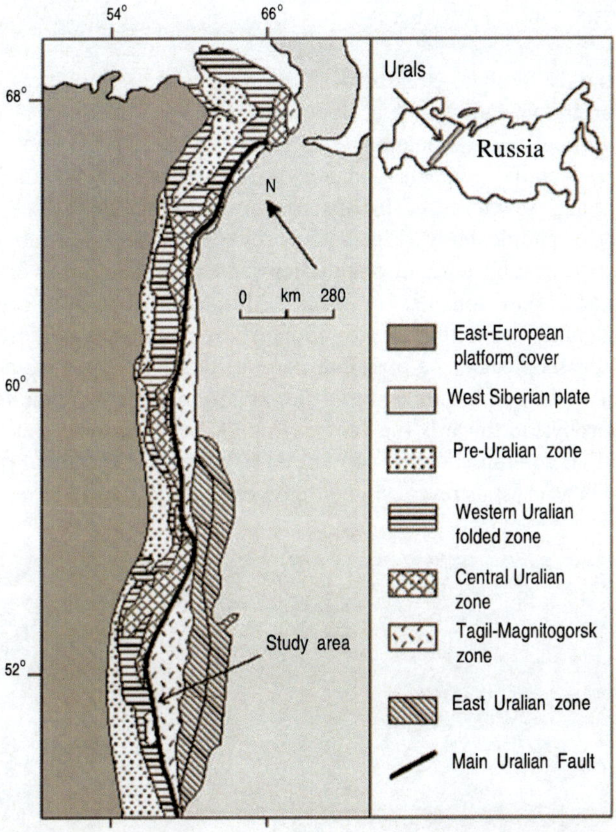

Fig. 29.1 Simplified geological map of the Urals showing the location of the study area. The location of the Urals in the Russia is shown in the inset. The map has been modified from Puchkov (2000)

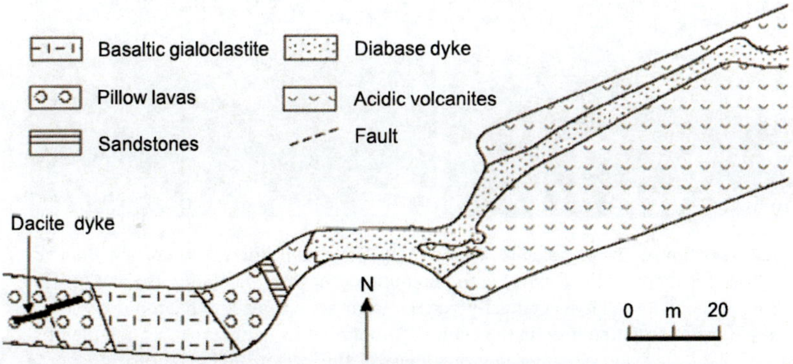

Fig. 29.2 Simplified field sketch of outcrop showing the studied dacite dyke that cuts Devonian pillow lavas of the Baimak-Buribai formation, Southern Urals, Russia. The thickness of the dacite dyke is not to scale. The map has been modified from Seravkin et al. (2003)

Petrography and Field Relations

In the study area the dacite dyke cuts the basaltic to andesitic pillow lavas of the Baimak-Buribai formation and strikes at about 70° NE (Fig. 29.2). It is traced along a distance of about twenty meters and varies in thickness from 5 to 20 cm. The dyke has sharp parallel contacts with the host lavas and shows no apparent field or petrographic evidence of *in situ* contamination by these rocks. The dyke appears to be a simple body (Fig. 29.3a) since both field and laboratory petrographic studies reveal no internal boundaries such as chilled zones or comb layers (e.g. Platten and Watterson, 1987; Platten, 2000). Macroscopically, the dyke is composed of aphanitic rocks that do not contain any visible intratelluric phenocryst phases. Microscopically, the marginal parts of the dyke are characterized by a non-crystalline microporphyritic texture (Fig. 29.3b) that changes gradually to crystalline microporphyritic towards the centre (Fig. 29.3c). The dyke consists of fine-grained rocks that contain about 35–40 vol.% of prismatic crystals of plagioclase (0.1–0.2 mm). The crystals reveal slight alignment with respect to its contacts;

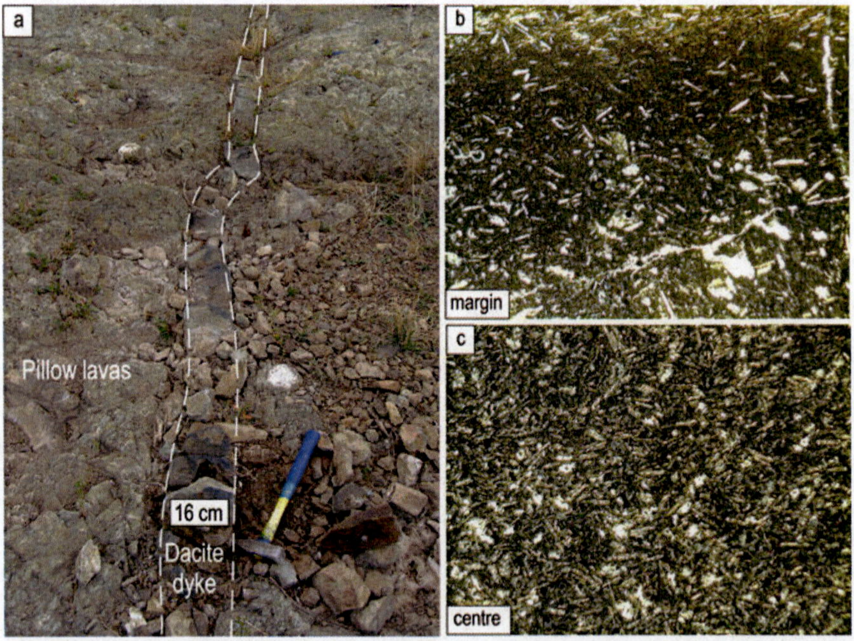

Fig. 29.3 Field photograph (**a**) of a dacite dyke cutting Devonian pillow lavas of the Baimak-Buribai formation (Southern Urals, Russia) with photomicrographs of its margin (**b**) and interior (**c**). The texture of the dyke tends to gradually change from non-crystalline microporphyritic at margins to crystalline microporphyritic in the centre. Throughout its entire cross section the prismatic crystals of plagioclase are set either in groundmass of fully devitrified glass (margins) or in groundmass of fine-grained aggregates of quartz, fibrous amphibole and chlorite that likely represent alteration products of pyroxenes (interior). Long dimension of photomicrographs is about 3 mm. Photomicrographs are in polarized light

they increase somewhat in size and abundance towards the dyke centre (Fig. 29.3b, c) and tend to be partially to completely sericitized. The interstices between plagioclase laths in the marginal parts of the dyke are occupied by fully devitrified groundmass. In the centre of dyke, the groundmass is mostly composed of plagioclase laths, small grains and aggregates of quartz (0.1–0.2 mm), as well as fibrous amphibole and chlorite that likely represent alteration products after pyroxenes. In addition, randomly distributed amygdales filled with quartz and/or calcite tend to be evenly distributed across the section (~2 vol.%) and show some increase in size from margins inwards from 0.1–0.5 mm to 1–2 mm in diameter. Towards the centre quartz and calcite in these amygdales are often partly and rarely totally replaced by aggregates of chlorite and amphibole. Small sub- to anhedral magmatic Ti-magnetite grains are present (<1 vol.%) in the groundmass.

Internal Compositional Variations

Systematic whole-rock variations in major and trace elements, as well as in some petrochemical parameters, are found across the dacite dyke (Fig. 29.4). The dyke exhibits a steady inward decrease in normative Pl+Or content, whole-rock SiO_2, Na_2O, Ba and Sr, and an increase in normative An (100An/(An+Ab)), normative Opx, and whole-rock MgO, FeO and V. Other components (e.g. Zr, Cr, TiO_2) do not show such clear compositional tendencies across the dike (Fig. 29.4 and Table 29.1). It should be emphasized that the most pronounced variations in the above parameters occur across the marginal parts of the dyke, whereas the interior is characterized by much more modest variations. It is also noteworthy that the extent of the observed internal variations is very high for such a small, 16 cm thick, dyke. For instance, the centre-to-margin difference in SiO_2, FeO, Na_2O and An(norm) is about 5 wt%, 3.5 wt%, 1.5 wt%, and 17 mol.%, respectively (Fig. 29.4). These compositional trends can be interpreted as reverse, since they are opposite to that normally expected from inward fractional crystallization of magma in dykes.

Three aspects related to the compositional zonation of the dacite dyke permit some preliminary inferences regarding the origin of its internal zonation. Firstly, the cross-sectional distribution of major and trace elements in the dyke do not reveal *in situ* contamination in direct contact with host rocks (Fig. 29.4). Therefore, *in situ* contamination can not be a factor responsible for the origin of an internal zonation of dyke.

Secondly, the sum of normative plagioclase and orthoclase tends to decrease inwards (Fig. 29.4). This tendency is additionally emphasized by an inward decrease in whole-rock Na_2O, Ba and Sr that can be taken as a proxy for plagioclase content. These observations indicate that crystals of plagioclase did not accumulate towards the centre of the dyke upon magma emplacement (e.g. flow differentiation). Also, quartz-calcite amygdales do not reveal any preferred concentrations within the dyke based on petrographic observations. This indicates that distribution of both plagioclase and amygdales play no role in the origin of the systematic compositional trends within the dyke.

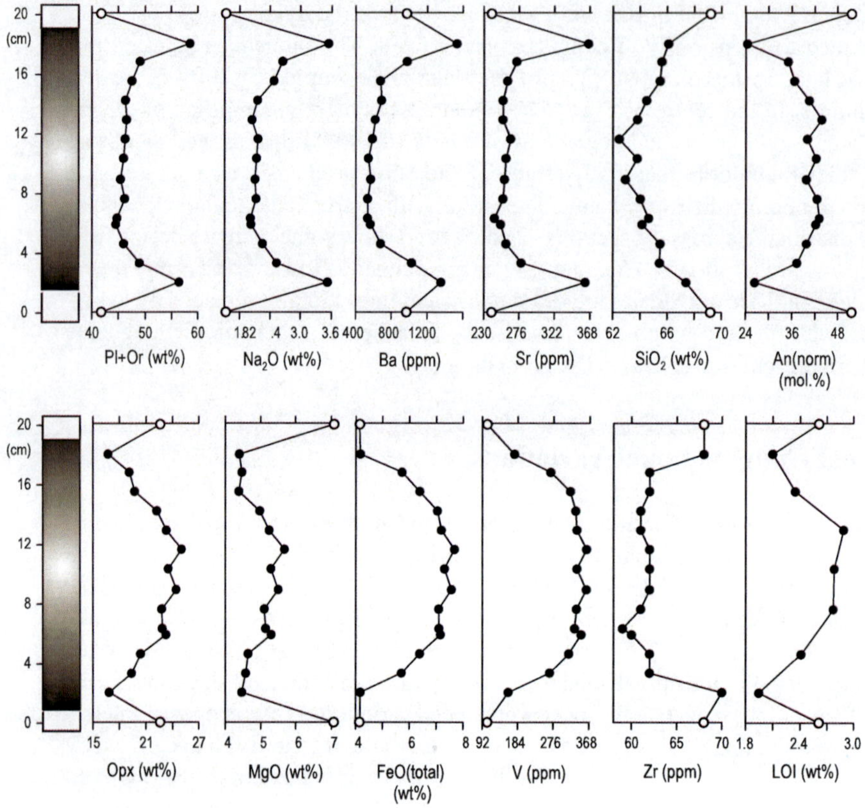

Fig. 29.4 Compositional variations (**a**) in normative Pl+Or, normative An, whole-rock Na_2O, Ba, Sr, and SiO_2 and (**b**) in normative Opx and whole-rock MgO, FeO (total), V, Zr and Mg number across dacite dyke from Southern Urals, Russia. Distance is shown from the north-west contact of the dyke. Compositional points for host rocks are indicated by open circles. An(norm) $= 100^*An/(An + Ab)$; Mg number $= 100^*Mg/(Mg + Fe_{total})$. Note that in terms of all geochemical indices the dyke is reversely zoned. See text for further discussion

Thirdly, the extent of postmagmatic alteration, as implied by variations in loss on ignition (LOI) does not correlate with the compositional trends in Ba, Sr, Rb, and K_2O (Fig. 29.4 and Table 29.1), as is usually the case with dykes in which internal zonation is due to secondary processes. Together with the uncontaminated nature of the dykes and their surprisingly systematic compositional trends, this implies that the observed internal zonation can be safely regarded as of primary magmatic origin, not related to postmagmatic modification of rocks.

Discussion

Reverse compositional trends observed in the dacite dyke (e.g. an inward increase in MgO and An(norm)) are not what is expected from closed system crystallization of magma in a dyke. Similarly, one cannot attribute them to filling of the dyke with magma that was changing its composition with time, a process that has been invoked in our previous studies to explain zonation in small mafic dykes (Chistyakova and Latypov, 2010). This process is characterized by a systematic inward decrease in MgO and An(norm), not an increase as is the case with the dacite dyke. These trends in the dacite dyke are identical, however, to those observed in the central parts of mafic dykes (Chistyakova and Latypov, 2009a; b; 2010). The trends were previously ascribed by us to *in situ* cumulate growth against dyke sidewalls from flowing magma. This process results in a progressive inward increase in the proportion of cumulus phases, such as plagioclase and pyroxenes. We believe therefore that the same mechanism can be applied to explain the origin of the compositional profiles in the dacite dyke. In this model, the internal zonation of this dyke can be considered as an inward progression from chilled liquid at margins to cumulate in the interior of the dyke. The significant chemical difference between the margin and centre of the dyke is likely a reflection of the large difference between dacitic parental liquid and corresponding noritic cumulate. The idea is schematically illustrated on the ternary diagram Fo-An-Qtz (Fig. 29.5). The most pronounced compositional variations occur across the marginal parts since the interior composed of noritic cumulate forms the major portion (about 80–85%) of the dyke (Fig. 29.4). The inward formation of cumulates takes place owing to effective removal of an evolved liquid boundary layer (rejected solute) from *in situ* growing crystals by magma flowing along dyke margins (Fig. 29.6).

One possible objection to this interpretation is that the rocks of the dacite dyke exhibit a fine-grained texture that is not characteristic of normal cumulate rocks in

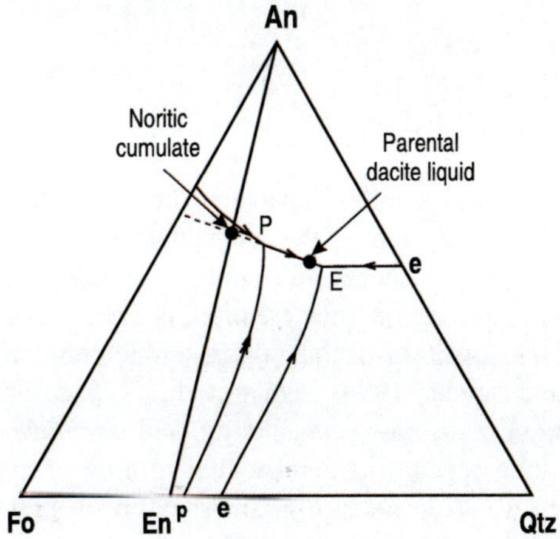

Fig. 29.5 Ternary diagram Fo-An-Qtz illustrating an idea that the significant compositional difference between chilled margins and the interior of the dacite dyke is due to the large compositional difference between dacitic parental liquid lying on the cotectic line An+En+L and its corresponding noritic cumulate lying on the connode An-En. Phase diagram is from Andersen (1915). E – *Qtz* + *En* + *An* = *L*; P – *En* + *An* = *L* + *Fo*. Cotectic and reaction lines are indicated by one and two *arrows*, respectively. *Arrows* point toward decreasing temperature. Hereafter in figures and text: E – eutectic point; P – peritectic point; L – melt

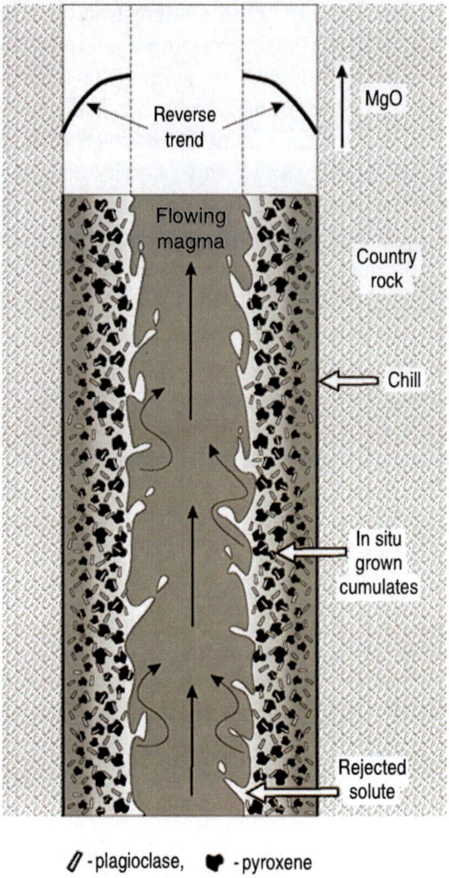

Fig. 29.6 Cartoon schematically showing the formation of a dacite dyke by *in situ* cumulate growth against dyke sidewalls. The reverse compositional trends in the dyke emerge as a result of a progressive inward increase in the proportion of crystallising cumulus pyroxene and plagioclase. The inward formation of cumulates takes place owing to effective removal of an evolved liquid boundary layer (rejected solute) from *in situ* growing crystals by magma flowing along dyke margins. See text for further discussion

large plutonic bodies. One should mention, however, that fine-grained cumulates are not uncommon in layered intrusions. They constitute, for instance, intraplutonic quench zones at the bases of megacyclic units. Although these zones are fine-grained, suggesting rapid nucleation, they are often impoverished in incompatible elements so that their compositions frequently resemble cumulates (Mathison and Booth, 1990; Wiebe and Snyder, 1993; Tegner et al., 1993; Miller and Ripley, 1996). This raises a question as to how such fine-grained cumulates are formed. In resolving this problem, one needs to remember that cooling of magma by cold country rocks is not the only reason for fine-grained texture in plutonic rocks. It is important to note that the fine-grained cumulates in layered intrusions do not commonly show skeletal and dendritic crystals that are expected for magma crystallizing with sufficient supercooling (e.g. Lofgren, 1974). The solution to this problem has been given by Tegner et al. (1993). They noted that experimental studies on natural basalts by Kouchi et al. (1986) have shown that stirring, or shear flow, during crystallization, dramatically changes both the nucleation rate and the crystal morphology of plagioclase compared with experiments under static conditions. At supercooling above 45°C, stirring increased the nucleation rate by about 10 times and skeletal crystal morphologies were never observed. According to these experiments, continuous magma flow along the cooling base of magma chambers or the cooling margins of a dyke, can account for a considerable increase in the nucleation rate resulting in formation of *in situ* grown fine-grained rocks.

Continuous magma flow during *in situ* growth also gives a simple and logical answer to the occurrence of fine-grained cumulates at the cooling floor of large plutonic intrusions and cooling margins of dykes. This question has already been touched upon by Snyder and Tait (1995). From their thermal analysis, no part of the magma flowing along the floor is expected to cool below the temperature of the ambient liquid during the formation of the fine-grained zones (supercooling in the range of 50–100°C). Since no minerals whose stability field lies below the ambient temperature will form in this layer, it will be impoverished in the components that constitute the low temperature phases. In this process, fine-grained cumulates may form if evolved liquid that forms a liquid boundary layer around growing crystals is continuously removed by flowing magma. In other words, cumulates at cooling margins originate when the cooling effect of country rocks favoring formation of "chilled" rocks is significantly overwhelmed by the effect of removal of evolved liquid by flowing magma favoring the formation of cumulates. Such "chills" or fine-grained rocks are thus very poor approximations of the injected liquid compositions. This is why liquid compositions estimated from such zones are too high in MgO or picritic or both (Berg, 1980; Tegner et al., 1993).

Coming back to the dacite dyke in this study, we conclude that, despite superficially looking like "chills", these very fine-grained rocks are still cumulates and cannot therefore be used to estimate parental magma composition. Perhaps, only 1–2 cm thick margins of this small dyke reflect the composition of the parental magma. It is quite conceivable that this situation may be common for small dykes of any composition worldwide.

Conclusion

Several important conclusions can be drawn from this study. Firstly, it indicates that internal chemical zonation in small dykes is not restricted to those from the Åland-Åboland dyke swarm, Finland (Chistyakova and Latypov, 2009a; b; 2010), but also occurs in small dykes from the Southern Urals and therefore may represent a quite common feature for small mafic dykes worldwide. Secondly, the fact that internal chemical zonation is found in small dyke of dacitic composition (62–68 wt.%, SiO_2) as well as of basaltic composition (48–52 wt.%, SiO_2) (Chistyakova and Latypov, 2009a; b; 2010) suggests that this phenomenon may be common for small dykes irrespective of their chemical composition. Thirdly, the study clearly demonstrates that even the very fine-grained textured small dykes can still be cumulates, not rapidly quenched liquids. The margin-to-center increase in whole-rock MgO and normative An in these dykes takes place from progressive inward increase in the proportion of crystallising cumulus pyroxene and plagioclase. The reverse compositional trends in dykes thus likely reflect a tendency of rocks to become more cumulate inwards owing to crystallization of flowing magma under progressively less supercooled conditions. This means that even small dykes may provide distorted information on compositions of parental magmas and should be used for such purposes with a great caution. In general, we believe that small dykes of any composition deserve more attention from igneous petrologists and geochemists. These dykes appear to exhibit many interesting puzzles that may help to shed light on processes of magma differentiation operating in larger plutonic bodies.

Acknowledgements We thank Evgeny Pushkarev for organizing fieldwork in the Urals as well as help and good advice with many aspects of this study. Richard James is greatly thanked for encouraging comments on the paper as well as editing that led to improvement in the wording of the text. We are indebted to Jouni Vuollo as an official reviewer of this paper for his helpful comments and suggestions. The authors are also very grateful to Rajesh K. Srivastava for the invitation to contribute to this IDC-6 volume. The research was supported by grants from the Finnish Academy of Science and the Renlund Foundation, Finland.

References

Andersen O (1915) The system anorthite-forsterite silica. Am J Sci 39: 407–454

Berg JH (1980) Snowflake troctolite in the Hettasch Intrusion, Labrador: Evidence for magma-mixing and supercooling in a plutonic environment. Contrib Mineral Petrol 72: 339–351

Brouxel M (1991) Geochemical consequences of flow differentiation in a multiple injection dyke (Trinity ophiolite, N. California). Lithos 26: 245–252

Chistyakova SYu, Latypov RM (2009a) Fine-scale chemical zonation of small dolerite dykes, Kestiö Island, SW Finland. Geol Mag 146: 485–496

Chistyakova SYu, Latypov RM (2009b) Two independent processes responsible for compositional zonation in mafic dykes of the Åland-Åboland Dyke Swarm, Kestiö Island, SW Finland. Lithos 112: 382–396

Chistyakova SYu, Latypov RM (2010) On the development of internal chemical zonation in small mafic dykes. Geol Mag 147: 1–12

Govindaraju K (1994) Compilation of working values and sample descriptions for 383 geostandards. Geostandard Newletter 18: 1–158

Hoek JD (1995) Dyke propagation and arrest in Proterozoic tholeiitic dyke swarms, Vestfold Hills, East Antarctica. In: Baer G, Heimann A, (eds) Physics and chemistry of dykes. Balkema, Rotterdam: 79–93

Kouchi A, Tsuchiyama A, Sunagawa I (1986) Effect of stirring on crystallization kinetics of basalt: Texture and element partitioning. Contrib Mineral Petrol 93: 429–438

Kretz R, Hartree R, Garrett D (1985) Petrology of the Grenville swarm of gabbro dikes, Canadian Precambrian Shield. Can J Earth Sci 22: 53–71

Lofgren G (1974) An experimental study of plagioclase crystal morphology: Isothermal crystallization. Am J Sci 274: 243–273

Mathison CI, Booth RA (1990) Macrorhytmically layered gabbronorites in the Windimurra gabbroid complex, western Australia. Lithos 24: 171–180

Miller JD, Ripley EM (1996) Layered intrusions of the Duluth Complex, Minnesota, USA. In: Cawthorn RG, (ed) Layered Intrusions. Developments in Petrology 15. Elsevier Science B.V., 257–301

Platten IM (2000) Incremental dilation of magma filled fractures: Evidence from dyke on the Isle of Skye, Scotland. J Struct Geol 22: 1153–1164

Platten IM, Watterson J (1987) Magma flow and crystallization in dyke fissures. In: Halls HC, Fahrig WF (eds) Mafic dyke swarms. Geological Association of Canada Special Paper 34: 65–73

Puchkov VN (2000) Paleogeodynamics of the Southern and Middle Urals. Ufa, Dauria (in Russian)

Seravkin IB, Kosarev AM, Znamenskii SE, Salixov DN, Minibaeva KR, Zaikov VV, Zdanovich GG, Ankusheva NN (2003) Guidebook to the field excursions. The second All-Russian symposium on volcanology and paleovolcanology "Volcanism and Geodynamics", Ekaterinburg, 59 p

Snyder D, Tait S (1995) Replenishment of magma chambers: Comparison of fluid-mechanics experiments with field relations. Contrib Mineral Petrol 122: 230–240

Spadea P, Kabanova LYa, Scarrow JH (1998) Petrology, geochemistry, and geodynamic significance of Mid-Devonian boninitic rocks the Baimak-Buribai area (Magnitogorsk zone, Southern Urals). Ofioliti 23(1): 17–36

Tegner C, Wilson JR, Brooks CK (1993) Intraplutonic Quench Zones in the Kap Edvard Holm Layered Gabbro Complex, East Greenland. J Petrol 34: 681–710

Wiebe RA, Snyder D (1993) Slow, dense replenishments of a basic chamber: The layered series of the Newark Island layered intrusion, Nain, Labrador. Contrib Mineral Petrol 113: 59–72

Chapter 30
Primary and Secondary Chemical Zonation in Mafic Dykes: A Case Study of the Vochelambina Dolerite Dyke, Kola Peninsula, Russia

S.Yu. Chistyakova and R.M. Latypov

Introduction

Mafic-ultramafic dykes often show internal modal or compositional zonation (Ragland et al., 1968; Steele and Ragland, 1979; Ross, 1983, 1986; Ross and Heimlich, 1972; Ernst and Bell, 1992; Halama et al., 2002; Kretz et al., 1985; Brouxel, 1991). The study of this internal zonation is important for the understanding of emplacement mechanism and processes of magmatic differentiation operating in dykes and plutonic bodies as a whole. Dykes are particularly suitable for such investigations because they are comparatively simple geologically, which allows the identification of individual processes of magmatic differentiation and assessment of their contribution to the compositional profiles. As an illustration, our recent study of compositional variations across several mafic dykes highlighted the role of important magmatic processes such as progressive changes in the composition of inflowing magma, *in situ* crystal growth against dyke sidewalls, and flow differentiation (Chistyakova and Latypov, 2009a, b, 2010). Care should be taken, however, in the interpretation of internal zonation in dykes because systematic compositional variations can also result from secondary, postmagmatic processes. Attempts to interpret zonations may therefore result in misleading conclusions with regard to magmatic differentiation processes. Here we report a study of a mafic dyke that reveals well-developed compositional variations that are assigned to postmagmatic alteration of the primary igneous mineral assemblage. The primary internal zonation of this dyke can only be identified after removal of the compositional effects of secondary processes. The study highlights the importance of detailed petrographic examination to avoid the misinterpretation of whole-rock geochemical data.

R.M. Latypov (✉)
Department of Geosciences, University of Oulu, FIN-90014 Oulu, Finland
e-mail: rais.latypov@oulu.fi

S.Yu. Chistyakova
Department of Geosciences, University of Oulu, FIN-90014 Oulu, Finland

R.K. Srivastava (ed.), *Dyke Swarms: Keys for Geodynamic Interpretation*,
DOI 10.1007/978-3-642-12496-9_30, © Springer-Verlag Berlin Heidelberg 2011

Samples and Methods

We present here whole-rock major and trace element data from a section across the Vochelambina dolerite dyke, Kola Peninsula, SW Russia (Fig. 30.1a, b). A mini-drill was used to collect 23 samples across the dyke and host rocks, with the distance between individual samples ranging from 4 to 20 cm (Fig. 30.2). One sample (D-1/12) has been excluded from consideration because of its high degree of secondary alteration. Major and trace elements were determined by X-ray fluorescence (XRF) using a Siemens SRS-303AS spectrometer with Rh tube at the University of Oulu, Finland. Calibration curves were determined by analysis of the geochemical rock standards NIM-N, NIM-P, NIM-D, BE-N, BIR-1, W-2, and MRG-1; data for each have been compiled by Govindaraju (1994). Precision and accuracy for all major elements were better than 5% and for most trace elements analyses within 15%. Detection limits are 12.1 ppm (Cl), 3.6 ppm (Sc), 2.6 ppm (Cr), 7.1 ppm (V), 3.7 ppm (Ni), 0.8 ppm (Cu), 1.5 ppm (Zn), 2.2 ppm (Ga), 1.6 ppm (Rb), 1.1 ppm (Sr), 1.5 ppm (Y), 1.5 ppm (Zr), 9.8 ppm (Ba), 5.9 ppm (Ce), 3.3 (Nd), 1.9 ppm (Pb). FeO was determined by titration and Fe_2O_3 by difference. Table 30.1 presents whole rock major and trace elements as well as CIPW normative compositions of rocks from the section across the Vochelambina dolerite dyke (Figs. 30.1b and 30.2). Plagioclase, clinopyroxene and mica compositions (Tables 30.2–30.4) were determined at the University of Oulu using electron microprobe JEOL JXA-8200. Operating conditions were 15 kV acceleration voltage, 10 nA cup current, 10–40 s counting times, and 10 μm beam diameter.

Geological Background and Field Relations

The Vochelambina dolerite dyke belongs to a mafic dyke swarm in the central part of the Kola Peninsula, Russia (Fig. 30.1a, inset), that cuts the Archaean to Proterozoic basement. The swarm consists of about ten dolerite dykes that range in width from several centimetres up to 15 m. The length of the dykes is difficult to determine because of poor exposure. They strike roughly E-W and are close to vertical. Chemically the dykes are tholeiitic basalts (Fedotov and Amelin, 1998). The isotopic age of the dykes, determined by two different techniques, is 1160.8±12.9 Ma ($^{40}Ar/^{39}Ar$ method) and 1176±28 Ma (Rb-Sr method) (Arzamastsev et al., 2006). This indicates that the dykes were mostly likely emplaced during the initial stage of the orogeny in the southern part of the Fennoscandian Shield (Gorbatschev et al., 1987). The studied dolerite dyke is exposed in the Vochelambina area within an Archaean greenstone belt (Mitrofanov and Pozhilenko, 1991; Daly et al., 1993). The dyke cuts the country rocks at almost right angles and can be followed along strike for a distance of about 350 m with a maximum width of 5–8 m (Fig. 30.1a). The dyke has sharp contacts with country rocks and shows no apparent field or petrographic evidence of *in situ* contamination by them. Locally, xenoliths of

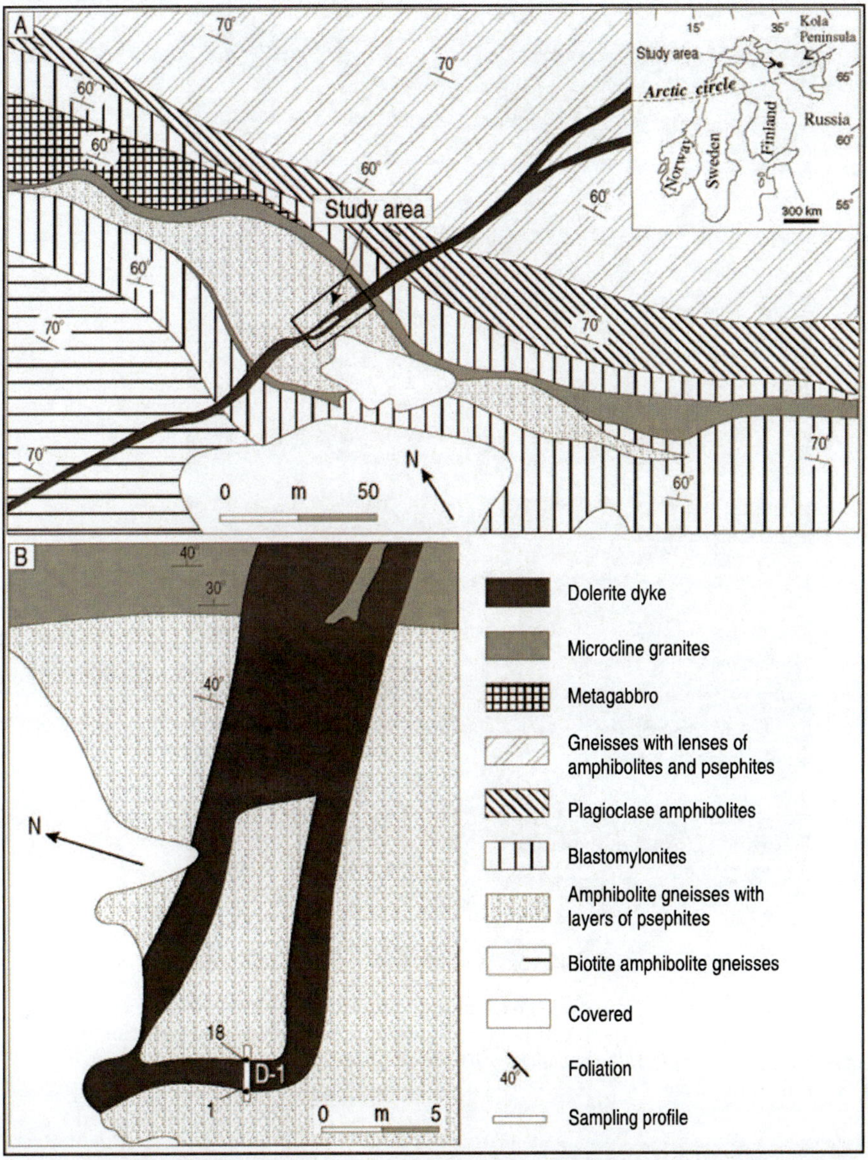

Fig. 30.1 Simplified geological map of the Vochelambina area (Kola Peninsula, Russia) showing the location of the studied dolerite dyke (**a**) and a field sketch of the outcrop where sampling of the dyke was carried out (**b**). The location of the Vochelambina area in the Fennoscandian Shield is indicated in the inset. The geological map has been modified from Mitrofanov and Pozhilenko (1991)

Fig. 30.2 Field photograph of the Vochelambina dolerite dyke (Kola Peninsula, Russia) cutting basement rocks. Both east and west contacts of the dyke with host rocks are very sharp. Minidrill sampling sites from D-1/1 to D-1/23 are indicated. The dyke is 140 cm thick. The length of pencil used as a scale is about 12 cm

country rocks are observed within the dyke in the form of angular, sharp-edged fragments with different sizes (Fig. 30.1b) that macro- and microscopically have undergone little change. In some places the dyke splits into two or more thinner sub-dykes, one of which is 140 cm thick and is the topic of this paper (Fig. 30.1). The dyke in this section appears to be a simple body: Neither field nor petrographic studies have revealed internal contacts, chill-zones or comb-layers (e.g. Platten and Watterson, 1987; Platten, 2000).

Petrography

Macroscopically, the dyke is aphanitic and does not contain any visible intratelluric phenocrysts. Microscopically, the dyke changes from microporphyritic at the margins to subophitic in the centre. The dyke contains about 20–25 vol.% of microphenocrysts of plagioclase, clinopyroxene and olivine (Fig. 30.3a, b), which reveal no preferred alignment and are uniformly distributed across the dyke (Pl, 1–2 vol.%; Cpx, 12–16 vol.% and Ol, 6–7 vol.%). The microphenocrysts become larger towards the centre of the dyke, probably as a result of post-emplacement growth. An about 20 cm-thick marginal zone of the dyke (along the western contact) is composed of 2–3 cm-thick alternating bands with contrasting grain sizes (from 0.01 to 0.35 mm).

The groundmass consists of small crystals of plagioclase, clinopyroxene and olivine that have apparently grown *in situ.* They show no compositional zonation and tend to increase in size from 0.01 to 0.5 mm from the dyke margins inwards. Groundmass plagioclase and clinopyroxene are compositionally more evolved than the microphenocrysts (Tables 30.2 and 30.3). Groundmass plagioclase varies between An_{77} and An_{64}. Groundmass clinopyroxene is compositionally augite with lower mg#, more TiO_2 and less Cr_2O_3 than the microphenocrysts. In addition, minor secondary biotite and sericite locally replace plagioclase crystals (Table 30.4). Fe-Ti oxides (< 1 vol.%), represented by magnetite, always occur as small anhedral crystals disseminated within the groundmass. From the margins the oxides increase in size inwards from 0.01 to 0.1 mm.

The plagioclase microphenocrysts form prismatic crystals and tend to increase in length from about 1 to 3 mm from the margins inwards. The microphenocrysts usually show normal zonation, with core compositions of about An_{80} grading into margins of about An_{70} (Table 30.2). Most plagioclase crystals are sericitized (Fig. 30.3c). The clinopyroxene microphenocrysts form equidimensional grains and aggregates of grains which increase in size from about 1.5 to 2.5 mm from the margins inwards. Clinopyroxene is commonly fresh and are compositionally endiopside or augite. The clinopyroxene microphenocrysts are normally zoned with mg# of 79–87 in cores and 72–86 in rims (Table 30.3). The size of olivine microphenocrysts gradually increases from 0.5 to 1 mm towards the centre of the dyke. All the olivine grains are completely altered to serpentine, talc and iddingsite.

Compositional Variations

The dolerite dyke reveals an unusual transverse compositional zonation. On the one hand, it shows an inward increase in normative An (100An/(An+Ab)), whole-rock Mg# (100 Mg/(Mg+Fe_{total})), MgO, Ni and Cr, together with an inward decrease in TiO_2, that are the opposite to what is expected for closed system differentiation of basaltic magma. On the other hand, it exhibits systematic inward increases in

Table 30.1 Major, trace and CIPW normative compositions for rocks from dyke Vochelambina

Samples		23	22	1	2	3	4	5	6	7	8	9
Rock type		Gneiss	Gneiss	Dyke	Dyke	Dyke	Dyke	Dyke	Dyke	Dyke	Dyke	Dyke
Distance (cm)		0	35	45	49	54	59	64	68	74	83	93
SiO_2	wt%	54.07	51.00	49.24	49.27	49.18	49.52	49.52	49.59	49.87	49.49	49.42
TiO_2	wt%	0.48	2.82	0.81	0.80	0.80	0.79	0.79	0.79	0.78	0.78	0.78
$Al2O_3$	wt%	15.78	12.79	14.48	14.20	14.33	14.41	14.51	14.59	14.57	14.57	14.48
Fe_2O_3	wt%	7.07	6.86	3.07	2.82	2.95	3.00	2.54	2.53	2.59	2.70	2.81
FeO	wt%	2.14	8.09	7.38	7.48	7.55	7.52	7.84	7.78	7.63	7.66	7.60
MnO	wt%	0.10	0.17	0.16	0.16	0.17	0.17	0.17	0.18	0.17	0.18	0.18
MgO	wt%	3.57	7.20	9.65	10.00	10.15	10.49	10.44	10.37	10.56	10.51	10.52
CaO	wt%	11.07	7.13	11.98	11.60	11.51	11.07	11.01	10.86	10.48	10.15	10.04
Na_2O	wt%	1.59	1.99	1.45	1.39	1.36	1.29	1.27	1.25	1.22	1.22	1.20
K_2O	wt%	3.43	1.35	0.28	0.45	0.35	0.39	0.51	0.66	0.90	1.09	1.14
P_2O_5	wt%	0.10	0.78	0.06	0.05	0.06	0.06	0.06	0.06	0.06	0.06	0.06
LOI	wt%	1.31	1.84	1.10	1.07	1.32	1.44	1.39	1.54	1.61	1.74	1.82
Cl	ppm	93	1628	117	138	116	113	121	112	98	111	103
Sc	ppm	19	34	38	41	38	38	37	38	40	39	39
V	ppm	222	246	266	260	260	259	262	261	256	255	261
Cr	ppm	47	179	598	694	623	611	623	632	642	670	670
Ni	ppm	48	108	191	189	189	188	190	187	191	192	196
Cu	ppm	28	169	104	96	103	99	99	100	96	102	100
Zn	ppm	41	137	72	75	71	69	65	69	62	66	63
Ga	ppm	29	19	15	14	14	14	14	14	13	14	14
Rb	ppm	61	73	22	37	29	33	44	59	80	99	103
Sr	ppm	902	234	84	83	85	82	84	88	94	101	103
Y	ppm	21	41	17	17	17	18	18	17	18	18	18
Zr	ppm	167	277	62	61	61	61	59	59	59	63	61
Ba	ppm	1301	407	50	65	47	55	62	76	107	124	119
Ce	ppm	53	59	28	28	19	19	21	18	8	16	17
Nd	ppm	18	41	4	14	7	13	5	7	11	9	4
Total	wt%	100.99	102.39	99.82	99.47	99.89	100.33	100.23	100.38	100.62	100.33	100.24
q	wt%	10.31	12.12	1.21	0.83	1.03	1.54	0.83	0.80	0.71	0.13	0.25
or	wt%	22.83	9.59	1.77	2.88	2.21	2.51	3.26	4.17	5.72	6.93	7.30
pl	wt%	44.24	46.75	48.09	46.42	47.02	46.68	46.20	45.91	44.87	44.56	44.19
cpx	wt%	21.61	7.97	23.42	22.68	21.54	19.24	18.89	18.33	17.23	16.40	16.25
opx	wt%	0.00	23.57	25.51	27.19	28.19	30.03	30.82	30.79	31.46	31.98	32.01
Mg#	at. %	42.81	47.37	62.93	64.03	63.95	64.66	64.79	64.77	65.41	65.03	64.95
An(norm)	mol.%	64.41	55.16	71.17	71.43	72.42	73.65	73.97	74.16	74.21	73.91	74.11

K_2O, Rb, Ba and Sr which are commonly considered as normal fractionation trends (Fig. 30.6) because they are consistent with fractional crystallisation of basaltic magma. Thus the compositional zonation of the dyke is normal in terms of some geochemical indices but reverse in terms of others. Three other features are noteworthy. Firstly, variations in K_2O, Rb, Ba and Sr are closely correlated with loss on ignition. Secondly, a strong and systematic increase in normative An across the 20 cm-thick marginal zone is not accompanied by a corresponding increase in K_2O and shows no correlation with variations in mineral grain size (Fig. 30.5). Thirdly, variations highly incompatible elements Zr (Fig. 30.5) and REE (Table 30.1) are fairly constant across the dyke.

Table 30.1 (continued)

Samples		10	11	13	14	15	16	17	18	19	20	21
Rock type		Dyke	Dyke	Dyke	Dyke	Dyke	Dyke	Dyke	Dyke	Gneiss	Gneiss	Gneiss
Distance (cm)		104	116	148	163	169	174	179	184	189	209	229
SiO_2	wt%	49.45	49.27	49.09	49.72	49.88	50.23	49.90	49.97	64.17	65.68	64.09
TiO_2	wt%	0.77	0.76	0.78	0.80	0.79	0.80	0.82	0.82	0.65	0.57	0.68
Al_2O_3	wt%	14.57	14.60	14.54	14.55	14.43	14.64	14.75	14.24	14.11	14.01	14.33
Fe_2O_3	wt%	2.82	3.16	3.92	3.26	2.35	2.25	2.26	2.41	2.93	4.19	2.87
FeO	wt%	7.57	7.09	6.60	7.27	8.02	8.13	8.26	8.08	4.01	2.99	4.30
MnO	wt%	0.18	0.18	0.17	0.18	0.18	0.17	0.18	0.17	0.07	0.06	0.06
MgO	wt%	10.49	10.53	10.02	10.15	10.29	10.17	10.07	10.10	4.09	3.59	4.21
CaO	wt%	9.97	9.75	9.68	10.11	10.20	10.89	10.95	11.03	4.60	5.40	4.94
Na_2O	wt%	1.20	1.18	1.19	1.27	1.29	1.39	1.46	1.43	3.83	3.83	3.35
K_2O	wt%	1.26	1.38	1.49	1.07	1.02	0.52	0.39	0.42	1.68	1.47	1.62
P_2O_5	wt%	0.05	0.05	0.06	0.06	0.06	0.06	0.06	0.05	0.14	0.17	0.15
LOI	wt%	1.85	2.09	2.33	1.85	1.66	1.22	1.06	1.07	1.52	0.95	1.47
Cl	ppm	113	110	103	132	161	155	180	187	970	880	1514
Sc	ppm	40	39	38	40	40	38	39	41	16	21	16
V	ppm	264	260	260	260	256	256	260	261	132	132	135
Cr	ppm	663	701	628	584	610	590	527	560	110	98	114
Ni	ppm	201	204	192	185	184	182	178	175	53	44	50
Cu	ppm	101	111	109	102	96	98	105	102	20	14	28
Zn	ppm	68	65	80	69	66	66	68	69	65	58	72
Ga	ppm	15	13	14	14	15	13	16	14	18	18	18
Rb	ppm	115	125	136	94	89	41	29	32	93	68	61
Sr	ppm	108	108	132	111	107	85	83	83	305	302	327
Y	ppm	18	19	19	18	18	18	17	18	19	24	21
Zr	ppm	62	60	66	62	63	60	60	63	176	169	177
Ba	ppm	126	139	149	119	106	70	58	70	362	284	565
Ce	ppm	23	10	16	16	18	18	25	16	42	59	49
Nd	ppm	7	9	10	8	8	6	11	6	21	31	27
Total	wt%	100.38	100.24	100.06	100.48	100.36	100.64	100.31	99.96	102.05	103.12	102.40
q	wt%	0.00	0.21	1.35	1.43	0.34	1.10	0.71	1.26	20.46	24.10	22.03
or	wt%	8.06	8.85	9.78	6.86	6.48	3.23	2.47	2.66	10.51	9.21	10.08
pl	wt%	44.00	44.10	44.30	45.19	44.38	46.57	47.69	46.26	51.59	51.99	50.24
cpx	wt%	16.03	15.35	15.69	16.52	17.01	18.37	18.37	20.00	4.70	7.49	3.58
opx	wt%	31.85	31.50	28.87	30.00	31.79	30.73	30.75	29.82	12.74	7.20	14.07
Mg#	at. %	64.90	65.40	63.84	63.96	64.41	64.12	63.57	63.77	52.34	48.66	52.19
An(norm)	mol.%	74.05	74.16	73.79	72.99	72.48	72.15	71.30	70.93	32.27	32.68	39.11

Mg# = 100 Mg/(Mg + Fetotal), An(norm) = 100An/(An + Ab), q – quartz, or – orthoclase, pl – plagioclase, cpx – clinopyroxene, opx – orthopyroxene, LOI – loss on ignition

Discussion

A key question to be addressed is why the Vochelambina dolerite dyke reveals an internal chemical zonation that combines features of both reverse and normal fractionation trends. It should be noted that reverse trends are quite common in relatively small dykes whereas normal trends are very rare (Chistyakova and Latypov, 2009a; b; 2010). There are several physico-chemical mechanisms that can operate in narrow dykes that are able to produce reverse compositional trends. In contrast, there is only one mechanism that can give rise to normal compositional trends: fractional crystallization of magma from the margins inward. However, this is inhibited

Table 30.2 Selected analyses of plagioclase from dyke Vochelambina

Sample	1-3.2	1-3.6	1-3.7	1-3.8	1-3.9	3-2.1	3-2.2	3-2.3	3-2.6	3-2.7	3-2.8	5-2.1	5-2.2	5-2.7	5-2.9	7-1.2
Position	Matrix	Matrix	Matrix	Matrix	Rim	Matrix	Matrix	Matrix	Core	Core	Rim	Rim	Matrix	Matrix	Matrix	Matrix
Oxide wt%																
SiO_2	50.99	51.56	50.81	0.25	0.67	53.60	52.61	0.18	0.27	0.23	0.65	0.09	0.28	0.21	53.30	52.06
TiO_2	0.02	0.02	0.01	0.02	0.07	0.03	0.07	0.09	0.00	0.04	0.00	0.00	0.00	0.00	0.03	0.01
Cr_2O_3	0.00	0.00	0.00	0.00	0.00	0.03	0.00	0.00	0.00	0.00	0.00	0.02	0.00	0.00	0.00	0.04
Al_2O_3	30.52	30.36	30.26	51.32	50.61	28.74	29.31	52.07	49.62	50.38	49.18	53.32	52.20	52.05	29.44	28.13
FeO	0.73	0.80	0.94	0.77	1.26	0.99	0.97	0.83	0.52	0.50	0.86	0.90	0.76	0.78	0.85	0.92
MnO	0.00	0.00	0.01	0.00	0.05	0.05	0.04	0.04	0.01	0.01	0.01	0.01	0.00	0.06	0.00	0.00
MgO	0.33	0.25	0.30	30.61	29.76	0.05	0.10	30.10	32.16	31.57	31.33	29.45	30.39	30.04	0.09	0.24
CaO	15.19	14.54	14.60	14.88	14.33	12.91	13.06	13.70	15.66	15.67	14.90	13.05	14.29	14.35	13.21	14.25
Na_2O	2.54	3.15	3.19	2.58	2.60	4.00	3.92	3.55	2.09	2.42	1.90	3.72	2.82	3.21	3.93	3.39
K_2O	0.06	0.08	0.12	0.14	0.53	0.14	0.14	0.10	0.48	0.16	0.74	0.14	0.09	0.05	0.13	0.11
ZnO	0.03	0.00	0.00	0.10	0.01	0.00	0.00	0.08	0.00	0.00	0.06	0.00	0.00	0.00	0.00	0.04
NiO	0.02	0.01	0.07	0.01	0.07	0.00	0.00	0.00	0.01	0.09	0.00	0.07	0.00	0.00	0.04	0.00
V_2O_3	0.00	0.00	0.00	0.00	0.00	0.00	0.00	0.00	0.00	0.00	0.00	0.05	0.03	0.01	0.05	0.03
Cl	0.00	0.00	0.00	0.01	0.00	0.00	0.00	0.01	0.00	0.00	0.00	0.04	0.02	0.00	0.01	0.00
Total	100.43	100.76	100.31	100.69	99.96	100.54	100.20	100.74	100.82	101.06	99.63	100.83	100.87	100.76	101.07	99.22
Cations/charges								20(O)								
Si	9.277	9.345	9.278	0.067	0.184	9.699	9.565	0.050	0.073	0.063	0.180	0.024	0.075	0.058	9.600	9.589
Ti	0.002	0.003	0.001	0.003	0.010	0.004	0.009	0.013	0.000	0.005	0.000	0.000	0.000	0.000	0.004	0.002
Cr	0.000	0.000	0.000	0.001	0.000	0.005	0.000	0.000	0.000	0.000	0.000	0.002	0.000	0.000	0.000	0.006
Al	6.546	6.487	6.514	9.308	9.291	6.131	6.281	9.428	9.018	9.123	9.057	9.617	9.420	9.423	6.250	6.106
Fe	0.111	0.121	0.144	0.117	0.193	0.149	0.147	0.126	0.080	0.075	0.133	0.135	0.114	0.119	0.127	0.142
Mn	0.000	0.000	0.002	0.000	0.008	0.007	0.006	0.006	0.001	0.002	0.001	0.001	0.000	0.009	0.000	0.000
Mg	0.089	0.066	0.082	6.544	6.439	0.013	0.026	6.423	6.890	6.738	6.801	6.262	6.464	6.410	0.024	0.066
Ca	2.962	2.824	2.857	2.891	2.819	2.503	2.545	2.658	3.050	3.041	2.939	2.523	2.764	2.783	2.549	2.813
Na	0.897	1.107	1.128	0.909	0.927	1.402	1.381	1.247	0.736	0.848	0.679	1.300	0.988	1.127	1.371	1.211
K	0.015	0.018	0.027	0.031	0.124	0.032	0.033	0.022	0.111	0.038	0.173	0.031	0.020	0.012	0.030	0.026
Zn	0.003	0.000	0.000	0.013	0.001	0.000	0.000	0.010	0.000	0.000	0.008	0.000	0.000	0.000	0.000	0.006
Ni	0.002	0.002	0.011	0.002	0.010	0.000	0.000	0.000	0.001	0.012	0.000	0.010	0.000	0.000	0.006	0.000
V	0.000	0.000	0.000	0.000	0.000	0.000	0.000	0.000	0.000	0.000	0.000	0.008	0.004	0.001	0.008	0.004
Cl	0.000	0.000	0.001	0.002	0.000	0.001	0.001	0.003	0.001	0.000	0.000	0.012	0.005	0.000	0.004	0.000
Total	19.904	19.971	20.043	19.888	20.005	19.946	19.993	19.985	19.961	19.946	19.969	19.925	19.855	19.941	19.973	19.970
An	76.76	71.84	71.70	76.09	75.26	64.09	64.82	68.06	80.57	78.19	81.23	65.99	73.66	71.17	65.02	69.90

Table 30.2 (continued)

Sample	7-1.3	7-1.4	9-1.7	9-1.8	9-1.9	11-1.7	13-2.2	13-2.3	15-1.5	15-1.6	15-1.7	18-1.6	18-1.7
Position	Matrix	Matrix	Core	Matrix	Matrix	Matrix	Matrix	Matrix	Matrix	Matrix	Matrix	Matrix	Matrix
SiO_2	52.23	53.17	49.99	51.36	52.71	53.27	51.76	52.56	51.02	52.20	51.40	50.45	52.71
TiO_2	0.00	0.04	0.00	0.02	0.03	0.07	0.06	0.02	0.00	0.05	0.03	0.00	0.05
Cr_2O_3	0.00	0.00	0.01	0.01	0.00	0.01	0.01	0.00	0.01	0.00	0.00	0.00	0.00
Al_2O_3	27.88	27.14	29.36	28.66	27.88	27.72	28.47	27.63	29.22	27.57	28.18	28.86	27.60
FeO	0.89	1.06	0.51	0.68	0.90	0.82	0.77	0.78	0.65	0.86	0.79	0.78	1.19
MnO	0.03	0.01	0.00	0.00	0.00	0.00	0.01	0.00	0.03	0.03	0.02	0.06	0.04
MgO	0.19	0.14	0.32	0.25	0.18	0.14	0.18	0.17	0.34	0.20	0.28	0.30	0.36
CaO	14.29	13.41	16.05	15.04	13.87	13.22	14.54	13.81	14.81	14.14	14.89	15.47	13.63
Na_2O	3.35	3.86	2.22	3.08	3.66	3.69	3.17	3.52	2.67	3.24	2.69	2.49	3.40
K_2O	0.09	0.10	0.07	0.08	0.12	0.11	0.07	0.07	0.06	0.14	0.06	0.05	0.12
ZnO	0.02	0.00	0.00	0.00	0.00	0.00	0.03	0.00	0.00	0.00	0.07	0.02	0.03
NiO	0.10	0.07	0.00	0.08	0.12	0.00	0.06	0.00	0.10	0.03	0.00	0.06	0.13
V_2O_3	0.00	0.00	0.00	0.03	0.00	0.00	0.02	0.01	0.00	0.00	0.05	0.02	0.00
Cl	0.00	0.01	0.02	0.00	0.01	0.00	0.02	0.00	0.02	0.02	0.01	0.00	0.01
Total	99.08	99.02	98.54	99.29	99.48	99.06	99.15	98.57	98.93	98.48	98.47	98.57	99.27
Cations/charges							20(O)						
Si	9.631	9.794	9.291	9.465	9.673	9.775	9.538	9.713	9.417	9.673	9.536	9.375	9.696
Ti	0.000	0.006	0.000	0.003	0.004	0.010	0.008	0.003	0.000	0.007	0.005	0.000	0.007
Cr	0.000	0.000	0.001	0.002	0.000	0.001	0.002	0.000	0.001	0.000	0.000	0.000	0.000
Al	6.061	5.892	6.431	6.226	6.032	5.997	6.184	6.018	6.357	6.022	6.162	6.322	5.984
Fe	0.138	0.163	0.079	0.105	0.139	0.126	0.119	0.121	0.101	0.134	0.123	0.122	0.184
Mn	0.005	0.002	0.000	0.000	0.000	0.001	0.001	0.000	0.004	0.005	0.004	0.009	0.006
Mg	0.052	0.039	0.088	0.070	0.049	0.039	0.048	0.048	0.093	0.055	0.077	0.084	0.099
Ca	2.824	2.647	3.196	2.971	2.727	2.600	2.870	2.734	2.930	2.808	2.959	3.080	2.687
Na	1.197	1.378	0.799	1.100	1.304	1.314	1.131	1.260	0.955	1.166	0.967	0.899	1.213
K	0.022	0.024	0.017	0.018	0.027	0.026	0.017	0.016	0.015	0.033	0.014	0.011	0.028
Zn	0.003	0.000	0.000	0.000	0.000	0.000	0.004	0.000	0.000	0.000	0.010	0.003	0.004
Ni	0.015	0.011	0.000	0.012	0.018	0.000	0.009	0.000	0.015	0.005	0.000	0.009	0.019
V	0.000	0.001	0.000	0.004	0.000	0.000	0.003	0.002	0.001	0.000	0.008	0.003	0.000
Cl	0.000	0.002	0.007	0.001	0.002	0.000	0.006	0.000	0.007	0.006	0.003	0.000	0.002
Total	19.948	19.958	19.909	19.977	19.975	19.887	19.939	19.913	19.896	19.914	19.868	19.917	19.928
An	70.23	65.77	80.00	72.97	67.65	66.42	71.74	68.44	75.41	70.67	75.37	77.41	68.89

Core – centre of microphenocrysts, rim – edge of microphenocrystal, matrix – in situ grown crystals, An = Ca/(Ca+Na+K).

Table 30.3 Selected analyses of clinopyroxene from dyke Vochelambina

Sample	1.1-9	1.1-10	1.1-11	3.2-1	3.2-2	3.2-3	3.2-4	5.1-1	5.1-2	5.1-3	5.1-4	7.2-2	7.2-3	7.2-4	7.2-5	9.1-1	9.1-2	9.1-3
Position	Core	Core	Matrix	Core	Rim	Rim	Rim	Core	Rim	Rim	Matrix	Core	Rim	Matrix	Core	Core	Rim	Matrix
Oxide wt%																		
SiO_2	52.22	52.62	52.68	52.24	52.68	53.55	54.19	52.84	54.12	52.87	51.12	53.62	53.20	54.31	52.20	53.27	53.50	52.41
TiO_2	0.29	0.34	0.49	0.29	0.28	0.22	0.25	0.18	0.16	0.17	0.69	0.19	0.34	0.27	0.27	0.21	0.32	0.39
Cr_2O_3	1.17	0.38	0.31	1.29	0.42	0.72	0.40	0.97	0.48	0.68	0.21	0.77	0.34	0.20	1.36	0.80	0.09	0.24
Al_2O_3	3.73	3.98	3.63	3.61	3.72	2.74	2.02	2.57	2.08	2.96	4.84	2.43	3.25	2.01	4.12	2.29	1.31	3.74
FeO	5.40	6.54	8.23	5.20	6.12	5.62	7.54	5.19	6.07	5.54	7.33	6.23	6.95	9.03	5.43	5.77	10.84	7.48
MnO	0.08	0.24	0.28	0.11	0.19	0.18	0.19	0.17	0.27	0.16	0.15	0.14	0.23	0.28	0.13	0.16	0.26	0.20
MgO	17.84	18.17	18.67	17.57	18.95	18.88	21.47	18.91	19.23	18.49	15.40	19.99	18.21	20.88	17.41	18.56	18.14	17.31
CaO	18.94	17.26	15.52	19.29	17.68	18.49	14.10	18.98	17.96	19.17	20.19	16.91	17.59	13.55	19.49	18.37	15.81	18.41
Na_2O	0.17	0.21	0.16	0.21	0.22	0.19	0.09	0.22	0.17	0.16	0.19	0.15	0.12	0.08	0.19	0.19	0.16	0.19
K_2O	0.00	0.03	0.00	0.02	0.00	0.00	0.00	0.01	0.00	0.01	0.00	0.02	0.01	0.02	0.02	0.01	0.01	0.01
ZnO	0.05	0.00	0.07	0.00	0.00	0.00	0.02	0.00	0.04	0.07	0.00	0.00	0.02	0.01	0.05	0.00	0.03	0.00
NiO	0.06	0.03	0.04	0.00	0.12	0.11	0.00	0.05	0.00	0.06	0.00	0.00	0.05	0.13	0.06	0.11	0.00	0.08
V_2O_3	0.07	0.08	0.04	0.05	0.06	0.05	0.08	0.06	0.00	0.06	0.11	0.04	0.05	0.00	0.08	0.00	0.04	0.04
Cl	0.00	0.01	0.00	0.00	0.00	0.00	0.01	0.01	0.00	0.00	0.02	0.00	0.01	0.01	0.00	0.02	0.01	0.00
Total	100.03	99.87	100.10	99.88	100.45	100.75	100.36	100.15	100.58	100.40	100.25	100.48	100.36	100.77	100.82	99.74	100.51	100.48
Cations/charges								4(O)										
Si	1.903	1.916	1.919	1.907	1.907	1.932	1.952	1.922	1.955	1.920	1.879	1.936	1.931	1.958	1.892	1.944	1.965	1.911
Ti	0.008	0.009	0.014	0.008	0.008	0.006	0.007	0.005	0.004	0.005	0.019	0.005	0.009	0.007	0.007	0.006	0.009	0.011
Cr	0.034	0.011	0.009	0.037	0.012	0.021	0.011	0.028	0.014	0.020	0.006	0.022	0.010	0.006	0.039	0.023	0.003	0.007
Al	0.160	0.171	0.156	0.155	0.159	0.117	0.086	0.110	0.089	0.127	0.210	0.103	0.139	0.086	0.176	0.098	0.057	0.161
Fe	0.165	0.199	0.251	0.159	0.185	0.170	0.227	0.158	0.183	0.168	0.225	0.188	0.211	0.272	0.165	0.176	0.333	0.228
Mn	0.003	0.007	0.009	0.003	0.006	0.006	0.006	0.005	0.008	0.005	0.005	0.004	0.007	0.008	0.004	0.005	0.008	0.006
Mg	0.969	0.986	1.014	0.956	1.023	1.016	1.153	1.025	1.035	1.001	0.844	1.076	0.985	1.122	0.941	1.010	0.993	0.941
Ca	0.740	0.673	0.606	0.754	0.686	0.715	0.544	0.740	0.695	0.746	0.795	0.654	0.684	0.523	0.757	0.718	0.622	0.719
Na	0.012	0.015	0.011	0.015	0.015	0.013	0.007	0.016	0.012	0.012	0.014	0.011	0.009	0.006	0.014	0.014	0.011	0.014
K	0.000	0.001	0.000	0.001	0.000	0.000	0.000	0.000	0.000	0.001	0.000	0.001	0.001	0.001	0.001	0.000	0.001	0.000
Zn	0.001	0.000	0.002	0.000	0.000	0.000	0.001	0.000	0.001	0.002	0.000	0.000	0.001	0.000	0.001	0.000	0.001	0.000
Ni	0.002	0.001	0.001	0.000	0.004	0.003	0.000	0.002	0.000	0.002	0.000	0.000	0.001	0.004	0.002	0.003	0.000	0.002
V	0.002	0.002	0.001	0.002	0.002	0.002	0.002	0.002	0.000	0.002	0.003	0.001	0.002	0.000	0.002	0.000	0.001	0.001
Cl	0.000	0.001	0.000	0.000	0.000	0.000	0.001	0.001	0.000	0.000	0.001	0.000	0.000	0.001	0.000	0.001	0.001	0.000
Total	3.998	3.992	3.991	3.996	4.006	3.999	3.995	4.012	3.996	4.008	4.001	4.001	3.989	3.993	4.000	3.998	4.003	4.001
mg#	85.48	83.20	80.17	85.76	84.65	85.70	83.54	86.67	84.95	85.61	78.94	85.12	82.37	80.48	85.11	85.15	74.89	80.49

Table 30.3 (continued)

Sample Position	11.1-3 Rim	11.1-4 Core	11.3-2 Matrix	13.1-2 Core	13.1-3 Rim	13.1-4 Core	13.1-5 Matrix	15.1-1 Core	15.1-2 Rim	15.1-3 Core	15.1-4 Matrix	18.1-1 Rim	18.1-2 Core	18.1-3 Matrix	18.1-4 Core
SiO_2	52.44	54.35	51.33	52.59	53.98	52.44	51.86	52.59	52.42	51.10	51.11	52.96	52.18	53.15	51.90
TiO_2	0.44	0.13	0.49	0.26	0.27	0.42	0.61	0.25	0.39	0.64	0.67	0.21	0.24	0.33	0.29
Cr_2O_3	0.05	0.72	0.23	1.08	0.32	0.19	0.01	1.37	0.40	0.33	0.05	0.36	1.20	0.50	1.32
Al_2O_3	2.64	1.73	3.42	3.15	2.05	3.27	2.92	3.43	3.71	3.98	3.53	3.42	3.66	3.00	3.81
FeO	11.94	5.44	7.87	5.74	7.10	7.56	10.69	4.81	6.59	7.66	10.96	5.60	5.60	7.90	5.07
MnO	0.33	0.15	0.23	0.18	0.22	0.21	0.27	0.11	0.19	0.23	0.25	0.16	0.16	0.21	0.11
MgO	17.47	19.52	16.35	18.37	19.41	17.46	15.77	17.35	17.61	16.23	14.97	18.32	18.82	19.76	17.60
CaO	14.40	18.21	18.47	18.66	16.90	18.54	18.03	20.04	18.19	18.90	18.30	18.67	17.19	14.50	18.90
Na_2O	0.13	0.13	0.11	0.22	0.18	0.24	0.17	0.19	0.22	0.15	0.19	0.20	0.21	0.13	0.18
K_2O	0.01	0.01	0.01	0.00	0.01	0.03	0.00	0.00	0.01	0.01	0.01	0.00	0.02	0.00	0.01
ZnO	0.00	0.00	0.00	0.00	0.00	0.02	0.01	0.00	0.00	0.00	0.07	0.04	0.00	0.00	0.00
NiO	0.05	0.07	0.02	0.00	0.01	0.03	0.05	0.03	0.12	0.00	0.08	0.00	0.05	0.06	0.00
V_2O_3	0.10	0.07	0.05	0.05	0.05	0.00	0.09	0.04	0.05	0.13	0.03	0.07	0.04	0.12	0.02
Cl	0.01	0.00	0.00	0.00	0.02	0.01	0.01	0.02	0.01	0.01	0.02	0.02	0.00	0.00	0.00
Total	99.99	100.52	98.58	100.29	100.50	100.43	100.49	100.23	99.89	99.36	100.22	100.02	99.36	99.66	99.22
Cations/charges							4(O)								
Si	1.939	1.960	1.915	1.912	1.954	1.916	1.919	1.913	1.915	1.893	1.903	1.924	1.907	1.935	1.904
Ti	0.012	0.003	0.014	0.007	0.007	0.012	0.017	0.007	0.011	0.018	0.019	0.006	0.007	0.009	0.008
Cr	0.001	0.020	0.007	0.031	0.009	0.006	0.000	0.039	0.012	0.010	0.001	0.010	0.035	0.014	0.038
Al	0.115	0.074	0.151	0.135	0.087	0.141	0.127	0.147	0.160	0.174	0.155	0.146	0.158	0.129	0.165
Fe	0.369	0.164	0.245	0.175	0.215	0.231	0.331	0.146	0.201	0.237	0.341	0.170	0.171	0.241	0.156
Mn	0.010	0.005	0.007	0.005	0.007	0.007	0.009	0.003	0.006	0.007	0.008	0.005	0.005	0.006	0.003
Mg	0.963	1.050	0.909	0.996	1.047	0.951	0.870	0.941	0.959	0.896	0.831	0.992	1.025	1.073	0.962
Ca	0.571	0.704	0.738	0.727	0.656	0.726	0.715	0.781	0.712	0.751	0.730	0.727	0.673	0.566	0.743
Na	0.010	0.009	0.008	0.016	0.013	0.017	0.012	0.014	0.015	0.011	0.014	0.014	0.015	0.009	0.013
K	0.000	0.000	0.000	0.000	0.001	0.001	0.000	0.000	0.000	0.001	0.001	0.000	0.001	0.000	0.001
Zn	0.000	0.000	0.000	0.000	0.000	0.001	0.000	0.000	0.000	0.000	0.002	0.001	0.000	0.000	0.000
Ni	0.001	0.002	0.001	0.000	0.000	0.001	0.002	0.001	0.004	0.000	0.002	0.000	0.002	0.002	0.000
V	0.003	0.002	0.001	0.002	0.001	0.000	0.003	0.001	0.002	0.004	0.001	0.002	0.001	0.004	0.001
Cl	0.001	0.000	0.000	0.000	0.001	0.001	0.001	0.001	0.000	0.001	0.001	0.001	0.000	0.000	0.000
Total	3.995	3.993	3.996	4.005	3.998	4.009	4.005	3.994	3.996	4.002	4.008	3.999	3.998	3.987	3.993
mg#	72.27	86.47	78.74	85.09	82.98	80.45	72.45	86.54	82.65	79.07	70.90	85.37	85.70	81.68	86.09

Core – centre of microphenocrysts, rim – edge of microphenocrystal, matrix – in situ grown crystals, mg#=Mg/(Mg+Fe).

Table 30.4 Selected analyses of secondary mica from dyke Vochelambina

Sample	1-3.3	1-3.4	3-2.5	3-2.9	5-2.3	7-1.5	9-1.5	9-1.6	11-1.9	13-2.1	15-1.4	18-1.5
Oxide wt%												
SiO_2	49.83	1.16	1.88	2.24	2.07	49.44	49.71	49.83	50.56	49.09	49.59	49.65
TiO_2	0.00	0.00	0.00	0.00	0.00	0.02	0.00	0.00	0.03	0.00	0.00	0.00
Cr_2O_3	0.01	0.00	0.01	0.00	0.05	0.03	0.00	0.03	0.00	0.00	0.00	0.00
Al_2O_3	30.62	49.74	49.97	50.47	49.59	28.87	29.22	30.10	31.28	30.45	30.03	30.62
FeO	1.45	1.13	1.95	0.89	1.19	2.17	1.13	1.20	0.90	1.36	2.11	1.31
MnO	0.01	0.10	0.02	0.05	0.03	0.05	0.08	0.10	0.13	0.10	0.08	0.12
MgO	1.63	32.03	30.33	30.13	30.19	2.60	1.89	1.66	1.29	1.27	1.48	1.94
CaO	0.04	0.08	0.06	0.07	0.03	0.05	0.07	0.10	0.23	0.04	0.11	0.08
Na_2O	0.09	0.11	0.08	0.08	0.04	0.04	0.05	0.13	0.75	0.07	0.30	0.05
K_2O	10.84	10.72	11.11	10.68	10.86	10.61	11.13	10.98	10.38	11.00	10.76	10.89
ZnO	0.00	0.00	0.03	0.00	0.00	0.01	0.04	0.05	0.00	0.00	0.00	0.00
NiO	0.06	0.00	0.01	0.00	0.00	0.00	0.00	0.00	0.12	0.04	0.00	0.06
V_2O_3	0.01	0.00	0.00	0.00	0.03	0.04	0.02	0.00	0.02	0.00	0.00	0.02
Cl	0.00	0.01	0.01	0.01	0.04	0.00	0.02	0.02	0.02	0.03	0.03	0.03
Total	94.58	95.07	95.47	94.62	94.09	93.93	93.34	94.18	95.71	93.43	94.48	94.76
Cations/charges						20(O)						
Si	9.690	0.333	0.543	0.645	0.602	9.727	9.808	9.738	9.690	9.680	9.701	9.648
Ti	0.000	0.000	0.000	0.000	0.000	0.003	0.000	0.000	0.005	0.000	0.000	0.000
Cr	0.002	0.000	0.001	0.000	0.008	0.005	0.000	0.005	0.000	0.000	0.000	0.000
Al	7.018	9.595	9.673	9.770	9.691	6.696	6.797	6.933	7.066	7.077	6.924	7.013
Fe	0.236	0.182	0.316	0.144	0.194	0.356	0.186	0.196	0.145	0.225	0.346	0.213
Mn	0.002	0.016	0.004	0.009	0.004	0.009	0.014	0.016	0.021	0.016	0.013	0.019
Mg	0.472	7.283	6.922	6.876	6.955	0.763	0.555	0.483	0.368	0.373	0.431	0.562
Ca	0.008	0.016	0.012	0.015	0.005	0.011	0.014	0.021	0.048	0.008	0.023	0.017
Na	0.032	0.041	0.031	0.029	0.015	0.015	0.021	0.049	0.279	0.027	0.112	0.019
K	2.690	2.639	2.744	2.638	2.707	2.663	2.802	2.738	2.538	2.767	2.687	2.700
Zn	0.000	0.000	0.005	0.000	0.000	0.002	0.005	0.007	0.000	0.000	0.000	0.000
Ni	0.009	0.000	0.002	0.000	0.000	0.000	0.000	0.000	0.018	0.006	0.000	0.009
V	0.001	0.000	0.000	0.000	0.005	0.006	0.003	0.000	0.003	0.000	0.000	0.003
Cl	0.000	0.002	0.003	0.003	0.013	0.000	0.007	0.007	0.007	0.008	0.010	0.010
Total	20.161	20.105	20.255	20.129	20.199	20.255	20.210	20.193	20.706	20.187	20.247	20.214

in small mafic dykes due to their rapid crystallization. The simultaneous occurrence of both reverse and normal compositional trends in the same dyke is puzzling since it apparently implies the operation of mutually exclusive processes.

Origin of Reverse Compositional Trends

There are four principle mechanisms that may result in the formation of such trends in narrow mafic dykes. These are *in situ* contamination, flow differentiation, *in situ* growth of crystals on dyke walls and changes in composition of inflowing magma (see a review in Chistyakova and Latypov, 2009b). Our previous studies have demonstrated that *in situ* contamination is effective only at distances of 1–5 cm from dyke contacts (Chistyakova and Latypov, 2009a, b) and this process cannot be responsible for an internal zonation that is developed throughout the 140 cm-thick dyke. This inference is also strongly supported by the distribution of TiO_2 that increases towards the margins of the dyke, while the country rocks have much lower concentrations (Fig. 30.4).

Flow differentiation causes reverse compositional trends due to the concentration of phenocrysts in the inflowing magma towards the center of a dyke (Bhattacharji and Smith, 1964; Bhattacharji, 1967; Simkin, 1967). However, the uniform distribution of phenocrysts of plagioclase, clinopyroxene and olivine across the dyke does not support the action of flow differentiation. The trace element data support this conclusion. The mechanism implies a decrease in concentration of highly incompatible elements, such as Zr and REE, towards the centre of the dyke due to a decrease in the amount of interstitial liquid. This is not the case with the studied dyke that shows quite constant Zr across its entire width (Fig. 30.5).

Another mechanism for the generation of reverse compositional trends is progressive inward increase in the proportion of liquidus phases crystallizing on the dyke walls (e.g., Ross, 1983; Miller and Ripley, 1996; Chistyakova and Latypov, 2009b). This process implies that the amount of liquidus minerals should increase inwards in response to crystallization under progressively less supercooled conditions. As a result, all compositional changes would be expected to closely correlate with an inward increase in mineral grain size. A systematic increase in normative An across the 20 cm-thick marginal zone of the dyke is not, however, correlated with mineral grain size variations that change here drastically and irregularly (Fig. 30.5).

The only option that appears to be left is the filling of the dyke with magma that becomes increasingly more primitive with time (e.g., Morse, 1979, 1981; Chistyakova and Latypov, 2009a, b). The dyke most likely records temporal changes in composition of magma rising from an underlying magma reservoir. This is quite an old idea in petrology that makes sense from the point of view of melting at the mantle source or magma staging at any intermediate level. It implies the emplacement of leading pulses of relatively evolved magma, produced by fractional melting without complete mixing in the mantle source, fractional crystallization of magma flowing through a conduit, or fractional crystallization and magma stratification in a subjacent chamber.

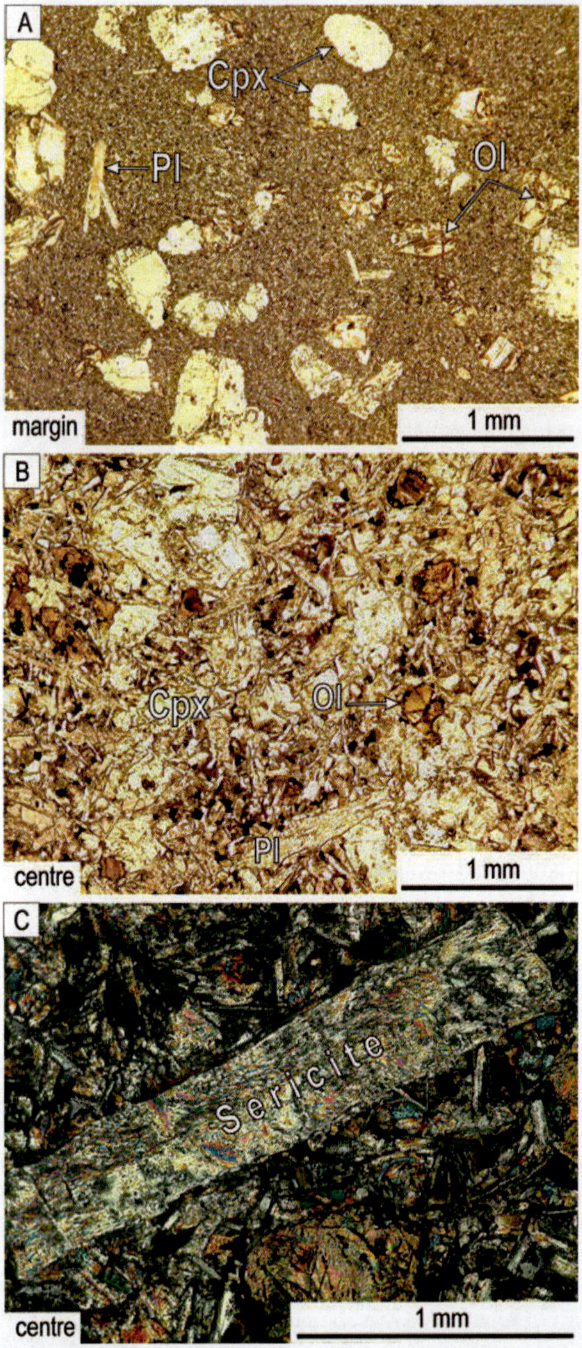

Fig. 30.3 Photomicrographs of the margin (**a**) and interior (**b**) of the Vochelambina dolerite dyke (in plane-polarized light). The texture of the dyke changes from microporphyritic at the margins to subophitic in the centre. (**c**) Close-up from the interior of the dyke illustrating fine-grained aggregates of sericite that have replaced phenocrystic and groundmass plagioclase (cross-polarized light)

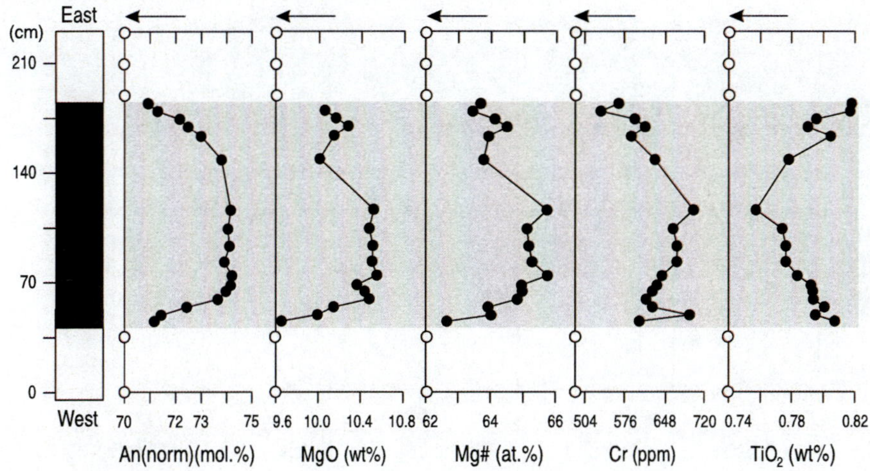

Fig. 30.4 Compositional variations in normative An, whole-rock MgO, Mg#, Cr and TiO_2 across the Vochelambina dolerite dyke (Kola Peninsula, Russia). Note that in terms of these geochemical indices the dyke is reversely zoned. Mg# = 100 Mg/(Mg+Fe_{total}); An(norm) = 100An/(An+Ab). The compositions of the host gneisses are indicated by open circles. Arrows show where these compositional points are off scale. The values for these points can be found in Table 30.1. See text for further discussion

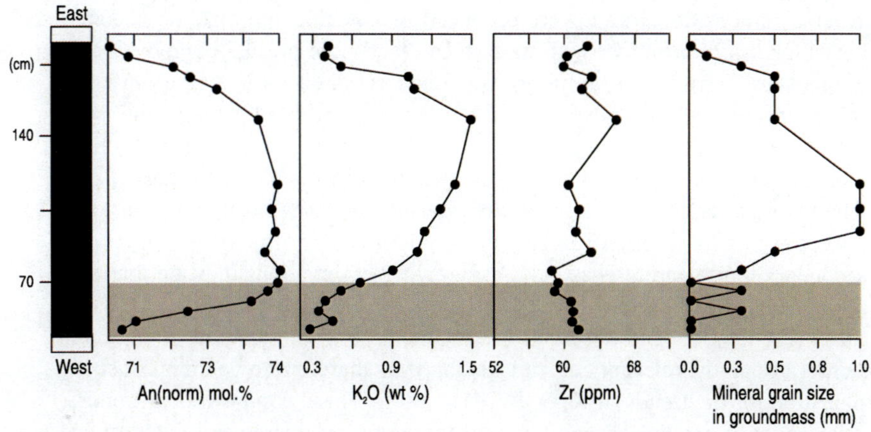

Fig. 30.5 Compositional variations across the Vochelambina dolerite dyke (Kola Peninsula, Russia) showing that systematic changes in normative An along the lower margin (grey field) do not correlate with grain size variations and K_2O, taken as proxy of secondary sericite. Also note that incompatible Zr has a fairly constant concentration through the dyke. See text for further discussion

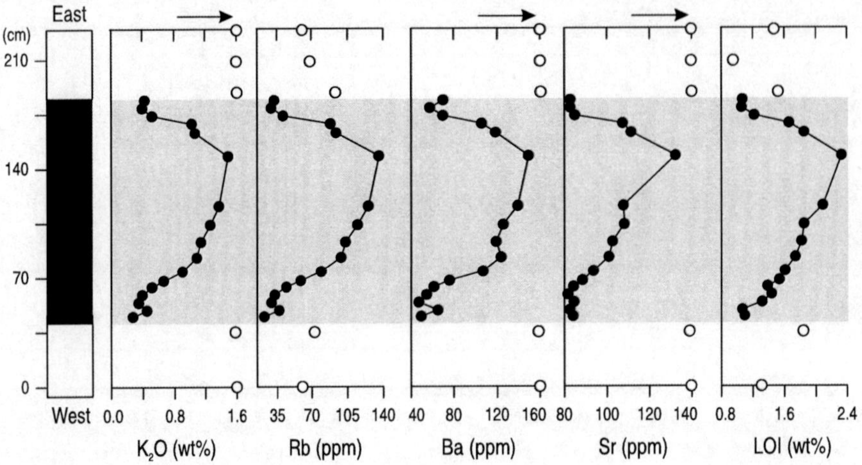

Fig. 30.6 Compositional variations in K_2O, Rb, Ba, Sr and LOI across the Vochelambina dolerite dyke (Kola Peninsula, Russia). Note that in terms of trace elements the dyke is normally zoned. Also note that all the observed compositional trends closely correlate with variations in loss on ignition (LOI). The compositions of the host gneisses are indicated by open circles. Arrows show where these compositional points are off scale (see Table 30.1 for values). See text for further discussion

Origin of Normal Compositional Trends

If the reverse compositional trends are produced by the inflowing magma becoming increasingly more primitive with time, then clearly the normal compositional trends cannot result from closed system fractional crystallization. There must be some other process, probably postmagmatic, that is responsible for their origin. In an attempt to resolve this paradox, we note that the inward increases in K_2O, Rb, Ba and Sr are closely correlated with loss on ignition. Since hydrous primary magmatic phases, such as amphibole or biotite, are lacking, then an increase in loss on ignition towards the centre of dyke can only be related to the development of some secondary fluid-bearing phase(s). Petrographic observations indicate that the most likely candidate is sericite that forms fine-grained aggregates after primary plagioclase (Fig. 30.3a). The observed inward increase in the amount of sericite, that incorporates all the relevant constituents, appears therefore to be the most simple explanation for the trends of K_2O, Rb, Ba and Sr. These compositional trends, along with variations in loss on ignition, are therefore considered as related to an increasing extent of postmagmatic alteration towards the centre of dyke.

Why does the extent of postmagmatic alteration increase towards the center of dyke? The mostly likely answer is that upward-migrating postmagmatic fluids had a tendency to concentrate in the centre of dyke that was the last portion to solidify. Although no specific studies have been undertaken to define the nature of these fluids, we believe that these were derived from the dyke itself, not from any external sources. If they were external, then it would have been more natural to expect most intensive alterations along the dyke margins, rather than the center. A question may arise as to what extent the postmagmatic alteration has affected other primary magmatic features of the rocks, such as their normative An-contents. The point is relevant since one can question whether an inward-increase in normative An is a primary feature, rather than related to the postmagmatic development of sericite aggregates after plagioclase. Since a systematic increase in normative An across the 20 cm thick marginal zone is not accompanied by a corresponding increase in K_2O that may be taken as a proxy of sericite in the rocks (Fig. 30.5), we believe that these variations are primary.

To conclude, the observed reverse compositional trends (Fig. 30.4) can be regarded as primary magmatic zonation that resulted from crystallisation of the dyke from magma that became increasingly more primitive with time. In contrast, the

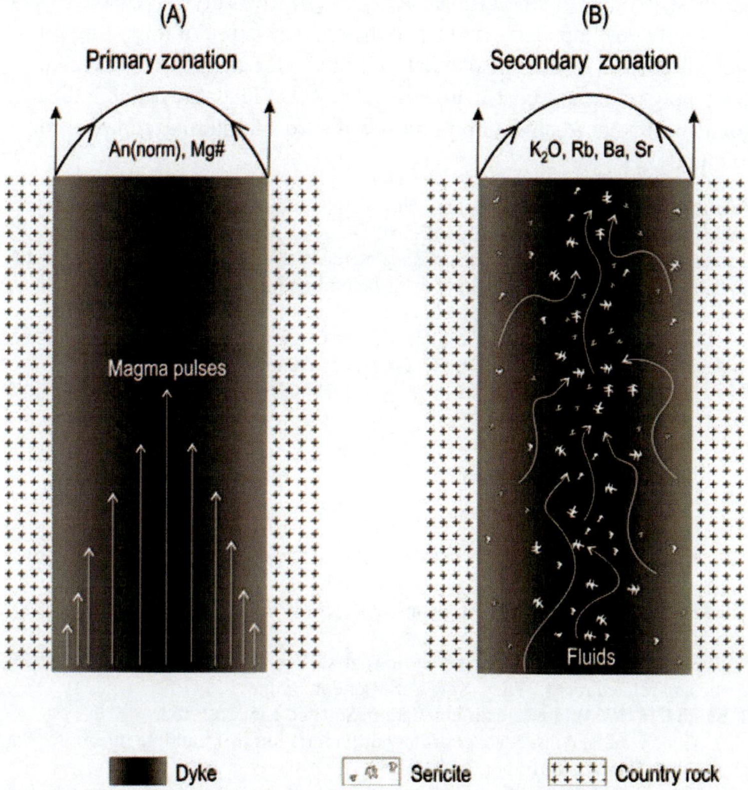

Fig. 30.7 A cartoon illustrating the successive formation of internal zonation of the Vochelambina dolerite dyke (Kola Peninsula, Russia) by two principle mechanisms: (**a**) successive changes in composition of inflowing magma that resulted in a primary magmatic zonation (e.g. an inward increase in normative An and whole-rock Mg#) and (**b**) alteration of the rocks by post-magmatic fluids producing fine-grained aggregates of sericite from plagioclase that resulted in secondary zonation (e.g. an inward increase in K_2O, Rb, Ba and Sr). See text for further discussion

normal compositional trends are a secondary feature (Fig. 30.6) that resulted from post-magmatic alteration superimposed on the primary magmatic zonation. The idea is schematically illustrated in Fig. 30.7.

Conclusions

A systematic increase in incompatible elements towards the centre of mafic dykes is commonly taken as evidence for their compositional zonation having formed by the fractional crystallization of basaltic magma. The present study shows, however, that this is not always the case. The studied dolerite dyke shows a clear inward increase in K_2O, Rb, Ba and Sr which is unrelated to primary processes of magmatic differentiation. The trends have resulted entirely from the postmagmatic replacement of primary plagioclase crystals by fine-grained sericite. The inwardly-increasing extent of alteration probably developed in response to the concentration of migrating fluids in the centre of the dyke. The chemical zonation of the dyke thus combines features of both primary magmatic and post-magmatic processes. The study thus indicates that care should be taken when interpreting the nature of internal compositional zonation in mafic dykes.

Acknowledgements We are grateful to Vladimir Pozhilenko and George Fedotov for help and good advice with many aspects of this study. Michail Dubrovskii, Nikolay Kudryashov, Pavel Pripachkin, Alexander Yefimov and Alexey Delenitsin are greatly thanked for assistance with field-work. We are indebted to N. V. Chalapathi Rao for his helpful comments and suggestions, which contributed to improving the manuscript. The authors are also very grateful to Rajesh K. Srivastava for the invitation to contribute to this IDC-6 volume. The research was supported by grants from the Finnish Academy of Science and the Renlund Foundation, Finland.

References

Arzamastsev AA, Montero P, Travin AV, Arzamastseva LV, Bea F, Fedotov GA (2006) Evidence for Sveconorwegian (Grenvillian) magmatic activity in the Northeastern Baltic Shield. Transactions. Doklady of the Russian Academy of Sciences/Earth Science Section 410(3): 461–465

Bhattacharji S (1967) Scale model experiments on flowage differentiation in sills. In: Wyllie PJ, (ed) Ultramafic and related rocks. Wiley, New York London Sydney: 69–70

Bhattacharji S, Smith CH (1964) Flowage differentiation. Science 145: 150–153

Brouxel M (1991) Geochemical consequences of flow differentiation in a multiple injection dyke (Trinity ophiolite, N. California). Lithos 26: 245–252

Chistyakova SYu, Latypov RM (2009a) Fine-scale chemical zonation in small mafic dykes, Kestiö Island, SW Finland. Geol Mag 146(4): 485–496

Chistyakova SYu, Latypov RM (2009b) Two independent processes responsible for compositional zonation in mafic dykes of the Åland-Åboland dyke swarm, Kestiö Island, SW Finland. Lithos 112(3-4): 382–396

Chistyakova SYu, Latypov RM (2010) On the development of internal chemical zonation in small mafic dykes. Geol Mag 147(1): 1–12

Daly JS, Mitrofanov FP, Morozova LN (1993) Late Archaean Sm-Nd model ages from the Voche-Lambina area: Implications for the age distribution of Archaean crust in the Kola Peninsula, Russia. Precambrian Res 64: 189–195

Ernst R, Bell K (1992) Petrology of the great Abitibi dyke, superior province. Can J Petrol 33: 423–469

Fedotov GA, Amelin YuV (1998) Post-Svecofennian dolerite dykes of the Kola region: Dual nature of cratonic magmatism. Vestnik MGTU 1(3): 33–42 (in Russian)

Gorbatschev R, Lindth A, Solyom Z, Laitakari I, Aro K, Lobach-Zhuchenko SB, Markov MS, Ivliev AI, Bryhni I (1987) Mafic Dyke Swarms of the Baltic Shield. In: Halls HC, Fahrig WF (eds) Mafic dyke swarms. Geological Association of Canada Special Paper 34: 361–372

Govindaraju K (1994) Compilation of working values and sample descriptions for 383 geostandards. Geostandard Newletter 18: 1–158

Halama R, Waight T, Markl G (2002) Geochemical and isotopic zoning patterns of plagioclase megacrysts in gabbroic dykes from the Gardar Province, South Greenland: Implications for crystallization processes in anorthositic magmas. Contrib Mineral Petrol 144: 109–127

Kretz R, Hartree R, Garrett D (1985) Petrology of the Grenville swarm of gabbro dikes, Canadian Precambrian Shield. Can J Earth Sci 22: 53–71

Miller JD, Ripley EM (1996) Layered intrusions of the Duluth complex, Minnesota, USA. In: Cawthorn RG, (eds) Layered Intrusions. Developments in Petrology 15. Elsevier Science BV: 257–301

Mitrofanov FP, Pozhilenko VI, (eds) (1991) The Archaean Vochelambina geodynamic area of the Kola Peninsular. Kola Science Center, Russia, Apatity: 196 p (in Russian)

Morse SA (1979) Kiglapait geochemistry I: Systematic, sampling, and density. J Petrol 20: 555–590

Morse SA (1981) Kiglapait geochemistry IV: The major elements. Geochim Cosmochim Acta 45: 461–479

Platten IM (2000) Incremental dilation of magma filled fractures: Evidence from dyke on the Isle of Skye, Scotland. J Struct Geol 22: 1153–1164

Platten IM, Watterson J (1987) Magma flow and crystallization in dyke fissures. In: Halls HC, Fahrig WF (ed) Mafic dyke swarms. Geological Association of Canada Special Paper 34: 65–73

Ragland PC, Rogers JJW, Justus PS (1968) Origin and differentiation of Triassic dolerite magmas, North Carolina, USA. Contrib Miner Petrol 20: 57–80

Ross ME (1983) Chemical and mineralogic variations within four dikes of the Columbia River Basalt Group, southeastern Columbia Plateau. Bull Geol Soc Am 94: 1117–1126

Ross ME (1986) Flow differentiation, phenocryst alignment, and compositional trends within a dolerite dike at Rockport, Massachusetts. Bull Geol Soc Am 97: 232–240

Ross ME, Heimlich RA (1972) Petrology of Precambrian Mafic Dikes from the Bald Mountain Area, Bighorn Mountains, Wyoming. Bull Geol Soc Am 83: 1117–1123

Simkin T (1967) Flow differentiation in the picritic sills of north Skye. In: Wyllie PJ, (ed) Ultramafic and related rocks. Wiley, New York London Sydney: pp 64–69

Steele KF, Ragland PC (1976) Model for the closed-system fractionation of a dyke formed by two pulses of dolerite magma. Contrib Mineral Petrol 57: 305–316

Index

DOI 10.1007/978-3-642-12496-9,